THEOLOGIES OF THE BODY:
HUMANIST AND CHRISTIAN

Benedict M. A

THE POPE JOHN CENTER

i

Nihil Obstat:
    The Rev. Robert F. Coerver, C.M., S.T.D.
    Censor Deputatus

Imprimatur:
    The Most Rev. Edward J. O'Donnell, D.D., V.G.
    Archdiocese of St. Louis

March 29, 1985

The Nihil Obstat and Imprimatur are a declaration that a book or pamphlet is considered to be free from doctrinal or moral error. It is not implied that those who have granted the Nihil Obstat and Imprimatur agree with the contents, opinions, or statements expressed.

Cover and dedication page design by the Rev. Larry D. Lossing

Copyright 1985
by
The Pope John XXIII Medical-Moral
Research and Education Center
186 Forbes Road,
Braintree, Massachusetts, 02184

Library of Congress Cataloging in Publication Data
Main entry under title:

Theologies of the Body

    Includes bibliographies and index.
    1. Body, Human (Christian theology)
    2. Body, Human. I. Title.
BT741.2.A84   1985   233'.5   84-15031
ISBN 0-935372-15-6

**TO MARY
THE NEW EVE
BY THE SPIRIT
MOTHER OF GOD'S SON
FOR US MADE
FLESH
DEAD ON THE CROSS
RISEN AND PRESENT
IN THE BREAD
COMING AGAIN
IN GLORY**

This icon of the Eastern Church symbolizes the theme of this book: Mary typifies the virgin Universe, Eden, the New Eve, the Chosen People, the Temple, the Church, the New Jerusalem. In her by the Holy Spirit the Word of God, as the New Adam, takes on our bodily humanity.

Contents

FOREWORD

I am acutely conscious that in this book I have ranged over many fields in which I have no claim to be a specialist. I have been compelled to do this by the excessive compartmentalization of thought that characterizes theology at the present time. I hope readers will overlook defects in detail and find some profit in my broader, synthetic purpose, namely to bring data and insights from history, science, philosophy and theology to enrich our understanding of what it is to be and to live as human bodies that think and choose.

My interest in this project ante-dates but has been greatly stimulated by my work in medical ethics under the auspices of the Institute of Religion and Human Development of the Texas Medical Center, Houston; the Catholic Health Association and the Pope John XXIII Medical-Moral Research and Education Center, both of St. Louis. I particularly want to thank Fr. Albert Moraczewski, O.P., first President of the Pope John Center, who first encouraged me to write this book and who has read it and made many suggestions, and the Rev. William Gallagher, the present President, the Rev. Larry Lossing, Msgr. Orville Griese and other members of its staff for their great generosity in sponsoring and preparing this work for publication. I owe much also to Fr. Kevin O'Rourke, O.P., the Director of the Center for Health Care Ethics of the St. Louis University Medical Center for his encouragement.

I also want to recall with gratitude the late Fr. William H. Kane, O.P. and the Albertus Magnus Lyceum of River Forest, Ill. which he founded and its members to whom, during the existence of the Lyceum, I owed my introduction to many of the topics discussed in this book. Most warmly I want to thank Fr. Ralph Austin Powell, O.P. who read the manuscript and to whom many of its best philosophical and theological insights are due, although on some points he might not be in full agreement. Finally I want to thank my Dominican brothers of the Province of St. Albert the Great, Aquinas Institute of Theology, and St. Louis Bertrand Priory for their love and support.

Rather than a bibliography, which in so wide-ranging a book would have been too large to be useful, I have included in the notes not only the references necessary for documentation of the text, but also certain other books and articles which have proved useful to me or to which I would refer readers who wish to explore some of the important topics more fully and from other points of view.

Benedict M. Ashley, O.P., S.T.M.

CAN WE CREATE OURSELVES?
"Humanist and Christian Theologies of the Body

Benedict M. Ashley, O.P.

Part I
SCIENCE, THE BODY AND THE
HUMANIST THEOLOGY

Chapter 1

Can We Create Ourselves?

I: We Are Bodies

I exist and live as a body in a world of bodies. We bodies contact each other and by this contact we create the world-space. To be a body, however, is to exist only in process: forming and decaying, appearing and disappearing, recurring and undergoing modification, development and growth. Although at times we seem as inert as sleeping dogs or rocks, we are never simply there, but are always becoming or perishing. As a living body I will die and my body will be indistinguishable from the dirt that has never lived, but may someday live. Yet even dirt has its history of chemical composition and decomposition.

As we bodies create space, so we create world-time, the sequence of comings and goings, events and pauses, occurrences, reoccurrences, failures to make it. Nowhere and at no time do I find myself to be anything but my body. Nowhere round me do I meet anything but bodies. Even the wind, even the empty space beyond our earthly atmosphere, turn out to be somehow bodily, filled with bodily activities, waves of energy vibrating and spreading, bodies precipitating out of pools of energy, or dissipating again into waves.

We are bodies. Or are we, perhaps, only fluid parts of one world-body, drops in an ocean of energy? In any case I experience myself always contacting what seem to be alien bodies, other objects or persons, which do not seem to be me, and yet which form the world of which I am now a part — for a time separated from other parts. Yet I know I have come to be myself out of this chaos of bodies that were not me and that I will pass away into bodies that will be me no longer.

Thus any questions that arise for me about my own self, or about the other selves and other things that make up my world must be questions about *body* — or to speak more generally — about *matter*, about these space-time beings that are only becomings. Why do I hurt? Why am I pleased? Will the other bodies around me give me pleasure or pain? What will become of us? Are there other bodies, persons or things, beyond the circle of my present contacts? Shall I try to explore this vast world whose horizon ever recedes? What will I find? Only more bodies?

All these questions are raised when I meet other bodies and thus become aware of myself as a body bumping into others. In fact any question I know how to ask concerns bodies, since even if something exists that is not bodily, I will know it only if somehow it contacts me as I am a body. Therefore, the puzzle of my body-self is a *universal* question, conditioning every other question I may ask. To ask any question I must question my body, and if I am to answer any question, I must somehow answer questions about my body also. It is my body that puzzles, questions, searches for answers in a world of bodies, or in the world-body which differentiates itself into parts only to reabsorb at last even us human, questioning parts into itself once more.

My puzzlement, however, is not to be satisfied by merely theoretical answers. As a body-self I find myself empowered to make a practical, active response to my situation in the world by taking control over the parts of my own body and over the bodies that environ me. I was created out of the world-chaos by processes that I find still going on within me. I can myself engage in these processes and in the face of many resistances in some measure direct their flow. In the flood of change I am a gate which can swing open or shut to redirect the current. As I emerge from the processes which have created me, I find myself able to be a creator. We humans came into the world as novelties, unique animals whose special newness is this capacity to share consciously and freely in creative processes by which we were ourselves produced. In the blind torrent of creative change I awaken, and I find you opening your eyes. As our eyes meet we realize

4

that we can and must take charge together of a world in process. We must recreate the world which has created us.

II. Can We Create Ourselves?

Today from many very different points of view thinkers seem to be converging on a new definition of the human being: *We are animals who create ourselves*. Walter Odajnyk expresses the view of Marxism when he writes:

> Man is now God, because he creates himself — he makes his own essence, and he also decides what values will be placed upon his life and his actions and even upon the life and actions of others. It is no wonder that anguish comes to haunt him in this work, for he must make his decisions alone without reference to heaven or to any stable norms; and he knows that he alone bears the entire responsibility for his decisions. He is a God, but a God chained; he is condemned to make these decisions. He has not chosen to play God, and yet no matter what he does he cannot escape his role.[1]

Jean Paul Sartre, the existentialist, says:

> Man, the creator and recreator of his own and all human life, the creator of meaning and values, the only factor actively managing history, increasingly conquering an existing situation to his objectivization, is man's future, as well as his immediate present. Man is man's problem, but he is also his neighbor and brother.[2]

It is not only anti-theological thinkers that sponsor this new definition. Radical theologians such as Thomas J. Altizer are convinced that moderns can no longer believe in a transcendent God. Nevertheless, they say, we may still be able to believe in Jesus as the Christ, as God become immanent in our world. Through the incarnation of God in a human being, we at last become fully responsible for ourselves and our world, and hence no longer dependent on a transcendental heavenly Father.

> It is precisely because the movement of the Incarnation has now become manifest in every human hand and face dissolving even the memory of God's original transcendent life and redemptive power, that there can no longer be either a truly contemporary movement to transcendence or an active and living faith in a transcendent God.[3]

Even theologians who continue to insist on the transcendence of God, such as the "process theologians" (followers of Whitehead or Teilhard de Chardin[4]) are convinced that we moderns can only accept a "God who is man's future,"[5] that is, a God who places in our hands the responsibility for ourselves and the world in its process of becoming.

No matter how paradoxical this new definition of ourselves as creative animals may seem, it certainly is rooted in our twentieth century experience. For most of our two or three million years on earth the human race was immersed in the cycle of nature. Humans were already distinguished from other animals by the human power of inventive intelligence, yet the pattern of human life remained largely determined by biological necessity. Only a few thousand years ago humans finally emerged from this enslavement to nature and began to live in villages, towns, and cities, in an environment which they themselves created. At last they could do something about their food and shelter, but still they could not do much about themselves, their own bodies. The mark of civilization is that we "naked apes"[6] put on clothes, but although we might cover our bodies, we could not improve them. Our bodies remained the same archaic, savage bodies that evolution had given us.

Now in this century we have at last taken a giant step, greater than our step onto the moon. We have begun to remake not only our environment, but our very body-selves. Daily we read of new successes in our efforts to patch up our bodies with transplanted and artificial organs. More prophetic is our growing ability to regulate our life processes through vitamins, hormones, and new drugs. Still more radical is the promise of the cracking of the genetic code by which the very origin and formation of our bodies is programmed. As we can now synthesize chemical substances not found in nature, so in the not too remote future we may be able to synthesize human beings having novel characteristics. We will make Supermen and Superwomen or perhaps the Superandrogyne.[7]

Nor does there seem to be any absolute obstacle to our producing humans who never grow old but live perpetually, exempt from any other death than through some freak accident. Since we now know we are essentially our brains, our ultimate problem is to remake our brains and keep them ever-living and ever self-improving.[8]

Of course there are serious doubts currently raised to cool our enthusiasm. Sober scientists insist we are still a long way from this degree of control. The complexity of the human genetic code (the combinations and permutations of perhaps 10,000 genes) and of the human brain (a network of over 25 billion neurons) is even more awesome than the immensity of astronomic space.[9] We have far to go before we know enough to program systems of such complexity.

6

Furthermore, many writers, such as Jacques Ellul,[10] warn us that our advance in technological control has not been matched by social and ethical control, so that technology instead of liberating is more and more enslaving us through environmental pollution, the "population explosion," "the energy crisis," iatrogenic disease,[11] totalitarian social domination, and the threat of nuclear obliteration. Perhaps the new definition of the human being should be not "the animal who creates itself," but "the animal who self-destructs."

While these doubts and warnings about the coming Age of Humanity give us fair notice that human creativity has its limits and its terrible risks, they do not prove either that technology must be abused, or that we can achieve our full humanity without technology. Pre-technological societies usually condemned most men and women to slavery, serfdom, or a narrow peasant existence burdening them with constant competition for the scarce necessities of life and with the stultifications of superstition and tribal conformism. We must, therefore, as creative animals, accept the responsibility of our creative powers. We must develop a new ethics, adequate to this responsibility.

III. The Body and Ethics

An ethics of creativity or a creative ethics, however, is a highly problematic notion. All the traditional systems of ethics were, in various ways, based on the assumption that the ultimate ethical norms are derived from "the nature of things." Yet human nature, as I have already noted, is first of all evident to us as our body-selves which exist in the context of Nature as the universe of bodies. Thus in eastern cultures ethics is based on some conception of cosmic law, for example, the Chinese *T'ao*.[12] In Judaism and Islam it is based on revealed commandments of God the Creator, the *natura naturans*.[13] In western culture under Greek influence all ethical systems rest on the notion of human nature as the stable and universal structure common to all members of the human species.[14] Christian theology accepted this conception and used it to develop and systematize its own biblically based morality.[15] The Protestant Reformation raised serious questions about this Hellenistic type of ethics, but did not find it easy to replace.[16] In the Enlightenment period various attempts were made to develop a new basis for ethics (for example Rousseau's ethics of naural feeling, Kant's ethics of duty, Bentham's utilitarian ethics of rational calculation, etc.[17]) but each of these in one way or another retained the notion of a universal, stable human nature. In fact the Enlightenment was characterized by its simple faith in this basic principle, and in the self-evidence of the "natural moral law."[18] It was only with Marx, and his

7

view that "man can create himself," that this long tradition began to come to an end.[19]

If human nature is not something fixed and given, but something which is itself problematic, ours to remake, then it cannot be the norm of our behavior, but rather the product and goal of that behavior. Human nature becomes our chief problem, not the principle by which our problems can be solved. This is the dilemma with which the twentieth century has confronted us and which makes necessary a rethinking of our conception of what it is to be human, and hence what it is to be a human body-self, a creative animal, a creative body among the countless bodies we call created Nature.

IV. A Hermeneutical Method

1. The Interpretative Dilemma

The universality of bodiliness which I described in section I of this chapter means that questions about body and matter in general must be approached by an inter-disciplinary, or perhaps trans-disciplinary method. Usually when we speak of such universal questions we call them "philosophical." Today many professional philosophers insist that philosophy deals with no special body of facts, principles or methods which define it as a field, since it deals not with realities as such, but with *language* about reality. Such philosophers renounce any ambition to present a universal view of reality, any system, or "metaphysics." They modestly confine themselves to clarifying the languages of the special disciplines by referring the technical languages of these disciplines to the ordinary language of life. Other philosophers, however, continue to defend the traditional view that philosophy deals not only with universal problems of language, but also with universal aspects of reality, with Being and its attributes.[20]

I do not want to enter into this debate here. It is enough to grant that "body" and "matter" are terms which enter into many, probably all, disciplines and are named in every language, however specialized. I will be free, therefore, to use sometimes the language of physics, sometimes of biology, or the behavioral sciences, as well as the more metaphorical and affective poetic and rhetorical modes of discourse. Because we humans are bodies that speak, all forms of human speech are needed to express our bodily existence. thus the inter-disciplinary range of the inquiry I have set myself would justify me in titling this book "A Philosophy of the Body."

Nevertheless, I have chosen to title it "Theologies of the Body: Humanist and Christian". The Greeks who invented the term "philosophy" also called the kind of universal inquiry I have undertaken "theology," because for them the Divine (*theos*) was a general term for the All, whatever its nature might be, and for many of them this All was universal matter.[21] The common distinction between philosophy as an investigation based on reason and theology as one based on revelation is of Christian origin and is of little relevance to any but Christians, and, perhaps, to such Jews and Muslims as have been influenced by the Christian body-soul dichotomy.[22] When used today by non-Christians the sense of the distinction is too often polemical. Since the eighteenth century Enlightenment the *philosophes* have employed it to make their point that while philosophy is founded on the secure rock of reason, subject to public, objective verification, theology rests only on the unstable sands of private, subjective, unverifiable faith.[23]

Today sociologists of religion and specialists in comparative religion have found it helpful to drop any attempt to define religion in terms of content, and prefer purely *functional* definitions.[24] They have found useful the definition invented by a theologian, Paul Tillich, for whom religion is "ultimate concern,"[25] that is, anyone's world-view and value-system, whether these are theistic or atheistic, dogmatic or relativistic, worldly or mystical, sacral or secular in content. The value of so vague a definition is precisely its functionality, since it makes it possible to compare and contrast on an even footing all the various world-views and value-systems within which human beings live, without labeling one "religious" and the other "irreligious," or one "rational" and the other "irrational." By using it we also avoid committing ourselves at the outset to any evolutionary scheme by which "religion" is seen as a feature of the past, and "secularity" an achievement of the present and a predictable characteristic of the future. If religion is defined in this functional way, then reflection on the content of particular religons and their comparative study ought to be called "theology" rather than "philosophy." It is in this sense that I have called this inquiry "Theologies of the Body." However, if any reader objects to this usage, he or she can translate "theology" as "philosophy," unless the context limits it to Christian theology as distinguished from a Christian's philosophy.

The method of any inter-disciplinary investigation is inevitably interpretative or hermeneutical because it requires the translation of one special language into another or into ordinary language. In the first half of this century the Logical Positivists supposed they could solve this difficulty by fashioning a "unified language of science," but this proved impossible even for the physical sciences. Today British and American philosophers confine themselves largely to the clarification of ordinary language; while continental European philosophers attack the problem on the still broader basis of a phenomenological, hermeneutical approach which deals with the epistemological and ontological conditions of the human thinking expressed in language.[26]

In this hermeneutical method the basic principle is that of the *hermeneutical circle*. When we try to understand the meaning of a text (or of any system of symbols) we can arrive at the meaning of the whole only by analyzing the whole into its parts, yet we must also take into account that these parts themselves have meaning only in their functional relation to the synthetic whole. Thus analysis and synthesis are mutually interdependent, so that interpretation is always an ongoing process of reading and re-reading in which the possibility of a still more adequate uncovering of the inherent meaning always remains open.[27]

What is true of the text is also epistemologically true of the interpreter. The knower as subject and the text as object are only an example of the general situation of knower and known in which subject and object form a mutually interdependent whole in which meaning or truth cannot be achieved except by exploring both terms of the relation in their distinction and identity. I cannot be objectively open to reality unless I become aware of my own subjectivity and the way it orients me to the objects I am trying to understand; yet I can know myself as subject only as I am intentionally directed to the objects of my world. Thus it is hopeless to attempt to understand myself by pure introspection, because as a bodily self I exist only in relation to a bodily world which demands my attention.[28]

Moreover, deeper even than the epistemological circle is the ontological circle of human living of which thinking is itself only a manifestation. We live in the stream of history out of which we have emerged as bodies which think, a history of the universe of matter, of human culture with the artificial incarnate structures which it produces, and of the symbolic systems which ultimately refer to bodily objects and events.[29] This stream of history is not effectively infinite, but exists for us within a *horizon* formed by the causal limits of what impinges on our lives and what we influence. My world, which forms me and which I in turn help to form, fades away at a finite horizon, and it does not perfectly coincide with your world. Each of us has unique experiences, understandings, ways even of using our common language. Because we are bodies living in the relativity of space and time, our communication with each other, with our pasts and our futures, raises problems of translation which are rooted in our very mode of being.

Thus the only method available to me for developing a theology of the body is one which is consciously and confessedly hermeneutical. I will not try violently to break out of the hermeneutical circle in which my inquiry is inevitably enclosed. I have to submit to the fact that the language used will have to labor under that vagueness from which any inter-disciplinary, non-technical discourse suffers. I will have to beg the reader to interpret and evaluate the parts of my argument in view of the plausibility of the whole panorama which I an attempting to lay out. Furthermore, I have to by-pass for the moment taking either the side of epistemological subjectivism with idealist thinkers or of objectivism with the realists. And finally I cannot commit myself at the outset either to a historicist or a metaphysical view of reality. Instead I must frankly acknowledge that I view my bodily self and the bodily world which is mine from within the encircling horizon of my life stance as a Roman Catholic Christian living in twentieth century, technological United States of America, looking certainly toward personal death in the not too distant future and perhaps to the rapid decadence of the American culture which has been my world. The hope that appears within this narrow, anguished world will be credible only if it emerges from the right reading of that world, from a careful exegesis rather than a forced eisegesis.

2. An Expanding Horizon

How then, if I stand within the horizon or hermeneutic circle of my own Christian commitment, will I be able to make use of that wealth of information and analysis developed by the modern sciences and arts which have been largely achieved within a secularized world-view? Has not Christianity shown itself singularly fearful of the body and all the material world — the world of sinful flesh? How is it possible for me to be open to all that other theologies have to say about the body, and especially to what is said by those "secular" theologies which more and more dominate contemporary culture?

Hans-Georg Gadamer, a leading hermeneutical philosopher, has shown that communication between those living in different worlds and speaking different languages can become possible through what he calls "the fusion of horizons."[30] Human worlds can overlap through the process of translation and interpretation and thus come to include each other. Is not this effort to transcend my own horizon, to use my imagination to translate another's world into the symbols native to me so as to enrich and even transform my familiar homeland, one of my strongest claims to be "a creative animal?" Poets who have mastered their own language may prove the best interpreters and translators of the works of alien poets. Just because I am the creator of my own world, the worlds that other animals like myself have created are opened to me by the creativity that is common to us all.

This is a process which Christians recently have vividly experienced through the ecumenical movement. For centuries Protestants, Catholics, and Orthodox Christians lived in different worlds from which the faith of others could only be understood as lies, ultimately inconsistent and unintelligible, false to the truth of authentic religious experience. By a remarkable historical development[31] experienced by us Christians as a deepening action of the Holy Spirit, we have begun to perceive that these diverse worlds can be interpreted positively to each other. The Biblical text common to all Christians now appears susceptible of different readings which formerly seemed falsifications, and the various theological traditions which once seemed contradictory, now begin to look complementary. The result of this hermeneutical process is not a conversion of one world into another, nor is it a mere overlapping in some least common denominator, but it is a *convergence*, or fusion achieved by a deepened and expanded understanding by each party of its own world, a common world which admits of varied expressions.[32]

To do justice to the problem of the human body it is necessary, I believe, to engage in such an ecumenical hermeneutic, not only to break through the barriers of the academic disciplines, but even more to break through the walls between the world-religions which seem to differ radically in their understanding of the human body, as on every other ultimate question about our human nature and destiny. To formulate the problem on any narrower basis runs the risk of failing to understand our bodily existence in its historicity, that is, in terms of all that the human race has experienced in its physical and cultural context and of its common projects for the future.

I do not imagine myself, however, capable of so vast a task. In spite of the progress in recent years in the comparative study of religions, we have only begun to enter into dialogue with so-called "primitive" religions, with the great eastern religions, or even with the growing world of Marxism. While I will make frequent references to what such studies can at present bring to the development of a Christian theology of the body, I acknowledge this can only be by way of suggestion.

Even within our western world since the eighteenth century the Christian world view has coexisted with the Humanism of the Enlightenment and with Marxism. In the United States of America Marxism has never had a deep influence. Therefore, I have of necessity chosen in this book to deal primarily with two theologies of the body (as its title indicates): the Christian one to which I am committed and the Humanist one in which I was born and with which I have been throughout my life in daily contact. I say "Humanism" rather than "Secular Humanism" as some do, to avoid any pejorative connotations and will explain its significance in chapter 3.[33] The expansion and fusion of horizons at which my hermeneutical method aims will be a convergence of these two worlds which historically have more and more diverged from each other. I hope to achieve this expansion not by stepping outside my own Christian world, which I believe to be rooted in ultimate reality, but by deepening my own understanding of the common human experiences out of which every world view has grown. By cultivating my own garden, as Voltaire the great Humanist advised,[34] I hope to discover that Eden which is the home of all who are human.

In undertaking this difficult task in a very tentative and sketchy manner, I want to avoid two pitfalls. One is the temptation of "futurism," that is, the windy prophesying of future human triumphs, without facing the painful task of laying the foundations on which such triumphs must be built. The other temptation is to indulge in the paradoxes in which many existentalist and process philosophers and theologians seem to revel. Every philosophical or theological exploration begins in a paradox (that is, in puzzlement), but a philosopher or theologian has the obligation to do what he or she can to resolve the apparent contradictions by the use of sober analysis and non-metaphorical language.

In the rest of this book I will propose a solution to these problems in the following steps. In the remainder of this Part I, I will first outline in Chapter 2 what science has shown us about ourselves as bodily beings in dynamic relation to the whole of Nature; and then in Chapter 3, I will describe the various "theologies" which Humanism since its origin in about 1700 has devised to interpret these scientific findings and give them ethical application. Then in Part II I will compare with these Humanist theologies the various Christian theologies of the body developed before the rise of modern science and after it, concluding by showing the need today for a more adequate Christian theology of the body. In Part III I will undertake a philosophical interpretation of the scientific facts reported in Chapter 2, an interpretation which I believe to be more adequate to these facts than that yet given by either Humanists or Christians. Finally in Part IV, I will propose a Christian theology of the body congruent with this philosophical interpretation of the scientific facts. In particular in Chapter 10, I will attempt to show how such an understanding of science can provide the ground for an ethics by which we can escape the dilemmas that I have described in this present Chapter 1, and in the the following chapters 11 to 14 I will show the implications this has for our understanding of God and of the eternal life of the resurrected body and transformed universe.

Notes

1. *Marxism and Existentialism* (Garden City, N.Y.: Doubleday Anchor Books, 1965), p. 13.

2. *Ibid.*, p. 139 from the 1957 Polish version of an essay which in revised form serves as a preface to Sartre's *Critique de la raison dialectique* (Paris: Librairie Gallimard, 1960). I have not been able to locate the passage in the French or in the translation *Critique of Dialectical Reason* (Atlantic Highlands, N.J.: Humanities Press, 1976).

3. Thomas J. Altizer, *The Gospel of Christian Atheism* (Philadelphia: Westminster Press, 1966), p. 136.

4. For a review of this movement see Gene Reaves and Delwin Brown "The Development of Process Theology" in a book edited by them, *Process Philosophy and Christian Thought* (Indianapolis: Bobbs-Merrill, 1971), pp. 21-64; and Charles Hartshorne, "The Development of Process Philosophy" in Ewert H. Cousins, ed., *Process Theology* (New York: Newman, 1971), pp. 47-66.

5. "The new concept of God — that is, faith in the One who is to come, in 'the wholly new One' who provides *us* here and now with the possibility of making human events into a history of salvation through an in-ward re-creation which makes us 'new creatures' dead to sin, thus radically transforms our committment to make a world more worthy of man, but at the same time it reduces to only relative value every result which has been so far achieved." Edward Schillebeeckx, O.P., *God the Future of Man* (New York: Sheed and Ward, 1968), p. 186. See also Jurgen Moltmann, *Theology of Hope* (New York: Harper and Row, 1965), and Carl Braaten and Robert Jensen, *The Futurist Option* (New York: Newman Press, 1970).

6. Desmond Morris, *The Naked Ape* (New York: McGraw-Hill, 1967).

7. See D.S. Halacy, *Cyborg: Evolution of the Superman* (New York: Harper and Row, 1965) and Carl Sagan, *The Dragons of Eden: Speculations on the Evolution of Human Intelligence* (New York: Ballantine Books, 1977).

8. J. F. Wilkinson, "The Deep Freeze Scheme for Immortality", *True*, Oct. 1967: 52-55, 79-82, reporting the work of Robert C. W. Ettinger and his pioneering book *The Prospect of Immortality*, 1964. George M. Martin, "Brief Proposal: On Immortality: An Interim Solution" in *Perspectives in Biology and Medicine* 14 (Winter, 1971), 339-340, proposes a system of cryobiological preservation of the central nervous system as an interim solution to death, and then, as an ultimate solution, the replacing of human beings by "a family of posthumanoid 'postsomatic' bioelectrical hybrids," capable of contributing to cultural evolution at "rates far exceeding anything now imaginable."

9. Monroe W. Strickberger, *Genetics*, 2nd ed. (New York: Macmillan, 1976), p. 542, estimates that for the human species the number of genes is in excess. of 10,000.

Charles R. Noback and Robert J. Demarest, *The Human Nervous System* (New York: McGraw-Hill, 1975), pp. 108, estimate there are more than 25 billion neurons in the brain (exclusive of the cerebellum in which there are 100 billion more), about 10 billion of which are in the cerebral cortex.

10. Jacques Ellul, *The Technological Society* (New York: Knopf, 1965) and *The Meaning of the City* (Grand Rapids, Michigan: Eerdmans, 1970). See also the discussion in William Barrett, *Irrational Man* (Garden City, N.Y.: Doubleday Anchor Books, 1958), pp. 184 ff.

11. See Ivan D. Illich, *Medical Nemesis*, New York: Pantheon, 1976.

12. Yu Lan Feng, *A History of Chinese Philosophy* (Princeton, N.J.: Princeton University Press, 1952-3), 1, 177-180. On the parallel of Jewish and Chinese thought see H.H. Rowley, *Prophecy and Religion in Ancient China and Israel*, New York: Harper, 1956, pp. 50-73, and Joseph P. Schultz, *Judaism and Gentile Faiths* (East Brunswick, N.J.: Associated University Presses, Fairleigh Dickinson University Press, 1981). Chapter 1: "The Way in Confucian and Rabbinic Traditions," pp. 35-69.

13. On Islamic ethics see Louis Gardet, *Mohammedanism*, vol. 143 of *Twentieth Century Encyclopedia of Catholicism* (New York: Hawthorne Books) Chapter V, "Man's Acts and His Destiny," pp. 114-128, and G. H. Bousquet, *L'ethique sexuelle de Islam*, vol. 14 of *Islam d'hier et d'aujourd' hui* (Paris: G.P. Maisonneuve et Larosse, 1957), pp. 2-28, on whether

Islam has an "ethic". On the relation of Islamic ethics to those of Judaism see Abraham Geiger, *Judaism and Islam* (New York: KTAV, 1970).

14. See Jacques Maritain, *Moral Philosophy: An Historical and Critical Survey of the Great Systems* (London: Geoffrey Bles, 1964) and Vernon J. Bourke, *History of Ethics*, 2 vols. (New York: Doubleday Image Books, 1968) are two histories of ethics in which this natural law foundation and reactions to it are very well brought out.

15. See Maritain, *Moral Philosophy*, pp. 71-91.

16. See Roger Mehl, *Catholic Ethics and Protestant Ethics* (Philadelphia: Westminister Press, 1971) and James M. Gustafson, *Protestant and Roman Catholics Ethics* (Chicago: University of Chicago Press, 1978).

17. Maritain, *Moral Philosophy*, pp. 92-118, and Vernon J. Bourke, *History of Ethics*, 1: 161-251; vol. 2: pp. 1-76.

18. Carl Becker, *The Heavenly City of the Eighteenth Century Philosophers* (New Haven: Yale University Press, 1932), pp. 33 ff., shows how the Newtonian conception of natural law was extended to moral law and sociology. For some qualifications of Becker's much discussed thesis see Raymond O. Rockwood, *Carl Becker's Heavenly City Revisited* (Ithaca, New York: Cornell University Press, 1958), pp. 33-36.

19. "But since for the socialist man the *entire so-called* history of the world is nothing but the creation of man through human labour, nothing but the emergence of nature for man, so he has the visible, irrefutable proof of his *birth* through himself, of his *genesis*. Since the *real existence* of man and nature has become evident in practice, through sense experience, because man has thus become evident for man as the being of nature, and nature for man as the being of man, the question about an *alien* being, about a being above nature and man — a question which implies the admission of the unreality of nature and of man — has become impossible in practice. *Atheism*, as the denial of this unreality, has no longer any meaning for atheism is a *negation of God*, and postulates the *existence of man* through this negation; but socialism as socialism no longer stands in any need of such mediation. It proceeds from the *theoretically and practically sensuous consciousness* of man and of nature as the *essence*. Socialism is man's *positive self-consciousness*, no longer mediated through the abolition of religion, just as *real life* is man's positive reality, no longer mediated through the abolition of private property, through *communism*.", Karl Marx, *Economic and Philosophical Manuscripts of 1844, Karl Marx and Frederick Engels, Collected Works,* (New York: International Publishers, 1975), vol 3; pp. 305-6. The editors note that for Marx at this point the term "communism" was used for the transition phase from capitalism to "socialism", but later he reversed the terms so that "communism" named the final stage (see editors' note 85, p. 603).

20. P. F. Strawson in his well known *Individuals: An Essay in Descriptive Metaphysics* (New York: Doubleday Anchor Book, 1963), says, "Metaphysics has been often revisionary, and less often descriptive. Descriptive metaphysics is content to describe the actual structure of our though about the world, revisionary metaphysics is concerned to produce a better structure." Strawson defends both types, and then writes, "The idea of descriptive metaphycis is liable to be met with scepticism. How should it differ from what is called philosophical or logical, or conceptual analysis? It does not differ in kind of intention, but only in scope and generality. Hence, also, a certain difference in method. Up to a point, the reliance upon a close examination of the actual use of words is the best, and indeed the only sure, way in philosophy. But the discriminations we can make, and the connexions we can establish, in this way, are not general enough and not far-reaching enough to meet the full metaphysical demand for understanding" (pp. XIII-XIV).

"Analytic philosophy recognizes that the question is not metaphysics or no metaphysics, but what kind of metaphysical thinking we will do." John A Hutchinson, *Living Options in World Philosophy* (Honolulu: University Press of Hawaii, 1977), p. 65.

21. See Werner Jaeger, *The Theology of the Early Greek Philosophers* (Oxford: Clarendon Press, 1947), pp. 31-34, and A. J. Festugiere, O.P., *La revelation d'Hermes Trismegistus,* 3 vols. (Paris: J. Gabalda, 1949), 2: 598-605.

22. Medieval Islamic and Jewish philosophers were as dependent on the Greek philosophical classics, especially Aristotle and the Neoplationists as were Christian philosophers. See Etienne

Gilson, *History of Christian Philosophy in the Middle Ages* , (New York: Random House, 1954), pp. 181-234.

23. The manifestation of this objective-subjective dichotomy in the organization of the modern university was brilliantly described by C. P. Snow in *The Two Cultures, And a Second Look*, 2nd ed. (Cambridge: Cambridge University Press, 1964).

24. John Edward Sullivan, O.P., "The Idea of Religion" in *The Great Ideas Today* (Chicago: Encyclopedia Britannica, Inc.), Part I, 1977, pp. 206-276, and Part II, 1978, pp. 22-276; finds five types of usage for the term "religion": *ceremonial, moral, mystical, revealed,* and *secular*. Comte and Feuerbach are typical of "secular religion," but the Marxists reject the term. Sullilvan concludes: "Putting together the dialectical concept of the religious object with the dialectical concept of religious activity (and the religious relationship) in one brief statement, we claim that *reverence for transcendent goodness* is an adequate formulation of religion in general" (p. 267). Robert D. Baird, *Category Formation and the History of Religions* (The Hague: Mouton, 1971), however, argues that the notion of "transcendence" is unnecessary, because he advocates a "functional" rather than an "essential-intuitional" mode of definition. Baird strongly defends Paul Tillich's "Religion is ultimate concern," pointing out that "ultimate" in this definition implies no particular metaphysical position. "By 'ultimate' I am referring to a concern which is *more important than anything else in the universe for the person involved"* (p. 18). For historical inquiry this means that (a) the study of religion is a study of man rather than God; (b) it can be pursued either on the level of ideal or observable realities; (c) it studies individual behavior first but also social behavior.

Baird poses and answers the objections that (a) some persons have no ultimate concerns; (b) the definition is too broad; (c) it assumes all persons are religious. For other discussions of Tillich's definition see John B. Magee, *Religion and Modern Man* (New York: Harper and Row, 1967), pp. 19-40; John A. Hutchinson, *Paths of Faith* (New York: McGraw-Hill), pp. 1-9; and Ugo Bianchi *The History of Religions* (London: Brill, 1975), whose thesis is that the definition of "religion" is necessarily analogous not univocal. See also J. Milton Yinger, *The Scientific Study of Religion* (New York: Macmillan, 1970), pp. 1-23, who formulates the following "functional" definition: "Religion, then, can be defined as a system of beliefs and practices by means of which a group of people struggles with these ultimate problems of human life. It expresses their refusal to capitulate to death, to give up in the face of frustration, to allow hositlity to tear apart their human associations. The quality of being religious, seen from the individual point view, implies two things: first a belief that evil, pain, bewilderment and injustice are fundamental facts of existence; and, second, a set of practices and related sanctified beliefs that express a conviction that man can ultimately be saved from these facts" (p. 7). Yinger, however, quite unconvincingly proposes to separate ethics from religion (pp. 41-46). C.Y. Glock and R. Stark, *Religion and Society in Tension* (Chicago: Rand McNally, 1965), pp. 3-17 ask whether "religion" means "every philosophy of life" or "only those views which include a reference to some 'Super-empirical reality'."

In my opinion the notions of "ultimate" or "totality" are necessarily non-empirical and in some sense "transcendental." For others who seem to agree with this broad definition of "religion" and "theology" see Joachim Wach, *The Comparative Study of Religions* (New York: Columbia University Press, 1958), pp. 27-58 (who, however, speaks of "pseudo-religions"); Edward J. Jurji, *The Phenomenology of Religion,* (Philadelphia: Westminister Press, 1963), (who speaks of "modern rivals of living faith" pp. 261-294; Clifford Geertz, "Religion as a Cultural System" in M. Banton, ed., *Anthropological Approaches to the Study of Religion*, (London: Tavistock Pub., 1968) (definition of religion, p. 4). Ninian Smart, *The Religious Experience of Mankind* (New York: Scribner's, 2nd ed., 1976), pp. 499-556, treats Humanism and Marxism as world religions, as does David Martin: "Marxism, Functional Equivalent of Religion" in his *The Dilemmas of Contemporary Religion* (London: Blackwell, 1978), pp. 75-90.

25. Paul Tillich, *Systematic Theology*, 2 vols. (Chicago: Unniversity of Chicago Press, 1951), 1: 8-15. Tillich says that the biblical basis of his concept of "ultimate concern" is the *shema, Deuteronomy* 6: 4-6. Ernest Becker wrote, "Society itself is a codified hero system, which means that society everywhere is a living myth of the significance of human life, a defiant creation of meaning. Every society thus is a 'religion' whether it thinks so or not: Soviet

'religion' and Maoist 'religion,' no matter how much they may try to disguise themselves by omitting religious and spiritual ideas from their lives . . . It was Otto Rank who showed psychologically this religious nature of all human cultural creation; and more recently the idea was revived by Norman O. Brown in his *Life Against Death* and by Robert Jay Lifton in his *Revolutionary Immortality."* *The Denial of Death* (New York: The Free Press, 1973), p. 7 f.

26. Cornelius A. van Peursen, *Phenomenology and Analytical Philosophy* (Pittsburgh: Duquesne University Press, 1971).

27. On the history of the notion of the hermeneutical circle see Richard E. Palmer, *Hermeneutics: Interpretation Theory in Schleiermacher, Dilthey, Heidegger and Gadamer* (Evanston, Ill.: Northwestern University Press, 1969), pp. 25-26. Schleiermacher was the first to develop the notion (pp. 86-88), but attributed the basis of his ideas to Friederich Ast (1778-1841) (p. 77, note 8). The concept was further developed by Dilthey (pp. 118-121). See also Martin Heidegger, *Being and Time* (New York: Harper and Row, 1962), pp. 194 f. (I, 5, 32, p. 153 German) and Hans-Georg Gadamer, *Truth and Method.* (New York: Seabury, 1975), pp. 234-245; and Wolfhart Pannenburgh, *Theology and the Philosophy of Science* (Philadelphia: Westminster Press, 1976), Chapter 3, "Hermeneutic: A Methodology for Understanding Meaning", p. 156-224. "The hermeneutic circle" (or spiral as we would like to call it) is essentially a very general model of the development of knowledge through a *tacking* procedure or dialectics . . . To express it in a slogan, 'No development of knowledge without foreknowledge'. In the process of developing knowledge, more and more of the hidden assumptions become known and articulated. Hence a *spiral coil* or an *oscillation with decreasing amplitude* is a suggestive *picture* or metaphor for this process. We can never make all the assumptions explicit. But something is won if we are aware of the fact there must be such assumptions. In analytic philosophy, Godel's results — his formal proof that we cannot prove the logical consistency of even a logical system of a simple type without utilizing an axiom from outside the circle — show an interesting parallelity with the idea of a hermeneutic circle — If what is asserted is at variance with our foreknowledge (so far as we are aware of it) this will, if we are fairly sure of the asserted hypotheses, force us to check that part of the foreknowledge which is itself based on the idea of a circle. (Note that the dialectics of worldpicture-hypotheses and scientific theories is also covered by the said figure (*Denkfigur*). The tacking process is *infinite* at both ends, so the speak." Gerald Radnitzky, *Contemporary Schools of Metascience*, 2nd ed. 2 vols. in one (New York: Humanities Press, 1970), p. 23 and 25.

28. Wilhelm Luijpen, *Phenomenology and Metaphysics* (Pittsburgh: Duquesne Press, 1965), Chapter 2, "Traditional Metaphysics and Realism", pp. 99-119, argues that, far from being idealistic, phenomenology is the salvation of philosophical realism. The disputes among the disciples of Husserl over this question show this compatibility of phenomenology and realism is not so clear. See Herbert Spiegelberg, *The Phenomenonlogical Movement: A Historical Introduction*, 2 vols. (The Hague: Martinus Nijhoff, 1960), 225 f., on Roman Ingarden's refusal to follow his master's idealism.

29. Erich Neumann, *The Great Mother: An Analysis of Archetypes* (New York: Pantheon Books, 1955), Chapter 4: "The Central Symbolism of the Feminine", pp. 39-54, shows how the archetypes are rooted in the symbolism of the human body. See also Carl G. Jung, "Psychological Aspects of the Mother Archetype" in *The Archetypes of the Collective Unconscious, Collected Works*, vol. 9 (New York: Bollingen Series, Pantheon Books, 1959), pp. 75-112.

30. Gadamer's term for "fusion of horizons" is *Horizontverschmelzung, Truth and Method* German original *Warheit und Methode* (Tubingen: Mohr, 1965), pp. 286-290. For a discussion of the notion in current hermeneutics, see Anthony C. Thiselton, *The Two Horizons* (Grand Rapids, Mich., Eerdmans, 1980).

31. Herbert Waddams, *The Struggle for Christian Unity* (New York: Walker and Co. 1968) (Anglican); Methodos Fouyas, *Orthodox, Roman Catholicism and Anglicanism* (London: Oxford University Press, 1972) (Orthodox); Charles Bouyer, S.J., *Le mouvement oecuménique: les faits — le dialogue* (Rome: Presses l'Universite Gregorienne, 1976) (Roman Catholic) trace the history and significance of the ecumenical movement from different points of view.

See also Paul M. Minus, Jr. *The Catholic Rediscovery of Protestantism: A History of Roman Catholic Ecumenical Pioneering* (New York: Paulist Press, 1976).

32. The possibilities of ecumenical convergence are neatly illustrated in the three essays of Alexander Schmemann (Orthodox), Jean Danielou (Roman Catholic) and Heiko Oberman (Lutheran) in Elmer O'Brien, S.J., *The Convergence of Traditions* (New York: Herder and Herder, 1966), written at the conclusion of Vatican II. For the new problems this raises for theological methodology see David Tracy, *Blessed Rage for Order: The New Pluralism in Theology* (New York: Seabury Press, 1975).

33. Of course the term "Humanism" is used in many ways, including "Christian Humanism". Recently in the form "Secular Humanism" it has become a polemical term often favored by ultra-conservatives. As will become clear in Chapter 3, I use it to designate, without prejudice or pre-evaluation, the world-view and value-system which originated with the Enlightenment as an alternative to Christianity and the other world religions. The humanism of the Renaissance period was entirely a different matter. I note that Susan Budd, *Varieties of Unbelief: Atheists and Agnostics in English Society* 1850-1900 (London: Heinemann, 1977), uses the term in my sense.

34. Last line of *Candide* in Voltaire, *Romans et Contes*, Bibliotheque de la Pleiade (Bourges: Gallimard, 1954), p. 237.

Chapter 2

What Does Science Say We Are?

I: We Are Thinking Bodies

1. The Scientific Vision

The split between the Christian and Humanist world-views, between Humanism and Marxism, and within Humanism between the sciences and the humanities has been the result of numerous historical forces which I will consider in Chapters 3 to 6, but the most obvious factor has been the rise of modern science which has forced every theology or philosophy to assimilate, more or less consistently and creatively, an immense fund of new data. Christian theology has often tended to see science as an inimical power, closely allied to Humanism, and has either rejected it or attempted to neutralize it. Few Christian theologians have viewed it positively as an important element of God's revelation of himself to us.

Yet we who live in a technological culture built on science cannot seek self-understanding only through myth, philosophy, or mysticism. We make our living, seek our enjoyments, by using the chemicals, machines, transportation, and communication science provides. Young people may seek a counter-culture in the desert, but to pass the time they take along paperbacks of science-fiction. Who can forget what science has taught us about ourselves — we have evolved from star-dust?

The vision of science is not that of the oriental sages or of Plato who defined us as spiritual selves exiled in matter. It is rather that of Aristotle and Darwin who saw us as always actively related to our environing world. I can discover myself only by discovering my world, and that world, before it is a cultural world is a natural world. In the light of science what are we?[1]

2. We are Communicators

Sociology and anthropology, the sciences of our total human behavior, acknowledge that all of us existing humans belong to a single species, but hesitate to speak of "human nature." The biological fact that all humans have essentially the same bodily structure and easily interbreed is largely overshadowed by the vast variations in human behavior due to cultural differences. Even masculinity and femininity turn out to be largely cultural rather than genetic.

Human culture cannot be reduced to inborn patterns like the instinctual behavior of other animals. We can invent, transmit and accumulate our ways of living through institutions, of which even the family is itself a highly varied invention. This unique characteristic of the human species makes it possible for it to adapt to many different environments without losing its genetic identity. Insects achieve adaptation by evolving into many species each with its special innate instincts. We humans remain genetically one species while diversifying our behaviors through culture.[2]

Fundamental to a culture are the tools used to modify the environment. We are born with the remarkable tools we call "hands," but we also use other objects for tools, and most remarkably we *invent* tools.[3] Of all our invented tools the most effective are symbols, including our many languages. By using symbols, words, we can accumulate our experiences, sort them out, rearrange and adapt them so as to invent new behaviors. We can manipulate symbols more rapidly and more accurately than we can manipulate the things they stand for. Moreover, language makes it possible for us to share our experience and our inventions socially through communication.

Human societies based on language differ strikingly from other animal societies.[4] The "languages" of bees and monkeys, as far as we now know, contain only signs for things and actions and lack syntactical signs ("and," "therefore," "not") which stand for *relations between symbols*. It is our human capacity to deal with symbolic relations which gives us our power to invent and use symbols as tools, and through them to invent other kinds of tools.

We exist, therefore, not merely as members of an abstract class but as an historical community. To understand any item of the behavior of any human we must specify the individual in his or her social relations at a particular time of life in a particular society at a unique period of its history. The crucial problem of the social sciences is that scientists must transcend the limits of their own cultures to enter into the system of symbols, meanings, values of other cultures, classes, epochs than their own. They have a *hermeneutical* or *translation* problem.[5]

No translation can be perfect, but the fundamental fact is that every human language can be translated into any other. We humans are one species not only because we are biologically an interbreeding population, but because we belong culturally through space and time to an intercommunicating society.

In summary: *Humans are a population which survives because of its capacity to control its environment by inventing and transmitting diversified cultures. Fundamental to these cultures are invented tools, especially symbol-systems or languages by which, in a measure, individual and cultural isolation can be transcended through communication and translation.*

3. We are Self-Aware

Psychology, which studies individual human development without neglecting human social environment, increasingly recognizes the strange interplay between our self-awareness and self-concealment. On one level our unique human sociality, inventiveness, and power of symbolic expression mark each of us as a supremely self-conscious, self-controlled, wide-awake animal. Individual human development builds up an ego, a clear sense of autonomous self.

On another level psychology has uncovered a wider realm of unconscious, concealed activity in us which conditions and energizes our conscious behavior.[6] Experimental psychologists such as Pavlov and Skinner explain this in terms of stimulus-response conditioning. Depth psychologists, like Freud and Jung, explain it in terms of the specification, sublimation, and integration of primary libidinal and aggressive drives. Structural psychologists, such as Piaget, see it as a step-by-step actualization of organic potentialities.[7] Perhaps these are not contradictory, but complementary theories.

All agree that each conscious self, the ego and its activities, are rooted in a wider, deeper self formed by cultural development and, Jung would add, in our history and pre-history (the "collective unconscious").[8] Only when we are consciously in touch with our total selves can we have full use of our distinctively human, creative powers. When the ego and the unconscious are severely incongruent, we become neurotic or psychotic. We regress to more primitive, rigid, irrational ways of dealing with our environment. We have less ability to communicate with other persons, to adapt, invent, survive.

Thus, we humans as individuals achieve maturity and sociability only by developing the total self, conscious ego and wider unconscious self, in a suitable family and societal environment where we can achieve harmony between personal autonomy and inter-personal communications.

II: Our Bodies Serve Our Brains

1. We are Brains

The more aware we become that our behavior is only partially controlled by direct insight, the more we begin to explore the inner workings of our body, which is, as it were, our *opaque* self. Biology and its auxiliaries study this human body, trying to understand why it surpasses other animal bodies in its adaptive power.

The ancients thought the "heart of man," the core of our being, is the physical heart, because we feel it beating in fear, excitement, or tenderness.[9] Anatomists soon discovered this core is not the heart but the brain and central nervous system, although we do not *feel* our brains at all. My brain is my very ego, myself, observed objectively from without rather than introspectively. By this central organ I can receive information from my body and also from outside and I can integrate this data into a unified pattern in terms of which I can then give a bodily response.

To survive I must exactly control countless bodily activities. My responses to be fully human must be not only adaptive but also "intelligent," "creative," "social." They are not only knowing-responses, but feeling-responses. These drives, needs, emotions also are rooted not in my heart but in my brain. The invention and use of symbolic tools takes place in its convolutions.

The human brain is composed of perhaps twenty billion (20×10^9) nerve cells (neurons) intricately inter-linked.[10] The analogy of the brain to a central switchboard is helpful, but the brain does far more than connect peripheral stations. It integrates and modifies information so that I can respond to my situation not part-to-part but whole-to-whole. As an ego, self-aware, self-controlling, personally communicating myself to other selves, I must think what I am doing. I must reflect on my own actions and synthesize them in consistent patterns and processes. Though much of what I do lies below the level of fully conscious control, somehow I must act as an individual body-person distinct from, yet appropriately related to my social and non-human environment. My nervous system is assisted in this by the endocrine system intimately linked to it. Together they form the "coordination system" of my body.[11]

The central nervous system is a vast array of elaborately interconnected neurons, whose pathways are still very imperfectly known. It is clear, however, that the outswelling of the upper-end of the spinal chord, the brain-stem, is the body's most essential vital center. The lower part of the brain stem is necessary to coordinate vital or metabolic functions on which the brain itself depends for energy. If these centers are seriously injured, we die. The upper-end of the brain-stem is needed for conscious life, both thought and feeling, and when it is injured we suffer perceptual and emotional defects.

The great size of the human brain compared to that of other primates is due, however, not to the brain-stem but to the enormous expansion of the cerebrum and cerebral cortex which overlays and surrounds it. Much of the cerebral cortex can be removed without destroying the conscious ego, but it is necessry for us to perform highly controlled, refined activities. When it is injured responses may become approximate, mechanical, ill-adapted, less flexible and less creative.[12]

Probably evolving humanoids at first had a brain only a little bigger than other primates, yet sufficiently better integrated so that they had the beginnings of true language. This new capacity, even in rudimentary form, greatly enhanced human ability to survive.[13] These humans now could act socially in hunting, food-sharing, defense, and invention. Natural selection thus favored rapid expansion of the human brain, especially the cerebrum. Humans could then refinc their uses of intelligence and speech so as to depend less on strength and instinct, more on their ability to live socially through invented, transmitted, progressive culture. The human family, with its primary bond between parents and children (unique as regards the role of the father[14]) and pair bonding between the male and the female, developed and became the basis of wider social groupings.

Therefore, *our remarkable human capacities to think, create, and communicate arise from the elaborate development of our central nervous system which enables us to integrate information received from the environment and live socially in a culture shared with others.*

2. We are Homeostatic Systems

The coordinating system of nerves and endocrine glands is the chief, but not the only organ-system of our bodies. There also are (1) the skeletal-muscular system which supports the body, protects its parts, helps it to move; (2) the integumentary system which clothes it with protective yet sensitive skin; (3) the alimentary system which processes food and drink;

(4) the respiratory system which takes in oxygen and removes some gaseous wastes; (5) the circulatory system which carries oxygen, food and endocrine secretions throughout the whole body and removes wastes; (6) the execretory system which eliminates these wastes; (7) the reproductive system by which parents produce offspring similar to themselves yet each of which is individualized by a new combination of traits.[15]

All these systems are necessary for the brain to live and function. Without their help the brain dies in minutes. Yet they, too, need the brain to coordinate all their activities in relation to the environment.

The whole body forms a *homeostatic* or self-regulating system, with a dynamic stability in which a constant in-put of matter and energy is balanced by a constant out-put of activity and wastage by devices which adjust the flow to meet environmental variations.[16] The body of an adult probably retains not a single atom from childhood. Yet each of us remains the same individual retaining the memories of past experiences in minute detail. Each of us is a river, but unlike a stream that washes away all traces, we retain our accumulating pasts, the continuity of our unique personal histories. It is this tenacious memory that makes it possible for us to use language. Isaac Asimov estimates that in a lifetime the human brain can store 10^{15} "bits" of information.[17]

Although all living organisms are homeostatic systems, we humans are the climax of the evolutionary trend toward the *internalization* of the environment. Because our bodies provide so delicately regulated an internal environment, our central nervous systems and especially our brains can work in undisturbed peace, yet remain in constant communication with the external world. Archimedes said, "Give me a proper fulcrum and I can move the world!" The fulcrum or "still point" from which a modern Archimedes might control the world would be his peaceful yet attentive brain.

It is helpful to compare this homeostatic system of our bodies with modern *cybernetic* self-regulating electronic machines (robots, electric-brains, computors).[18] Already we are building machines to simulate many kinds of human behavior which seem adaptive or purposeful by using the "feed-back principle" which enables machines to respond not "mechanically" but flexibly, adjusting to variations in the environment. Eventually we may even produce machines that build themselves, simulating the power of organisms to repair themselves, grow, learn, even reproduce.

Probably we can duplicate living systems to the extent that we understand them. At present, however, we are very far from producing a homeostatic system as complexly integrated as even the simplest living cell, let alone the human brain integrated out of billions of cells. Indeed, every cell of my aging body contains the "genetic code" according to which I was built from a single cell. This code was composed by millions of years of evolutionary processes.

This evolution in producing man did not render other living things obsolete. Dolphins have brains as remarkable in some ways as man's.[19] Insects have an "intelligence" built on a very different model than our own. More important is the fact that we humans depend for the energy by which we live on other living things. With them we form an ecological system (ecosystem) which has its own kind of homeostasis. Astronauts can survive on the lifeless moon only in "life-support systems" that simulate our ecosystem.

The development of this ecosystem or biosphere of which we are a part has taken some three billion (3×10^9) years. It is now formed by some three million species of organisms, most of which are relatively new.[20] However, the main types or (phyla) were established early and exemplify the main options of body-plan and life-style. Three radiations of life each in its own turn occupied most of the ocean water-ways of the earth before the dry land became habitable.[21]

The first radiation produced tiny organisms (bacteria, algae) isolated or in loose colonies.[22] Some developed a central organ or *nucleus*. Some got their energy by fermentation, chemicosynthesis, or parasitism, but others more efficiently used solar-energy by photosynthesis, and some (protozoa) achieved greater mobility by feeding on the photosynthesizers.

The second radiation produced much larger, multicellular creatures (metazoa) with bodies differentiated into tissues, organs and organ-systems. They could locate food at a distance by *sensation* and were mobile to pursue it. These were true animals, fully developing the protozoan style of life. Some (sponges and parasite worms) still had little structure, but the radiata (jelly-fish, etc.) had beautiful, hollow, radially symmetrical bodies with a single opening for mouth and anus. They had genuine organs of sensation and the movements of their body-parts were correlated by a nerve network.

Since round animals are not well adapted for rapid movement, the third radiation produced bilaterally symmetrical animals with a true gut-tube opening in a mouth at one end and an anus at the other, which gave them an internal, physiological environment independent of the one in which they moved. To mature such an elaborate structure necessitated *larval* phases of growth. These well-developed animals branched into two great types: (1) proterostomes, which retain the larval mouth in the adult; (2) deuterostomes (we are deuterostomes) who perform an embryological flip-flop, so that the larval mouth becomes the adult anus, and a new mouth opens to mark the "head" of the animal.

The proterostomes, which include segemented worms and molluscs, culminate in the arthropods, especially in the insects, countless in species and wonderfully adapted to all environments by elaborate instincts. The worms have only a crude ladder-like nervous sytem, but the molluscs and arthropods have large masses of nerve cells (ganglions) associated with such excellent sense organs as the eyes of the squid and the bee. Yet the "head" of the bee is by no means so centralized an organ as that of the bird or man.

The proterostomes include starfish who have regressed to resemble the radiata, and also the chordates whose nervous system is integrated into a central chord. The simplest of these are not impressive, but the vertebrates attain a genuine central nervous system and true brain, the fundamental pattern of our human body.

All these main phyla of animals (possibly excepting the vertebrates) originated more than 500 million years ago when fossil records begin, probably at the point when competition for survival in the oceans required the protection of bone and shell.[23] About 360 million years ago life began to expand to dry land, but not until the first vertebrates, the fishes, filled the seas. First on land were arthropods, then vertebrate amphibians.

Probably at the same time true plants, without which a terrestrial ecosystem would have been impossible, originated. Living, like the algae, by photosynthesis, they had to remain immobile to expose large, leafy surfaces to sunlight for hours at a time, so they required neither sense organs nor a nervous system. The cells of a plant body do not communicate easily with others because of their cellulose walls which support and protect it in its immobility.[24]

Plants formed a series of ecosystems with animals as the earth and its climate slowly changed. The ancient epoch of landlife (Paleozoic) saw plants evolve from the simplest forms to mosses and then to great forests sheltering insects and amphibians. Then came conifer (gymnosperm) forests where reptiles dwelt. When the epoch ended 225 million years ago with the rise of the Appalachian and other mountains, terrestrial life was fully established with insects, reptiles and conifers completely adapted to land, independent of the mothering seas.

The Mesozoic epoch which lasted until 65 million years ago when the Rockies, Andes, Alps and Himalayas were built, saw the great dinosaurian reptiles flourish and decline. The seed-plants (angiosperms), including hard-wood trees, partly replaced the conifers, because seed-plants protect their seeds in "ovaries" and become dormant in winter. Thus, they can live in more varied climates than the conifers. Birds appeared in this age, and then the mammals.

26

While amphibians and reptiles cannot protect their nervous systems from great temperature changes, birds and mammals are warm blooded. Their highly internalized environment permits them to develop better nervous systems, acute senses, rapid, coordinated movements.[25] Birds, still reproducing by eggs, live by rigid instincts. Their wings exclude the development of fore-paws or hands to manipulate the environment. Mammals, however, retain their young for a long development, first in the womb, then in the breast. They can give the young time to form an intricate brain, and then to develop it by prolonged interaction with its mother. Their nervous system is freed from elaborate instinctual programming, and their behavior becomes more flexible and adaptive.[26]

We live in the epoch (the Cenozoic) since the last great mountain building, with a stable ecosystem, although during this time the land mass has divided into continents that have drifted apart.[27] It has been the time of great mammalian evolution. The first mammals still laid eggs. Then the marsupials bore young still very undeveloped. Finally the perfect mammals appeared who bear well-developed young. These branched into four great groups:

1. Carnivores (dogs, cats, etc.) modified to eat other animals, and herbivores (ranging from elephants to deer, cattle and horses) to eat plants.
2. Whales and porpoises modified to live in the sea.
3. Rabbits and rodents, small animals adapted to tight living-situations.
4. Tree-living creatures like shrews, bats, sloths, anteaters, but especially the primates or man-like creatures such as monkeys.

This last group has fore-limbs adapted to grasp branches and fruits. They tend, therefore, to erect posture, a head free to move about, eyes in front, so that front-paws and eyes can be coordinated. In a period of great glaciation two or three million years ago, a drier climate in Africa forced primates out of the trees to become the first hominoids, the australopithecenes. These small primates walked erect, used their hands, and invented tools, but it is not clear that this invention also implies the invention of true language.[28]

As the selective advantage provided by this better brain favored its rapid evolution, Java and Peking man (homo erectus) appeared, then modern man (homo sapiens) along with some side-branches (Neanderthal man) which probably resulted from periods of isolation produced by glaciations. Relative isolation also accounts for racial divisions of the human species which probably have existed throughout most of our history.[29]

By the end of the glaciations, between 50,000 and 10,000 years ago well-developed cultures, as evidenced by the cave paintings, etc., had been created by humans. By 8,000 years ago they had achieved remarkable control of their environment. By domestication of animals, cultivation of cereals and other foods, they were able to live in villages, then in cities where culture can accumulate and be transmitted through stable institutions and written records. Meanwhile, humans were changing their ecosystems, destroying great herds of mammals, clearing forests, diverting streams.

Our creative human brain is served by excellent sense organs and by a mobile body with dexterous hands and nimble tongue, so we are able to receive information, communicate socially, and manage our external environment. Such a brain is possible only because the whole human body is constructed on a plan, selected from many models (multicellular vs. acellular, plant vs. animal, radially vs. bilaterally symmetrical, deuterostomic vs. proterostomic, vertebrate vs. invertebrate, mammal vs. reptile or bird, primate vs. carnivore, whale or rodent), *so that it provides a finely regulated internal environment well-supplied with energy derived from our ecosystem.*

3. We are Genetic Codes

The evolutionary construction of our homeostatic, brain-controlled body can be traced to crucial events within the smallest life-unit, the cell, some 75×10^{12} of which compose the adult.[30] As our body grows from a single cell by some 45 series of cell-divisions, there is a differentiation into diverse types of cells to form bone, muscle, nerves, skin, and blood each in its proper quantity, shape and position. Each newly differentiated group of cells influences differentiation of still newer types.[31] While this goes on, the basic vital functions of the whole organism must be maintained. The business of living cannot wait till the final structure is complete.

This embryological process is regulated by a *genetic code* contained in minute living threads called chromosomes within the nucleus of the fertilized cell, and every subsequent cell descended from it.[32] Chromosomes in the fertilized cell are paired, one in each pair provided by the male, and one by the female parent. Each contains a definite series of genetic units or genes. Because each member of a pair has a different evolutionary history, one may have a dominant influence, so that the actual organism produced (the *phenotype*) does not fully manifest all the genes (the *genotype*) it may transmit to its descendants. In any population of interbreeding individuals, there exists a gene pool or fund of possible traits which may appear in some combination in an actual individual.

Each individual (except perhaps identical twins) is a unique selection of such traits.

Genes are relatively stable chemical entities duplicated from generation to generation. In about every 50,000 to one million duplications a gene is affected by an influx of high-energy radiation or other accident occurring during duplication or assemblage in the chromosome.[33] Such accidental modifications are called *mutations,* and they introduce new genes into the gene pool, removing older ones. Thus, occasionally individuals emerge who not only combine old traits but new ones to transmit to their descendants.

Therefore, in the course of many generations any interbreeding population, or *species,* consists of a very considerable range of types. At some stages of their histories living species exist only as quite small populations. Then a "sampling error" may result in "genetic drift," i.e., certain traits will become common, others will diminish or be lost. Hence, no species remains completely stable, but gradually shifts its average characteristics.

Species-change, however, is not merely a chance affair.[34] Organisms react with their environment in the struggle to adapt, survive and reproduce. Only those that survive to reproduce effect the traits of the next generation. Traits which in a certain environment give even some slight advantage in this contest tend more and more to dominate the species. This process is *natural selection* which slowly alters a given species, adapting it to its environment, eliminating disadvantageous traits, universalizing the advantageous. "Advantage" is purely relative to a given environment. If the environment changes, the effect of natural selection also changes, rendering what was fit for survival unfit, and favoring former under-dogs. Environmental changes open new avenues to life, and living things in their dynamism seize their opportunities.[35]

Natural selection acting on the variations in the genetic code produced by mutations, genetic drift, and other factors slowly produces an evolution of new types of life. If, however, these new types continued to interbreed, the result would be merely a widening range of types, but environment includes factors tending to divide and isolate portions of the population into pockets, each of which develops in its own direction under different influences. Ultimately these isolated groups may be so different they can no longer interbreed effectively, even if the environmental barriers which isolated them are removed.

Thus, at any moment of history the kingdom of living things is divided into distinct species, groups that cannot interbreed but which form an ecosystem adapted to the range of environments. They are a community that is not only competitive but complementary.

*We humans, marvelous as we are, are only one such species pro-
duced through natural selection of chance genetic mutations through
interaction with a changing environment.*

III: We Are Matter-Energy

1. We are Matter

Whence the genetic code? Life is based on interaction between two
kinds of substances which compose any organism. One consists of relatively simple materials built up of carbon, hydrogen, nitrogen, oxygen,
phosphorus, and small traces of many other elements. The other substance,
built of the same elements but much more highly organized, is DNA (with
its related RNA).[36] It is a macromolecule made up of two strands, each
a polymer or long string of similar but not identical parts. The variations
of these parts form the genetic code, in a sequence broadly characteristic
of each species, but unique to each individual organism. This two-stranded
structure is the basis of all living reproduction, since, when the strands
separate, each of the pair can produce an identical partner.

DNA alone could not constitute a living organism. It must react in
a long sequential process with the simpler living materials to produce proteins of various kinds, each determined by the genetic code. These proteins are enzymes which act as chemical catalysts to speed up chemical
reactions. Without such a speed-up complex molecules would never be
constructed, but would be obliterated by simpler chemical reactions. By
aid of such enzymes the constructive process goes on and culminates in
the production of DNA itself, so that the organism can reproduce. This
cycle of processes catalyzed by enzymes is *metabolism* or "life" itself,
the controlled homeostasis of a living system.

Metabolism produces three types of organic substances:[37] (1) *proteins*
(including the nucleic acids of which DNA and RNA are examples) or
catalysts regulating living processes; (2) *carbohydrates* which store energy
in readily available form; (3) *fats* which store reserve energy in less space
than carbohydrates. These basic materials are built up into macromolecules
suitable to construct filaments, membranes, granules, liquid or viscous
colloids which can be obtained to form tissues and organs.

Metabolism also releases controlled energy needed for nutrition, synthesis, growth, reproduction, movement — the basic life activities, all
rooted in the simplest unit of the organism, its cells. Each cell is a minute
drop of plasma (the simpler living materials), which contains many structures or organelles of which the most important are the filaments of DNA
and the *ribosomes* where the protein enzymes are constructed under the

guidance of DNA and the mediation of "messenger" RNA. In most organisms the DNA and ribosomes are centralized in a nucleus.

The origin of the first living things on earth is still little understood, although we can experimentally produce simple proteins and nucleic acids from non-living matter. We know, too, that some organic compounds exist in outer space, and could antecede the earth itself.[38] It seems plausible that in the earth's early stages such materials collected in small pockets along seashores. The high energy supplied by solar ultra-violet radiation and by electric storms may have built up out of these materials rudimentary plasmas and DNA-like proteins. Separated from the chemical "soup" by membranes, formed also chemically, these droplets were homeostatic systems.

Once formed such units were subject to natural selection. Droplets able to maintain homeostasis persisted and took up available materials, until the less stable droplets starved and perished. Gradually a definite pattern of controlled chemical processes was established so that a genetic code was transmitted as each unit grew and subdivided. This was then subject to mutation, and biological evolution was underway.

Thus, the problem of the origin of life is a problem about the establishment of a controlled use of energy. *We humans are the most complexly integrated systems of energy-transformations in our solar system.*

2. We are Energy

The world of living and non-living is matter in motion, that is, matter and energy. Energy manifested in motion is called *kinetic.* Manifested statically as *inertial mass* or matter, it is resistant to a change of state of motion and rest. Matter and energy are mutually convertible so that we observe "matter" only as it is manifested in some energy-transforming event. Sometimes kinetic energy is decreased with a proportionate increase of mass, or vice versa, in accordance with Einstein's formula $E = mc^2$. Since c (the velocity of light in a vacuum) is 186,000 miles per second, a very small mass yields enormous kinetic energy. This velocity of light seems the natural measure of all motion, beyond which no entity can be accelerated.[39]

Energy undergoes constant transformation which at present we can reduce to four fundamental forces:[40]

1. A strong nuclear force.
2. An electromagnetic force, only about 1/137th as strong as the first.
3. A weak nuclear force, only about 10^{-13} as strong as the first.
4. A gravitational force, only about 10^{-38} as strong as the first.

31

Strong and weak nuclear forces have very short range, about 10^{-13} cm. Electromagnetism and gravitation extend indefinitely, decreasing as the square of the distance. Since electromagnetism has a positive and negative form which tend to neutralize each other in a limited region, only gravitation is significant for cosmic distances.

These forces are not disembodied, but are best conceived as "exchange forces," transformations of energy in a spatial field. They can also be conceived as particles definitely located in this field and moving through it with definite velocities, i.e., with momentum. Some particles are "matter" since they have inertial mass. Others have no mass, but still have "momentum" (defined in this case as E/c) and can be converted into matter.

During the period 1930-1960 after the invention of the cyclotrons by which sub-atomic particles could be accelerated to great speeds and used to probe the inner structure of the atoms, in place of the positively charged protons and the negatively charged electrons which physicists had for some time supposed were the ultimate building-blocks of chemical substances, a bewildering multitude of "elementary" particles were discovered. In the last twenty years, however, it has become possible to classify these particles and to begin to understand their relations to the four fundamental types of interaction just mentioned. [41]

First are the heavy-weights or *baryons* to which the *protons* belong, as well as the neutrons which are also found in atomic nuclei.

Second are the middle-weights or *mesons,* some of which also exist in the atomic nuclei not in the relatively stable manner of the proton and neutron, but as "virtual particles", i.e. as a process of energy exchange.

The baryons and mesons together form the class of *hadrons* (from Greek for "strong") because they are subject to those interactions which constitute the strong nuclear force. It is now thought that the thousands of hadrons which can exist outside the atom for brief, often very brief periods, can all be elegantly explained by supposing that they are composed of *quarks* (a name derived humorously from a phrase in James Joyce's *Finnegan's Wake*) of which there are at least five and probably six types. The strong interaction so closely binds these quarks together within a hadron that they never appear as independent particles.

Some physicists believe that quarks are the "rock-bottom" constituents of matter, but we must also allow for a third class of quanta, the light-weights or *leptons,* which include especially the *electron* or negatively charged particle which in the atom orbits around the nucleus, and the photon which is altogether massless. Electromagnetic radiation or light is a stream of such photons outside the atom, but within the atom they exist as "virtual particles" by the interchange of which the electromagnetic force is constituted. The most surprising of the leptons is the *neutrino,* which is without electromagnetic charge and is either massless or of extremely low mass, and which, once emitted from an atom, travels through matter with only very rare interactions and thus fills the entire cosmos in countless numbers. If in fact they have some small mass, then neutrinos probably constitute 90% of the mass of the universe.

Recently, this classifactory scheme has been completed by the concept of gluons (binding particles) of four types which are ultimate bearers of the four fundamental forces. The *"colored" gluons* (the term is purely arbitrary) bind the quarks together in the hadrons and thus correspond to the strong nuclear force. The *photons* constitute the electromagnetic force binding the orbital electrons to the atomic nucleus. The *weakons* account for the weak nuclear interactions and also for that spontaneous decay of nuclei which we know as radioactivity and which results in the transformation of the chemical "elements" into simpler elements. It has been discovered, also, that it is possible to unify in a single "field theory" the electromagnetic force with the weak nuclear force, and possibly with the strong nuclear force. The eventual goal of such a unified field theory is to include the gravitational force as well. It is probable that at very high energies, such as existed at the very beginning of the cosmic expansion (which I will describe later,) all four forces converged in a single type of interaction which differentiated only at lower energies. Finally it must be noted that each of the types of particles comes in pairs, e.g. the electron is paired by a positron, which constitute *anti-matter.* When a particle collides with its anti-particle they "annihilate" each other with the production of new hadrons and electrons.

Most hadrons, because of the weak force, are unstable, existing only briefly before spontaneously "decaying" into smaller particles and photons. The electron, photon, proton, and neutrino are relatively stable, but they can be transformed into other particles. Even the proton, the most stable of all, has recently been shown to be subject to decay in something like 10^{31} years. The relative stablity of the proton and electron, however, makes possible the existence of atoms as permanent (but not everlasting) entities. It may be that some of the actual atoms of hydrogen and helium in the atmosphere around us have existed as individuals since the first hours of our cosmos.

In the simplest kind of atom, hydrogen, a single proton acts as a nucleus around which moves one electron in such an "orbit" that its momentum prevents it from falling into the nucleus under the electromagnetic force which holds them together.[42] The hydrogen nucleus, however, can also contain, along with a proton, one or two neutrons that have no electromagnetic charge. Such atoms with the same nuclear charge and number of electrons but with nuclei of different masses, are called *isotopes* and differ little in chemical properties, since such properties depend primarily on the number of orbital electrons.

When sufficient energy is available the nuclear forces tend to build up more and more complex nuclei, and these with their corresponding orbital electrons, constitute some 100 "elements", each having several isotopic forms. Some of these kinds of atoms, due to the weak interactions, are radioactive, spontaneously decaying at various average rates. Others are stable. When atoms collide with other atoms of the same or different kinds they may form *chemical bonds,* either transiently as in gases or liquids, or more permanently in crystalline solids. [42] These bonds may be so loose that the atoms remain distinct units, or so close that they are restructured into a highly unified molecule. In crystals these molecules may also remain loosely bound, or may unite to form a macromolecule. The construction of all chemical substances requires only the same four fundamental forces found within the atom, although at this level of organization it is the electromagnetic force which predominates.

We can actually photograph in a "cloud-chamber" the event in which an electron and its anti-particle collide, annihilate each other, and produce a photon of light-energy. This event occurs at a definite point in space and time. Yet the photon as it moves off can no longer be exactly located. We can only calculate the probability it will enter into some other observable event in a rather large volume of space in a given stretch of time. The photon acts more like a wave spreading in space than like a projected particle.

Not only massless photons, but all particles exhibit this wave-aspect in some measure. When particles travel in large numbers as a beam, they exhibit wave phenomena such as diffraction and interference. We cannot accurately predict the path of any individual particle, but only give statistical laws of average behavior. Hence, the ultimate natural laws governing the fundamental forces of nature are statistical not absolute. They obey a *principle of indeterminacy.* [43].

The relation of particle and wave aspects of energy-transforming events is expressed by De Broglie's formula $\lambda = h/p$, which states that the wave-length associated with a moving particle is inversely proportional to its momentum according to a constant h. This leads to Heisenberg's *principle of uncertainty.* If we imagine or arrange experimentally any situation in which we can locate an individual particle, then the more exact the determination of that location, the wider the range of its possible moments. Conversely, the more exactly we determine the momentum, the wider the range of its possible locations.

This uncertainty is negligible in predicting the behavior of the large bodies of ordinary experience, but for sub-atomic particles it is so great that it becomes impossible to apply unambiguously to them the standard concepts used in describing macroscopic bodies. It seems that this principle of uncertainty does not merely express the imperfection of our knowledge or observational techniques but indicates that matter and energy at the subatomic level have structures that are only analogous to those of macroscopic bodies. Nevertheless, it is possible for us to come to some understanding of them by the use of analogous "models" derived from the macroscopic level, such as the models of "particles" and of "waves" provided that we do not take them too literally. Thus sub-atomic particles are not like small billiard balls nor are they spread out in space like waves, but are energy-transforming states in a spatio-temporal field which under certain circumstances of observation behave like particles and in others like waves and are knowable by us only through these different modes of observation. Moreover, the behavior of such sub-atomic entities can be predicted only by *statistical* laws which give us the probability that this or that transformation of energy will take place at a given time in a given place.

Thus, we are brought to realize that *we humans are systems of macromolecules composed of atoms, and these of protons, electrons and other particles, all the result of four fundamental and inter-related forces which govern energy transformations according to statistical laws.* Such laws are open to surprises!

IV: We Have Evolved And Will Perish In The Cosmic Epoch

1. We Have a Cosmic History

The element of surprise or novelty implied by the statistical character of natural law in turn implies that time, unlike space, has a direction, an *anisotropy* or lack of symmetry.[44] We cannot go backwards into past time, but only forward, because a little of the unpredictable inevitably enters every event. The universe must move in one time direction; it has a history. Of course, it is possible our universe never "began," but always existed in much the same state as now. This "steady-state hypothesis" is still favored by some scientists, but the weight of evidence now favors the view that our universe began about 19 billion years ago with a "Big Bang," something like a hydrogen bomb explosion.[45]

At the moment of the Bang the fireball consisted of matter and anti-matter in equal proportions with a vast excess of photons and at a temperature far exceeding 10^{11} C. At 1/10 second of history the temperature had fallen to 10^{11} C. and there were already a large number of electrons and neutrinos with their anti-particles and a much smaller number of protons. At 14 seconds the temperature was down to 3×10^9 and the percentage of protons was increasing. At 3.5 minutes the temperature had fallen to 10^9 C. and the hydrogen and helium nuclei of the first atoms were beginning to form in the proportion of 74% to 26% as they still are. The other elements came into existence much later. For some 100,000 years of further cooling down to 3000 C. the hydrogen and helium atoms were forming. The heavier elements were synthesized much later, probably in the stars or in their explosion as supernovae.

At first the more massive particles were so agitated by jostling photons they could not collect; but as the universe expanded, cooled, became relatively empty, gravitation began to bring them together to crystallize into fine dust, most of which still remains scattered through space. In these vast clouds of gas and dust, turbulence produced denser patches, attracting still more atoms, gradually snow-balling. A condensation could survive, however, only if it were as large as 10^{40} grams in mass. In time only island condensations remained in a great thin sea of space, although photons and gravitons along with the stray particles of very high velocity we call "cosmic rays," still voyaged from island to island.

The chaotic islands continued to move apart, but within each island gravitational contraction replaced the explosive expansion. This happened about 30 million years ago when the islands were already at average distances about 100 times their average diameters. The universe then had cooled to "room temperature," was about 10^{24} grams per cubic centimeter in density, and was almost totally dark.

The islands, within the survival-range already mentioned, were of all different sizes.[46] Some were spinning, hence pulled out in ellipsoidal or flattened forms. Others had little spin and hence were spherical in form. All were composed of cool, dark gases, which streaming along with each island's rotation, broke up into turbulence in a hierarchy of big eddies containing smaller and smaller eddies. Gravitation condensed these further till they began to glow hot through the mutual bombardment of particles in the confined space, to emit light, and finally to initiate chain-reactions of nuclear fission and fusion as in a hydrogen bomb, reminiscent of the original "Big Bang."

Throughout each island stars began to shine, until in the order of 100 million years (1×10^7) each became a vast collection of rotating dust and gas filled with innumerable stars, which we call a *galaxy*, such as the average one we live in with its 100 billion (1×10^{11}) stars spread out in a disc 100,000 light years in diameter.

The stars we now see in the night sky are at different stages of development. A star is at first diffused, of low temperature and red color. By contracting it becomes hotter, more brilliant, and blue-white. At its center the gases are so packed and heated that the atoms are stripped again to nuclei and free electrons as in the primordial state of the universe. Thermonuclear reactions begin again, maintaining a constant output of energy as in a vast, controlled atomic reactor, until most of the hydrogen is exhausted. Then the star contracts further, rapidly loses its energy, and becomes a "white dwarf," perhaps no bigger than our earth, ending in a cold, burned-out mass of dust.

In its decline a star may undergo violent phases. In our galaxy and others we observe unstable stars in pulsing regular rhythms, or racked by explosions which eject huge quantities of matter. Some (novas or supernovas) actually blow up. Moreover, there are sources of energy (quasars, pulsars) whose exact nature and even location are still a puzzle. It is even suspected that some stars suffer total gravitational collapse and become "black-holes" or energy-sinks.

Our galaxy is between 1 and 2 x 10^{10} years old, our sun about 4.6 billion years old. The sun is largely composed of hydrogen, being rapidly converted at its heart into helium, but it contains small amounts of almost all the other heavier elements. This probably means it is a second or third generation star composed of materials previously forged in interiors of other stars born with the galaxy, but since exploded. Our sun is an average star in mass and brilliance, consuming itself at a rather slow, stable rate, yet troubled by cycles of change (sun-spot cycle). It may continue this way for another 5 x 10^9 years and then begin to expand and engulf some of its planets, as it passes through various unstable states on its way to death.

Probably around each star in its prime there circulates considerable material held by gravity, yet not pulled into its furnace, perhaps replenished by ejaculations from the star. Perhaps in half the stars this matter forms a second star (binary) or several (multiple stars) sufficiently massive to be self-luminous. But for others, like our sun, this material (less than 1 per cent in our case) is not sufficient to form a binary system. It seems likely (though not yet proved) that all single stars have a planetary array like our sun.

Much of the material about our sun was originally diffuse hydrogen gas. Along with small quantities of other heavier elements it condensed into small crystals which, as they orbited the sun, collided and collected.[47] By a kind of natural selection this constant hail of small bodies produced the inner planets (Mercury, Venus, Earth, Mars) composed of rocks and metals. Beyond Mars the planetoids are perhaps the remains of this process. The inner planets lost their hydrogen, but Earth and Venus developed considerable atmospheres. The outer planets (Jupiter, Saturn, Uranus, Neptune, Pluto) remained largely gaseous. Some planets captured moons, while the debris remains in the form of comets, meteors, and clouds of dust.

The earth, when the sun was itself just beginning to heat up and glow more brightly, accumulated layer by layer from cold materials.[48] Under the impact of meteors, from the energy of its own radioactive elements, and through the immense pressures of gravitational contraction it warmed up. It developed a molten core of iron and nickel, and over this a mantle of porous rock and a lighter crust. Through the porous mantle lava escaped to remake this crust repeatedly, while gases escaped to form our complex atmosphere from which water rained down to form oceans. As the sun grew hotter it supplied the earth with the constant flux of energy that produced our weather and seasons.

Our earth is not perfectly stable in its inner structure or even the orientation of its poles in space. It has suffered great cycles of geological formation in which mountains were built, then eroded as seas advanced to deposit sediments. Then these troughs of sedimentation, structurally weaker than the older rocks, were again folded up into mountains. The present continents broke off and drifted away from one great landmass. The sun's energy probably varied, and there were cycles of climactic change. All these changes, most very gradual, contributed to develop the complex balance of forces necessary to successive ecosystems.

The more we know about the other planets in our solar system, the more unique this delicate balance on earth appears. However, Harlow Shapely "conservatively" estimated that in our universe there are now at least 10^{17} planetary systems. Of these 10^{14} have a planet of the right temperature; 10^{11} of these are of the right size, and 10^8 at a stage of development ripe for the origin of life. Of these, he guessed, 10^5 have developed human life, and perhaps civilization surpassing ours.[49] Such estimates are guesswork, but to some scientists they suggest that the same forces are at work through the whole universe, so that our earth is cosmically multiplied.

We may conclude that *we humans could not exist unless the delicately balanced environment of the earth had been produced by a very long cosmic history, but this historical process possibly has and will continue to produce many regions similar to ours.*

2. We are Upstream Swimmers

In tracing the transformation of energy from the atom to the marvel which we humans are because of our super-marvelous brains, we should not distort the scientific world picture by thinking that all energy-transformations tend to greater and greater order. A basic scientific principle governing all energy-transformations is the Second Law of Thermodynamics which states that in any closed system the all-over trend of energy transformation is toward maximum disorder or *entropy*.[50] In every energy-transforming event, there is always an increase in the percentage of the total energy which is in the form of random, disorderly motion ("heat") no longer useful to produce orderly processes or structures.

Evolutionary trends described in this chapter are only upstream eddies in a more universal current of energy-transformations running downhill. Evolution is parasitic on remains of higher forms of energy which were at their maximum in the Big-Bang, but which are rapidly dissipating in expanding space. We know of no process that will "put Humpty-Dumpty together again," so as to return the universe as a *whole* to its original high-energy state.[51]

In the Big-Bang energy was so concentrated and chaotic that no organized structures or processes were possible. Every trace of order or pattern that appeared was promptly blasted by colliding particles. As matter and energy spread out through space, regions and epochs appear in which there is enough free energy to form complex structures, ranging from atoms to men, and yet empty enough to give the peace and time needed for gradual evolution. The time is inevitably approaching, however, when in each solar-system, each galaxy, and the whole universe, energy will have so dissipated and the level of entropy so increased that every structure and process will again be dissolved.

Freeman J. Dyson points out this breathing space for evolution results from a precarious balance, from certain "hang-ups" which hold back the entropic process from proceeding too rapidly:[52]

1. *Size* hang-up: the Big-Bang scattered the primordial matter sufficiently to slow down gravitational collapse back into the original state.
2. *Spin* hang-up: the rotations of galaxies, solar-systems, and stars slow down their gravitational collapse.
3. *Thermonuclear* hang-up: burning-out of the stars is offset by the predominance of hydrogen-fuel for thermonuclear reaction, and this proceeds slowly because it goes by *weak-interactions*. If the stars contained more heavy hydrogen which would burn by strong-interactions (as in a hydrogen bomb) it would be too rapid.
4. *Opacity* hang-up: the cooler surface gases in stars and the cold mantle of the earth prevent the rapid dissipation of heat from the hotter cores of these bodies.
5. *Nuclear surface-tension* hang-up: the surface tension of the nuclei of radioactive elements slows down their spontaneous loss of energy.

These are only some of the counter-forces in the universe which slow down its entropic decline. They no way contradict the Second Law of Thermodynamics which does not state the *rate* at which the universe will run down, but only that it must eventually do so.[53]

What is true for the universe, is true for the biosphere on our earth.[54] Thus, the atmosphere holds back the energy of the sun from destroying life, the ecological balance makes possible the slow evolution of species, etc. Nevertheless, life also gradually poisons itself with its own waste products and entropically down-grades the environment on which it feeds. The ecological problem produced by modern technology is only a speeding up of the process by which life eventually destroys itself by entropic pollution.

Thus we can conclude that *in a universe where the ultimate tendency seems to be toward maximum entropy, the evolutionary trends toward order, including our own human history and our progresive mastering of our environment through scientific intelligence, are only local and temporary, but the date of the entropic doomsday which must finally overcome our human existence and the ordered universe out of which it has arisen is not set by any law yet know to us.*

V: A Scientific Definition Of The Human Person

From this summary of the current scientific understanding of how we humans have come to be, what makes us distinct from other objects in the universe, and what is likely to become of us, it is apparent that the old definition "man is a rational animal" has not been falsified, provided we understand that definition in such a way that our "rationality" is seen as profoundly conditioned by our "animality." It is necessary, moreover, to expand the terms "animal" and "rational" so that they include all that rich detail that scientific investigation has so far uncovered, and leave them open indefinitely to further enrichment.

For the present purposes, in the light of the scientific understanding so far achieved we can formulate, in broad outline, the following definition of what it is to be human:

We human persons are creative, communicating, socially intelligent animals, motivated by unconscious and conscious emotions and purposes, capable of achieving scientific knowledge through which we can control our own behavior, our environment and our own evolution. At any moment of our historical development we are limited by our past, by barriers that divide our human community, yet we continue to strive to transcend these limits and so to build a novel future.

We think and feel with our brains and nervous systems in bodies which supply our brains with the energy to function and with information about ourselves and the world around us. Our bodies execute the orders by which we explore, control, and modify our world. We have evolved by natural selection from our environment and can know and control it only by interacting with it.

This interaction and evolution take place by a sequence of energy transformations involving the fundamental forces of gravitation, electromagnetism, strong and weak nuclear forces, transmited by elementary wave-particles which sometimes exist stably in the highly integrated forms of atoms, molecules, organisms. These in turn exist in larger but less integrated systems such as the earth's ecosystem, the solar system, the galaxy, and the expanding universe.

At some future date, which we cannot securely predict, either this human race will destroy itself, or it will be destroyed by changes in the solar system. Even if, in the manner of science fiction, we so master space as to be able to escape the demise of our earth and journey beyond the solar system, at some further date the universe as a whole will die and the present mode of existence of intelligent beings come to an end in the dispersal of all matter and energy.

Notes

1. I take it for granted that this is a question worth asking, unlike Bertrand Russell who wrote, "I still think that man is cosmically unimportant, and that a Being, if there were one, who could view the universe impartially, without the bias of *here* and *now*, would hardly mention man, except perhaps in a footnote at the end of the volume" *My Philosophical Development* (London: Allen and Unwin 1959), p. 213.

2. "There is no fossil evidence of existence at any one time of more than a single human or human-like species, except, possibly in the case of Australopithecinae . . . Mankind preserved its specific unity through its evolutionary development during Pleistocene times, although it always was, and still is, subdivided into races. Human evolution never led to differentiation of a single species into a group of derived species, some of which might have become lost and others survived." Theodosius Dobzhansky, *Evolution, Genetics, and Man* (New York: John Wiley and Sons, 1955), p. 333. Carleton Coon, *The Origin of Races* (New York: Alfred A. Knopf, 1962), attempted to refute this but was refuted in turn by Ashley Montagu, *Man's Most Dangerous Myth: The Fallacy of Race* (New York: Columbia University Press, 1945). See also L. Lieberman, "The Debate Over Race" in Ashley Montagu, ed., *Race and IQ* (New York: Oxford University Press, 1975), pp. 18-41, and G. Ledyard Stebbins, *Darwin to DNA, Molecules to Humanity* (San Francisco: W. H. Freeman, 1982) (a well-reviewed synthesis of current scientific thinking on human evolution to which I will refer frequently in these notes) who points out that, although the human species differs from the chimpanzee genetically by only about 2.5%, this amounts to some 2,500 gene pairs (p. 320) and the biochemical differences between the modern human races are superficial (p. 345).

3. See Theodore Dobzhansky, *The Biology of Ultimate Concern* (New York: New American Library, 1967), pp. 54 ff., on the significance of tools and language in human evolution and G. Ledyard Stebbins, *Darwin to DNA*, pp. 321-330. On tool-invention by chimpanzees see Jane van Lawick-Goodall, *In the Shadow of Man* (New York: Houghton Mifflin, 1971). Sheo Dan Singh, "Urban Monkeys," *Scientific American*, 221, 1 (July, 1969): 108-115, has studied the question as to whether urban monkeys can develop a culture and concludes: "We have observed that the urban way of life causes monkeys to change their feeding and sleeping habits, alters their behavior toward one another, increases their aggressiveness, makes them highly responsive and manipulative in their approach to novel or complex features of the environment and in general enhances their psychological complexity, but it does not advance their intelligence, although their behavior may appear to exhibit a high degree of shrewdness," p. 114. Of course the urban environment itself was made by men, not by

monkeys. V. Reynolds, *The Apes* (New York: E. P. Dutton, 1967), referring to the observations of Jane Goodal, writes that chimpanzees share with man "the ability to use sticks and stones as tools and even as weapons, to kill and eat animals and even share the meat, to dance, to drum, to roister, and to celebrate abundance of good food with comunity gathering and sexual activity" (p. 112). They make twigs into tools, use leaves for toilet, use leaves as a cup, build nests (p. 123), but they have no speech (219 ff.) and seem to exhibit no social cooperation in problem solving (p. 215); yet they do solve problems that require foresight of the consequences of their actions (p. 197 f.).

4. G. Ledyard Stebbins, *Darwin To DNA,* compares man and the apes on the basis of reproductive strategy, aggressiveness, social structure and communication, and on the last of these writes, "With respect to language . . . scientific opinion has undergone a revolutionary change during the past decade. The old belief that symbolic language is a peculiarily human trait no longer commands credibility" (p. 326). Stebbins recognizes "three distinctly novel human characteristics — artisanship, conscious time binding, and imaginal thinking" (*ibid.*). The evidence, however, which he cites for this "revolutionary change" of opinion has been very severely criticized. See Winthrop P. Kellogg, "Communication and Language in Home-Raised Chimpanzee," *Science*, 162, 5 (Oct., 1968): 423-427; David Premack, "Language in the Chimpanzee?", *ibid.*, 172, 21 (May, 1971): 80-822; and especially Thomas A. Sebeok and Robert Rosenthal, *The Clever Hans Phenomenon: Communication with Horses, Whales, Apes and People* (New York: New York Academy of Sciences, 1981).

Victoria Fromlan and Robert Rodman, *An Introduction to Language*, 2nd ed. (New York: Holt-Rinehart and Winston, 1978), Chapter 3, "Animal 'Languages'", p. 39-53, sum up as follows: "If language is defined merely as a system of communications, then language is not unique to humans. There are, however, certain characteristics of human language which are not found in the communication systems of any other species. A basic property of human language is its creative aspect — a speaker's ability to string together *discrete units* to form an *infinite* set of "well-formed" novel sentences. Also, children need not be taught language in any controlled way; they require only linguistic input to enable them to form their own grammar." (p. 51). "Birds, bees, crabs, wolves, dolphins, and most other animals communicate in some way. Limited information is imparted, and emotions such as fear, and warnings, are emitted. But the communication systems are fixed and limited. This is not so of human language. Experiments to teach animals more complicated language systems have a history of failure. Recently, however, some primates have demonstrated an ability to learn more complex systems. It is possible that the higher primates have the limited ability to be taught *some* more complex rules. To date, however, language still seems to be unique to humans" (p. 50). "No animal language or communication system has developed that is remotely as complex as human language, even with the intervention of human teachers. If other species have the ability equivalent to the human language ability, one wonders why it has never been put to use. It thus seems that the kind of language learned and used by humans remains unique to the species." (p. 51). But if Sebeok and Rosenthal are right, these primates are not really communicating in the languages which they have been so painfully taught but only responding to cues unconsciously supplied by their human teachers. Eugene Linden, "Endangered Chimps in the Lab", *New York Times Magazine*, May, Dec. 19, 1982, p. 77 ff., shows some of the ethical dilemmas that have resulted from this controversy.

5. The universal translatability of human languages has raised for Noam Chomsky (*Reflections on Language*, New York Pantheon, 1975) and others the old question about innate "structures" underlying all languages. See Jean Piaget, *Structuralism* (New York: Basic Books, 1970), pp. 74 ff. Piaget opposes both Chomsky's innatism and the positivism of other linguists and prefers the middle hypothesis according to which the child "constructs" its language by using its innate capacities to interact with its cultural environment.

6. The fullest account of the development of the understanding of our "wider self" including the unconscious psyche is to be found in Henri F. Ellenberger, *The Discovery of the Unconscious* (New York: Basic Books, 1970). The influence of the Romantic Idealists and of Nietzsche's critique of bourgeois morality on the discovery of the unconscious was of prime importance. See especially, Chapter 4, "The Background of Dynamic Psychiatry",

pp. 254 ff. See also Lancelot Law Whyte, *The Unconscious Before Freud* (New York: Basic Books, 1960).

7. Jean Piaget, *Structuralism* (note 5 above).

8. For Jung's mature thought on this question see his "Approaching the Unconscious" in a collection edited by him, *Man and His Symbols* (New York: Dell Paperbacks, 1964), pp. 1-94, and the essay by M. L. Franz, "Science and the Unconscious", *ibid.* pp. 575-588. For an interesting comparison, in favor of Freud, see Walter A. Kauffman, *Discovering the Mind* (New York: McGraw-Hill, 1980), 3 vols., vol. 3, *Freud Versus Adler and Jung.* In the first two volumes Kauffman traces some of the philosophical antecedents of Freud in Goethe, Kant, Hegel, and Nietzche.

9. Cf. Edwin Clarke and C. D. O'Malley, *The Human Brain and Spinal Cord: A Historical Study illustrated by Writings from Antiquity to the Twentieth Century* (Berkley: University of California Press, 1968), Chapter I, "Antiquity and Medieval Period," pp. 1-26, and Richard B. Onians, *The Origins of European Thought* (New York: Arno Press, 1973 [1951]).

10. I have used the following: Isaac Asimov, *The Human Brain* (Boston: Houghton Mifflin, 1964); C. U. M. Smith, *The Brain, Towards an Understanding* (New York: G. P. Putnam's Sons, 1970); Steven Rose, *The Conscious Brain* (New York: Alfred A. Knopf, 1973); Charles R. Noback and Robert J. Desmond, *The Human Nervous System* (New York: McGraw-Hill, 1975) and Carl Sagan, *The Dragons of Eden: Speculations on the Evolution of Human Intelligence* (New York: Ballantine, 1977). "The central nervous system is more than a transmission system which merely integrates bodily functions. It becomes, above the reflex level, a system which is flexible in response, modifiable because of previous experience, and retentive of that past experience. To provide for these more complex functions the central nervous system has increased in complexity. This complexity involves multiple branchings of alternate pathways, excitatory and inhibitory mechanisms, and expansion and elaboration of cortical association areas on which learning, reasoning, and imagination depend so heavily;" Loyd S. Woodburne, *The Neural Basis of Behavior* (Columbus, Ohio: Charles E. Merrill, 1967), p. 4., Smith, *The Brain,* p. 28, notes that a typical insect brain contains 1 or 2 x 10^3 neurons, while John von Neumann says that "The largest automaton which we know how to plan could consist of something of the order of 10^4 units; see Warren S. McCulloch, *Embodiments of Mind* (Cambridge, Mass: MIT Press, 1965), p. 139. See also F. W. Went, "The Size of Man," *American Scientist,* 56, 4, 1968, 400-413, who shows that our human size is fundamental to our power to develop a technology which requires a midpoint between micro- and macrocosmic physics; a condition which the insects with their small body size, limited by their special physiology and anatomy, cannot meet.

11. On inter-relation of nervous and endocrine systems see C. Donnell Turner, *General Endocrinology,* 4th ed. (W. B. Saunders, Philadelophia, 1966), pp. 1-18.

12. See Wilder Penfield, "Speech, Perception and the Uncommitted Cortex," in John C. Eccles, ed., *Brain and Conscious Experience* (New York: Springer Verlag- New York, 1966), pp. 217-237.

13. "With the evolution of *Homo*, emphasis will be placed on the dominant rôle of language and the associated imagery and conceptual abilities in bringing about the extraordinarily rapid expansion of the neocortex . . . There is an alternative explanation that cerebral development was associated with the increased demand for fine motor control in the construction and use of tools. Undoubtedly this is a contributory factor, particularly in relation to the imagery and planning necessary for good performances, but in the human cerebral cortex the linguistic areas are much larger than those concerned in movement control . . .In searching for an explanation (of the very rapid evolution of this big brain) we have to look for extraordinarily strong selection pressures that could induce such amazing developments . . . a big factor must have been the growing importance of linguistic communication in hunting, in warfare and in social organization . . . Another factor would be the continual improvement in tool culture with its demand for skilled movement control, both in the manufacture of stone tools and their use." See also Sir John C. Eccles, *The Human Mystery,* The Gifford Lectures, University of Edinburgh, 1977-1978 (New York-Heidelberg-Berlin: Springer International, 1979), pp. 94-95; See also Lecture 6, "Cultural Evolution with Language and

Values, The Human Person", pp. 98 ff. Eccles suggests that polygyny in early cultures favored the offspring of men who were effective leaders.

14. Peter J. Wilson, *Man the Promising Primate* (New Haven: Yale University Press, 1980), p. 65; see whole argument, pp. 50-82.

15. Paul B. Weiz, *The Science of Biology*, 3rd ed. (McGraw-Hill, New York, 1967), pp. 161 f. See also William T. Keeton, *Biological Science*, 3rd ed. (New York: W. W. Norton, 1980), p. 187-554.

16. See Ludwig von Bertalanffy, *General Systems Theory: Foundations, Development, Applications* (New York: George Braziller), Chapter 5, "The Organism Considered as a Physical System", pp. 120 ff.

17. Isaac Asimov, *Human Brain*, (note 10 above) shows the close connection of speech ability with memory; "It is estimated that in a life-time, a brain can store 1,000,000,000,000,000 (a million billion) "bits" of information." (p. 33). "One can therefore argue that all the abstruse facets of modern mathematics and physical science are but reflections of those facets of the physical universe which are simpler in structure than the human mind. Where the limit of understanding will be, or whether it exists at all, we cannot well predict, for we cannot measure as yet the complexity of either the mind or the universe outside the mind. However, even without making measurements, we can say as an axiom that a thing is equal to itself, and that therefore the human mind, in attempting to understand the workings of the human mind faces us with a situation in which the entity that must understand and the object to be understood are of equal complexity." (p. 340).

18. For a detailed development of the brain-computor analogy in non-technical language see Dean E. Woodbridge, *The Machinery of the Brain* (New York: McGraw Hill, 1963). For difficulties with the model see Hubert L. Dreyfus, *What Computors Can't Do: A Critique of Artificial Reason* (New York: Harper and Row, 1972). For a recent argument that computors *are* intelligent see the Nobel Prize winner Herbert Simon, "Is Thinking Uniquely Human?," *University of Chicago Magazine*, Fall, 1981; 12-22. This issue is discussed at length in Chapter 8.

19. John C. Lilly, *Man and Dolphin* (New York: Pyramid Communications, 1974), argued that since an 8 ft. dolphin has a brain 350 grams larger than a 6 ft. man, dolphins may be more intelligent than we! William N. Tavologa, et al., *Development and Evolution of Behavior* (San Francisco: W. H. Freeman, 1970), showed that dolphin "whistles" are signals not coded speech (p. 291). The animal's very large brain is probably related to its remarkable ability to navigate by underwater "sonar".

20. "The grand total is 1,492,000 described species. This is likely to be perhaps only one-half of the species actually in existence. The total number of species that ever lived is even harder to estimate. Simpson (1960) gives 50 million to 4 billion as the outside limits, with about 500 million as a reasonable guess." (p. 24 f.) "The destiny of species in time varies from preservation with little change to tranformation, to extinction. Extinction is in the long run the most probable fate; according to Simpson (1953,1960), not a single species known in fossil condition in Cambrian times has living descendants alive on our time level. Nevertheless, the total organic diversity, as measured by the number of species has increased with time." (p. 25). Theodosius Dobzhansky, "On Some Fundamental Concepts of Darwinian Biology" in T. Dobzhansky, M. K. Hecht and W. C. Steere, *Evolutionary Biology* (New York: Appleton-Century Crofts, 1968), 2, 1-34.

21. The schematization of "three radiations" is from Weisz, *Science of Biology* (note 15 above). Classifications even of the phyla vary with different authors; see Keeton, *Biological Science* (note 15 above), pp. 912-915, for a comparison of different systems. Keeton himself prefers a five-kingdom classification: 1. *Monera* (one-celled forms lacking a nucleus); 2. *Protista* (one-celled forms with a nucleus); (3) *Plantae*; (4) *Fungi* (differing from plants in lacking the power of photosynthesis and in lacking perfect separation of cells); (5) *Animalia*, each divided into many phyla.

Can we speak of hierarchical order of these life forms? Sir Julian Huxley, *Evolution: the Modern Synthesis*, (New York: Harper and Brothers, 1942), defined progress as increase in dominance of organisms over their environment. George Gaylord Simpson in *The Meaning of Evolution* (New Haven: University Press, 1949), took issue with this definition and maintained there has been no universal trend that can be objectively defined as progress. To Simpson, evolutionary progress could only be defined with reference to some particular goal selected by the human observer. In my own opinion the basic fact is that evolution has not been merely adaptive radiation of similar forms of life into different habitats or niches, but organisms have repeatedly evolved new ways of exploiting the same or similar environments. In doing so, their bodies have evolved new levels in the hierarchy of complexity from the macromolecular level to organelles, cells, tissues, organs and organ systems. Achieving these levels required the accumulation of new genetic information, concerned largely with the integration of development and metabolism and with regulating the translation of genetic information into form and function. Thus objectively from the standpoint of life as a whole, the achievement of these new levels of complexity can be designated as progress, just as Huxley thought. See also G. Ledyard Stebbins, *The Basis of Progressive Evolution* (Chapel Hill, N. C.: University of North Carolina Press, 1969), p. 29. In the work cited in note 2 above, p. 141, Stebbins defines evolutionary progress as "directional change toward more complex organisms."

22. Eccles, *Human Mystery*, note 13 above, p. 51, says: "In this search (for signs of early life) the first successes were fossils of algae . . . with a radioactive dating of 2 billion years ago. More recently fossils of presumed organisms 'Eobacterium,' have been found . . . which are dated up to 3.4 bilion years ago." (p. 51).

23. Dates from Stebbins, *Darwin to DNA* (note 2 above), p. 105. Eccles, *Human Mystery* (note 13 above) agrees on the end of Paleocene at 65 million years, but dates end of Paleozoic at 285 million years.

24. David Hawkins, *The Language of Nature* (New York: Doubleday Anchor Books, 1967), p. 329, says: "The cell wall is the end of a chapter and the beginning of a new one. It is an incipient sense organ." It would be better to say that the cell wall was the prototype *either* of the insensitive protective "skin" of plants *or* the sensitive communicative skin of animals. R. D. Preston, "Plants without Cellulose," *Scientific American*, 218, 6 (June, 1968), 102-108, shows that some algae do not have cellulose cell-walls yet can photosynthesize. He says: "All we can conclude is that plants on dry land need in their cell walls a balance between the rigidity necessary to hold the plant upright and extensivity — perhaps even a controlled breakdown — that will allow the plant to grow. Perhaps this balance can be achieved only in a wall based on cellulose" (p. 108).

25. "Here, then, we have the mammal's formula for success — a more efficient, more sensitive body, a complex brain, constant high temperature, athletic limbs, fewer offspring with longer dependence on their parents and, above all, *activity and intelligence*," Desmond Morris, *The Mammals: A Guide to the Living Species* (London: Hodder and Stoughton, 1965), p. 11. Morris believes that hair increased temperature-control but also brought development of numerous sweat-glands to balance heat-retention. The mammaries evolved from such glands. "It would seem, therefore, that mammals first stood on their own feet as a distinct group, not as model parents feeding their young with their own bodies, but as animated thermostats," p. 11. Birds, however, may surpass mammals in certain problem-solving abilities, which seem to require other parts of the brain than the cerebral cortex; see L. J. Stettner and K. A. Matyniak, "The Brain of Birds," *Scientific American*, 218, 8 (June, 1968), 64-77. Moreover, bird instincts develop through a process of "in-printing" and learning; they are not simply ready-made; see Jack P. Hailman, "How an Instinct is Learned," *ibid.*, 221, 6 (Dec., 1969), 98-108.

26. On the evolution of the nervous system see the attractive work of H. Chandler Elliott, *The Shape of Intelligence; The Evolution of the Human Brain* (New York: Charles Scribner's Sons, 1969).

27. See Kent C. Condie, *Plate Tectonics and Crustal Evolution* (New York: Pergamon Press, 1976), pp. 175-199, for the general theory, William N. McFarland, F. Harvey Pough, Tom J. Code and John B. Herser, *Vertebrate Life* (New York: Macmillan, 1979), for the effect of continental drift on evolution.

28. Just when tool making originated is debatable. See C. F. Hockett and R. Ascher, "The Human Revolution," *Current Anthropology* 5 (June, 1964): 135-168, and the differing view of Donald L. Wolberg, "The Hypothesized Osteodontokratic Culture of the Australopithicinae:

A Look at the Evidence," *ibid.*, 11 (Feb., 1970): 23-27. Tools seem most clearly associated with *homo habilis* which has often been classified as one species of Australopithecus (Stebbins, *Darwin to DNA*, note 2 above, p. 335). *Homo habilis* had a brain some 50% larger than other australopithicinae (Eccles *Human Mystery,* note 13 above, p. 77). Steven M. Stanley, *The New Evolutionary Timetable* (New York: Basic Books, 1981), discusses the problem of the accepted gradualist version of evolution and the new revisionists "punctualist" theory which holds that evolution is not continuous but episodic. According to Stanley the earliest humanoid species, *australopithecus africans*, persisted virtually unchanged for a million and half years, coexisting for the latter parts of its career with a later species, *australopithecus robustus*. This latter lasted also with little change for another million years, during the latter half of which it coexisted with *homo erectus* who suddenly appeared about 1.5 million years ago and lasted unchanged for a million years. Finally *homo neanderthalensis* (100,00 years ago) and *homo sapiens* (40,000 years ago) appeared suddenly and coexisted with *home erectus* for about 5,00 years. According to Jeffrey T. Laitman, "Ancestors," *Natural History* 93 (August. 1984): 20-28 recent studies indicate none of the hominoids before *homo sapiens* (c.400,000 years ago) were anatomically capable of speech.

29. See references in note 2 above.

30. "The entire body, then, contains about 75 trillion cells" 25 trillion of which are red blood cells. Cf. Arthur C. Guyton, *Textbook of Medical Physiology* (Philadelphia: W. B. Saunders, 1981), p. 2.

31. See Keeton, *Biological Science*, (note 15 above), pp. 727-748.

32. On the nature of the genetic code see F. H. C. Crick, "The Genetic Code: III", *Scientific American*, 215, 4 (Oct. 1966), 55-63. That identical twins are not completely genetically identical seems evident from studies reported in Amram Scheinfeld, *Twins and Supertwins* (Philadelphia: J. P. Lippincott, 1967). See also Eccles, *The Human Mystery* (note 13 above), p. 60-73.

33. "A rate of one mutation per gene locus in every 100,000 cells is a conservative estimate. Because all higher organisms contain at least 10,000 gene loci, and most of them contain more, we can conservatively say that one individual out of ten carries a newly mutated gene at one of its loci . . . From various experimental studies we can arrive at a conservative estimate of the proportion of useful mutations as one in a thousand. On the basis of these estimates we can calculate that in any species about one in ten thousand individuals in each generation would carry a new mutation of potential value in evolution. Using conservative values of 100 million as the total number of individuals per generation and 50,000 as the number of generations in the evolutionary life of the species, we would expect that at least 500 million *useful* mutations would occur in this life span. We do not know how many new mutations are needed to transform one species into another, but five hundred is a reasonable estimate. On this basis, only one in a million of useful mutations or one in a billion of all mutations which occur need to be established in a species population to provide the genetic basis of observed rates of evolution." Cf. G. Ledyard Stebbins, *Processes of Organic Evolution* (Englewood Cliffs, N. J.: Prentice-Hall, 1966), p. 30.

34. See Ernest Mayr, *Population, Species and Evolution* (Cambridge, Mass: Harvard University Press, 1970), for the biological notion of species, pp. 10-21. "Species are groups of interbreeding natural populations that are reproductively isolated from other such groups," p.12. For the neo-Darwinian concept of "natural selection" see George Gaylord Simpson, *The Major Features of Evolution* (New York: Columbia University Press, 1953). For current views see Stebbins, *Darwin to DNA* (note 2 above). For the revisionist or "punctuational" view see Stephen J. Gould, *Ever Since Darwin* (New York: Norton, 1977). For the philosophical aspects see John N. Deely and Raymond J. Nogar, eds., *The Problem of Evolution* (New York: Appleton-Century-Crofts, Meredith Corp., 1973) with ample bibliographies. Note that "natural selection" does not proceed merely by competition." Pugnacity and aggressiveness are often less conducive to biological success than is inclination 'to live and let live' and to cooperate with other individuals of the same and other species. The fact is that both competition and cooperation are observed in nature. Natural selection is neither egotistic nor altruistic. It is, rather, opportunistic: life is promoted now by struggle and now

by mutual help." Cf. Theodosius Dobzhansky, *Evolution, Genetics and Man*, (note 2 above), p. 113.

35. "Modern theories of evolution consider the environment to be the directive force in the evolutionary process. But the environment does not change the organism from without, as some theorists of the past believed. The environment furnishes the challenge, to which a living species may respond by adaptive transformations of its gene pool. Whether the species does or does not respond to the challenge depends on the presence in the gene pool of the proper raw materials, mutated genes and gene combinations." *Ibid.* p. 123. The mutuality of this process is well illustrated by the fact that there was probably no free oxygen in the atmosphere until it was produced by green plants; see P. Cloud and A. Gibor, "The Oxygen Cycle," *Scientific American*, 223, 3 (Sept., 1970), 110-123.

36. See note 32 above and Stebbins, *Darwin to DNA* (note 2 above). pp. 32-37; Eccles, *The Human Mystery* (note 13 above), pp. 60-69. DNA has the remarkable power to repair itself when it suffers mutations. If this power were lacking, stable species and life itself would be impossible; but if DNA always *perfectly* repaired itself, evolution would become impossible for lack of inheritable variations. See P. C. Hanwalt and R. H. Haynes, "The Repair of DNA", *Scientific American*, 216, 2 (Feb., 1967), 36-43.

37. See William T. Keeton, *Biological Science*, (note 15 above), pp. 46-64. Nucleic acids are a fourth class of living materials, but these are closely related to proteins.

38. See Gosta Ehrensvard, *Life: Origin and Development* (Chicago: University of Chicago Press, 1962). On the possible extra-terrestrial origins of life see Elso S. Barghoorn, "The Oldest Fossils," *Scientific American*, 224, 5 (May, 1971), 30-53. On the first stages of evolution see Lynn Margulis, "Symbiosis and Evolution", *ibid.*, 225, 2 (August, 1971), 48-61. For more recent views see Stebbins, *Darwin to DNA* (note 2 above), pp. 174-189, and Eccles (note 13 above), pp. 60-65. Sir Fred Hoyle, "The World According to Hoyle", *Spectrum* 22 (Nov., 1982), 9-13, defends his "steady-state" theory by arguing that the world must always have existed much as it is now, because it has required a great deal of time for life to evolve by sheer chance and to have spread through the whole cosmos!

39. See Cecil B. Mast, "Matter and Energy in Scientific Theory," in Ernan McMillan, *Concept of Matter* (Notre Dame, Ind.: Notre Dame University Press, 1963), pp. 595 ff. For possible qualifications of this principle see Gerald Feinberg, "Particles that Go Faster than Light", *Scientific American*, 222, 2 (Feb., 1970), 68-81. Such particles have not yet been observed.

40. On the four fundamental forces see Henry Semat and John R. Albright, *Introduction to Atomic and Nuclear Physics*, 5th ed. (New York: Holt, Rinehart and Winston, 1972), pp. 584 ff., and the fascinating popular account of recent advances toward a unified field theory by Timothy Ferris, "Beyond Newton and Einstein," *New York Times Magazine*, Sept. 26, 1982, pp. 36-47. On such a unification as the goal of modern science see Carl Friederich von Weizäcker, *The Unity of Nature* (New York: Farrar, Straus, Geroux, 1980), and Benjamin Gal-Or, *Cosmology, Physics and Philosophy* (New York: Springer Verlag New York, 1981).

41. Heinz R. Pagels, *The Cosmic Code: Quantum Physics as the Language of Nature* (New York: Simon and Schuster, 1982), presents an up-to-date and non-mathematical statement of the present phase of the very rapid development of atomic physics, which I follow here. See also, G. F. Chew, Murray Gell-Mann, and Arthur H. Rosenfeld, *Scientific American*, 210, 1 (Jan, 1964), 82ff.; Henry W. Kendall and Wolfgang Panofsky, "The Structure of the Proton and Neutron," *ibid.*, 22, 4 (June, 1971), 60 ff.; R. and S. van der Meer, "The Search for Intermediate Vector Bosons," *ibid.*, 246, 3 (March, 1982), 48-60, and "Science and the Citizen: Most Wanted List", *ibid.* pp. 75 f.; E. D. Bloom and G. J. Feldman, "Quarkonium", *ibid.*, 246, 5 (May, 1982), 66-77.

42. On the current theory of chemical bonding see Francis A. Carey and Richard J. Sundberg, *Advanced Organic Chemistry* (New York: Plenum Press, 1977), vol. 1 (Part A), pp. 1-38. It should be noted that while it is customary to speak of the "92 elements" only 85 occur naturally, but a number of short-lived elements of higher atomic number have been artificially produced. There are some 287 istopes of which 16 are radioactive but some of these have half-lives comparable to the age of the earth.

43. See Max Jammer, *The Philosophy of Quantum Mechanics* (New York: Wiley Interscience, 1974); Gerald Feinberg, "The Philosophical Implications of Contemporary Particle

Physics" and Clifford A. Hooker, "The Nature of Quantum Mechanical Reality," in Robert G. Colodny, ed., *Paradigms and Paradoxes: The Philosophical Challenge of the Quantum Domain* (Pittsburgh: University of Pittsburgh, Press, 1972), pp. 33-46 and 67-302. Also Enrico Cantore, *Atomic Order* (Cambridge, Mass: MIT Press, 1969). An especially clear discussion of the relation of indeterminacy to causality is found in Ernest Nagel, *The Structure of Science* (New York: Harcourt, Brace and World, 1961), pp. 277-235. The more common interpretation, following Werner Heisenberg, *The Physicist's Conception of Nature* (London: Hutchinson, 1958), pp. 32 ff. is that the uncertainty of measurement is a subjective consequence of the objective indeterminacy of the behavior of the particle. Nagel is concerned to show that this subatomic indeterminancy does not lead to macroscopic indeterminism. See also Norwood Russell Hanson, "The Copenhagen Interpretation of the Quantum Theory" in A. Danto and S. Morgenbesser, *Philosophy of Science: Readings* (New York: Mentor Books, World Pub., 1960), pp. 450-470. For a defense of the determinist view of Einstein and Bohm see Gal-Or, *Cosmology* (note 40 above).

44. J. J. C. Smart, *Philosophy and Scientific Realism* (London: Routledge and Kegan Paul, 1963), attacks the notion of time as transitory as a kind of anthropocentricity. In a book edited by him, *Problems of Space and Time* (New York: MacMillan, 1964), several essays indicate the many difficulties about giving time (and hence motion) reality in the modern scientific world-view. See in the same volume especially the extreme position of F. H. Bradley, "The Unreality of Space and Time," pp. 132-138; also W. V. Quine, "Time", pp. 370-374, who treats time as "space-like;" and Adolf Grünbaum "Time, Irreversible Process, and the Physical Status of Becoming," pp. 397-426, who shows the connection of time with irreversible processes of which entropy is not the only one. Grünbaum, however, in defending the objective reality of time considers that only the time-intervals are objectively real and that "the present" is purely psychological. See his essay, "The Status of Temporal Being" in Richard Gale, ed., *The Philosophy of Time* (Garden City, N.Y.: Doubleday, 1970), pp. 322-353, and "Are Physical Events Themselves Transiently Past, Present, and Future" in *British Journal for Philosophy of Science*, 20 (1969), 146, and the criticism of his position by Frederick Ferre, "Grünbaum on Temporal Becoming: A Critique" in *International Philosophical Quarterly*, 12, 3 (Sept. 1972), 426-445. See also Harold F. Blum, *Time's Arrow and Evolution*, 2nd ed., (New York: Harper Torchbooks, 1962); T. Gold and E. L. Schumacher, eds., *The Nature of Time* (Ithaca, New York: Cornell University Press, 1967); J. T. Fraser, F. C. Haber, and G. H. Müller, *The Study of Time* (New York: Springer Verlag New York, 1972); Benjamin Gal-Or, "The Crisis about the Origin of Irreversibility and Time Anisotropy," *Science*, 176, 3 (April 7, 1972), 11-17, concludes, "Thus the problem of the irreversibility of nature, which is intimately coupled to the very concept of time and initial conditions, incorporates in it issues that are as far beyond our reach now as they were in the early days of thermodynamics" (p. 16). See also Robert G. Sachs, "Time Reversal," *ibid.*, 176, 5 (May 12, 1972), 587-597.

45. See George Gamow, *The Creation of the Universe*, rev. ed. (New York: Viking Press, 1961), and *Matter, Earth and Sky* (Englewood Cliffs, N. J.: Prentice-Hall), c. 23, "Cosmology", pp. 582 ff.; Wiliam H. McCrea, "Cosmology Today," *American Scientist*, 58 (Sept-Oct, 1972), 521-527; Steven Weinberg, *The First Three Minutes* (London: Andre Deutsch, 1977); Michael Rowan-Robinson, *Cosmology* (Oxford: Clarendon Press, 1977), and Eccles, *Human Mystery* (note 13 above) pp. 14-24. See also the article by Sir Fred Hoyle (note 38 above) in which he contends that the steady-state hypothesis is still alive and well.

46. On the dimensions and development of the galaxies see Martin J. Rees and Joseph Silk, "The Origin of the Galaxies," *Scientific American*, 222, (June, 1970) pp. 26-31 and Eccles, *Human Mystery* (note 13 above) pp. 33-48. "Before the Sun was formed, almost every, stellar temperature, density and life history had been experienced somewhere in our Galaxy. The residues are the many kinds of atoms in the universe. Some of the processes that synthesize one element will destroy others; the enormously complicated composition of our bodies, for example, requires that many different stars contributed atoms for the raw material out of which our Sun, the planets, and ourselves were born.", J. L. Greenstein, "Stellar Evolution and the Origin of the Chemical Elements," *American Scientist*, 49, 1 (Dec., 1961), 449-473, p. 458.

47. Eccles, *Human Mystery* (note 13 above) pp. 33-48.

48. John Verhoogen, et al., *The Earth: An Introduction to Physical Geography* (New York: Holt, Rinehart and Winston 1970). See also Kent C. Condie, *Plate Tectomics* (note 27 above).

49. Shapely's speculations can be found in his *Of Men and Stars* (Boston: Beacon Press, 1958), and *The View from a Distant Planet* (New York: Basic Books, 1963). Eccles, *Human Mystery*, (note 13 above) p. 43 f., takes the opposite view and quotes S. S. Kumar, "Planetary Systems" in W. Salaw, and K. Jacobs ed., *The Emerging Universe* (Charlottesville: University Press of Virginia, 1972) who argues that single stars are relatively uncommon and that planetary systems are much rarer than is customarily claimed: "In our galactic system with 10 stars, there are probably not more than 10 planetary systems, and there may be only one!" (p. 44).

M. Rowan-Robinson, *Cosmology* (note 45 above) writes, "The anti-geocentric viewpoint which encourages us to believe the cosmological principle, leads naturally to the idea that we are not unique in the universe. Calculations of the probability of other inhabited planets in our Galaxy are rather meaningless at this stage of our knowledge of the origin of life. But in the framework of the cosmological principle we should assume that there is at least one inhabited planet per galaxy. Naturally it is intriguing to think that somewhere else all the controversies outlined in this epilogue have been resolved. Yet we should not expect to learn these solutions through some intergalactic broadcasting service, for in the framework of evolutionary cosmology, in which the galaxies form simultaneously and evolve in parallel, the light from even the nearest comparable galaxy, M31, set off a million years ago, long before Andromeda-man would have appeared on Andromeda-earth. We have to solve these problems on our own" (p. 145). The "cosmological principle" is simply the supposition that all regions of the universe are similar. It is a heuristic not an empirical principle.

50. For a discussion of the concepts of "entropy", "negentropy," and "information" see Albert L. Lehninger, *Bioenergetics: The Molecular Bases of Biological Energy Transformations* (New York: W. A. Benjamin, 1965) and Harold F. Blum, *Time's Arrow and Evolution* (note 44 above).

51. On the theory of the Big Crunch (i.e. the reverse of the Big Bang) see J. A. Wheeler, "Genesis and Observership" in R. Butts and J. Hintika, eds., *University of Western Ontario Series in the Philosophy of Science* (Boston: Reidel, 1977). According to Wheeler, cosmic expansion will quit at 18.9 billion light years diameter and then will collapse to a minimum expansion at 59 billion years from the Big Bang: "With gravitational collapse we come to the end of time. Never out of the equations of general relativity has one been able to find the slightest argument for a "re-expansion" or a "cyclic universe" or anything other than an end." M. Rowan-Robinson, *Cosmology* (note 45 above), p. 78: "It is not known if the (oscillating) universe can go through more than one cycle." "Another question we could ask pertains to the forces which caused the initial expansion of the universe, and to the state of contraction which was the starting point of all our discussion. Mathematically we may say that the observed expansion of the universe is nothing but the bouncing back which resulted from the collapse prior to the zero point of time a few billion years ago. Physically, however, there is no sense in speaking about that 'Prehistoric state' of the universe, since indeed during the stage of maximum compression everything was squeezed into the pulp, or rather the ylem, and no information could have been left from the earlier time if there ever was one," George Gamow, "The Origin and Evolution of the Universe," *American Scientist*, 39 (July, 1951), 393-406, p. 405 f. See also K. S. Thorne, "Gravitational Collapse," *Scientific American*, 217, 5 (Nov. 1967), 88-102, and Roger Penrose, "Black Holes", *ibid.*, 226, 5 (May, 1972), 38-55.

52. Freeman J. Dyson, "Energy in the Universe," *Scientific American*, 224, 3 (Sept. 1971), 50-59.

53. Blum, *Time's Arrow* (note 44 above), pp. 29 ff.

54. Statistically the outstanding feature of life is the continuous creation of material structures of an extremely high degree of orderliness. A man, a fly, a fungus, and even bacteria are highly ordered, and, therefore, most improbable, structures. This means that they are very unlikely to be created by shaking the requisite number of different kinds of atoms in a test tube. However, in order to continue living and to grow, these structures have to

metabolize and, in doing so, they raise the total entropy to a larger degree than corresponds to the growth of orderliness which they achieve. The phenomenon of life presents questions which we have so far been unable to formulate but a violation of the second law of thermodynamics is not one of them." Cf. K. Mendelssohn, "Probability Enters Physics," *American Scientist*, 49 (March, 1961), 37-49. The balance of the ecosystem produces cycles which slow down the rate of entropy (pollution) but never quite offset it.

Chapter 3

Humanist Theologies of the Body (1700-2000)

I: The World-View of Humanism

1. The Humanist Interpretation of Science

Since about 1700 modern science has no longer developed under the patronage of Christianity but under the increasing cultural dominance of Humanism. Some suppose that this is because there is an inherent incompatibility between the *facts* discovered by modern science and the Christian world-view. The possibility should be considered, however, that the scientific data described in the last chapter is susceptible both to a Christian and a Humanist interpretation. In Part II of this book I will relate the history of the efforts of Christians to develop a world-view which is both honestly scientific and Christian.

In this concluding chapter of Part I, I will attempt to present the Humanist interpretation (with its variations) of what modern science has been able to tell us about ourselves in our world. In doing so I must range through a distracting number of topics which may seem to have only a remote relation to the scientific account of the human person as a body in the world of bodies. Yet this is necessary because one of the principal problems of Humanism, as we shall see, has been how to overcome the dualism of Mind and Matter proposed in the seventeenth century by René Descartes, a proposal which is commonly said to have initiated "modern" thought. In order to solve this problem Humanists have either reduced the whole of human reality to Matter, or conversely the whole of physical reality to Mind, but today the former reduction predominates. Consequently it is necessary to show how all the principal aspects of the Humanist world-view and value-system are related to the way it interprets the scientific findings about the human body in the world of bodies.

Before launching on this attempt to describe the Humanist world-view, one caution is in order. It would be prejudicial to the dialogue between Humanism and Christianity which I want to promote if we were unthinkingly to identify Humanism with "modernity" or "secularization" as it often done. The advance of science has conditioned most of us to assume that what is new is also "improved", that the historically more recent world-view must automatically be the truer since it rests on a refutation of the older by the discovery of new data. If this were a law of inevitable progress then the Humanist interpretation of science and the "secularization" which seems to be the usual result of the adoption of scientific technology would be of unquestionable authority. We have seen, however, that the empirical data of science do no evidently manifest any such "law of progress." Such a law, if there is one, is itself a feature of a theological interpretation and is a function of the theology to which we are committed.

We often hear that our modern culture is characterized by increasing "secularization", but the term is ambiguous. Peter Berger defines it as "the process by which sectors of society and culture are removed from the domination of religious institutions and symbols."[1] If, however, religion is defined *functionally* (as I argued in Chapter 1 it must be if we are to compare radically different world-views) then this process of secularization is not just the decline of "religion," but is the rise of some new functional equivalent to the old declining religion. Nor does it help to speak, as do some, of a "desacralization," since what is "sacred" and what "profane" is a function of each particular religion.[2]

What Berger really has in mind by "secularization" is the decline of Christian religious institutions and symbols in economically advanced countries and parallel processes in the Jewish community, in Islamic, Hindu, and other cultures, that is, the replacement of the older world religions by some new functional religions. He comes to the conclusion that the "carrier" of secularization is the modern scientific, technological, industrial economic system. While this is certainly true, we should not make the mistake of assuming that "modernity" is necessarily connected with secularization. It may be true that modernity is a favorable climate for secularization, but it is possible that all the ancient religions may not only survive, but even be strengthened by modernization. We must not forget that while the United States is among the most technologically modern of countries, it remains considerably less secularized than countries like Sweden, France, or Italy by the usual criteria, such as belief in God, church membership and attendance, etc.[3]

If, then, we cannot assume that secularization is the decline of religion or that it is identical with modernity, what is it? Thomas Luckman[4] sees it as the growth of what he calls "the Invisible Religion," i.e., of private religions, entirely individual or shared by small, loosely structured, and highly mobile groups. So amorphous and pluralistic a secular religion would defy any characterization. Undoubtedly Luckmann has pointed out an important phenomenon coincident with secularization, but he has not demonstrated the identity of secularization and the growth of religious privatism. How could such private world-views ever fulfill the social function of religion to give meaning to the life not only of isolated individuals, but of the community as a whole? In fact, modern individuals do not each invent their own world-views or value systems, but derive them largely from the community of which they are parts. What, then, is this public "secular" religion which is tending to replace Christianity and other traditional religions in societies whose culture is highly technological or "modern"?

Robert Bellah and others[5] have tried to answer this question by a theory of "civil religion" which, in the United States, is supposed to provide a functional religion for the nation by the use of symbols of Protestant origin which have been now so modified as to be acceptable to all the main segments of American society. Whatever the merits of this theory as an explanation of the peculiar church-state relations in the United States, it does not explain how the secularized sector of our society interprets these symbols in a way congruent with its own world-view. Obviously this world-view differs from the Christian world-view not only in negative but also in *positive* ways.

In Chapter 1, I suggested that this new religion which is the *terminus ad quem* of the so-called secularization process can properly be called "Humanism" in the western democracies and "Marxism" in the Soviet bloc and developing world. Like the older world religions these two new world religions have already become denominationalized, yet each exhibits an underlying unity just as do Christianity, Islam, or Buddhism. This becomes evident when we ask how the Humanist culture of our times understands human personhood in its bodily existence and what is the value system which it draws from this self-understanding.

53

2. Contemporary Humanism and the Human Body

In formulating this world-view we are assisted by those humanists who accept this title and have themselves drawn up a "set of common principles that can serve as a basis for united action — positive principles relevant to the present condition. They are a design for a planetary society on a planetary scale."[6] While the *Humanist Manifesto II* (1973) was only subscribed by a small group of thinkers, it is, as I will show later, a very fair statement of views much more widely held which have their own inner logic. It can be summarized (keeping close to the original wording) as follows:

A: Religion
1) In the best sense, religion may inspire dedication to the highest esthetic ideals. The cultivation of moral devotion and creative imagination is an expression of genuine "spiritual" experience and aspiration.

    We believe, however, that traditional dogmatic or authoritarian religions that place revelation, God, ritual, or creed above human needs and experience do a disservice to the human species. Any account of nature should pass the tests of scientific evidence; in our judgment the dogmas and myths of traditional religions do not do so. Even at this late date in human history, certain elementary facts based upon the critical use of scientific reason have to be restated. We find insufficient evidence for belief in the existence of a supernatural; it is either meaningless or irrelevant to the question of the survival and fulfillment of the human race. As nontheists, we begin with humans not God; nature not deity. Nature may indeed be broader and deeper than we know; any new discoveries, however, will but enlarge our knowledge of the natural . . . We appreciate the need to perserve the best ethical teachings in the religious traditions of mankind, many of which we share in common. But we reject those features of traditional religious morality that deny humans a full appreciation of their own potentialities and responsibilities.
2) Promises of immortal salvation or fear of eternal damnation are both illusory and harmful. They distract humans from present concerns, from self-actualization and from rectifying social injustices . . . Science affirms

that the human species is an emergence from natural evolutionary forces. As far as we know, the total personality is a function of the biological organism transacting in a social and cultural context. There is no credible evidence that life survives the death of the body. We continue to exist in our progeny and in the way that our lives have influenced others in our culture . . . Traditional religions are surely not the only obstacles to human progress . . . Purely economic and political viewpoints, whether capitalist or communist, often function as religious and ideological dogma. Although humans undoubtedly need economic and political goals, they also need creative values by which to live.

B: **Ethics**

1) We affirm that moral values derive their source from human experience. Ethics is *autonomous* and *situational,* needing no theological or ideological sanctions. Ethics stems from human need and interest. . . . Human life has meaning because we create and develop our futures. Happiness and the creative realization of human needs and desires, individually and in shared enjoyment, are continuous themes of humanism.

2) *Reason and intelligence* are the most effective instruments that humankind possesses. There is no substitute: neither faith nor passion suffices in itself. The controlled use of scientific methods . . . must be extended further in the solution of human problems . . . Nor is there any guarantee that all problems can be solved or all questions answered. Yet critical intelligence, infused by a sense of human caring, is the best method that humanity has for resolving problems. Reason should be balanced with compassion and empathy and the whole person fulfilled.

C: **The Individual and the Democratic Society**

1) *The preciousness and dignity of the individual person* is a central humanist value . . . We believe in maximum individual autonomy consonant with social responsibility. Although science can account for the causes of behavior, the possibilities of individual *freedom of choice* exist in human life and should be increased.

2) In the area of sexuality, we believe that intolerant attitudes, often cultivated by orthodox religions and

puritanical cultures, unduly repress sexual conduct. We do not approve of exploitative, denigrating forms of sexual expression, neither do we wish to prohibit, by law or social sanction, sexual behavior between consenting adults. The many varieties of sexual exploration should not in themselves be considered "evil."

3) To enhance freedom and dignity the individual must experience a full range of civil liberties in all societies . . . We are committed to an open and democratic society. We must extend *participatory democracy* in the true sense [to all levels of society] . . . The separation of church and state and the separation of ideology and state are imperatives . . . Human societies should evaluate economic systems not by rhetoric or ideology, but by whether or not they *increase economic well-being* for all individuals and groups . . . *The principle of moral equality* must be furthered through elimination of all discrimination based upon race, religion, sex, age, or national origin . . . We believe in the *right to universal education.* Everyone has a right to the cultural opportunity to fulfill his or her unique capacities and talents . . . We deplore the division of mankind on nationalistic grounds. We have reached a turning point in human history where the best option is to *transcend the limits of national sovereignty* and to move toward building a world community in which all sectors of the human family can participate. Thus we look to the development of a world law and a world order based upon transnational federal government. This would appreciate cultural pluralism and diversity . . . This world community must renounce *the resort to violence and force* as a method of solving international disputes . . . [It] must engage in *cooperative planning* concerning the use of rapidly depleting resources. The planet earth must be considered a single *ecosystem* . . . The problems of *economic growth and development* can no longer be resolved by one nation alone . . . *Technology is a vital key* to human progress and development . . . Technology must, however, be carefully judged by the consequences of its use . . . We must expand communication and transportation across frontiers . . . We thus call for full international cooperation in culture,

science, the arts, and technology *across ideological borders* . . . What more daring a goal for humankind than for each person to become, in ideal as well as practice, a citizen of a world community!

Christians can agree, of course, with most of what is positive in this statement, but the Christian is convinced that human beings, bodies that they are, are made in God's image and are called to share in the life of God's Son, who became a human in bodily existence, died, and rose to immortal *bodily* life, not by the power of human technology, but by the power of God's Spirit. Such a dogma seems to the Humanist not only an illusion, but a dangerous one, because it distracts humankind from its real tasks. It seems illusory, not precisely because it is logically contradictory to the scientific world-picture I have described in the last chapter, but because its verification exceeds the limits of scientific reason, that "critical intelligence, infused by a sense of human caring," which the *Manifesto* believes "is the best method that humanity has for resolving problems."

How many people today live within the horizon of the world-view thus formulated by the *Humanist Manifesto*? Empirical studies so far have been done mainly in terms of the secularization process understood negatively as a decline in committment to traditional religions. However, if we take as a rough measure the number of people who no longer believe in human survival after death, we find that in the western democracies from 20 to 50% of the population are "secularized" and thus presumably have adopted either the Humanist or the Marxist world views available to them as functional religions alternative to Christianity or Judaism.[7] We also know that this secularization is especially pronounced among the educational elites, and even more among those specializing in the humanities and social sciences than among natural scientists.[8]

Although I am not acquainted with any thoroughly empirical attempt to discover the positive views of these elites, all the evidence in current writing, both literary and scientific, lead me to believe that the world-view and value-system of these elites are pretty well formulated by *Humanist Manifesto II*. Through the various levels of popularization in the media this point of view is constantly presented to the American and European publics, so that in public policy debates today most people regard it as out of place to introduce arguments based on Christian revelation, while arguments based on Humanist assumptions are considered altogether proper.

It is important to note that while *Humanist Manifesto II* sounds politically and ethically "liberal," its basic principles are common to both liberals and conservatives in the elites of the western democracies. The differences between left and right in the twentieth century is no longer between Humanists or Marxists and Christians, but between Humanists and Marxists, or (as in the United States) between Humanists who stress "social welfare" and Humanists who stress "free enterprise."[9]

3. What Separated Humanism from Christianity?

The Humanistic world-view, including its special understanding of the human person as body, has much in common with the Christianity out of which it originated in dialectical opposition. To understand Humanism and its relation to Christianity as regards their two visions of the human body, it is necessary to ask why this separation and opposition occurred. Why could not Christianity retain within itself the questions and insights of Humanism?

Three explanations often given seem inadequate. The first of these is that the rise of modern science with Galileo and Newton made it impossible for intellectuals any longer to believe the Christian world-view and that the masses were later swept into this same loss of Christian faith by the rise of a culture based on scientific technology. But historians of science have shown conclusively that while science began its rapid modern expansion only in the seventeenth century its roots lay deep in the Middle Ages.[10] The chief seventeenth century founders of modern science such as Copernicus, Galileo, Kepler, Harvey and Boyle were Christians who were convinced their scientific discoveries would bolster the Christian faith, not undermine it. In England, which became the central focus both of scientific progress and the rise of Humanism, it was the Puritans who were especially active in promoting scientific research![11] The Galileo case and similar battles between "science and religion" in Protestant lands need not have been decisive, but might simply have been developmental episodes in the history of Christian culture.

A second explanation is that the renaissance of Greek learning in the fifteenth and sixteenth centuries which shifted the focus of human attention from clerical interests in the things of God to the interests of the laity in Man had so restored confidence in our own power to create our own world that Humanism, as the name implies, was the inevitable result. Today, however, most students of the Renaissance do not consider it as anti-Christian or a revival of "paganism."[12] Christianity had met the challenge of classical "paganism" in the patristic age and found principles by which to assimilate Greek and Roman culture. This process had continued through the High Middle Ages. The Renaissance was only a further step by which, after the medieval assimilation of Greek philosophy, Christian culture assimilated the literary, artistic, and mathematical achievements of the ancients. Most of the leaders of the Renaissance were Christians profoundly convinced that they were "despoiling the Egyptians" and, indeed, that they were strengthening the Christian Church by purifying it through a "return to the sources" of the Bible and the Church Fathers.

A third explanation is that the novelty of the materials from the ancient past together with the new scientific discoveries and geographical explorations stimulated the rising educated middle-class to rebel against the narrowness of scholastic theology and its authoritative enforcement by the Church, resulting in the spread of a sceptical, critical attitude subversive of Christian faith and ecclesiastical authority. Some proponents of this explanation see the Reformation as a first step toward this "freedom of thought" by its emphasis on the freedom of the individual conscience, or by the advocacy of the so-called "Protestant ethic." This explanation, however, is not only unfair to Protestantism which, in true Renaissance style, always regarded itself as the restoration of the original Christianity, but it assumes as evident that Christianity is incompatible with freedom of thought. Such assumptions are useless in comparing world-views, because every religion accuses the others of lacking freedom of thought, since the concept of "freedom" is a function of each world-view. Thus Marxists charge Humanists with lacking "freedom of thought" on the grounds that the views of Humanism are dictated by the class-struggle inherent in capitalism.

Studies of the rise of "scepticism" or "Pyrrhonism" in the sixteenth and seventeenth centuries have shown that for the most part this scepticism was not anti-Christian, but anti-intellectual.[13] Luther, Montaigne, and Charron attacked the contradictions of the theologians and philosophers not to destroy Christian faith, but to prove its necessity in a world that had become an intellectual Babel. This scepticism was really *fideism*. Faith was exalted not only at the expense of scholasticism, but also as a bulwark against "enthusiasm," that is, the sectarian and individualistic excesses of mysticism and fanaticism. Moreover the rise of the "critical attitude" was not only the work of non-Christians like Spinoza, Hobbes, or Hume, but of convinced Christians such as Descartes, Mersenne and Gassendi who courageously faced the most radical questions of epistemology.[14] Again it is a polemical assumption to speak of someone else's world-view as "uncritical," since the criteria of truth are a function of each world-view.

What then were the fundamental factors in the rise of Humanism, to which the foregoing may have been contributory but could not have been decisive? Arnold Toynbee[15] is surely right in arguing that the chief factor in the rise of "secularization" (or, as I have argued, secular Humanism) in dialectical opposition to Christianity was not the Reformation as such, but the religious wars, persecution, and bitter intolerance of the sixteenth and seventeenth centuries, notably the devasting Thirty

Year War of Catholic and Protestant nations (1618-1648), and the Catholic-Calvinist conflict in France culminating in the revocation of the Edict of Tolerance of Nantes (1685). These ferocious, fanatical wars of Christian against Christian made the claims of Christianity as a religon of faith and love appear incredible and called forth the new religion of Humanism, just as at an earlier time the Trinitarian and Christological quarrels in the patristic Church had called forth Islam.

Of course the famous thesis of Max Weber[16] contended that it was the theology of Calvinism, with its emphasis on the worldly vocation of the Christian and the so-called "Protestant work-ethic" which paved the way for modern capitalism and with it modern scientific technology. The Marxists, on the other hand, attribute these changes to the overthrow of the "feudal" aristocracy by the rising middle-class. H. R. Trevor-Roper in his *The Crisis of the Seventeenth Century* disputes these theories and argues that the revolutions of 1620-1660 which swept away the social system of the Renaissance and shifted the European cultural centers from southern Catholic to northern Protestant lands require a different explanation.[17]

Trevor-Roper shows that the capitalist entrepreneurs and scientific innovators in Catholic countries were driven out into northern lands by the increasingly top-heavy, over-centralized bureaucracies built up by the great Renaissance princes and reinforced by the Counter-Reformation Church. Protestant states suffered from the same bureaucratization, but their internal religious divisions, especially in Calvinist countries, gave more room to economic and technological change and underwent more radical revolutions, especially in England. Calvinism did not directly advocate these changes, but its severe doctrine of the sovereignty of God provoked more vigorous religious dissent and thus undermined the old order more radically.

Furthermore, Trevor-Roper explains the great witch craze and the anti-semitism of this period (which originated in the Catholic Inquisition but were taken up in Protestant lands) as the last great effort to maintain Christian unity by defending what he calls "the pseudo-Aristotelian" world-view. (What he is really talking about is the medieval renaissance synthesis of Christianity with Greek philosophy, in which Neo-Platonism with its demonology, rather than Aristotelianism, dominated, as I will show in Part II).

Thus to many thoughtful people of the seventeenth century, and particularly to those interested in economic and scientific progress, Christianity seemed hopelessly confused, warring against itself, and precariously maintaining a semblance of unity only through a superstitious, paranoid suppression of thought imposed by an obsolescent, bloated bureaucratic tyranny of altar-and-throne in the name of a tyrant God. The political revolutions of 1620-1660 reflected a profound shift in world-view usually called the Enlightenment but which I understand as the first phase of the new religion of Humanism that no longer believed in the Christian creed which neither the Renaissance nor the Reformation had questioned.

The reaction against the Christian world-view, however, would not have resulted in a successful new religion, unless there was also some strong *positive* factor working in its favor. This positive factor, it is generally admitted, was the rise of the new science along with its technological applications which made possible a new type of capitalistic economy. For Humanism, as we have seen, our human hopes rest not on God but on "the controlled use of scientific methods" which "must be extended further in the solution of human problems . . . Nor is there any guarantee that all problems can be solved or all critical questions answered. Yet critical intelligence, infused by a sense of human caring, is the best method that humanity has for resolving problems."

Since modern science is an empirical study of the material world and therefore views human beings as bodies among the various bodies that make up the material universe, Humanism, positively based as it is on science, had to develop it's own philosophy (or theology) of the human body as an interpretation of the findings of science and as the foundation of our human self-understanding.

II: The Empiricist Development Of The Humanist World-View

1. The Newtonian World-View and Deisms

The overthrow of Ptolemaic astronomy and Aristotelian physics by Copernicus (d.1543) and Galileo (d.1641), and the project of extensive empirical research for practical, technological purposes put forward by Francis Bacon (d.1626), gave rise to a first general revision of the scholastic philosophy which had served Christianity, a revision that was principally the work of René Descartes (d.1650) and which remained within the context of Christianity. Although Cartesianism was at first opposed by the Jesuits, in that period the intellectual leaders of the Catholic clergy, it was eventully promoted by them and up to the end of the eighteenth century dominated the training of priests and of many Protestant ministers.[18]

Descartes' system, which I will discuss further in its Christian context in Chapter 6, had the effect of reviving in radical form Plato's model of the human being as a *mind-in-a-body* and of the body as a *mechanism,* a model which Arthur Koestler, somewhat unfairly, has called "the ghost

61

in the machine."[19] This mechanistic understanding of the human body and of the visible cosmos long survived Cartesianism as a system and came to dominate the whole development of modern science. Consistent with this development was the tendency to conceive of God as the Divine Mind who had designed and constructed the world-machine, wound it up and left it to run itself, which came to be called "Deism,"[20]

For men and women scandalized by the Tyrant God in whose name Christians were piously killing each other and refusing to accept the progress of scientific investigation, Deism was an attractive alternative. According to the Deists God makes himself known to us not through a historic intervention in the first century Jew Jesus of Nazareth, but to the reason of all human beings in all times and places through the manifest order of the world-machine and through our own human instincts built into us by that same Watch-Maker. This God requires of us nothing but the acknowledgement of his wisdom and the observance of natural laws evident from universal human instinct and confirmed by scientific research, liberated from imposed, irrational, traditional notions and customs. The great scientific success of the world-view of Isaac Newton (d.1727) supplied this deistic theology with a very convincing way of manifesting this natural order, a way superior to Christian Cartesianism because it avoided a-priorism and grounded itself firmly on empirical data. "I make no hypotheses," Newton said.[21]

That Newton was an Englishman was important because it was in England, weary with its religious struggles, in reaction to the fierce Calvinism of the Puritans, freed of the Stuart despotism by the Revolution of 1689, and in the process of building a vast commercial empire through which seafarers came into contact with global humanity in all its religious diversity, that the conditions for the success of Humanism were best satisfied. Herbert of Cherbury (d.1648) had already sketched the philosophy of Deism. Thomas Hobbes (d.1679) had proposed a thoroughly mechanistic account of the world, John Locke (d.1704) had put Deism on a securer basis by a strictly empiricist psychology (in contrast to Lord Herbert and Hobbes who still accepted Cartesian innate ideas). It was Voltaire (d.1788) who popularized Deism in France where it took on a still more radical character and whence it spread to Germany and all of Europe.[22]

Deism, however, was only a transitory step in the rise of Humanism. In France after Voltaire, the *philosophes,* in their famous *Encyclopedie,* although they had rejected Descartes in favor of Newton, argued from Descartes' dualism of Matter and Mind that in view of the Newtonian demonstration of the nature of Matter as a self-sufficient mechanical system it should be possible to dispense both with the Divine Mind and then with the human mind as anything more than the working of the brain. La Mettrie (d.1751) in his well-titled *L'homme machine* and Diderot (d.1784) went on to argue that if external human behavior can be explained by the mechanisms of the body, and mind is, as Locke had shown, nothing more than bodily impressions and "reflections" on these impressions, then the awkward Cartesian dualism is unnecessary and the human person is simply a body thinking and operating by mechanical laws. Furthermore, Diderot, Buffon (d.1788) and Maupertuis (d.1759), without yet having any convincing evidence, put forth the hypothesis of transformism or evolution, arguing that life could have arisen from inanimate matter by natural processes and then been differentiated into all the present living species, including the human. Frederick II the Great of Prussia (who ruled 1740-1786), after spending part of his youth in France, gathered many of the *philosophes* at his court, including Maupertuis as head of his Prussian Royal Academy. As the first of the "enlightened despots"[23] Frederick demonstrated in a very practical way how this French version of the Enlightenment could become a powerful political ideology, but he probably did not foresee, anymore than did Voltaire, that this Humanism would eventually lead to the modern "democratic" state for which it would provide a civil religion.

Thus by about 1750 the whole world-view of Humanism had been proposed along with its chief ethical implications, namely, that humans should courageously accept responsibility for their own destiny without concern for any supernatural powers. Peter Gay in his brilliant and sympathetic study of the Enlightenment says:[24]

> The men of the Enlightenment united on a vastly ambitious program, a program of secularism, humanity, cosmopolitanism, and freedom, above all, freedom in its many forms—freedom from arbitrary power, freedom of speech, freedom of trade, freedom to realize one's talents, freedom of esthetic response, freedom in a word, of moral man to make his own way in the world . . . Kant saw the enlightenment as man's claim to be recognized as an adult, responsible being.

The eighteenth century, however, did not end before David Hume (d.1776) had raised serious questions about the great confidence of the *philosophes* in the power of human reason. Hume showed that empiricism does not lead us to a knowledge of universal laws inherent in the physical world nor even moral laws inherent in the human self, but only to the observation of *regularities.* The principle of casuality on which natural laws must rest, which had been accepted by the founders of modern science from Galileo to Newton,[25] is not, Hume argued, either self-evident or empirically verifiable. Nor can we be sure that there is a human "self" behind the stream of our consciousness. Hence we must be content with probable knowledge only, which for practical purposes suffices. The sceptics of the Renaissance had come to adopt the same probabilism, but they used it as an argument for blind trust in Christian revelation. Hume, however, concluded (in very English fashion) that we must just "muddle through." Thus the pragmatic criterion of truth became dominant not only over the correspondence criterion but even over the consistency criterion.

Thus the theologies of Humanism have usually not only identified the human person with the body but have believed this body to be reducible *objectively* to a mechanism, albeit a marvelously complex mechanism, and *subjectively* to a phenomenal stream of consciousness, thus eliminating any noumenal "self." Our self-understanding, therefore, becomes limited to the pragmatic, and this suffices, since the purpose of scientific research is conceived primarily in terms of the power to control ourselves and our world, not to rise to the contemplation of some transcendent God.

2. Darwin and the Evolutionary View of Nature

During the nineteenth century, physicists and chemists continued to develop the Newtonian world-view of inanimate nature, although more and more they realized how great can be the gap between a mathematical model of the world and its physical reality. Consequently they thought that scientific knowledge consists only in more or less warranted hypotheses whose justification is their usefulness in guiding further research and technological invention. In the first half of the century, however, biology and psychology still had established no well-warranted, over-arching hypothesis to explain the realm of animate nature, since the transformist or evolutionary hypothesis still remained without solid empirical verification. To most scientists it still seemed necessary to appeal to the intervention of a Creator to explain the origin of life and human intelligence.

In the second half of the century this dilemma was overcome by Charles Darwin (d.1882) and Gregor Mendel (d.1884) by their convincing development of the theory of evolution through the natural selection of variations produced by genetic mutations. This theory, in spite of many difficulties, has seemed capable of explaining in a reasonable way the origin of all living things through natural processes reducible to the fundamental physical forces of Newtonian cosmology. Nevertheless, Darwinian evolution introduced into Newtonian cosmology a new way of seeing things, quickly exploited by such philosophers as Herbert Spencer

(d.1903). For Newton the cosmos was a static whole whose cyclical processes were governed by universal permanent laws excluding all real novelty. With the success of the Darwinian theory of evolution, however,
scientists began to think of the universe not merely in terms of its lawful
processes but also in terms of its *history*. They no longer explained the
world as a repetition of natural events, but as the unfolding of a sequence
of unique, never to be repeated events. From the evolution of life forms,
this historical perspective was extended both backward and forward (as
Kant in his "nebular hypothesis" of the origin of the solar system had
already anticipated) to the evolution of the whole universe.

Spencer and many others understood this evolutionary view of the
world in the sense of an inevitable progress from chaos to ever higher
forms of order, but this interpretation, at first congenial to those who were
uncertain whether to desert Christianity for a full-fledged Humanism, in
the long run proved, like Deism, an uncomfortable compromise between
the two world views. If evolution implied teleology, then teleology seemed
to imply that evolution had to be interpreted as the work of a sovereign
Creator, a conclusion which clear-headed humanists found oppressive. Yet
it was not until the present century that the discovery of the genetic code
and the development of molecular biology provided a way to understand
biological evolution in a completely non-teleological way. As Jacques
Monod has argued,[26] the apparent purposefulness found in living
organisms and their evolutionary origin can be satisfactorily explained
in terms of systems that are self-directing through "negative feed-back"
after the analogy of servo-mechanisms.

Thus, since even such high levels of organization as those found in
higher organisms can now be explained by evolutionary processes which
are the result of the operation of *necessity* (natural laws) and *chance*
(history) without any imposed or inherent purpose (teleology), humanists
believe we have finally arrived at the fulfillment of the Enlightenment program of giving a purely mechanical explanation of the world and even
of our own human bodily existence, which frees us to give our own human
meaning to that in itself meaningless world.

3. Einstein and the Relativistic World-View

As Hume had brought a tone of moderating scepticism to the
Enlightenment, so at the beginning of the twentieth century Albert Einstein (d.1955) with his relativity and quantum theories eliminated the last
vestiges of the metaphysical aspects of Newtonianism. The general and
special theories of relativity give us a cosmos which can be expressed purely mathematically, but the mathematics used no longer corresponds to

our human experience of space and time. One of the great themes of the Enlightenment had been that nature is completely intelligible to universal human reason, even to the man in the street, if it is properly explained. This conviction at first made some hesitate to accept Newton's new physics, because of his notion of "action at a distance" and the abstruseness of his mathematics, but in the long run it had been possible for Voltaire and other to popularize Newtonianism. With the theory of Einsteinian relativity, however, it became necessary to admit that empiricism does not mean that science can remain limited to the use of models that can be seen or at least imagined. In order to arrive at invariant laws of nature it becomes necessary to understand the world in terms of mathematical relations of a highly abstract character, remote from ordinary sense experience.

This new realization was reinforced by the success of the quantum theory. The ultimate particles of matter, which any mechanistic cosmology supposes to be the ultimate realities, turned out to be unimaginable because, as we have seen, they behave as particles in some experiments and as waves in others. Moreover, the laws which determine their behavior are only probable or statistical, not merely (as Hume thought) because of the defects of our powers of knowing, but somehow because of their own mode of existence in relation to us. In such a theory the reality of "matter" in any ordinary sense seems to disappear and to be replaced by a space-time *field* modified by various energy-states.

Thus our familiar human bodies turn out to be realities inaccessible to our imaginative understanding in terms of our sensible experience of these bodies. To say "I am my body" really means that I am a field through which waves are passing in a period of time. Moreover the "field," the "waves," and the "time" turn out to be no more than mathematical relations justified by ingenious experiments which can be interpreted by several theories. The significance of these mathematical theories about my body is not that they have any experiential meaning for me, but that they promise a better control over my body. The "events" of which my body is constituted are experimentally observable and, within limits, predictable and controllable, since the very acts by which I observe my body are physical acts capable of objectification and manipulation.

The growing realization that the scientific method, although capable of giving us ever greater control over the environment and our own bodies, can never hope to give us any kind of intuitive understanding of ourselves and our world except the kind expressible in mathematical terms, led at the beginning of this century to that ultimate form of positivism called "logical" or "logical empiricism." The logical positivists, of which the leaders were from the Vienna Circle,[27] rejected all forms of alleged cognition, religious, metaphysical, esthetic or moral, which cannot be reduced to the kind of terms used in the sciences, as nothing but subjective feeling-states. Accepting the view of Russell (d.1970) and Whitehead (d.1947) that mathematics can be reduced to logic, they believed it should be possible to express all objective knowledge in a "unified language" of logical symbols and to organize it into a universal axiomatic system. Of course at least some of the symbols in this language had to be interpreted in terms of experimental data, but this data itself could be restricted to quantitative measurements, "pointer-readings."

It was not long, however, before it became evident that this ambitious program could not succeed. After the mathematician Gödel[28] had demonstrated that no formal axiomatic system capable of application even to simple arithmetic can be proved to be self-consistent, it was admitted that no system of purely formal language can function without being interpreted in the terms of *ordinary* language which is culturally conditioned, constantly shifting in its meanings, and inescapably ambiguous. Thus ultimately all scientific observations and experiments rest on the bodily sensations of the human beings who observe, experiment, theorize, and debate their findings in the common, non-technical language of the streets derived from ordinary human experiences. Such commonsense descriptions are somewhat clarified by commentary in technical and symbolic languages but cannot be replaced by them.

Consequently, today in Humanist circles where empiricism is most valued, especially in English-speaking lands, the dominant philosophy of Humanism has become "ordinary language" or, more properly, "analytic" philosophy.[29] The analysts modestly content themselves with attempting to clarify the terms used both in the sciences and public discourse in order to eliminate as much confusion, fallacious reasoning, and illegitimate rhetoric as possible, while honestly recognizing that theirs is a task like that of Penelope's.

Their critics, however, point out that this wholly admirable aim is conditioned in the work of most analysts by a commitment to the moderate scepticism of Hume, the elder philosopher whom they most admire. Thus many analysts not only are sceptical about transcendental realities, as are all Humanists since Kant, but they attempt to show that "God-language" is "meaningless", thus rendering ecumenical discussions with Christians or other theists impossible. They have also generally espoused the Identity Theory of the relation of mind and body (i.e. that mind is nothing but brain states), sometimes in such extreme form as to cast doubt on whether such terms as "consciousness" or "introspection" have any meaning, thus committing themselves to a radical behaviorism in their account of human personhood and to much puzzlement over whether we can even speak of the continuous identity of you or me.[30]

Obviously this kind of scepticism tends to weaken the Humanist faith in the autonomy of human persons as the ultimate realities in our universe.

III. The Idealist Critique Of The Empiricist Interpretation Of Humanism

1. The Kantian Epistemological Revolution

No doubt every world-view when embodied in an actual community of believers tends to polarize. At one pole is a dominant position which adheres unquestioningly to the creed in its simplest and most literal interpretation. At the other in the course of historical development arises a "loyal opposition," persons who remain faithful adherents of this world-view, but are sensitive to the narrowness of its official formulations which they see neglect certain undeniable aspects of human experience, and thus tend to discredit the faith. This loyal opposition also tends to be more ecumenical in the sense that it seeks to mediate between the world-view to which it remains faithful and other world-views to which the first stands in dialectical opposition. Within Christianity a persistent polarization between an unworldly or dualistic and a worldly or incarnational understanding of the teachings of Jesus has again and again manifested itself. Buddhism also has experienced the polarization of the Hinayana and Mahayana traditions, Islam is polarized into Sunnites and Shiites, etc.

In the historical development of the theologies of Humanism it is evident that a counter-current exists within the main stream, a current of thought critical of the "objectivism" of Empiricism and its apparent inability to deal with human *subjectivity*, i.e. with the human "spirit" as well as the human body. Present from the beginning, this critique fully emerged only toward the end of the eighteenth century, after the first triumphs of the Enlightenment and the American and French Revolutions, in the form of *Romanticism*. Romanticism (a term which has been defined in a thousand ways[31]) can be characterized for my present purposes as an effort *within* Humanism[32] to do justice to human subjectivity in the face of the objectivism of the natural sciences and to acknowledge the reality not only of "facts" but of "values." Certainly the empiricists had struggled heroically to develop a science which was "value-free," but, queried the romantics, "Are not values just as real or even more real for us human beings as are mere facts?" A world-view, if it is to function as a religion, must found a value-system, an ethics, and it seems impossible for a value-free science, especially a positivistic empiricism, to provide such a foundation.

The romantic reaction in philosophy, literature, art and music to the scientism of the Enlightenment began in France with Rousseau (d. 1778) (not to mention the Enlightenment leader Diderot himself), and then spread to England and Germany. This reaction climaxed in the first third of the nineteenth century in the general conservative turn against the excesses of the French Revolution. While in some respects it was simply reactionary and led not a few of its adherents (e.g., Chateaubriand, Lacordaire, Friedrich Schlegel) back to Christianity, for the most part the romantics remained firmly within the Humanist world-view in strong opposition to Christianity and firmly dedicated to Englightenment ideals. Indeed, they greatly enriched, strengthened and furthered these ideals, especially the ethics of human *freedom* which, as Peter Gay wrote in the quotation given above, seems the very essence of the Enlightenment and, I would add, of Humanism as a functional religion.

While it was undoubtedly the Swiss Rousseau writing in French who first effectively protested against the exaggerated rationalism of the Enlightenment and strove to make a place within Humanism for feeling, imagination, and intuition, it was the German Immanuel Kant (d.1804), an admirer of Rousseau,[33] who provided the romantic tradition with its fundamental critique of both rationalism and empiricism, that is, of objectivism. Descartes had raised the epistemological problem in full force and had split reality into Mind and Matter. Locke had interpreted Mind in terms of our empirical consciousness, and Hume had concluded that the human mind cannot arrive at any truth other than pragmatic probabilities. Kant, who deserves to be considered the central philosopher (I would say "theologian") of Humanism, its Augustine or Aquinas, saw clearly that it was necessary somehow to integrate the Newtonian world-view of science based on objective natural laws with the subjective human world of purposes, values, and feelings; to bring together Locke, Newton, Hume with Rousseau. Since Locke and Hume were primarily moralists, Kant correctly believed that there were good grounds in their own thought for reconcilation.[34]

Kant based his synthesis on what he called his "Copernican revolution" in which the human person as subject (i.e., as conscious, purposeful, free) faces the world as object (i.e., as determined by mechanical value-free natural laws) and forms a unifed world-view and value system not by seeking to conform the mind to objective reality, but by fitting objective reality to the categories natural to the human imagination and reason and the goals chosen by the will under the guidance of reason. In other words, Kant abandoned the classical notion of truth as "conformity of the mind to reality," in favor of the notion of truth as consistency and pragmatic efficacy. He believed that by this revolution in the concept of truth he had overcome the threat posed by Hume's scepticism to Newtonian science, as well as the atheism and amoralism of the more radical French *philosophes*. It seemed to him that Newton's laws were secure because they rested not merely on empirical facts but on the *a priori synthetic* activity of the human mind obeying the laws of its unchangeable structure. Moreover, for Kant the moral law was also secure because, although he agreed with Hume there could be no theoretical proof of God's existence or of the immortality of the human soul, yet these beliefs on which the moral law is grounded are *practical* truths presupposed by all rational human life and society and thus pragmatically justified.

Kant's philosophy (for all its many loose-ends which critics were quick to note) has remained the most influential of all Humanist theologies because it provides a defense of science and of morality independent of any metaphysics. While Kantianism has dominated all subsequent continental European thought, it has been less successful in British and American lands, always suspicious of grandiose system-building, where a Humean scepticism tends to predominate.

Yet it cannot be denied that most modern philosophies of science including logical positivism, are built on Kantian foundations. Few really question the basic Kantian split of the world into the objective and subjective realms (in place of Descartes' Mind-Matter dualism), nor Kant's elimination of any metaphysics of super-sensible being. Humanists generally suppose that Kant definitively showed that the human mind can never attain to transcendent realities (not even to the God of Deism) without falling into paradox and illusion. But most humanist philosophers today are less convinced than was Kant that the existence of God and the immortality of the soul are pragmatically necessary assumptions. Thus Kai Nielsen[35] thanks Kant for exploding the classical arguments for God's existence and then goes ahead to argue against Kant for a "morality without religion," i.e., without God or immortality.

In Kant's philosophy the material world and the human body as part of that world are governed by deterministic natural laws and have no inherent teleology. The functional purpose which we seem to see in our bodies and the beauty which shines out from this apparent purposefulness in organisms, in crystalline minerals, and in the order of the heavens, Kant explains as a projection of our own subjective sense of purpose and value.[36] Of course this projection is not arbitrary, since the data of our experience can be given this meaning in a consistent way accessible to all human subjects; but, nevertheless, we attribute that meaningfulness to the world out of our own subjectivity, we do not receive it from the world. Thus it is quite understandable that although Kant affirmed, in a pragmatic sense, the existence of God and a moral law, he refused to pray to God, a refusal which expresses the fundamental creed of Humanism, i.e., we must solve our own problems.[37]

2. Idealism and Historicity

The subject-object dualism with which Kant had replaced the matter-mind dualism of Descartes obviously failed to provide a completely unified concept of human existence, yet this dualism has persisted throughout the history of Humanism to date in the form of what C. P. Snow called "The Two Cultures," the scientific culture of facts, and the more humanistic culture of values.[38] The major attempt to overcome this Kantian dualism has been the idealism of the German romantics, chiefly Schelling (d. 1854) and Hegel (d. 1831) who sought to achieve this unification by replacing the notion of "nature" (so basic to ancient and medieval philosophy and still dominant in the Enlightenment phase of Humanism) by that of "history" or "development," "transformism," "evolution," or "process."

But isn't this historicism incompatible with the fundamental article of the Humanist creed, namely, human autonomy? If humanists accept the biblical conception of linear, teleological time, of historical progress toward an *eschaton,* are they not logically obligated also to admit the sovereignty of Divine Providence over history? A way out of this dilemma was soon found by substituting for a transcendent, providential God an immanent evolutionary *élan* culminating in the emergence of human intelligence and human self-creation through culture. This conception of historical progress has proved an important feature of many Humanist theologies.[39]

For Schelling and Hegel Nature is the immanent Spirit's objectivication of itself, which results in the dialectical opposition of Humanity as subject to Matter as object. This opposition overcomes itself through the development in time of human cultures, which are themselves the dialectical embodiment of creative human thought, culminating finally in a total human self-consciousness in which the immanent Spirit itself comes to full actuality. Thus the creative human mind, as the highest exemplification of Spirit, becomes creative of the whole world, not merely in interior subjective thought, but in the objective expression of thought in culture, in art, politics, and technology. This idealist theory of history is susceptible both of a Christian and a Humanist interpretation, and undoubtedly for Schelling and Hegel themselves it was an overcoming of the split between these two opposed world-views. In Chapter 9, I will inquire whether this hope was mistaken or not, but for the present it is enough to say that idealism provided for the Humanists, discontent with the empiricist theologies, a way to reincorporate many of the values of the Christian tradition into Humanism without, however, succumbing once again to Christian faith.[40]

Although romantic idealism was unsuccessful in its many attempts to develop a "philosophy of nature" that could compete in detail with the scientific method of empiricism, nevertheless, its basic approach to nature seemed powerfully confirmed by the fact that empiricism itself eventually took the steps already described from a static "natural law" view of the world to a historical, evolutionary view with Darwin, and to a relativistic, probabilistic view with Einstein.

Moreover idealism, if it was not very effective in the natural sciences, proved much more fruitful in the *Geisteswissenschaften,* the historical or cultural disciplines. Through the nineteenth and into the twentieth century the development of historicism, of cultural relativism, and of the social sciences showed that it is possible to work out a scientific understanding of the individual, the unique, the subjective, the historical. While ancient and medieval thought had accepted the Aristotelian dictum that "science is of the universal," it now became possible for historical, comparative, and statistical methods to find intelligibility in the singular. The result for some Humanists was a reconciliation with the past that the Humanists of the Enlightenment would have found shocking, but for others it led to a radical critique of Humanism itself and the rise of still another new religion, that of Marxism — but that is another story.

It was among the idealists also that the notion of "unconscious mind," which Cartesians would have considered a contradiction in terms, developed.[41] One way of overcoming both the mind-matter and the subject-object dichotomies is to attribute mind-like aspects to matter, i.e., panpsychism. Then it becomes easy to understand how the human mind has emerged through the evolution of matter. Such a primitive mind-like aspect of matter is necessarily "unconscious," i.e., potentially or virtually conscious, and the development of the human psyche is seen as the emergence of fully conscious mind from unconscious but not mindless matter.

Sigmund Freud (d. 1940) was committed to a thoroughly empiricist, mechanistic view of the world, yet through his clinical experience came to believe that this notion of "unconscious mind" could be used to explain human behavior in its biographical singularity as a product of hidden energies struggling into self-consciousness and freedom.[42] Carl Gustav Jung (d. 1961), more directly influenced by idealism, showed that this same notion might be used to explain the whole realm of the subjective, of religion, art, and literature in analogical symbolic terms rather than in the univocal symbols of mathematics, as projections of a psyche seeking its own individuation and integration.[43]

72

Thus Humanist "realism" which tended to dominate the latter half of the nineteenth century in reaction to the Romanticism which had dominated its first half, in the twentieth has itself been somewhat discredited, and Humanists have sought to reconcile these two trends of thought with each other by reinterpreting the whole realm of subjectivity and of values as the expression of the unconscious aspect of human personality which underlies and energizes the rational, conscious aspect. Science itself is now seen as a creative enterprise rooted in subjective motivations in the scientist's psyche. Freud, while remaining a scientific positivist and materialist whose only faith was in the scientific method of solving human problems, came to admit that his own drive to achieve objective scientific knowledge might be a manifestation of hidden sexual frustrations, of boyish curiosity about the facts of life.[44]

This new emphasis on the irrational led in continental European philosophy to the development of " life-philosophies" which emphasized the primacy of action and will over thought, following a lead already given by the idealists Fichte (d. 1814) and Schopenhauer (d. 1860) and reaching expression in twentieth century American pragmatists such as William James (d. 1910) and John Dewey (d. 1952), the latter probably the principal author of *Humanist Manifesto I* (1933). The most incisive thinker of this line was Nietzsche (d. 1900) who came at last to attack historicism by proposing a return to the cyclical view of time ("the doctrine of the eternal return") as the only way to escape theism, while at the same time announcing the coming of "the death of God" and of "Superman" who will be his own God. His exaltation of the Dionysian against the Apollonian spirit, i.e., of spontaneous creativity as against reflective analysis, has been very influential in many recent writers who have urged us to escape the "cerebralism" of modern life and return to "living in one's whole body."[45]

3. Humanism, Phenomenology and Hermeneutic Philosophy

The irrationalism of the life-philosophies provoked the rise of the phenomenological movement headed by Edmund Husserl (d. 1938), originally a mathematician, who was deeply dedicated to the development of philosophy as "a rigorous science without presuppositions," and hoped to return to Descartes without falling into the fatal Cartesian mind-body, subject-object dualism. Husserl, however, was also influenced by British empiricism and by William James.[46]

Two basic notions guided Husserl's work. The first (derived from neo-scholasticism[47]) was that of "intentionality," i.e., the understanding of human consciousness not as a collection of ideas or copies of external realities, in the manner of Descartes and Locke, but as direct awareness of objects (essences) real or mental. Hence the conscious human subject does not exist in dualistic isolation from the world or from other human subjects, but always as intentionality, i.e., in conscious relation to them.

The second notion was that of "the phenomenological reduction," i.e., a rejection of the "natural attitude" of mind taken by the sciences in favor of a critical attitude in which the objects (essences) of consciousness can be described just as they appear (as *phenomena*) apart from any judgment about their concrete conditions of existence. Thus viewed these objects or essences can be the foundation of genuine scientific certitude. Moreover in this light such essences are no longer seen in a dualistic fashion as simply other than the subject but as intrinsically related to but not identified with the subject. In this way it appears to phenomenologists that the subject-object dualism may at last be overcome.

Nevertheless, the movement immediately split into an idealistic and a realistic wing, when Husserl moved toward idealism by introducing the notion of a "transcendental ego" which constitutes all phenomena by giving them their meaning.[48] Husserl, however, did not deny that the "hyletic data" or material out of which the ego constitutes its objects are derived passively through the sense-organs of the body. Moreover, in the last phase of his thought, he developed the notion of the *Lebenswelt* or the world of ordinary lived experience in its concrete historicity as the source out of which the subject derives the empirical content of consciousness.

This concern for the concrete life-experience of the subject led to an important development in the phenomenological movement known as Existentialism. By "existence" existentialists mean much what Husserl meant by "intentionality," i.e., that our human mode of being is "being-in-the-world" except that they understand this not only of cognition but of the totality of human living. Hence existentialists dwell more on ethical themes than did Husserl whose interests were primarily epistemological. Jean Paul Sartre (d. 1980) argued that the greatest human value is freedom, thus professing a fundamental article of the humanist creed.

The meaning of "freedom" is a function of each world-view. Mortimer Adler in his useful study *The Idea of Freedom*[49] has surveyed the use of this term in western literature and distinguished three principal senses: (1) *psychological* freedom, i.e., the capacity to choose without predetermination between alternative means to a goal; (2) *moral* freedom, i.e., the capacity to choose rationally in view of long range and universal goals, rather than for immediate satisfactions and merely individualistic goals; (3) *social* freedom, i.e., the capacity to do what one chooses without social coercion.

Christians tend to make moral freedom primary, to see psychological freedom as the condition of moral freedom (which freedom they believe to be severely restricted by original sin) and to see social freedom as an important value but one limited by the requirements of morality and the common good. Humanists, on the other hand, tend to make social freedom a central value on the grounds that the ultimate measure of morality is the maximum satisfaction of the individual according to individual preferences, whether these are "rational" by some objective standard or not, provided that the social freedom of other members of the society is not infringed.

Humanists tend to see the notion of "moral freedom" as simply an excuse for the imposition by one group of its own values on others. Humanists of the liberal persuasion are convinced that expert opinion and public authority must be used to promote social progress. Those of the conservative or "libertarian" tendency reject such social control and favor an extreme individualism within the broad limits of law and order. Nevertheless, both wings of Humanism agree that the goal of social policy, whether "libertarian" or progressive, is to establish those conditions which they believe will foster the highest degree of individual freedom in the sense of freedom from social coercion.

As for psychological freedom, humanist philosophers, for the most part, tend to understand this only as a determinism of motivation consistent with the view that human behavior is ultimately reducible to the same natural forces as that of animals. Thus the free individual is one who is motivated by realistic goals and is not held back from attaining these goals by unconscious compulsions or inhibitions, or by traditional prejudices, yet the behavior of such "free" individuals, just as much as that of the unfree, is ultimately determined by genetic, environmental, and cultural factors. Sartre, however, so valued human freedom that he argued for atheism on the grounds that an omnipotent God would be incompatible with our freedom, and he rejected every type of determinism, contending that no one can excuse him or herself on any grounds from full moral responsibility.

For Sartre this world into which we are thrust is without inherent purpose or value in itself.[50] It is "absurd," since there is no God, and its "facticity" nauseates us like forced-feeding. This fate of ours, however, does not excuse us from moral responsibility but demands that by our own free choice we give our own lives the only kind of meaning they can ever have. A difficulty of this view, heroic as it may be, is that it seems to reassert the Cartesian dualism of human spirit struggling against its own body and the world of matter, which phenomenology hoped to overcome.

Maurice Merleau-Ponty (d.1961), a friend of Sartre's, attempted to meet this difficulty by giving an emphasis to the role of the human body in knowledge greater than that given it by most phenomenologists.[51]

Merleau-Ponty was deeply interested in the scientific researches of modern psychologists, especially those of the Gestalt school, but, along with other phenomenologists, he was strongly critical of modern scientists for their "natural attitude" leading to a neglect of the role of the subject in knowledge. At the same time he criticized Husserl for placing the ultimate source of knowledge in the clear light of the conscious ego. Rather our subjectivity has its roots in the dark pre-conscious life of the body through which we are first present to the world. Only by a long process in which sense perception plays the mediating role does our subjectivity emerge into the light of self-consciousness. Hence, although Merleau-Ponty until late in life shared Sartre's rejection of the existence of God in order to defend human freedom, he always believed that this freedom is limited, not absolute. We are free only within the limits placed by our historical situation and by the bodily conditions which that involves. Indeed our "body" is not only our biological organism, but the whole weight of the past, of our environment and our culture and our personal biography which our freedom enables us to transcend, but only to a degree. Thus Husserl's late emphasis on the *Lebenswelt* is further underlined by Merleau-Ponty .

The consequences of this trend are still more striking in the greatest disciple of Husserl, Martin Heidegger (d. 1976) who is perhaps the most widely influential philosopher of the latter half of this century.[52] Heidegger, in Hegelian fashion but with a very different result than Hegel, rewrote the whole history of western thought to show that philosophy as it has been understood from Plato to Nietzsche has at last come to its fated end where it must yield to some more profound view of Being, perhaps in union with eastern thought.

For Heidegger the historicity of human existence implies that in every age and culture our world-view (Being as it reveals itself to us) cannot be understood simply as a human creation, but is rather our *fate*. We are born into a culture, a language, and a system of ideas which are of course man-made and yet which have their own inner logic that we who have made them cannot escape. Even modern science, for all its claims to objectivity, is only a subjective interpretation of reality, an expression of our human "will to power" as Nietzsche made evident. The history of the West has been the "forgetting of Being," that is a progressively increasing preoccupation with "things," with our own will to create our own world. But thus to forget Being is to forget ourselves, since authentic human existence is not only the courageous facing of our own inevitable death (our historicity), as the existentialists and the early Heidegger himself thought, but also, just because we are fated to die, "the shepherding of Being."

Heidegger's implication is that we are destined by history in the twentieth century to return to some more contemplative attitude toward the universe in which we dwell, its heaven and earth, its "gods" and mortals — an attitude more poetic than "philosophical" as that term has been understood since Plato and Aristotle, yet demanding a new type of "thinking" more radical than philosophy.

This conclusion, however, is ambiguous with an ambiguity which Heidegger scrupulously maintains in view of future historical developments which must simply remain open. On the one hand his thought can be taken as an obituary of Humanism and a return to a Christian or other theistic, or at least transcendental, view of human reality. On the other (although this was probably not Heidegger's personal intention[53]) it can be understood as an atheistic way of answering all the criticisms against Humanism raised by previous romantics. If Being is not something transcendent to human experience, but simply human experience itself in its historical development, in which development to date Humanism has been the fated outcome, then should we not simply achieve authenticity by accepting our fate, with the realization that in time Humanism will be replaced by some new world view just over the horizon? Thus, does it not seem that Heidegger (already anticipated by Oswald Spengler, Arnold Toynbee, and other philosophers of history) has given us a kind of Götterdamerung of the Enlightenment with a stoicism free even of the illusory sense of responsibility with which Existentialism had invested it?

Out of this Heideggerian twilight has emerged the hermeneutic philosophy of which I have already written in Chapter 1. Just as an analytic philosophy gave to the empiricist interpretation of Humanism an emphasis on the clarification of language in the hope of relating the scientific worldview with ordinary human experience, so the romantic tradition has come with Heidegger to an emphasis on language as the expression of the relation of the human person to the world. Heidegger in his late period more and more stressed that Being comes to light in us through language which both uncovers and conceals meaning. As an anti-metaphysical thinker he abandoned traditional philosophical terminology for a language which he believed was more closely connected with fundamental human experiences.

The developers of hermeneutic philosophy, of whom Hans-Georg Gadamer is representative,[54] stress that we all live in different worlds by reason of our cultural histories and that we will remain fatefully imprisoned in these worlds unless we find a way to transcend these communication barriers which separate us. One of the latest styles in philosophy, Structuralism,[55] attempts to find deep structures in language which make it possible to discover points of contact between diverse linguistic expressions underneath their surface differences. These deep structures may be fixed in our genetically determined neural circuits and may thus insure that all members of our species have in common certain ways of seeing the world, somewhat like Descartes' innate ideas or Kant's *a priori* categories. Or it may be that they are rooted, as Jung thought, in a collective unconscious which is somehow the depository of the historic memory of the human race. Or they may themselves be subject to human restructuring either by genetic engineering or by new cultural forms. In any case as humans we have both possibilities of sharing the same global world and possibilities of ending in Babel.

In the midst of these diverse Humanist philosophies it has become very clear in this century that the scientific language of the empirical sciences may have a privileged position in a technological world under human control, but it cannot be the only language. Our human bodily existence, our human historicity, our human creativity, limited as Merleau-Ponty showed, by our biological and cultural past, can close us to one another and to our common future unless we find a way to be open to other world-views, other theologies than our own. That is as true for Humanism as for Christianity. Consequently Humanism to be true to itself must assimilate a hermeneutic point of view, and thus open itself to dialogue even with the Christianity against which it revolted and with the Marxism which revolted against it.

4. The Human Body in Art and Literature

It might seem that the most direct, concrete evidence of the Humanist attitude to our bodily condition might be found in art and literature rather than in philosophy, just as Christian art often reveals more of the Christian world-view than the lucubrations of theologians. Three broad trends within Humanist culture can be noted.

The first of these is the influence on art and literature of the Humanist effort to develop a democratic society on the basis of an industrial economy. In such a society there is strong pressure for every citizen to become a productive worker in industry, a part of the machine. Consequently the rococo style of the early eighteenth century still dominated by the aristocracy soon gave way to middleclass sobriety and good sense.[56] Class distinctions in costume became less pronounced. Men came to dress in "business suits," while women's styles, although more and more commercialized and hence constantly shifting, more and more simplified until today they tend to the unisex. Similarly, architecture has progressively become more and more functional and, after passing in the nineteenth century through a series of experiments with historical styles (the "Gothic

revival," the "Greek revival," "romanesque," etc.) has settled down (until recently) to the international style of unornamented boxes of glass and steel.[57] In line with this trend the human body is thought of as a mechanism to be kept in good running order by drugs and surgery. The modern medical center in its stark architecture and its industrial processing of the human body swathed in sheets or hospital gowns or nakedly exposed to the knife or the x-ray probe seems to sum up this understanding of the human person as machine.[58]

The second trend which influences art and literature is the intense Humanist concern to "humanize" this mechanistic culture. In the visual arts the results have been surprising. The American and French Revolutions at first attempted to express republican ideas by still another classical revival in art, but this soon gave way in the nineteenth century to an emphasis on "realism," i.e., the noncommittal (or occasionally propagandistic) depiction of daily life, free of transcendental symbolism.[59] This trend gradually triumphed over the "historical painting" of the academies, but very quickly was itself replaced by impressionism, symbolism, expressionism, and surrealism, so that by the end of the nineteenth century the visual arts had become centered on the subjectivity of the artists, their reaction to their world, rather than on the objectivity of the world.

This development went through many stages but is marked by the gradual loss of interest in the human figure as an *object*. Renaissance and baroque art had concentrated on the human figure and glorified the nude body, male or female, but in Humanist culture we at first see a growing interest in the environment in its own right. Thus landscape paintings, architectural scenes, still-lifes were created for themselves not as mere settings for the human figure.[60] The nude body in the nineteenth century was largely left to academic historical painting, and when it began to appear in realism it was that of the female studio model as in Courbet's *The Artist's Studio* (1855) or Manet's *Lunch on the Grass* (1863). Soon it was reduced to an occasion for the impressionic study of light as in Renoir's nudes or of proto-cubistic composition as in Cezanne's.[61] In surrealistic art the body is dismembered and becomes occultly symbolic. This Humanism centers itself not on the body as seen objectively, but as felt from within. At the beginning of the present century with the rise of cubism the visual arts became more and more non-objective, non-figurative, "abstract," and when these abstractions retained an objective reference this tended to be surrealistic, that is to refer to unconscious realms rather than to every-day life.[62].

The famous statement of the nineteenth century poet and critic Theodore Gauthier that "all modern art approaches music" seems fully verified in our times in the sense that as music expresses moods and feelings without specifying the objects which give rise to them, so non-objective art by patterns of color and shape and by textures and lines that embody human gestures expresses not objects but subjective events.[63] Yet modern music, divorced from the context of church or court, after passing through its romantic phase of greater and greater emotional expressiveness, has today itself become largely a kind of abstract expression not so much of emotion as of the composers's mental computations. Consequently, Humanism, which hoped to find in the fine arts a secular equivalent of Christian contemplation, toward the end of our century is faced with the difficult problem of making art once more a part of democratic life.[64].

It is in response to this unsolved problem of humanizing industrialized society, that a third trend has been a constant feature of Humanist culture, namely the search for an anthropocentric mythology to replace the symbol system of Christianity. The most obvious way this can be done is through *science fiction* in which human beings live together with robots in artificial worlds and themselves may be robotized by the insertion of artificial organs or ruled over by computor lords. Myths of this type are popular in fiction and in film and have influenced modern art from the time of the Italian Futurists.[65] A more subtle solution was that of Art Deco, the Bauhaus and other schools of design which enthusiastically accepted machine-living but sought to give it an esthetic value by the artistic designing of its instruments and environments. If a Greek vase can be beautiful, why not a steam radiator, or a laundromat? The ultimate in this approach is provided by those city-planners who attempt to construct a total living-space combining gardens and advanced technology, all designed with a view both to function and beauty. Such efforts, however, demand a degree of social control which a democratic Humanism finds difficult to achieve.[66]

Perhaps more effective than the effort to redesign our environment and our instruments has been the effort to change the way we *perceive* our modern world by the use of the media of communication. The Humanist evaluation of the world is communicated in ways never before available to any functional religion through advertising, political propaganda and fiction which reach into every home not only in printed material but more directly by radio and television. In this way the esoteric formal inventions of the visual arts and of music are popularized and penetrate the mass consciousness.

The literary genre which has been most characteristic of Humanism (as the epic and the verse play were characteristic of older cultures) has been the novel which has also formed the basis of much that appears in the cinema and television.[67] The novel which achieved its modern form in the eighteenth century is largely free of myth, yet it is functionally equivalent to it. The novel does not ordinarily take place in mythic time or place but in our contemporary world or in a definite historical setting, and it deals not with gods nor even demi-gods but with ordinary human beings in earthly settings. Generally its narratives deal not with transcendental themes nor even with fundamental moral decisions but with the interplay of human relations viewed primarily in *psychological* terms. In this way the world-view and value system of Humanism with its conception of what it is to be human, especially its emphasis on human historicity, is communicated throughout society, so that the novelist, or the cinema director who dramatizes the novel, becomes the effective preacher of Humanism, replacing the Christian Bible narratives and the pulpit homilist who commented on them.

In the twentieth century novel since Joyce and D. H. Lawrence, the Humanist understanding of our bodily condition and especially of human sexuality has become very explicit, but it has been present in all the great novelists. A Dickens, a Flaubert, or a Tolstoy have presented characters in the stream of history whose physiognomy, bodies, gestures, voices, even their smell has sunk into our imaginations and help to form our own image of ourselves as earthly beings. Perhaps no art form has ever been so *physical* and so *historical* as the novel, and it is in the novelistic picture of life that Humanism has been most successful in overcoming the subject-object dualism that has plagued its more formal theology.

Cinema and television, because of a flexibility which far surpasses the theater, have been able to dramatize the novel and remove it from the realm of the active imagination to the passive realm of seeing and listening. They have also influenced and profited from the developments of modern art so as to adopt techniques not only of realism, but of impressionism, expressionism, surrealism, cubism, and abstractionism. In recent science fiction movies we find ourselves in a world that transcends the space and time of earth and enters a world of light and darkness where the form of the human body takes on protean shapes and the distinction between nature and machines vanishes. The objective world seems totally absorbed into the subjectivity of the human inventor and yet that subjectivity gives birth to nothing but robots engaged in star wars.

Recent literary criticism has been much influenced by hermeneutic philosophy and has influenced it. One radical trend is the "Deconstructionism" of Jacques Derrida and others.[68] Borrowing from Heidegger the notion of the "destruction" of classical philosophies by re-reading them to uncover their deeper meaning, Derrida has come to the conclusion that every effective critical reading of a literary text is actually the creation of a new work of art which may be radically different from what the author consciously intended or what other readers have found in it. Moreover, he has extended this notion to a general epistemological theory according to which all of "objective reality" is a "text" which can be known only by deconstructive interpretations. Heidegger, I believe, would see this as another forgetting of Being i.e. of the will to power of our technological culture.

IV: Resolution Of The Inner Contradictions Of Humanism

1. The Marxist Critique

The romantic criticism of Humanism has been from within Humanism itself, and has been intended to enrich, not destroy Humanism. If taken too seriously it seems to prove that in a technological world dominated by science, dehumanization is inevitable. If taken too lightly, the humane, intuitive, and ethical aspects of Humanism are reduced to mere wishful thinking unsupported by empirical reality. A vital Humanism must give full weight to the dimensions of reality pointed out by Romanticism, without in any way undercutting the usefulness of the scientific world picture as a firm ground for human control over the world. As long as romantic Humanists criticize modern science and technology in a reactionary way, they seem to fall back into the irrationalism and obscurantism of which they accuse Christianity. A Humanist fideism is as incredible as a Christian one.

In the middle of the nineteenth century it had become apparent that Romanticism, for all its stress on human freedom, might become a support for reaction since its historicism could be interpreted as traditionalism. Hegel, the great philosopher of history, was interpreted by the Right Hegelians as a defender of the Prussian *status quo* and by the Left Hegelians as a proponent of permanent revolution.[69] In turning to the left Marx said he was "turning Hegel on his head" by reasserting materialism (as the only philosophy consistent with scientific progress) against idealism, but at the same time he opposed mechanistic materialism and advocated a "dialectical" or "historical" materialism. Marx admitted that human history has been determined by the laws of Darwinian evolution and of economics. Human thought has been rigidly conditioned by the class interests it was developed to serve, and has been largely a rationalization of the material aims of the oppressing classes. Marx, however, also believed that the advent of modern science and technology, for which he believed his own philosophy would serve as a critical foundation, would make it at last possible for us to come to full humanity, to overcome our alienation from ourselves, and to achieve a genuine freedom to control and remake the world and even our own human nature.[70]

Marxist Humanism, therefore, while it rests on a scientific and antitheistic world view, does not regard "values" as merely subjective as we have seen Humanism proper tends to do. For Marx the ethical values of bourgeois society were a deceptive ideology masking the greed of the oppressors and its religion was merely "an opiate of the people," but he believed that in the coming communist society ethical values will arise from fundamental human needs objectively determined by the scientific study of our human biology and its natural environment.[71] The humanly created values of culture — art, music, literature, recreation — will no longer express the values of the oppressor, but will become the expression of the feelings and aspirations common to productive human beings and will promote their human solidarity and comradeship.

Marxism soon separated itself from Humanism to become a distinct and antagonistic world-view precisely because it saw in the empiricist-romanticist polarization within Humanism an internal contradiction which prevented Humanism from ever achieving the real goals of the Enlightenment.[72] Marxism has always attacked the idealist tradition of continental philosophy as a hidden form of Christianity used by the capitalist classes to prevent the logical development of modern science, and has argued that the outcome of Romanticism has been Fascism and National Socialism with their appeal to obscurantist traditionalism. On the other hand Marxism also criticizes empiricism for asserting that science must be value-free, thus leaving the whole area of history and ethics to private subjectivism beyond political commitment.

Humanists have not been able to deny the seriousness of this Marxist criticism and have been aroused to assimilate many elements of Marxist thought into their own world-view, especially into its more liberal version. Thus the *Humanist Manifesto II* (1973) explicitly advocates some type of socialism as its political program. Humanists, however, for their part also criticize the Marxist solution of the object-subject dilemma as reactionary, since it has been purchased at the price of a dogmatism which has to be enforced by a totalitarian state, obviously contrary to the Enlightenment advocacy of free inquiry and debate. Moreover, Humanists point out the apparent lack of creativity in Soviet science and art as evidence of the obscurantist effects of this dogmatism.

2. Process Philosophy as a Revision of Humanism

Within Humanism recently new attempts are being made to overcome the subject-object dichotomy without falling into that romantic denigration of science (such as we have noted in the phenomenological movement) which would undermine the very foundations of Humanism. In the United States the most influential of these attempts today is what is called "process philosophy." This is characterized by an effort to develop a metaphysics or ontology (as opposed to analytic philosophy's anti-metaphysical attitude) which will, however, be grounded in modern science (as opposed to phenomenology's by-passing of science). It seeks to understand modern science not merely as a set of mathematical symbols whose only significance is in technological application, but as a valid description of the cosmic reality which provides a secure ground for a value-system.

The initiator of this movement was Henri Bergson (d. 1941)[73] who attacked the reductive character of scientific intelligence as interpreted by positivism, but at the same time accepted the scientific evolutionary view of the world and of the human person as the starting point of his philosophy. Bergson believed that the world revealed by science is not really one of rigid mechanism but a process of "creative evolution" which reaches its climax in human consciousness and freedom. This consciousness operates in the scientific or technological mode as analytical intelligence, but also in a deeper intuitive mode which makes it possible for us to understand the creative flow of natural processes and human history and to transcend determinism as conscious participants in that same creativity. Bergson, by birth Jewish, was led by this philosophy into Christianity, and it was a Jesuit priest, Teilhard de Chardin (d. 1955), who gave to the Bergsonian world-view the most concrete elaboration.[74] Nevertheless, it is obvious that such a process view, which has not a little in common with Marxism, need not be Christian, but holds possibilities for Humanism since it centers on the concept of "creativity" understood in an immanent rather than a transcendent sense.

This tendency received classical form in the process philosophy of Alfred North Whitehead (d. 1947) and his most prominent American follower Charles Hartshorne.[75] Whitehead's system, which he entitled a "cosmology" but which he was also quite willing to call a "metaphysics," was theistic, but in a way which transforms the traditional Christian notion of God even more radically than had Deism, since it makes God dependent on the world for his own actualization and self-knowledge.

Whitehead's principal aim, however, was not to develop a theodicy or even an anthropology, but to meet the crisis caused by Einsteinian physics which had at last overthrown the mechanistic view of space and time. Under the influence of Leibnitz and Bergson, Whitehead hoped to find a new way of interpreting modern science that would not reduce it to mere mathematicism or pragmatism. He did not depart

84

epistemologically from the accepted method of modern empiricism, namely, the validation of a set of hypothetical axioms by deducing from them empirically verifiable conclusions. But instead of hypothesizing that the fundamental explanatory principles of all empirical phenomena are matter and energy as classical mechanism had taught, he accepted Einstein's new view that matter itself is reducible to energy. Yet he proposed that we conceive this "energy" not in the purely mathematical way that Einstein had done, but as "creativity." For Whitehead energy is the fundamental principle of cosmology. It is not a mechanical force but a creative impulse (Bergson's *élan vital,* Teilhard's "radial energy"), best understood on the analogy of our artistic or esthetic impulses and responsible for the interiority or subject aspect of all objejctive reality, coming to full consciousness at last in the human psyche. He believed that such a hypothetical first principle could be justified by its superior ability to "save the phenomena" by overcoming the dualism between the objective and subjective aspects of our experience for which mechanism had failed to account.

Whatever is to be thought of these efforts of the process philosophers (whose views I will discuss more fully in Chapter 8), it is evident that if Humanism is to remain viable as a world religion it must be able to interpret the results of modern science in ways that do not lead to a reductionism by which the creative purposefulness of human existence is made to appear a vain illusion. Since for Humanism human creative freedom is the supreme value, if the scientific world-picture cannot be understood in such a way as to permit us to believe both in freedom and in science as the chief instrument of freedom, Humanism will die as certain other world-religions have already died.

3. The Strength of Humanism

Although I am a Christian, I do not think that Humanism will soon die, but rather that it will continue for a long time as one of the great world religions alongside Christianity, Judaism, Islam, Hinduism, and Buddhism. Its present crisis, caused in part by the attacks of its child and rival Marxism, and its own internal contradictions (its difficulty in relating facts and values, objectivity and subjectivity, and in controlling the abuses of technology resulting from the instability of its value-system), seems no deeper than other crises through which each of the great surviving world religions has passed. Humanism's strength lies in the values expressed by the lines of the *Humanist Manifest II* which I once more recall:

> Nor is there any guarantee that all human problems can be solved
> or all questions answered. Yet critical intelligence, infused by a

sense of human caring, is the best method that humanity has for resolving problems. Reason should be balanced with compassion and empathy and the whole person fulfilled.

Humanism's goal is the freedom and autonomy of individuals who each define what "fulfillment" means for them. Humanism recognizes, however, that human cooperation is necessary to achieve this goal. We must help each other to acquire education and we must use government to protect us in our property, health, and freedom from coercion. We can do this only if we all can participate in political decision-making to see that our rights are thus protected.

Furthermore, Humanism is convinced that this freedom for individual fulfillment is gounded in the continued search for objective truth, i.e., truth arrived at by free discussion and by the critical standards of modern science. Only such progressive search for truth and its free communication among all members of society can guarantee political freedom, the protection of rights, the advance of human control over the environment and social change, and the individual's control over his or her own body and psyche.

Hence, in opposition to both Christianity and Marxism, Humanism rejects the notion that human knowledge can ever be more than probable ("Nor is there any guarantee that all problems can be solved or all questions answered"), although it takes as *pragmatically* certain that free inquiry according to the scientific method and public debate in political matters must be protected against every form of "dogmatism" or "censorship." Consequently, it fosters freedom for other religions, including Marxism and Christianity, provided that dogmatic religions which claim certain truth confine their proselytism within limits that do not threaten the dominance of the basic principles of Humanism.

Finally, Humanism's emphasis on the fulfillment of the individual as self-defined means that it is greatly concerned with "compassion and empathy," with promoting individuals' subjective sense of well-being, their freedom from suffering and from morbid psychological states, their ability to relate honestly and intimately with others. Humanism vigorously resists attempts to sacrifice individual happiness to abstract moral ideals, or to the aims of institutions and governments. Hence it rejects the Marxist willingness to sacrifice present generations to some future utopia, just as it agrees with Marxism in rejecting the Christian promises of heaven. Humanism seeks happiness for all in earthly life.

Following this creed, Humanism has contributed much to human culture. First, it has been largely responsible for the remarkable advance of science and technology which in the last two hundred years has given us an increasing fund of reliable knowledge and of technical control over our human destiny. Second, it has produced the modern democratic order with its concerns for human rights which has abolished slavery, inaugurated universal literacy and franchise and promulgated the concept of fundamental human rights. Third, it has produced a vast wealth of philosophical reflection, literature, art and music, along with a development of depth psychology which has made it possible for us to understand ourselves more fully and express authentically our real needs to each other and to overcome our irrational guilt and fears.

These strengths of Humanism as a world-view and a value system which seem to guarantee it a long future as a world religion, rest in large measure on Humanism's straightforward acceptance of the bodily character of human existence. Beginning with the Enlightenment, the notion that we human beings are souls whose presence in the body and in the material world is an unfortunate accident began to fade. Humanists are convinced with at least a pragmatic certainty that we humans are bodies and only bodies, although bodies endowed with remarkable brains that enable us to acquire scientific knowledge and to empathize with one another so as to be able to transcend in a measure the determinism of our environment and our common and personal histories. Because we are such thinking, feeling bodies, our freedom and autonomy are fundamental values. Since, for Humanists, we probably have only one life to live, it is imperative that we have maximum control over that life, free from external coercion or the imposition on the individual of values or goals other than those he or she has chosen. Yet even our own body-selves should not be regarded as simply given, since my self is not a static substance but a stream of consciousness, so that what counts for me is how I experience that stream of events and control it for my own purposes.

Humanism seeks to create a system of values in a world which of itself is without purpose and with a body which is the product of blind evolutionary forces. This value system exists in the perspective of the inevitable death of each individual and finally of the human family, yet it is possible for some men to attain the humanist ideal which Peter Gay ascribes to one of the chief architects of Humanism, David Hume:

> He was so courageous that he did not have to insist on his courage. He followed his thinking where it led him, and he provided through his own life . . . a pagan ideal to which many aspired but which few realized. He was willng to live with uncertainty, with no supernatural justifications, no complete explanation, no promise of permanent stability, with guides of merely probable validity; and what is more, he lived in his world without complaining, a cheerful Stoic. Hume, therefore, more decisively than many of his brethren

in the Enlightenment, stands at the threshold of modernity and exhibits its risks and possibilities. Without melodrama but with sober eloquence one would expect from an accomplished classicist, Hume makes plain that since God is silent, man is his own master; he must live in a disenchanted world, submit everything to criticism, make his own way.

V. Conclusion Of Part I

In Chapter 1, I posed the problem that modern technology promises to give us humans the power to re-create ourselves by a profound alteration of our bodies, thus undermining all of the classical ethics or value-systems which were based on the supposition that "human nature" is the fixed standard by which values can be measured as natural or unnatural, reasonable or irrational, good or bad. What standard now is to measure the values according to which we are to remake ourselves? Thus the technological malleability of our bodies poses for us an inescapable problem about who we are and what we ought to become.

Since this new technological power is the fruit of modern science, in Chapter 2, I inquired what kind of self-understanding can we obtain from current scientific study of the origin and structure of the bodies we now are? What are the materials out of which the superpersons of the future must be reconstructed? From this necessarily sketchy account of scientific anthropology two especially important facts emerged. First that modern science no longer attempts to explain the world and human emergence only in terms of natural laws which govern the ever-repeating cycles of natural events, but has become increasingly given to explaining the present world *historically* as the result of a unique series of events in which chance plays a determining role.

Therefore, any modern effort to achieve human self-understanding by the help of natural science must interpret modern science not as purely objective truth independent of history, but as a part of humanity's historical effort to understand its own total experience through critical reflection. As relativity physics has exposed the myth that the scientific observer is totally outside the world observed as his or her object, so today we must understand the results of all scientific inquiry as the activities of human subjects, namely, the scientists who engage in these inquiries and the human community to which they interpret their findings. In such a context the "science of material bodies" is an activity of *human* bodies reflecting on themselves and their world, human bodies who think, feel, create.

Obviously this scientific truth as it exists in the lives of any actual human community is too tentative and too full of gaps to stand alone as a world-view adequate enough to be the basis of the value-system by which that community acts together. The empirical findings of science require interpretation in terms of their human relevancy, an interpretation which, if it can be done at all, seems to be the task of philosophy. The technological application of these scientific findings, because it involves a pattern of life for the community and its members, also requires a *theology* critically explicating the "faith" by which this community justifies its life commitments. Consequently in Chapter 3, I have described the interpretation supplied by the world-view and value-system which I have called Humanism, because Humanism in fact dominates the life of the "free world" just as Marxism does that of much of the rest of our globe. Although modern science originated within the Christian community, it has been fostered and developed chiefly under the patronage of the Humanist religion, and it is the Humanist interpretation of the results of modern science which has won the widest acceptance by scientists themselves. I tried to show as sympathetically as I could for a religion which rivals my own, the great service which Humanism has performed in promoting the development of a scientific anthropology and in applying it ethically to the development of the modern concept of human freedom and human rights.

At the same time I showed that Humanism has from its outset been faced by an internal "contradiction" which it must overcome if it is to remain a viable world religion in the face of Marxism and the growing resurgence of older world religons, especially in the Third World. This conflict arises from the fact that modern science, as it is usually interpreted, is "value free." If then, as Humanism believes, science is the most objective and reliable kind of truth available to us, is not the whole realm of values by which the world becomes humanly meaningful relegated to mere arbitrary, irrational subjectivity? Does not this undermine the possibility of the public debate and realistic consensus on which the existence of any free human community depends? What then becomes of the principle article of the Humanist creed, the conviction that human beings can work together to make for themselves a better world? To evidence the seriousness of this problem I showed how the romantic wing of Humanism has made many attempts, most of them modifications of Kant's idealistic notion of truth, to reconcile facts and values, but without complete self-satisfaction.

Two outcomes of this effort by Humanists "to get their own act together" appeal to me as promising to provide common grounds between Humanists and Christians for living and working together for a more human world. The first is a result of the increasing historical character of scientific explanation already mentioned. Phenomenology, and especially hermeneutic philosophy, by establishing the inevitability of philosophical pluralism have given Humanists a justification from within their own horizon to be open to dialogue with other world-views. If the historicity of all human truth, even scientific truth, is granted, then Humanists no longer need, as did their fellow Humanists of the Enlightenment and even of the nineteenth century out of loyalty to their commitment to scientific empiricism to dismiss out of hand the older "religious" world-views as necessarily irrational and anti-scientific. Modern Humanists are able consistently to view their own Humanism as one among many ways of interpreting human experience, all of which, at least for purposes of dialogue, stand on an equal footing or at least are mutually tolerable, provided that those who make each of these interpretations are willing to deal seriously and not merely selectively with the data of experience. With this proviso, none of these interpretations should be labeled in advance as irrational, anti-scientific or anti-modern.

The second outcome of the intra-Humanist efforts at self-consistency which to me seems favorable for dialogue is the interest which process philosophy shows in the development of a cosmology or philosophy of nature. In the Christian tradition, as in many of the older world religions, the cosmos is *inherently* purposeful or meaningful. We do not merely impose human meaning on the world, but rather we discover that meaning and enrich it. Because we are part of that meaningful cosmos by reason of our bodily existence, at least one of the clues to the meaning of life is the meaning and purpose inherent in the structure and function of our own bodies. The philosophy of Kant which has so dominated Humanist thinking as a way of overcoming its internal polarization has seemed to close off this way of understanding the results of science, although many voices from the romantic wing of Humanism have been raised against this Kantian door-slamming. Process philosophy, however, once again opens the way to an interpretation of the scientific world-view which would permit us to conceive human meaning as emerging out of the material cosmos and the human body as the ground of the creative human psyche.

Perhaps, however, there is also an obstacle to this dialogue on the part of the Christian tradition arising from the fact that although Christians profess to find meaning in matter, yet throughout their history they have been troubled by their own "inner contradiction", since they have also seemed to reject matter and the body in favor of a one-sided concern for "the salvation of the soul." It is this problem to which I now turn in Part II of this book.

Notes

1. Peter L. Berger, *The Sacred Canopy: Elements of a Sociological Theory of Religion* (New York: Doubleday Anchor Book, 1969), p. 107.

2. "The word *secularization* began as an emotive word, not far in its origin from the word *anticlericalism*. Sometimes it meant a freeing of the sciences, of learning, and of the arts, from their theological origins or theological bias. Sometimes it meant the declining influence of churches, or of religion in modern society. Then the sociologists, heirs of Comte, aided by certain historians and anthropologists, did a service by showing how deep-seated religion is in humanity and in the consensus which makes up society. They, therefore, made the word unemotional; a word used to describe a process, whatever that process was, in the changing relationship between religion and modern society, a process arising in part out of the industrial revolution and the new conditions of urban and mechanical life, in part out of the vast growth in new knowledge of various kinds. Modern sociologists (and others) have not always kept the word in this unemotional plane. Sometimes they have used it as a word of propaganda. This is because this study is like trying to write the history of the Reformation in the year 1650 — everyone is still committed." David Martin, *A General Theory of Secularization* (Oxford: Blackwell, 1978), p. 264.

Some Protestant theologians, e.g. Friedrich Gogarten, have hailed "secularization" as "desacralization of the world" which they claim has a positive biblical value since only God is sacred. Joseph Comblin, "Secularization: Myths and the Real Issue," in Roger Aubert, ed., *Sacralization and Secularization, Concilium* vol. 47 (New York: Paulist Press, 1969), pp. 121-136, attacks this view as an identification of faith with the philosophy of idealism. In every society a balance must be struck between the sacred and the secular, passivity and technology, myth and demythologization, law and freedom. "In short we might well say that two things are taking place in the world right now: (1) the rearranging of the frontiers between the sacred-mythic and the profane-scientific, prompted by the new, urban society; (2) the demise of Roman society and the role which religion played in it." (p.131). "Let us not interpret the present changes in mythical terms. We are not confronted with the decline of the world, or the birth of a new humanity. One civilization, the Greco-Roman civilization of the past, is disappearing — and with it, the role which religion and the clergy exercised in that civilization. A new urban and industrial civilization is coming to birth, causing a new displacement in the sacred and mythic elements. But there is no reason to enmesh all this in the framework of idealist philosophy. To create a theory of secularization is to echo the dream of Joachim of Fiore in modern terms. It is to ask that history purify Christianity and perform the task allotted to each new generations of Christians." (p. 132). Enrique Dusell, "From Secularization to Secularism: Science from the Renaissance to the Enlightenment", *ibid.* pp. 93-120, (useful bibliographical note p. 93 f.), distinguishes "secularism" (my "Humanism") from secularization which he seems to identify with "modernization."

The predominantly Catholic authors collected by Albert Schlitzer C.S.C. ed., *The Spirit and Power of Christian Secularity* (Notre Dame, Ind.: Notre Dame University Press, 1969), seem to understand by "secularity" simply a positive attitude toward earthly life as contrasted to the medieval *contemptus mundi*. On the other hand David L. Edwards, *Religion and Change* (London: Hodder and Staughton, 1969), argues that "Secularization occurs when supernatural religion — that is religion based on 'belief in God or a future state' — becomes private, optional, and problematic." (p. 16). "To sum up 'the coming of age of mankind,' the sense that man is 'on his own now,' may be seen from a religious angle in two ways. First, it is the sense, so widespread in our time, that men can do what previously was impossible; *men can celebrate the power and responsibility of adulthood.* Second, it is the sense so noticeable in the smaller company of the sensitive, that men have to reconcile themselves with the fact that, because they cannot do everything, everything will not be done; *men must accept tragedy maturely."* (p. 38).

Richard D. Brown, *Modernization: The Transformation of American Life* 1600-1865 (New York: Hill and Wang, 1976), writes, "Christianity itself, with its emphasis on the linear advance of history toward judgement day, and its demand that people change themselves and society according to the divine will, is fundamentally modern. Insofar as it permeated the mentality of European society after the decline of the Roman Empire, Europe was experiencing the process of modernization."(p. 15). He describes the "modern model" as dynamic, technological, with shifting economic and political structures. It does not legitimate its political order by reference to the past. It stimulates popular participation, develops bureaucracy, opposes explicit elitism, emphasizes rational analysis rather than supernatural revelation, and promotes a personality type of individual autonomy and initiative.

Thomas G. Sanders, *Secular Consciousness and National Consciousness in Southern Europe*. A Report from the Center for Mediterranean Studies, Rome, Italy (Hanover, New Hampshire: American Universities Field Staff, 1977), attempts definitions which distinguish "modernization" from "secularization." Modernization according to all analysts includes two motives (a) the economic, social, and political process of differentiation or specialization; (b) a world view centered on science and technology (p. 10). "Secularization" on the other hand "is a process by which sectors of society and culture are detached from the domination of religious institutions and symbols" (p. 15). Sanders adds that this second motive may mean "that secularization as an aspect of modernization in developed countries has merely substituted *covert* religious influence and symbols for *overt* religious influence and symbols in economic, political and social institutions" (p. 17). He also notes that "secular consciousness" i.e. "an individual or collective psychological attitude by which the meaning of life and basic decisions are determined by non-religious interpretations" (p. 19) may not coincide with "secularization".

Larry Shiner "The Meanings of Secularization" in James F. Childress and David B. Harned, *Secularization and the Protestant Prospect* (Philadelphia: Westminister, 1970), pp. 30-42, distinguishes five meanings of the term: (1) decline of religion; (2) conformity to "the world", (3) desacralization of "the world"; (4) disengagement of society from religion; (5) transposition of beliefs and behavior from "religious" to "secular" sphere. See also Peter E. Glasner, *The Sociology of Secularization: A Critique of the Concept* (London: Routledge, Kegan Paul, 1977), and Alan D. Gilbert, *The Making of Post-Christian Britain: A History of the Secularization of Modern Society* (London and New York: Longmans, 1980).

3. Sanshiro Shirakashi, "Religious Attitudes and Ways of Life of the World's Youth," The 1972 International Youth Survey of the Office of the Prime Minister of Japan, *Journal of Church and State,* 18 (3) 1976, pp. 523-536, studied youths of 18-24 in Japan, U.S.A., UK, West Germany, France, Switzerland, Yugoslavia, India, Philippines and Brazil and found youths of the U.S.A. second in church attendance. The U.S. Department of Commerce, Office of Federal Statistical Policy and Standards, Bureau of the Census in *Social Indicators,* 1976, p. 555, found that in the U.S.A. 86% considered religion "important" or "fairly important" compared with France where only 55% rated it similarly, or the Scandinavian countries only 45%. *Religion in America: The Gallup Opinion Index,* 1977-78, (Princeton, N.J.: The American Institute of Public Opinion), found that 70% Americans are church members and 60% say their religious beliefs are "very important" in their lives.

Hans Mohl, ed., *Western Religion,* A *Country by Country Sociological Inquiry* (The Hague: Mouton, 1972), p. 553, concluded that two-thirds of the total population of the U.S.A. are church members (close to four-fifths of adults) and about 45 of all adults attend church during a typical week, with many more attending at less frequent intervals. In Sweden, on the other hand, "a general high rate of religious participation occurs only in relation to those services which approximate to 'rites of passage' " (p. 494), in France only 20% go regularly to Sunday Mass (p. 179) and in Italy 28%.

See also Lee Sigelman, "Review of the Polls: Multi-Nation Surveys of Religious Beliefs", *Journal of the Scientific Study of Religion,* 16 (1977): 289-294. Those who do not believe in life after death ranged from 48% in France and West Germany to 20% in the U.S.A. and 17% in sub-Saharan Africa. Martin E. Marty in *The Modern Schism: Three Paths to the Secular* (London: SMC Press, 1969), distinguishes between "utterly secular countries" (France and Germany), "merely secular countries" (England), and "controlled" or "ambiguously" secular countries (U.S.A.).

4. Thomas Luckman, *The Invisible Religion* (New York: Macmillan, 1967).

5. See Robert N. Bellah, *The Broken Covenant: American civil religion in a time of trial* (New York: Seabury, 1975), and Bellah and Phillip E. Hammond, *Varieties of Civil Religion* (San Francisco: Harper and Row, 1981).

6. "Humanist Manifesto", *The New Humanist* 6 (May/June) 1933, reprinted in *The Humanist* 33 (Jan./Feb.), 1973: 13-14, and "Humanist Manifesto II", *ibid.* pp. 4-9. For commentary see Corliss Lamont, *The Philosophy of Humanism,* 5th ed. (New York: Frederick Ungar, 1965), esp. his view of "The Humanist Tradition" pp. 30-80. See also "The Paris Statement of 1966 on Ethical Humanism" in Paul W. Kurtz and Albert Dondeyne, eds., *A Catholic/Humanist Dialogue* (Buffalo: Prometheus Books, 1972), p. 3 f.; Paul W. Kurtz ed., *The Humanist Alternative: Some definitions of Humanism* (Buffalo, N.Y.: Prometheus Books, 1973) and *In Defense of Secular Humanism* (same publisher, 1983). For a Humanist critique of Marxism see Kurtz, *The Fullness of Life,* (New York: Horizon, 1974). For the historical origins of non-theistic Humanism in the United States in Unitarianism see Stow Persons, *Free Religion: An American Faith* (New Haven: Yale University Press, 1941). For an incisive criticism of the optimistic assumption of some humanists by one who is nevertheless, it would seem, essentially a humanist, see David Ehrenfeld, *The Arrogance of Humanism* (New York: Oxford University Press, 1978).

7. See note 3 above.

8. See Evert C. Call, Jr. and Seymour M. Lipset, *The Divided Academy,* Carnegie Foundation for the Advancement of Teaching (New York: McGraw-Hill, 1975), p. 166, found only 23% of teachers of Christian background at elite colleges and universities attend church once a week, and 41% attend once a year or less. According to Appendix C, p. 344-45, teachers of all religions and all fields of study 32% attend once a year or less. The highest rate of church attendance is in teachers in the professional schools, lowest in Social Psychology (30%) and Anthropology (37%). Note that rates for these elites are approximately the reverse for the general population but similar to those for the general population of west Europe.

Michael Argyle and Benjamin Beit-Hallahmi, *The Social Psychology of Religion,* rev. ed. (Boston: Routledge and Kegan Paul, 1975), pp. 86-93, and Harry E. Smith, *Secularization and the University* (Richmond, Virginia: John Knox, 1968). *The Connecticut Mutual Life Report on American Values in the 80's* has compared the "values" of 1,500 leaders in different fields, such as business, military, law, government, news media, religion, science, voluntary associations and education with 2,000 members of the general public and found a wide gap between the two groups, the leaders generally favoring moral views typical of liberal Humanism.

9. Thomas P. O'Neill, *The Rise and Decline of Liberalism* (Milwaukee: Bruce, 1953) clearly demonstrated that both the "liberal" and "conservative" tradition are heirs of what in Europe is called "liberalism" based on the Enlightenment. George H. Nash in his excellent study *The Conservative Intellectual Movement in America since 1945* (New York: Basic Books, 1976), distinguishes three intellectual trends that accept the term "conservative" (1) "classical liberals" or "libertarians" resisting the threat of the ever expanding state to liberty, private enterprise, and individualism"; (2) "new conservatives" or "traditionalists" who "urged a return to traditional religious and ethical absolutes"; (3) militant, evangelistic anti-communists (p.xiii). I would note that although the "traditionalists" include defenders of Christianity, they resemble in many respects the "romantic humanists" in that what they admire about Christianity is simply that it is traditional, or "absolute", not that it is true. As for the other two types, the opposition to Communism is primarily due to the fear of its anti-individualism rather than to its atheism as such. Adam Smith and Edmund Burke, rather than the Church Fathers or the Bible, are at the roots of all these types of conservatism.

10. For example, A.C. Crombie, *Augustine to Galileo,* 2nd ed. (Cambridge, Mass.: Harvard University, 1961), 2 vols. in one; Marshall Clagett, *The Science of Mechanics in the Middle Ages* (Madison, Wis.: University of Wisconsin Press, 1959); Annaliese Maier, *Studien zur Naturphilosophie der Spätscholastik,* 5 vols., (Rome, 1949-1958); James A. Weisheipl O.P., *The Development of Physical Theory in the Middle Ages* (New York: Sheed and Ward, 1959), and William A. Wallace, O.P.,*Causality and Scientific Explanation*, 2 vols., (Ann Arbor, Michigan: University of Michigan Press 1972-4).

11. See Robert K. Merton, *Science, Technology and Society in the Seventeenth Century,* (New York: Howard Fertig, 1970) and Charles Webster, *The Great Instauration: Science, Medicine and Reform 1626-1660* (New York: Holmes and Meier, 1976). For modifications of this theory see James R. Jacob and Margaret C. Jacob, "The Anglican Origins of Modern

Science: The Metaphysical Foundations of the Whig Constitution," *Isis* 71 (June, 1980) 251-267, and Lotte Mulligan, "Puritans and English Science: A Critique of Webster", *ibid.,* 71 (258) (Sept., 1980) 456-469. See also Eugene M. Klaren, *Religious Origins of Modern Science* (Grand Rapids, Mich.: Eerdmans, 1977), on the role played by the doctrine of creation in early modern science, especially in the spiritualism of Helmont and the mechanism of Boyle.

12. The view of the Renaissance as pagan became popular through the great work of Jakob Burkhardt, *The Civilization of the Renaissance in Italy* (New York: Harper Torchbooks, 1958) 2 vols.; but see the criticism of M.I. Bush, *Renaissance, Reformation and the Outer World* (New York: Humanities Press, 1967), pp. 148 ff. "Not only the northern humanists centering upon Erasmus but also the Italian humanists remained firmly within the Christian framework." (p. 15). Students of the French Renaissance especially have debated this question; see Henri Busson, *Le pensée religieuse de Charron à Pascal.* (Paris: J. Vrin, 1933); René Pintar, *La libertinage érudit dans le première moitié du xvii siècle* (Paris: Bonvin, 1943). See also Don Cameron Allen, *Doubt's Boundless Sea: Scepticism and Faith in the Renaissance* (Baltimore: John Hopkins Press, 1964); Franklin L. van Baumer, *Religion and the Rise of Scepticism* (New York: Harcourt Brace, 1960); and Richard H. Popkin, *The History of Scepticism from Erasmus to Spinoza* (Berkley: University of California Press, 1979), "Layman's religion", pp. 230-292.

13. See references in note 11 above.

14. On Spinoza's relation to scepticism see Popkin, *History of Scepticism* (note 12 above) pp. 229-348; on Hobbes and Hume, Popkin, *The High Road to Pyrrhonism* (ed. by Richard A. Watson and James E. Force) (San Diego: Austin Hill Press, 1980), pp. 11-37; 103-132; on Gassendi, Olivier René Bloch, *Le philosophie de Gassendi* (La Haye: Martinus Nijhoff, 1971), pp. 474-481, and on Mersenne, Robert Lenoble, *Mersenne ou la naissance du mécanisme,* 2nd ed. (Paris: J. Vrin, 1971).

15. Arnold Toynbee, *An Historian's Approach to Religion,* 2nd ed., (Oxford University Press, 1979), especially pp. 165-192. Toynbee identifies as the two great evils of modern history the intolerance of Christianity and nationalism, the latter of which he considers to be the result of the Greco-Roman worship of the state.

16. Max Weber, *The Protestant Ethic and the Spirit of Capitalism,* translated by Talcott Parsons with a foreward of R.T. Tawney (London: Unwin University Books, 1970, originally published 1930).

17. H.R. Trevor-Roper, *The Crisis of the Seventeenth Century: Religion, the Reformation and Social Change* (New York: Harper and Row, 1968).

18. Paul Mouy, *Le développement de la physique cartésienne,* 1646-1712 (Paris: J. Vrin, 1934), traces the influence and decline of Cartesianism. Adam Vartanian, *Diderot and Descartes* (Princeton, N.J.: Princeton University Press, 1953), pp. 21-40, discusses the relation of Descartes to religious thought and p. 41 dates the beginning of the decline of his influence in the 1730's.

19. *The Ghost in the Machine* (New York: Macmillan, 1968).

20. The three great classics of deism are Locke's *The Reasonableness of Christianity* (1965); Hume's *Dialogue Concerning Natural Religion* (1750, first published posthumously 1779) and Kant's *Religion Within the Bounds of Mere Reason* (1793). See Ernest C. Mossner, article "Deism" in Paul Edwards ed., *The Encyclopedia of Philosophy,* 2:326-366; E. Graham Waring, *Deism and Natural Religion: A Source Book* (New York: Frederick Ungar, 1967), and Peter Gay, *Deism: An Anthology* (Princeton, N.J.: Van Nostrand, 1968). Also see Norman L. Torrey, *Voltaire and the English Deists,* (New Haven: Yale University Press, 1930); Adolf G. Koch, *Religion of the American Enlightenment* (New York: Thomas Y. Crowell, 1968, orginally 1933); Paul Hazard, *European Thought in the Eighteenth Century* (Cleveland: Meridian Books, World Pub. Co., 1964), pp. 402-415; Frank E. Manuel, *The Eighteenth Century Confronts the Gods* (Cambridge, Mass: Harvard University Press, 1959); R.J. White, *The Anti-Philosophers: A Study in the Philosophes in Eighteenth Century France* (London:MacMillan-St. Martin's Press, 1970). The last two works detail the transition from deism to agnosticism to atheism.

The post-Cartesian trend toward atheism has been traced recently in Hans Küng's, *Does*

94

God Exist? (Garden City: New York, Doubleday, 1980), but he only retraces ground more adequately covered by Cornelio Fabro, *God in Exile: A Study of the Internal Dynamic of Modern Atheism from its Roots in the Cartesian Cogito* (Westminster, Md.: Newman Press, 1968), and by James Collins, *God in Modern Philosophy* (Chicago: Regnery, 1959).

21. Wallace, *Causality,* (note 10 above), 1:200 ff.

22. For a general overview (not always very accurate) of this development see Ira O. Wade, *The Intellectual Origins of the French Enlightenment* (Princeton, N.J.:Princeton University Press, 1971), Also R.J. White, *The Anti-Philosophers* (note 20 above).

23. See Roger Wines, ed., *Enlightened Despotism: Reform or Reaction* (Boston: Heath, 1967) and Leonard Krieger, *Kings and Philosophers 1689-1789* (New York: W.W. Norton, 1970).

24. Peter Gay, *The Enlightenment,* 2 vols. (New York: Alfred Knopf, 1967) 1:3. See also Leroy E. Loemker, *Struggle for Synthesis:The Seventeenth Century Background of Leibniz's Synthesis of Order and Freedom* (Cambridge, Mass: Harvard University Press, 1972), for a somewhat different interpretation of the origins of the Enlightenment.

25. William A. Wallace, *Causality* (note 10 above), 1:194-210.

26. Jacques Monod, *Chance and Necessity* (New York: Knopf, 1971). See also John Lewis, ed., *Beyond Chance and Necessity, A Critical Inquiry into Prof. Jacques Monod's Chance and Necessity* (Atlantic Highlands, N.J.: Humanities Press, 1974).

27. According to Victor Kraft, *The Vienna Circle: The Origin of Neo-Positivism* (New York: Philosophical Library, 1953) the Circle was agreed that "The traditional concerns of philosophy fall into three groups: first, questions about empirical facts; their solutions must come out of the factual sciences. Secondly, questions concerning representation, language: these questions must be settled through clarification of concepts and propositions. And thirdly, there were the metaphysical questions; such questions cannot be answered at all, since they cannot even be formulated in the language of science, in terms of scientific concepts." (pp. 191 f.).

"The sort of conception of philosophy which the Vienna Circle has advocated is by no means a radical innovation. Kant already limited philosophy, considered as a cognitive discipline, to epistemology, and the positivists already left all responsibility for factual knowledge to the special sciences. What makes, however, the Vienna's Circle's conception superior to the positivist conception is the idea of the unity of science. For thus the problems of a unified world-view, which constitutes a primary concern of the philosophies of the past, remain as scientifically legitimate problems of a unified system of scientific knowledge. And further it was in the Vienna Circle that the method of epistemology was precisely defined as logical analysis of language. Knowledge is based on designation, representation, on language, and therefore the analysis of knowledge must be performed by means of linguistic analysis." (p. 191). On the school's contributions see Peter Achinstein and Stephen F. Barker, eds., *The Legacy of Logical Positivism* (Baltimore: John Hopkins Press, 1969) and J.O. Urmson, *Philosophical Analysis: Its Development between the Two World Wars* (Oxford:Clarendon Press, 1956).

28. See Ernest Nagel and James R. Newman, "Gödel's Proof" (New York: New York University Press, 1958).

29. On the history of analytic philosophy see G.J. Warnock, *English Philosophy since 1900* (London: Oxford University Press, 1955) and Barnard Williams and Alan Montefiore, *British Analytical Philosophy* (New York: Humanities Press, 1960), Introduction, pp. 1-16.

30. For discussion of this controversy see P.F. Strawson, *Individuals* (New York: Doubleday Anchor Books, 1963) and Amélie Oksenberg Rorty, Ed., *The Identities of Persons* (Berkley:University of California Press, 1976).

31. See Arthur O. Lovejoy, "On the Discrimination of Romanticisms", in *Essays in the History of Ideas* (New York: Putnam's Sons, 1960) on the difficulty of defining Romanticism; see also the articles on "Romanticism" in Philip P. Weiner, ed., *Dictionary of the History of Ideas,* vol. 4, especially Franklin L. Baumer, "Romanticism: c.1780-1830" pp. 204-205. Further analysis will be found in Leonard M. Trawick, ed., *Backgrounds of Romanticism* (Bloomington, Ind.: Indiana University Press, 1967), pp. vii-xxl; Paul van Tieghem, *Le romantisme dans la littérature européene* (Paris: Albin Michel, 1969), who analyzes the interior

and exterior elements of the movement, pp. 221-289, and the recent work of Hugh Honour, *Romanticism* (New York: Harper and Row, 1979).

32. See Geoffrey Clive, *The Romantic Enlightenment* (New York: Meridian Books, 1960), and H.G. Schenk, *The Mind of the European Romantics* (New York: Frederick Ungar, 1966), pp. 3-48 on the nature of Romanticism, and pp. 49-80 on its religious tendencies, and pages 88-120 on its relations to Christianity. Paul Hazard, *The European Mind: 1680-1715* (Cleveland: World Pub. Co., 1964), pp. 378-392, shows how the elements of Romanticism are already present in Diderot. See Jacques Barzun, *Classic, Romantic and Modern* (Garden City, N.Y.: Doubleday, 1961), for a defense of Romanticism p. 155-168, and also G.A. Borgese, "Romanticism", *Encyclopedia of the Social Sciences,* New York, 1934, vol xiii (vii). "In truth, the Romantic Movement was as much a revolution as a counter-revolution. It may, indeed, be thought of as the first protest against the "modern world," that is, the rational scientific civilization, which had begun in the seventeenth century, and which assumed major proportions in the eighteenth century. But it protested in the name of a new modernity. Some of the romantics thought of themselves as "modern" in the sense of Christian, and anti-classical in artistic taste. But they were modern in still another sense. We might say that they were more modern than they knew, especially in their exploration of the night-side of life, of dreams and the unconscious, and in providing the theoretical basis for modern nationalism. In these, among other areas, the romantics obviously set in motion waves of thought that were not to have their full impact until the twentieth century." Baumer "Romanticism" (note 31 above), p. 268.

Note that the sense in which romantics contrasted "Christian modernity" to "classical antiquity" was that they considered that with Christianity "subjectivity" entered culture as opposed to classical "objectivity," a notion related to the Protestant conception of Christianity. Isaiah Berlin, section "Counter-Enlightenment" of article "Enlightenment" in *Encyclopedia of Philosophy,* 2:100-113. See also Ernest Cassirer, *The Philosophy of the Enlightenment* (Boston: Beacon Press, 1951), p. 104 ff. for the way that the Enlightenment rationalism shifted from pure rationalism to an emphasis on human *affectivity* as a determinant in judgment, thus preparing the way for Kant's exaltation of the practical reason as against theoretical reason. See also, M.H. Abrams, *Natural Supernaturalism: Tradition and Revolution in Romantic Literature* (New York: Norton, 1971) who contends that the romantics "Secularized" Christian principles and values.

33. See Herman J. de Vleeschauwer, *The Development of Kantian Thought* (London: Thomas Nelson and Sons, 1962), pp. 28, 39, 40-43. Also Dieter Heinrich, "Some Historical Presuppositions of Hegel's System" in Darrel E. Christensen, *Hegel and the Philosophy of Religon,* The Wofford Symposium (The Hague: Martin Nijhoff, 1970), pp. 25-60.

34. On Kantian ethics see H.J. Paton, *The Categorical Imperative* (Chicago: University of Chicago Press, 1948) and Jacques Maritain, *Moral Philosophy,* (London: Geoffrey Bles, 1964), especially Maritain's criticism of Paton, pp. 97-101; and H.H. Schroeder, "Some Misinterpretations of the Kantian Ethics, " *The Philosophical Review,* 49 (1940), 424-426. Also A.E. Teale, *Kantian Ethics* (Westport, Conn.: Greenwood Press, 1975).

35. *Ethics without God* (Buffalo: Prometheus Books, 1973). On the deliberate efforts to develop substitutes for the Christian religion see D.G. Charlton, *Secular Religions in France* (London: Oxford Press, 1963).

36. *The Critique of Judgement,* trans. by J.C. Meredith (Oxford: Clarendon Press, 1952; 1st ed. of German, 1790), Introduction, sections 6-9, pp. 26-39. See H.W. Cassirer, *A Commentary on Kant's Critique of Judgement* (New York: Barnes and Noble, 1938), pp. 131-154.

37. Kant admitted the value of prayer as adoration of God with the purpose of motivating moral life, but rejected prayer in the fundamental sense of petition because he thought this to be an effort to persuade God through flattery, an effort unworthy of human dignity and insulting to God. See Jean-Louis Bruch, *La philosophie religieuse de Kant* (Aubier: Editions Montaigne, 1968), pp. 189-218, and Michel Despland, *Kant on History and Religion* (Montreal: McGill-Queen's University Press, 1973), pp. 112, 175, 212.

38. C.P. Snow, *The Two Cultures, a Second Look,* 2nd ed. of *The Two Cultures and the Scientific Revolution* (Cambridge: Cambridge University Press, 1964).

39. "Thus it was the New Men of Europe, the merchants and traders and manufacturers, the owners of mines and mills and of banks, and their technicians and managers

and doctors and clerks, whose experience tallied with the new philosophy, and whose needs called forth the new science. They had little regard for privileges of birth or for divine rights, but required instead rewards for merit, freedom of contract, protection of property and defenselessness of labour, and in the end they won, for their economy was the more efficient. In the two or three centuries before the 1680s or 1690s, their advance and the growing acceptance of their values were sufficiently marked to make a doctrine of progress seem plausible to them." Sidney Pollard, *The Idea of Progress* (New York: Basic Books, 1968), p. 17. Pollard, p. 14-17, shows the economic basis of the idea which was first clearly expressed by Bacon (1620) and clearly affirmed by Descartes. See also pp. 148-181 on the "doubters and dissenters" to the notion.

40. An excellent discussion of the nineteenth century struggle of Christianity and Humanism through which Humanism came to dominate the twentieth century can be found in Owen Chadwick, *The Secularization of the European Mind in the Nineteenth Century* (Cambridge University Press, 1975).

41. See Henri F. Ellenberger, *The Discovery of the Unconscious* (New York: Basic Books, 1970).

42. A useful introduction to Freud's thought is Gerald Levin, *Sigmund Freud* (Boston: Twayne World Author's Series, 357, Twayne Pubs., 1975), with a select bibliography pp. 161-167; see especially the conclusion p. 146-151 on Freud's cultural significance. A philosophical analysis of his basic concepts, especially that of the Unconscious will be found in Roland Dalbiez, *Psychological Method and the Doctrine of Freud* (London and New York: Longmans Green, 1941), 2 vols. A severe negative criticism from the empirical standpoint is found in Hans J. Eysenck and Glenn D. Wilson, eds., *The Experimental Study of Freudian Theory* (London: Methuen, 1973), and from the phenomenological point of view in Gerald N. Izenberg, *The Existentialist Critique of Freud,* (Princeton, N.J.: Princeton University Press, 1976). A balanced, sympathetic Christian view is given by Gregory Zilboorg, *Sigmund Freud* (New York: Charles Scribner's Sons, 1951), and by Jacques Maritain, "Freudianism and Psychoanalysis: A Thomist View", in Benjamin Nelson, ed., *Freud and the Twentieth Century* (Gloucester, Mass.: Peter Smith, 1974) which contains a number of other interesting evaluations by very different specialists. See also Hans Küng, *Freud and the Problem of God* (New Haven: Yale University Press, 1979), and Chapter 2, note 7. Paul C. Vitz, *Psychology as Religion* (Grand Rapids, Mich.: Eerdmans, 1977), argues that most of these modern forms of therapy are really pseudo-religions of the Humanist type.

43. A useful introduction to Jung's thought will be found in Jolande Jacobi, *The Psychology of C.G. Jung* (New Haven:Yale University, 1962). Carl G. Jung with M.L. Franz, J.L. Henderson, J. Jacobi and A. Jaffé, *Man and His Symbols* (Garden City, New York: Doubleday, 1964) is a summing up by Jung and his disciples. A severe negative evaluation of Jung in comparison with Freud is Edward Glover, *Freud or Jung* (Cleveland: World Publishing Co., Meridian Books, 1956). Glover argues that Jung abandoned Freud's basic discovery, the conflictual unconscious and presents only a psychology of the conscious self.

44. "This instinct for "knowledge and research" cannot be counted among the elementary components, nor can it be classed as exclusively belonging to sexuality. Its activity corresponds on the one hand to a sublimated manner of obtaining mastery, while on the other hand it makes use of the energy of scopophilia. Its relations to sexual life, however, are of particular importance, since we have learned from psychoanalysis that the instinct for knowledge in children is attracted unexpectedly early and intensively to sexual problems and is in fact possibly first aroused by them." Sigmund Freud, *Three Essays on the Theology of Sexuality,* II, "Infantile Sexuality" 5, *Collected Works,* James Strachey, ed. (London: Hogarth Press, 1953), 7:194. Note that Freud *does* allow for a non-sexual source for curiosity, but emphasizes only the sexual source.

45. See Norman O. Brown, *Love's Body* (New York: Random House, 1966), and *Life Against Death,* (Wesleyan University Press, 1970); Herbert Marcuse, *Eros and Civilization* (New York: Vintage Books, 1962).

46. For the influence of William James on Husserl see Herbert Spiegelberg, *The Phenomenological Movement,* 2 vols. (The Hague: Martinus Nijhoff, 1960), 1:111-116.

47. See *ibid.*, pp. 39-41, and J.N. Mohanty, "Husserl's Concept of Intentionality" in Anna-Teresa Tymieniecka, ed., *Analecta Husserliana,* (Dordrecht, Holland: D. Reidel, 1970), 1:100-132.

48. Joseph J. Kockelman, "Husserl's Transcendental Idealism" in the collection edited by him, *Phenomenology* (Garden City, N.Y.: Doubleday, 1967), pp. 183-193, and "World-Constitution, Reflections on Husserl's Transcendental Idealism" in Tymieniecka, *Analecta* (previous note) 1:11-35. For the criticisms by Husserl's disciples of his idealism see Roman Ingarden, *On the motives which led Husserl to Transcendental Idealism* (The Hague: Martinus Nijhoff, 1975).

49. Mortimer Adler, *The Idea of Freedom* (Garden City, New York: Doubleday, 1958-61), 2 vols.

50. See his novel *Nausea* (New York: New Directions, 1959).

51. Maurice Merleau-Ponty, *Phenomenology of Perception* (New York: Humanities Press, 1962, French 1945).

52. The standard exposition of Heidegger's thought is William J. Richardson, S.J., *Heidegger:Through Phenomenology to Thought* (The Hague: Martinus Nijhoff, 1963). See also Marjorie Grene, *Martin Heidegger* (London: Bowes and Bowes, 1957) and Thomas Langan, *The Meaning of Heidegger* (New York: Columbia University Press, 1961). Some of its implications and recent criticism of it can be found in Michael Murray, ed., *Heidegger and Modern Philosophy* (New Haven: Yale University Press, 1968). An especially helpful discussion of its hermeneutical aspect is Anthony D. Thiselton, *The Two Horizons* (Grand Rapid, Mich.: Eerdmans, 1980). Perhaps no one has expounded better Heidegger's approach to "Being" than John MacQuarrie, *Martin Heidegger* (Richmond, Virginia: John Knox Press, 1978) when he writes. "If it appeared at first that man was to be made the measure of all things, this trend is reversed by the dialectic which finally subjects man to Being. Man remains unique as the existent, the place of openess among all the beings. But the notion of autonomous man yields to the notion of man as responsible steward, and perhaps this is by far the more mature notion, and one that has great relevance as technological man increasingly subjects the world to his control and has its resources more and more at his disposal. Is he the absolute master, or is he rather the one to whom Being has graciously entrusted itself and on whom it has conferred an almost frightening responsibility? Heidegger's answer is clearly given: 'Man is not the lord of beings. Man is the shepherd of Being.' (Brief über den Humanismus, p. 29)'', p. 50.

53. See John R. Williams, *Martin Heidegger's Philosophy of Religion* (Canadian Corporation for Studies in Religion, 1977). It seems that Heidegger's attitude was that religious questions so transcend philosophy that it is a distortion of the religious quest to attempt to raise the God-question philosophically; but he did not intend to close the way to a mystical approach to God.

54. Hans-Georg Gadamer, *Truth and Method* (New York: Seabury Press, 1975), pp. 235-245.

55. David Robey, *Structuralism: An Introduction* (Oxford:Clarendon Press, 1973), and Philip Pettit, *The Concept of Structuralism: A Critical Analysis.*(Berkley: University of California Press, 1975).

56. Robert Rosenblum, *Transformations in Late Eighteenth Century Art* (Princeton, N.J.: Princeton University Press, 1967) (Berkley: University of California Press, 1975).

57. See Henry Russell Hitchcock and Phillip Johnson, *The International Style* (New York: Norton, 1966) and Hitchcock, *Modern Architecture:Romanticism and Reintegration* (New York: Hacker Art Books, 1970); Charles Jencks, *Architecture 2000* (New York: Praeger, 1971), pp. 117-123, and Reyner Banham, *The New Brutalism:Ethic or Aesthetic?* (New York: Reinhold Pub. Corp., 1966).

58. See Ivan D. Illich, *Medical Nemesis* (New York: Pantheon, 1976), for the argument that modern medicine is an application of the "industrial model" to the problem of health which results in an increasingly unhealthy population in spite of rising life expectancy.

59. Linda Nochlin, *Realism,* (Harmondsworth, Sussex: Penguin Books, 1971).

60. Kenneth Clark, *Landscape into Art* (Boston 1949: Beacon Press) traces the rise of landscape as an independent *genre*.

61. Kenneth Clark, *The Nude* (New York: Random House-Pantheon, 1956), pp. 248-310.

62. See Wylie Sypher, *Rococo to Cubism in Art and Literature,* (New York: Vintage Books, 1963), who argues, to my mind convincingly, that cubism is the principal style of twentieth century art.

63. See my article, "Significance in Non-Objective Art" *Proceedings of the American Catholic Philosophical Association* (Washington, D.C.: Catholic University of America, 1965), pp. 156-165.

64. John Dewey's, *Art as Experience* (New York: G.P. Putnam's Sons: Capricorn Books, 1958) (originally published in 1934) is a powerful argument for the "democratization" of art. Leon Jacobson, "*Art as Experience* and American Visual Art Today," *The Journal of Esthetics and Art Criticism,* 19 (1960), 117-126, attacks Dewey on the grounds that he proved to be a poor prophet of progressive trends in art.

65. See Joshua C. Taylor, *Futurism* (New York: Museum of Modern Art and Garden City, N.Y., Doubleday, 1961); Marianne W. Martin, *Futurist Art and Theory, 1909-1915,* (Oxford: Clarendon Press, 1968) and Rosa T. Clough, *Futurism* (New York: Greenwood Press, 1968). The pro-Fascist, anti-democratic political tendency of this artistic movement is evidence that progressivism in art styles is not always identified with social liberalism.

66. The struggle of our culture to develop livable cities is recounted in Charles N. Glaab and A. Theodore Brown, *A History of Urban America,* 2nd ed., (New York: Macmillan, 1976).

67. Lionel Trilling, *The Liberal Imagination* (Garden City: NY, Doubleday, 1953).

68. See Jacques Derrida, *Speech and Phenomena: and other essays on Husserl's theory of sense* (Evanston: Northwestern University Press, 1973); *Writing and Difference* (Chicago: University of Chicago Press, 1978); *Positions* ((Chicago: University of Chicago Press, 1981); *Of Grammatology* (Baltimore: John Hopkins Press, 1976), and essays on Derrida in Mark Krupnick, ed., *Displacement: Derrida and After* (Bloomington, Ind.: Indiana University, 1983). Hegel's *Aufhebung* emphasized the transcendence of the past by its sublation in a new synthesis. Heidegger's "destruction" of the past is also a sublation, or a "retrieval." But Derrida emphasizes *discontinuity.* For him the most significant re-readings of a "text" are those which "displace" or "de-center" its traditional reading so as to give it a new and disturbing meaning.

69. David McLellan, *The Young Hegelians and Karl Marx* (London: Macmillan, 1969), studies this development in detail.

70. See Chapter 1, note 19.

71. George L. Kline, "Was Marx an Ethical Humanist?", *Studies in Soviet Thought,* 9 (June, 1969), 91-103, concludes that Marx had a humanist ideal for the future ("fallacy of deferred value") but failed to develop humanist principles to guide future life; but see Eugene Kamenka, *The Ethical Foundations of Marxism,* New York, 2nd ed., (London: Routledge and Kegan Paul, 1972), and Erich Fromm, *Marx's Concept of Man* (New York; Frederick Ungar, 1961).

72. Henry J. Koren, *Marx and the Authentic Man,* Pittsburgh: (Duquesne University Press, 1967) treats of Marx's place in the modern search for "authenticity" of life and the exposure of "bad faith."

On the antecedents of Bergson see Newton P. Stallknecht, *Studies in the Philosophy of Creation, with special reference to Bergson and Whitehead* (Princeton, N.J.: Princeton University Press, 1934), and Ben-Ami Scharfstein, *Roots of Bergson's Philosophy,* (New York: Columbia University Press, 1943), pp. 128 ff. Bergson's chief works are *Matter and Memory* (London: George Allen and Co., 1913), *Creative Evolution* (New York: Macmillan, 1938), and *The Two Sources of Morality and Religion* (Garden City, New York: Doubleday, 1935). They were subjected to a penetrating yet sympathetic criticism by Jacques Maritain in *Bergsonian Philosophy and Thomism,* rev. ed (New York:Greenwood Press, 1968).

73. Teilhard's works are published in a series of 13 volumes, *Oeuvres,* Editions du Seuil, Paris. *The Phenomenon of Man* (New York: Harper and Row, 1959); *The Divine Milieu* (New York: Harper and Row, Harper Torchbooks, 1960); *Hymn of the Universe* (New York: Harper and Row, Colophon Books, 1965); and *The Heart of the Matter* (New York: Harcourt Brace Jovanovich, 1979) are among his chief works. His biography can be found in Mary and Ellen Lukas, *Teilhard,* new and annotated edition (New York: McGraw-Hill Paperbacks, 1981), and

a good bibliography of his works and works about in him Ursula King, *Towards a New Mysticism: Teilhard de Chardin and Easter Religions* (New York: Seabury Press, 1981), pp. 293-300. The relation between Bergson and Teilhard is discussed by M. Barthelemy-Maudale, *Bergson et Teilhard de Chardin* (Paris: Edition du Seuil, 1963).

For severe criticisms of his thought from a Catholic point of view see Phillipe de Trinité, O.C.D. *Teilhard de Chardin:Etude critique,* 2 vols. (Paris: La Table Rond, 1968), and Jacques Maritain, *The Peasant of the Garonne* (New York: Holt, Rinehart and Winston, 1968), "Teilhard and Teilhardism", pp. 116-126 and appendices, who said: "From the start one mixes everything together, since, scanty as it may have been in such and such an age, philosophy of nature, metaphysics, theology, natural mystique, even touches of supernatural mystique, all of which are made to contaminate and corrupt one another in a powerful high-soaring lyrical flight — unnatural and deceptive because it is pseudo-angelic.", p. 115; and from a scientific point of view P.B. Medawar, "Critical Notice of *The Phenomenon of Man*" *Science,* Series 2, 70 (Jan. 1961), 99-106, with the comments of Bernard Tower, "Scientific Master Versus Pioneer: Medawar and Teilhard", in his *Concerning Teilhard and Other Writings on Science and Religion* (London: Collins, 1969), pp. 89-99. Also Gaylord G. Simpson, *Scientific American,* April, 1960, and Paul Chauchard, *La Pensée scientifique de Teilhard* (Paris: Éditions Universitaires, 1965).

For the defense see Henri du Lubac, S.J., *The Religion of Teilhard de Chardin* (New York: Desclée, 1967), and Robert North, S.J. *Teilhard and the Creation of the Soul* (Milwaukee: Bruce, 1966). An excellent introduction to the thought of Teilhard from a Catholic point of view is Robert L. Faricy, S.J. *Teilhard's Theology of the Christian in the World* (New York: Sheed and Ward, 1967).

74. Whitehead's chief systematic work is *Process and Reality: An Essay in Cosmology,* (New York: Macmillan and the Free Press, Paperback, 1969) (original edition 1929). The books I have found most useful in understanding this difficult work are William A. Christian, *An Interpretation of Whitehead's Metaphysics* (New Haven: Yale University Press, 1959), which attempts to show the consistency of Whitehead's mature system; Robert W. Palter, *Whitehead's Philosophy of Science,* (Chicago: University of Chicago Press, 1960), which deals with the mathematical and scientific origins of the system (often neglected by theological admirer's of Whitehead); Victor Lowe, *Understanding Whitehead* (Baltimore: John Hopkins Press, 1962), which treats the development of his thought and its sources; and Donald W. Sherburne, *A Key to Whitehead's Process and Reality,* (New York: Macmillan, 1966), which tries to make the central line of argument in PR more perspicacious. Lowe, p. 186 ff. stresses the ways in which Whitehead's thinking differed from the outset from that of his co-worker Bertrand Russell who more and more approached the standpoint of logical positivism or empiricism, while Whitehead always opposed Hume and the positivist tradition.

On the influence of Bergson on Whitehead and their differences see Lowe, pp. 193 ff. and 257 ff; also Robert M. Palter, "Bergson's Influence on Whitehead", *The Personalist,* 35 (1955): 250-257. For Charles Hartshorne see especially his *The Divine Relativity: A Social Conception of God* (New Haven: Yale University Press, 1964 edition), and *A Natural Theology for Our Time* (Lasalle, Ill.: Open Court, 1967), and the bibliography in Ewert H. Cousins, ed., *Process Theology* (New York: Newman Press, 1971), pp. 351 ff. In Whitehead the notions of creativity and process are primary, while for Hartshorne *relativity* (i.e. relationality) seems more fundamental. I will discuss this question more at length in Chapters 7 and 8. An interesting discussion of the relation of Teilhard to Whitehead is Ian G. Barbour, "Teilhard and Whitehead", an appendix in Cousins, *Process Theology,* pp. 323-350. For the history of the whole movement see Gene Reeves and Delwin Brown, "The Development of Process Theology" in *Process Philosophy and Christian Thought,* edited by them (Indianapolis: Bobbs-Merrill, 1971), pp 21-64.

75. *The Enlightenment* (see note 24 above), vol. 1, pp. 418 f. For another summary of the Humanist attitude see Robert Anchor, *The Enlightenment Tradition* (Berkley: University of California Press, 1979), p. 144: "It is not necessarily true, *of course,* that a godless universe and a silent nature deprive human life of meaning. It is only true that, in a meaningless universe, men must bear responsibility for giving value to what they create and add to a world which they did not make and which will outlive them. A consciousness of this task is the lasting legacy of the Enlightenment" (italics are mine).

PART II
CHRISTIAN THEOLOGIES
OF THE BODY

Chapter 4

The Platonic Christian Theology

I: Biblical Themes

1. The Body is a Tomb

The first phase of the development of a Christian theology of the body was dominated by the influence of a dualism deriving from the philosophy of Plato. Adopting a Pythagorean saying that reflected the widespread ancient belief in the transmigration of souls, *sōma sēma*, "The body is a tomb," Plato and his followers expressed their deep conviction that the true human self is the spiritual soul and that the soul's earthly existence in the body is a kind of death or exile or imprisonment. The body is only a garment of which the soul must divest itself, a sleep from which it must awaken as from a dream to its true life.[1]

This conception has something in common with the Christian vision of what it is to be human, namely, that our inner spiritual life somehow transcends our outward bodily life. But more fundamentally it contradicts our belief that "the Word became flesh" (John 1:14) and "If the dead are not raised, then Christ was not raised; and if Christ was not raised, your faith is worthless" (I Cor. 15:16-17). The body is not a tomb but the spouse of the soul in the wedding celebration of eternal life.

Yet undeniably this ancient Platonic notion profoundly colored for many centuries the way that Christians read their bible and that Christian thinkers struggled to give consistent expression to the biblical message concerning human salvation. On the one hand it assisted these thinkers in developing a profound appreciation (not easy to achieve) of human spirituality, and on the other it obscured the development of a corresponding theology of the body. In this chapter I will trace in broad outline the influence of this dualistic conception of the nature of the human person up to its final rejection by the great scholastic theologians of the Middle Ages. Its influence on popular Christianity still remains to be overcome.

2. Bible and Body

Historically the Christian world-view and value-system exists in the communal praxis and self-understanding of the Church, a community universal in space and time, continuous with the ancient community of Israel. This self-understanding came to unique expression in a literature called the Bible which the Church recognized as the work of the prophetic spirit and, therefore, under the guidance of the same Spirit, canonized as the *normans non normata* of its faith and life. This Bible, however, requires to be interpreted under the guidance of that same prophetic Spirit who continues to animate the Church whose faith the Bible both regulates and expresses.

Thus Tradition, as the living, historically continuous and self-consistent faith of the Christian community; and Scripture, as an inspired and historically unique expression of that faith in its formative phase, are mutually confirmatory and constitute an hermeneutical circle.[2] The Church to be self-consistent in its own witness must remain faithful to the Scriptures, and the Scriptures must be understood through the historically developing life of the Church out of which they arose and which they essentially express.

The real, but still imperfect, unity of the Christian community, has been obscured by differences in the understanding of this relation of Scripture to Tradition. To put it very broadly, the eastern churches understand Tradition as fixed by the first seven ecumenical councils and refuse to accept as binding the subsequent developments of Tradition in the western church defined by the Bishop of Rome or by councils approved by him as ecumenical.[3] The churches of the Reformation, on the other hand, accuse both eastern and western churches of deserting the norm of the Scriptures, so that for Protestants the continuity of Tradition is problematic. At Vatican II Roman Catholics arrived at a clear statement of the principle that Tradition involves a historic development of the understanding of revelation congruent with the apostolic witness of the Scriptures.[4]

This fundamental consistency of Tradition and Scripture is guaranteed to the Church by the abiding power of the Spirit, whose witness is infallibly expressed by definitive doctrinal decisions of the popes and ecumenical councils approved by them. Their "infallibility," however does not exclude various kinds of human defects in the formulation of such decisions, provided these defects do not nullify the definitive character of the Church's witness to the essential elements of the faith, just as the "inerrancy" of Scripture does not exclude such defects, provided they do not nullify its normative character as communicating the Word of God.

Consequently, to discover what the Word of God can tell us about our human bodily existence, we must first of all look to the Bible, but we should not claim to be able to arrive at a correct theological understanding of its message exclusively by some "objective" scientific method independent of the Tradition of the believing Church. Modern historical criticism, by techniques designed to minimize prejudice, anachronism, and eisegesis, greatly contributes to our discovery of the true sense of the Scriptures, but Tradition must be our ultimate guide to that true sense, since it judges by the Spirit what the Spirit is saying to the churches (Rev. 2:7).

Thus, rather than beginning my exposition of the Christian understanding of our bodily existence by reviewing the opinions of modern critical exegetes,[5] I prefer first in the next three chapters to trace in broad outline the historic development of the Church's understanding of scriptural teaching on the nature of the human body-person, leaving problems of current exegesis to be discussed at various points in the other, constructive chapters of this book. To introduce this outline I will first survey in highly schematized fashion and with minimal documentation the principal scriptural themes which have been the main topics of theological reflection throughout Christian history relating to our human existence as living, dying, thinking, feeling, striving bodies.

3. Liberation and Resurrection Theme

The biblical teaching which casts the strongest light on our bodily existence is the New Testament's witness to Jesus' resurrection to eternal life and his ascension to the right hand of the Father by the power of the Spirit. By the Spirit He shares in the eternal "I am" of the Father in a spiritualized life in which whatever was vulnerable, weak or transitory in his bodily existence has been overcome, yet He remains not just God's Son, but his *incarnate* Son, "the Word . . . made flesh" (John 1:14).

Jesus' own preaching was centered not so much on himself as on the coming of the Reign of God, yet clearly He implied that this Reign had begun in his own person, and would be manifested in power in himself as Risen Lord. St. Paul, therefore, speaks no longer of the Reign of God, but of the crucified and risen Christ who, although ascended to the Father, remains present to his community, the Church, in his Spirit. The Church itself, because it makes the Risen Lord still visible and audible to the world is, according to St. Paul, his very *body* into which Christians have been *incorporated* by faith, baptism and the eucharist. This Pauline teachng is reinforced by the Johannine writings which dwell on Jesus' promise to send his Spirit to be the undying life of his community and to unite it in faith and love in a unity like that of the fruit-bearing branches with the vine of which they are organic members.

This resurrection theme of the New Testament had its antecedent in the central theme of the Old Testament: the liberation of the Hebrews by the Exodus and their formation as the people of God by the Covenant. Later, at the time of the exile, the Deutero-Isaiah reinterpreted the Covenant as a new creation or rebirth of the people through their second liberation from captivity, a captivity understood as just punishment for their many breaches of the Covenant under faithless kings and false prophets and for their refusal to heed the warnings of the true prophets.[6]

The failure of those Jews who returned from the Exile fully to realize the ideals of this renewed Covenant led the later Old Testament writers to look forward with longing to this perfect realization of the Reign of God under a future messianic king, a descendent of David. In the Maccabean period, oppression by the Greeks brought this expectation of the Jews to the boiling point in the apocalyptic writings in which the expected Liberator was envisioned no longer as a mere earthly king, but as a savior from heaven. The death of the Maccabean martyrs as witnesses of the One God and His Law in resistance to the hellenizing efforts of the Seleucids, convinced the Jews, who had previously been sceptical of the pagan myths about the future life, that after all a just God would certainly resurrect these faithful witnesses to everlasting life in His Kingdom.[7] Finally, the Roman oppression provoked the political rebellion of the Zealots and furthered that identification of the Reign of God with the liberation and triumph of the Jewish nation manifest in the thought of the Pharisees and Essenes. It was at this historic crisis that Jesus came preaching his own universalist interpretation of the Reign of God which this resurrection theme summarizes.

4. Creation or Image Theme

Because both the Old and New Testament are centered on the historic event of the liberating Covenant sealed by the blood of Jesus, the theme of creation is subordinate to that event, yet provides its necessary context.[8] Thus *Genesis* was prefixed to the Torah as an explanatory introduction to the account of the Exodus and the gift of the Law at Sinai. The

Covenant between God and humankind is at least as old as Noah (9:9) and re-establishes that concord which existed between the Creator and Adam and Eve before they broke His commandment not to eat of the tree of the knowledge of good and evil.

Adam and Eve were created in God's image, that is, the human race corporately and individually were created in personal relation to God, able to know Him, to praise His wisdom in creation, and to obey Him as stewards of that creation. They might have shared His deathless life, but by violating the order of creation, they became vulnerable to all the ills to which they were liable as bodily beings created out of the dust of earth. No doubt if they had not sinned against this creation it would have attained its consummation in an event at which *Genesis* only hints when it refers without further explanation to the "tree of life" in the garden, namely, God's gift of everlasting life to humankind.[9]

Because of their sin, this consummation was delayed until the coming of Jesus Christ, who according to the Pauline writings, is the new and sinless Adam whose redeemed bride is the Church (Romans 5:12-19; Ephesians 5:22-33). In Him and His Body, the Christian community, the original image of God in humankind, is at last restored. Christ's body has become the new temple where God is present on earth in an infinitely fuller sense than He was present to Adam and Eve in the garden, since "In Christ the fullness of deity resides in bodily form" (Col. 2:9).[10] His rising again to bodily life also effects the transformation of the world, the incorporation of all humanity and indeed of the whole cosmos (Col. 1:15-20) in One who is not merely "god-like" (as the serpent had falsely promised Eve she might become) but who is in the strictest sense, the Son of God, the First-Born (Col. 1:15)

5. Sin-Death-Cross Theme

While the biblical message is essentially positive, a gospel or good news founded on the promises of God, yet this good news is given to those "who sit in darkness and the shadow of death" (Luke 1:79). From this negative angle the Bible is the shocking history of human failure, beginning with the folly of Adam and Eve, through the account of the decay of the human race before the flood, the wanderings of the patriarchs whose lives are clouded by various morally dubious acts of survival, the enslavement in Egypt, the bloody conquest of the promised land, the apostasy of the kings of Judah and Israel, the near extinction of the exile, the disappointments of the return and restoration, the oppression by the Greeks and Romans leading to a new nationalism and a more rigid piety, and ending in the rejection of the way of peace offered by Jesus and the destruction of Jerusalem. It seems to confirm the sad, wise words of Qoheleth (1:2), "Vanity of vanities! All things are vanity!"

Yet the Bible, reflecting on such events which it never flinches from depicting in all their horror and banality, refuses to blame God for these disasters. They are all alien to God's original creative purpose and to His created image in humanity. Death has its origin in the sin of free and responsible creatures, whether these be superhuman powers or mere humans. These creatures have freely chosen to deny their true selves, by denying their origin in the gift of God in favor of an idolatrous autonomy. Rather than share with God and with each other God's reign over the cosmos, they have set themselves up as gods independent of him and of each other, each seeking tyrannical supremacy. The ironic result has been their enslavement to delusion and to each other, and even to the worship of beasts and lifeless objects. Thus the creation has been reduced by sin to the chaos out of which God brought it into orderly and beautiful existence. It has become, indeed the vanity of vanities, the apparent cancelling of God's purposes by the Cross.[11]

The resurrection, image, and death themes are the principal ones though which the Bible interprets our bodily existence in its original beauty, its ruin, and its glorious restoration. There are also, however, several secondary themes, closely related to the primary ones, which can be schematized as follows.

6. Wisdom Theme

As already noted, a whole class of Old Testament writings is called "wisdom literature." On the one hand, this literature dwells on the splendor of God's created order; and, on the other, on the ever recurrent problems of earthly life.[12] It differs from other types of biblical writing in its concern for the permanent order and the cyclical events of human life rather than for salvation history. Yet it is not, as sometimes asserted, a-historical, since by stressing this permanent and cyclical cosmic order it provides a framework within which the unique events of history take on their special meanings. Wisdom writings declare the original intentions of God and His ultimate purposes which are to be realized through salvation history.

The Torah itself culminates in *Deuteronomy* which is understood by some exegetes as a reinterpretation of older Torah traditions under the influence of those scribal circles from which the wisdom writings also issued.[13] *Deuteronomy* calls the people to discern God's purposes in the world and to choose resolutely between the way of death and the way of life. They are called to cooperate with God by following the way of wisdom which is conformity to the order of His creation, and warned that the way of folly leads to ruin and the return of the primeval chaos. Later wisdom writings explicitly relate the creation theme to the salvation history theme, praising God both in the works of creation and the saving events of history.

Thus in a non-philosophical way the Hebrew scriptures deal with the epistemological problem usually considered the concern of the philosophical Greeks, namely, how can we fallible humans ever attain to truth? For the Bible true wisdom is *both* a gift of God and the fruit of long reflections on human experience. While for *Proverbs* and *Qoheleth* this wisdom is directed to instructing us how to live our lives with moderation and accept old age and death with resignation; for *Job* it goes beyond this to more profound questions about the sufferings of the innocent, crying out in faith to God for ultimate justice; and for the *Book of Wisdom* this vindication is confidently expected beyond the horizon of this mortal life.

Finally, Jesus as the One Teacher reveals a higher wisdom, which St. Paul says is "the folly of the Cross" (I Cor. 1:18-25). This wisdom is both a final reinterpretation of the Torah, as in the Sermon on the Mount, and the historical person of Jesus himself, in his acts and his suffering.

7. Stewardship Theme

True wisdom requires us to obey the wise God, to live in accordance with His cosmic plan, but at the same time it guides us to use His gifts well, cultivating the garden of the world. Adam was given dominion over the earth, but a dominion which was a stewardship, accountable to God. Hence the wisdom writings praise human skill and prudence, but also warn against idolatry and the pride manifested in the Fall, the building of the Tower of Babel, the unrighteous rule of the kings of Judah and Israel, and the perverse, unnatural practices of the gentiles.

The Remnant of God, those remaining faithful to the Covenant, are pictured as the righteous poor (*anawim*) who, rather than share in unjust riches and tyrannical power, accept poverty for righteousness' sake. At the same time eschatologically the coming of the Reign of God is pictured as an entrance into a promised land of plenty, of milk and honey, prosperity and peace.[14] In the Beatitudes Jesus announces the coming of the time when the poor will at last inherit the land, and the hungry will be satisfied. In the Parable of the Talents (Matt. 25:14-30) he assures his disciples that if they are to enter into the joy of the Kingdom they must be good stewards of God's gifts and return God a handsome profit on his investments!

8. The Fertility Theme

The Old Testament teaches that God created humanity male and female in view of the expansion of the human community, since we are social animals. Yet sexual union is not merely a biological function since it promotes intimate human love expressed in bodily union and common life by which man and woman complete each other. The fruitfulness of this love enables them to cooperate with the Creator in transmitting life through space and time to form God's people and to possess the earth. The image of God is found in man and woman alike, and according to the prophets, marital love is the image of the Covenant between God and his people. Unlike the gods of the heathens, Yahweh has no other wife than his people to whom he remains ever faithful, even after this people has proved unfaithful and followed false gods. Jesus announces the renewal of this covenant of love and identifies himself as the bridegroom who will celebrate his wedding with the Church in the eternal banquet of the Reign of God.[15]

In the wisdom writings marriage is regarded as a great gift of God to be valued especially for its fertility which ensures the just a kind of immortality in their children and the survival and growing power of God's chosen people. Barrenness is a curse. The Law protects marriage and family ties with many regulations, yet permits a man to take several wives and to divorce one and to take another, especially for the sake of posterity. Sexual license is severely condemned as a kind of folly that leads to destruction, but the joy of sexual love in marriage is praised, especially in the *Song of Songs*, as one of the great consolations of our troubled life. At the same time sexuality is seen to be so linked with human physical and moral frailty, to our mortality, as to be a source of ritual impurity like contact with a dead body.[16]

In the New Testament Jesus works his first miracle to complete the celebration of a marriage at Galilee. On another occasion he abolishes the Law's permission of the husband's unequal right to divorce and forbids remarriage after divorce, thus restoring the original order of creation. He himself, contrary to Jewish custom, remains celibate and teaches that in the resurrection marriage will cease, since sexuality pertains to this passing life of birth and death. St. Paul follows Jesus both in living as a celibate and in urging Christians to accept the same mode of life in view of the shortness of the time, while at that same time denouncing those who condemn marriage as sinful. Paul shows special pastoral concern to teach married people that their love for each other should be modeled after that of Christ for the Church. He and the other New Testament writers take great pains to instruct families in their duties of love and fidelity to

110

each other and to inculcate a sobriety and modesty of domestic life that some modern critics find somewhat too bourgeois.[17]

Thus for the New Testament marriage is a Christian and sanctifying institution, but compared with the Old Testament stress on its necessity for the survival of the chosen people, its value is relativized, because the growth of the Church as the New Israel depends not on physical birth but on spiritual rebirth through baptism. The Church inaugurates not a temporal but an eternal kingdom for which married love is only a metaphor. The resurrection and the abolition of death put an end to the absolute value of fertility, unless we speak of the church as the spiritually fruitful bride of Christ, the messianic king and father of all nations.

9. The Suffering Servant Theme

The wisdom which makes Jesus a teacher of the whole human family as well as its spiritual father, also make Him its lord and king, but a king who does not seek to dominate. Rather Jesus presents himself in humility as the servant of the people, especially the most despised and neglected. This servitude wins Him nothing but death on the Cross as an outcast and a criminal.

The meekness of Moses and Job were praised in the Old Testament, as well as the clemency of King David, but their patience and humility pale before the hidden splendor of the Lord who washes his disciples feet. By refusing to take vengeance on his enemies and praying for their forgiveness, He brings to an end the holy wars of the Old Testament and restores the peace of Eden, symbolized in *Genesis* by the vegetarianism of Adam and Eve[17] and broken by Cain's murder of Abel and the ever increasing violence of later ages, a violence which Yahweh himself is depicted as commanding in the service of the survival of his chosen people.

Jesus' kingdom is not to be built by the sword but only by the cross of martyrdom and the patient care of the poor, the sick, and the rejected in whom He will remain always present to be served by his true followers. By thus emptying himself, the Son of God restores all things and presents them at last to his Father in an act of redemption, reconciliation, and recreation. By surrendering the glory of His Sonship and becoming a human being of flesh and blood, subject to all the weariness of life, Jesus has achieved his victory, even over the spiritual forces of evil and ensured the final transformation of the whole of creation as God's Kingdom.

All these diverse themes concerning our bodily existence are summed up in the eucharistic words of Jesus which are enshrined in the church's central act of worship.[19] These words clearly state the theme of the Suffering Servant who is to rise from the dead, "This is my blood, the blood of the covenant, which is poured out on behalf of many. I solemnly assure you I will never again drink of the fruit of the vine until the day when I drink it new in the reign of God" (Mark 14:25); and they are associated with the other themes, since this meal is the eschatological wedding feast (the fertility theme); it is a eucharist or thanksgiving for the gifts of the earth (the Jewish prayers on which it was based praised the Creator for these first fruits); and it is a wisdom which reveals the meaning of all of history, "Do this in memory of me" (I Cor. 11:25).

II. Eastern Platonic Christian Theology

1. From the Jewish to Greek World-View

Christianity at first appeared as a particular sectarian interpretation of Judaism and, as we see in the New Testament, its self-understanding was expressed in various "theologies" such as the more conservative theology of *Matthew* and the *Epistle of James* or the more radical theology of St. Paul, yet all of them, even that of the Johannine literature,[20] remained within the thought forms typical of Jewish culture. For some time theologies of this type, called by Jean Daniélou "Jewish Christian",[21] prevailed at least in parts of the Church. One of the characteristics of Jewish Christian thought was the retention of Jewish expectation of the establishment of the Kingdom of God in earthly existence for "a thousand years" ((*millenialism,* Rev. 20:2). In such perspectives there seemed little room for a body-soul dualism. Yet it is probable that already among heterodox Jewish sects, especially among the Samaritans, the moral dualism common in Jewish apocalyptic and in Essene writings, had begun to take on a cosmological form in which the body and matter were identified with moral evel and the soul with moral good.[22]

The Church quickly found itself confronted with the problem of translating the Gospel, not merely into the Greek language as is already the case in the New Testament, but into the thought forms of the Greco-Roman world. The problem of this transition should not be exaggerated by setting a "Hebrew mentality" off sharply against a "Greek mentality." Not only did Jew and Greek come out of a much older common Mediterranean culture, but Palestine had been hellenized since the conquest of Alexander the Great. Moreover, Jewish thought preceding the New Testament, as we see clearly in the Septuagint translation of the Old Testament, in Josephus, and especially in Philo, had already faced up to this kind of problem quite effectively. Nevertheless, for Christians the transition to Greco-Roman culture proved no small difficulty.[23]

The first effort to meet this challenge and one which came close to absolute disaster, was that Christian theology we know as *Gnosticism*, which for a time in the second century threatened to overcome all the others.[24] Gnosticism is especially significant for the theme of this present book because it interpreted Christianity as a radical cosmological dualism. Although it made use of certain Jewish texts, especially the first chapters of *Genesis*, it reinterpreted them so as to reject the Old Testament God as an evil or at least ignorant being responsible for creating the material world with its human miseries. Into this world a spiritual Christ descends not to be incarnated, but to reveal to certain more spiritual humans the secret knowledge (*gnosis*) of their true origin in the spiritual world, so that they may be liberated from their bodies and return to their true home. In spite of their insistence on this radical opposition of spirit to matter, it is doubtful that the Gnostic theologies presented a very advanced conception of the metaphysical character of the spiritual order, since they spoke of it in mythical rather than ontological terms. One form of Gnosticism, Manicheism, which arose in Persia in the third century and took to itself remnants of the ancient dualistic religion of Zoroastrianism, was to last into the fifth century and to reappear again in the Middle Ages as Catharism, Albigensianism, etc., although it is perhaps better rated a new religion than a Christian heresy.

The Christian Church, after a hard struggle, rejected Gnosticism as radically inconsistent with some of the basic themes I have mentioned, especially the doctrines of the Creation, Incarnation, and Resurrection. The most effective opponent of Gnosticism was St. Irenaeus (fl. 177) who taught that the world and human beings are the free creation of the One Holy God of the Old Testament acting through His only begotten Son, the Word.[25] Adam and Eve were created innocent, and evil came into the world through their freely chosen sin of disobedience. Salvation comes through a true incarnation of the Word who came to restore the creation.

> When He became incarnate and was made man, He recapitulated in himself the long history of man, summing up and giving salvation in order that we might receive again in Christ Jesus what we had lost in Adam, that is, the image and likeness of God.[26]

"Recapitulation" (*anakephalaiosis*), a term derived from the Pauline *Ephesians* 1:10, was Irenaeus' fundamental principle in constructing a Christo-centric, salvation-historical theology. As Word, Jesus Christ is the beginning of history and as Incarnate Word its summing up. As incarnate He is the New Adam, and his mother Mary is the New Eve who by her obedience repairs Eve's disobedience and provides the Son of God with a true human nature. That human nature as first created in Adam and Eve is "in the image and likeness" of God (Genesis 1:26):

113

There are three things out of which . . . the complete man is composed — flesh, soul, and spirit. One of these does indeed save and form — this is the spirit; while as to another it is united and formed — that is the flesh; that which is between these two — that is the soul, which sometimes indeed, when it follows the spirit, is raised up by it, but sometimes it sympathizes with the flesh and falls into carnal lusts.[27]

God's "image" is found in both our body and our soul by which we resemble God in having intelligence and freedom, but we are said to be in His "likeness" only by reason of the presence of the Holy Spirit. Because this Spirit was lost through Adam and Eve's misuse of their freedom, their descendants have been left in the fallen state of unlikeness, open to the temptations of the disordered impulses of the flesh. Only in the New Adam is this likeness restored, so that we can come to share in the divine immortality and incorruptibility.

A special feature of Irenaeus' theology, however, is that for him this recapitulation in the New Adam is something more than a mere restoration, because he understands *Genesis* 2:25, "The man and his wife were both naked, yet they felt no shame," to mean that the original innocence of Adam and Eve was a *child-like* state.[28] They could have arrived at spiritual adulthood only through a process of growth which is now achieved in Christ, the New Adam who has suffered evil and who thus not only restores creation but brings it to final completion. In this way Irenaeus refuted Gnostic dualism by providing the Church at the very beginning of its theological development with a remarkable anthropology in which the human body is portrayed as (1) essentially good; (2) an integral part of the human person; (3) a source of sin only when human free will destroys the relation of that person in the Holy Spirit to God; (4) a potential state which reaches its full perfection only in the risen Christ and those united in the Spirit to Him. Yet Irenaeus' thought still remains largely within biblical thought-forms, working with typology and using such vague distinctions as those between body, soul, and spirit, terms common to the Bible and Greek thought.

2. Anthropology of Origen and the Cappadocians

The first successful effort to give the Christian world-view a thoroughly Greco-Roman expression is what we may call the Platonic Christian Theology whose anthropology is dominated by the view that the human person is a spiritual soul which uses a body but is not substantially identified with it.

As Nietzche with his famous distinction between the Apollonian and Dionysian tendencies in Greek culture pointed out[29], this culture had both its rational and irrational side; but by the fourth century B.C. the rational tendency had led to the establishment of well defined, highly systematic schools of philosophy, deriving their concern for logical rigor from the unique discovery by the Pythagoreans of scientific mathematics.[30] These schools all attempted to reform popular Greek religion,[31] and by their influence on rhetoric or the art of public discourse, which was a central feature of Greco-Roman culture, their ideas had penetrated daily life. The Christian Church which had to enter into this public discourse by its preaching, apologetics, and polemics was forced to reinterpret its own biblical message in this philosophical language, at least in its popular rhetorical form.[32] The only question was which of the diverse schools of philosophy would best suit its purposes. One solution was to avoid the problem by eclecticism. This path was followed by most Christian authors who had little knowledge of or concern for the nice distinctions between one philosophical system and another; but the deeper minds among them could hardly be content to be so superficial and were keenly aware that choices had to be made if Christian thought was to become intellectually respectable and pastorally effective.

St. Justin Martyr (d.c.165), a Palestinian of Greek descent, who engaged in debates with the Jews (*Dialogue with Trypho*[33]) and the Greeks (*Apologies I* and *II*[34] and *On the Resurrection*) recounts his own efforts as a young student of philosophy to make such a choice:

> When I first desired to contact one of these philosophers, I placed myself under the tutelage of a certain Stoic. After spending some time with him and learning nothing new about God (for my instructor had no knowledge of God, nor did he consider such knowledge necessary), I left him and turned to a Peripatetic who considered himself an astute teacher. After a few days with him, he demanded that we settle the matter of my tuition fee.[35]

The Pythagorean to whom he went next required Justin first to learn mathematics before dealing with existential questions. Impatiently Justin turned to a famous Platonist:

> Under him I forged ahead in philosophy and day by day improved. The perception of incorporeal things quite overwhelmed me and the Platonic theory of ideas added wings to my mind, so that in a short time I imagined myself a wise man. So great was my folly that I fully expected immediately to gaze upon God, for this is the goal of Plato's philosophy.[36]

At last Justin met a truly wise man who by engaging him in a dialogue in the Socratic-Platonic style exposed the weakness of Platonism itself, especially in its doctrine of the pre-existence and transmigration of the human soul, and led him to faith in Christ the True Philosopher.[37] Thus Justin, contemporary with Irenaeus, established a model for Christian theology which was ready to use the rational, demythologizing method of Greek philosophy, while remaining sharply critical of some of its doctrines.

Why did most Christian theologians, like Justin, prefer the Platonic formulation of the Greek world-view to those of other schools? The answer is clear enough if we look at the alternatives available. Scepticism, whether as the system of Pyrrho (d.c.270 B.C.) and Sextus Empiricus (fl.c.250 A.D.) or as the disillusionment into which Greek philosophy had fallen by the time of Cicero (d.43 B.C.), had little appeal to young, enthusiastic Christianity, although the Apologists did not disdain to use its arguments to expose the contradictions of pagan thought and, as we will see later, it ultimately came into its own at the time of the Reformation as a support for Christian fideism.

Nor was there much appeal for Christians in the philosophy of Epicurus (d.270 B.C.) which in the eighteenth century reaction to Christianity was to be revived by Humanism as its classical prototype, except, perhaps, in Epicureanism's ethical exaltation of friendship and the security of rural retreat from the hubbub of civic engagement (values later revived by the romantic wing of Humanism).[38] Epicurus, following the pre-Socratic Democritus, taught that the universe, including the remote and mortal gods, is nothing more than a chance arrangement of an infinite number of atoms falling through an endless void. The human person is only a collection of such atoms which will be dispersed again at death. Hence we have nothing to fear after death, but also nothing to hope for except to live on earth the most serene and pleasant life we can manage with the help of our friends.[39]

Far more plausible for believing Christians was Stoicism, founded by a thinker of semitic origin, Zeno of Citium (d.264 B.C.), and this system did in fact have considerable influence on the development of Christian theology, in part because it synthesized many of the best points of the other Greek schools, and was probably the most popular of them all in the Roman Empire.[40] Stoicism, like Epicureanism, was a materialism, but it conceived matter not mechanistically as unchangeable atoms moving randomly in a void, but *dynamistically* as denser stuff energized by a "spirit" or "fire" (which was itself only a more tenuous stuff) constantly forming the world and then destroying it according to a purposeful and providential *logos* or *natural law* which may be worshipped as "Nature" or even as "God." The human person is a product of this plastic energy; the soul being a spark of the divine Fire giving life to the denser body. Hence at death the human soul does not perish but is reabsorbed into the universal Fire or Logos of the universe, as the body returns to the fund of denser material. Human intelligence and freedom are simply a participation in and submission to this universal energy into which the individual mind will be reabsorbed.

Thus for the Stoics morality was an uncomplaining submission to natural law. By this submission the philosopher can be saved from any internal conflict of emotions, hopes or fears, and in this way arrive at that same *ataraxia* or serenity which the Sceptics sought to attain by withdrawing from philosophical controversy and the Epicureans by retirement from the friction of public life and religious quarrels. The Stoics, however, believed that a true philosopher can remain serene even when fulfilling public offices in the service not only of his country, but even in the service of all mankind and of the cosmic order. Much of this could be interpreted in a Christian sense, but no Christian could accept Stoicism's impersonal God and impersonal immortality. Moreover, how could the passion of Christ be explained in a system in which the true philosopher is indifferent to suffering?

Hence the choice of a system which might be useful in constructing a Christian theology fell by elimination on those of Plato (d.348 B.C.) and his great pupil Aristotle (d.322 B.C.) both of whom had taught the existence of one personal, spiritual God, the immortality of the human intelligence, and an ethics whose goal was the contemplation of God, prepared by a life of moral discipline and the fulfillment of social obligations.[41]

Yet to Christians Aristotle appeared chiefly as the favorite pupil of Plato, who had greatly developed his teacher's logic, natural science, and ethics, but who, unfortunately, had not been altogether true to Plato's lofty theology. Plato was supposed (erroneously) to have taught in the *Timaeus* that the world had a beginning, and he certainly did argue strenuously for personal immortality and portrayed his own teacher Socrates as the ideal of virtue in the midst of poverty. Aristotle, on the contrary, was accused (correctly) of teaching the eternity of the world and (incorrectly) of denying divine providence and the immortality of the soul. Moreover he had unedifyingly argued that the life of virtue requires at least a modicum of material possessions![42] Hence Christian writers, although they made some use of Aristotle's logic, natural science, and ethics, believed that Plato was a better theologian.

At the same time the Christians from the beginning, as we have seen in Justin's case, were acutely aware that on at least three points Plato's thought could not be made to square with the Bible: (1) even if (which was doubtful) Plato had taught that the world was created, he did not understand creation as a *free* act of God but as a kind of emanation as necessary as the sun's shining; (2) he taught the immortality of the soul,

117

but conceived this as its liberation from the body, thus excluding the resurrection; (3) he conceived human beatitude as the contemplation of God, but taught that this could be achieved by the soul's own efforts to return to its original home with God, not as a gift of grace. To overcome these difficulties, while employing Plato's method, language, and major philosophical insights to construct a systematic Christian theology intelligible and persuasive to men and women raised in the Greco-Roman world, required an enormous hermeneutic effort. How could the lofty spirituality of Plato be joined to the Christian conviction that the material world and especially the human body because they are the free creations of God are not only essentially good but are capable of being elevated by the grace of God to a share in divine and eternal life?

In developing this Platonic Christian Theology the Alexandrian school of Clement (d.before 216), Origen (d.253) and St. Athanasius (d.373) played the leading role.[43] Clement of Alexandria attempted to meet the Gnostics half-way, but opposed their dualism, especially by his defense of marriage. Like his Jewish predecessor in Alexandria Philo, Clement argued that Greek philosophy had been derived from the wisdom of Moses. Yet for Clement Christ was the supreme philosopher. Hence the use of Greek philosophy in theology was not the introduction of something alien into Christianity but a kind of redemption of truth from its pagan captivity. Origen, also following Philo's lead, used allegorical methods of biblical exegesis, which the Greeks themselves had already employed to reconcile Homeric mythology with philosophy.[44] Moreover, he produced a systematic theology in his *On Principles*[45], boldly attempting, while deferring to the Bible and Church Tradition, to reinterpret the Christian world-view in terms of Plato's system or rather of Middle Platonism, that is, the eclectic form of Platonism which had developed under Stoic influences.

According to Origen's system while God is the free Creator, all created intelligences have pre-existed forever with God. To be intelligent is to be free, hence such spirits are always liable to sin, and many have in fact fallen. God in His goodness has to find a way to lead them back to Himself by a purgative and educational process. For this reason God created our material world as a place to discipline and redeem fallen souls. He has created this material world through the Logos, or first-created Son of God, and this Son of God has become incarnate in Jesus Christ in order to lead all fallen souls under the guidance of his Holy Spirit back to the Father. All fallen souls, perhaps only after many reincarnations and purgatories after death, will eventually return to God (*apokatastasis*). Obviously the great difficulty of this scheme is to fit the Resurrection (which Origen strongly affirmed) into it in any really meaningful way.[46]

Moreover, the pre-existence of souls, the possibility of their falling from heaven again and again, and the conception of the Son as subordinate to God the Father, were all divergences from the Tradition as this came to be defined by the Church councils after Origen's time. Nevertheless, Origen had put forward a powerful synthesis of Scripture and Platonic theology which later theologians tended to take as their starting point, while attempting to remove the difficulties which eventually led to the Church's condemnation of "Origenism" as a heresy. Origen's influence was especially great on the ascetic, spiritual tradition of the hermits and the cenobites as the monastic life began to be institutionalized in the Church at the beginning of the fourth century.

Through St. Athanasius of Alexandria the Church at the Council of Nicaea (325) took the decisive step of meeting the problem which both Gnosticism and this early Platonized theology had raised about the Incarnation by defining the divinity (i.e. the equality of the Son with the Father) of Christ, while at the same time affirming, against every gnostic tendency, his genuine humanity, thus giving a firm foundation to the anthropological principle that human beatitude is the divinization (*theosis*) of the human person through grace without the destruction of the bodily character of human existence. Thus Athanasius sometimes argues from our need to be divinized to the union of divinity and humanity in Christ:

> The reason, therefore, that the Son assumed a created human body was so He might in Himself deify what He had created and renewed, and might thus lead all of us who share His likeness into the Kingdom of heaven. For humans would not have been deified by being united to the Son if He were merely a creature and not truly God; nor would they have been admitted to the Father's presence unless the Son was the Father's natural, true Word who had put on a body.[47]

The defect in Athanasius' Christology is not with regard to Christ's body, whose reality he strongly affirmed, but with regard to His human soul of which he scarcely speaks, taking St. John's "the Word became flesh" (John 1:14) as if "flesh" simply meant "body." For Athanasius (who does not follow Irenaeus in distinguishing "image" from "likeness"), Christ is the Image of the Father in the sense of the Father's coequal, and we were created and will be divinized by Christ according to that Image; but the notion of the image of God in Christ's humanity is passed over.[48]

The full development of the Platonic Christian Theology in its earlier phase, however, is best seen not in these Alexandrians, but in the Cappadocian Fathers, Gregory Nazianzen (d.c.390), Basil the Great (d.379) and his brother Gregory of Nyssa (d.394), who had the considerable advantage of knowing Platonism, not in the somewhat confused form of Middle Platonism, as had the early writers, but in the form of the Neo-

Platonism of Plotinus (204-270) a contemporary and perhaps a fellow student of Origen, yet an opponent both of Gnosticism and orthodox Christianity. Plotinus, no doubt in reaction to Christianity, gave to Platonism a systematic and consistent form not to be found in the dialogues and epistles of Plato himself, but which today is generally admitted to be a genuine development not a distortion of Plato's deepest thought.[49] Plotinus managed to find in Plato's works remarkable parallels to many of the best features of the Christian world-view, while himself remaining strictly faithful to Plato's basic assumptions in sharp contrast to those of the Bible. With this advantage of knowing this highly systematized form of Platonism, the Cappadocians were able to develop a magnificent cosmology and anthropology, especially in Basil's *Hexameron* or commentary on *Genesis 1* and the completion of this work by Gregory of Nyssa in his *The Making of Man.*[50]

St. Basil interprets *Genesis 1* with the help of the science of his time, especially that of the Aristotelians and Stoics, but insists that all such knowledge is only probable, since it is only orthodox faith in scriptural revelation which can uncover the truths relevant to our salvation.[51] What *Genesis 1* shows us is that the One God is the creator of all things visible and invisible. His wisdom is revealed in the interdependence (*syndesmos*) and harmony (*sympatheia*) of all things — Stoic themes found in Philo.[52] This world under the firmament (Gen. 1:6) is the place of time and change which reflects the spiritual world but is ever fading away.

The human race was created part of this world of time, but the human soul was made to share God's eternal existence. Freed by the grace of Christ, it first rises above the instability and darkness of this sinful world to the pure realm of the first heaven (the visible firmament). Then, as it attains to wisdom, it becomes fit to dwell in the second heaven of the angels, whose bodies are pure light and whose mode of existence is between time and eternity. Finally, it rises to the third heaven (of which Paul spoke in II Cor. 12:2) of deification where the soul is united in eternity to God. Thus for Basil, while the fundamental ontological division is between Creator and creature, this duality is reflected in the created world by the division between the intelligible realm of light, and the sensible world created in the image of the realm of light yet subject to the darkness of sin and corruption. The human person is a microcosm sharing in both worlds, but destined ultimately for life with God. Therefore, Basil's dualism is not one of opposition, but of dynamic unity in which the lower realities reflect the higher.

Gregory of Nyssa in *The Making of Man* completed Basil's scheme with a more detailed and philosophically subtle anthropology.[53] God's creative act is beyond time, and the six days of creation are only symbolic. First God created Adam as part of the intelligible world of spirits in the second heaven. Adam then had a body of pure light which in no way dimmed his simple unity as an image of the One God. Yet as a created being having free will, Adam was liable to sin, and, in view of this liability to fall, God gave him a second, provisional mode of existence (the "second creation") as a member of the darker, sensible world of matter in which he could lead a temporal life of discipline and probation through which he might learn to return to God forever. Thus Adam or the human person in sensible existence is divided into male and female, in order that the human race dwelling in a world of change might still be preserved, and is also endowed with the other bodily organs and instincts (passions) necessary for getting on in such a world.

After this foreseen fall of Adam and Eve took place, God in His mercy sent His Son to become human and to share our suffering human condition in order to lead us back to our original home in heaven by the grace of His Holy Spirit. Gregory in his very influential spiritual works also goes on to show that beyond the second heaven the soul ascends through grace by faith and love to the third heaven of the Trinity.[54] It is in this union with God that soul is deified, yet it remains distinct from God whose essence is absolutely inaccessible to created minds. Only through His uncreated *energies* does God manifest himself to the souls whom He has drawn to himself in Christ, his only coequal Image.

Obviously Gregory is very close to Origen in much of this teaching, although he rejects Origen's notion of the eternal pre-existence of souls as contrary to the doctrine of creation. Yet Gregory does go so far as to accept Origen's theory that hell is only a temporary purgatory out of which all created spirits will ultimately return to God. This theory was rejected by later orthodox theologians, but the rest of Gregory and Basil's anthropology profoundly influenced the whole subsequent history of Christian thought. This Cappadocian anthropology retains the doctrine of the original goodness of the body and strongly affirms its resurrection, but it seems that it conceives this body not as our terrestrial body but as a body of pure light pertaining to the first creation. Did not St. Paul say, "If there is a natural body, be sure there is also a spiritual body . . . the first man was of earth, formed from dust, the second is from heaven" (1 Cor. 15:45, 48) ?

3. Anthropology of the Pseudo-Dionysius and St. Maximus

The development of this Platonic type of theology did not end with the Cappadocians. Christian Neo-Platonists always contended that they had found ways to reconcile Aristotle with Plato, and in the work of Nemesius, Bishop of Emessa, *On the Nature of Man* (c.400), significant elements of Aristotelianism were introduced into Christian anthropology.[55] More important was the reworking of this whole tradition by St. Maximus the Confessor (d.662) who developed what is today more and more recognized as the most balanced and classic version of eastern theology, performing for that tradition somewhat the same service as did St. Thomas Aquinas for western theology.[56]

St. Maximus even intensified the Platonism of this eastern tradition by making a thorough critique of the Origenism of the Cappadocians with its eclectic, Middle Platonic elements, and by shifting to a more consistent adherence to the systematic New Platonism of Plotinus and Proclus (d.485).[57] He was assisted in this shift by the writings of an unknown author (probably a Syrian monk) for whom Maximus by his sponsorship obtained orthodox acceptance and vast influence. This unknown writer who wrote under the name of a disciple of St. Paul, Dionysius the Areopagite (Acts 17:34), produced a Christian theology so close to that of the pagan Plotinus that one recent author has argued that he was not a Christian but a Greek philosopher attempting to evade the censorship of the Emperor Justinian.[58] The contribution of this Pseudo-Dionysius to an understanding of our bodily existence is to be found chiefly in two themes already present in the writings of Gregory of Nyssa[59] whose thought may have influenced him, but to which he gave much more systematic form.

The first of these is the concept of the human being as the microcosm or midpoint of an immense cosmic *hierarchy* (the word was introduced by Dionysius[60]) in which all the perfections of any substantial being are contained more perfectly and in more unified form in its immediate superior on up to the Absolute One in whom all being and perfection is unified. This hierarchy, however, is not static, since it is produced by the cascading overflow (by necessary emanation in Plotinus, but by free creation in Dionysius) of Being from the One to the many, and by the return of the many to the One through contemplation (for Plotinus a human effort, for Dionysius the work of grace as well). Dionysius elaborates this hierarchical scheme by a speculative angelology and an ecclesiology in which the earthly hierarchy of the Church images the linear hierarchy of the angelic kingdom. In this hierarchy matter is simply the manifestation of remoteness from the One, the dispersion of Being like light scattering in the darkness. Thus the human body reflects the imperfect unity of the human soul which requires to be reunified through contemplation —what has been scattered must be *recollected.*[61]

The second of these contributions is Dionysius' attempt to explain the dynamism of this cosmic hierarchy through its relation to the One above all Being and Non-Being. The One (who is also the ineffable Trinity) can be reached only by a negative or *apophatic* theology achieved through contemplative *ecstasis* in which all objective knowledge is overcome;[62] while Non-Being is the principle of plurality and of evil (since evil is the mere lack of Being). In this scheme matter tends to become identified with evil, not in the dualistic sense of Gnosticism, but in the sense of "distance" from the One. Matter is the fading away of Being into nothing; *ontologically* in the descent from the One, and *morally* in the failure of creatures to turn back through contemplation toward the One because of undue concern with lesser realities. Thus humanity as a microcosm and midpoint of the hierarchy tends to move toward the One by reason of its spirituality, but toward nothingness by reason of its corporeality. The human body is good to the degree that its beauty and order, and the beauty and order of the visible world revealed to our bodily senses, awaken the spirit to seek the Source of beauty and order through contemplative recollection; but evil to the degree that it becomes a burden to this return by drawing the spirit downward in pursuit of transitory things.

St. Maximus makes full use of this Dionysian system, but strives to correct its dangerous tendencies, as he also joins Gregory of Nyssa in correcting the dangers of Origenism. Maximus achieves this correction by insisting that since God freely creates by means of the Divine Ideas or *logoi,* He not only wills the return of all things to unity, but He also has eternally decreed their differentiation. Applied to anthropology this principle means that the composition of the human person from body and soul is not transitory, but permanently necessary for human completeness. Hence, St. Maximus is pre-eminent among eastern theologians in giving a clear reason for the resurrection. Against the tendency of Monophysitism, Monoenergism, and Monothelitism[63] to absorb the distinct humanity of Christ into his Divinity, Maximus gave his life in witness to the Chalcedonic doctrine of the permanent distinction of the two natures in Christ. Similarly, Maximus defended the positive reality of the human body distinct from the soul and complementary to it, thus opposing the Neo-Platonic tendency to describe the body chiefly in negative terms as a sign of the distance of the human soul from the Creator. At this point we are very close to an Aristotelian solution to the problem of the unity of the human composite, but Maximus went no further. For him the reason the human soul has a body is only that the Creator has so willed it. Why? Maximus did not ask.[64]

The ultimate refusal of the Eastern Church to be deceived by the extreme dualism to which it was tempted by Platonism was most decisively manifested in its resistance to the Iconoclasm which Emperor Leo III (717-740) tried to impose on it with persecutory zeal. Influenced by the

Old Testament commands against the worship of "graven images," but also undoubtedly by Platonic dualism, Leo tried to abolish the veneration of the icons and the crucifix. He would have reduced Christian liturgy to the style of the Jews, Moslems, and Calvinist Protestants. St. John of Damascus was himself one of the victims of this persecution, which was finally brought to an end by the seventh ecumenical council (the last recognized by the Eastern Church) Nicaea II in 787.[65] Although pagan Platonism and Neo-Platonism had not rejected the use of idols in popular worship, they interpreted all such visible material objects as mere *signs* pointing to a transcendent spiritual realm of purely intelligible realities. The Council saw clearly that this Platonic distrust of images which fostered Iconoclasm would lead to a docetic understanding of the Incarnation according to which the human Jesus would be considered a mere apparition manifesting the Second Person of the Trinity, rather than the hypostatic or *substantial* presence of the Word made flesh. The veneration of icons is thus an appropriate way of proclaiming the truth of the Incarnation and the Resurrection. The Eastern Church, however, did not go as far as the Western Church in permitting the liturgical use of three-dimensional sculpture, but restricted itself to two-dimensional images.[66]

During the Middle Ages the one severe crisis within eastern theology was the controversy over the views of St. Gregory Palamas (d.1359) who sought to defend the great tradition of eastern mysticism against attacks by the first Byzantines to become acquainted with Latin theology as developed in the medieval universities of the west.[67] In this defense Gregory developed a theory, based on certain texts of the Cappadocian Fathers, according to which the essence of God must remain forever hidden from human contemplation, but nevertheless the mystical union is possible through a vision of the Divine Light or "Uncreated Energies" of God. This "Light of Tabor" can be seen even with human physical eyes, since it was seen by the disciples in the Transfiguration of Jesus on that mountain. To theologians formed in the western scholastic mentality this talk of seeing God with bodily eyes seemed to deny the Divine Simplicity, the immediacy of the beatific vision, and its wholly spiritual character. A synod of the Eastern Church, however, accepted Gregory as an orthodox defender of the eastern tradition against western rationalism. The significance of this controversy for a theology of the body is that Gregory made a unique if not wholly successful effort to show how the glorified body is made participant along with the soul in the vision of God.

III: The Western Platonic Christian Theology

1. The Anthropology of St. Augustine

The Latin tradition of theology had the same roots as that of the East, and its first works were largely apologetic and anti-heretical. The most important Latin writers were Africans raised in the Roman rhetorical tradition whose philosophical interests were superficial and eclectic such as Tertullian (d.c.220), Cyprian (d.258), and Lactantius (fl.c.317), along with certain Roman writers of similar stamp such as Minucius Felix (fl.c 200), Hippolytus (d.235, who wrote in Greek) and Novatian (fl.250). Of these the chief was the earliest, Tertullian, who was more evidently under Stoic than Platonic influences. Although Tertullian was certainly no pantheist and taught the free creation of the world, yet he used the term *corpus,* body, for every *substance,* including God who, he says "is a body, although a spirit" (*corpus, etsi spiritus*);[68] and he believed that the human soul is a kind of shoot (*tradux*) pruned from the souls of the parents.[69] His conception of original sin was, therefore, very physical. Against the Gnostics of his day, such as Marcion, he defended the reality of the human nature of Christ and consequently the reality of the bodily presence of Christ in the Eucharist. It is perhaps more consistent with this materialism than it seems at first sight that Tertullian accepted the Montanist emphasis on the continuation of prophecy in the Church and a certain moral and ascetic rigorism, because he longed for a religion which was concrete, tangible, and visible in opposition to the all too "spiritual" myths of the Gnostics.[70]

The Platonic influence first becomes evident in the West in the anti-Arian works of Hilary of Poitier (d.367) who had spent some time in exile in Asia Minor where he became acquainted with Alexandrian theology. In his effort to defend the divinity of Christ, Hilary favored the view of Clement of Alexandria that while the Son of God was truly man, nevertheless his body was a heavenly one (*corpus caeleste*), i.e. already glorified as it was in the Transfiguration and when he walked on the water, and hence he was free of all bodily needs and passions. Christ's passion, therefore, was a voluntary surrender of his inherent glory.[71]

This Platonizing tendency is also very manifest in St. Ambrose of Milan (d.397) who was much influenced by Philo and Origen, although in Ambrose's moral works direct Stoic influence also is in evidence.[72] Ambrose, however, is much clearer than Hilary on the humanity of Christ and is generally very careful to correct the more dangerous aspects of Platonism. In his *Hexameron* and *On Paradise*[73] Ambrose transmitted much of St. Basil's thought on the first chapters of *Genesis* to Latin theology. For him the soul (or *mens,* the highest part of the soul or "the

125

inner man") is the human person whose body is only an outward instrument. Humanity was created in the image of Christ who in turn is the Image of God in the sense that by grace Christ's mind was united to the Holy Spirit and his body or outer man was in harmony with his mind, in one "spirit." After Adam had fallen through his own free choice, the total human person was alienated from God, is no longer in His image, and can be spoken of as "flesh." The image is restored by grace, but the conflict between the inner and outer man, between spirit and flesh, continues to show that we are still on the way to the final restoration of the paradisal peace in the Resurrection at the end of history.

Platonism is still more evident in the works of the African rhetorician Marius Victorinus (d. after 326), well acquainted with Greek philosophy both directly and through Cicero, and came to its climax in the great St. Augustine of Hippo (354-430), who fixed this tradition for the whole subsequent history of the Western Church. While Tertullian was familiar with Greek literature, Augustine, born only a little more than a hundred years after Tertullian's death, probably did not know Greek well and depended for his knowledge of the eastern theological tradition chiefly on translations. He was, however, in correspondence with St. Jerome (d.420) and Rufinus of Aquileia (d.410), Latin monks living in the east and well acquainted with the works of Origen. Moreover, Augustine, although he may have known very little of Plato directly, had read in translation some of Plotinus and much of Plotinus' pupil Porphyry.[74]

In his own spiritual odyssey Augustine had passed from the popular, non-intellectual Christianity of his childhood, through the influences of Cicero and the Latin rhetoricians with their rich Stoic, Epicurean, and Platonic borrowings, through Manicheism with its Gnostic dualism, and then through Neo-Platonism to the Scriptures as interpreted by Ambrose, to his own conversion to Christ. His own theology is powerfully original, yet it remains in its basic form a Platonic Christian Theology very similar to that of the eastern tradition. For my purposes, therefore, it is sufficient to note only the ways in which his anthropology moved beyond that tradition.

To understand Augustine's unique contribution I must say something about the issue of the subject-object duality which has so much occupied modern Humanist thought. Certainly the whole Platonic tradition is focused on the self-conscious subject, on the human soul which seeks truth by entering into itself rather than by turning outward to the objects of the visible world which at best serve it only as reminders of the inner reality. We have seen how all of eastern theology is occupied with spiritual subjectivity in this sense. For Plato himself, however, the goal of this inner search is not the subject itself, but the Ideas which are innate in the soul and need only to be remembered. These Ideas form a realm transcendent to the individual soul and are the *objects* of its contemplation. Thus the original Platonism is, in this special sense, objectivizing.

In Neo-Platonism, however, there is a transition to a new interest. Here there is not so much a concern with the Ideas, which exist in the Nous (or Second God) but are transcended in the One (or First God), as in the *process* by which all reality flows out of the One by emanation and returns to it by contemplation and ecstasy. This Neo-Platonic interest in process was very probably a result of the influence of the dynamism of Stoicism, back of which lies Aristotle's dynamic, as opposed to Plato's static, mathematicist, view of reality. Hence Neo-Platonism becomes interested in subjectivity, self-consciousness, in what characterizes spiritual existence as such.[75]

In Augustine this interest in subjectivity as such becomes what today we call *existential,* that is, it thematizes the drama of the soul faced with the crises of *free* decision. In Neo-Platonism the notion of "conversion" plays a central role in the sense of the turning back of the soul toward the One from which it originally emanated, but this conversion is conceived almost as an inevitable result of the soul's search for self-knowledge through reflection. The Christian theologians before Augustine were, of course, well aware that this conversion to God must be a free, moral act made possible for us only by the grace of Christ, but they dwelt more on the ascetic discipline by which this conversion results in spiritual growth, than on the converting decision itself.

Augustine's focus, however, is on the existential decision, that drama of freedom of choosing God in preference to self or tragically, self in preference to God. Hence it is in Augustine's thought that the modern concern with subjectivity as such was first clearly enunciated, and with it the problem of the *will*, never prominent in Greek philosophy.[76] Of course *love* is a prominent theme in Plato's dialogues and Augustine is faithful to Plato in conceiving God pre-eminently as the Lovable (the Good and Beautiful) and of human persons as drawn by love toward union with Him, yet Augustine's emphasis, unlike that of the Greeks, is on the *freedom* both of God and creature, i.e. on the grace which makes it possible for us to respond freely to God's free love for us.

Among Christian theologians it was Augustine who made the mystery of human freedom through grace the very heart of his theology. While by the humanity given us in creation we have the power of free choice between alternatives, this power cannot be actualized in conversion toward God or in continuing growth in obedience to God except by God's freely given help. Inevitably, without God's grace, burdened as we are by the effects of our sinful origin as members of the human race alienated from

God from the time of our first parents, we will come to choose ourselves and other creatures for ourselves, not for God. The choice of God is fully actualized free choice or *liberty* (*libertas*); while the choice of self and creatures is a free choice of enslavement. The grace which makes our liberty possible comes only through the Liberator, Jesus Christ. [77]

Augustine remained true to this basic insight, which he had experienced in his own life, no matter into what theological dilemmas it plunged him. The chief of these dilemmas was how to make this doctrine of grace coherent with the facts of human existence, personal, pastoral, and social which he was too realistic to ignore. Augustine was unable to turn from these cruel realities toward the ideal heaven of contemplation as easily as the Eastern Fathers seem to have done. To him it seemed only too obvious that the evident evil in the world is hard to reconcile with the absolute goodness and mercy of God revealed in Jesus. It is because Augustine faced this crucial theological problem of evil more squarely than any previous theologian that he has gained the reputation of pessimism, and even of crypto-Manicheism. It would be more just to say that he at last unflinchingly confronted the problems which Gnosticism had raised for Christianity and which earlier theologians had too easily dismissed without deeper reflection.

The centrality of the will and of love in Augustine's theology is also evident in another very important and original contribution to Christian anthropology, Augustine's theory of the Trinitarian image in the human soul. [78] Many of the Fathers before Augustine had used various analogies taken from nature and from human psychology in their attempts to make the doctrine of the Trinity more understandable, but Augustine went beyond this to argue that since the creation reflects the Creator and since the Creator is the Triune God, creation must somehow show traces (*vestigia*) of the Trinity. He believed that these are to be found in the *existence, truth,* and *goodness* found in every creature, manifesting respectively the Father, Son, and Holy Spirit.

Furthermore, following the exegesis of many of the Fathers, Augustine understood the words of *Genesis* 1:26, "God said, Let us make man in our image and likeness" because of the plural "us" to mean the human person is an image of the Trinity. Augustine was the first to teach that although the body like every creature bears a *trace* of the Trinity, the human soul alone is it *image.*

First, the soul in knowing itself images the Trinity when the soul knows itself, because this self-knowledge is a word distinct from the soul yet perfectly expressing it, just as the Son or Divine Word is distinct from the Father yet perfectly reveals Him. Moreover, the word of self-knowledge, although distinct from the soul, is united to it by the act of self-love proceeding from the soul through this word of self-knowledge, just as the Holy Spirit proceeding from the Father through the Son unites them in the unity of the Godhead. For Augustine the soul as the origin of self-knowledge and self-love is called the "memory", i.e. the soul's abiding psychic life which distinguishes it from corporeal things which pass away.

Second, when the soul knows and loves itself, it only imperfectly images the Trinity, but when it remembers, knows, and loves God this imperfect image is perfected through the power of the Holy Spirit by which the soul transcends itself in a faithful and loving search for God. What is especially striking and original in these speculations of Augustine is that he seems to have been the first theologian to identify the Holy Spirit with the act of Divine Love, as previous theologians had identified the Son with Divine Knowledge because the Scriptures speak of Him as "Word". It is frequently said today that Augustine's psychological theory of the Trinity is only a theory, insufficiently grounded in Scripture, but it would be more just to say that it is a profound theological effort to make explicit what is implicit in the scriptural language by which the Son is called "Word" and the Spirit is called "Gift".

Augustine's stress on human and divine love and on the freedom of the will did not, however, lead him (as it did some later medieval Augustinians, notably Luther) into anti-intellectualism or fideism. Indeed, from the very earliest phase of his thinking as a Christian, Augustine was deeply concerned with a question to which the Greek Fathers had given little attention: the *epistemological* problem. How can a firm faith in God dwell in the human mind if the human mind is incapable of truth?

Augustine knew well, through Cicero, that Platonic thought in its middle phase before the rise of Neo-Platonism had declined into a disillusioned scepticism. In his *Against the Academics,*[79] Augustine strove to defend the capacity of human reason, once it has been healed by the grace of faith, to attain to truths that are both objectively and subjectively certain. In refuting scepticism Augustine formulated the famous anticipation of Descartes' *cogito ergo sum* with his own *si enim fallor sum* (if I am mistaken, I must exist), i.e. the very fact that I doubt shows with certitude that I exist at least as a doubter.[80] The interesting thing about this notion is that it founds objective certitude on human subjectivity, i.e. on the indubitable experience of self-consciousness.

Augustine was also convinced that there are many truths, such as those of mathematics (here his Platonism is evident), which cannot be reasonably doubted or thought to be otherwise than they are. Such truths are not variable but eternal. How then is it possible for my mind, which obviously is not eternal, but variable and doubting, to attain such unvarying, eternal truths? And what is the ground of the eternity, the infallible objective certitude of these truths? Certainly it cannot be my human experience of this sensible world in which all is passing. The only answer must be that these truths originate in some One Eternal Truth which both grounds all particular objective truths and also our capacity as fallible creatures to know these truths with subjective certainty. It is by God's illuminating action, "by His light we see light," (Psalm 36:10), that we are enabled to attain certain knowledge, not of our empirical experiences, but of the eternal truths about the natures of things, including both the truth about God and of our human selves in relation to God.

We might suppose that this very Platonic emphasis on eternal truths would have led Augustine to a very a-historical understanding of human nature as it had Plotinus who would never talk about his personal biography to his disciples[81], but among early Christian thinkers there is none who equals Augustine in his keen understanding, very much in line with biblical mentality, of the *historicity* of human existence. His *Confessions* is the first autobiography in the modern sense of a self-revelation (although in the Cappadocians' orations on the death of relatives and friends and in some of the letters of the Fathers this specific Christian note is already struck, with a precedent in the epistles of St. Paul).[82] The *Confessions* also concludes with profound reflections on creation and time. Moreover, Augustine's vast *City of God* is the first Christian attempt at a theology of history, other than Irenaeus' somewhat vague recapitulation theory already mentioned.[83] Finally, in Augustine's works against the Donatists, we find important reflections not merely on the spiritual but on the institutional and therefore historical character of the Church. Why this concern with human historicity? Modern existentialist thought has enabled us to see that Augustine is, in a sense, the first existentialist because his interest in the crisis of free, moral decision brought him to see human life as a series of crucial events not as a pattern determined by a static human nature.

It is this historical understanding of anthropology which forced Augustine to give a more systematic theory of original sin than any of the Eastern Fathers had attempted.[84] Thus the Cappadocians had understood original sin primarily as the loss to humanity of its original immortality, with the consequence that the human race must now be maintained by sexual generation and suffers from many miseries, so that we need baptism to recover the gift of immortal life. For Augustine these miseries of the human condition were rooted in that loss of inner liberty which manifests itself in an enslaving, restless desire (*concupiscentia*) for those passing created goods which can never make us permanently happy. He was never sure exactly how this condition of unfreedom should be understood, whether as a contamination of the soul because it is an offshoot of the souls of its parents (Tertullian's theory already mentioned), or as a contamination of the soul directly created by God through its relation to a body produced by sinful parents (the theory of the Cappadocians).[85]

The weakness of Augustine's theology with regard to original sin, a weakness which was to have tragic results in the theology of the Reformers and the Jansenists, was his failure to distinguish clearly (as Aquinas and the Council of Trent later did in correcting him[86]), in the use of the term "original sin" between "sin" taken in the proper sense of an evil, free and responsible moral decision committed by Adam and deserving of the punishment of damnation, and "sin" taken in an analogical sense as the *effects* of Adam's sin on his descendents, which deprived them of the grace which would have otherwise come to them through him as head of the human family. This deprivation leaves all humankind naturally open to the internal conflict of reason and sensual desire, to suffering and death, but it does not make infants "sinners" in the proper sense of persons morally responsible for their own sins and thus deserving of damnation, but rather makes them victims of the sins of others. The Augustinians of a later time, such as the Reformers and the Jansenists, were to draw the conclusion that sin in the proper sense does not presuppose freedom of the will, a conclusion which completely nullified the very foundation of Augustine's anthropology, namely, his conviction that to be truly human is to be in God's image, and since God is free, Love Itself, to be truly human is to be free.

Today Augustine is often attacked for what is asserted to be his crypto-Manichean, dualistic view of human sexuality.[87] Actually his theology of sexuality made an important positive advance over that of the Eastern Fathers for whom, as we have seen, God created the sexes only in view of the foreseen fall of humanity into sin. Augustine, on the contrary, insisted that sexual differentiation was part of the original order of creation independent of whether sin would have occurred or not and that in Eden the multiplication of the human race would have been through sexual intercourse freed from any compulsiveness that would diminish its perfect freedom.

Nevertheless, because of his stress on liberty as characteristic of humanity as God originally created it, Augustine was puzzled by the undeniable common experience that we are driven by our greed for created goods (concupiscence), a greed that is physically manifest in the case of sexual desire. Such desire has a compulsive instinctual power independent of our rational will and often resistant to it. Was not this erotic madness, symbolized by the goddess Aphrodite or Venus, one of the greatest themes of pagan literature? Today it is commonplace to accuse Augustine of fearing sexual freedom; but for him the ambiguity of sexuality was precisely that in our actual fallen condition sex is never experienced as truly free, but always imposes its unfreedom on us who were created by God to be free.

Hence, although Augustine was the first theologian to write a systematic treatise in defense of marriage,[88] he believed that it was simply realistic to admit that sexual relations even in Christian marriage are always tainted with the unfreedom of selfishness, although this inevitable sinfulness is rendered excusable because marriage continues to perform its original God-given and noble purposes of procreation and mutual love of the partners, and in our fallen condition lessens temptations to extramarital indulgence. Certainly this theology of marriage failed to do full justice to the spiritual value of the physical expression of love which Augustine might have developed on the basis of such biblical texts as the *Song of Songs,* but it must be admitted that this theme is not only absent from Christian writers before Augustine, but that it would have been distasteful even to the great Stoic philosophers who dominated pagan culture. This applies even to Plato, who praised sexual love (chiefly in its pederastic form) only as an awakening to the power of Eros, whose earthly form must be entirely transcended in favor of a heavenly Eros beyond all physical expression.[89]

In summary it can be said that Augustine's immense contribution to a Christian anthropology was that, while retaining Platonism's profound understanding of human spiritual *interiority,* he gave to this interiority an existential, historical sense by emphasizing the centrality of will, freedom, and love in the relation of humanity to God and God to humanity. The acceptance of human historicity can only lead to a more positive understanding of the relation of the body to the soul, and eventually to an overcoming of Plato's error in identifying the human person with the human soul.

Augustine's thought quickly came to dominate western theology, although it remained largely unknown to the Eastern Church until the thirteenth century.[90] This is illustrated by the way in which Faustus of Rietz (d.after 485) in Stoic manner maintained that the soul, because it is localized in the body, must be material. He was answered by Claudianus Mamertus (d.c.474) in his *On the State of the Soul* and Cassiodorus (d.c.570) in his *On the Soul,* both closely following Augustine's anthropology.[91]

2. Medieval Theology Before the Rise of the Universities

The patristic age ended in the West with Pope St. Gregory I the Great (d.604) who, although he had lived six years in Constantinople, knew no more Greek that had Augustine.[92] Not a very original or speculative mind,

Gregory was very largely dependent on Augustine's thought, but gave his own teaching a more practical, pastoral form in his very influential *Moralia in Job*. Making use also of the tradition of the Desert Fathers through Cassian (d.c.430), Gregory stressed the psychological aspects of the interior life and showed a profound sympathy for the frailty of the human condition. He explained at length the seven capital sins as the roots of this frailty and described the struggles between the aspiring soul and the weary and troublesome body which weighs down the soul. Unlike Cassian, however, Gregory wrote not only for monks but for all Christians.[93]

After Gregory, the seventh century and the Carolingian Revival in the eighth, added little to the Augustinian theology until the remarkable John Scotus Erigena (d.877) who (having somehow learned Greek in Paris where nobody else seemed to have known it!) translated the works of the Pseudo-Dionysius, the *Making of Man* of Gregory of Nyssa, and parts of the commentaries of St. Maximus the Confessor on the works of the Pseudo-Dionysius. Thus the most advanced form of Platonic Christian Theology was made available to the Latins. Although in the form given it by Erigena this theology was condemned by the Latin Church for its paradoxical, pantheistic sounding formulas, it continued to exercise a profound influence throughout the Middle Ages. Erigena gave to the theology of Dionysius a very radical form, but remained orthodox in his Trinitarian doctrine, and in spite of appearances, did not really succumb to pantheism.[94] He taught that God the One creates the eternal ideas of all things in the Word, in this sense "creating himself", i.e. manifesting His hidden essence which is Non-Being in the sense of that which is beyond all Being, i.e. all multiplicity. These Eternal Ideas then become the source of the whole of the spiritual and corporeal creation by a process of hierarchical descent through multiplication, each lower member of the hierarchy being, as it were, a negation of the superior from which it proceeds, until individual things, including the least and most corporeal entities, are reached.

The Eternal Idea of Humanity is one and sexless; but it is manifested in temporal human beings who are fallen and in whom sinful sexuality appears, along with the other conflicting drives of the human condition. Through Christ, the Incarnate Word, all things will eventually return to God. Lower creation returns to God through being known by human beings, for the human being is a microcosm reflecting both the corporeal and the spiritual ranges of the created hierarchy. Humanity begins to return to God by contemplation and then by the liberation of the soul from matter at death. At last, in the Resurrection, the body will itself attain a spiritual condition through its reunion with the soul, so that all matter will return to God through the spiritualized human person. Yet this return will not be a pantheistic reabsorption into God, since the multiplicity of Ideas in the Word is eternal. Erigena admits eternal damnation but not eternal corporeal punishment; because for him damnation is simply the willful ignorance of God of those souls who have rejected Him.

Erigena's system, based largely as it was on Greek theology, still fitted into the Augustinian tradition, yet did not immediately affect that tradition as is evident from the works of the great Anselm of Canterbury (d.1109), the first figure in the "Twelfth Century Renaissance". Anselm remained firmly in the line of Augustine's Platonism, teaching the knowledge of eternal truths through illumination, the psychological image of the Trinity in the human soul, and the distinction between "freedom of choice" which remains in the sinner, and "liberty" which is lost by sin. Dionysian Platonism, however, (together with a variety of other Platonisms, as Fr. Chenu has shown,[95]) is apparent in the thought of two great clusters of twelfth-century theologians, the School of Chartres and the School of St. Victor, along with the Cistercians.

Bishop Fulbert of Chartres (d.1028) founded a school which has been brilliantly described for us by John of Salisbury (d.1180).[96] What is manifest in the writers of this school is the strong influence of the cosmic view of Plato's *Timaeus* (known through the Latin translation of a fourth century, probably Christian writer Chalcidius), in which the process of world production is described and the harmony between the spiritual and the visible order which imitates it is extolled.[97] William of Conches (d.c.1154) underlines the dynamic character of this cosmic order by adopting the view that the Holy Spirit is the World-Soul, like the Stoic Logos and the Plotinian Third God. Moreover, through the *Timaeus* these medievals became acquainted with the remarkable mathematical, mechanistic, and atomistic cosmology which Plato had derived from the pre-Socratics, especially Pythagoras and Democritus.

The Victorines and the Cistercians were more concerned with the problems of the spiritual life, but they too added other types of Platonism to Augustine's. Hugh (d.1141) and Richard (d.1173) of St. Victor and others of this school[98] proposed a mystical cosmology in which the soul ascends in contemplation to God by intellectual as well as ascetical discipline, as also did the Cistercians St. Bernard of Clairvaux (d.1153) and Aelred of Rievaulx (d.1166). Isaac of Stella (d.c.1167) especially developed the theme of the human person as a microcosm. It is by knowing ourselves that we come to know God and in His light all creation. In his *Letter on the Soul* addressed to his brother Cistercian Alcher, Isaac writes.

> There are then three realities, — the body, the soul, and God;
> I profess that I do not know their essence, and that I understand
> less what the body is than what the soul is, and less what the
> soul is than what God is.[99]

Nevertheless, the interest in the body was growing as is exemplified by William of St. Thierry in his *The Nature of the Body and the Soul* in which he borrows his treatment of the soul largely from Gregory of Nyssa's *The Making of Man,* but prefaces this with a very detailed discussion of the body derived not only from Nemesius, but from Constantine the African (d.1087) who had translated many Arabic works on medicine going back to the Galenic tradition. We have thus a clear indication of the turn toward the serious, scientific interest in the material universe which Christian thought was about to take in the thirteenth century.

If we now look back over Platonic Christian Theology as it had developed in east and west by the thirteenth century, we see that both in the east with such thinkers as St. Maximus the Confessor and in the west with the Twelfth Century Renaissance, theologians in this tradition had gradually worked to overcome the dualistic tendencies inherent in Platonism by recognizing that the human body is not simply the prison of the soul, but somehow its necessary instrument and expression, even in eternal, resurrected life. At the same time Christian thought continued to move within the Platonic epistemology in which the senses played only the role of awakening the soul to turn inward to an interior truth and upward to a Truth inexpressible in earthly terms. In such a theology the Bible had to be read primarily in other-worldly terms, and its celebration of earthly, bodily existence was suppressed.

Notes

1. See E.R. Dodds, *The Greeks and the Irrational* (Berkley: University of California Press, 1951), p. 135-156 who argues in Chapter 5, "Greek Shamans and the Origin of Puritanism", that the notion of metempsychosis originated in shamanism and the experience of the soul leaving the body in trance states: "As an example of the [archaic guilt-culture], we have seen how the Classical Age inherited a whole series of inconsistent pictures of the "soul" or "self" — the living corpse in the grave, the shadowy image in Hades, the perishable breath that is spilt in the air or absorbed in the aether, the daemon that is reborn in other bodies," p. 179.

It would seem that all of these notions are also reflected in the Old Testament. Note also that as Henry Chadwick, *Early Christian Thought and Christian Tradition; Studies in Justin, Clement, and Origen* (New York: Oxford University Press, 1966) says: "The Platonic idea of the relation between spirit and matter was capable of being interpreted either in an optimistic or a pessimistic way. It could be construed to mean that the visible world mirrors the glory of the supra-sensible world. It could also be taken (as by the Gnostics) to justify a radical rejection of the material order as an accidental smudge, resulting from a mistake", p. 113; and C.J. de Vogel, "The *Sōma-Sēma* Formula: Its Function in Plato and Plotinus compared to Christian Writers," in H.J. Blumental and R.A. Markus, *Neoplatonism and Early Christian Thought* (London: Variorum Publs., 1981).

2. "Tradition" here, of course must be distinguished from merely human "traditions" (Colossians 2:8). See Yves Congar, *Tradition and Traditions* (New York: Macmillan, 1967) for an extensive discussion of this distinction.

135

3. See John Meyendorff, *Byzantine Theology* (New York: Fordham University Press, 1974), pp. 4-11. Meyendorff, however, also stresses that Orthodoxy, because of its strong mystical tradition, holds that "Revelation, therefore, was limited neither to written documents of Scripture nor to conciliar definitions, but was directly accessible, as a living truth, to a human experience of God's presence in His Church" (p. 9). He seems to limit "development" to personal experience rather than doctrinal declaration: "In Jesus Christ, therefore, the fullness of Truth was revealed once and for all. To this revelation the apostolic message bears witness, through written word and oral tradition; but in their God-given freedom, men can experience it to various degrees and in various forms" (p. 10). See also Martin Jugie, *Theologia Dogmatica Christianorum Orientalium,* 5 vols. (Paris: Letouzey et Ané, 1926-1935), vol. 1, 642-669.

4. *Dogmatic Constitution on Divine Revelation,* Nov. 18, 1965, especially nos. 7-13 in Austin Flannery, O.P., ed., *Vatican Council II: The Conciliar and Post-Conciliar Documents* (Collegeville, Minn.: The Liturgical Press, 1975). For commentary see Abbott Christopher Butler, "The Constitution on Divine Revelation" pp. 43-53 and Paul S. Minear, "A Protestant Point of View" pp. 89-98 in John H. Miller ed., *Vatican II: An Interfaith Appraisal* (Notre Dame: University of Notre Dame, 1966), and James Gaffney S.J., "Scripture and Tradition in Recent Catholic Thought," pp. 141-169, Cornelius Ernst, O.P., "The Ontology of the Gospel", pp. 170-181, and Thomas Camelot, O.P., "Tradition", pp. 182-195, in Anthony D. Lee, ed. *Vatican II: The Theological Dimension* (Washington, D.C.: The Thomist Press, 1963).

5. See Werner G. Kümmel, *Man in the New Testament* (London: Epworth Press, 1963, rev. ed.) and Hans Walter Wolff, *Anthropology of the Old Testament* (Philadelphia: Fortress Press, 1974) for an introduction to the results of current exegesis for biblical anthropology. Also Leo Scheffczyk, *Man's Search for Himself* (New York: Sheed and Ward, 1966), and Albert Gelin, *The Concept of Man in the Bible* (London: Geoffrey Chapman, 1968). A useful summary is Claude Tresmontant, *A Study of Hebrew Thought* (New York: Desclée, 1960), pp. 83-114.

6. On the theology of Deutero-Isaiah see John L. McKenzie, *Second Isaiah,* Anchor Bible (Garden City, New York: Doubleday, 1968), "Introduction" pp. 6, 1vi-1xvii.

7. H.C.C. Cavallin, *Life After Death: Paul's Argument for the Resurrection of the Dead,* (Lund: Gleerup, 1974) vol. 1, "An Inquiry into the Jewish Background" treats of all the pre-Christian evidence (see especially the discussion of *II Maccabees* pp. 111-115, the Sadducees, pp. 193-196, and the author's synthesis pp. 198 ff.). He concludes that among the Jews there were many views about the after-life, but none was securely established as orthodox until after the beginning of the Christian era.

8. "The thesis of 'Creative Redemption' in the Book of Consolation (Isaiah 40-55) can be enunciated most briefly: from the new redemption of Israel, to the creation of the entire world of Israel; from the creation of the entire world of Israel, to the creation of the entire world *simpliciter*; from the creation of the entire world, to the redemption of this world" (p. 237). "The undeveloped and somewhat inconsistent form of Deutero-Isaiah's idea of creation places him *theologically inferior* to P [Priestly author of *Genesis* 1] and for that reason presumably *chronologically prior* as well. While P achieved a fuller theological presentation of *creation as such,* Deutero-Isaiah's attention centered on the *person* of Yahweh the *creator.*" Carroll Stuhlmueller, C.P., *Creative Redemption in Deutero-Isaiah* (Rome: Biblical Institute Press, 1970), p. 56.

9. Many scholars (e.g. Gerhard von Rad, *Genesis: A Commentary,* The Old Testament Library (Philadelphia: Westminster, 1961), pp. 76-78) see the lack of a clear explanation of the "tree of life" in the text of *Genesis* as evidence that it is an addition to the original story of J from another source; but it seems essential to the story as we now have it. St. Maximus the Confessor taught that both trees were *good,* but that the "tree of knowledge" signifies a knowledge of God by our own powers, and the "tree of life" a knowledge by the grace of faith. Thus the serpent spoke a half-truth in saying that the "tree of knowledge" would *deify* those who ate it. His deception was in concealing the fact that we can be deified only if we first believe God and through faith in Him come to know in a god-like way. Many

of the Fathers also identified the "tree of knowledge" with the "tree of life" in the paradoxical symbol of the Cross; cf. Lars Thunberg, *Microcosm and Mediator* (Lund: Gleerup, 1965), pp. 175-177.

10. See Yves Congar, *The Mystery of the Temple* (Westminister, Md.: Newman, 1962) for an extensive discussion of this theme.

11. Protestant theology has often accused Roman Catholic theology of not "taking sin seriously" because of the Catholic rejection of such terms as "total depravity." The mutual misunderstanding seems to result from the fact that Catholic theology tends to take an *ontological* point of view and to insist, therefore, that human sin cannot totally destroy the good of God's creation, since that would amount to annihilation; while Protestant theology stresses the *moral* point of view and insists that the sinner is totally incapable of fulfilling the Divine Law by his own efforts. Of course Catholic teaching agrees with this last assertion, while not denying that the sinner can perform some good actions which, however, cannot merit justification; a point which the Reformers also admitted when they granted that sinners can perform good "civil acts."

12. On the origin and nature of the Wisdom Literature see R.B.Y. Scott, *Proverbs and Ecclesiastes,* The Anchor Bible (Garden City, N.Y.: Doubleday, 1965), Introduction, pp. xv-li, and William McKane, *Proverbs, The Old Testament Library* (London: SCM Press, 1970), Part I, "International Wisdom," pp. 51-210; Roland E. Murphy, "Wisdom: Theses and Hypotheses," pp. 25-33 and Hans-Jürgen Hermisson, "Observations on the Creation Theology in Wisdom", pp. 35-42 in John G. Gammie, *et al., Israelite Wisdom: Theological and Literary Essays in Honor of Samuel Terrien* (Missoula, Mont.: Scholars' Press for Union Theological Seminary, New York, 1978).

13. McKane, *Proverbs* (note 12 above), pp. 51-210.

14. See Albert Gelin, *The Poor of Yahweh* (Collegeville, Minn.: Liturgical Press, 1964) and Jacques Dupont, *Les Béatitudes,* new ed., 3 vols. (Bruges: Abbaye de Saint André, 1958-1973) vol. 2, pp. 19-53.

15. Jesus seems to have compared himself to the bridegroom at a wedding feast in replying to the objection that unlike the Pharisees and John's disciples, Jesus' disciples did not fast (Mark 2:18-22). He also gave to this metaphor an eschatological thrust in predicting the time of sorrow to come. This theme is then reflected in the parable of the wedding banquet which the king gives for his son (Matt 22:1-14; cf. Luke 14:7-11) and in the parable of the wise and foolish virgins (Matt. 25:1-13; cf. Luke 12:35-40) and is then taken up in Ephesians 5:22-33 and Revelations 21-22.

16. On the ritual impurity resulting from contact with the dead and on the general notion, see Ronald de Vaux, O.P., *Ancient Israel* (London: Darton, Longman and Todd, 1965), pp. 56-61 and 460-464. "A corpse defiled one who touched it (for seven days. . .), or anything on which a corpse was placed. If one wanted to engage in any holy activity elaborate decontamination from death-pollution was necessary. Secretions from the body were, as it were, dead (though this reason seems not to have operated on their minds) and they defiled automatically, with different effects in each case. Semen, menstrual blood, urine, faeces, spittle . . . defiled. The touch of a menstruant defiled. Sexual intercourse defiled, and in theory the partners must immerse themselves immediately thereafter," J. Duncan M. Derrett, *Jesus's Audience* (London: Darton, Longman and Todd, 1973), p. 130.

17. See Rudolf Schnackenburg, *The Moral Teaching of the New Testament* (New York: Herder and Herder, 1965), pp. 296-306 (Paul); 359-364 (James); 365-371 (Peter). For a feminist criticism of the *"Haustafel* trajectory" in the New Testament see Elisabeth Schüssler Fiorenza, "Discipleship and Patriarchy: Early Christian Ethos and Christian Ethics in a Feminist Perspective," *The Annual of the Society of Christian Ethics, 1982,* Society of Christian Ethics (Perkins School of Theology, Southern Methodist University, Dallas, Texas), pp. 131-172. Fiorenza contrasts what she calls the "submission ethic" of the *Haustafeln* (household codes) with the "co-equal discipleship ethic" which she considers central to the New Testament. I will discuss this problem further in the later chapters of this book.

18. Even the animals in Eden were vegetarian! "God also said, "See, I give you every seed-bearing plant all over the earth and every tree that has seed-bearing fruit on it to be

your food; and to all the animals of the land, all the birds of the air, and all the living creatures that crawl on the ground, I give all the green plants for food'" (Gen. 1:29-30). But after the flood God concedes to Noah (evidently because "The desires of man's heart are evil from the start" (8:21) that "'Every creature that is alive shall be yours to eat; I give them all to you as I did the green plants. Only flesh with its lifeblood still in it you shall not eat'" (9:3-4). In the early Church vegetarianism was sometimes associated with celibacy. Thus the Syrian Church held that in Eden there was neither meat eating nor sexual intercourse and it appears that this notion was derived from Jewish thought, since Rabbi Rab held there was no sex in Eden. Consequently in Syrian Christianity celibacy was required after baptism. See David Edward Aune, *The Cultic Setting of Realized Eschatology in Early Christianity* (Leiden: E.J. Brill), pp. 215-219. On the Jewish origin of these ideas see p. 216, note 4.

19. See Louis Bouyer, *Eucharist: Theology and Spirituality of the Eucharistic Prayer* (Notre Dame, Ind.: University of Notre Dame, 1968), p. 15-135.

20. Raymond E. Brown, *The Gospel According to John,* The Anchor Bible (Garden City, New York: Doubleday, 1966), vol. 1, Introduction, pp. 1iii-1xvi.

21. Jean Danielou, *A History of Early Christian Doctrine,* vol. 1, *The Theology of Jewish Christianity* (London: Darton, Longman, Todd, 1964).

22. See Leon Morris, *Apocalyptic* (Grand Rapids, Mich.: Eerdmans, 1972), "Dualism", pp. 47-49.

23. "From about the middle of the third century B.C. *all Judaism* must really be designated *'Hellenistic Judaism'* in the strict sense." Cf. Martin Hengel, *Judaism and Hellenism,* 2 vols. (London: SCM Press, 1974), 1, p. 104. Hengel, pp. 17-217, shows in detail how Greek ideas and thought forms affected the later books of the Old Testament Canon and the extra-canonical literature. James W. Thompson, *The Beginnings of Christian Philosophy: The Epistle to the Hebrews,* Catholic Biblical Quarterly Monographs, no. 13 (Washington, D.C.: Catholical Biblical Association of America, 1982), has shown convincingly that within the New Testament itself the reformulation of the Gospel in Middle Platonic terms has already begun; cf. especially pp. 17-14, and the conclusion pp. 152-162.

24. See Pheme Perkins, *The Gnostic Dialogue* (New York: Paulist Press, 1980), especially the discussion pp. 16-19 of the Jewish origins of Gnosticism. Perkins mediates between the position of Hans Jonas who points out the anti-Jewish character of much of the gnostic literature and George McRae who stresses its extensive use of Old Testament symbols by arguing that it arose within Judaism from heterodox sects in bitter conflict with orthodoxy. Such sectarians more easily became Christian because Jewish Christianity was itself a sect within Judaism.

25. On Irenaeus and Justin Martyr see Daniélou, *History of Early Christian Doctrine* (note 21 above), vol. 2, *The Gospel Message and Hellenistic Culture,* pp. 157-196; Johannes Quasten, *Patrology,* 5 vols. (Westminster, Md.: Newman, 1951), vol. 1, p. 196-219; H.B. Timothy, *The Early Christian Apologists and Greek Philosophy* (Assen: Van Gorcum, 1973), pp. 23-39; Eric Osborn, *The Beginning of Christian Philosophy* (Cambridge: Cambridge University Press, 1981); John Lawson, *The Biblical Theology of St. Irenaeus* (London: Epworth Press, 1948); D.E. Jenkens, "The Make-up of Man according to St. Irenaeus," *Studia Patristica* (Berlin: Akadamie Verlag, 1962) vol. 6, pt. 4, pp. 91-95; Michel Aubineau, "Incorruptibilité et divinisation selon s. Irenée", *Recherches Patristiques* (Amsterdam: A.M. Hakkert, 1974), pp. 197-224; E.P. Meijering, *God Being History: Studies in Patristic Philosophy* (Amsterdam: North-Holland, 1975) is largely devoted to studies on Irenaeus.

26. *Adversus Haereses,* III, 18, 1. My translation from the Latin translation in which the complete work is preserved, from text ed. by Adelin Rousseau and Louis Doutreleau, *Sources Chrétienne* (Paris, 1974) tom. 2, pp. 342 f. On "recapitulation" see Lawson, *Biblical Theology of St. Irenaeus* (note 25 above) pp. 140-188.

27. *Ibid.* 5, 9, 1, pp. 106 f. My translation.

28. *Adversus Haereses,* III, 22, 4, pp. 438 ff.; IV, 38, 1 and 3, pp. 943 ff. and *Demonstration of the Apostolic Preaching,* 12, French trans. by L.M. Froidevaux, *Sources Chrétiennes* (Paris: Cerf, 1959), p. 50 f. On this subject see N.P. Williams, *The Ideas of the Fall and Original Sin* (London: Longmans Green, 1927), pp. 189-199, who says that this conception of the imperfection of the first parents is found also in Theophilus and Tatian.

138

29. Friederich Nietzsche, *The Birth of Tragedy from the Spirit of Music,* vol. 1. *Complete Works* ed. by Oscar Leng (New York: Russell and Russell, 1964), pp. 21-28. Nietzsche characterizes the contrast between the Apollonian and the Dionysian world as that between "dreamland and drunkenness."

30. It is debatable whether the understanding of mathematics as an axiomatic science really goes back to Pythagoras or originated in the Platonic Academy. See C.J. De Vogel, *Pythagoras and Early Pythagoreanism* (Assen, The Netherlands: Van Gorcum, 1966), pp. 192-217; J. A. Philip, *Pythagoras and Early Pythagoreanism* (Toronto: University of Toronto Press, 1966), pp. 76-109; Erich Frank, *Plato und die sogennanten Pythagoreer* (Tübingen: Max Niemeyer Verlag, 1970); W.A. Heidel, "The Pythagoreans and Greek Mathematics" in David J. Furley and R.E. Allen, *Studies in Pre-Socratic Philosophy* (London: Routledge and Kegan Paul, 1970), vol. 1, 350-381, and Walter Burkert, *Lore and Science in Ancient Pythagoreanism* (Boston: Harvard University Press, 1972),

31. See A.J. Festugière, O.P., *L'idéal religieux des Grecs et l'Évangile* (Paris: J. Gabalda, 1932), for the different approaches of each of these schools to religious reform. That the Church Fathers in their turn did not accept pagan philosophy without penetrating critical revision is well shown by Claude Tresmontant, *La métaphysique du Christianisme et la naissance de la philosophie chrétienne* (Paris: Seuil, 1961), especially on the theme of the freedom of creation as against Greek determinism, pp. 190-195.

32. Danielou, *History of Early Christian Doctrine* (note 21 above), vol. 2, *The Gospel Message and Hellenistic Culture,* and vol. 3, *The Origins of Latin Christianity,* treats of this transition in great detail.

33. PG 6, cols. 471-800. For commentary see J.C.M. Van Winden, *An Early Christian Philosopher: Justin Martyr's Dialogue with Trypho,* Chapters 1 to 9 (Leiden: Brill, 1971). On Justin see references in note 25 above and H. Chadwick's treatment in A.H. Armstrong, ed., *The Cambridge History of Later Greek and Early Medieval Philosophy* (Cambridge: Cambridge University Press, 1967), pp. 158-167; M.J. Lagrange, O.P., *Saint Justin, philosophe, martyr,* 2nd ed. (Paris: Gabalda, 1914); and E.R. Goodenough, *The Theology of Justin Martyr,* (Jena, 1923; Reprint Amsterdam: Philo Press, 1968). The most recent English translation of his works is by Thomas B. Falls in Ludwig Schopp, ed., *The Fathers of the Church* (New York: Christian Heritage, Inc., 1948), *Trypho:* pp. 147-368.

34. *Ibid., First Apology,* pp. 33-114, *Second Apology,* pp. 119-138; PG. 6, cols. 327-440; 441-470.

35. *Trypho,* 2, p. 149; PG 6, cols. 476-477.

36. *Ibid.,* p. 150 f.

37. *Ibid.* 3-8, pp. 151-161; PG 6, cols., 477-494.

38. *Festugière, L'idéal religieux* (note 31 above), pp. 59-65, and *Epicurus and His Gods* (New York: Russell and Russell, 1955). Tertullian wittily speaks of "the nobility (*honor*) of Plato, the force (*vigor*) of Zeno, the balance (*tenor*) of Aristotle, the stupidity (*stupor*) of Epicurus, the melancholy (*maeror*) of Heraclitus, and the madness (*furor*) of Empedocles", in *De Anima,* 3, PL vol. 2, pp. 647.

39. Festugière, *Epicurus and His Gods,* (note 38 above) pp. 8-39, on the concept of "serenity" (*ataraxia* or *untroubledness*) as the goal of the philosophic life.

40. Festugière, *L'idéal religieux* (note 31 above), pp. 66-72, and *La révélation d'Hermès Trismegiste,* 4 vols. (Paris: Lecoffre 1944-45) 2, pp. 260-340, on Stoicism, and M. Spanneut, *Le stoicisme des Pères de l'eglise,* 2nd ed. (Paris: J. Duculot, 1969).

41. It is sometimes argued that because Plato was known to Christians principally through the Middle Platonists and Neoplatonism that his thought was wrongly understood. The tendency of modern scholarship, however, has been to stress the continuity of the Platonic tradition; see R. Klibansky, *The Continuity of the Platonic Tradition* (London: Warburg Institute, 1939, reprinted 1950); H.A. Wolfson, *The Philosophy of the Church Fathers* (Cambridge, Mass.: Harvard University Press, 1956); Philip Merlan, *From Platonism to Neoplatonism* (The Hague: Martinus Nijhoff, 1960) (argues p. 3 f. that even Aristotle is a Neo-Platonist!); C.J. De Vogel, *Philosophia, Part I, Studies in Greek Philosophy* (Assen: The Netherlands: Van Gorcum, 1969); Jean Daniélou, *History of Early Christian Doctrine*

139

(note 21 above), vol. 2, Chapter 4, "Plato in Christian Middle Platonism", pp. 107-128; John Dillon, *The Middle Platonists* (Ithaca, N.Y.: Cornell University Press, 1977); John N. Findlay, "The Neoplatonism of Plato" in R. Baine Harris, ed., *The Significance of NeoPlatonism* (Norfolk, Va.: Old Dominion University, 1976), pp. 23-40.

In an interesting study, E.N. Tigerstedt, *The Decline and Fall of the NeoPlatonic Interpretation of Plato* (Helsinki; Commentationes Humanarum Literarum, Societas Scientiarum Fennica, 1974), t.52, shows that St. Augustine originated the theory of an "esoteric" systematic doctrine of Plato (as contrasted to the unsystematic exoteric doctrine of the dialogues), which still has some supporters, but argues that "systematic" Platonism was the work of the Middle and Neo-Platonists based not on an esoteric tradition but simply on their interpretations of the dialogues. See also pp. 54-63 on the Protestant objections to the Platonic influences in the Catholic tradition, and for a modern evaluation of the Platonism of Melancthon.

42. Festugière, *L'idéal religieux,* pp. 220-263 (note 31 above) deals with Aristotle's reputation in antiquity and shows that some of these accusations are based on the spurious *De Mundo;* see also Festugière's discussion of this work in *La révélation* (note 40, above), 2, 555-572.

43. Quasten, *Patrology* (note 25 above), 2,1-120; Armstrong, *Cambridge History* (note 33 above), pp. 137-194 on Alexandrines. Cf. also Joseph C. Mc Lleland, *God the Anonymous: A Study in Alexandrian Philosophical Theology* (Philadelphia: Philadelphia Patristic Foundation, 1976) who stresses the danger of "classical theism" stemming from Philo which assumed an "optimistic rationalism", conceiving God as the "Impassible" and neglected other aspects of Greek theology thus provoking the recent development of process and death of God theologies, and other efforts to show that God becomes actual only in the world of time; E.F. Osborne, *The Philosophy of Clement of Alexandria* (Cambridge: Cambridge University Press, 1957) and Elizabeth A. Clark, *Clement's Use of Aristotle* (Lewiston, New York: Edwin Mellen Press, 1977).

44. See Jean Daniélou, *Origen* (New York, Sheed and Ward, 1955) and Chadwick, (note 1 above). On the relation of Origen to Plotinus see Franz Heinrich Kettler, "War Origenes Schüler des Ammonios Sakkas?" in J. Fontaine and Charles Kannengiesser, ed., *Epektasis: Melanges patristiques offerts au Cardinal Jean Daniélou* (Paris: Beauchesne, 1972), pp. 330-334. On Origen's influence on the history of Christian spirituality see Louis Bouyer, F. Vandenbroucke and J. Leclercq, *A History of Christian Spirituality,* 3 vols. (New York: Desclée, 1969) 1, pp. 276-302, 369-394.

45. PG 11: cols. 115-414, H. Crouzel and M. Simonetti eds., *Sources Chrétiennes,* no. 252 (Paris: Cerf. 1978-80). Translation by G.W. Butterworth with an introduction by Henri de Lubac (New York: Harper Torchbooks, 1966). For Origen's use of philosophy see Henri Crouzel, *Origène et la philosophie* (Paris: Aubier, 1959) who concludes (p. 177) that Origen's uncritical use of Platonism reduces to two errors (1) pre-existence of souls; (2) Christological subordinationism. See also Crouzel, *Théologie de l'image de dieu chez Origène* (Paris: Aubier, 1956) especially on the theme of "image and likeness", pp. 217-246.

46. Apparently (it is not altogether clear because of the fact we must depend chiefly on the Latin translation of Rufinus of the *On First Principles)* Origen thought that eventually the resurrected bodies of the blessed would attenuate to nothingness. See *On First Principles,* Book III, Chapter vi, pp. 245-255, with notes and p. lvi of the introduction by Butterworth. See also Jacques Dupuis, S.J., *"L'esprit de l'homme": Étude sur l'anthropologie religieuse d'Origene* (Paris: Desclee de Brouwer, 1967). See also Henry Chadwick, "Origen, Celsus and the Resurrection of the Body," *Journal of Theological Studies* 48 (1947): 34-9.

47. *Contra Arianos* II, 70. PG 25, col. 296.

48. See J.N.D. Kelly, *Early Christian Doctrines,* rev. ed. (New York: Harper and Row, 1978), pp. 284-289, and Quasten, *Patrology* (note 25 above), 3, 72-76, who concludes that although Origen had clearly affirmed the existence of a human soul in Christ, Athanasius never does, but also never explicitly denies it.

49. See note 41 and 44 above. On the attitude of the Fathers to Plato see E.P. Meijering, *God Being History* (note 25 above), esp. "Wie Platonisierten Christen?" p. 133-146, against H. Dörrie, "Was ist spätantiker Platonismus?", *Theologische Rundschau* (N.F.) 36 (1971): 285-302 who argues that the Christian creed of the Fathers was absolutely inconsistent with Platonism which was more the cause of heresies than an influence on Christian orthodoxy.

50. Basil's *In Hexameron* PG. 29, cols. 3-308. Edition and French translation by S. Giet, *Sources Chrétienne,* no. 26 (Paris: Cerf, 1949), English translation by B. Jackson, P. Schaff and H. Wace eds, *Nicene and Post-Nicene Fathers,* 2nd. Series (Grand Rapids, Mich.: Eerdmans, 1952, reprint of 1892 edition) 2nd series, vol. 8, pp. 52-107. The authenticity of Homilies 10 and 11, usually regarded as spurious, is defended in the edition and French translation by A. Smets and M. Van Esbroeck, *Sources Chrétienne,* no. 160, 1970. Gregory of Nyssa *De hominis opificio,* PG 44, cols. 125-1256; edition and French translation by J. Laplace, *Sources Chrétienne,* no. 6, 1943 and its completion in the *Explicatio apologetica in Hexameron,* PG. 44, cols. 61-124. The former work was Englished by H.A. Wilson, *Nicene and Post-Nicene* Fathers, 2nd series, vol. 5, pp. 387-427. Quasten, *Patrology* (note 25 above) 3, 217, points out that Basil's exegesis is not allegorical but literal. Laplace in his introduction to Gregory's work shows that in distinction to Basil's Platonism, Gregory introduces the Aristotelian notion of *physis* as composed of matter and form.

51. Excellent syntheses of the philosophical thought of St. Basil and St. Gregory of Nyssa are to be found in the two essays, John F. Callahan, "Greek Philosophy and Cappadocian Cosmology" pp. 29-58 and Brook Otis, "Cappadocian Thought as a Coherent System" pp. 96-124 in *Dumbarton Oak Papers,* no. 12, (Cambridge, Mass.: Harvard University Press, 1958). Brook Otis, p. 94 note 1 argues for the dependence of Gregory of Nyssa on the thought of Basil and Gregory Nazianzen although he is usually credited with being the more original thinker See also Jean Louvras, *Les idées anthropologique et l'activité morale de saint Basile.* (Lille: University of Lille, 1958: Diplome).

52. See Werner Jaeger, *Nemesios von Emesa* (Berlin, 1914) and *Two Rediscovered Works of Ancient Christian Literature: Gregory of Nyssa and Macarius* (Leiden: Brill, 1954), pp. 25-26 on the notion of *syndesmos.* On *syndesmos* and *sympatheia* see Max Pohlenz, *Die Stoa,* (Gottingen: Vandenhoeck and Ruprecht, 4th ed. 1970), I, p. 43 f, II, p. 58 with references, who believes it was Posidonius who made these terms central to Stoic cosmology. They are found in Basil, *In Hexameron,* Homily 6, 10, Giet (note 50 above) pp. 380 f. and in Nemesius; see commentary by William Telfer, *Library of Christian Classics* (Philadelphia: Westminster Press, 1955), vol. 4, I, cc. 2-3, pp. 230-232. This inter-relation and linkage of things was explained by the Stoics as evidence of a World Soul. Humanity was seen as the mid-point of the chain of being, the microcosmos reflecting the macrocosmos and binding together the higher and the lower order of things. On Posidonius see John M. Rist, *Stoic Philosophy* (London: Cambridge University Press, 1969), pp. 210-218. On the World Soul see Joseph Moreau, *L'âme du monde de Platon aux stoiciens* (Hildesheim: George Olms, 1965).

53. I have used Jean Daniélou, S.J., *Platonisme et théologie mystique; Doctrine spirituelle de Saint Grégoire de Nyssa,* rev. ed. (Paris: Aubier, 1944); "La résurrection des corps chez G.", *Vigiliae Christianae,* 7 (1953): 154-170; *"L'apocataste chez G.",* *Recherches des science religieuse,* 30 (1940): 328-347; "Notes sur trois texts eschatologique de Saint G." *ibid.* 348-356; *L'être et le temps chez G.* (Leiden, Brill, 1970) (especially Chapter x on the apocatastasis); "Metempsychosis in G." in David Neiman and Margaret Schatkin, *The Heritage of the Early Church,* Orientalia Christiana Analecta, no. 195 (Rome: Pont. Inst. Studiorum Orientalium, 1973), pp. 227-243. Also see Harold Cherniss, *The Platonism of G.* (Berkley: University of California, 1930); A.H. Armstrong, "Platonic Elements in St. G.'s Doctrine of Man," *Dominican Studies* 1 (1948), 113-126; Hans Urs van Balthasar, *Présence et Pensée* (Paris: Beauchesne, 1942); Roger Leys, *L'image de Dieu chez saint G.* (Paris: Desclée de Brouwer, 1951); Jérome Gaith, *La conception de liberté chez G.* (Paris: Vrin, 1953) (critical of Balthasar); Gerhart B. Ladner, "The Philosophical Anthropology of G." in *Dumbarton Oak Papers,* no. 12 (Cambridge, Mass. Harvard University Press, 1958); 29-58; David L. Balàs, *Metousía Theou: Man's participation in God's perfections according to G.* (Rome: Herder, 1966); E. Corsini, "Plérôme humain et plérôme cosmique chez G." in Marguerite Harl, ed., *Écriture et culture philosophique dans pensée de G.* (Leiden, Brill, 1971), pp. 111-126, and "L'harmonie du monde et l'homme microcosme dans le *De hominis opificio* in J. Fontaine et C. Kannengieser, *Epektasis* (see note 44 above) pp. 455-462; John P. Cavarnos, "Relation of Soul and Body in the Thought of G.", in *Gregor von Nyssa,* International Colloqium Freckenhorst, Germany, 1972 (Leiden: Brill, 1972), pp. 61-78; Reinhard M. Hübner, *Die Einheit des Leibes Christi bei G.* (Leiden: Brill, 1974); David L. Balàs, "Eternity and Time in *Contra Eunomium", ibid.,* 128-155, and Monique Alexandre, "L'exégese de Gen. 1: 1-2a in l'*In Hexameron de G.;* Deux approaches du problème de la matière" and discussion, *ibid.* pp. 159-192.

141

54. See articles of Otis and Callahan in note 51 above.

55. See note 52 above and Boleslaw Domanski, *Die Psychologie des Nemesius, Beiträge sur geschichte der philosophie des mittelalters,* bd.iii, heft 1, and E. Skard, article "Nemesios", Pauly-Wissowa, *Realencyklopädie der Classichen Altertumswissenschaft* (Stuttgard: J.U. Metzlersche, 1940), Suppl. vii, pp. 561-566.

56. See Hans Urs Von Balthasar, *Liturgie Cosmique: Maxime le Confesseur,* trans. from German by L. Lhaumet and H.A. Prentout (Paris: Aubier, 1947) and Lars Thunberg, *Microcosm and Mediator: The Theological Anthropology of Maximus the Confessor* (Lund: Gleerup, 1965). In English we have Polycarp Sherwood, O.S.B. *The Earlier Ambigua of St. Maximus the Confessor,* Studia Anselmiana no. 36 (Rome: Herder, 1955) and by the same translator, *The Ascetic Life* and *The Five Centuries of Charity* in *Ancient Christian Writers* no. 21 (Westminster, Md.: Newman Press, 1955) with a valuable introduction pp. 3-102.

57. Daniélou, *Platonisme* (note 53 above), pp. 13-28 identifies the four principle elements in Maximus' thought as (1) a Christian Platonism derived from the Cappadocians and the Areopagite; (2) an Aristotelian philosophical rigor and emphasis on the value of the natural world derived through Leontius of Byzantium; (3) a practical mysticism derived from Evagrius Ponticus and hence from Origen; (4) a firm opposition to Monophysitism in Christology. Balthasar stresses the notion that the Aristotelian element so modified the Platonic basis of his thought as to bring it close to the thought of Thomas Aquinas, in its respect for the natural order of things while exalting the distinction between nature and the divine order.

58. Ronald F. Hathaway, *Hierarchy and the Definition of Order in the Letters of Pseudo-Dionysius* (The Hague: Martinus Nijhoff, 1969), p. xxi.

59. "We have a whole body of evidence seemingly obliging us to place 'Dionysius' in the direct line of Gregory of Nyssa, not without indisputable references to Evagrius [Ponticus, a friend of the three great Cappadocians]. But this comes back to making him a disciple and a faithful continuator of the great Christian Alexandrines, however removed from them he may have been in time and place" Bouyer, *History of Christian Spirituality* (note 44 above) 1, 400.

60. Cf. Dominic J. O'Meara, *Structures hierachiques dans la pensée de Plotin* (Leiden: Brill, 1975), pp. 1-18, especially note 1.

61. See Rèné Roques, *L'Univers Dionysien* (Paris: Aubier, 1954) and his articles "Denys L'Areopagite (Le Pseudo-)" in the *Dictionnaire de Spiritualité* (Paris: Beauchesne, 1957), t.3, pp. 243-286, and articles by various authors on his influence, pp. 287-427.

62. "The negative (apophatic) way attempts to know God not in what He is (that is to say, in relation to our experience as creatures) but in what He is not. It proceeds by a series of negations. The Neo-Platonists and India use this way too . . . But outside Christianity, it only ends in the depersonalization of God, and of the man who seeks Him . . . A Gregory of Nyssa or a Pseudo-Dionysius the Areopagite . . . does not see, in apophaticism, revelation but the receptacle of revelation: they arrive at the personal being of a hidden God. For them the negative way is not resolved in a void where subject and object will be reabsorbed: the human person is not dissolved but has access to a face to face encounter with God, a union without confusion according to grace." Vladimir Lossky, *Orthodox Theology: An Introduction* (Crestwood, N.Y.: St. Vladimir's Seminary Press, 1978) p. 32.

63. On these terms see G. Owens, "Monothelitism" in *New Catholic Encyclopedia,* 9, pp. 1067-1068.

64. Thunberg, *Microcosm and Mediator* (note 56 above), pp. 100-119.

65. *Fons Scientiae,* II *De haeresibus,* PG. 94, cols. 681-683 translated by F.H. Chase, Jr., in *Writings, The Fathers of the Church* ed. by R.J. Deferrari (New York: Fathers of the Church, Inc., 1958), p. 113. On this period I have consulted Martin Jugie, *Theologia Dogmatica* (note 3 above), 1, pp. 404-481, and Han-George Beck, *Kirche und Theologische Literatur im Byzantischen Reich* (Munich: C.H. Beck, 1959); J.M. Hussey, *Church and Learning in the Byzantine Empire, 868-1185* (New York: Russell and Russell, 1963); John Meyendorff, *Byzantine Theology* (note 3 above), and Klaus Oehler, *Antike Philosophie und Byzantisches Mittelalter* (Munich: C.H. Beck, 1969), especially "Aristoteles in Byzanz", pp. 272-286, with a good bibliography, and "Die Dialektik des Johannes Damaskenos", pp. 287-299.

142

66. See Leslie W. Barnard, *The Greco-Roman and Oriental Background of the Iconoclastic Controversy* (Leiden: Brill, 1974); Léonide Ouspensky, *The Meaning of Icons,* (Crestwood, N.Y.: St. Vladimir's Seminary Press, 1982).

67. John Meyendorff, *A Study of Gregory Palamas* (London: Faith Press, 1964); Vladimir Lossky, *The Mystical Theology of the Eastern Church* (London: J. Clarke, 1957). A much harsher view is the older Roman Catholic study of Jugie, *Theologia Dogmatica* (note 3 above) vol. 1, pp. 431-489, and vol. 2, pp. 47-183, especially conclusions on pp. 182 f.

68. *Adversus Praxean,* 7, 8; PL 2: 153-196, col. 162. Edited by A. Kroymann and E. Evans, *Corpus Christianorum, Series Latina* (Turnholt, Belgium, 1934), vol. 2, pp. 1159-1205, p. 1166 f. On Tertullian's thought see Timothy D. Barnes, *Tertullian* (Oxford: Clarendon Press, 1971) and H.B. Timothy, *Early Christian Apologists* (see note 25 above).

69. *De Anima,* 27 (also 9, 19, 20, 22, 36) PL 2: 642-746, col. 694 ff. Also in *Corpus Christianorum, Series Latina* edited by J. Waszink, vol. 2, pp. 781-869, pp. 822-824; translated by E.A. Quain, *Fathers of the Church* no. 10, 1950, *Apologetic Works,* pp. 179-312, cf. pp. 242-245. On Tertullian's traducianism and its influence see Quasten, (note 25 above) II, pp. 287-290 and A. Michel, "Traducianisme" in the *Dictionnaire de Théologie Catholique,* t. 15 (1946) cols. 1350-1366.

70. See Jean Daniélou, *History of Early Christian Doctrines.* (note 21 above), vol. 2. *The Origins of Latin Christianity,* 1977, pp. 214-231 on Tertullian's materialistic realism and Michel Spanneut, *Tertullian et les prémiers moralistes africain* (Gembloux: J. Duculot, 1969).

71. For Clement of Alexandria see *Stromata,* 6, c.9; PG. 9: 292, English translation by A. Roberts and J. Donaldson, *Ante-Nicene Fathers* (New York: Scribner's, 1900), vol. 2, pp. 346-567, pp. 496 ff.; for Hilary see *De Trinitate* 10, 18: PL 10:556 f., translated by Stephen McKenna, *Fathers of the Church,* vol. 25 (1954) p. 411. On Hilary's notion of the impassibility of the Incarnate Word see Paul Galtier, S.J., *Saint Hilaire de Poitiers,* (Paris: Beauchesne, 1960), pp. 131-141.

72. See especially Angelo Paredi, *Ambrose: His Life and Times,* (Notre Dame, Ind.: University of Notre Dame, 1964), especially pp. 264-275 on the literary character of the *In Hexameron;* J.C.M. Van Winden, "St. Ambrose's Interpretation of the Concept of Matter," *Vigiliae Christianae,* 16 (1962): 205-215 and "Some Additional Observations on St. A.'s Concept of Matter", *ibid.* 18 (1964): 144-5, Wolfgang Seibel, *Fleisch und Geist beim heiligen Ambrosius* (Munich: Karl Zink, 1958), especially the summary pp. 195-7 where he shows that for Ambrose the soul or rather the *mens* is the true man, the body its garment and instrument, while "flesh" is the total person of fallen man; and Goulven Madec, *Saint Ambroise et la Philosophie* (Paris: Etudes Augustiniennes, 1974), who shows that while Ambrose used philosophy, he saw it as a rival of Christian wisdom which is the intellectual and spiritual ideal.

73. *In Hexameron* and *De Paradiso,* ed. by C. Schenkl, PL 14, 131-289 and 289-331; *Corpus Ecclesiasticorum Latinorum,* Vienna, vol. 32, 1, pp. 261 and 265-336. Translated by John J. Savage in *Fathers of the Church,* vol. 42, pp. 3-355.

74. See Pierre Courcelle, *Les confessions de saint Augustin dans la tradition littéraire: Antécedénts et posteérite.* (Paris: Études Augustiniennes, 1963). Note especially pp. 27-88 on the Neo-Platonic sources of Augustine's thought and pp. 461-511 on Augustine's influence on the Romantics. See also Courcelle's *Connais-toi toi-même de Socrate à saint Bernard,* 2 vols. (Paris: Études Augustiniennes, 1974); Henri Marrou, *St. A. et la fin de la culture antique,* 4th Ed. (Paris: Boccard, 1938) and *Saint Augustin et l'Augustinisme* (Paris, Seuil, 1959).

75. S.E. Gersh in *Kinesis arkinetos: A study of spiritual motion in the philosophy of Proclus* (Leiden: Brill, 1973) shows how in Neo-Platonism all of reality is conceived as a process, although this process consists in eternal logical relations which, however, are essentially *dynamic* in the way that thought is dynamic.

76. The notion of will was, of course, not absent from Greek thought, but not as clearly defined as a faculty as was the intelligence. See Anthony Kenny, *Aristotle's Theory of Will* (London: Duckworth, 1979), and André Jean Voelbe, *L'idée de volonté dans le stoicisme,* (Paris: Presses Universitaires de France, 1973), pp. 162-190.

77. On the notion of *libertas* in Augustine see Charles Boyer, S.J., *Essais anciens et nouveaux sur la doctrine de saint A.* (Milan); Mary T. Clark, *Augustine: Philosopher of Freedom* (New York: Desclee, 1958) and E. Bailleux, "La liberté augustinene et la grace," *Mélanges de science religieuse* 19 (1962), pp. 30-48; M. Huftier, *Libre arbitre, liberté et péché chez S.A.* (Louvain: Nauwelaerts, 1968). I have found the following especially useful: N. Kaufman, "Les éléments aristotéliciennes dans la cosmologie et la psychologie de saint A.", *Revue Néoscolastique de Philosophie,* 2 (1904), 140-156; Paul Henry, "Augustine and Plotinus", *Journal of Theological Studies,* 38 (1937) 1-23; A. Pegis, "The Mind of St. A.", *Medieval Studies,* 6 (1944) 1-61; Vernon J. Bourke, *Augustine's Quest of Wisdom* (Milwaukee: Bruce, 1944); G. Verbeke, *L'évolution de la la doctrine du pneuma du stoicisme à S.A.* (Paris: Desclée du Brouwer, 1945) (it was the Christians not the Platonizing Stoics who transformed the materialistic *pneuma* into a spiritual principle); Rita Marie Bushman, "St. A.'s Metaphysics and Stoic Doctrine," *New Scholasticism* 26 (1950): 283-302. M. Anna Ida Gannon, "The Active Theory of Sensation in S.A.", *New Scholasticism* (1956): 154-180; Henri I, Marrou, *S. Augustin et la fin de la culture antique* (note 74 above) 1958 (A. little interested in cosmological but only in anthropical questions, unlike Greek Fathers); Eugéne Portalie, S.J., *A Guide to the Thought of Saint A.* (Chicago: Regnery, 1960); Étienne Gilson, *The Christian Philosophy of St. A.* (New York: Random House, 1960); James F. Anderson, *St. Augustine and Being* (The Hague: Martinus Nijhoff, 1965) (A. not an "essentialist", pp. 66 ff.); Robert J. O'Connell, *St. Augustine's Early Theory of Man,* 386-391 (1968); *St. Augustine's Confessions: The Odyssey of the Soul,* (1969) and *Art and the Christian Intelligence in St. A.* (1978), all three (Cambridge, Mass.: Harvard University Press); and Eugene Teselle, *Augustine the Theologian* (New York: Herder and Herder, 1970). Also the excellent collection of articles in *Augustinus Magister* (Paris: Études Augustiniennes, 1954); 3 vols, especially G. Verbeke, "Spiritualité et immortalité de l'âme chez S.A." vol. 1; 329-334, and the valuable series of St. Augustine Lectures held at Villanova University (Villanova, Penn.); Vernon J. Bourke, *St. Augustine's View of Reality,* 1963, pub. 1964; A.H. Armstrong, *St. Augustine and Christian Platonism,* 1966, pub. 1967 (for A., unlike Plotinus, soul is mutable and hence not divine); John A. Callahan, *Augustine and the Greek Philosophers,* 1964, pub. 1967; Mary T. Clark, *Augustine's Personalism* 1966, pub. 1970. John Edward Sullivan, O.P., *The Image of God: The Doctrine of St. A. and Its Influence* (Dubuque: Priory Press, 1963) gives one of the most complete treatments of Augustine's anthropology in relation to the "image and likeness" theme.

78. *De Trinitate,* PL 42, translated by Stephen McKenna, *Fathers of the Church,* vol. 45. See Eugéne Portalie, *Guide* (note 77 above), pp. xi, 129-135, and Michael Schmaus, *De Psychologische Trinitatslehre des heiligen Ausgustinus* (Münster im Westfallen, Aschendorff, 1966).

79. PL 32: 905-958; CSEL 63: 1-81 ed. by Pius Knöll, 1922, pp. 1-85; translated by John J. O'Meara, *Ancient Christian Writers,* no. 12, edited by J. Quasten and J.C. Plumbe (Westminster, Md.: Newman, 1950).

80. Actually Augustine's own expression is *Si non esses falli omnino non posses,* "You could not be mistaken at all if you did not exist," *De Libero Arbitrio* III, 7; PG 32: 1222-1310, p. 1243; CSEL 74: 1-81, II 20, line 10, p. 42, ed. William M. Green, 1956; Translated by R.R. Russell, *Fathers of the Church,* 1968, vol. 59, pp. 72-241, p. 114. See also G. Lewis, "Augustinisme et Cartésianisme" in *Augustinus Magister* (note 77 above) 2, 1087-1104. On Augustine's epistemology see Bourke, *Augustine's Quest* (note 77 above) pp. 112-117, 236 ff.; I. Quiles, ":Para una interpretacion integral de la illuminacion agostiniana," *Augustinus* 3 (1958), 255-68 who distinguishes four types of metaphysical experience: (1) cognitive experience of one's own soul; (2) cognition of being, truth, goodness; (3) cognitive experience of God; (4) immediate cognition of the *rationes aeternae;* C.E. Schetzinger, *The German Controversy on St. Augustine's Illumination Theory* (New York: Pageant Press, 1960) and Bruce Bubacz, *St. Augustine's Theory of Knowledge: A Contemporary Analysis* (Lewiston, N.Y.: Edwin Mellen Press); Étienne Gilson, *Études sur le rôle de la pensée médiévale dans la formation du systeme cartésien* (Paris: J. Vrin, 1967), "Le cogito et la tradition augustinienne", pp. 191-201.

81. See Robert J. O'Connell, *St. Augustine's Confessions: The Odyssey of a Soul* (Cambridge, Mass.: Harvard University Press, 1969), p. 108. "Plotinus, the philosopher of our times, seemed ashamed of being in the body. As a result of this state of mind he could never bear to talk about his race or parents or native country and he objected so strongly to sitting to a painter, or a sculptor that he said to Amelius, who was urging him to allow a portrait of himself to be made, 'Why really is it not enough to have to carry the image in which nature has encased us, without you requesting me to agree to leave behind me a longer-lasting image of the image, as if it was something genuinely worth looking at?' " *Porphyry's Life of Plotinus* in Plotinus' *Enneads,* edited and translated by A.H. Armstrong (Loeb Library, Cambridge, Mass.: Harvard University Press, 1966), vol. 1, c.1, p. 3.

82. See for example St. Gregory Nazianzen's funeral orations on his brother St. Caesarius, his Sister St. Gorgonia, and his friend St. Basil, PG 35: 755-815, 494-606, translated by Leo P. McCauley in *Fathers of the Church,* vol. 22, 1953.

83. See John Edward Sullivan, *Prophets of the West: An Introduction to the Philosophy of History* (New York: Holt, Rinehart and Winston, 1970), pp. 3-20, and John O'Meara, *Charter of Christendom: The Significance of the City of God,* St. Augustine Lecture, Villanova University, 1961 (New York: MacMillan, 1961) (Augustine was not hostile to the state as such but to the polytheism of the pagan state).

84. See Gerard Phillips, *La raison d'être du mal d'après saint Augustin* (Louvain: Ed. du Museum Lessianum, 1927) (originality of Augustine in teaching that God would not have permitted sin of the reprobate except for the good of the elect); J. Chené, *La théologie de saint A.: Grace et Prédestination* (Le Py-Lyon: Ed. Xavier Mappus, 1961), especially pp. 84 ff.; Gerald Bonner, *St. Augustine of Hippo: Life and Controversies* (Philadelphia: Wesminister Press, 1963); F. Floeri, "Remarques sur la doctrine Augustinienne du péché original," *Studia Patristica* (Berlin: Akademie Verlag, 1966), vol. 9, Part III, pp. 416-421; A. Sage, "Péché original: Naissance d'un dogme", *Revue des études augustiniennes* 13 (1967): 211-248; Robert F. Evans, *Pelagius: Inquiries and Reappraisals* (New York: Seabury Press, 1968), Gerald Bonner, *Augustine and Pelagianism in the Light of Modern Research,* St. Augustine Lecture, 1970 (Villanova, Penn.: Villanova University Press, 1972).

85. Antoine Slomkowski *L'état primitif de l'homme dans la tradition de l'église avant saint Augustin* (Paris: Gabalda, 1928).

86. *Summa Theologiae* I-II, q. 81, a.1. and 82, a.1. The Council of Trent, Session V, canon 5 declared that in the *baptized* the remaining concupiscence is not sin in the proper sense but is "an effect of sin and an inclination to sin," it did not define that the unbaptized infant is in a proper sense "a sinner" but only that it has suffered the deprivation of grace or "death of the soul" through the sin of its forebearers. See Henri Rondet, *Original Sin: the patristic and theological background* (Staten Island, N.Y.: Alba House, 1972), pp. 160-167; 259-277.

87. For an excellent discussion see Peter Brown, *Augustine and Sexuality,* Center for Hermeneutical Studies in Hellenistic and Modern Culture, 46th. colloquy, 22 May, 1983 (Berkley: Graduate Theological Union and University of California, 1983). Brown shows that Augustine's development was always in the direction of a more positive view of sexuality, and in his last works he grants that in Eden there would have been a legitimate enjoyment of sexual pleasure free of sin. A.H. Armstrong, *St. Augustine and Platonism* (note 77 above), pp. 11 f. compares Augustine's sexual teaching to that of the earlier Fathers and writes: "Augustine . . . is often more balanced and positive — and not, as sometimes seems to be assumed, more unbalanced and negative — in his attitude to the body, sex and marriage than most of his Christian contemporaries. He made two advances of special importance towards a more positive and constructive way of thinking about these matters. By his clear-cut insistence that the cause of sin lies in the will, not in the body (e.g. *City of God,* 14, 3), he did a great deal to banish from Western Christian thinking the shadow of the Pythagorean-Platonic belief in the dark, recalcitrant element which is a necessary constituent of the material world and the source of evil to the soul which comes into contact with it; a belief which persists in Plotinus, though the later pagan Neoplatonism abandoned it. And by his rejection of the doctrine of that other great Christian Platonist, St. Gregory of Nyssa, which persisted in later Greek Christian thought, that the division of the human

race into sexes was made *ratione peccati,* with a view to procreation only after the Fall, and was no part of the original creation in the image of God, and his insistence that there would have been begetting and birth of children in Paradise *(City of God,* 14, 22 ff), Augustine took at least the first step toward a positive, Christian valuation of sexuality." See also Michael Müller, *Die Lehre des heilegen Augustinus von der Paradiesehe und ihre Auswirkung in der Sexualethic des 12 und 13 Jahrhunderts bis Thomas von Aquin,* (Regensberg: F. Pustet, 1954) and *Grundlagen der katholischen Sexualethik,* same place and publisher, 1968. Also John J. Hugo, *St. Augustine on Nature, Sex and Marriage* (Chicago: Sceptor Press, 1968), and William A. Alexander, "Sex and Philosophy in St. Augustine," *Augustinian Studies,* 5 (1974): 197-208.

88. *De Bono Conjugali,* PL 40, cols. 373-394; CSEL. 41, 188-231, ed. J. Zycha; translated by Charles T. Wilcox, *Fathers of the Church,* 27, 9-54.

89. Plato, *Symposium,* 210-212, the conclusion of the speech of Diotima; see also John J. O'Meara, "Virgil and Saint Augustine. The Roman Background to Christian Sexuality", *Augustinus* 13 (1968): 307-326.

90. According to E. Dekker, "Traductions grecques des écrits patristiques latins", *Sacris Erudiri* 55 (1953): 192-233, from the third to the fifth century some Latin works of Tertullian, Cyprian, Novatian, Ambrose, Rufinus, Jerome and Cassian were translated into Greek. In the following centuries some juridical and hagiographical works were also translated, mainly for use by Greeks in southern Italy and Rome. In the thirteenth and fourteenth centuries some patristic opuscula, mainly of Augustine and Boethius were known to erudite Byzantines. The translations were often by public notaries rather than by scholars. Pierre Courcelle, *Les confessions de saint Augustin dans la tradition littéraire: Antécedents et postérité,* (Paris: Études Augustiniennes, 1963), pp. 202 f. referring to this article of Dekkers and one by B. Altaner, "Augustinus in der griechen Kirche bis auf Photius", *Historisches Jahrbuch* t. 71 (1952), pp. 52 ff., says that little of the works of Augustine were known to the Greek Church until about the thirteenth century, and there is no proof his *Confessions* were read there until modern times.

91. Claudianus, PL 53: 697-778; CSEL 11, edited by A. Engelbrecht, 1885; see P. Courcelle, *Late Latin Writers and their Greek Sources,* (Cambridge, Mass.: Harvard University Press, 1969), pp. 238-251. Cassiodorus: PL 70: 1279-1308; see Courcelle, *Les confessions de Saint Augustin* (note 90 above) pp. 334-409. Cassiodorus was answering Faustus of Rietz whose motto was *omne corpus fugiendum,* cf. p. 250 f.

92. F.H. Dudden, *Gregory the Great* (London: Russell and Russell, 1967), vol. 1, p. 153. The *Moralia in Job,* PL 75: 540-1162 and 76: 1-782. Critical edition by Robert Gillet and André de Gaudemaris, *Sources Chrétienne,* 1952. The only English translation (in 5 vols.) is by John Henry Parker, Oxford, 1844. The *Dialogues* have been edited with an important introduction by Adalbert Vogüe, *Sources Chrétiennes,* 1978.

93. See the Introduction by Adalbert de Vogüe to his edition of the *Dialogues, Source Chrétiennes,* vol. 1, (Paris: Ed. du Cerf, 1978), and the article by Robert Gillet "Gregoire le Grand", *Dictionnaire de Spiritualité* (Paris: Beauchesne, 1964), t. 6, cols. 872-910, especially on the *Dialogues,* cols. 878 f.

94. See Stephen Gersh, *From Iamblichus to Eriugena: An Investigation of the Prehistory and Evolution of the Pseudo-Dionysian Tradition* (Leiden: Brill, 1978), and John J. O'Meara, "The Present State of Eriugenian Studies" in *Studies in Medieval Culture,* ed. by J.R. Sommerfeld and E.R. Elder (Kalamazoo, Mich.: The Medieval Institute, Western Michigan University, 1973-1974), pp. 8-9, 15-18; and John J. O'Meara and Ludwig Bieler, eds. *The Mind of Eriugena,* Dublin Colloqium July, 1970 (Dublin: Irish University Press, 1970), especially I.P. Sheldon-Williams, "Eriugena's Greek Sources", pp. 1-15 and Werner Beierwaltes, "The Revaluation of John Scotus Eriugena in German Idealism, pp. 190-199.

95. M.D. Chenu, "The Platonisms of the Twelve Century" in his *Nature Man, and Society in the Twelfth Century,* (Chicago: University of Chicago Press, 1968), pp. 49-98. See also Klibansky, *Continuity of Platonic Tradition* (note 41 above). Tullio Gregory, *Platonisme medievale: Studi e richerche* (Rome: Instituto Storico Italiano per il medio evo, 1958), and Werner Beierwaltes, ed., *Platonismus in der Philosophie des Mittelalters* (Darmstadt: Wissenschaftliche Buchgesellschaft, 1969), especially Josef Koch, "Augustinischer und Dionysicher Neoplatonismus und das Mittelalters" (1957), pp. 317-342.

96. Some of the members of the School were Thierry (d.c. 1155), his brother Bernard (d.c. 1130), Gilbert de la Porée (d. 1154), Bernard Silvestris (d. 1153), William of Conches (d.c. 1154) and Clarebaud of Aras (d.c. 1170).

97. *Timaeus a Calcidio translatus commentarioque instructus,* ed. J.H. Waszink in Plato *Latinus,* ed. R. Klibansky (Leiden: Brill, 1962); see also J. Parent, *La doctrine de la création dans l'école de Chartres: études et textes,* (Paris: J. Vrin, 1938).

98. Others of the Victorines were Godfrey (d. 1194), Walter (d.c. 1180) and Alan of Lille (d. 1202).

99. See Bernard McGinn, ed., *Three Treatises on Man: A Cistercian Anthropology,* with introduction by the editor (Kalamazoo, Mich.: Cistercian Publications, 1977) which includes William of St. Thierry, *The Nature of the Body and Soul,* pp. 101-152 (text in PL 180: 695-726); Isaac of Stella, *Letter on the Soul,* pp. 153-178; (PL 194: 1875-1890); and Alcher of Clairvaux (?), *Treatise on the Spirit and the Soul,* pp. 179-288 (PL 40: 779-832). Translations are by B. Clark, B. McGinn, and E. Leiva and Benedicta Ward respectively. See also McGinn, *The Golden Chain: A Study in The Theological Anthropology of Isaac of Stella* (Washington, D.C.: Cistercian Publications; Consortium Press, 1972).

Chapter 5

The Aristotelian Christian Theology

I. The Aristotelian Alternative

1. Aristotle in the Eastern Church

An alternative to the Platonic dualism was available for the service of Christian theology from the beginning but its value was long unrecognized. Both in the Greek Church where the works of Aristotle were generally available and in the Latin Church where they were only partially known, the Stagirite was thought of as a pupil of Plato whose chief importance was as a logician, although his scientific and ethical works were recognized as filling out details in the cosmology which Plato had only sketched and which had not much interested the Neo-Platonists.

This Aristotelian logic used as an instrument of Christian theology proved a two-edged sword. On the one hand it provided the exegetes of the Antiochene School, Diodorus of Tarsus (d.c. 394) and his followers St. John Chrysostom, Theodore of Mopsuestia, Theodoret of Cyrus and the heresiarch Nestorius, with a sober method of reading the Bible with due care for its literal meaning as contrasted to the sometimes fantastic allegorical reading employed by the Platonic Alexandrian School of Origen and his followers.[1] On the other it spawned the Nestorian heresy and the subsequent doctrinal controversies, one of the causes of which was the Aristotelian effort to reduce analogical to univocal language, a reductionism inevitably disastrous for a theology of transcendent mysteries. Another example of this reductionism is to be found in the work of John Philoponus (fl.c.500-530), perhaps the most important of early Christian commentators on Aristotle, who was led by it into Monophysitism, the opposite error to Nestorianism.[2]

The danger of Aristotle's dialectical method in theology manifested in such episodes (later repeated in the Western Church) made his philosophy suspect until Orthodox theologians began to see how this method, when employed with moderation, might be very helpful for clarifying the confusions that produced heresies. This became evident in the work of so thorough a Platonist as St. Gregory of Nyssa, of Leontius of Byzantium (d.543), a converted Nestorian and vigorous opponent of Monophysitism, and especially in the thought of St. Maximus the Confessor whom Dolger[3] does not hesitate to call "a penetrating thinker of the Aristotelian school", although, as was shown in the last chapter, his theology is rooted in the Neo-Platonism of the Pseudo-Dionysius. The finest product of this methodology is the *Fount of Knowledge* of St. John of Damascus (d.c. 749) which is a veritable scholastic *summa* of theology, the first part of which is devoted to a treatment of Aristotelian dialectics which is then used to systematize the second, doctrinal part, thus constituting the first real manual of Orthodoxy.[4]

Nevertheless, in subsequent Eastern theology the typical doctrines of Aristotelianism, including its conception of the relation of soul to body, never played a dominant role.[5] When the University of Constantinople was re-established in the ninth century by Caesar Bardas, the great scholar Photius (d.897) gave to its studies not so much a philosophical as a humanist and antiquarian emphasis; and when this university experienced a renewal in the eleventh century its leading philosopher Michael Psellus (d. after 1078) regarded Aristotle's works only as a subject of study preliminary to that of Plato, Plotinus and Proclus who were the authorities in metaphysics leading up to theological studies. Psellus' pupil John Italus, who from his Italian birth-place was acquainted with western scholasticism, was condemned as a heretic for his iconoclasm and his belief in reincarnation, a condemnation which permanently discouraged philosophical studies in Byzantium. In the fourteenth century there were debates over whether Plato or Aristotle was the superior philosopher, as witnessed by the dialogue *Florentius* of Nicephoras Gregoras (d.1359) who took the side of Plato even to the point of defending the notion of a World-Soul. At the very end of Byzantian history Gemistus Plethon (d.1464), who is often called the "last Byzantine philosopher" and who attended the Council of Florence in 1438, was so ardent a Platonist that he persuaded Cosimo de Medici to form a Platonist Academy and is reported to have recommended, in view of the divisions within the Christian Church, a return to pagan Greek religion![6]

2. Aristotle in the Latin Universities

The fortune of Aristotelianism in the Western Church, however was very different from its relative neglect in the East, although in the West at first it was viewed with great suspicion. In the Carolingian revival of the ninth century, along with the Neoplatonism of Erigena there had also been something of a revival of logical studies based on the versions of some of Aristotle's logical works made by Boethius (d.526), with the outcome for theology of the rise of disturbing controversies over predestination and the Eucharist. Again in the twelfth century certain dialecticians such as Abelard (d.1142) aroused the ire of St. Peter Damian and St. Bernard of Clairvaux by their theological rationalism. The result might well have been, as it was in Byzantium, that emphasis on the philosophical aspect of theology would have remained under ecclesiastical suspicion, with the further result that the distinction between theology based on the interior illumination of faith from above and philosophy based on human empirical experience from below might never have become seriously operative.

The essential difference of Platonism and Aristotelianism does not consist simply in a methodological difference, but in epistemology. For Plato true knowledge (*episteme*) can never come from the senses which attain only contingent, transitory reality, but only from an innate insight into eternal truth (the Ideas). For Aristotle, on the contrary, true knowledge can only arise out of sense experience insofar as that experience is fully actualized by the human intelligence. An Aristotelian Christian theology, therefore, while admitting that the gift of faith gives to humans revealed insights which surpass anything that their intelligence can discover in the data of the senses (as well as anything like the innate insights of Platonism), must grant to purely human knowledge a genuine autonomy. Moreover, it must show how the realms of faith and reason can be harmonized in the unity of the human person as knower and believer.

The gradual entry of Aristotle's complete works into the curriculum of the medieval universities, the tardy acknowledgment of the radical epistemological differences between his thought and that of Plato, the conservative resistance of theologians to these innovations, and finally the slow awakening of the theologians and the bishops to the possible orthodoxy and value of an Aristotelian type of theology is a story on which much research has been done and it need not be repeated in detail here.[7] The battle-line of the conflict in the schools was between the theology and the arts (i.e. the liberal arts, sciences, and philosophy) faculties. As a result of the Aristotelian epistemology the liberal arts for the first time in the history of Christian culture found an adequate justification for their genuine autonomy as secular disciplines. It was now realized that "philosophy" (which included all that today we would call the natural, life, and behavioral sciences) had a method and validity of its own, not merely as an instrument of theology, but as clearly differentiated and independent fields of knowledge. The arts faculties wished to maintain this autonomy even at the risk of accusations of heresy. Some theologians also began to see that these disciplines could not really be of service to theology unless they developed according to their own proper principles.

Canon Van Steenberghen[8] has given us what is probably the most balanced analysis of this medieval development. He points out that the earliest scholastics such as William of Auxerre (d.1231), William of Auvergne (d.1249), Phillip the Chancellor (d.1236) at the University of Paris, and the great Robert Grosseteste (d.1253) at Oxford were all faithful followers of the Platonic Christian Theology of St. Augustine, but already were making use of philosophical views which can best be described as an eclectic, Neo-Platonizing Aristotelianism. This new interest in Aristotle was the result of these scholars experiencing that the works of St. Augustine did not supply them with the developed, systematic philosophy for which they felt a need in university teaching to students who had already completed their studies of the liberal arts, and which the works of Aristotle amply supplied. These Aristotelian works, however, were often interpreted with the help of Arab Neo-Platonizing commentators who read them in a manner which for some time concealed the difficulties inherent in this eclecticism.

We can get a good notion of the approach to anthropology of these eclectic theologians from two examples. William of Auvergne in his *De Anima* and *De Immortalitate Animae* (parts of his great *Magisterium Divinale*) argued that the human person is not the soul, but the composite of body and soul, yet he dealt principally with arguments to show the superiority of the soul to the body and the possibility of its action independent of the body, essentially treating the body as an instrument to be used or discarded.[9] Robert Grosseteste, on the other hand, showed an intense interest in scientific investigations of the material world and established the scientific tradition of medieval Oxford which led quite directly to the rise of modern science,[10] but his approach to science remained primarily Pythagorean and Platonic in its stress on the abstract mathematical order of the cosmos, rather than (as for Aristotle) on its physical, material, dynamic character. Grosseteste made special use of the so-called "light metaphysics" already present in the School of

Chartres and among the Victorines in which the substance of the physical world was reduced to light (energy). In such a cosmology the human body is thought of as a lamp which both radiates and obscures the flame of the soul.

The full development of scholastic anthropology came only after the entry of the Dominican and Franciscan Friars into the universities in a partnership that soon developed into a rivalry. This divergence can be shown by comparing four generations. In the first were the Dominican Albert the Great (c. 1280) and the Franciscan Roger Bacon (c. 1220 to after 1292).[11] Albert the Great for the first time fully expounded the whole range of Aristotle's works, stressing two themes: (1) these works are not a complete encyclopedia of human knowledge, but require further research for their completion; (2) the method of these sciences must rest on sense experience (*experimentum*). On these points Roger Bacon fully agreed with Albert but added his own stress on the importance of mathematics and languages as necessary instruments of the advancement of knowledge. Bacon's epistemology, however, in spite of this special appreciation of sense experience, remained essentially Platonic, because for him the first principles of natural science did not arise from sense knowledge but from an interior illumination. Albert was more Aristotelian in his epistemology, but also hesitated wholly to abandon Platonism. Indeed by his commentaries on the works of Pseudo-Dionysius, Albert gave impetus to the Neo-Platonic mysticism which was to flourish in the next century.[12]

Thomas Aquinas (1225-1274), a pupil of Albert the Great, must be credited with the first thorough going use of Aristotelian philosophy in theology, and if we can speak of an Aristotelian Christian Theology it is to be found in its most unqualified form in his work. Cornelio Fabro[13] has forcefully argued that while Aquinas was certainly an Aristotelian, it must also be recognized that he accomplished in a unique way that synthesis or reconciliation of Aristotle with Plato which had been the great ambition of the Neo-Platonists.

Thoroughly Aristotelian in epistemology, Aquinas also accepted an Aristotelian cosmology based on the four "causes" (matter, form, agency, finality) and a metaphysics in which the existence of spiritual as well as material substances was defended and analyzed in terms of act and potency. He saw clearly, however, the source of those features of Aristotle's worldview which had made this view unacceptable to previous Christian thinkers, namely, Aristotle's failure to show how all substances, both material and spiritual, depend *totally* on the One God not only for their actualization (essence) as this or that kind of being but even for their very existence. If we adhere simply to the letter of Aristotle's text we are left with the impression that matter and the many spiritual intelligences not only coexist eternally with the Prime Mover but that their existence is as necessary as His.[14] Such a metaphysics, while not incompatible with some of the more primitive views of God found in the Bible, is radically incompatible with the full biblical conception of a God who creates *ex nihilo*.

Aquinas saw how this fundamental difficulty could be overcome by borrowing from Plato the notion of "participation", a notion which Aristotle knew but did not much use because of its association with the Platonic "separated forms" (Ideas) which were incompatible with Aristotle's epistemology, involving as they did a confusion between the real and the ideal orders. Aquinas, however, following Augustine, placed the Ideas in God (exemplarism) and thus saw all *essences* as participating in the One Divine Essence. So far he was in line with Platonic Theology; but his unique contribution was to distinguish sharply between the line of *essence* and that of *existence* which in creatures are really distinct but in God identical.[15]

For Aquinas creatures are completely contingent, yet they truly participate in God's existence in the sense that God freely causes them to be with a real autonomy not identical with but analogous to His own autonomous existence. Moreover, Aquinas integrated this doctrine with the Aristotelian doctrine of act and potency by showing that in creatures their existence is related to their essence as act to potency. God is Pure Act and the cause of actuality in creatures by which they each in various analogical ways participate in his Actuality.

Arthur Little and others[16] go further than Fabro and argue that Aquinas is *more* a Platonist than an Aristotelian, since his whole thought is unified by this Platonic participation. It is clear, however, that Aquinas accepts Aristotle's fundamental project, which was not to overthrow the Platonic world-view, but to revise it critically by eliminating the "separated ideas" known through an innate intuition, and to replace these by an intellectual abstraction of the essences of things known in sense experience. In pursuing this goal Aristotle, (at least in the works that have come down to us,[17]) stopped short at certain crucial points and it is precisely these hesitations which the Neo-Platonists attempted to press further, and which persuaded Christian theologians that Neo-Platonism was a more useful instrument for their purposes. What Aquinas accomplished was to tackle these same difficulties in a consistently Aristotelian manner and solve them more profoundly than had the Neo-Platonists. In doing so he was able to assimilate to Aristotelianism the fundamental Platonic insight of "participation", which Aristotle had not rejected but to which he had been unable to do full justice.

153

For a theology of the body this revised Aristotelianism presented by Aquinas was a major advance, because it provided a way to break once and for all with the Platonic dualism in anthropology. Aquinas fully accepted the Aristotelian theory of the human soul as the substantial form of the human body and excluded all the various medieval systems of a "plurality of forms" in the human person. At the same time Aquinas not only rejected the obviously more Neo-Platonic interpretations of Aristotle, but also the rigid Aristotelianism of the Arabian Averroes (Ibn Rochd, d.1198) which had become the standard interpretation of these texts in the arts faculties.[18]

Aristotle insisted against Plato that rational thought depends for its content on our bodily sense organs, yet conceded to Plato that reason, since it attains the universal and essential in our particular, concrete experience, cannot itself be the act of a bodily organ. Aristotle, however, never made clear what this spiritual "agent intellect" might be.

The Averroists interpreted Aristotle as teaching that there is only a single cosmic Intelligence (the Third God or World Soul of Neo-Platonism) which acts in each human body. For Augustinian theologians this notion was interpreted to mean that God is the Agent Intellect illuminating every individual mind. Aquinas' interpretation, certainly more consistent with Aristotle's basic principles, was that each human being has its own intellect as the highest power of its spiritual soul which is the form or actualization of its own body. Thus Aquinas was able to defend the proper autonomy of the individual human person, yet also to maintain the spirituality and immortality of the human soul, which, however, requires union with its own body at the resurrection to become again fully a human person.[19] At the same time he maintained the total dependence of the human creature on God even in the creature's most intimate and free acts, because the agent intellect of each human is kept in act only by the constant concurrence of God, the Pure Act who energizes and elevates the whole of each human person to a participation in the divine knowledge and freedom both at the level of nature and at the level of grace.[20]

Aquinas was also deeply concerned to understand the concrete relation of the mind to the body through the internal and external senses and the bodily appetites. These faculties were understood by him as the dynamic forms of definite physical organs, so that the whole body was viewed as a system of inter-related instruments of the human life bestowed on the body by the soul. Aquinas rejected the Platonic doctrine that sensation is a spiritual activity which merely uses the sense organs as channels, since he regarded sensation as the reception of the forms of external reality through a physical change in the sense organs, a change, however, which is not mechanical but vital. He understood the constitution of the sense organs and of the whole body as the result of physiological ("vegetative") functions in an organic system of nutrition and growth. Finally, these physiological functions were analyzed as chemical changes of compound substances, ultimately resolvable into the fundamental elements (earth, water, air, and fire) characterized by resistance and inherent temperature.

These elements into which the human body and all earthly realities are resolved (Aquinas hypothesized with Aristotle that the heavens are formed of a fifth element, which accounts for their permanence) are not permanent atoms, but are constantly subject to transmutation into each other, so that the substrate of all change, which never exists without at least elemental form, is prime matter or pure potentiality, i.e. the capacity for unlimited change. Because all earthly things are in process, they are each divisible into parts and thus form bodies with quantitative dimensions. Aquinas accepted Aristotle's refutation of the Platonic "empty space", and held that space is nothing but contiguous bodies of various densities which form a finite universe. The processes of change are the result of the primary forces in the world and attain a kind of coordinated unity through the action of the heavenly bodies and ultimately through the outer sphere of the universe which is kept perpetually in motion by God (the Prime Mover).[21] Such a coordinated universe, however, is not perfectly unified by some World Soul, as the Platonists had thought, but is substantially pluralistic. The human person, by reason of its spiritual faculties of intelligence and will is free and transcends the cosmic physical order, since the heavenly bodies cannot directly act on the intelligence and will, but only God can do so.

The theological applications of these philosophical theses are very extensive. In later chapters I will indicate some of them. Here it suffices to point out one such application. For Aquinas (*Summa Theologiae* III, q. 46) the Incarnate Word effected human salvation by every one of his earthly acts but supremely by his passion. Since in this life the human person cannot perform any spiritual act that is not also a bodily act, it follows that Jesus saved us through his body as the conjoined instrument of his soul. In the sacrament of the Eucharist Jesus is truly present bodily (although in a miraculous mode). The other sacraments are, as it were, extensions of this bodily eucharistic presence, and they convey grace to us by a physical action (e.g. baptism by washing the body). They are instruments of Christ's own body which is the instrument of his soul. As a painter paints with his hand (a conjoint instrument), his hand paints with a brush (a separated instrument); so Christ saves us through his body, which saves us through the sacraments. This thoroughly "physical" explanation of sacramental causality differs radically both from a Platonic explanation according to which a sacrament is simply a symbol manifesting a hidden reality, and from the later Nominalist explanation according to which God acts directly so that a sacrament is merely the occasion on which he has committed himself to act.

This radical Aristotelianism of Aquinas was not very well understood and certainly not generally accepted. The Archbishop of Paris in 1277 (after Aquinas' death) condemned Averroism and other errors, among which were included some of the key points of Aquinas' novel interpretation of Aristotle. Although Aquinas' orthodoxy was later guaranteed by his canonization in 1323, both at Paris and Oxford the traditional "Augustinianism" remained dominant, and was given a more developed form. Thomism itself sunk into comparative obscurity until its revival by John Capreolus (d.1444).[22]

Averroism itself, although it lost all influence in theology, survived in the faculty of arts in Paris and especially in the universities of Northern Italy and made important contributions to the development of the natural sciences, although the Averroists became ever more intransigent in their defense of the Aristotelian texts to become finally the chief opponents of the innovations of Galileo.[23]

Nevertheless, the scientific interests stirred up by Aristotelianism began to make some headway outside Thomist and Averroist circles. The Franciscan Francis de Marcia, then John Buridan, Albert of Saxony, Nicholas Oresme and Marsilius of Inghen recast the Aristotelian theory of motion by the hypothesis of "impetus" which was eventually to lead to that of "force".[24] Thomas Bradwardine (d.1349) and his followers at Oxford introduced into this study of motion the use of mathematical theories of the measurement of velocity and acceleration. "Bradwardine's goal, namely to embrace all motions, terrestrial and celestial, and all variations of velocity, in a single mathematical formula, was not even reached by Galileo; it had to wait for Newton to find temporary realization."[25] Bradwardine was philosophically a Platonist in the line of his great predecessor at Oxford, Robert Grosseteste, and it was Platonic Pythagoreanism, i.e. the mathematical understanding of nature, which was the positive link of the Middle Ages with the science of the Galilean epoch, not Nominalism, for which this credit has often been claimed.[26]

This same mathematical tendency was also evident in the more Aristotelian work of the Dominican Theodoric of Freiberg (d.1310), a disciple of St. Albert the Great, and the first to propose an approximately correct mathematical theory of the rainbow.[27] Thus the fourteenth century was slowly preparing the ground for modern science, and the type of scientific thinking which it fostered was more and more *objectivizing*, i.e. a study of natural objects in their external relations as measurable, rather than precisely as natures, i.e. as having internal principles of action.

156

3. The Survival of Platonic Dualism

In the theological faculties of the universities, however, after the condemnation of Averroistic Aristotelianism in 1277 in which the thought of Aquinas was thought by many to be also implicated, the Platonic Augustinian tradition continued to dominate. It had previously found its purest expression in the Franciscan contemporary of Aquinas St. Bonaventure (d.1274).[28] His thought has a fully Augustinian flavor, but is above all a *symbolic* theology in which the Book of Nature is shown to reflect the Trinity just as do the Book of the Bible and the Book of Life (Jesus Christ known contemplatively). Bonaventure compares Aristotle to Plato unfavorably by saying that Aristotle tended to follow the way of science which turns downward to creatures and remains at their level, while Plato followed (though less perfectly than Augustine) that way of wisdom which never remains content with knowing the forms of creatures, but always moves upward toward the exemplar ideas in God from which those forms are derived. Thus Bonaventure seems unsympathetic to the Thomistic extroversion to the visible world as a way to discover ourselves and our God and favors an introversion to the image of God in our souls, knowable only through an inner illumination radiating from God. It should not be forgotten, however, that St. Francis whose biographer Bonaventure was, brought into Christian culture a fresh new vision of the beauty and poetry of the visible world as the mirror of God, a vision which inspired Dante and Giotto and the Renaissance of literature and art in fifteenth century Italy. St. Bonaventure is the supreme theologian of this Franciscan praise of God in nature.

It was another Franciscan Duns Scotus (d.1308) who attempted to construct a theology which would incorporate Aristotelian learning, but which would avoid the errors of Averroes and Aquinas.[29] In deciding whether Scotus' theology should be considered fundamentally Platonizing or Aristotelian in the sense that I have been using these terms the crucial question is epistemological. At first sight it appears that on this criterion Scotus, for all his "realism" as regards the famous problem of universals, is an Aristotelian, since he resolutely abandons the Augustinian illumination theory supported by St. Bonaventure and most of the Franciscan masters and adopts Aristotle's abstraction theory.[30] Nevertheless, Scotus modifed this abstraction theory in a way that permitted him, in the last analysis, to remain essentially Platonic, although in a very original manner.

For Aristotle and Aquinas the unity of soul and body in the human person is required by the fact that, although the human intellect is spiritual (since only a spiritual power can penetrate to the essences of existing realities), yet it is dependent on the bodily senses not only for the data from which this essential knowledge is abstracted but also for determining with certitude the existence and uniqueness of singulars, which singulars are the only extra-mental realities. Scotus, however, avoided having to deny to the human intellect the capacity to know singular existents as such, by claiming that although this intellect requires sense data for abstractive knowledge, it also has a power to know singular existents as such by an immediate intellectual *intuition* for which sense knowledge is a mere condition, not a principal cause.[31] This ingenious and novel doctrine of the intellectual intuition of singular existents became accepted as a fundamental thesis of the Franciscan School of theology and was to play that role in Ockham's own anti-Scotistic Nominalism.

Although Scotus accepted the hylomorphic composition of all material things from a potential matter and actualizing form, he radically reinterpreted this hylomorphism in line with his new epistemology by insisting that it is absurd to conceive matter as pure potentiality (as Aristotle had done) because this would seem to deprive it of all intelligibility, since only that which is in act is intelligible. Consequently, for Scotus, matter has a minimal but positive actuality so that, at least by divine power, it can exist without form.[32] Moreover, the individuation of the members of any species of material things is not to be attributed to their matter, but to a "thisness" (*haecceitas*) which is neither matter nor form, yet pertains to the unique actuality of the thing, and which can thus be known by an intellectual intuition. As for the forms of things, which are known by abstraction, Scotus conceived them as an ordered set of really distinct but inseparable formalities, e.g. the vegetative, animal, and rational functions in the human person. Thus Scotus rejected Aquinas' doctrine of the unicity of the human soul actualizing the matter of the body and the actualization of this composite essence by a unique act of existence distinct from the essence, in favor of a Platonic realism in which the human person is composed of a set of distinct forms, along with a minimally actual matter and an actual *haecceitas* which he does not call form, but which cannot be pure potentiality either.

Thus for Scotus the human soul, while it is the form of the body with which it constitutes an *unum per se*, an existent singular, nevertheless, does not satisfy the body's need to have a form of its own which the body retains even after the departure of the soul at death, so that what the soul gives to the body is not its existence as such, but only its *esse vivum*, its vitality. Rather paradoxically, in spite of this doctrine which favors Platonic dualism, Scotus was led by his insistence on Aristotelian logical rigor, to deny the possiblity of a certain demonstration of the immortality of the human soul (which Aquinas by a benign interpretation of Aristotle had retained[33]). Scotus was content with merely probable philosophical arguments for the soul's immortality, although faith assured him of its certainty. He was able, moreover, to agree with Aquinas that the separated soul is not the human person, yet (and this is the crucial point) he did not admit that the soul is united to the body for the sake of the soul, but only for the sake of the composite nature willed by God. Scotus' reason for this position was that for him the proper object of the human intelligence was not, as it was for Aristotle and Aquinas, sensible material being, but Being as such, comprehending not only material but even immaterial being (and the infinite God Himself) under a single *univocal* concept.[34] If this were not the case, Scotus thought, metaphysics and a philosophical knowledge of God in this life and the beatific vision in the next would be impossible.

At the same time, Scotus was able to admit that *de facto* our knowledge in this life is restricted to what can be known through sense experience either by intuition or abstraction. He did not attribute this restriction to the nature of the human intelligence, however, but to original sin or some other unknown reason caused or permitted by God.[35] Thus for Scotus, as for Plato and Augustine, the spiritual soul in its own nature is independent of the body. While going very far in adopting much of Aristotle's system, Scotus did so with that "subtlety" for which he was famous and by means of which he was able to retain the most characteristic feature of the Platonic Christian Theology, i.e. the intrinsic independence of the spiritual element in man from any necessary relation to the material body.

Scotus, by raising serious doubts about whether many of the metaphysical demonstrations of the scholastics could really meet the rigorous requirements of Aristotelian logic, loosened the relation between theology and philosophy and consigned to theology and to faith many of the fundamental questions of anthropology. In theology Scotus' great concern seems to have been to defend the absolute freedom of God against the deterministic tendencies of Averroistic Aristotelianism. While Scotus was not an extreme voluntarist[36], he constantly maintained that the created order is unconditionally dependent on the will of an infinite God whose purposes are largely hidden from us. The human person is especially the image of this hidden God in our participation in His freedom by which we are open to loving Him who is Love itself. Thus Scotus by his moderate voluntarism provided a justification for the Franciscan tradition (of which St. Bonaventure remains the supreme master) that stresses love as the ultimate human act, rather than, as Aquinas taught, knowledge.[37] Hence, at the end of the great age of scholastic theology an essentially Platonic Christian Theology for which the human body was conceived primarily as a *limitation* on the human spirit was still dominant in the Western as in the Eastern Church.

In the fourteenth century other theologians attempted a return to an even more radical Augustinian Platonism; for example that forerunner of the Reformation John Wycliff (d.1384) at Oxford. Others following the Franciscan emphasis on affectivity rather than cognition developed an anti-scholastic, anti-intellectual and mystical piety which reached its culmination in the famous *devotio moderna* associated with the Brothers of the Common Life (Gerard Groote and Thomas à Kempis). As Étienne Gilson has shown,[38] the *Consolation of Theology* of John Gerson, Chancellor of the University of Paris, in 1418 called for a simplified theology, free of metaphysics, scripturally based and directed toward pastoral care.

Most radically Platonic of all was the Dominican Meister Eckhart (d.1327)[39] who followed the lead not of Aquinas but Albert the Great in Albert's commentaries on the Pseudo-Dionysius and produced a speculative mysticism according to which union with God is to be achieved by a total "letting go" (*gelassenheit*) of all images and concepts derived from the visible creation, in order that in the soul thus made completely void the Word of God might be born in silence. Without questioning Eckhart's personal orthodoxy, a papal commission condemned many of his paradoxical formulas, but his Dominican disciples John Tauler (d.1361) and Henry Suso (d.1365), as well as the secular John Ruysbroek (d.1381)[40] and the unknown authors of the famous *Theologia Germanica* and *The Cloud of Unknowing,* gave to this Dionysian spirituality a more clearly orthodox expression.[41] Tauler and Suso colored Eckhart's very abstract spirituality with a much more imaginative, incarnational devotion to the Humanity of Christ in his passion that brought it closer to the *devotio moderna*. It should be noted, however, that in Italy during this fourteenth century a much more out-going and socially oriented spirituality influenced by Thomism developed, of which Dante and St. Catherine of Siena are notable examples,[42] which was to flower in the Renaissance.

II: Nominalism and the Shift from Nature to Law

1. Nominalism and Voluntarism

David Knowles has characterized the fourteenth century as a time in which every intellectual position was pushed to its ultimate extremes.[43] Nominalism, which came to dominate the universities during this century and down through the Renaissance of the next, was such an extreme. It was a result of pushing Aristotelian logic to its critical limits in the service, paradoxically, of the Platonic Augustinian tradition and in behalf of the radical party of Franciscan Spirituals led by the Englishman William of Ockham (d.c.1350), an intransigent critic of the system of his fellow Franciscan Dun Scotus.[44] It had the effect of demonstrating the futility of the long effort which I have been tracing, to cover over the radical difference between a Platonic and an Aristotelian epistemology; and it radically shifted, as we will see, the whole focus of theological discussion on Christian anthropology from the theme of human *nature* to the theme of *law*.

Ockham sincerely motivated by the Franciscan ideal of poverty, a thinker of great orginality, highly gifted as a logician, rigorously applied the Aristotelian theory of scientific demonstration to all areas of philosophy and theology while working out the ultimate implications of Scotus' doctrine of the intellectual intuition of singular existents. As a result Ockham led the movement to abandon the moderate realism to which most of the older scholastics had subscribed. He proposed a Nominalism (Terminism, Conceptualism) according to which the essences of material realities are not merely difficult to know, as Aristotle and Aquinas had admitted,[45] but are totally inaccessible to human intelligence. Thus our intellectual intuition of singular existents tells us what each is in itself only *phenomenally* as observed by our senses or (as regards our self-knowledge) by introspection.

Ockham even anticipated David Hume in denying the validity of the principle of causality, as Nicholas of Autrecourt in 1346 was condemned for doing.[46] We can of course observe that certain events regularly follow others, but since we do not know the essences of the things in question, we cannot know with certitude that one thing causes the other. Thus Ockham was led not only to deny the possiblity of proving philosophically that God is the cause of the world, but even that the cause of human actions is a spiritual soul. Certitude about such noumenal matters must be derived from faith not from reason.

Moreover, Ockham, while retaining the Aristotelian hylomorphism and the reality of those accidental qualities of bodies accessible to empirical observation, reduced all the other traditional philosophical categories to names that signify nothing more than extension (quantity) or merely mental relations (space, time, causality, etc.) Thus for the Nominalists, who quickly came to dominate all university theology, no ontological understanding of nature was possible but only a phenomenalistic description of isolated existents. Paradoxically, Ockham, starting from Scotus' Platonic intuition of singulars, had managed by his use of Aristotelian logic to undermine both Plato and Aristotle and to come very close to the empiricism of Hume which was to be the starting point of the Humanist world-view which we described in Chapter 3.

161

Although the noted physicist and historian of science Pierre Duhem thought that Ockham and his Nominalist followers, by reason of their empiricism, were forerunners of modern science, more recent research has shown that their chief philosophical interests were in logic (where again they approached very close to the modern standpoint) and their chief theological interest was in establishing the sovereign freedom of God.[47] As a "spiritual" Franciscan Ockham was himself actively involved in the reform of his Order and the Church, and his theology reflects a wide-spread fear that academic theology had become rationalistic and dangerous to piety. This anti-rationalism, fueled by a sense of guilt over the increasing gap between theological technicalities and Christian living, led to *fideism*. God's freedom precludes human reason from understanding God's purposes. Hence the Law of God requires our blind obedience. God's free creation, the universe, no longer appears intelligible, but simply confronts us in its sheer facticity, to be observed and explained only conjecturally.

The Nominalists were fond of showing the vast scope of God's power (his *potentia absoluta*) and contrasting it with the creation which he has in fact freely chosen to institute (by his *potentia ordinata*) which order is known to us only by revelation or by empirical observation.[48] Thus undoubtedly the Nominalists prepared the way for the rise of a non-ontological, non-telelogical natural science, not by directly motivating this inquiry, which their sceptical tendencies could only have discouraged, but by discrediting the possibility of an Aristotleian philosophical understanding of nature.

Having eliminated the possiblity of an ethics based on the exploration of human nature, the Nominalists began to develop a deontological, voluntaristic, legal theory of morality. By his *potentia absoluta* God could have decreed that what he has in fact forbidden (*potentia ordinata*) in the Ten Commandments would be morally good, and what he has commanded would then be morally bad.[49] Moreover, God could have willed to grant us heaven as a reward for acts which are within our natural power; in which case Christian ethics would be indistinguishable from a purely philosophical ethics. From this followed novel conclusions about the role of grace in human life.

Although it is often asserted that the nominalists were Pelagians, this was not the case. How could a system rooted in Franciscan Augustinianism have accepted the heresy of Augustine's chief foe? Pelagianism is the doctrine that we can be saved by our own moral efforts with the aid only of such external graces as the guidance of the Scriptures and preaching. Semi-Pelagianism is the doctrine that while interior grace is necessary for

162

salvation it *presupposes* a naturally good will in the recipient of grace. The nominalists, however, followed Augustine in teaching that no one can merit saving faith by *true* merit (*de condigno*). They argued, however, that since the sovereignly free God could by his absolute power grant salvation as a reward for purely natural acts, it may be that in the order which He has actually established by his *covenant* with humanity, He has freely willed to grant faith to any unbeliever who by his own natural powers does such good as he can (*facere quod in se est*), in which case he is said to "merit" not in a strict, but a conditional sense (*de congruo* or by a certain appropriateness), *in view of the covenant.*

To the still Augustinian Nominalists this seemed a happy solution to the old objection that Augustine's theology of grace made God seem unjust because he distributes his grace as much to the very wicked as to those striving for a certain moral goodness. Moreover, it explained why it made theological sense to urge sinners to strive to obtain the grace they still lacked. The reason that this approach seemed to make sense in the fourteenth century, when it had been firmly rejected in the thirteenth by Aquinas and others,[50] was because the Nominalists were beginning to think of grace not in ontological terms as a transformation of the human person elevating that person from the natural order to a share in the supernatural life of God, or as a restoration of the divine image, but as an extrinsic, legal (forensic) relation between God and man, dependent simply on the free favor of God who could either grant it unconditionally or on certain conditions which he had freely established. This notion of conditions established by God seemed to have biblical justification in the notion of the Old and New Covenants, and was eventually to be used by the Puritans in their "covenant" or "federal theology."[51]

2. From Nature to Law

The significance for the theology of the body of this new voluntarism in ethics may not at first be apparent. Up to this point, however, Christian anthropology had been dominated by the Biblical theme of the image of God and that image had been understood in accordance with Greek philosophy as reflected in human *nature* i.e. in the intrinsic principles of human behavior. For Platonic minded theologians this image was revealed as the soul freed itself from the body in order to take flight to God. For Aristotelian theologians it was also found in the soul but in that soul as it was naturally wedded to the body through which alone it could come to know the world and its Creator.

Moreover, for both Platonists and Aristotelians nature is *teleological* or goal-oriented, purposeful not in the sense that every nature is consciously self-determining, but in the sense that it is determined by its nature to tend to act in a regular way that maintains it in existence against opposing forces and enables it to develop to its full actuality. Now the moderate voluntarism of Scotus was pushed by Ockham and the Nominalists to the point that the relation of humanity to God could no longer be conceived in terms of an ontological similarity between human nature and the Divine Nature, but only in terms of human conformity to God's Will, i.e. in terms of conformity to the Divine Law.

In some respects this voluntaristic shift from *nature* to *law* was congenial to the Biblical basis of theology, since law is one of the fundamental categories of Hebrew thought. Yet in the Bible (and generally in ancient thought) the Law or Will of God was conceived as embodied in the ontological structure of things.[52] The Word or Law of God is creative, and therefore the creation reflects that Law in its internal structure, its nature, and in the activities rooted in that nature. What developed in the fourteenth century, as a result of the turmoil of the times and the failure of the theologians to resolve the conflict over epistemology and anthropology, was a profound scepticism about the possibility of knowing the nature of things and of human nature in particular as a key to the nature of God, and a new reliance on the biblical revelation of the Divine Will as an order and meaning imposed on the world not from within but from without. In such a legalistic perspective the importance of the body as furnishing a teleological basis for ethical norms dwindles, and the natural world is viewed more and more *objectively* as a collection of things to be used for their utility rather than to be contemplated as a mirror of the Creator.

This objectification of the material world fitted in with the increasing use of mathematics in natural science and the tendency to understand natural processes in terms of measurable *forces* rather than teleologically. It also intensified the tendency we have already noted in the spiritual writers of the period to stress the interiority and *subjectivity* of human existence in which cognition was secondary to affectivity, to the conscience and the will. The human person was seen more and more as a self-determining subject isolated in a world of alien objects that had to be controlled and dominated by force of will, but a will accountable to the sovereign power of the Divine Will. Surely this new self-understanding of the individual reflects the social situation of the times which saw the rise of powerful national states centralized under absolutist monarchs like Phillip IV (the Fair) in France, opposed by a similarly absolutist Pope Boniface VIII.[53]

III: The Renaissance and the Platonic Idealization of the Body

1. The Ideal Body

The Renaissance was a re-birth of culture not after the Dark or Middle Ages, as so often imagined in popular literature, but after the disintegration and gloom of the fourteenth century. In many ways it was the continuation and completion of the earlier Renaissance of the twelfth and thirteenth centuries. It saw a renewal of both the Platonic and Aristotelian traditions after their humiliation by Nominalism. The new Platonism, however, was marked by a more positive attitude to the human body as a result of the new enthusiasm for Greek and Roman art and literature. The body was still regarded only as the garment or tomb of the soul, but now it was realized that this garment and tomb might be a magnificent expression of the soul it clothed or memorialized. Such an expression of the soul demanded, however, that the body be *idealized,* that is, that it reflect the eternal forms of beauty, freed of the defects of perishable matter. Had not St. Paul himself said, that in the Resurrection we shall receive back our bodies but they will be spiritualized, "A natural body is put down and a spiritual body comes up" (I Cor. 15:44)? For the Christian Platonists of the Renaissance, art could picture the human body in its spiritual ideal as it left the hand of the Creator and as once again restored and glorified it will live in eternity.

This renewal of Christian hope was encouraged in the fifteenth century by the rise in Italy of a prosperous middle class. At the same time the end of the Byzantine Empire in the East flooded Italy with emigrés bringing to the west the literary culture of the Greeks. The university schools of philosophy and theology, however, remained largely conservative and still dominated by Nominalism, with Averroistic Aristotelianism still retrenched in the liberal arts and medical faculties. Through the patronage of the religious orders, nevertheless, Scotism flourished and Thomism began to be revived, especially through the work of Thomas de Vio, Cardinal Cajetan (d.1534), although his was a Thomism not altogether free of Averroistic influences.[54]

The vitality of this Renaissance, however, was not centered in systematic theology but in Humanism in the older sense of that word[55], that is, in literary studies, first of the Latin heritage, then of the Greek, carried on not so much in the universities as by independent scholars, sometimes grouped in "academies" patronized by nobles or by the newly rich capitalists whom they served by adding luster to their courts and providing them with secretarial and propaganda services.[56] The Humanism of this

Renaissance shifted the whole mode of thought from the dialectical method of the clerical schoolmen with its logical precision directed toward theory building and transcendental contemplation, toward the rhetoric of practical moral persuasion. Cicero, not Aristotle or Plato, was its model and literary beauty rather than intellectual rigor its ideal. Thus the voluntaristic and legalistic tendencies of the fourteenth century were implemented by the persuasive powers of the rhetoricians.

As I noted in Chapter 3, the Burkhardtian theory of the Renaissance as a time of the emergence of the free, secularized individual, has now been largely abandoned by historians.[57] . The leading humanists were, for the most part, sincere Catholic Christian laymen, but untrained in the scholasticism of the universities and repelled by its dry, overly technical language and its clerical ethos. They shared the tendency of the *devotio moderna* to look for a Christian piety that was scriptural, affective and oriented toward practical morality rather than monastic contemplation. What they added to such piety, however, (and this is of the greatest importance for our study of a theology of the body) was the Platonic theme of *beauty*. As Courcelle has shown[58] for one of the first humanists, Petrarch (d.1374), this theme was derived from Plotinus and Augustine. It had, of course, reappeared in the Middle Ages, especially in the twelfth century Victorine Platonists and had been richly expressed in Gothic art,[59] but concern for sensible beauty had always been kept in check by the fact that literature was dominated by writers who were celibate clerics, who emphasized asceticism and the danger of fascination with the beauty of a passing world. Certainly this other-worldly asceticism had been a legitimate interpretation of Platonism but the humanists shifted the emphasis in a this-worldly direction.

For the humanists, beauty of form both in language and in the plastic arts became a legitimate rhetoric through which *virtue* could be taught. The theory of "art for art's sake" was not typical of the Renaissance, but the rhetorical or moralistic justification of beauty was constantly stressed. Thus a reformer like the Dominican and Thomist Savonarola (d.1498) could urge the burning of licentious pictures, but also encouraged artists to use their art magnificently in the production of works which made virtue attractive.[60] Kenneth M. Clark in his work *The Nude: A Study in Ideal Form,*[61] has contrasted the Gothic view of nudity which always considered nakedness as an *indignity* to the human person, as when the fallen Adam and Eve realized their shame or when the Crucified was exposed to mockery and death, with the Renaissance idealization of the body as the revelation of human dignity in its pristine state.

The Gothic nude is painfully realistic, meager or gross, exposed to the weather and to scornful or lustful eyes, while the Renaissance nude is ideal, god-like. Clark also shows how this ideal implies Platonic dualism in its iconography of the earthly and the heavenly Venus. It is the heavenly beauty, the absolute ideal, which gives to earthly, visible beauty its *raison d'être*. We have a right to enjoy beauty-in-the-flesh, in its sensuous perfection, precisely because it is a genuine exemplification of a spiritual beauty. Thus the Platonism of the Renaissance, while remaining Platonism, assimilated something of the Aristotelian notion that earthly realities are not mere shadows of the ideal, but existential embodiments of the ideal. Nevertheless, one has only to look at one of the great Renaissance nudes, whether in painting or sculpture, to realize that they are not realistic portrayals of the model but efforts to discover the ideal in the real.

As Kristeller[62] and Trinkhaus[63] have brilliantly shown, the Christian Platonic Theology (with assistance from the Byzantine contacts so important in the enlargment of humanism from Latin to Greek literature), was revived in this fifteenth century with a strong emphasis on the theme of God's creation of humanity "in our image and likeness" (the title of Trinkhaus' study). This thematization has often been misunderstood as an anti-theistic "glorification of man," but in fact the humanist point of view remained strongly theistic and was based on the notion that because we are so conscious of the miseries of life and our own guilt, we tend to despair and therefore we need to be aroused to hope in God's mercy and in the real possibility of regaining our original dignity or "virtue." Hence the humanists believed it was their tasks to celebrate human dignity by their rhetorical "amplification" of this theme of man as God's image.

Trinkhaus shows that Petrarch, who set the pattern for these Renaissance panegyrics on "the glory of man", worked with themes from patristic literature, such as:

> . . . man's creation in order to rule over the rest of creation which was given him by providence; the beauties and utilities of the world at man's disposal; the capacities of the human mind, soul, and body, for the work of ruling the subhuman universe; the erectness of man's posture pointing to his heavenly goal; the immortality of the soul; the gift of the Incarnation which divinely honoured man; the consequent beatification of man and the resurrection of the body; man's ascent in dignity beyond the angels. Thus, whether it was a question of man's position in this world or of his ultimate end and destiny, with a variety of arguments and variation of emphases, beginning with Petrarch, the Italian humanists combined and elaborated upon these old theological themes. Although no one of these ideas was original with them, their combination of these ideas into a literary form

made it their invention. Later in the Quattrocento, a more secular emphasis on man's worldly accomplishments, inventiveness, love of beauty and pleasure would be woven into the more traditional theological frame.[64]

Nevertheless, to the very end of the Renaissance this Christian theological framework remained intact. In Michelangelo's Sistine Chapel frescoes, dating from the middle of the sixteenth century, we move from the magnificent creation of Adam in his noble nudity to the terrific Last Judgment in which frantic naked bodies tumble down into hell, but over which the prophet Jonah turns in righteous indignation toward God, only to be rebuked for his failure to understand God's unfailing mercy.[65] For Catholic humanists the dignity of man remained a glorification of the Creator and a ground for hope in God's redemptive forgiveness. Consequently, in Renaissance art the Savior is pictured not merely as the suffering servant or wise teacher as in medieval art, but as the New Adam, heroically beautiful. What the Reformers were to label the Pelagianism of the great humanist Erasmus (d.1536) can better be explained not as a denial of our human dependence on the grace of God, but rather as the praise of God for creating man in His own image, a likeness which pleads with God for grace.[66].

The Renaissance also moved toward a more *universalistic* view of humanity, in contrast to the Augustinian particularism which stressed salvation only through an *explicit* faith in Christ. Humanists justified their admiration for the wisdom of the ancient pagans by the notion of "the ancient theology"[67] as this had been accepted by the early Christian Apologists, relying on the teaching of the *Gospel according to St. John* that the Logos "enlightens all men" (1:9). Thus humanists could argue that Greek and Egyptian wisdom were derived from Moses or even from some pre-Mosaic revelation such as that given (supposedly) to Hermes Trismegistus.[68] Hence, when the New World and China were discovered, it was possible to argue that these pagans derived their evident virtues from some primitive revelation of Divine Truth. Later this notion was to be taken up by the secular humanists of the Enlightenment, but it also has a strictly Christian version, as is evident from the ecumenism of the Second Vatican Council.

Among the elements of this ancient theology which Renaissance thinkers believed compatible with the Gospel was the *cosmic religion* found in the Hermetic literature[69] (which of course was also acceptable to both Platonists and Aristotelians because actually it had been eclectically derived from their systems), along with the notion of the *anima mundi* (World Soul), often identified with the Christian Holy Spirit. This cosmic religion seemed to the humanists to be supported and enriched by the *Kabbala,* itself a Jewish attempt to achieve a similar mythical interpretation of the Hebrew Bible. Marsilio Ficino (d.1499) and the Florentine Academy, in close relation to Savonarola and other Dominicans of San Marco, typify this effort to find a synthesis of Christianity and the universal cosmic religion, often with a fascination for astrology and "natural" magic, in which strongly Stoic elements were blended with Neo-Platonism.[70] In the thought of Giordano Bruno (d.1600) and Tommaso Campanella (d.1639), both Dominicans, this universalism reached its extreme and in the case of Bruno (who ended on the Inquisitor's pyre) actually converted Christianity into pantheism.[71]

Similarly on the basis of Plato's *Timaeus,* which had so fascinated the theologians of the twelfth century, along with eclectic borrowings from Stoic and Epicurean sources, such thinkers as Cardano (d.1576), Telesio (d.1588) and Patrizzi (d.1597) developed pseudo-scientic cosmologies which at least served to stimulate more genuinely scientific efforts in the Italian universities to revise the Aristotelian physics still being taught there with Averroist and Nominalist variations.[72]

2. The Mathematization of Natural Science

The Platonic idealization of natural forms found in the arts also had its effects in the sciences. I have already shown that efforts to extend the ancient Ptolemaic application of mathematics in the study of the heavens to earthly phenomena, especially by the Oxford Mertonians,[73] had already begun in the fourteenth century. J. H. Randall[74] has shown how these efforts were taken up with enthusiasm in Italy in the universities of Padua and Bologna. These schools, unlike those of France and Germany, had no theological faculties but were noted for medical faculties which were continuously concerned with the revision of Aristotelian physics and Galenic medicine throughout the fifteenth and sixteenth centuries. The considerable progress which they made culminated in the work of Galileo (d.1642), so that by the beginning of the seventeenth century the logic of the modern *scientific method* which has made possible our modern technological world was thoroughly worked out. At the same time, as Koyré has shown,[75] the Pythagorean-Platonist tradition in astronomy was undergoing a remarkable new development in the work of the Pole Copernicus (d.1543), the Dane Tycho Brahae (d.1601), and the German Johannes Kepler (d.1630).

Nevertheless, according to Randall, these Platonic astronomers essentially belonged to the old line of Ptolemaic thought since they still conceived science as a mere "saving of appearances." Nor did the naive empiricism of the English Francis Bacon (d. 1626) with its faith in "induction" from sense experience really contribute much to laying the theoretical foundations of modern science.[76] Rather it was the Paduans who developed in a medical and non-mathematical way the logic of hypothetical reasoning and verification on an Aristotelian basis (of which the demonstration of the circulation of the blood by William Harvey in about 1615 is a classical example).[77] These Paduans brought together hypothetico-deductive reasoning, mathematization of observation, and empirical verification to form the modern scientific method. Recent research has shown that Galileo's thought was fundamentally Aristotelian, but an Aristotelianism revised and partly undermined by his use of Italian advances in mathematics (based on newly discovered Greek texts) and of new instruments of observation and measurement.[78] To these factors should be added, as Randall argues, the inventiveness of Leonardo da Vinci (1452-1519), who developed the use of mechanical models and new methods of experimentation, and the initiative of Bacon, who in place of the scholastic universities or the humanists' literary academies with their concentration on the study of ancient texts, proposed the foundation of scientific academies devoted to experimental research and technological application.[79]

These Paduan developments took place not so much in a Platonic atmosphere, such as in Florence with its strong concern for religious reformation, but in an Aristotelian milieu. This Aristotelianism was at first Averroistic, but with increasing knowledge of the older Greek commentaries on Aristotle (such as those of Alexander of Aphrodisias, third century A.D. and John Philoponus, sixth century A.D.) the Thomistic refutation of the Averroistic notion of a single world-intellect for all humans came to be accepted and this notion was replaced by an emphasis on the objective public mind and a naturalistic conception of the human person. This new trend finally led to the denial of the natural immortality of the soul by the Averroist Pietro Pomponazzi (d. 1530), a view promptly condemned by the Fifth Lateran Council.[80] Even the great Thomist Cajetan (d. 1534) was led into this radically anti-Platonic position.[81] Later Thomists repudiated this thesis of Cajetan, without, however, accepting Platonic dualism. Pomponazzi did not reject the Christian doctrine of eternal life through grace, nor did he deny that probable philosophical arguments could be brought to support this doctrine, but he was concerned to defend the autonomy of the natural goals of human nature, and to defend a frankly naturalistic interpretation of miracles and of the origin of religion.

The Renaissance interaction of religious and scientific interests was strikingly summed up in the thought of the learned Cardinal Nicholas of

Cusa (d.1464)[82] who had studied law and mathematics at Padua but claimed nothing more than a *docta ignorantia*, a "learned ignorance" derived from the negative theology of John Scotus Erigena and Meister Eckhart.[83] Unlike Eckhart, however, Cusa was greatly interested in the cosmological, scientific aspects of the Platonic tradition. He anticipated Copernicus in rejecting the Aristotelian notion of the earth as the fixed center of the universe, and defended the view (quite foreign to all Greek thought) that the universe is spatially infinite. This seemed to make our earth and human life upon it trivial to anyone but ourselves, and thus to contradict the common Renaissance theme of human dignity, but also to reinforce the religious emphasis on human *subjectivity*.

For Cusa, God was the *coincidentia oppositorum*, the paradoxical or dialectical transcendence of all the polar oppositions of thought and experience which so enriched Renaissance culture — a synthesis beyond any conceptual expression. Such mystical negations joined with the theological fideism of the Nominalists to produce the "New Pyrrhonism" (scepticism) of Michael de Montaigne (d.1592) which left all "metaphysical" questions to blind faith.[84] On the other hand the *Realpolitik* of Machiavelli (d.1527)[85] reduced the ethical voluntarism and legalism of the Nominalists to the sheer will to power. In this cultural climate where the "metaphysical" conception of human nature proposed by Plato and Aristotle had so repeatedly been called into question, frequently humanists, lawyers, and scientists still respectful of Greek thought turned to the less metaphysical philosophy of Stoicism because it seemed to provide a secular ethic compatible with the Christian faith which could be accepted pragmatically.[86]

Thus we see that the real concern of Renaissance Humanism was not the development of a systematic theology into which could be integrated the many new insights arising from the increased appreciation of the literature and art of antiquity or the new discoveries of the world through exploration and the revision of ancient science, but rather the development of an ethics and politics, not incompatible with the Gospel, but more relevant to the life of the laity in a rapidly changing society than had been the clerical spirituality of the Middle Ages. Nor, as the spiritual writings of Erasmus and the great religious art of the High Renaissance demonstrate, was there any reason that this Christian Humanism need have undermined the Christian world view and its understanding of the human person in its bodily existence. Rather it was a needed and promising development in the Christian tradition which helped to correct false dualistic tendencies stemming from an excessive other-worldliness by a new interpretation of Christian Platonism which made more room for the Aristotelian appreciation of the natural world and the human body as an epiphany of the ideal. This new synthesis, however, because it was more eclectic than profound, was unstable.

Underlying all these developments were the pragmatic concerns of the laity. Increasing scientific research was often associated with an equal interest in what to us seem the pseudo-sciences of astrology, alchemy and magic.[87] What was sought was not just knowledge, but power, whether technological or magical. Thus the late Middle Ages and Renaissance were a time when the supreme goal of life was no longer contemplation, as Plato and Aristotle and the monastic tradition had declared it to be, but the active life of politics, technology and art. No wonder then that God Himself appeared more as Sovereign Lord and Lawgiver than as "Thought Thinking Itself,"[88] and human beings were seen not so much as contemplators and adorers of God as clients seeking His favor for their projects.

The men of the Renaissance in southern lands, in spite of their frequently Machiavellian cynicism, for a time seemed confident of their Lord's favor. They gloried in the gifts He had lavishly poured out on them. Yet they also trembled at the warnings of a Savonarola that this favor could be as quickly withdrawn as it had been given. In northern lands where the chilly mists of the Middle Ages had lingered longer, these doubts, this sense of guilt before the Sovereign's dread eyes were more disturbing. To such troubled consciences the superficiality of much Renaissance theology, such as that of Erasmus, was only an irritant.

IV: The Body and the Sacraments

1. The Reformation Subordination of Sacrament to Word

For Protestants generally, Luther (d.1546) is a heroic, prophetic figure who broke with the abuses by which the papacy had deformed the Gospel into a Pelagian system of salvation through "good works." He achieved this by returning to the plain message of the Bible, especially to St. Paul's proclamation of the Gospel of justification by faith, as against Catholic tradition which had become a religion of Law. But ecumenically minded Catholics find it easier to accept Luther as a true reformer if he is seen as proposing an original synthesis of a number of reforming tendencies already active in the Catholic Church since the fourteenth century.[89]

Especially significant for understanding the Protestant contribution to the development of a theology of the body are the following elements which were combined in Luther's original synthesis. *First*, while there can be no question that Luther's effort to be strictly faithful to the Bible and thereby to purify Christian preaching of all alien, even if traditional elements, was his greatest strength; yet it must also be acknowledged that this insistence that all theology must be biblically rooted was shared by all the great medieval theologians.[90] Luther had the advantage, however, of the new concern of the humanists for a return to the text of the Scriptural writings in their original language and thought forms. Although in many respects the Reformation was a reaction against Renaissance worldliness, in their biblicism Luther and Calvin were typical humanists scholars for whom the reverence for the original text of the classics was the proper point of departure for all learning. Consequently, for the Reformation, theology became in the fullest sense a theology of the Word, the word printed and the word preached.[91]

No one, however, reads a text without pre-understandings. One such is the *second* element of the Lutheran synthesis, namely, the Augustinian tradition in which Luther, like any medieval university student, and more particularly one educated like himself as a member of the Augustinian Order, lived and breathed.[92] Augustine's Platonic dualistic outlook and his urgent concern for the problem of grace raised by Pelagius had been somewhat balanced in medieval theology, as we have seen, by other influences stemming from the Eastern Fathers and Aristotelianism. These balancing factors, however, were of little interest to Luther, for whom the anti-Pelagian problematic oriented his whole reading of the Bible and shaped his famous notion of a "canon within the canon," according to which those passages of St. Paul which directly answer the problem of justification become the key to the interpretation and even the canonicity of the rest of the Bible.[93]

Third, the specific form of Augustinian theology in the context of which Luther struggled with the problem of justification, was the Nominalism derived from Ockhamist voluntarism.[94] Although Luther had some acquaintance with the works of Scotus, Aquinas, and the thirteenth century scholastics, it was only as followers of the *via antiqua*, while he walked in the *via moderna* of the Nominalists who had been his masters. Hence, he shared in the Nominalists' assured scepticism about the possibility of knowing the intrinsic nature of things by reason. Consequently, he also shared in their way of posing theological problems not in terms of the analogy between the Divine and human natures (i.e. in terms of the *imago Dei* of the Fathers and thirteenth century scholasticism) but in the voluntaristic and legal terms of the Sovereign Will and the obedient or rebellious human will. Although Luther came to oppose the Nominalists' solution to these problems by invoking St. Augustine's reading of St. Paul, he never broke from their voluntaristic way of formulating theological questions.

Fourth, the intensely experiential, interior and subjective character of Luther's spirituality, which undoubtedly accounts for so much of his contemporary and subsequent influence, could hardly have been derived from the dry dialectics of Nominalist academe, but had its roots in the *devotio moderna* and the Rhineland mystics. Very directly Luther was influenced by the *Theologia Germanica* and John Tauler, and through them by Meister Eckhart.[95] Luther is strongest as a religious teacher when he writes passionately of the *theologia crucis* according to which the hidden God is most truly revealed in the Cross and the sincere believer approaches God most nearly when sharing in the darkness of humiliation and absolute faith. This spirituality is authentically Eckhartian, although Luther strove to give it a purely biblical expression, freed from every trace of Neo-Platonic metaphysics and psychology.[96]

What is especially striking is how Luther's rejection of the scholastic view of grace as an intrinsic transformation of the soul and his adoption of the view that the reborn Christian is holy only by the *alien* holiness of Christ echoes Eckhart's teaching that the Christian is sanctified not by the infusion of virtues, but by an emptying (*kenosis*) of self to make room for the birth of the Word. Paradoxically this kenotic conception of grace is reinforced in Luther's theology by the very different Nominalistic *forensic* (legal and external) conception. For Eckhart grace is alien but interior, for the Nominalists alien but exterior. No doubt recent Lutheran scholarship is correct in rejecting the common Catholic accusation that Luther had a purely forensic conception of grace.[97] Certainly for Luther God's grace, although it is God's free declaration of forgiveness and favor, unites the forgiven sinner to God in a profoundly interior way, since for Luther faith is a rebirth of the "inner man" (II Cor. 4:16). Yet in this interior union the believer remains *simul justus et peccator*, a sinner himself but by faith made heir to the holiness of Christ.

Fifth, consistent with this late medieval type of spirituality was Luther's subordination of the sacramental life of the Catholic church, even the Eucharist, to the preaching of the word. In the thirteenth century the Dominican Friars Preachers sometimes urged the laity to prefer to hear preaching than to attend Mass when a choice had to be made, on the grounds that the sacraments were unfruitful for those who were not instructed in the faith.[98] In the fourteenth century, writers of the *devotio moderna* much more strongly deprecated excessive reliance on the *ex opere operato* of the sacraments in favor of a more interior, purely spiritual piety. As the Reformers were to do, they reinforced this anti-sacramentalism with a vigorous criticism of the frequent unworthiness of the clergy who purveyed the sacraments. Luther himself, for all his attacks on sacramental abuses, remained conservative in liturgical matters. Not only did he defend infant baptism, but also the objective Real

174

Presence of the Body and Blood of Christ in the Eucharist against Zwingli (d.1531) and others who reduced the Lord's Supper to a merely symbolic rite, and also against the still more radical sectarians who wished to replace the visible rite by a purely spiritual communion.[99]

Yet even for Luther the Word seems to have taken the central place which for Catholics the Eucharist occupies. He understood the sacraments primarily as devotional acts by which the faith of the believer in the forgiveness of sins declared in the Scriptures and in preaching can be more personally realized. The Eucharist was viewed by Luther more as Holy Communion for the edification of the individual recipient than as the Church's incorporation in the Body of Christ offered once for all on the Cross. Hence Luther rejected the Catholic doctrine of the *Sacrifice* of the Mass because this seemed to him just another Pelagian effort on the part of human beings to effect their own salvation by works.[100]

The French Reformer John Calvin (d.1564) took an intermediate position on the Real Presence by speaking of a "virtual" presence by which the heavenly Christ is present to the believer through His spiritual power, but in time Calvinists tended to the Zwinglian view. Since Calvinism came to dominate much of the Protestant world and to affect deeply even the Lutheran churches, the sacraments played a very much reduced role in the Reformation. Although the Lutheran churches made very important contributions to musical culture, as in the great works of Telemann, Bach, Handel and Mendelssohn, and Protestants have greatly enriched Christian literature, Calvinism took a negative view on the use of the arts in worship because this had no clear warrant in the New Testament and seemed unspiritual.[101] Hence the plastic arts and the Renaissance idealization of physical beauty ceased to flourish in Protestant culture, lost their sacramental character, and became secularized. Hence, although the Reformation had surely been right in attacking the neglect of preaching and in calling for a closer union of Word and Sacrament, the practical result was that the sacraments were largely replaced by preaching.

For many Protestant theologians this de-sacramentalization of Christian life did, and perhaps still does, seem progress toward a more spiritual understanding of the Gospel. External rites were replaced by interior experience. But as Luther himself pointed out to those who tried to dissuade him from his belief in the Real Presence, this kind of excessive spiritualism undermines the fundamental Christian belief in the Incarnation. Thus Luther echoes the Seventh Ecumenical Council (Nicaea II, 787) which condemned Iconoclasm on the ground that to call the veneration of icons idolatry is to deny worship to the Humanity of Christ.[102] One of the Zwingli's chief arguments against the Real Presence was the text, "It is the spirit that gives life; the flesh is useless" (John 6:63); but Luther rightly replied that "spirit" and "flesh" here are not the Platonic "soul" and "body", but the Biblical concepts of "grace" and "humanity without grace."[103]

It must be admitted, however, that Luther himself was not altogether consistent in applying this profound insight. Not only did he reject the sacramental character of marriage, but he proposed as a solution to the classical theological problem of the relation of Church and State his unfortunate Two Kingdom theory. According to this doctrine, the Church is an interior community of faith, while the State (provided it maintains that minimum of law, order, and freedom required for the Church to preach the Gospel) should be given complete control over public affairs.[104] This reduction of the Gospel to the purely private sphere, powerless to guide public life, and the reduction of marriage to a secular institution seem to be very difficult to reconcile with the incarnational and sacramental character of the Gospel and with the concept of the Kingdom of God in both the Old and New Testaments.

Calvin, a lawyer by education, while liturgically more radical than Luther, was closer to the Catholic tradition in his ethical thinking. Luther resisted antinomianism, but for him the Law had only two uses for the Christian. First it was necessary to maintain that external order without which the State was impossible. Second it was necessary for the preacher to arouse in his hearers the acknowledgement of their own sinfulness, so that when the Good News of God's mercy was announced it might effect a genuine conversion. But the believer is no longer under the Law. He requires no other motive for obedience to God than gratitude for God's mercy; and no other instruction in God's will except the inward guidance of the Spirit. Although Luther as a peacher often gave excellent moral instruction, he distrusted any attempt to develop a systematic ethics as a dangerous concern for self-righteousness and a return to the yoke of the Law.[105]

Calvin, although he too believed in the internal guidance of the Spirit, admitted a third use of the Law to instruct the Christian in the will of God. He also taught in closer conformity with Catholic tradition that the sanctification of the believer is not only a relation to the holiness of Christ, as Luther declared, but a positive growth in personal character through the action of grace.[106] Moreover, although he entirely agreed with Luther on justification by faith alone, yet he stressed more emphatically that good works are a sign of the sincerity of the faith of which they are necessary fruits. Following St. Augustine, Calvin says in the *Institutes*[107] that only two things are really worth knowing: God and oneself. True self-knowledge reveals our "total depravity" through Adam's sin and our own and drives us to throw ourselves entirely on God's mercy. Calvin dwells on this corruption rather than on human dignity, as did Renaissance writers, not because he denies this dignity given us by the Creator, but because, he said, we easily admit our dignity but stubbornly deny our sinfulness.[108] Nevertheless, Calvin argued that Christians need not be depressed by the knowledge of their own corruption, because faith in Christ gives absolute assurance not only of forgiveness but also of the grace to grow in character.

Both Luther and Calvin believed it essential to maintain that faith gives the true believer *subjective* certitude of salvation, but Calvin went further than Luther in grounding this certitude in the shocking doctrine of double predestination.[109] Is not this Calvinistic conception of God as the Sovereign Will who creates some for salvation and some for damnation to manifest His mercy and justice the terrifyingly logical outcome of the voluntaristic tradition in theology? Nevertheless, believers certain that they were predestined could not rest on that assurance, because if they failed to produce good works this raised serious doubts whether their faith was genuine, or merely hypocritical pride. The result was that remarkable "Protestant ethic" to which Max Weber attributed the rise of modern capitalism. Weber's thesis has been shown by historians to be a half-truth, but that half is of no little importance.[110]

Thus Calvin, avoiding the Lutheran tendency to leave social problems to the State, attempted at Geneva, as Savonarola had done at Florence, to establish a Christian commonwealth. Throughout its history, Calvinism, although maintaining the separation of Church and State, has attempted to exert a powerful moral influence on society.[111] Like the Lutherans, Calvinists refuse to call marriage a sacrament, but the Calvinistic Puritans developed a highly spiritual almost mystical, understanding of married love.[112] Nevertheless, many historians think that in the long run the social effect of Calvinism was strongly secularizing, because its religious dynamism depended so much on the doctrine of double predestination that when this doctrine was undermined by the accusation that it made of God an arbitrary tyrant, Calvinism collapsed into a moralism that was indistinguishable from a humanistic ethics.[113]

I have noted these ethical consequences of Reformation doctrine because the chief concern of this book is the grounding of ethics in human nature and especially in that nature as it is subject to manipulation and reconstruction by technology. The desacralization of the human body submits this body to reconstruction at the will of human power. Historically this desacralization began when the sacramental significance of the body was undermined, and this was largely the result of the voluntaristic conception of God and of Christian ethics which led to the shift from categories of nature to those of law. Although the Reformers sought to liberate the Christian conscience from the heavy burden of Law, they actually reinforced this voluntaristic theology by posing the problem of salvation in legalistic terms. Their solution to this problem was an excessive spiritualization and interiorization of religion and a consequent secularization of earthly existence.

It would be a mistake, however, to suppose that this secularization was immediate. If it had been, the Protestant countries would have quickly become Humanist. Instead they developed a vigorous Christian culture which entered into a long bitter stalemate with the Catholic culture. The result of this exhausting polarization was a disillusionment among intellectuals with Christianity itself. The theological support of this new Protestant culture was centered in the universities which had undergone the reform or been newly founded under the auspices of Protestant princes.

The reforming theologians, to answer Catholic accusations that the Reform was a mere babble of contradictory opinions, quickly began to develop systematic theologies for university teaching of the new clergy. Among Lutherans the leader in this trend was Luther's close friend Philip Melanchthon (d.1560) who favored a reconcilation of Platonism and Aristotelianism, but who used Aristotelian methodology (but not Aristotelian epistemology or metaphysics) in his great *Loci Communes* of 1521.[114] Thus arose the Lutheran Orthodoxy of the seventeenth century with such formidable systematists as Martin Chemnitz, John Gerhard, Abraham Calov, and George Calixtus. Calixtus, by promoting an "analytic" ordering of theology in place of the older "synthetic" ordering, gave Lutheran doctrine a thoroughly philosophical, scholastic formulation. Today this Lutheran scholasticism is regarded by many Lutheran theologians as a perversion of Luther's original insights, especially because it so stressed the literal intepretation of the Bible (the so-called "formal principle" of the Reformation) rather than its "material principle" of justification by faith.[115]

The aridity of this state-fostered orthodoxy encouraged the rise of Pietism which developed the more subjective, mystical elements of Luther's thought. Jakob Boehme (d.1624) was a visionary in the line of Rhineland mysticism. He proposed the notion which was to prove very influential that the dark, unconscious or even evil side of the human spirit reflects a certain polarity within God Himself.[116] Johann Arndt (d.1621), P.J. Spener (d.1705) and A.H. Francke (d.1727) and others worked to revive a more intense spiritual life for individuals and for sectarian communities.[117] The same intensification of piety was promoted by the anabaptist sects of the Radical Reformation and greatly influenced the present Baptist Churches and especially the Methodist movement which arose in the Angelican Church in the eighteenth century. American Protestantism on the whole has been molded more by such pietistic tendencies than by Lutheran or Calvinist Orthodoxy.[118]

In the Calvinist tradition (out of which the Baptists and Methodists were to evolve) there were developments parallel to those in Lutheranism. The doctrine of double predestination was given systematic formulation not by Calvin himself, but by his successor at Geneva Theodore Beza, influenced, it would seem by the ex-Dominican Thomist Martin Bucer.[119] It became the keystone of a very scholastic type of Calvinist Orthodoxy. Calvin had been influenced by Stoicism.[120] Aristotelian methodology, however, was used at first by Calvinist systematizers, but was very soon replaced by a new, explicitly anti-Aristotelian method called "Ramism" after its inventor Pierre de la Ramée (d.1572). This methodology, of rhetorical rather than logical inspiration, encouraged an extremely mechanical method of classifying ideas by rigid dichotomization and proceeded by a purely deductive form of reasoning, which encouraged a frigid rationalism in theology.[121]

Orthodox Calvinism was given systematic formulation by J. Wollebius (d.1629), G. Voetius (d.1676) and especially by Francois Turrentin (d.1687). It received official approval by the Synod of Dort in 1619 and the Helvetic Confession of 1675. Nevertheless, it was soon succeeded by the Federal or Covenant Theology first proposed by a friend of Zwingli, Johann Bullinger (d.1575) and Caspar Olevianis, one of the authors of the Heidelberg Catechism of 1563, but which was given its full exposition by Coccejus (Johann Koch, d.1609) and finally enshrined in the Westminister Confession of 1647.[122]

This Federal Theology (which I have already shown was in some measure anticipated by the Nominalists) sought to overcome the apparent arbitrariness of Calvin's doctrine of grace and predestination by conceiving the relation of God to humanity as a covenant initiated by God but in which humanity can also play its part by obeying God. Yet the Reformation doctrine of justification by faith was saved by maintaining that the "Covenant of Works", ratified with Adam but broken by him and his descendants, has been replaced by the "Covenant of Grace" ratified in Christ. More radical was the outright defense of the freedom of the will against Luther and Calvin made by the French School of Saumur (Moses Amyraut and others) and then by Jacob Arminius (d.1609) who had been a pupil of Theodore Beza. Arminianism, a term which came to shelter all attempts to soften Reformation orthodoxy, gradually made deep inroads and (after the Puritan dominance of the seventeenth century which combined Federal Theology with pietistic asceticism) controlled most of the Reformed Church and even tended to Unitarianism.[123]

Thus by the opening of the eighteenth century, the Reformation movement, both Lutheran and Calvinist, had markedly declined, partly because of the rigid systematization of orthodoxy, [124] partly because of the anti-intellectual reactions and tendency to pietistic sentimentalism in the face of this aridity, and above all as a result of the devastation of the religious wars of the seventeenth century with its policy of *cuius regio eius religio.* This caused constant superficial shifts in religious allegiance. It was out of these fratricidal struggles that the new religion of Humanism, with its conception of the human person as autonomous, free of dependence on God, yet in its bodily condition subject to all the determinisms of the laws of physics, was to emerge. Reformed Christianity, which had arisen in protest of Renaissance worldliness, by freeing the Word from the sacraments, by interiorizing and privatizing religion, had made way for a secularization of culture which was in fact a new Religion of Man.

2. The Counter-Reformation Reduction of Sacraments to Law

The Catholic Counter-Reformation shared with the Reformation the same voluntaristic, legalistic conception of the relation of God to man that both equally had inherited from late medieval Nominalism, but the Counter-Reformation glorified the Sacraments as much as the Reformation glorified the Word. In response to Protestant criticisms the Council of Trent urged a revival both of preaching and a more intense sacramental practice, but it was sacramental devotion that proved the greater strength of the Catholic renewal.

When the Council of Trent closed in 1565 the Age of the Baroque began, during which, over against the Protestant North, Spain with its immense empire and France with its powerful centralized monarchy were able to establish a certain stability and prosperity that favored a splendid triumph of artistic and literary creativity. The Catholic baroque style quickly became international and even in Protestant countries literary artists such as Shakespeare, Milton, and Dryden, or composers such as Bach and Handel, or painters like Rembrandt were inspired by its themes and techniques. The image of man and woman in the art of this time is vital, sensuous, affirmative of life, and usually posed dramatically in some grandiose landscape or architectural vista. The glorification of humanity that had characterized High Renaissance art now became less that of idealized form and more that of *dramatized movement,* heroic actions triumphant over forces that are in retreat but still dangerously resistant. These heroic acts take place in the theater, as it were, of the cosmic court over which the Sovereign Will surrounded by His angelic courtiers and pages presides from

His throne. The court of the Sun King, Louis XIV at Versailles was a reasonably colored facsimile of this baroque vision of heaven.[125]

This Baroque Style passed through several phases, beginning with that of Mannerism (a transition style of deliberately shocking strangeness inspired especially by Michelangelo), flowering in the full exuberance of the style in such artists as Bernini and Rubens, moderating itself by a more sober neo-classicism with Poussin, and ending in the lightened miniaturization of the Rococo. Underlying this whole range was the fundamental Counter-Reformation transformation of the Renaissance spirit as is apparent from the fact that baroque art used the same themes and classical motifs as the Renaissance but gave to them a very different interpretation. The Renaissance in its effort to develop a culture that remained Christian yet provided for the temporal concerns of the laity had linked the spiritual and the material realms by an idealization of the human body. The Baroque followed the same path but avoided the mere dreaminess into which such idealization so easily slips, by introducing the element of dramatic conflict (and often of incongruity, humor, and madness) into the ideal. This was made theologically possible by the new Counter-Reformation attitude to the sacraments.[126]

In the Middle Ages the actual sacramental practice of the people had been very lax. Although reverence for the Mass was high, the reception of communion was infrequent. Popular religion tended to be centered in secondary symbols and ritual practices rather than in the Eucharist. As a result of Trent and particularly through the vigorous efforts of the great new form of religious order, the Clerks Regular, of which the Society of Jesus founded by St. Ignatius Loyola (d.1556) was the leader, a systematic campaign was successfully carried out to encourage the regular attendance at Mass on Sundays and great feasts and the relatively frequent reception of the sacraments. The key to this new discipline was the sacrament of the confessional in which the priest instructed the penitent on how the moral law applied to his or her actual duties in life and enforced their observance by assigning penances and in some cases refusing absolution and communion. The development of written catechisms which included precise moral instructions furthered this discipline for the literate laity.[127]

Thus the sacraments were no longer viewed primarily in the context of the contemplative experience of praising and thanking God but rather (as for the Lutherans) in the context of the forgiveness of sin; and sin was seen not so much as a turning away from union with God as a breaking of His laws for which forgiveness and restoration to His sovereign favor must be sought through the confessor who sat in judgment in His place. In this way the correspondence between the Heavenly

Court and the earthly court of the Church and the State was felt by the laity as a reality and reflected in the art and literature which they admired. Because this conception was in terms of a cosmic order maintained by the power of the Sovereign Will, the sacraments became instruments of control, symbols of a power struggle, and this power struggle was manifested in the baroque style which is conflictual, dramatic, and triumphant.

In Protestantism the spirituality of this age was interior, anguished, subjective, as we have seen, — the fearful drama of despair and dramatic conversion and salvation. No less dramatic was the Catholic spirituality of the time which reached its classic expression in the writings of the great Spanish Carmelite mystics St. Teresa of Avila (d. 1582) and St. John of the Cross (d. 1591). These saints who, like Luther, were influenced by the Rhineland mystics,[128] developed a profoundly experiential and psychological phenomenology of the ways in which the grace of God purifies the Christian through the "dark nights of the soul and the spirit", transforming the human person by the indwelling of the Holy Spirit. It is only this Holy Spirit who can elevate the faithful Christian into a higher way of prayer and life which transcend the human mode of action. St. Ignatius Loyola out of his own personal experience of that transformation developed in his famous *Exercises* a precise method of directing both clergy and laity in this perilous journey.[129] This highly developed psychology of mysticism necessarily entailed an interest in the physical phenomena of the interior life, so that the bodily conditions which promote or hinder meditative concentration began to receive careful attention. The concern was no longer for the sometimes extravagant asceticism of earlier times, but rather on a severe but sanely moderate discipline of the body compatible with the individual's social and occupational situation.[130]

The same forces that in Protestant universities led to Lutheran and Calvinist Orthodoxy as a means of stabilizing and moderating the explosive tendencies of the Reformation were also at work in Catholic universities. The result was a decline of Nominalism and a revival of the more balanced systems of the High Middle Ages. Scotism was still influential in the Franciscan Order which experienced new vitality at this time, and most striking of all was the remarkable revival of Thomism which for the first time, partly because of the success it had enjoyed at the Council of Trent, began to assume hegemony. The most important of all the opponents of Luther was Thomas de Vio, Cardinal Cajetan, whom I mentioned earlier as a typically Renaissance figure; but in his wake in Spain came Francis of Vitoria (d.1546), Dominic Soto (d.1560), Melchior Cano (d.1560), Bartholomew of Medina (d.1581) and Dominic Bannez (d.1604), Dominicans who considerably elaborated the style and expanded the scope of the Thomistic tradition, particularly as regards natural science and political theory. This opened the way to a more political conception of human nature and raised questions about the transcultural character of human nature as a result of increasing Spanish contact with the native peoples of the New World and the Far East.[131]

Since St. Ignatius Loyola had advised his order to educate its members as Thomists, the Jesuits soon produced a freer and more independent Thomism through the work of a number of philosophers and theologians, culminating in that of Francis Suarez (d.1617) who attempted to synthesize Scotism and Thomism, another case in the long history of attempts to overcome the Platonic-Aristotelian dichotomy.[132]

Yet it would be a serious mistake to think that the unique Aristotelianism of Aquinas at last had achieved a definitive triumph. Baroque or "Second Scholasticism"[133] was characterized not by a creative capacity to construct a unified theological system like those of the High Middle Ages, but by a concentration on certain "special questions" disputed among the schools. Nor did it progress very far in its limited attempts to advance into the new fields being opened up by natural sciences and political theory. Its energies were too much drained by involvement in two great theological disputes of which I must give some brief account in order to justify my contention that the Counter-Reformation, just like Reformation theology, was based on an anthropology which was less ontological than legalistic.

The first of these controversies concerned the same problem that had so much preoccupied Luther and Calvin: predestination versus free will. In the famous Commission on Grace (*Congregatio de Auxiliis*) set up by Pope Clement VIII, the Dominicans for nine years (1598-1607) defended a strong view of predestination without accepting the Calvinist double predestination, while many Jesuits adopted the new system of Luis de Molina (d.1600). He refuted Calvin's determinism by arguing that God predestines to salvation only those whom he foresees (in view of their character and the life situations in which His providence will place them) will cooperate with His grace. For the Dominican Thomists of whom Dominic Bannez was the leader, this "Molinism," just as the Nominalistic theory of grace before it, tended to Pelagianism; but they also rejected it as logically implying the very same determinism which was so objectionable in Calvin's system.[134]

The second controversy, not unrelated to the first, was over "Probablism" or the rules of Christian prudence which govern difficult moral decisions.[135] The Jesuits who were so zealous and skillful in the pastoral ministry of the confessional wrote their vast *Institutiones* or manuals of moral theology no longer on the plan of the Christian virtues but on that of the list of the commandments of God and of the Church, thus shifting from the theme of sanctification to that of obedience to law. Significantly the Protestant treatises on Christian ethics adopted the same legal pattern. [136]

Of course handbooks for confessors had been common throughout the Middle Ages, [137] but these post-Tridentine manuals, comprehensive in scope and exhaustive in detail, began to shape theological language systematically. They focused attention no longer on the development of the virtues of Christian character, but on the moral dilemmas of the Christian laity living in an increasingly secularized culture. In such dilemmas Dominicans and more conservative Jesuits, still influenced by the older teleological modes of thought, counseled penitents to follow what appeared to be the "more probable" opinion, because anyone wishing to reach a goal can only reasonably take the path which seems more likely to lead to it. The new view, first proposed by the Dominican Bartholomew of Medina, but developed by Jesuit authors, was that in such cases a penitent can reasonably follow any course of action which is probably moral, even if less probable than some alternative. The argument for this more lenient approach (which gave greater latitude to the laity to solve the dilemmas of life in the world without disrespect to the law of God or of the Church) was that the burden of the obligation to promulgate the law clearly was on the lawgiver, not on the subject to discover the lawgiver's hidden will. [138]

The Counter-Reformation popes were at first very suspicious of this novel moral methodology, but they eventually came to tolerate it with certain restrictions. Protestant theologians on the contrary denounced it as sophistical, pharisaic, and lax. From within the Catholic camp the Jansenists in France carried on a bitter war against the Jesuits, as in the famous *Provincial Letters* of Pascal, [139] primarily on the unjust charge that they were promotors of moral laxism. The Jansenists and the Protestants agreed that it was the Christian's duty to follow the strictest opinion (rigorism) out of respect to the Divine Sovereignty. [140]

Today this controversy over "moral systems" seems misguided, especially when we see that most of the probable opinions which the Jesuits defended would now seem extremely strict, but the real issue was not the proposed norms themselves but the attitude toward Divine Law which these seemed to reflect. The Rigorists did not want to give any opening to a softening of Augustinian pessimism about the moral condition of fallen humanity; while the Probabilists sought to avoid Pharisaism in the application of the moral law to a struggling humanity. In Chapter 10, I will show how this problem is still current and that it has its historic roots in the legalistic conception of morality which results when ethics is cut lose from its foundation in human nature as manifested in our bodily as well as our spiritual needs.

184

The profound influence which these problems of pastoral counseling also had on what today is called "political theology" is evident in the great work of the Jesuit already mentioned, Francis Suarez, the *De Legibus* in which he attempted to reconcile Thomism with the voluntarism of Scotus. Suarez and his fellow Jesuit Cardinal Bellarmine (d.1621) developed the thought of the Dominican Francis of Vitoria into a remarkable system of political theory on international relations, the relations of church and state, and on human rights, including the right of resistance to tyranny. [141] Thus the medieval theories of natural law were reshaped in the Counter-Reformation in terms of "the rights of man," a theme which was also taken up by Protestant legalists such as Hugo Grotius (d.1645) and Samuel von Pufendorf (d.1694). [142]

The Jesuit theologians, made cautious by the Church's experience with the quietism and "enthusiasm" which had too often been the fruits of Rhineland mysticism and which seemed to savor of Jansenism or of the Lutheran theories of passive justification, also developed a theory that Christian santification normally (especially for the laity) proceeds by an "ordinary way" of active obedience to the Commandments of God and the Church and of vocal and meditative prayer. Only a few, therefore, are called by a special grace of God to a passive sanctification by mystical prayer, and even they cannot safely travel that road except under obedience to a spiritual director. [143] Thus the exuberance of the Baroque Age was held always in dramatic tension by an elaborate system of social and religious controls.

3. The Desacralized Body

If the results of this survey of the development of theology from 1300 to 1600 are summarized, the following negative and positive effects for the development of the Christian understanding of the body are noteworthy.

First, the Platonic dualism inherited primarily from St. Augustine and secondarily from the Eastern Fathers, especially the Pseudo-Dionysius, was basic to the theology of the medieval universities and was reinforced in the Renaissance and Reformation periods by the humanists' first-hand acquaintance with Platonic and Neo-Platonic texts and their enthusiams for the rhetorical mode of the writings of Augustine.

Second, the important place given to Aristole in the medieval universities, although it resulted in a full-blown scholastic methodology and in an increasing interest in the scientific study of the material world, never succeeded in completely replacing the Platonic dualistic anthropology and epistemology. St. Thomas Aquinas who, while by no means neglecting the great contributions to theology of the Platonic tradition, was the first scholastic to break with Platonic dualism in anthropology and to adopt a radically Aristotelian epistemology, was not understood because his position was confused with that of the Averroist Aristotelians who denied the doctrine of creation and personal immortality.

Third, the introduction of Aristotelian methodology into the Platonic tradition of the schools resulted with the work of William of Ockham in the rise and dominance of the universities by Nominalism and with it a shift in anthropology from an understanding of the relations between humanity and God no longer in terms of teleological nature but in voluntaristic, legal categories. At the same time a continuation of the development of the natural sciences, stimulated by Aristotelianism, but taking the Pythagorean-Platonic road of mathematization, began to *objectify* the physical universe and human body as the realm of mechanical forces. Over against this, the spirituality of the *devotio moderna* and the Rhineland mystics promoted a heightened awareness of human *subjectivity.* Thus Platonic dualism was taking on a new form as the subject-object opposition.

Fourth, the Renaissance saw a renewal of the Platonic tradition outside the universities, and a much more favorable attitude toward the human body as the expression of the inner spirit; but it achieved this only by an idealization of the body through art.

Fifth, the Reformation, reacting against this idealization as if it were anti-Christian, achieved a new synthesis of the various tendencies present in late medieval thought and the Renaissance rhetorical mode, to produce a Theology of the Word, which minimized the traditional sacramental system. The physical world and the human body thus no longer appeared as sacraments through their sensuous beauty revealing the spiritual world, but were desacralized. The whole visible order of bodily existence and the physical world studied by science thus began to lose their religious significance and were assigned to the secular sphere, while religion became ever more private, and subjective.

Sixth, the Counter-Reformation in opposition to Protestant tendencies reinforced the Catholic sacramental system, but, remaining within a voluntarist theology, it understood the sacraments not contemplatively but as a means of moral discipline and social control, as became evident in the irresolvable controversies over grace and the "moral systems." These tensions within Catholic culture conceived as a balance of power were expressed in the rich art of the Baroque which affirmed the human body but chiefly as a manifestation of power and of the will to conquer and control.

186

It was in this world of a Christianity conceived in terms of power that the terrible religious wars of the seventeenth century were fought, wars that gave rise to the new religion of Humanism which Christians were shocked to see arising within their own territory.

Notes

1. See Henri De Lubac, *Histoire et Ésprit: L'intelligence de l'Écriture d'apres Origène* (Paris: Aubier, 1950), for a corrective of the oversimplified dichotomy often set up between these two schools, and J. Emile Pfister, S.D., *St. Gregory of Nyssa: Biblical Exegete* (Woodstock, MD.: Woodstock College, 1964, abstract of diss.) with its bibliography.

2. See Walter Böhm, ed., *Joannes Philoponus: Christliche Naturwissenschaft im Ausklang der Antike, Vörlaufer der modern Physik, Wissenschaft und Bibel: Ausgewahlt Schriften,* edited, translated and commented (Paderborn: Schönigh, 1967).

3. F. Dolger, "Byzantine Literature", c. xxvii of *The Cambridge Medieval History* (Cambridge: Cambridge U. Press), vol. 4, pt. 2, pp. 207-264, p. 244.

4. See references in Chapter 4, note 65.

5. Basil Tatakis, *La philosophie byzantine,* 2nd ed. (Supplementary fascicle no. 11 to Emile Brehier, *Histoire de le Philosophie)* (Paris: Press Universitaires de France, 1949), and Klaus Oehler, *Antike Philosophie und Byzantisches, Mittelalter* (Munich: C. H. Beck, 1969), especially pp. 272-286, on the influence of Aristotle in Byzantium.

6. On Pletho see F. Masai, *Pléthon et la Platonisme de Mistra* (Paris, 1956); Deno J. Geankopolis, *Greek Scholars in Venice* (Cambridge: Harvard University Press, 1962), pp. 85-86; p. 327 ff. and Steven Runciman, *The Last Byzantine Renaissance* (Cambridge: Cambridge University Press, 1970), pp. 77-80. Pletho's *De Platonicae et Aristotelicae philosophiae differentia* is printed in PG 160, cols. 889-934.

7. See Étienne Gilson, *History of Christian Philosophy in the Middle Ages* (New York: Random House, 1954), pp. 235-245; also Daniel A. Callus, "Introduction of Aristotelian Learning to Oxford," *Proceedings of the British Academy,* vol. 29, 29-155, and Paul Moraux, *D'Aristote à Bessarion. Trois exposés sur l'historie et la transmission de l'aristotélisme grec,* Conferences Charles De Koninck, No. 2 (Quebec: Laval University Int'l. Book Service, 1970). For the role of the Arabs see Frances E. Peters, *Aristotle and the Arabs* (New York: New York University, 1968), and Majid Fakhry, *A History of Islamic Philosophy* (New York: Columbia University, 1970).

8. *Aristotle in the West: The Origins of Latin Aristotelianism* (Louvain: E. Nauwelaerts, 1955), and *La philosophie au xiii^{e}siècle* (Louvain: Publications Universitaires, 1966).

9. Gilson, *History* (note 7 above), pp. 250-58. The *Opera Omnia* of William of Auvergne (Paris, 1674) are available in a Minerva Reprint, Frankfurt-am-Main, 1974; note especially vol. 2 Supplementum, *II De Anima,* Pars 24, "Quid corpus animae profit?" pp. 150-152.

10. See A. C. Crombie, *Robert Grosseteste and the Origins of Experimental Science, 1100-1700* (Oxford: Clarendon Press, 1953) and Daniel A. Callus, *"Robert Grosseteste as Scholar"* in *Robert Grosseteste: Scholar and Bishop: Essays in Commemoration of the Seventh Centenary of his Death* (Oxford: Clarendon Press, 1955), pp. 1-69, and James McEvoy, *The Philosophy of Robert Grosseteste* (Oxford: Clarendon Press, 1982), especially the conclusion pp. 450 f.

11. The major portion of Roger Bacon's works are available in Robert Steele et al. ed., *Opera hactenus inedita Rogeri Baconi,* 16 vols. (Oxford: Clarendon Press, 1940-1941). On his life and works see A. G. Little, "On Roger Bacon's Life and Works in collection edited by him, *Roger Bacon: Essays Contributed by Various Writers on the Occasion of the Commemoration of the Seventh Centenary of His Birth* (Oxford: Oxford University Press, 1914); Stewart C. Easton, *Roger Bacon and His Search for a Universal Science* (New York: Columbia University, 1952), and Thomas Crowley, *Roger Bacon: The Problem of the Soul in His Commentaries* (Louvain, Éd. de l'Institute Superieur de philosophie, 1950). An evaluation of his scientific achievements will be found in A. C. Crombie, *Medieval and Early Modern Science,* 2 vols. (Garden City, N.Y., 1959). The *Opera Omnia* of St. Albert the Great are

available in the old edition by Auguste Borgnet, 38 vols. (Paris: Vives, 1890-99), and are being critically edited in the Cologne Edition by the Institutum Alberti Magni Coloniense (Münster im Westfallen: Aschendorff, 1951 seq.). On his life and works see James A. Weisheipl in the collection edited by him, *Albertus Magnus and The Sciences: Commemorative Essays*, 1980 (Toronto: Pont. Inst. of Medieval Studies, 1980), pp. 13-52, and on his notion of science my essay, "St. Albert and the Nature of Natural Science", *ibid.*, pp. 73-102, An up to date bibliography is on pp. 585-616. See also James A. Weisheipl, "Albertus Magnus and the Oxford Platonists," *Proceedings of the American Catholic Philosophical Association* (Washington, D.C.: The Catholic University of America, 1958), pp. 124-139.

12. Martin Grabmann, "Der Einflus Alberts des Grossen auf das mittelalterlichen Geistesleben," in *Mittelalterliche Geistesleben* (Münster: Max Hueber, 1926), Bd. II, pp. 324-412, and W. Stammler, "Albert der Grosse und die deutsche Volksfrömmigkeit des Mittelalters," in *Freiburg Zeitschrift der Philosophie und Theologie,* 3 (1956): 287-319. On his philosophical influence see G. Meersseman, ed., *Geschichte des Albertismus II*, (Institutum Historicum Fr. Praedicatorum, Rome, Paris: R. Haloua, 1933; Bruges: Beyaert, Rome, 1935).

13. Cornelio Fabro, C.P.S., *La nozione metafisica di participazione secondo S. Tommaso d'Aquino* (Turin: Societa Internazionale, 1950), "L'obscurassement de l'esse dans l'école thomiste", *Revue Thomiste,* 63 (1958): 443-472; *Participation et Causalité selon S. Thomas d'Aquin* (Louvain: University of Louvain, 1961); "Dall' 'essere' di Aristotele all' 'esse' di S. Tommaso" in *Mélanges offerts a Étienne Gilson* (Toronto: Pont. Institute of Medieval Studies, 1959), pp. 227-247; "The Transcendentality of 'Ens-Esse' and the Grounds of Metaphysics,' *International Philosophical Quarterly* 6 (1966), 389-427; "The Problem of Being and the Destiny of Man," *ibid.*, 1966, 407-436; "The Intensive Hermeneutics of Thomistic Philosophy: The Notion of Participation," *Revue of Metaphysics*, 27 (1973-74), 449-491. Also G. Lindbeck, "Participation and Existence in the Interpretation of St. Thomas, *Franciscan Studies* 17 (1957), 1-22, 107-125, and Sister M. Annice, "Historical Sketch of the Theory of Participation," *New Scholasticism*, 26 (1952), 49-79. The most authoritative work on the life and dating of the works of Aquinas is James A. Weisheipl O.P., *Friar Thomas D'Aquino* (Garden City, N.Y.: Doubleday, 1974); the best introduction to reading him is M. -D. Chenu O.P., *Toward Understanding St. Thomas* (Chicago: Regnery, 1964). On the present topic see N. A. Luyten O.P., ed., *L'anthropologie de St. Thomas* (Fribourg: Publications Universities, 1974).

14. See my "Aristotle's Sluggish Earth: The Problematics of the *De Caelo*", *New Scholasticism,* 32 (Jan, 1958), 1-31 and (April, 1958), 202-234, for discussion and references. On the history of this problem see Leon Baudry, *Problème de l'origine et de l'éternité du monde dans la philosophie grecque de Platon à l'ère chrétienne* (Thesis, Paris, 1931).

15. Étienne Gilson, *History of Christian Philosophy* (note 7 above), pp. 368-383, especially the summary on pp. 382 f., on the centrality of the act of existence in the thought of Aquinas.

16. Arthur Little, S.J., *The Platonic Heritage of Thomism* (Dublin: Golden Eagle Books, 1949), reviewed by W. Norris Clarke, S.J., *Review of Metaphysics,* 8 (1954), 105-124; Robert J. Henle S.J., *St. Thomas and Platonism* (The Hague: Martinus Nijhoff, 1956); K. Kremer, *Die neuplatonische Seinsphilosophie und Ihre Wirkung auf Thomas von Aquin* (Leiden: Brill, 1971) (often inaccurate); Cornelio Fabro, "Platonism, Neo-Platonism, Thomism," *New Scholasticism,* 44 (1970), 69-100; James A. Weisheipl O.P., "Thomas' Evaluation of Plato and Aristotle," *ibid.*, 48 (1974), 100-124; J. Moreau, "La platonisme dans la Somme Théologique" in *Tommaso D'Aquino nel suo settimo centenario* (Naples: Ed. Domenicane Italiane, 1974) 1: 238-247; A. Von Ivanka, "St. Thomas platonisant," *ibid.* 1, 256-260 (Aquinas platonizes in ontology but not in epistemology); M. D. Phillipe, "Analyse de l'être chez saint Thomas", *ibid.*, 6, 9-28 (Aquinas is Aristotelian even in ontology).

17. On the present state of our knowledge of Aristotle's lost works see Anton-Hermann Chroust, *Aristotle: New light on his life and some of his lost works* (Notre Dame, Ind: Notre Dame University, 1973) vol 2; Paul Moraux, *Der Aristotelismus bei den Griechen* (Berlin: de Gruyter, 1973), p. 3-44, and Jonathan Barnes, M. Schofield and R. Sorabji, *Aristotle: A Selective Bibliography* (Oxford: Oxford University Press, 1977).

18. See the works of Van Steenberghen, note 8 above for a full account of "Latin Averroism."

188

19. "The soul is part of human nature, and hence, although it can exist apart from the body, because it still retains the nature of a part which can be reunited [with the body], it can not be called an individual substance which is a hypostatis or first substance; just as neither can a hand or any other part of man; and thus neither the name nor the definition of person pertains to it." *Summa Theologiae* I, q.29, a.1 ad 5, my translation. Cf. also I, q.75, a.4 ad 2; *Sent.*III, d. 5, 3 ad 2; *Pot.* 9, 2 ad 14.

20. "Only God illuminates humans, enkindling in them the natural light of the active intellect, and still further the light of grace and glory, but the active intellect illumes the images [in the human mind] as a light received from God," *Quaestio Disputata de Spiritualibus Creaturis*, a.10 ad 1, my translation.

21. On the cosmology of Aquinas see John H. Wright, *The Order of the Universe in the Theology of St. Thomas Aquinas* (Rome: Gregorian University, 1957), and Thomas Litt, *Les corps céleste dans l'univers de Saint Thomas d'Aquin*, Philosophes Médiévaux, 7 (Louvain: Publications Universitaires, 1963). Aquinas' views on the Prime Mover are often inaccurately presented as a result of a failure to consult his Aristotelian commentaries (cf. Weisheipl, *Friar Thomas* (note 13 above), pp. 281 ff. for a defense of the importance of these commentaries in the development of Aquinas' thought. For an accurate analysis see Vincent E. Smith, *The General Science of Nature* (Milwaukee: Bruce, 1958) pp. 336-383. Aquinas correctly understands the Prime Mover not only as the final cause of the motion in the universe but also as the efficient cause of its motion and its existence. Many modern commentators on Aristotle wrongly separate his treatment of the question in the *Metaphysics* made primarily in terms of finality from the presupposed treatment in the *Physics* in terms of efficiency. For a defense of this reading of Aristotle as against the common opinion of modern authors see Venant Cauchy, "La causalité divine chez Aristote,", *Mélanges à la memoire de Charles De Koninck* (Quebec: Laval University, 1965), pp. 103-114.

22. John Capreolus was born at Rodez, France. His great work, published in parts between 1408 and 1433 is *Libri defensionem theologiae divi Thomae de Aquino* and its current edition by C. Puban and T. Pègues was published at Tours, 1900-1907. See M. Grabmann, "Johannes Capreolus O.P., der 'Princeps Thomistarum' (d. 1444) und sein stellung in der Geschichte der Thomisterschule", *Mittelalterliches Geistesleben* (Münster, Max Huber, 1956), 3: 370-410. On the followers of Aquinas who preceded Capreolus see Frederick J. Roensch, *The Early Thomistic School* (Dubuque, Iowa: Priory Press, 1964).

23. See Gilson, *History of Christian Philosophy* (note 7 above), pp. 521-527, on these Averroists of whom John of Jandun (d. 1328), Marsilius of Padua (d. before 1343), Paul of Venice (d. 1429), Cajetan of Thien (d. 1463), Jacob Zabarella (d. 1589) were the chief ones. Pietro Pomponazzi (d. 1525) was chief of the "Alexandrists" who were also Aristotelians but favored the commentator Alexander of Aphrodisias rather than Averroes.

24. For a bibliography of the works of Annaliese Maier the chief expert on the "impetus theory" see her posthumous *Ausgehendes Mittelalter* (Rome: Edizioni di Storia e Letteratura, 1977) vol. 3, pp. 617-626, prepared by A. P. Bagliani. Cf. also E. J. Dijksterhuis, *The Mechanization of the World Picture* (Oxford: Clarendon Press, 1961).

25. See James A. Weisheipl, O.P., *The Development of Physical Theory in the Middle Ages* (New York: Sheed and Ward, 1959), p. 81.

26. *Ibid.*, pp. 62-88.

27. William A. Wallace, O.P., *The Scientific Method of Theodoric of Freiberg* (Fribourg, Swizerland: The University Press, 1959).

28. On Bonaventure see Étienne Gilson, *The Philosophy of St. Bonaventure*, rev. ed. (Paterson, N.J.: St. Anthony Guild Press, 1965); Ewert H. Cousins, *Bonaventure and the Coincidence of Opposites* (Chicago: Franciscan Herald Press, 1978) (Chapter 8, pp. 229-268 has a particularly interesting discussion of the relation of B.'s thought to modern philosophy); Joseph Ratzinger, *The Theology of History of St. B.* (Chicago: Franciscan Herald Press, 1971); and John Francis Quinn, *The Historical Constitution of St. Bonaventure's Philosophy* (Toronto: Pontifical Institute of Medieval Studies, 1973). Quinn (p. 1-100) reviews in detail the various interpretations of Bonaventure's philosophy given by Mandonnet, De Wulf, Gilson, Van Steenberghen, Patrice Robert, Ratzinger, van der Laan and others. He also makes careful comparison of Bonaventure with Aquinas and shows that although their systems are very

different, their conclusions tend to converge. He maintains that B. holds (contrary to Gilson's interpretation) for a formal distinction of philosophy from theology and that his philosophy is that of a Christian but is not "Christian philosophy." He also denies that B.'s thought can be reduced to "Aristotelizing Augustinianism" or to "Platonizing Aristotelianism" (Van Steenberghen). Rather it is a unique philosophy in its own right which makes appropriate use of many sources, of which Aristotle and Augustine are major. On the fundamental question of epistemology, Quinn sums up by saying, "Although the influence of Augustine is predominant in St. Bonaventure's solution to the problem, and the influence of Aristotle is predominant in the solution of St. Thomas, even so, St. Bonaventure's solution owes a lot to Aristotle, as the solution of St. Thomas owes much to St. Augustine."

In the final analysis, however, the two solutions to the problem depend essentially on the different principles of metaphysics and of knowledge supporting the personal conceptions by St. Bonaventure and St. Thomas of the universe and its relation to God" (p. 886). But are not their metaphysics determined by their epistemological starting points? On the degree to which Bonaventure assimilated Aristotelianism see Léon Elders, "Les citations d'Aristote dans le 'Commentaire sur les Sentences' de S.B." in A. Pompei, ed., *San Bonaventura Maestra di Vita Franciscana e di Sapienza Cristiana* (Rome: Pontificia Facolta' Teologica "San Bonaventura": 1976) 1: 831-842; Angelo Marchen, "L'Atteggiamento di S.B. di fronte al pensiero di Aristotele," *ibid*. pp. 843-859; and José Oroz-Reta, "Aristotelismo y Augustinismo en la doctrina de S.B." *ibid*., pp. 861-881. Note that Bonaventure did *not* accept the Platonic *sōma sēma* doctrine: "Since we see that the soul, no matter how good, does not want to be separated from the body . . . which would be strange if it did not also have a natural inclination to the body, as to its companion, *not as to its prison*." II Sent. 18, q.2, Conclusion 3, *Opera Omnia* (Quaracchi, Collegio S. Bonaventurae, 1185); my translation.

29. See Étienne Gilson, *Jean Duns Scot: Introduction à ses positions fondamentales* (Paris: J. Vrin, 1952); Efrem Bettoni, *Duns Scotus: The Basic Principles of His Philosophy* (Washington, D.C.: Catholic University of America, 1961). His works are published in the critical edition edited by C. Balić, *Joannis Duns Scoti Opera Omnia,* Commissio Scotistica (Vatican: Vatican Press, 1950).

30. On Scotus' theory of knowledge see Gilson, *Jean Duns Scot* (note 29 above), pp. 511-573; Bettoni, *Duns Scotus* (note 29 above), pp. 93-131, and Cyril L. Shircel, O.F.M., *The Univocity of the Concept of Being in the Philosophy of Duns Scotus* (Washington, D.C.: Catholic University of America, 1942),

31. See Gilson, *Jean Duns Scot* (note 29 above), pp. 545-555; Bettoni, *Duns Scotus* (note 29 above), pp. 121-123, and Sebastian J. Day, *Intuitive Cognition, A Key to the Significance of the Later Scholastics* (Saint Bonaventure, N.Y.: The Franciscan Institute, 1947). Day (whom I have followed in my account) differs somewhat from Gilson's interpretation. According to Gilson for Scotus this intuition of the singular is a knowledge of its existence not of its singularity as such in the present state of the human intellect. Thus Scotus is not so far from the view of Aquinas as it might seem, since both deny a scientific knowledge of the singular as such. Nevertheless, the concern of Scotus to maintain the essential independence of the intellect from sense remains a fundamental difference in their two approaches, a difference indicative of Scotus' fidelity to the Augustinian and Platonic tradition. On Aquinas' insistence that we have intellectual knowledge of singulars "only indirectly and as it were by a certain reflexion on the phantasm," see *Summa Theologiae* I, q.86, a.1 and *Quaestio Disputata De Anima,* a.20; also Rudolf Allers, "The Intellectual Cognition of Particulars", *Thomist* 3 (1941): 95-163, and Kenneth C. Clatterbaugh, "Individuation in the Ontology of Duns Scotus," *Franciscan Studies* 32 (1972): 65-73.

32. See Gilson, *Jean Duns Scotus* (note 29 above), 432-444; Bettoni, *Duns Scotus* (note 29 above), pp. 48-52.

33. See A.C. Pegis, *St. Thomas and the Problem of the Soul in the Thirteenth Century* (Toronto: Pontifical Institute of Medieval Studies, 1934), and *At the Origins of the Thomistic Notion of Man,* The Saint Augustine Lecture, 1962 (New York: Macmillan, 1962) on the way in which Aquinas interpreted and adapted Aristotle's doctrine on the separated intellect.

34. See Shircel, *Univocity* (note 30 above), pp. 57-73, 521-522; Bettoni, *Duns Scotus* (note 29 above), p. 27-46. The latter sums up as follows: "The pre-occupation of guaranteeing man's capacity to arrive at God inspires the Scotistic theory of the proper object of the human mind, just as it inspired all the theories of intellectual illumination worked out by the thirteenth-century Augustinians. This theory, strictly linked up to the doctrine of univocity [of being] is, in my opinion, Duns Scotus' most significant doctrine. It is this, first, because in it we see the solution of the controversy between the Augustinian and Aristotelian schools, which is the scope of all his philosophic endeavors; secondly, because this doctrine offers the key to a full understanding of the Scotistic synthesis both in its spirit and in its intellectual achievement." p. 46.

35. Firmata est autem illis legibus [sapientiae] quod intellectus noster non intelligat pro statu isto nisi illa quorum species relucent in phantasmate, et hoc sive propter poenam peccati originalis, sive propter naturalem concordiantiam potentiarum animae in operando, secundum quod videmus quod potentia superior operatur circa idem circa quod inferior, si utraque habeant operationem perfectam. Et de facto ita est in nobis, quod quodcumque universale intelligimus eius singularis actu phantasiamur. Ista tamen concordiantia, quae est de facto pro statu isto, non est de natura intellectus unde intellectus est, nec etiam unde in corpore; quia tunc in corpore glorioso necessario haberet similem concordiantiam, quod falsum est. "It has been established by those laws of Wisdom, that our intellect in the present state may know only those things whose species are illuminated in a sensible image, and this is so either as a punishment of original sin, or because of the natural concord of the faculties of the soul in functioning, according to the way we see a superior faculty functions in regard to the same object as does the inferior, if both together are to function perfectly. So it is *de facto* in us, so that whatever universal we know we must also in that same act imagine a singular instance. Nevertheless, that concord which *de facto* exists in our present state, is not due to the nature of our intellect as it is an intellect, nor even because it is in the body; for then in the glorified body there would have to be the same kind of concord, which is false." *Opus Oxiensis Ordinatio* I, dist. 3, pars 1, q.3, a.4, no. 187: *Opera Omnia,* tom. 1. pp. 351-352.

36. See Bernardine M. Bonasea, O.F.M., "Duns Scotus' Voluntarism" in John K. Ryan and B.M. Bonasea, eds., *John Duns Scotus 1265-1965* (Washington, D.C.: Catholic University of America), 1965) pp. 83-121; and Robert Prentice O.F.M., "The Voluntarism of Duns Scotus seen in his Comparison of the Intellect and Will," *Franciscan Studies* 28 (1968): 63-104.

37. See Josef Pieper, *Happiness and Contemplation* (New York: Pantheon, 1958) and *About Love* (Chicago: Franciscan Herald Press, 1974) for an attractive presentation of the Thomistic position. Also in Thomas Gilby, O.P., ed., Aquinas' *Summa Theologiae,* (New York: McGraw Hill, 1969), see Gilby's Appendices 10, vol. 1, and 4 and 5 in vol. 10, for a discussion of the relation of the Thomist and Franciscan traditions on this problem.

38. On Gerson, see Étienne Gilson, *History of Christian Philosophy* (note 7 above), pp. 529-545 and Steven E. Ozment, *Homo Spiritualis: A Comparative Study of the Anthropology of Jean Gerson, Johannes Tauler and Martin Luther (1509-16)* (Leiden: E.J. Brill, 1969).

39. On Eckhart see the excellent introduction on his life and theology in Edmund Colledge, O.S.A. and Bernard McGinn, *Meister Eckhart: The Essential Sermons, Commentaries, Treatises and Defense,* The Classics of Western Spirituality, (New York: Paulist Press, 1981) with its bibliography p. 349 ff. for editions and translations. See also the articles from the Eckhart Symposium, *The Thomist* 42 (April, 1978) including my article "Three Strands in the Thought of Eckhart the Scholastic Theologian", pp. 226-239, and the extensive bibliography by Thomas O'Meara, pp. 171-182. Of special value in understanding Eckhart's thought are Bernard J. Muller-Thym, *The Establishment of the University of Being in the Doctrine of Meister Eckhart of Hockheim* (New York: Sheed and Ward, 1939); Vladimir Lossky, *Théologie Négative en Connaissance de Dieu chez Mâitre Eckhart* (Paris: Vrin, 1960) and Reiner Schürman, *Meister Eckhart: Mystic and Philosopher* (Bloomington, Ind.: Indiana University Press, 1978); C.F. Kelley, *Meister Eckhart on Divine Knowledge* (New Haven: Yale University Press, 1967). John D. Caputo, *The Mystical Element in Heidegger's Thought* (Athens, Ohio: Ohio University Press, 1978) makes a very helpful comparison with

certain trends in modern philosophy. A very full survey and bibliography of recent literature is Wolfram Malte Fuse, *Mystik als Erkenntnis: Kritischen Studien zur Meister-Eckhart-Forschung* (Bonn: Bouvier, 1981).

40. On Suso see the introduction to *The Exemplar,* edited by Nicholas Heller, translated by Sister M. Ann Edward, O.P., 2 vols. (Dubuque: The Priory Press, 1962), vol. 1, and the critical edition by Dominikus Planser, O.P. of the *Horologium Sapientiae* (Instituto Storico Domenicano: Rome, 1937).

On Tauler see *Spiritual Conferences,* translated and edited by Eric Colledge and Sister M. Jane, O.P. (St. Louis: Herder Book Co., 1961) with bibliography pp. ix-xii. A new translation with introduction is to appear in *The Classics of Western Spirituality.* On the influence of Tauler on Luther see Steven E. Ozment, *Homo Spiritualis* (note 38 above); also his "Homo Viator: Luther and Late Medieval Theology" in the collection edited by him, *The Reformation in Medieval Perspective* (Chicago: Quadrangle Books, 1971), and in the same volume Heiko Oberman, "Simul Gemitus et Raptus: Luther and Mysticism", pp. 219-252, and Ozment's "German Mysticism and Protestantism", *Thomist 42* (April, 1978): 259-280. Similar discussions will be found in Ivar Asheim, ed. *The Church, Mysticism, Sanctification and the Natural in Luther's Thought,* Third International Congress on Luther Research, Jarvenpaa, Finland, 1966 (Philadelphia: Fortress Press, 1967). Lutheran writers are anxious to emphasize Luther's originality, which is really not in question. The most evident element of this originality for me is his effort to free the exposition of Christian spirituality from its Neo-Platonic formulations for the sake of a more biblical statement. The continuity with late medieval spirituality, however, is also a historical fact which helps to explain Luther's tendency to reject the notion of sanctification as an *intrinsic* transformation and elevation of the Christian, thus excluding the whole Eastern tradition of *theosis* which the scholastics of the thirteenth century had labored to assimilate. This tendency is already present in Eckhart and Tauler. On Ruysbroek see E. Colledge, "Ruysbroek, Jan van B1." in *New Catholic Encyclopedia,* 12: 763-765 and Colledge's translation and commentary *The Spiritual Espousals,* (New York: Harper, 1953).

41. New translations with helpful introductions are available in the Paulist Press *The Classics of Western Spirituality* series, *The Theologia Germanica of Martin Luther,* trans. and presented by Bengt Hoffman (New York: 1980); *The Cloud of Unknowing,* trans. and presented by James Walsh (New York: 1981). Another translation of *The Cloud* is by Clifton Wolter (Hammondsworth/New York: Penguin Books, 1978). William Johnson, *The Mysticism of the Cloud of Unknowing* (New York: Harper's, 1961) provides a helpful commentary.

42. See *Catherine of Siena, The Dialogue,* introduction and trans. by Suzanne Noffke, O.P., The Classics of Western Spirituality (New York: Paulist Press, 1980) with a select bibliography pp. 367-320; *The Prayers of Catherine of Siena* (same editor-translator and publisher, 1983); *Le lettre di S. Caterina da Siena,* ed. by Pietro Misciatelli (Siena: Giuntini et Bentivoglio, 1913-1922) 6 vols.; *Epistolario di Santa Caterina da Siena,* ed. by E.D. Theseider (Rome: Tipografia del Senato, 1940) (a critical edition never completed). On Dante see Bruno Nardi, *Saggi di filosofia dantesca* (Milan: Società Dante Alighieri, 1930) and Étienne Gilson, *Dante the Philosopher* (London-New York: Sheed and Ward, 1949).

43. David Knowles, "A characteristic of the mental climate of the Fourteenth Century," in *Mélanges offerts à Étienne Gilson* (Toronto: Pontifical Institute of Medieval Studies, 1959), pp. 315-325. Also Georges de Lagarde, *La naissance de l' esprit laique au déclin du moyen age,* rev. ed. (Louvain: Nauwelaerts, 1956-1970) 5 vols. and Gordon Leff, *The Dissolution of the Medieval Outlook* (New York: Harper and Row, 1976).

44. Leon Baudry, *Guillaume d'Occam, sa vie, ses oeuvres, ses idées sociales et politique* (Paris: J.Vrin, 1950) and Gordon Leff, *William of Ockham* (Manchester: Manchester University, 1975). See also the Introduction of Philotheus Boehner, O.F.M. ed., *William of Ockham: Philosophical Writings* (Edinburgh: Thomas Nelson and Sons, 1957), pp. ix-xlix. Boehner emphasizes that Scotus initiated among the Franciscans a highly varied movement of independent thinkers of whom Peter Aureoli (d. 1321) and Ockham were the chief.

45. St. Thomas Aquinas, *In libros Posteriorum Analyticorum Aristotelis* II, lect. 13, n. 7. See also the excellent explanation of this passage by T. Zigliara in the Leonine edition of Aquinas, tom. I, p. 375 a-b. and my article, "Does Natural Science Attain Nature or Only

the Phenomena?" in Vincent E. Smith, ed., *The Philosophy of Physics* (Jamaica, N.Y.: St. John's University, 1961), pp. 63-82.

46. Julius R. Weinberg, *Nicholas of Autrecourt* (New York: Greenwood Press, 1969) (originally Princeton University Press, 1948). Weinberg notes that Nicholas anticipated much of Hume's argumentation against the principle of causality and shows that it is possible that Hume learned of these arguments through Malebranche (p. 228).

47. See Gordon Leff, *William of Ockham* (note 44 above), pp. 436-468, on the radical contingency of creatures, summarized in Ockham's saying, "Ontologically, however, man's capacity for laughter is as conditional upon God's will as Peter's salvation" (*Quodlibet* vi, q. 2).

48. On this distinction see Heiko A. Oberman, *The Harvest of Medieval Theology: Gabriel Biel* and *and Late Medieval Nominalism* (Cambridge: Harvard University Press, 1963), pp. 30-56, definitions on p. 473; and Leff, *William of Ockham* (note 44 above), pp. 15-18, and entry "God, absolute power, ordained power" in index on p. 658.

49. David W. Clark, "Voluntarism and Rationalism in the Ethics of Ockham, *Franciscan Studies* 31 (1971): 72-87 protests against classifying Ockham's ethics as a form of voluntarism, because it also contains a strong rationalistic element in that Ockham admits a natural law. However, Clark also admits that this natural law consists only in very broad principles of a purely formal character. Since for Ockham the ultimate root of such principles is the will of God which is free to alter the concrete rules of moral behavior, it seems to me Ockham's position is fairly classified as voluntarism, even a radical voluntarism. Clark and Boehner, *Philosophical Writings,* (note 44 above) p. xlix, both admit that Ockham taught that God by his absolute power could command us to hate Him and that to do so would then be an act of charity! Cf. *Sent.* II, q.19, III, q.12; and *Quodlibet* II, q.13.

On the other hand, the common charge that the Nominalists were Pelagians (*prima facie* improbable considering Ockham's extreme Franciscan Augustinianism) is well refuted by Francis Clark, "A New Appraisal of Late Medieval Theology", *Gregorianum* 46 (1965), 733-65, a strongly critical review of Heiko Oberman's work cited in note 48 above. The purpose of the Nominalists' famous *facere quod in se est* (God will grant grace to those who do what they can to obey Him) was not to support either a Pelagian or Semi-Pelagian claim that man can wholly or partially save himself but to exalt the gracious mercy of an utterly free God who has willed to reward His creatures not only for good actions that deserve reward in justice (*de condigno*), but even for actions which fall short of the Law but which have at least some tendency to obedience to the Law and thus can be *fittingly* encouraged (*de congruo*). They saw this as pastorally encouraging and as providing an answer to those who accused God of arbitrariness in granting His graces; a difficulty which had become acute because of their extreme emphasis on the Divine Freedom.

What is most significant in this controversy is the theological shift from thinking in ontological terms about grace as a transformation of the human person elevating it from the natural order to a share in the supernatural life of God or as restoration of the divine image (as both the Eastern and Western theological tradition had done right up through the High Middle Ages) to an understanding of grace as an extrinsic, legal (forensic) relation between the sovereign God and subject man, dependent simply on the free favor of God who could either grant His grace unconditionally or in certain conditions which he himself had freely established.

50. On the scholastic theories of grace and the infused virtues, see Odon Lottin, *Psychologie et Moral au XIIe ET XIIIe siècles,* 6 vols (Gembloux, Belgium: J. Duculot, 1957-1960), vol. I, pp. 520-523; vol. 3, pp. 197-252 and 460-535. Also Henri Rondet, *The Grace of Christ: A Brief History of the Theology of Grace* (Westminster, Md.: Newman, 1967), pp. 58-62, 72 f., 87 note 103, 261-264, 279 f., 292 f., 310 f., 365-377.

51. "Federal" or "covenant theology" was developed in the Calvinist tradition as a way to stress human ethical responsibility within the Reformation doctrine of justification *sola gratia.* Saved by grace and not by works, sinful human beings were still bound by covenants with God to obey his law, and God, in turn, was "bound" by his covenant agreement to show favor to the obedient. See references in note 122 below.

52. See Claude Tresmontant, *Christian Metaphysics* (New York: Sheed and Ward, 1965), for an extensive study of the philosophical themes contained in the Scriptures. Note, however,

that in this and his *A Study of Hebrew Thought* (New York: Desclée, 1960), Tresmontant exaggerates the contrast between biblical and Greek thought.

53. See C.W. Previte-Orton, *The Shorter Cambridge Medieval History* (Cambridge: Cambridge University Press), 2, 940-952. On the question of the continuity or discontinuity of the Renaissance relative to the Middle Ages see the essays, "The Concept of the Renaissance" by Federico Chabod, in his *Machiavelli and the Renaissance*. (Cambridge, Mass: Harvard University Press, 1960), pp. 149-200, with the extensive bibliography on the controversy, pp. 201-247.

54. On the problem of Cajetan's departures from the positions of Aquinas see Emilia Verga, "L'immortalità dell' anima nel pensiero de Cardinale Gaetano", *Rivista di filosofia neo-scolastica,* 27 (1935), pp. 21-46, especially pp. 33-35; T.A. Collins, O.P., *American Ecclesiastical Review,* 128 (1953), 90-100, and Étienne Gilson, "Cajetan et l'humanism théologique," *Archives d'Histoire Doctrinal et Littéraire du Moyen Age,* 22 (1955), 113-136.

55. See Laura Martines, *The Social World of the Florentine Humanists, 1390-1460* (Princeton, N.J.: Princeton University, 1963).

56. See M. Maylender, *Storia delle accademie d'Italia,* 5 vols. (Bologna: 1926-1930), and F.A. Yates, *The French Academies of the Sixteenth Century* (London: 1947), and also John E. Sandys, *A Short History of Classical Scholarship* (Cambridge: Cambridge University Press, 1915), pp. 189-192.

57. As Petrarch, the father of the humanists, said, "I certainly am not a Ciceronian, or a Platonist, but a Christian", M.L. Bush, *Renaissance, Reformation and The Outer World* (Chapter 3, note 12). See the same note for other references.

58. *Les confessions de saint Augustin dans la tradition littéraire: Antécédents et Posterité* (Paris: Études Augustiniennes, 1963). See also William J. Bowsma, "The Two Faces of Humanism: Stoicism and Augustinianism in Renaissance Thought" in Heiko A. Oberman with Thomas Brady, Jr., *Iternarium Italicum: The Profile of the Italian Renaissance in the Mirror of its European Transformations* (Leiden: E. J. Brill, 1975), pp. 3-60, who shows how for both Catholics and Protestants Stoicism was revived as a secular philosophy complementary to a renewed Augustinianism in theology. D. C. Allen, "The Rehabilitation of Epicurus and his Theory of Pleasure in the Early Renaissance," *Studies in Philology,* 41 (1944), 1-5; Eugene F. Rice, Jr. *The Renaissance Idea of Wisdom* (Cambridge, Mass.: Harvard University Press, 1958), and Allen G. Debus, *Man and Nature in the Renaissance* (Cambridge: Cambridge University Press, 1978), illustrate other aspects of the philosophical currents that influenced Renaissance Christianity.

59. See Edgar de Bruyne, *Études d'esthétique médiévale* (Brugge: De Tempel, 1946), 3 vols in one. See also the great works on theology and art of Hans Urs von Balthasar, *Herrlichkeit* 7 vols. (Einsiedeln: Johannes Verlag, 1961 —) and *Theodramatik,* 3 vols. (Einsiedeln: Johannes Verlag, 1976-1980) reviewed by Thomas F. O'Meara, O.P., "Of Art and Theology: Hans Urs von Balthasar's Systems", *Theological Studies,* 42 (1981) 272-276.

60. Ronald M. Steinberg, *Fra Girolamo Savonarola: Florentine Art and Renaissance Historiography* (Athens, Ohio: Ohio University Press, 1977), especially chapter 13, pp. 82 ff.

61. New York: Pantheon, 1956, pp. 3-29 and 308-347.

62. Paul Oskar Kristeller, *The Philosophy of Marsilio Ficino* (Gloucester, Mass.: Peter Smith, 1953), and *Medieval Aspects of Renaissance Learning,* especially 'Thomism and the Italian thought of the Renaissance," pp. 29-94.

63. Charles Trinkhaus, *In Our Image and Likeness: Humanity and Divinity in Italian Humanist Thought* (London: Constable, 1970), vol. 2, pp. 761-774.

64. *Ibid.,* vol. 1, pp. 192 f. See also the daring work of Leo Steinberg, *The Sexuality of Christ in Renaissance Art and in Modern Oblivion* (New York: Pantheon, 1984).

65. Charles de Tolnay, article, "Michaelangelo Buonarroti", *Encyclopedia of World Art,* vol. 9, pp. 862-914, in discussing the iconography of the Sistine Chapel (885-7) says that Jonah was chosen for its place over the "Last Judgment" because Jonah is a sign of the Resurrection. However, the fresco does not represent Jonah's escape from the whale, but the concluding incident of the book in which he complains to God because the vine which mercifully protected him from the sun has withered, and God reminds him that he also complained when God was merciful to the repentant Ninevites. Thus the direct meaning of the arrangement is that God's mercy is greater than His justice.

194

66. See Louis Bouyer, *Erasmus and the Humanist Experiment* (London: Geoffrey Chapman, 1959) for a nuanced evaluation of Erasmus' work and influence.

67. Daniel P. Walker, *The Ancient Theology: Studies in Christian Platonism from the Fifteenth to the Eighteenth Century* (London: Duckworth, 1970), note especially Chapter 2, "Savonarola and the Ancient Theology", pp. 42-62. On the artistic expression of this theology see Edgar Wind, *Pagan Mysteries in the Renaissance*, rev. ed., (London, Faber and Faber, 1968).

68. A. J. Festugière, O.P., *La révélation d'Hermès Trismégiste,* 4 vols. (Paris: Lecoffre, 1944-45).

69. *Ibid.*, vol. 2 "Le Dieu Cosmique."

70. See Kristeller, *Philosophy of Marsilio Ficino* (note 62 above); Ardis B. Collins, *The Secular is Sacred: Platonism and Thomism in Marsilio Ficino's Platonic Theology* (The Hague: Martinus Nijhoff, 1974), and Daniel P. Walker, *Spiritual and Demonic Magic from Ficino to Campanella* (London: University of London, 1958).

71. On Bruno see Francis A. Yates, *Giordano Bruno and the Hermetic Tradition* (Chicago: University of Chicago Press, 1964); on Campanella see Leon Blanchet, *Campanella* (Paris, 1920), reprinted by Bert Franklin, New York (no date); and Gisela Bock; *Thomas Campanella: Politisches Interesse und Philosophische Spekulation* (Tübingen: Max Niemeyer Verlag, 1974).

72. See Frederick Copleston, S.J., *A History of Philosophy* (Garden City, N.Y.: Doubleday Image Books, 1963), vol. 3, Part II, pp. 54-63.

73. See Weisheipl, *Development of Physical Theory* (note 25 above), pp. 72-88.

74. See John M. Randall, Jr., *The School of Padua and the Emergence of Modern Science* (Padua: Ed. Antenore, 1961); *The Career of Philosophy* (New York: Columbia University Press, 1962), vol. 1, pp. 256-307, and "Paduan Aristotelianism Reconsidered" in Edward A. Mahoney, *Philosophy and Humanism: Renaissance Essays in Honor of Paul Oskar Kristeller* (New York: Columbia University, 1976). The eminent physicist Carl Friedrich von Weizäcker sums up his opinion of the Galileo case as follows: "The late middle ages are in no way dark ages, they are a time of high culture, bristling with intellectual energy. They adopted Aristotle because of his concern about reality. But the main weakness of Aristotle was that he was too empirical. Therefore he could not achieve a mathematical theory of nature. Galileo took his great step in daring to describe the world as we do not experience it. He stated laws which in the form in which he stated them never hold in actual experience and which therefore cannot be verified by any single observation but is mathematically simple. Thus he opened the road to a mathematical analysis which decomposes the complexity of actual phenomena into single elements. The scientific experiment is different from everyday experience in being guided by a mathematical theory which poses a question and is able to interpret the answer. It thereby transforms the given "nature" into a manageable "reality." Aristotle wanted to preserve nature, to save the phenomena; his fault was that he made too much use of common sense. Galileo dissects nature, teaches us to produce new phenomena; and to strike against common sense with the help of mathematics." *The Relevance of Science* (London: Collins, 1946), p. 104. See also W. P. D. Wightman, *Science and the Renaissance*, 2 vols. (Edinburg-London, 1962).

75. Alexandre Koyré, *From the Closed World to the Infinite Universe* (Baltimore: John Hopkins University Press, 1957).

76. On the scientific value of Francis Bacon's work see A. C. Crombie, *Medieval and Early Modern Science* (Garden City, N.Y.: Doubleday Anchor Books, 1959), vol. 2, pp. 286-300.

77. *De Motu Cordis* was actually published in 1628. On Harvey see Gweneth Whitteridge, *William Harvey and the Circulation of the Blood* (New York: Elsevier, 1971); Sir Geoffrey Keynes, *The Life of William Harvey* (Oxford: Clarendon Press, 1966), and Walter Pagel, *Harvey's Biological Ideas* (Basel-New York: Karger, 1967).

78. William A. Wallace, O.P., *Galileo's Early Notebooks: The Physical Questions* (Notre Dame, Ind.: University of Notre Dame, Press, 1977).

79. See works of Randall cited in note 74 above.

80. On Pomponazzi see Frederick Copleston, S.J., *A History of Philosophy* (Garden City, N.Y.: Doubleday Image Books, 1953), vol. 3, Part II, pp. 27-31, who shows that Pomponazzi admitted that the human intelligence transcends matter, but considered that to demonstrate its immortality it was further necessary to prove that it was independent of the body in its acquisition of the materials of knowledge, a point no Aristotelian would want to admit; but which Aquinas for one did not consider a necessary premise in the proof. See also the introduction to the selection from Pomponazzi in E. Cassirer, P. O. Kristeller and J. J. Randall, eds., *The Renaissance Philosophy of Man* (Chicago: University of Chicago Press, 1948).

The condemnation in question is as follows: "We condemn and reject all who assert that the intellective soul is mortal, or that there is only one such soul for all humans, as well as those calling this into doubt, since that soul is not only truly *per se* and *essentialiter* the form of the human body, as is contained in the canon of our predecessor of happy memory Pope Clement V issued in the Council of Vienne (902 [48]), but it is truly immortal and multiplicable, multiplied, and to be multiplied individually for the multitude of bodies in which it is infused . . ." H. Denziger and A. Schönmetzer, *Enchiridion Symbolorum*, 32 ed. (Fribourg-Bresgau: Herder, 1963), 1440 [738] p. 353. While some theologians argue that although this condemns the denial of the supernatural immortality of the soul, it does not assert its natural immortality; this hardly agrees with the historical context of the canon, since Pomponazzi did not deny the supernatural immortality, but the demonstrability of the natural immortality.

81. See references in note 54 above.

82. Alexandre Koyré, *From The Closed World To The Infinite Universe* (note 75 above); Pauline Moffitt Watts, *Nicolaus Cusanus: A Fifteenth Century Vision of Man*, Studies in the History of Christian Thought, 30 (Leiden: Brill, 1982) (emphasizes how Cusanus synthesized scholastic, mystical, humanistic, and Platonist trends, with a stress on the divine transcendence and ineffability); Edmond Vansteenberghe, *Le cardinal Nicolas de Cues 1401-1464: l'action, la pensée* (Paris, 1920, reprint Minerva, Frankfurt am Main, 1963), and Jasper Hopkins, *A Concise Introduction to the Philosophy of Nicholas of Cusa* (Minneapolis: University of Minnesota Press, 1978), Introduction, pp. 1-43, bibliography, pp. 47-57, and an edition and translation of Cusa's *Trialogus de Possest* "On Actualized-Possibility".

83. Herbert Wackerzapp, *Die Einfluss Meister Eckharts auf die ersten philosophischen Schriften des Nikolaus von Kues 1440-1450*, Beiträge zur Geschichte der Philosophie und Theologie des Mittelalters (Münster, Westfalen: Aschendorff, 1962), Bd. 89, Heft 3.

84. See Richard A. Sayce, *The Essays of Montaigne: A Critical Exploration* (London: Weidenfeld and Nicolson, 1972) for an evaluation of the different interpretations of this ambiguous writer.

85. On the various interpretations of the thought of Machiavelli see Federico Chabod, *Machiavelli and the Renaissance* (note 53 above) with the interesting critical introduction by A. P. D'Entrèves.

86. Léontine Zanta, *La renaissance du stoicisme au XVIe siécle* (Paris: Champion, 1914). Discussing the influence of Stoicism on the Reformers, pp. 47 ff., Zanta shows that although undoubtedly Stoic ethics and the Stoic determinism influenced some of them, especially Zwingli and Calvin, yet Calvin took special care to reject Stoic materialism and other features that appeared to him incompatible with the Scriptures.

87. See Daniel P. Walker, *Spiritual and Demonic Magic*, (note 70 above).

88. Aristotle, *Metaphysics*, Book Lambda (XII), chapter 9, 1075a 1-10.

89. See Jared Wick, S.J., ed., *Catholic Scholars Dialogue with Luther* (Chicago: Loyola University Press, 1970), especially the masterly essay of Joseph Lortz, "The Basic Elements of Luther's Intellectual Style", pp. 3-34. Also Edwin Iserloh on Luther's development in *Reformation and Counter Reformation*, vol. 5 of Hubert Jedin and John Dolan eds., *The History of the Church* (New York: Seabury, 1980), pp. 3-96.

90. On the attitude of the Catholic scholastics to the Scriptures as the basis of theology see Beryl Smalley, *The Study of the Bible in the Middle Ages* (New York: Philosophical Library,

1952) and Thomas Gilby, "Appendix 11, "The *Summa* and the Bible" in the McGraw-Hill edition of St. Thomas Aquinas, *Summa Theologiae* (New York, 1963), vol. 1, pp. 133-139.

91. On the influence of Humanist thought forms on the Reformers see Trinkaus, *In Our Image* (note 63 above), 2, p. 770 f. on *theologia rhetorica*; Lewis W. Spitz, *The Religious Renaissance of the German Humanists* (Cambridge, Mass.: Harvard University Press, 1963); Quirinus Breen, *Christianity and Humanism* (Grand Rapids, Mich.: Eerdmans, 1968) (especially good on Calvin and relation of Reformation theology to the Renaissance interest in Roman law); Leif Grane, *Modus Loquendi Theologicus: Luthers Kampf um die Erneuerung der Theologie (1515-1518)* (Leiden: Brill, 1975). For parallels in Catholic circles see John W. O'Malley, *Praise and Blame in Renaissance Rome: Rhetoric, Doctrine and Reform in the Sacred Oratory of the Papal Court c. 1450-1521* (Durham, N.C.: Duke University Press, 1979).

92. On the Augustinian revival in the later Middle Ages see Heiko A. Oberman, *Masters of the Reformation* (Cambridge: Cambridge Univesity Press, 1981), pp. 64 ff. Oberman explains the preoccupation of the Reformers with Pelagianism as follows, "For Luther, the *modus loquendi* of the Apostle and his faithful intepreters remained valid *post Pelagium* because that fifth-century controversy belongs as much to Luther's present as the church's past. Since Pelagianism arises continually in the natural theology of the sinner's heart, an anti-Pelagian *modus loquendi* is the mode of speaking for all authentic theology." (p. 110). But are there not also other heresies that "arise continually in the sinner's heart" which an authentic theology must address e.g. anti-nomianism?

93. "John's gospel and St. Paul's epistles, especially that to the Romans and St. Peter's first epistle are the true kernel and marrow of all the books [of the Bible. There you will find] depicted in masterly fashion how faith in Christ overcomes sin, death and hell, and gives life, righteousness and salvation. This is the real nature of the gospel, as you have heard." (p. 361 f.). "In a word St. John's Gospel and his first epistle, St. Paul's epistles, especially Romans, Galatians and Ephesians and St. Peter's first epistle are the books that show you Christ and teach you all that is necessary and salvatory for you to know, even if you were never to see or hear any other book or doctrine. Therefor, St. James' epistle is really an epistle of straw, compared to these others, for it has nothing of the nature of the gospel about it." (p. 362) Luther, *Preface to the New Testament*, in *Luther's Works*, American Edition (Philadelphia and St. Louis: Muhlenberg Press, 1955), vol. 35, (afterwards LW), p. 357-65. "The epistle [to the Romans] is really the chief part of the New Testament, and is truly the purest gospel", Luther, *Preface to Romans, ibid.*, p. 365.

On the Protestant development of the *sola scriptura* doctrine see the classic work of J. A. Dorner, *History of Protestant Theology* (Edinburgh: T. and T. Clark, 1871), 2 vols.; and Robert D. Preuss, *The Theology of Post-Reformation Lutheranism* (St. Louis: Concordia Publishing House, 1970); on Luther's own views see William Jan Koolman, *Luther and the Bible* (Philadelphia: Mulenberg Press, 1961). On the question as to whether Luther really took into account the *whole* Bible see the essay of Lortz (note 89 above), pp. 29-31.

94. See Heiko Oberman, *Harvest of Medieval Theology*, (note 48 above); also his article "*Facientibus quod in se est Deus non denegat Gratiam*: Robert Holcot, O.P. and the Beginning of Luther's Theology" in Steven E. Ozment, ed. *The Reformation in Medieval Perspective*, (Chicago: Quadrangle Books, 1971), pp. 118-141 who argues that Luther broke with Nominalism as a student even before he began his professorship.

95. On the *Theologia Germanica* which Luther edited in 1518 see Spitz, *Religious Renaisance* note 91 above, pp. 240 ff. and note 41 above. On the influence of Tauler see Ozment, *Homo Spiritualis* (note 38 above). On the general question see Bengt Hoffman, *Luther and the Mystics* (Minneapolis: Augsburg Publishing House, 1976) and David C. Steinmetz, *Luther and Staupitz* (Durham, N.C.: Duke University Press, 1980). Protestant writers often use the term "mysticism" as if it necessarily implied a pantheistic absorption of the creature into the Creator, or at least a substitution of private revelations for the public revelation of the Gospel, and consequently deny that Luther could have approved mysticism. But the Catholic conception of mysticism is a union with God through faith, hope, and love; the creature always remaining a creature, and the faith a faith in the Gospel, not in some inner experience.

197

96. Ozment's argument (see previous note) against the importance of Tauler's influence on Luther turns on the assertion that for the Eckhartians the union of the Christian with God is mediated by the *scintilla animae* or most spiritual part of the soul, for Luther it is mediated by *faith* alone. It seems to me that this is a misunderstanding of Eckhart and Tauler. For them also the medium is faith alone, but they were interested in the psychology of the act of faith, which Luther was not, and hence they situated the act of faith in what is highest or deepest in man, the *scintilla animae*, but as orthodox Catholics they taught that man is incapable of faith without the gift of faith, just as Luther did.

97. "Between grace and gift there is this difference. Grace actually means God's favor, or the good which in himself he bears toward us, by which he is disposed to give us Christ and to pour into us the Holy Spirit with his gifts. This is clear from Romans, Chapter 5:15, where St. Paul speaks of 'the grace and gift in Christ,' etc. The gifts of the spirit increase in us every day, but they are not yet perfect since there remains in us the evil desires and sins that war against the spirit, as he says in Romans 7:5 ff. and Galatians 5:17 . . . Nevertheless, grace does so much that we are accounted completely righteous before God. For his grace is not divided or parceled out, as are the gifts, but takes us completely into favor for the sake of Christ our Intercessor and Mediator. And because of this, the gifts are begun in us." *Preface to Romans*, LW, vol. 35, p. 369 f. In texts like this Luther combines (a) grace as favor (forensic); (b) Uncreated Grace (the Holy Spirit); (c) the gifts of grace as an intrinsic tranformation of the believer which "increase in us every day." On the justification question see Stephanus Pfurtner, *Luther and Aquinas on Salvation* (New York: Sheed and Ward, 1964).

The Lutheran difficulty against Catholic doctrine remains that expressed by Gerhard Ebeling, *Luther: An Introduction to His Thought* (London: Collins, 1972), pp. 159-174, who thinks that the Catholic *fides formata caritate* theology takes away from (1) the passivity of man before God, since man becomes truly a person only through the gift of faith; (2) man's total dependence on God in temptation; and leads Catholics to trust in their own capacity to love, rather than in God's love for them, p. 174. In fact, however, Catholic doctrine teaches that our power to believe in God and to trust Him and our power to love Him are all equally His gifts. Moreover, while we can know if we have faith, we cannot know if we have charity. On this whole question see Erwin Iserloh, "Luther's Christ-Mysticism" in Wicks, *Catholic Scholars Dialogue With Luther*, (note 89 above), pp. 37-58 and Peter Manns, "Absolute and Incarnate Faith", *ibid.*, pp. 121-157.

98. "Christ heard Mass only once; there is no mention of His having gone to confession, but He laid great stress on prayer and preaching, especially on preaching', Humbert of Romans, O.P. quoted in Charles H. G. Smyth, *The Art of Preaching* (New York: 1940), SPCK, p. 16; "If you can only do one of these two things — hear Mass or hear a sermon — you should let the Mass go, rather than the sermon . . . There is less peril for your soul in not hearing the Mass, than in not hearing the sermon." St. Bernadine of Siena, O.F.M., *ibid.*, p. 16 f. Other examples are given by G. G. Coulton, *Five Centuries of Religion* (Cambridge: Cambridge University Press, 1923-50), 1, p. 124.

99. The primacy of the Word over the Sacraments as Lutherans understand it is well expressed by Eric W. Gritsch and Robert W. Jenson in *Lutheranism: The Theological Movement and Its Confessional Writings* (Philadelphia: Fortress Press, 1976) as follows: "Therefore Lutheran sacramental teaching is in the first place polemic. As we will see, 'word' and 'sacraments' are fundamentally but two inseparable aspects of the one event that Reformation theology calls '*the* Word.' But considered as religious phenomena, there is this difference between them: the performances we call "sacraments" are more *obviously* something *we* do than are verbal words; they are more directly to be grasped as religious works of the believers. It would be difficult to suppose that we are justified by the preacher's work at getting his point made; but it is easy to suppose that we are justified by attending the eucharist, or by reciting our sins in confession. Our impulse to works-righteousness, seeking to find security in our religiousness, therefore regularly seeks to separate the gospel promises from the sacraments, in order to claim the sacraments as its own. Lutheran sacramental doctrine is first of all an attack in advance on any doctrine or practice that makes sacraments and the gospel-promise rivals to each others." p. 80 f.

Luther's defense of the real, objective presence of the Body and Blood of Christ in the Eucharist against Zwingli and others is to be found especially in *That these Words of Christ "This is my Body" Etc. Still Stand Firm Against the Fanatics*, LW, vol. 37, 13-150; *"This is My Body" ibid.*, pp.   and *Confession Concerning Christ's Supper, ibid.*, pp. 161-372. On the history of the controversy see Herman Sasse, *This is My Body*, (Minneapolis: Augsburg, 1959). On Luther's eucharistic theology see Werner Elert, *The Structure of Lutheranism*, 2 vols. (St. Louis: Concordia, 1962) pp. 300-321; Paul Althaus, *The Theology of Martin Luther* (Philadelphia: Fortress, 1966), pp. 375-403 and Eric W. Gritsch and Robert W. Jenson, *Lutheranism* (reference above).

100. "If the mass is a promise . . . then access to it is to be gained, not with any works, or powers, or merits of one's own, but by faith alone" (p. 38 f.). "Therefore, just as distributing a testament or accepting a promise differs diametrically from a sacrifice, so it is a contradiction in terms to call the mass a sacrifice, for the former is something that we receive and the later is something that we give. The same thing cannot be received and offered at the same time, nor can it be both given and accepted by the same person, any more than our prayer can be the same thing as that which our prayer obtains, or the act of praying be the same as the act of receiving that for which we pray." (p. 54). "The Mass was provided only for those who have sad, afflicted, perplexed and erring consciences and they alone commune worthily" (p. 57). Luther, *The Babylonian Captivity of The Church*, LW vol. 36, pp. 3-126.

Thus for Luther the Eucharist seems to be purely a communion service, and since it is a gift of God to us it cannot be a sacrifice (i.e. a gift of ours to God). The weakness of this argument is twofold (a) in every sacrifice we can only give to God what He has already given to us, that is why it is a eucharist or thanksgiving; (b) according to traditional doctrine the sacrifice of the Mass is Christ's offering of himself to the Father through the ministry of the priest and the people's offering themselves with Christ. In this same work Luther agrees that the people should offer themselves and all they possess to God at Mass. What he hesitates to accept is the idea that the priest and people can *participate* in Christ's offering to the Father. For Luther Christ offers Himself for us and we can only receive the benefits of this offering but can not share in it as an act.

101. *Institutes of the Christian Religion*, I, 1,1, edited by John T. McNeill, translated by Ford L. Battles, 2 vols. *Library of Christian Classics*, vol. 20 (Philadelphia: Westminster Press, 1960), vol. 1, p. 35.

102. See Althaus, *Theology of Martin Luther* (note 99 above) pp. 394 ff. On Iconoclasm see Chapter 4, note 66. See also G. G. Coulton, *Art and the Reformation* (Oxford: Backwell, 1928) especially pp. 268-320; John Phillips, *The Reformation of Images: Destruction of Art in England 1535-1660* (Berkley: University of California, 1973) and Martin Marty, *Protestantism* (New York: Holt, Rinehart and Winston, 1972), pp. 227-256 on Protestant attitudes toward the religious use of the arts.

103. Luther repeatedly attacks the argument taken from John 6:63, "It is the spirit that gives life; the flesh is useless" by showing exegetically the different ways that "flesh" and "spirit" are used in Scripture and by arguing that in any case the Flesh of Christ is the source of our salvation, e.g. *Confession Concerning Christ's Supper* (note 99 above), pp. 245-248; *"This is My Body"* (note 99 above), pp. 78-101.

104. See John Toulsen, *The Church and the Secular Order in Reformation Thought* (New York: Columbia University Press, 1971). For a sympathetic presentation of Luther's Two Kingdom doctrine see Althaus, *Theology of Martin Luther,* (note 99 above), pp. 112-154; for modern Protestant criticism see Karl Barth, *Community, State and Church* (Garden City, N.Y.: Doubleday Anchor Books, 1960) and Dietrich Bonhoeffer, *Ethics* (New York: MacMillan, 1962). On the political influence of Calvin's thought, see Robert H. Kingdom and Robert D. Linder, eds., *Calvin and Calvinism: Sources of Democracy* (Lexington, Mass.: Heath, 1970).

105. "Oh it is a living, busy, active, mighty thing this faith. It is impossible for it not to be doing good works incessantly. It does not ask whether good works are to be done, but before the question is asked, it has already done them, and it is constantly doing them. Whoever does not do such works, however, is an unbeliever. He gropes and looks around for faith and good works, however, but knows neither what faith is nor what good works

are. Yet he talks and talks, with many words about faith and good works." *Preface to Romans* LW, vol 35, p. 370. Yet Luther's own practical moral instruction can be found in such works as *The Ten Sermons on the Catechism*, LW, vol. 57, pp. 137-193 and in his *Treatise on Good Works*, LW, vol. 44, pp. 15-114. See also his letters, *Luther: Letters of Spiritual Counsel*, ed. and trans. by Theodore G. Tapert, *The Library of Christian Classics* (Philadelphia: Westminster Press, 1955), vol. 18. See N. H. Søe, "The Three 'Uses' of the Law" in G. H. Outka and P. Ramsey, *Norm and Context in Christian Ethics* (New York: Scribners', 1968), pp. 297-324; Dietrich Bonhoeffer, *Ethics* (note 104 above), pp. 271-285, and Werner Elert, *Law and Gospel* (Philadelphia: Fortress Press, 1967) *passim*. The most complete presentation of Luther's ethical thought is Paul Althaus, *The Ethics of Martin Luther* (Philadelphia: Fortress Press, 1972).

106. On Calvin's conception of Christian ethics and sanctification see T. F. Torrance, *Calvin's Doctrine of Man*, rev. ed. (Grand Rapids, Michigan: Eerdmans, 1957); Georgia Harkness, *John Calvin: The Man and His Ethics* (Abingdon Press: Nashville, 1958); Ronald S. Wallace, *Calvin's Doctrine of the Christian Life* (Grand Rapids, Mich: Eerdmans, 1959) (esp. pp. 123-140 and 170-194 on the role played by the notion of "moderation" or "soberness"); R. Stauffer, *The Humanness of John Calvin* (Nashville, Tenn.: Abingdon, 1971); Lucien Joseph Richard, O.M.I., *The Spirituality of John Calvin* (Atlanta: John Knox Press, 1974).

107. Calvin: *Institutes of Christian Religion*, (note 101 above), vol. 1, I, chapter 1, pp. 33-35 and II, chapter 2, pp. 243-4. On the style of the *Institutes* see Francis M. Hignan, *The Style of Calvin* (Oxford: Oxford University Press, 1967), who shows that Calvin varied his style in various works to fit the audience but always strove for great sobriety and economy, yet followed a rhetorical rather than a scholastic mode.

108. "Nothing pleases man more than the sort of alluring talk that tickles the pride that itches in his very marrow. Therefore, in nearly every age, when anyone publicly extolled human nature in most favorable terms, he was listened to with applause. But however great such commendation of human excellence is that teaches man to be satisfied with himself, it does nothing but delight in its own sweetness; indeed, it so deceives as to drive those who assent to it into utter ruin . . . Whoever, then, heeds such teachers as hold us back with thought only of our good traits will not advance in self-knowledge, but will be plunged into the worst ignorance." *Institutes*, II, 1, 2, (note 101 above), vol. 1, p. 243. Calvin goes on in what follows to say that we should first consider the great gifts with which God has endowed us, and then our own utter incapability of living up to what God expects of us, and therefore our need to rely absolutely on God.

109. Paul Hacker, *The Ego in Faith: Martin Luther and the Origin of Anthropocentric Religion* (Chicago: Franciscan Herald Press, 1970), develops at length the thesis that Luther's emphasis on "salvation *for me*," i.e. on the assurance of individual salvation is the characteristic mark of the Reformation. See Werner Elert, *The Structure of Lutheranism* (note 99 above), pp. 68-90, for a defense of Luther against this charge of *Heilsegoismus*. Elert admits the shift to emphasis on the *pro me*, but argues that it is biblically justified. For Calvin's doctrine of "assurance" see *Institutes*, III, 14, 20 and 24, 4-10. John S. Bray, *Theodore Beza's Doctrine of Predistination* (Nieukoop: De Graaf, 1975) pp. 58-60, says that Calvin based the assurance of salvation of faith in Christ as Savior rather than in the certainty of being predestined, but secondarily admitted the help of the so-called *syllogismus practicus* ("I am bearing the fruit of good works, therefore I must be predestined to salvation"). Later Calvinists tended to shift the emphasis to this secondary sign of assurance.

110. Max Weber, *The Protestant Ethic and the Spirit of Capitalism*, translated by Talcott Parsons with a foreword by R. T. Tawney (1930) (London: Unwin University Books, 1970). For various views see Robert W. Green, *Protestantism and Capitalism: The Weber Thesis and its Critics* (New York: Heath, 1959); M. J. Kitch, *Capitalism and the Reformation* (New York: Barnes and Noble, 1967); S. N. Eisenstadt, *The Protestant Ethic and Modernization* (New York: Basic Books, 1968); and H. R. Trevor-Roper, *The Crisis of the Seventeenth Century* (New York: Harper and Row, 1968); with the judicious summing up of John Gilchrist, *The Church and Economic Activity in the Middle Ages* (New York: MacMillan), "The Weber-Tawney Thesis in Retrospect", pp. 122-139.

200

111. See André Biéler, *The Social Humanism of Calvin* (Richmond, Virginia: John Knox Press, 1964), for a very positive presentation of Calvin's social and political thought.

112. See Harkness, *John Calvin* (note 106 above), Chapter VII "Domestic Relations", pp. 127-156 and James T. Johnson *A Society Ordained by God: English Puritan Marriage in the First Half of the Seventeenth century* (Nashville: Abingdon Press, 1970). It is claimed the Puritans applied their covenant theology to marriage and reversed the Anglican view which put procreation first before help as the end of marriage. Milton went further and interpreted "mutual help" as *compatibility* so as to conclude that in case of incompatibility a couple by mutual consent could dissolve their marriage. To me the evidence cited for this reversal of ends seems somewhat weak.

113. See Frank Hugh Foster, *A Genetic History of New England Theology* (New York: Russell and Russell, 1963) for a case history of the way Calvinism tended to break down into Unitarianism and then Humanism.

114. See Lewis W. Spitz, "The Course of German Humanism" in Oberman and Brady, *Iternarium Italicum* (note 58 above), who shows how Melanchthon derived the notion of *loci* from the rhetoricians. The same transference was made by the Catholic Thomist Melchior Cano, O.P. (d. 1560) in his *Loci theologici* (1563). See *Melanchthon on Christian Doctrine: Loci communes,* translated and edited by Clyde L. Manschreck, introduction by Hans Engellard (New York: Oxford University Press, 1965). See also Carl E. Maxy, *Bona Opera: A Study in the Development of Doctrine in Phillip Melanchthon* (Nieukoop: De Graaf, 1980) and Peter Petersen, *Geschichte der Aristotelischen Philosophie im Protestantischen Deutschland* (Stuttgart — Bad Cannstatt, 1964, originally Leipzig, 1921: F. Frommann Verlag). The latter work shows Melancthon's influence on the rise of the Enlightenment.

115. See Robert D. Preuss, *Theology of Post-Reformation Lutheranism* (note 93 above), for an extensive attempt to answer such criticisms.

116. See Alexandre Koyré, *La philosophie de Jacob Boehme: Étude sur les origines de la métaphysique allemande,* (Paris: Vrin, 1929).

117. A. Rischl, *Geschichte des Pietismus,* 3 vols. (Bonn 1880-1886) is the most comprehensive work. See also August Lang, *Puritanismus und Pietismus* (Darmstadt: Wissenschaftliche Buchgesellschaft, 1972 (Reprint of 1941), who traces the movement back to Martin Bucer; F. Ernest Stoeffler, *The Rise of Evangelical Pietism* (Leiden: Brill, 1965); Friedrich Wilhelm Kantzenbach, *Orthodoxie und Pietismus (Evangelische Enzyklopädie)* (Gütersloh: Gerd Mohn, 1966) and Louis Bouyer, et al. *A History of Christian Spirituality* (New York: Desclée, 1965), vol. 3, pp. 169-183.

118. On Methodism see Bouyer, *ibid.,* vol 3. pp. 187-197.

119. On Beza see John S. Bray, *Theodore Beza's Doctrine of Predestination* (Nieukoop: De Graaf, 1975), Bray argues that Calvin, although he taught double predestination did not make it central to his theology as did the scholastics. Beza is a transition figure who tends more toward scholasticism but cannot be said to be the source of the scholastic emphasis on predestination as the central doctrine of "Calvinism". Bucer's *De Regno Christi* is translated with an introduction and bibliography by Wilhelm Pauck in *The Library of Christian Classics* (Philadelphia: Westminister, 1969). As to Calvin, Edward A. Downey Jr, *The Knowledge of God in Calvin's Theology* (New York, 1965) (whom Bray quotes with approval) says, "Indeed, if there is a persistent theme in Calvin it is that God's way and thoughts are incomprehensible to man without special revelation. His theology then is an expression of faith and complete trust in God, written by a man of faith to encourage and aid the faithful of God. As such the rational dimension is clearly subordinated to the religious. In this program theology is designed not to meet the demands of a rationally acceptable and defensible system but to assist the faithful in understanding God's revelation" (p. 3). Calvin, however, certainly believed in double predestination, cf. *Institutes* III 21, 7, but Downey says that in his writings this is an "isolated doctrine" (p. 217), nor did he ever discuss the question of supra- or infralapsarianism (pp. 55 f.). Bray attributes the development of scholastic Calvinism to two causes (1) the need to answer Catholic attacks; (2) the tendency to systematize Calvin's teaching around a central theme (pp. 137-143) and sees in this scholasticism the influence of Aristotelianism and Thomism.

120. Before his conversion Calvin wrote, *Commentary on the De Clementia of Annaeus Lucius Seneca,* ed. and translated by F.L. Battles and A.M. Hugo, (Leiden: Brill, 1969). See Zanta, *La Renaissance du Stoicisme* (note 86) and Charles Partee, *Calvin and Classical Philosophy* (Leiden: Brill, 1977).

121. See Walter J. Ong, S.J., *Ramus: Method and the Decay of Dialogue from the Art of Discourse to the Art of Reason* (New York: Octagon Books, 1974), and Neal W. Gilbert, *Renaissance Concepts of Method* (New York: Columbia University Press, 1960) on Ramus, pp. 129-220, and on Bartholomew Keckermann (d. 1609) an interesting Aristotelian opponent of Ramus, pp. 214-220.

122. On the history of the "federal theology" see Peter Toon, *The Emergence of Hyper-Calvinism in English Non-Conformity 1689-1765* (London: The Olive Tree, 1967) and Holmes Rolston, *John Calvin Versus the Westminister Confession* (Richmond, Virginia: John Knox Press, 1972), who argues that the distinction of the "covenant of works" from the "covenant of grace" is not found in Calvin's writings. For Calvin, Adam was created in a "covenant of grace."

123. On Amyraut see Brian G. Armstrong, *Calvinism and the Amyraut Heresy: Protestant Scholastisticm in Seventeenth Century France* (Madison: University of Wisconsin Press, 1969), Armstrong concludes (p. 269) that Amyraut was truer to the text of Calvin than his "scholastic" accusers. He was also a strong defender of the notion of the *natural law* in ethics (Appendix 1, pp. 273-275). On Arminius see Carl O. Bangs, *Arminius: Study in the Dutch Reformation* (Nashville: Abingdon, 1971).

124. On Protestant "scholasticism" or "orthodoxy" see Kantzenbach, *Orthodoxie und Pietismus* (note 117 above), pp. 28-56; Andrew L. Drummond, *German Protestantism since Luther* (London: Epworth Press, 1951), pp. 11-35, and Bray, *Theodore Beza's Doctrine* (note 119 above) pp. 137-143. For a defense of Lutheran Orthodoxy see Preus, *Theology of Post-Reformation Lutheranism* (note 93 above).

125. On the spirit of the Baroque Age see Heinrich Wölfflin, *Principles of Art History* (New York: Dover, 1950) *Renaissance and Baroque* (Ithaca, N.Y.: Cornell University Press, 1964); Victor Lucien Tapie, *Baroque and Classicisme* (Paris: Plon, 1957); Sir Sacheverell Sitwell, *Baroque and Rococo,* (New York: Putnam, 1967); A.C. Sweter, *Baroque and Rococo,* (New York: Harcourt Brace Jovanovich, 1972) and Frank J. Warnke, *Visions of Baroque: European Literature in the Seventeenth Century* (New Haven: Yale University, 1972).

126. See Jedin and Dolan, *History of the Church* (note 89 above), 5:562 f.

127. See John W. O'Malley, S.J., *The Jesuits, St. Ignatius and the Counter Reformation: Some Recent Studies and Their Implications for Today* (St. Louis: Studies in Spirituality of the Jesuits, 14 Jan., 1982) for some of the new insights on the role of the Jesuits in the Counter Reformation.

128. See Jean Orcibal, *Le rencontre du Carmel thérésien avec les mystiques du Nord* (Paris: Presses Universitaires de France, 1959) and *Saint Jean de la Croix et les Mystiques Rhéno-flamands* (Paris: Desclée de Brouwer, 1966). Also Giovanna dell Croce, O.C.O., *Johannes von Kreuz und die deutsch-niederländische Mystik.* Jahrbuch für mystiche Theologie, 1960 (1) (Vienna: Verlag Heiler, 1960.)

129. *The Spiritual Exercises of St. Ignatius of Loyola,* a new translation and a contemporary reading by David L. Fleming, S.J. (St. Louis: St. Louis Institute of Jesuit Sources, 1978). See also Paul Begheyn, S.J., *A Bibliography on St. Ignatius' Spiritual Exercises,* (St. Louis: Studies in the Spirituality of the Jesuits, n. 13, March, 1981).

130. For an analysis of the teaching of St. Ignatius of Loyola on the spiritual life see Hugo Rahner, *The Spirituality of St. Ignatius Loyola: An Account of its Historical Development* (Westminister, Md.: Newman, 1953), and Piet Penning de Vries, *Discernment of Spirits According to the Life and Teachings of St. Ignatius of Loyola* (New York: Exposition Press, 1973).

131. After John Capreolus (d. 1444) came Peter Niger (d. 1477), Paul Concina (d. 1494), Dominic of Flanders (d. 1500) Chrysostom Javelli (d. 1545), Francis Sylvester de Sylvestris Ferrariensis (d. 1528), and Thomas de Vio, Cardinal Cajetan (d. 1534), all Dominicans. For the Spanish Thomists see Carlo Giacon, S.J., *La Seconda Scolastica: I Grandi Commentatori de San Tommaso* (Milan: Fratelli Boca, 1944), vol. 1. On the important role of

Francis de Vitoria see James Brown Scott, *The Catholic Conception of International Law* (Washington, D.C.: Georgetown University Press, 1934). A selection of his legal works can be found in *The Classics of International Law*, no. 7 (Washington, D.C.: The Carnegie Institute, 1977).

132. Among these Jesuit Thomists are Francis Toletus (d. 1596, a pupil of Dominic Soto, O.P.), the Portugese group called the Conimbricenses of whom the chief was Peter Fonseca (d. 1599); Gabriel Vasquez (d. 1604), Gregory of Valentia (d. 1603), Leonard Lessius (d. 1623, Luis of Molina (d. 1600) and especially Francis Suarez (d. 1617). On the last see Karl Werner, *Franz Suarez und die scholastik die letzen jahrhunderts*, rev. ed. (Regensburg 1889), 2 vols., and the excellent discussion of his metaphysical position in Copelston, *History of Philosophy* (note 72 above), vol. 3, Part II, pp. 202-228.

133. On baroque scholasticism see works of Giacon (note 131 above) and Werner (note 132 above).

134. On this controversy see Jedin and Dolan, *History of The Church* (note 89 above), pp. 542-545.

135. The most extensive treatment of this controversy is J. J. I. von Döllinger and F.H. Reusch, *Geschichte der Moralstreitigkeiten in der römisch-katholischen Kirche*, 2 vols. (1889, reprinted (Munich: Scientia Verlag, 1968). See also Johann Theiner, *Die Entwicklung der Moraltheologie zur Eigenständigen Disziplin* (Regensburg: Verlag F. Pustet, 1970).

136. On the origin of the *Institutiones Theologiae Moralis* or "moral manuals" see Louis Vereecke, "Préface à l'histoire de la théologie morale moderne" in *Studia Moralia* (Rome: Academia Alfonsiana Institutum Theologiae Moralis, I, 1962), pp. 87-120. On Protestant "legalism" see the references in note 124 above.

137. See Pierre Michaud-Quantin, *Sommes de casuistique et manuels de confession au moyen age (xii-xvi siècles)* (Louvain: Nauwelaerts, 1962), and Thomas N. Tentler, *Sin and Confession on the Eve of Reformation* (Princeton, N.J.: Princeton University Press, 1977).

138. For the history of the Probabilism Controversy see Th. Deman, "Probablisme", *Dictionnaire Théologique Catholique*, tom., 13 (1), 417-619. For the present status of the theory see Marcellinus Zalba, *Theologie Moralis Compendium* (Madrid: Biblioteca de Autores Cristiana, 1958), vol. 1, pp. 379-406.

139. See Walter E. Rex, *Pascal's Provincial Letters: An Introduction* (London: Hodder and Staughton, 1977). Also Jan Miel, *Pascal and Theology* (Baltimore: John Hopkins University, 1969), and Léon Brunschwicq, *Descartes et Pascal: Lecteurs de Montaigne* (New York-Paris: Brentano, 1944).

140. On the Jansenist opposition to probablism see Jean-Robert Armogathe and Michel Dupy, article "Jansénisme", *Dictionnaire de Spiritualité* (Paris: Becauschesne, 1974), vol. 8, cols. 102-147.

141. See Heinrich A. Rommen, *The Natural Law: A Study in Social History and Philosophy* (St. Louis: B. Herder, 1948), pp. 34-74, and his *Der Staatslehre des Franz Suarez, S.J.* (M. -Gladbach: Volksverein Verlag, 1927), Selections from his legal works are published in No. 20 of *The Classics of International Law*, Carnegie Institute (Oxford: Clarendon Press, 1944). On Bellarmine, John Clement Rager, *The Political Philosophy of Bl. Cardinal Bellarmine* (Washington, D.C.: Catholic University of America, 1926), and Franz X. Arnold, *Die Staatslehre des Kardinals Bellarmin* (Munich: Max Hueber Verlag, 1934).

142. See Edward Dumbauld, *The Life and Legal Writings of Hugo Grotius* (Norman: University of Oklahoma, 1969); Charles S. Edwards, *Hugo Grotius: The Miracle of Holland* (Chicago: Nelson Hall, 1981); Leonard Krieger. *The Politics of Discretion: Pufendorf and the Acceptance of Natural Law* (Chicago: University of Chicago Press, 1965). Modern editions of major works by Grotius and Pufendorf can be found in J.B. Scott, ed., *The Classics of International Law*, Nos. 3 and 10, (Carnegie Institute, Washington, D.C. 1913-1934).

143. See for example the classical spiritual work of Alphonsus Rodriguez, S.J. (d. 1616) *Practice of Perfection and Christian Virtues* (Chicago: Loyola University Press, 1929), 5th Treatise, no. 4, "Two Sorts of Mental Prayer, pp. 289-291. Of course many Jesuit writers have not accepted this "two way" theory. On this controversy see Reginald Garrigou-Lagrange, O.P., *Christian Perfection and Contemplation* (St. Louis: B. Herder, 1944), pp. 23-42; Antonio Royo, O.P. and Jordan Aumann, O.P., *The Theology of Christian Perfection* (Dubuque, Iowa: Priory Press, 1962), pp. 184-195 and the summation by Charles Baumgarten, article "Contemplation", *Dictionnaire de Spiritualité*, tom. 2 (2): col. 2171-2193.

Chapter 6

Christian Theology Confronted
By Humanism

I: The Cartesian Transition

1. The Phenomenal Body

In my sketch of the history of the Humanist theologies of the body in Chapter 3, I dated the rise of Humanism from 1700, but indicated that it was during the 1600's that the wars of religion and the simultaneous rise of modern science negatively and positively prepared its way. I have now to go back and recall some of these same events from the viewpoint of the Christian theologians for whom Humanism was and is a formidable challenge. It was a challenge whose real scope they did not at first recognize, so preoccupied were they with some of the older controversies which I described in the last chapter.

My intention is to show that what characterized the intellectual life of Europe in this period was the step-by-step abandonment of an ontological conception of truth for a phenomenological conception and that this ideological shift is the typically modern way of devaluating the bodily aspect of our human existence, as Platonism was the ancient way of the same devaluation. In the last chapter I showed that this shift had already been prepared in the field of anthropology by the nominalists and mystics of the fourteenth century, the Renaissance humanists of the fifteenth, and the Reformers of the sixteenth by their replacement of a theory of human life in terms of a teleological human nature by a voluntaristic legal theory and by their turning from the object to the subject as the point of departure for knowing and acting. But this shift had not yet received the full support of the natural sciences which, although they had begun to undergo mathematization, were still mainly understood as an exploration of the nature of things in the ontological sense.

With the seventeenth century natural scientists came to abandon any serious thought of arriving at a knowledge of understanding natural phenomena in terms of the natures which they were once thought to reveal, and became content to study the phenomena as such, without any hope or even any need being any longer felt to penetrate to the ontological noumena which had been the goal of scientific investigation for both Platonists and Aristotelians. For any theology of the body this meant that just as the soul became, in the new way of thinking, simply "the stream of consciousness," so the body became "the process of evolution."

2. The Cartesian Christian Theology

The tentative first steps of modern science and technology (and the more detailed understanding of the physical world which it was to bring about) had already been taken in the later Middle Ages. But at the beginning of the seventeenth century natural science entered into a phase of rapid development of which the first great center was the University of Padua. Padua was still dominated by Averroistic Aristotelians such as Jacopo Zabarella (d.1594) who at the end of the previous century had made an important contribution to the logic of science, but these textual commentators were little concerned with new investigations. In the sixteenth century Thomistic Aristotelians such as Dominic Soto (d.1560) had actually advanced important new physical theories, but the last of the great baroque Thomists, John of St. Thomas (Jean Poinsot, d.1644), although also an important logician, was indifferent to more recent scientific developments.[1] In fact Catholic theologians were for the most part so deeply involved in the controversies over grace and the moral systems that they left these new developments to the philosophers.

Although many of the Paduan Averroists were so intransigently faithful to the letter if not the spirit of Aristotle, it was at Padua that Galileo (d.1642) initiated modern physics and from there that William Harvey (d.1657) returned to England to initiate modern biology. These two men were destined to overthrow the Aristotelian world-view that had been

205

accepted also by the Neo-Platonists and which had for so many centuries served Christian theology, but they both had been thoroughly trained in its methodology. Even before coming to Padua Galileo had been grounded in this method by his Thomistic Jesuit professors at the Collegio Romano.[2] What Galileo and Harvey added to this method were more practical methods of observation and controlled experimentation and, in Galileo's case, of measurement and mathematical theorization.

Aristotelian science had never denied the value of the use of mathematics but always considered it preliminary to physical demonstrations through the "four causes", matter, form, agent, and goal (*telos*).[3] The Pythagorean-Platonic tradition, on the other hand, had considered mathematical theories as superior to physical explanations, but even these they regarded as no more than a probable "saving of appearances," as in their application to astronomy, especially in the Ptolemaic geocentric astronomy which had proved so highly successful in accounting for the data of observation.

Moreover, as already noted in the last chapter, in the thirteenth and fourteenth centuries at Oxford Robert Grosseteste and Thomas Bradwardine had attempted seriously to extend this mathematical method to physics, but, lacking practical experimental methods, with only feeble success. Before Galileo, however, Copernicus (1473-1543) had revolutionized Ptolemaic astronomy by his mathematically simpler heliocentric theory, inspired by his readings of Cardinal Cusa and what he had learned about Pythagorean thought from his studies at Bologna under Dominicus Maria de Novaro and through reading the humanist Marsilio Ficino.[4] This heliocentric theory was perfected by Johannes Kepler (d.1630) and Tycho Brahe (d.1630) through whom it began to receive acceptance in Protestant circles.

It was Galileo, however, who succeeded in extending this mathematical approach from the heavens to earth as a universal method for natural science and combining it with the Aristotelian determination to go beyond "saving the appearances" to giving genuinely physical explanations. The glory and the tragedy of Galileo was precisely this discontent with a merely phenomenological science, because if science can reveal the world to us as it really is and not just as it appears, then it can also come into conflict with Christian revelation which also claims to attain reality, as the Inquisition before which Galileo was condemned for heresy rightly understood. We now recognize that these inquisitors employed a faulty hermeneutic in their reading of the Bible, and we also know that Galileo's scientific arguments were faulty, since empirical verification of heliocentrism was never satisfactorily achieved until the discovery of stellar parallax two hundred years later![5] Yet in principle the confrontation was genuine and not an illegitimate confusion of science and religion as is so often asserted.[6]

Galileo, however, not only added something to the Aristotelian notion of science, he also made a very significant subtraction. His concern to formulate his theories mathematically led him to abandon Aristotelian teleology, and to retain only arguments based on material, formal, and agent causes. The effect of this was that as physics developed in the seventeenth century it rapidly became *mechanistic,* reverting to the Democritean type of science which Plato had in part accepted, but Aristotle vigorously rejected,[7] and seeking to explain phenomena in terms of goalless forces acting on inert particles in empty space. In the Renaissance Cardano and Telesio had attempted to build this mechanical system of physics, relying on the Epicurean tradition of philosophy in which the thought of Democritus was best preserved, but with little success. Now in the seventeenth century this effort was renewed by René Descartes (1596-1650) and his friend and fellow student Marin Mersenne (d.1648), a Minim priest, and independently by Pierre Gassendi (d.1655).[8]

Descartes is commonly entitled "The Father of Modern Philosophy", but this is said from the perspective of Humanism which sees in him the beginning of the breaking free of philosophy from its long Christian tutelage to become a replacement for Christian theology rather than its handmaid. This claim is too much like that other claim which is also sometimes made that Luther was the inaugurator of the Humanist dogma of "free-thought." In fact Descartes was a serious Christian and his thought, however one evaluates it, is deeply embedded in the Christian tradition which we are tracing. In Chapter 3, I indicated how some elements of his thought contributed to the rise of Humanism but this does not make Descartes, any more than Luther, a proto-Humanist.

Like Galileo, Descartes had been educated by Jesuits in the baroque scholastic tradition and had also taken great interest in the Renaissance developments in mathematics.[9] But while Galileo remained a physicist who used mathematics as an instrument of his science; Descartes, who early in life made a fundamental mathematical invention, namely, analytical geometry, took mathematics as the model for all thought because of its wonderful deductive clarity and certitude. In fact he abandoned the Aristotelian concept of truth as the *correspondence* between theory and sense experience for a *consistency* theory based on logical coherence with self-evident axioms derived not from sense experience but from innate ideas. Thus Descartes belongs to the Pythagorean-Platonic tradition, as is evident from his conviction that his invention of analytical geometry would at last solve the problem of the continuum and of irrational numbers that had always been the scandal of that rationalist tradition.[10]

Descartes, however, early fell under a very different sort of influence, namely the revival of anti-scholastic Augustinianism sponsored by the priests of the Oratory.[11] How often we have seen a return to Augustine and with what various results! We have seen that scholasticism was itself profoundly Augustinian, but we have also seen that both the Renaissance humanists and the Reformers had found in his more rhetorical, more subjective mode of thought and writing an antidote to the dry objectivity of the Aristotelian logic of scholasticism. This particular early seventeenth century return to Augustine was prompted by weariness with the controversies of baroque scholasticism and the growing scepticism in France (typified by Montaigne, d.1592) with regard to Christian dogma as a result of the Catholic-Huguenot conflict. Avid for absolute certitude (which he identified with clarity), Descartes found comfort in the thought of Augustine who had so vigorously refuted the sceptics of his own time.[12] It was probably from Augustine that Descartes derived his own epistemological grounding in the famous *cogito ergo sum*, "I may doubt, but doubt is thought and if I think I must also exist."[13] For Aristotle the certitude of my existence in a real world is founded on my sense of touch and can always be verified by it. I feel my body and with my body I keep in contact with my world. But for Plato, Augustine, and Descartes rational thought must be its own guarantee of its own contact with reality; it cannot receive this certitude from anything outside its own self-evidence.

We have already seen how this centering on human subjectivity and its polar opposition to objectivity had been gradually intensifying since the late Middle Ages in the voluntarism of the Scotists and Nominalists, in the mysticism of the Rhineland and the *devotio moderna*, in the rhetorical style of the Renaissance humanists, in the conversion-experience theology of the Reformers. In each case reaction against the Aristotelian logic and the empirico-objective epistemology of university scholasticism had taken the form of a return to Augustine's rhetorical, autobiographical, Platonic interpretation of Christianity. Obviously, the sociological causes of this long-range trend toward the development of an intellectual elite seeking certitude for its Christian world-view within the individual consciousness must lie deep. The autobiography of Descartes, a layman working outside a religious community or a university, precariously supported by the patronage of the aristocracy, in contact with current thought mainly through correspondence, troubled by the bitter religious disputes of his time, fascinated by the new advances in the sciences, gives us some clue.[14]

What was most original about Descartes was not his revival of these traditional philosophical themes, but the way in which he applied them to the development of a mechanical physics. Paradoxically, what was to make his thought so vastly influential was not its grounding in self-verifying innate ideas nor in self-affirming self-consciousness, but rather that these assumptions seemed to receive a practical confirmation by their fruitfulness in the advance of science.

To Descartes, as to Galileo, the scholastic theory of matter as *potentiality* seemed unnecessarily obscure and unsuitable for mathematical formulation. Had not Descartes himself by his invention of analytic geometry neatly removed the last vestige of potentiality, i.e. the continuum, from mathematics and reduced all quantity to discrete numbers? Matter, argued Descartes, is nothing more than *extended* substance, substance having numerable parts. All its other attributes or qualities can be reduced to the arrangement of these parts. This of course was an old Epicurean notion, but Descartes hesitated to go all the way and reduce matter to unalterable particles moving in a void by some purposeless, inherent force. He retained the Aristotelian notion that the world is a material *plenum,* and then attempted to account for all the phenomena of physics in terms of *vortices* or self-perpetuating currents set up by the Creator in this fluid cosmic mass. Here we see the influence of Stoic conceptions with which many thinkers of the time were also experimenting.[15] Descartes does not seem to have realized the inconsistency of attributing the condensation and expansion demanded by fluid movement to a matter whose very essence was pure (and therefore rigid, motionless) geometry.

Having reduced the phenomena of this sensible world to the mechanical notion of purely geometrical substance (he even concluded that animals are mere automatons, and our bodies machines), Descartes was left with the problem of accounting for his own thought by which this new physics had been deductively constructed. Consequently over against material substance, he posited a second absolutely different kind of substance called "Mind," whose essence was thought. Matter was extension; Mind was thought which, since it is self-conscious and therefore altogether-in-itself, is unextended like a point. Matter is known by us always as an object, i.e. it is viewed from without, from its surface. Mind is known always as a subject, i.e. it is viewed from within, from its center. Descartes, using these necessary mathematical metaphors very literally, ended by flatly opposing subject and object, soul and body in a way that the scholastics would have regarded as naive, but which had a great appeal to a culture disillusioned with scholastic subtleties.[16] Descartes is one of those thinkers like the Pre-socratics Heraclitus, Parmenides and Zeno of Elea who have stirred up philosophy not by the reasonableness or the consistency of their outrageous paradoxes but because these paradoxes fascinate puzzle-solvers. What counts for our present purposes is that Descartes at last in the crudest possible terms exposed the Platonic dualism

209

between body and soul, matter and spirit which had so long troubled Christian theology. Of this dualism Cartesianism is the *reductio ad absurdum*.

Cartesianism immediately became a subject of controversy. Descartes' works were put on the *Index of Forbidden Books* by his own Church and also by the Calvinist Synod of Dort.[17] Yet gradually Cartesianism was adopted by many Jesuits as well as by their Jansenist opponents and together with the Ramist logic penetrated the teaching of philosophy in the seminaries of France, newly established to replace university education (or no education) for the clergy, so that by the time of the French Revolution the Cartesian philosophy was dominant in Catholic theology. It was so dominant that even the surviving scholastic traditions tended to be understood in a Cartesian, rationalistic manner. Its influence in France is seen especially in the Oratorian philosopher Malebranche (d.1715) who developed it in explicitly Augustinian form as *ontologism* (all things are known in God) and in the Jansenists Antoine Arnauld (d.1694) and the great Blaise Paschal (d.1662). Paschal, a mathematician of genius, a bitter opponent of the Jesuits, and a Christian apologete, tried to supplement Cartesian rationalism with an ardent fideism which in fact is quite compatible with it.[18]

Outside of France in the Netherlands Cartesian rationalism stimulated Baruch Spinoza (d.1677) to develop his own system which sought to overcome dualism by pantheism, combining the Neo-Platonism of Jewish cabalistic mysticism with a rationalistic attack on the inspiration of the Old Testament, which was to make an important contribution to the development of Humanism in the next century.[19] In Germany, G.W. von Leibnitz (1646-1716) a Protestant of strongly ecumenical motivation, a great mathematician and logician, gave to Cartesianism a new form in which the human mind is conceived as a kind of spiritual "atom" or monad associated with the unextended material atoms which form the body, all of which monads are casually independent of each other yet coordinated by the predestinating plans of their Creator.

The greatest influence of Leibnitz, however, was in his propagation of a rigidly rationalistic conception of philosophy and of theology in the German universities, deductive in method, and characterized by an epistemology of innate ideas and a metaphysics in which the absolute necessity of metaphysical reasoning is sharply contrasted with the contingency of all the empirical sciences.[20] This strong dichotomy between metaphysical necessity and empirical contingency is oddly reminiscent of the nominalist distinction between the *potentia absoluta* and *ordinata* of God. Leibnitz was in fact unusually well acquainted for a philosopher of this period with scholastic thought and his own method deeply influenced the later phase of Lutheran Orthodox Theology.[21] Leibnizian rationalism was made a standard for German university education largely through its schematization by Christian Wolff (1679-1754), the influence of which was soon felt in Catholic textbooks so that as late as the Thomistic revival at the end of the nineteenth century, much of what passed for "Thomism" was really Wolffianism.[22]

The great significance of this Cartesian episode is that it obliterated what remained of the Aristotelian tradition and substituted for it a simplified, essentially Platonic epistemology based on innate ideas and a deductive rationalism as a single method for all fields of knowledge. Yet the motivation of this Cartesian movement was not the overthrow of the Christian world-view but its defense against the rise of religious scepticism and materialism by a reconciliation of the new mechanistic physics with a pious fideism and a rationalistic orthodoxy. Yet in the Jewish Spinoza and the Protestant apologist Pierre Bayle (d.1706), another ardent admirer of Descartes, this rationalism extended to an attack on the inspiration of the Bible and even the very concept of revelation.[23]

In England, Cartesianism was less well received. Francis Bacon (d.1626), although his actual achievements as a scientist and logician were minimal, had made great propaganda for an empirical, inductive approach to science, and had fostered the development of the Royal Society as a center of research outside the scholastic traditions of Oxford and Cambridge, emphasizing the practical character of science as a technology in the service of government, the military, and business. The result was that when the philosophy of Descartes invaded the English universities its chief effect was to deal the death blow to Aristotelianism. Surprisingly, Calvinistic Puritanism seems to have contributed an important motivation to the advance of scientific technology because its millenarianism suggested the coming of a "new age" in which humanity would gain remarkable control over nature.[24]

Thomas Hobbes (d.1679), personally acquainted with Galileo and Paduan science, gave the new mechanism a frankly materialist interpretation in sharp opposition to Descartes' dualism. Moreover, Hobbes went back to the "realistic" political theory of Machiavelli (d.1527) and, under the influence of Calvinistic pessimism about the human condition and its Federal Theology, proposed a politics and ethics based on a contract theory.[25] He concluded that the sovereignty of the state is the only perservative of order in the face of biologically determined egoistic drives. This Hobbesian materialism was too radical to be widely accepted and was quickly succeeded by the better camouflaged views of John Locke (d.1704) who opposed the Cartesian innate ideas by means of the Aristotelian doctrine he had learned at Oxford that all human knowledge begins in sensation, but changed it radically in a materialist sense by denying any essential difference between sensation and intellection. [26]

This revival of materialism by Hobbes and Locke brought back into philosophy the position common to most of the Pre-Socratic Greek thinkers, attacked by Plato and Aristotle, but continued by the Stoics and Epicureans. It had always been rejected by Christians with the possible exception of Tertullian. Nevertheless, it is the contention of this book that the materialists have an important contribution to make to thinking which Christian theology has too much neglected.

We must not neglect to note that against this growing materialism there was considerable resistance in the universities, especially on the part of the so-called Cambridge Platonists (Ralph Cudworth, d.1688, Henry More, d.1687, etc.) who defended innate ideas and, turning against Calvinism, laid the ground for the Armininian latitudinarian theology in the English Church. [27]

II: Christian Reaction to the Rise of Humanism

1. The Rise of Empiricism

The main factors in the rise of empiricism were on the one hand the work of David Hume (d.1776) who derived from Locke's epistemology a thorough going pragmatic scepticism, and on the other the great scientific achievements in chemistry of Robert Boyle (d.1691) and in physics of Isaac Newton (d.1727). Boyle and Newton, in spite of Hume's corrosive scepticism, were able to demonstrate that at least in the field of the physical sciences by the painstaking pursuit of the Galilean methodology and mechanistic type of theory, it was possible to arrive at mathematically universal laws that were empirically verifiable in a way that to most minds amounted to the genuine certitude which Descartes had sought but not achieved. [28]

In Chapter 3, we saw that in England beginning with Herbert Lord Cherbury (d.1648) a deistic theology had begun to develop as an antidote to the religious controversies which had resulted in civil war. This Deism, absorbing earlier universalist tendencies such as that of Socinian Unitarianism within the Radical Reformation[29], provided a "natural religion" transcending all the bitter religious wars. We have seen too how this Deism was the first phase of the new religion of Humanism which separated itself definitively from Christianity, with Locke, Hume and Newton as its first "theologians". Because of the prestige given by Newton's remarkable scientific discoveries, Humanism passed by way of Voltaire to

France and then to Germany where, surprisingly it fused with the rationalism of Wolff.[30]

Thus by the beginning of the eighteenth century the Christian Churches, deeply divided against themselves and exhausted and discredited by these struggles, found themselves confronted with a new religion as formidable as Islam had been for the divided Church of the seventh century, a consistent world view in which revelation, grace, and ultimately God were no longer necessary, and in which man was encouraged to exploit to the full the wonderful new possibilities that Newtonian physics and the new technology were opening up.

It should not be forgotten, however, that the founders of this new science which was to provide the apologetic miracles for the new religion were not only Christian but often strongly motivated by the desire to defend through their science a theistic view of the world against the rising scepticism of the seventeenth century. It was only the blindness of Christian theologians, too much preoccupied with old quarrels, and their inability to integrate this new science into Christian theology, which made it possible for science to be so easily co-opted by Humanism. One of the difficulties of such a theological task was that it had to be faced without adequate philosophical tools. Scholasticism had been long ossified and the Ramism and Cartesianism which had taken its place proved hopelessly inadequate to this new task. As a result, throughout the eighteenth century Christian theology was engaged in a struggle merely to survive.

The dilemma presented by Humanism to Christianity was terrifying. On the one hand there was the scepticism of Hume which imperilled any possibility of certitude about a divine revelation made to humanity or of any miracles or prophecies witnessing it, and on the other was the mechanistic universe of Newton in which everything was determined by natural law, excluding human free will or revelatory events, but which could be known with a clarity that would have satisifed Descartes, not by ideas innate to a spiritual mind, but by a scientific cognitional process which could itself be reduced to sensation and that again to material mechanisms. The human person had been reduced at last to matter in motion.

2. The Experiment with Idealism

In Chapter 3 we saw how Immanuel Kant (1724-1804) provided an impressive solution to this apparent opposition of Humean scepticism and Newtonian science for the Humanist world-view, and thus made Humanism the dominant religion that it is today. Christian thought, therefore, had to face up to Kantianism and the many systems which sprang from it and to ask itself whether or not this new philosophy might be so modified as to be of service to the Christian world-view as had been the philosophies of Greek paganism.

Kant himself came of a pietistic Lutheran family and was educated in the text-book metaphysics of Christian Wolff. This Cartesian-Leibnitzian rationalism strongly influenced the work of German Lutheran theologians who found it not ungenial to the scholasticism of Lutheran Orthodoxy which I have previously described, such as that of J.G. Reinbeck (d.1741), Jacob Carpov (d.1768), I.G. Canz (d.1753). Peter Reusch (d.1757), and most influential of all, S.J. Baumgarten (d.1757). This rationalism also influenced German Calvinists such as Daniel Wyttenbach (d.1779), J.F. Sutpfer (d.1775), J. Chr. Beck (d.1785) and Samuel Endeman (d.1789).[31] Most of these theologians appear to us now hopelessly obscure, but they wrote useful textbooks and it is largely through textbooks that the pre-understandings we bring to reading the classics are usually formed. On the other hand, the very foundation of the Reformation began to be questioned by the rise of a historico-critical approach to the Scriptures already suggested by Spinoza, in the work of H.S. Reimarus (d.1768), G.E. Lessing (d.1781) and S.J. Semler (d.1791, often considered the "father of liberal Protestantism").[32]

During this same period Catholics continued a very learned scholastic theology, and also the exploration of patrology and church history by such groups as the Jesuit Bollandists and the Benedictine Maurists, and by isolated individuals like Richard Simon (d.1708) and Jean Astruc (d.1753), who helped to initiate the historico-critical study of the Bible. Among the theologians can be mentioned the Jesuit Dionysius Petavius (d.1652), who established a more historical approach to theology; Dominican Thomists such as Cardinal Gotti (d.1742) and Daniel Concina (d.1756), and Franciscan Scotists like Claude Frassen (d.1680), along with a host of not very effective apologists against the Humanist philosophies.[33]

In Germany too there were Protestant efforts to oppose rationalism by introducing pietistic elements into theology, by J.A. Bengel (d.1752), F. Ch. Oetinger (d.1782) and Ch. A. Crusius (d.1775), while in England similar but more philosophical efforts were made by Joseph Butler (d.1752), William Paley (d.1805) and especially by Bishop George Berkeley (d.1753) who in a very original way turned Locke's empiricism into a theistic idealism. Finally there was a real resurgence of a mystical, sometimes anti-intellectual enthusiasm through such diverse but charismatic figures as the founder of the Quakers, George Fox (d.1691), the visionary Immanuel Swedenborg (d.1777), the founder of pietistic communities Count Zinzendorf (d.1760), and the great evangelist and founder of Methodism within the Anglican Church John Wesley (d.1791).

Among Catholics too, there was much mystical writing both in orthodox and Jansenist circles, and great activity in the foreign missions.[34]

I have listed some of these Christian activities during the eighteenth century to make concrete the assertion that in spite of the very marked decline in the institutional church during this century and its apparent helplessness against the rise of Humanism, it was nevertheless a time in which there was important theological activity. Moreover, a fundamental transformation was beginning to occur in the conception of theology, namely, a realization of the *historical* character of biblical revelation and of church tradition. This transformation was taking place gradually under the surface of the prevailing, very unhistorical rationalism which was the legacy of Cartesianism. In time it would mean a new understanding of human bodily existence, because (as I will argue later in this book) it is by our bodies that we exist in the flow of historical time.

The direct confrontation of Christianity with Humanism, however, took place in terms of the new notion of *truth* which Kant introduced in what he called his "Copernican revolution." Although Descartes' rationalism had already tended to replace the medieval correspondence theory of truth with a consistency theory, yet this consistency theory depended on the notion of innate ideas which the British empiricists had discredited and which had ended in the dilemma between Hume and Newton which I have described. Kant sought to escape this dilemma by replacing innate ideas with innate categories which the human mind by its very structure is compelled to use in giving order to the data of sensation. Thus we interpret the world in terms of such causal laws as those discovered by Newton not because we can detect causal relations in the data of sensation (Hume had showed this to be impossible, Kant thought), but because all human minds are so constructed as inevitably to make this kind of causal interpretation.

According to Kant what we naively believe to be the truth about the objective world apprehended by us through our senses, including its location in space and time, is in fact constructed by us mentally by a kind of hermeneutic process out of the intrinsically unordered data of sensation. Nevertheless, the laws of Newtonian science remain valid because we all have the same kind of minds. Obviously, this means that the bodies that make up our universe and our own human bodies are known to us only phenomenally and not as they are in themselves (noumenally). But it also means that our souls are not known to us noumenally either, but only the stream of our thought as it actively constructs the world out of our sensations. Thus, although for Kant the two substances Mind and Matter which Descartes posited lose their substantiality, yet we do not need

215

to succumb with Hume to scepticism, since the natural laws of science are fully justified by the fit between the data and our mental categories.

Nevertheless, although Kant believed he had saved the truth of natural science, he contended that we must abandon any hope of arriving at objective truth about the noumenal world, about any substantial existent as such, and about entities which do not appear within the horizon of our sense experience. There is no way to know, he believed, that the law of causality or any other of the traditional "metaphysical principles" hold when applied to anything else but our sense data since our minds are structured to deal with only such data. Hence the existence of the human self, the person; the survival of the soul after death; and the existence of God become impossible to demonstrate. Even the existence of any teleology inherent in nature becomes uncertain, since science only establishes mechanical laws.[35]

Kant, however, realized that such a view of the world, even if it seemed to guarantee the validity of Newtonian science, was not sufficient to constitute the kind of rational philosophy of life which he and other Humanists were attempting to construct to replace the old traditional Christian revelation which had darkened the world with its fanatical wars. Consequently, he argued that although there cannot be an objective scientific justification for theism, for morality, and for the eventual triumph of good over evil, yet there can be a subjective and pragmatic justification of these beliefs as necessary for an orderly personal and social life, since no other justification in practical matters than a practical one is required.[36] Here Kant came rather close once more to Hume's own system of morals, since Hume also believed that humanity would act more reasonably if it ceased to try to ground the norms of behavior on revelation or some metaphysical natural law and based them instead on the commonsense of experience. Moreover, Kant came close to Rousseau, whom he admired (as Hume did not), since Rousseau grounded morality on universal human instincts unspoiled by a distorting culture and expressed socially as a "general will", somewhat as Kant himself grounded morality on the universality of practical reason.[37]

To Christian theologians this pragmatic moralism presented a dangerous temptation. Would it not be possible to save the Christian religion by giving up the attempt to justify it ontologically and objectively, and instead to defend it pragmatically? The trouble with this of course, as Kant had himself fully realized, is that pragmatically it is difficult to show that any one human religion which supports morality and social order has a privileged place. The privileged place of Christianity as the support of morality was precisely what Humanism had come to deny.

From all sides Kant's total system fell under severe criticism and to-day is defended in its original form by few writers; but its notion of truth, and its attempt to found a world-view in terms of our need for a moral order that gives meaning to our lives continues to have the widest influence. Because his philosophy remains the basis of philosophical instruction in Germany (as does that of Descartes in France, or Locke and Hume in England) and because the greatest centers of Protestant biblical and theological scholarship are German, Kantian presuppositions underlie much of Protestant theology today. Moreover, the rising influence of Germanic influence in Catholic theology since Vatican II is tending to a similar result for Catholic theology.[38]

In its original form Kant's system seemed inimical to Christian theology rather than an applicant for a job as its handmaid. Like the philosophies of Plato and Aristotle, it required considerable re-training before it was suitable for the task. The first step in this direction was taken by the idealist philosophers Fichte, Schelling, and Hegel, whose efforts were then put to direct theological use by the Protestant Friederich Schleiermacher (d.1834) and by some German Catholic theologians such as those of the Munich school led by Joseph Görres (d.1848) and George Hermes (d.1831) of the University of Bonn. As explained in Chapter 3, this idealist interpretation of Kant is a result of the polar opposition that exists within Humanism.[39] While Kant himself was especially concerned to give a philosophical justification of Newtonian science, the Romantic wing of Humanism to which Idealism belongs reacted strongly to the Newtonian mechanization of the world, although it was not antagonistic to the progress of the study of nature.[40]

It might appear that this new polarization between romantics and mechanists was simply the old opposition of Platonists and Aristotelians in a new form. No doubt there is an analogy between these two polarizations; but the older opponents shared a correspondence theory of truth, while the new opponents were agreed on the new Kantian conception of truth. Kant's critics very quickly pointed out that his sharp opposition between the knowing "subject" and the "object" known raises many difficulties. For Plato and Aristotle the "objects" of knowledge are intrinsically intelligible because they possess an orderly, teleological structure in themselves, in their noumenal reality. They reveal themselves to the human knower whose mind is of itself dark and formless, requiring to be illumined and formed by the object, although capable of seeing the object when it is presented. Yet neither Plato nor Aristotle were "naive realists", as they are sometimes accused of being. Both granted that this illumination of the mind by the object does not occur without a learning process in which the mind is actively cooperative. For Plato this process is a kind of awakening of the mind from forgetfulness to a recollected attentiveness. The mind turns inward to perceive the light that comes to it from the *ideal* objects already innately present in the soul.[41] For Aristotle this preparation requires an illumination of the sensed object by the "agent intellect" by which whatever is accidental in the sensed object is removed (abstraction) so that the intrinsic intelligibility of the essential noumenal object can be intellectually seen. This process is very gradual and is achieved only by a careful comparison and analysis of our sense experience, including the experimental exploration of the objective world in order to assist this process of separating the essential from the accidental.[42]

For Kant, on the other hand, the "object" as such is without intrinsic order or light, it is dark, opaque, meaningless, just "data." The phenomena apprehended by the senses in no way reveal the noumenal reality which we are spontaneously convinced must underlie them. Consequently all illumination must come from the human psyche, first at the level of perception and then at the level of thought, giving an intelligible order to the data. The "subject" as self-consciousness is, as Descartes had seen, of itself light and transparent, but it constantly finds itself opposed by the opaqueness of objects. Thus we seem faced with a dualism even more painful than that of Descartes' Mind and Matter, for Descartes at least was convinced that Mind can know Matter mathematically, while for Kant even the spatial and temporal order on which mathematics is based is also a construction of our act of sensing, not something in the object sensed.[43] The scientistic wing of Kantianism soon came to interpret this Kantian dualism in a *positivistic* way by claiming that all we really know are the data, while the theories by which we interpret them are more or less arbitrary constructs having only a pragmatic justification.[44]

The idealists, however, tried to overcome this dualism in the opposite way by claiming that it is the mind which generates the object and then reassimilates it to itself by encompassing it within its own illuminating horizon. In order to get over the paradox in Kant's system which results from the assumption that in order to guarantee the universality of Newtonian natural laws, all human minds must have the same structure (although our noumenal selves, our own and those of others, are hidden from us), the idealists went beyond the phenomenal human consciousnesses to an Absolute Mind in which our finite human minds participate. It is in this Absolute Mind that subject and object are opposed and then synthesized in an eternal dialectic.[45]

It was Fichte (d.1814) who proposed the tripartite thesis-antithesis-synthesis process by which the Absolute Mind creates the world by *willing* it to be and then thinking itself through to a perfect reunification.[46]

218

In Fichte the late medieval voluntaristic conception of God as primarily will and freedom thus reappears, reflecting both Kant and Hume's pragmatic tendency. Schelling (d.1854) developed this dialectic still further. For him Nature (the world studied by science, no longer conceived as Newtonian clock-work, but instead *organically* as informed by a World Soul) is the Absolute Mind's objectivization of itself. This objectification is necessary in order that the Absolute can become fully transparent to itself, fully self-conscious not just in a general way but in the concrete details of every possible realization of itself. We note here the influence on these idealists of Spinoza who had overcome the Cartesian dualism by considering Mind and Matter as two modes of the same Deity.[47]

Finally, it was Hegel (d.1831), building on the work of Schelling, who conceived this dialectic not just as some abstract, timeless process, but as human *history* in which Nature comes to self-consciousness in humanity, and the Absolute Spirit re-unifies all things in those human minds which achieve a share in this unified self-consciousness (subjectivity) by transcending every partial point of view and identifying with the Absolute Spirit. Hegel called his system *objective* idealism because he believed it was an exploration of the life of the subject which does full justice to objective reality, that is to the total stream of phenomena of nature and of history in their concrete, temporal actuality.[48]

Ancient and medieval philosophy and Cartesian rationalism as well had taught that "science is of the universal and necessary", so that a science of history, i.e. of the singular and contingent, was declared impossible. The Catholic philosopher Giovanni Battista Vico (d.1744) had attempted to refute this ancient opinion earlier in the century but he had little influence until later.[49] Consequently it was Hegel who revolutionized modern thinking about history, supported by other romantics of less powerful philosophical gifts as J. G. Herder (d.1803) and the Catholic Friederich Schlegel (d.1829), who gave to history that centrality which it now occupies in modern thought.[50] While Hegel did not include biological evolution in his scheme, the model which he provided of all history as a kind of organic process prepared the milieu in which in a few years after his death the scientific world-picture itself was to be transformed from the Newtonian natural law model to the evolutionary historical model.[51]

Hegel had begun his career as a student of Protestant theology and he sincerely regarded his philosophy as a Christian theology which culminated in the doctrine of the Incarnation and the Trinity. Although he considered philosophy superior to religion, he seems to have meant by this that the human spirit attains its full development not at the level of popular religion but at the level of reflective understanding of the con tent of popular religion. In other words, for Hegel, philosophy is Christian theology. The question is thus whether German idealism is Humanist as Kantianism clearly was, or rather an effort to reformulate Christian faith in a philosophy derived from Kantianism. Certainly many interpreters of Hegel understand his philosophy as Humanist, as a pantheism or even an atheism, since it is he who first said, "God is dead."[52]

Perhaps the best way to understand Schelling and Hegel is to say that like many of the romantic wing of Humanism, they desired to preserve for Humanism the immense historical values they appreciated in Christianity and which they believed the scientist wing of Humanism had unthinkingly rejected. They hoped by their comprehensive philosophical systems to include this heritage in a new view of reality in which the particularism of Christianity was overcome. Yet their systems (I do not speak of their personal faith on which the opposite judgment might be valid) remain essentially Humanist because they rest on human reason, not on faith in the Word of God. The content of Christian faith is included in these systems by an hermeneutic process, but the guarantee of its truth rests ultimately not on faith, but on its consistency (as re-interpreted) with a rational system. Of course this only amounts to saying that their systems are not theological but philosophies; but the point is that quite explicitly for Hegel his philosophy *is* a theology in the sense of the *ultimate* worldview. Evidently this claim for the ultimacy of reason is not a Christian, but a Humanist position. Nevertheless, it is clear that Hegel, by forging a philosophy of history, provided Christian theology with an instrument which could be adapted to an historical hermeneutic of the Bible which the older philosophies had failed to supply.[53]

To sum up: the German idealists, Hegel in particular, made a major and irreplaceable contribution to the Christian theological understanding of our bodily existence by showing that human nature can never be adequately understood if we investigate it only as the Greeks and medievals did in its abstract universality. With Vatican II, the Catholic Church has acknowledged that theology (to be true to the Scriptures and to the very nature of Tradition, and doctrinal development) must take full account of the fact that to be human is to live in history and to make history. Since it is because we are not merely minds, but minds that think and act through bodies and in a world of bodies, that we are in the flux of time and change (sharing its dialectic of determinism and freedom), therefore to recognize our essential historicity is, contrary to Plato and the whole idealistic tradition, to recognize that our bodies are not adjuncts of the self, but essential to it. Thus, paradoxically, the nineteenth century idealists in recognizing human historicity refuted their own Idealism.

III: Compromises

1. Protestant Approchement

In Protestant circles, the effort to use Idealism as the philosophical instrument for the reconstruction of a theology, a revision whose necessity grew more and more acute throughout the nineteenth century with the advance of the historico-critical study of the Bible, moved through a series of stages. These can be conveniently schematized by saying that Schleiermacher (d. 1834, of Calvinist and pietistic background, but a friend of Schelling) dominated the first phase of what has ever since been called "liberal Protestantism."[54] Friedrich H. Jacobi (d.1819) had already criticized Kant's dualism and attempted to refute it by arguing that it can be overcome by a kind of innate faith which assures us of the reality of the supersensible. Schleiermacher similarly hoped to overcome this dualism by an innate faith, a sense of human dependence on the Divine. He was content to leave to philosophy and science the whole range of concrete data, reserving to theology the realm of subjectivity, of feeling and intuition. This latter realm of piety was essentially the awareness of our total dependence on God. Through this sense of dependence we are aware of God not in some remote heaven, but as present in our empirical world, the very same that is explored by philosophy, science, and practical living. Thus theology and religion deal with no distinct subject matter, but with a dimension of the subject matter common to all other disciplines. This is the *ethical* or value dimension of human life. Schleiermacher, with great sensitivity, proceeded to explain the whole content of the Bible and of traditional Christian dogma in such a way as to understand it as an expression of this religious sense. For example, Christ becomes the man in whom this sense of total dependence on God becomes fully manifest to us and who awakens in us the same sense. Therefore, an ontological Christology loses all importance. Kant's substitution of morals for metaphysics is thus fully operative in Schleiermacher's theology.

A second phase of nineteenth century Protestant theology appeared with the sensational *Life of Jesus* of David Friederich Strauss (d.1873) in 1835, followed by the critical biblical studies of the so-called Left Wing Hegelians such as Bruno Bauer (d.1882) and Ferdinand Christian Baur (d.1860). These Hegelians interpreted Hegel in a Humanist sense, and in the light of his philosophy of history raised the question as to the reliability of our historical sources for the knowledge of Jesus, thus forcing liberal Protestantism more and more to see Jesus as a symbolic rather than a historical figure. At the same time, the revival of evangelical religion which took place in mid-century, was reflected in the "mediating theologians"

221

who in various ways attempted to reinterpret the old confessional formulae in the new idealistic terms.[55]

A third phase occurred in the latter part of the century when Darwinism and positivistic materialism once more got the upper hand. It began with the abandonment of the speculative system of the romantic idealists and a "Back to Kant" movement. The chief figure was Albrecht Benjamin Ritschl (d.1889), who proposed a strongly moralistic theology based on the concept of the Kingdom of God, resolutely rejecting every metaphysical concept in the interpretation of the Bible, even to the point of arguing that Luther had failed to free Christianity from metaphysics and proposing Kantian moralism as the true essence of Protestantism.[56] Thus by the end of the century it seemed that in Germany, the home of the Reformation, Protestant theology had become almost indistinguishable from Humanism. In such an anti-metaphysical theology, the spiritual subjectivity which idealism had attempted to perserve had been reduced to psychology, and psychology itself was beginning to be studied in scientific terms that reduced mind to matter in motion.

In striking opposition to these reductive tendencies was the work of Søren Kierkegaard (d.1855) whose greatest influence was to be felt in the twentieth century. Rejecting what he understood as Hegelian "rationalism", Kierkegaard argued that reason can only lead us to false gods, since, as Descartes saw, reason begins in doubt, in rebellion against God. God cannot be reached by abstract, universal, essentialist reason, but only by an existential faith, "a leap in the dark." The human will also, because it seeks happiness for itself, will inevitably turn us from God who can be reached only through suffering. Nor can we find God through the state or even organized religion, but only individually and in a purely interior way in which we stand before God in "fear and trembling." Kierkegaard's fideism certainly seems like a return to Luther's own stance, except that Kierkegaard is even more radically individualistic than Luther, because Kierkegaard no longer finds himself in a Church requiring reform but in a world no longer Christian.[57] In spite of Kierkegaard and the survival of Lutheran and Reformed Orthodoxy among theological conservatives, there is no doubt that "Liberal Protestantism" was in the ascendancy by the end of the nineteenth century, and for this type of theology the doctrines of Incarnation, Redemption, and Resurrection were embarrassments that had to be reinterpreted in a purely psychological and moral sense. Moreover, morality was grounded not in human nature but in the Kantian *a priori* of the Categorical Imperative. In such a perspective the human body has purely secular significance without any special theological meaning. Even for Kierkegaard its theological significance is simply as an element of human frailty before the Divine Imperatives.

2. Catholic Approchement

For Catholics also the nineteenth century was a time in which Humanism in a series of battles step by step took control of political and intellectual institutions and reduced Christian faith to a purely private status. This Humanist campaign was called *laicism* as if it were a battle of the laity against the clergy, but in fact it was the rise to power of a new spiritual elite. First there was the direct onslaught on the Church by the French Revolution. This was followed by Bonapartism, the Holy Alliance, etc., in which Humanists compromised for a time with the still largely Christian majority until their power might be consolidated. The revolutionary movements of 1848 brought a second reaction in France in the form of the Second Empire, but the "liberal" forces again gradually gained ground until the Third Republic (1870) saw its complete triumph. The basic reaction of French Catholics was the *Traditionalism* typified by the writings of Joseph de Maistre (d.1821) and Louis de Bonald (d. 1840). The brilliant Félicité de Lammenais (d.1854) gave this traditionalism a liberal pro-democratic and Rousseauan turn. Such Traditionalism can be understood as an attempt by Catholics to find an ally in the romantic wing of Humanism. Rejecting Cartesian rationalism in which it saw the seeds of the Enlightenment and Revolution and which it often blamed on the influence of Protestantism, the traditionalists argued that without the guidance of revelation, a revelation given primitively to Adam and passed down through the ages in the tradition of the Church, reason inevitably leads to blind irrationality.[58]

Reacting to Traditionalism, some French thinkers defended the Cartesian heritage, relating it to the Augustinianism of great spiritual writers such as Bossuet (d.1704) and Fenelon (d.1715) and the Ontologism of Malebranche.[59] Louis Bautain (d.1867), perhaps the most important theologian of this period in France, combined Traditionalism with a belief in an intuition of divine realities. He linked this with the German idealist distinction found in Kant between the lower discursive reason proper to science (*Verstand*) and the higher intuitive reason of self-consciousness (*Vernunf*) which Bautain believed, with some of the idealists, to be capable of reaching noumenal, supersensible reality. A similar but more cautious combination of Traditionalism and Ontologism came also to dominate the Catholic University of Louvain which was under even stronger German influence.

Other French thinkers, however, tended to draw away from traditionalism and to rely more strongly on the Cartesian and ontologistic tradition. In this they were influenced by the Italians Antonio Rosmini (d.1855) and Vincenzo Gioberti (d.1852), the latter of whom favored an outright Ontologism by which all truths are known in God. This Ontologism had wide influence, for example in America in the thought of Orestes Brownson (d.1876), a former Protestant and member of Emerson's Transcendentalist group. Various lay philosophers, notably Maine de Biran (1824) and Victor Cousin (d.1867), attempted to develop a "Christian philosophy" along Augustinian and frankly Platonic lines, but also clearly influenced by Kantianism and Idealism. Cousin, a great power in the University of Paris, opposed the orthodox Catholics, yet defended the Catholic tradition by a kind of Hegelianism open to Christian values.[60]

In Germany Catholics were also faced by the aggressive attitude of the state which more and more attempted to enforce a public conformity to secular (i.e. Humanist) standards. The universities of Germany in the second half of the nineteenth century became the most advanced and influential in the world because of their dedication to scientific and historical research. A few of them admitted Catholic faculties of theology along with the Protestant ones. Thus under the influence of Friederich Schlegel, Johann Sebastian von Drey founded the Catholic faculty at Tübingen in 1817, which produced a number of important theologians of whom Johann Adam Möhler (strongly influenced by Schleiermacher) is the most famous. This faculty was traditionalist and anti-metaphysical and sought to ground the philosophy it used for its theologizing in *Vernunft* conceived as a higher, supersensible intuition. Directly influenced by Schlegel, Schelling, and Hegel, it sought to develop a dynamic, dialectical theology of history and a theory of doctrinal development. In the work of this school the romantic notion of *organism* was extremely important, the Church and its faith being seen as undergoing organic growth through the inner life of the Spirit.[61]

At the University of Münster, on the other hand, Georg Hermes (d. 1831) took a more strictly Kantian line (semi-rationalism) by assigning to history the task of establishing the fact and content of revelation by purely scientific methods ("a positive theology"). He handed over the certitude of faith to the Categorical Imperative of Kant which obliges us to believe in God as a universal moral duty.[62]

Perhaps the most complete theological system developed in the nineteenth century was the work of the Viennese Anton Günther (d.1863) a friend of St. Clement Maria Hoffbauer the founder of the Redemptorist order.[63] Günther rejected Aristotelian scholasticism as it was still taught in Cartesianized form in the seminaries, because it remained on the level of mere *Verstand*. He wished to build on *Vernunft*, unitive self-consciousness, without, however, falling into the pantheism that seemed to have been the fate of the Protestant idealists. He thought that the source of this error had been the idealist conception of God as an impersonal Absolute. Instead it was necessary to go back to Descartes' *cogito ergo sum* and to understand this in a personalistic way as revealing the spiritual self-conscious "Ego-Idea," which develops into full consciousness by the principle of "contrapositional dualism", i.e. by dialectically opposing to itself a Non-Ego. Hence all of reality is to be understood not by a pantheistic obliteration of opposites, but as a dialectical interrelation of clearly distinguished opposites.

Thus the dialectical synthesis is not an Hegelian *Aufhebung* in which the thesis and antithesis are absorbed, but a dualism. The finite human Ego demands an Other, an infinite Non-Ego which is God. The human soul demands a body as its other. The human person is a mystery, "the world sphinx" in which body and soul are joined in a "marital union" that cannot be rationally explained, but only grasped in their unity in an intuitive way.

Although Günther accepted the Cartesian *cogito*, he did not accept Descartes' mechanical model of nature, but with the idealists conceived nature as an organic whole animated by a world soul. In order to give content to our intuitive notion of God, Günther resorted to Traditionalism. He thought that the first Adam had intuitive reason (*Vernunft*) in its pure state so as to be able to understand that God, as the Infinite Ego, must be Triune. This intuition was lost through sin, but restored in Christ, the New Adam. Thus Günther tended to close the gap between natural human knowledge and divine revelation, a cardinal point of Catholic orthodoxy.

The Holy See, concerned to protect two points of orthodoxy: (1) the distinction between divine revelation and human knowledge, and (2) the validity of natural human reason, rejected all these nineteenth century systems as either rationalistic or idealistic or both. Greogry XVI condemned the writings of Lammenais (1832-1834), Hermes (1835), and Bautain (1840); Pius IX the followers of Hermes (1846), Bonnetty (a follower of Lammenais, 1855); Günther (1857), and the system of Ontologism (1861). The opponents of the new systems, belonging to the older scholastic tradition at the universities of Strasburg and Mainz and at the Collegio Romano of the Jesuits (who were not so much scholastics as advocates of a positive theology) had little to offer when it came to meeting the needs of the times with any fresh formulation of orthodoxy. At the First Council of the Vatican (1869-1870) the bishops under Pius IX strongly reaffirmed the two points just mentioned but were not prepared by their theologians to speak in more positively inspiring and illuminating terms.[64]

Thus the confrontation with Kantian thought and the Humanism of which it was the leading theological justification had been a battle to the draw. In a hundred years war both Protestant and Catholic thinkers had struggled either to refute Kant or to find a way that his philosophy might be modified so as to be a suitable instrument of Christian theology. Why had they failed? Because an idealistic interpretation of Kant led them into an immanentism which called into question the distinction between Creator and creature; while a scientistic interpretation of Kant closed off the possibility of any rational defense of transcendent reality and left only a reduction of religion to morality.

This narrative may seem a long way from the theme of a theology of the body, but it has everything to do with it, at least in that it demonstrates that without an adequate way to deal with what modern science daily reveals to us about the physical world and our bodily existence in it, Christian theology in the nineteenth century was either stagnant or hopelessly adrift. The romantic flight from the objective realities with which science deals has resulted in the isolation of Christian theology from the modern world and its own best traditions.

IV: The Dialogue with Scientistic or Positivistic Humanism

1. The Thomistic Revival

In Chapter 3, I showed that after Darwin's proposal of a plausible theory of biological evolution a thoroughly materialistic world-view became entirely credible and the atheistic philosophies of thinkers like Feuerbach, Marx, and Nietzsche rapidly gained ground. Shortly after the beginning of the twentieth century Sigmund Freud would give Feuerbach's contention that God is only an illusory projection of human ideals (a view which after all was already implicit in Kant's system) a convincing clinical confirmation. In the face of this apparent triumph of scientific reductionism, the Romantic Idealism which Christian theologians had seized upon as their best friend in the Humanist camp was seen to be not an ally but a liability.

Among Protestants, except for a few isolated voices like that of Kierkegaard, Liberalism had largely surrendered to Humanism as far as the theologians, (but not the more conservative faithful) were concerned. Among Catholic theologians during the middle part of the century a consensus was growing that what was needed was a unified theological method firmly rooted in the Catholic tradition prior to the rise of the religious divisions of the Reformation and the new religions of the Enlightenment. This growing consensus was made official soon after the First Vatican Council when Leo XIII (Joachim Pecci) in his encyclical letter *Aeterni Patris* in 1879 required all Catholic colleges and universities to base their philosophical and theological instruction on the great scholastics of the thirteenth century of whom St. Thomas Aquinas was singled out as the safest guide.

Thomas J. A. Hartley in his *Thomistic Revival and the Modernist Era*[65] has traced the stages leading up to this decision. As Bishop of Perugia, Joachim Pecci with the assistance of his brother Joseph (the more scholarly of the two) set up at the seminary of the diocese a center of Thomistic studies whose professors included the Spanish Dominican Paolo Carbo and the Italian Thomas Zigliara of the same order. Joseph Pecci had learned the essentials of Thomism from a Jesuit Serafino Sordi who had studied theology at the seminary of Piacenza with Vincenzo Buzzetti (d. 1824). Both Sordi and Buzzetti had learned their philosophy at the Vincentian Collegio Alberoni which was also located in Piacenza. This Vincentian Collegio had a Thomistic tradition going back to a Francesco Grassi who had begun to teach there in 1751 as a result of the decision of the Vincentian Congregation to adhere to Thomism in educating its members. At the same time other groups were at work attempting to revive Thomism, notably certain Spanish and Italian Dominicans, of whom the most influential was Salvatore Roselli whose textbook on philosophy (1783) became a principle resource of the revival; at the Collegio Romano the Jesuits, including the Sordi brothers, Luigi Taparelli d'Azeglio (Joachim Pecci was one of his students), Matteo Liberatore and Curci the founder of the *Civiltà Cattolica* which from 1850 carried on a campaign against "the errors of the day" and in favor of Thomism; Gaetano Sanseverino (d.1865) and Salvatore Talamo (d.1875) of Naples who were the first important writers in the century on an authentic Thomism; and finally the noted German Jesuit Josef Kleutgen became in Rome the chief advisor of the popes in the condemnation of the idealistic theology of Günther.[66]

It must not be thought, however, that the efforts of these men and of Leo XIII led to a general acceptance or understanding of the thought of St. Thomas Aquinas. At the beginning of this revival there was serious lack of the historical research necessary to make clear exactly how Aquinas differed from the other great scholastics. The proponents of Ontologism had claimed Aquinas was a Platonist. Many of the Jesuits thought that Suarez was the most reliable interpreter of the Angelic Doctor. Even the Dominicans, who had a continuous tradition, often presented Aquinas under the distorting influences of Christian Wolff (and hence of Leibnitz and Descartes) in an effort to up-date him, not to mention the fact that they were not fully aware of the biases introduced into their tradition even by such able commentators as Cajetan and John of St. Thomas.[67]

Moreover, it was not long until a kind of textbook Thomism, lifeless at best and at its worst a caricature, became the view most people had of Aquinas.[67] Leo XIII had hoped that St. Thomas would be taken as the basis of a vigorous development of a modern theology able to assimilate all the good in modern culture. What probably did more than anything to harden the Thomist revival and to frustrate this hope was the rise of the Modernist heresy and its severe repression by St. Pius X in the decree of the Holy Office *Lamentabili*, and the encyclical letter *Pascendi*, both issued in 1907.[68]

The Modernist movement in theology of which the chief figures were Alfred Loisy (d.1940) and George Tyrrell (d.1909), had its origins in the advance of historico-critical methods in exegesis but also in the continued influence of German Idealism. In the nineteenth century Cardinal John Henry Newman (d.1890) had put forward a remarkable theory of doctrinal development.[69] While Newman's philosophical education had been in Lockian empiricism, his own conception of "experience" was much broader. He was convinced that we have an intuitive awareness of supersensible, spiritual reality and of its workings in the drama of history. In this respect his thought independently joined that of J. A. Möhler of the Catholic Tübingen school[70] and the interest in human historicity of Romantic Idealism, yet Newman's understanding of history is not Hegelian. Because of these insights Newman was able to conceive the development of doctrine in the history of the Church in a less rationalistic way than did those Thomists who attempted to work out a theory of development as if it were a purely logical process of deduction from the articles of faith.[71]

The Modernists, on the other hand, under Hegelian influences, did not conceive the development of doctrine in a truly historical way, but in what Walgrave calls a "transformic" manner.[72] That is, they granted a historical continuity of Christian doctrine in the sense of a continuity of "inspiration", but not in the sense of a growth rooted in the permanent truth of a Revelation or Word of God given once for all. Pius X rightly saw (however lacking he may have been in a thorough understanding of the scholarly controversy) that this notion of development revived in even more seductive form all the errors against which his predecessors in the papacy had been struggling for a hundred years and which the First Council of the Vatican had definitively rejected. The result of his condemnation, as necessary as it was, was a period of almost fifty years during which Catholic biblical scholars worked in "fear and trembling" to integrate modern historico-critical methods of biblical exegesis with the scholastic theology of Aquinas, although M. J. Lagrange, O.P. and other biblical scholars had pointed out that the Thomistic theory of biblical inspiration provided an opening for this historico-critical hermeneutic which, in fact, Leo XIII himself had already approved in his encyclical *Providentissimus Deus* in 1893.[73]

Clearly what was needed was a new understanding of Thomism which would repair what, it was becoming clear, was its main defect if it was to serve as the basis of a modern theological revival: it's failure to take full account of human *historicity*, restricted as it was by the Aristotelian dictum that "science is of the universal."[74] Not that this weakness was that of Thomism only, since it underlay the whole struggle of the romantics, especially Hegel, with the even more anti-historical rationalism of Descartes, Leibnitz, and Kant himself. Unfortunately the ideas of Giovanni Battista Vico, of Rosmini who was influenced by Vico, and those of Newman had failed to penetrate Catholic circles where they should have had a better reception. Nevertheless, during the latter part of the nineteenth and the first half of this century there was great activity in historical scholarship among Catholics. The facts were accumulating, but their implications for theology were only slowly being felt because the Thomists, limited by their non-historical methodology did not know what to do with them.

The result was that Thomism, although it advanced remarkably before the Second World War toward an accurate understanding of the thought of Aquinas based solidly on historical and textual studies and began to be applied to many new problems, tended to split into a number of disparate schools. There are various ways to characterize these developments. Helen James John, S.N.D., in *The Thomist Spectrum*,[75] divides them into three principal trends: (1) the Christian Philosophy controversy, under which she treats "Strict-Observance Thomism" typified by Reginald Garrigou-Lagrange, O.P., and also the more progressive work of Jacques Maritain and Étienne Gilson; (2) the Participation Theory controversy (i.e. the problem of the Platonic elements in Aquinas' thought), under which she treats Aimé Forest and André Marc with their stress on a return to the concrete, Pedro Descoqs' attack on the real distinction of essence and existence and a defense of the Suarezian version of Thomism, Cornelio Fabro's discovery of the role of participation, L. B. Geiger's defense of the role of essence, and Ferdinand De Raeymaeker's contribution to the philosophy of science; (3) the Maréchal Tradition which produced the "transcendental Thomism" of Josef Maréchal, S.J., and such followers as André Hayen, Emerich Koreth and Karl Rahner, to which can conveniently be added the independent "cognitional theory" of Bernard Lonergan, S.J. This third trend arose from a direct dialogue with the thought of Kant and sought to discover in Aquinas' epistemology a certain *a priori* element. It is noteworthy that not all, but the majority of these authors in all three trends were concerned to minimize the Aristotelianism of Aquinas and to uncover Platonic elements or anticipations of Kant's critical philosophy in Thomistic thought.

Similarly Gerald A. McCool in *Catholic Theology in the Nineteenth Century: The Quest for a Unitary Method*[76] distinguishes four stages of Neo-Thomism: In the first stage St. Thomas was presented through the eyes of the classic commentators as an attack on the "becoming" of the modernists and as a defense of "being" as the fundamental *concept* of metaphysics, although Maurice Blondel and Pierre Rousselot, S.J., appealed for a greater emphasis on *intellectus* as against *ratio* i.e. on intuition in Aquinas' epistemology. In the second stage, between the World Wars, there developed three Thomisms: Maritain was orthodox but creative in his application of the system to new problems; Gilson stressed Thomism as an "existentialist" philosophy; and Maréchal and numerous writers of the University of Louvain began the dialogue with Kantianism. In the third stage, after World War II came the so-called "new theology" which had a Maréchalian basis and which opposed the sharp dichotomy traditional in Thomism between the natural and the supernatural by developing Aquinas' teaching on the "natural desire for God." Finally, the fourth post-Vatican II stage saw the triumph of Maréchal, Lonergan and transcendental Thomism, along with a vindication of the Catholic Tübingen tradition with its stress on history and its Hegelianism to which has been added strong new influences from phenomenology and especially from the thought of Martin Heidegger and his analysis of human historicity. Thus not only has pluralism become powerful in Catholic theology, but this pluralism is justified both by the transcendental Thomists and by the German romantic tradition as inevitable because of the historicity of human existence. For some conservatives, alarmed at the results of Vatican II, this sounds as if the heresy of Modernism has taken over Catholic theology. It would be truer to say that the problems raised by the modernists are still with us, and Catholic theologians believe that Vatican II called them to attack these problems honestly and fearlessly.[77]

It is clear from McCool's account as well as that of John's that the central issue as regards Thomism is the conflict between the Aristotelianism of Aquinas and the post-Kantian notion of truth and most particularly the historicity of truth. Fabro, who has done so much to uncover the Platonic element in Aquinas, nevertheless has remained firm on the point that Aquinas' thought is essentially Aristotelian and that his assimilation of the Platonic theory of participation is simply another of the many historic attempts to reconcile Plato and his great pupil. Aquinas' way of effecting this reconciliation was not by Platonizing Aristotle, but by retaining the whole of Aristotle's epistemology of intellectual knowledge grounded in sense experience and his act-potency cosmology, while extending these doctrines so as to include the unifying notion of "participation in being" in metaphysics to fill a great lacuna in Aristotle's system.[78]

230

It is obvious, however, that Maréchal and his followers wish to break with the Aristotelian epistemology and to revise Thomism in view of Kant's "Copernican revolution" in the theory of truth, of course not without an acute criticism of Kant's own system. The theology of Karl Rahner, S.J., so powerfully influential at Vatican II and since, is the finest fruit of this transcendental Thomism.[79] In order to effectively incorporate the historicity of human existence and of God's dealings with humans, it seeks to find an *a priori* element in human knowledge, so that it is the continuity of human thought as a quest of the Infinite through history which gives structure to history and conditions the way in which God has historically revealed himself. Thus this type of theology finds itself in opposition to the Aristotelian tradition in Catholic theology and more sympathetic to the Platonic tradition in which the intuitional element of thought, prior to the stream of concrete objective data, was preserved. It might seem paradoxical that Plato who was even more a defender of eternal, timeless truth than was Aristotle, should be found more congenial by these historical minded thinkers. It is the centrality of the subject for Plato as for Descartes and Kant which makes Plato more "modern" and more congenial to this trend of thought, since for Transcendental Thomists, as Rahner has said, "theology is first of all anthropology"[80] and by "anthropology" they understand the self-exploration of the subject.

On the other hand the Thomists who have rejected the "transcendental" approach are generally agreed with Gilson and Fabro that the special contribution which Aquinas made to Aristotelianism was the real distinction of essence and existence and the primacy of the *actus essendi*, grasped not by an analysis of essential concepts, but by a judgment of *separatio*, i.e. the judgment that "being" is not exclusively material being. It should be noted that this way of defending the validity of metaphysics differs from that of the Maréchalians, who rest its validity on the *a priori*, the mind's reflection on its own action, although efforts are made by some Maréchalians (e.g. McCool[81]) to equate these two positions.

Among these non-transcendental Thomists, however, there is also an important split (very relevant to the theme of this book) which John and McCool do not take into account. This is the question of the relation between metaphysics and natural philosophy. One school of thought which for a long time dominated the Thomism of the University of Louvain regarded the Aristotelian philosophy of nature as obsolete, or (in the tradition of Christian Wolff) simply as a branch of metaphysics, and left all detailed study of nature to the positive sciences. A second, of whom Maritain was the leader, insisted on the present validity of a philosophy of nature, distinct from metaphysics, but restricted it to the task of establishing hylomorphism, and acknowledged the natural sciences as two new types of science (the *empirio-metric* and the *empirio-schematic*) not recognized by the medieval scholastics. A third, of whom Robert, Geiger, Kane, and Smith were spokesmen, held that the Thomistic phiosophy of nature remains valid today, although requiring revision in view of the factual discoveries in the sciences, and can be of great importance in resolving the confusions which plague current scientific thought in interpreting its own theories.[82]

Thus the first two schools maintain that Thomist metaphysics is totally independent of the philosophy of nature, although they do not deny that modern metaphysicians should be cognizant of the results of the sciences. The third school holds that although metaphysics is formally independent of the particular disciplines including the natural sciences, it is materially dependent on the establishment by natural philosophy (which is identical with the natural sciences critically understood) of the existence of non-material substantial entities as ultimate causes of the existence and processes of material entities and hence of a wider sense of *being* than the physical. This position defends the fundamental Aristotelianism of Aquinas and its present validity.[83]

The importance of this issue for our present purposes is that if Thomism is to have any relevance for a modern theology of the body some decsion has to be made on the relation between the metaphysics of Thomism and modern science. The attempts to by-pass this question which seem to underlie both the first and second of these forgoing positions are not very promising. As for the third position it must be confessed that so far it has found very little support, but I will examine its possible merits more critically in the next chapter.

The revival of Thomism can best be understood in retrospect as a Catholic effort to "return to the sources", to re-establish the continuity and rooted vitality of its own tradition, not as entrenchment in the past but as an assured beach-head for future deployments. Vatican II seems a confirmation of this understanding since the theologians who served it were for the most part well-grounded in Thomism, but determined to be open to new advances. Parallel to this Catholic "return to the sources" was the powerful theological movement somewhat misleadingly called Neo-Orthodoxy among the churches of the Reformation tradition. These parallel trends have tended to converge in the present concern on both sides for ecumenism, hermeneutics, and the social implications of the Gospel.

2. Recent Renewals

At the beginning of this century liberal Protestantism going back to Schleiermacher had taken a strong social, even socialist trend in the "Social

Gospel" of Troeltsch, Rauschenbusch, and Reinhold Niebuhr to such an extent that to many it appeared to be allowing itself to be coopted by Humanism. [84] In vigorous reaction Karl Barth (d.1968) a Swiss Reformed churchman, himself socially a progressive, became the prophet of Neo-Orthodoxy, although his thought showed the influence of Kierkegaard and the rising existentialist philosophies with their opposition to rationalism and cultural relativism.

Barth, confidently loyal to the Reformed tradition, rejected liberal Protestantism as a compromise with a religion of man, rather than obedience to the Word of God. He made war on any influence of philosophy on theology and on every kind of "natural theology", which he believed to be not an instrument of Christian theology (as the Thomists maintained) but a covert substitute for a theology based on naked faith. He even attacked every attempt, even that made by his colleague in the movement Emil Brunner [85] to find in the human person any point of insertion for the Word of God. For Barth humanity stands before God helpless, deaf and blind, just as the Reformers had contended. To God belongs the entire initiative and power of revelation and salvation. Barth's Catholic critics thought that he carried this doctrine so far that it raised serious doubts whether we really can hear the Word of God at all, thus opening the way to Feuerbach's atheism in which God becomes nothing more than a human projection. [86] As a result of such criticism, in his later writings Barth somewhat softened his language, but it is not clear that he ever admitted any real truth in these criticisms. [87]

In his vast *Church Dogmatics,* [88] Barth attempted to rethink and preserve the whole doctrinal content of Christian tradition in a thoroughly biblical way, free of philosophy, yet without accepting biblical fundamentalism, since he refused to identify the Bible with the Word of God, and without reverting to the orthodoxy of seventeenth century scholastic Protestantism. Indeed Barth's mode of thought, for all his opposition to conformism to Humanism and his rejection of philosophy, is recognizably Kantian in its presuppositions. Kant's influence is seen in Barth's rejection of the metaphysical aspect of theology, and in his tendency to reduce human faith to obedience and every theological formulation to a dialectical paradox impervious to reason.

Very different, but also influential, was the work of Paul Tillich (d.1965) who in his *Systematic Theology* [89] also attempted to save what he called the "Catholic substance" (matter or content) of Christian tradition by giving this tradition an interpretation in the light of the "Protestant principle" (form) according to which the content of faith is *symbolic*, so that no literal formulation of faith can be definitive. Although this sounds rather like the Thomistic view that all theological language is analogical, it seems rather to mean that these symbols are subject to changing interpretations through history which retain only an analogical continuity, a view resembling that of the Catholic modernists. Tillich was deeply influenced by the Romantic Idealism of Schelling and by the Rhineland mystics who had inspired Schelling.[90]

Dietrich Bonhoeffer (d.1945) from a Lutheran background came, as a result of the crisis of World War II, to criticize the traditional Lutheran Two Kingdom Doctrine and what he called "cheap grace" in favor of a more socially-engaged view of Christian discipleship.[91] At the same time he became radically critical of institutionalized religion for functioning as a refuge for the insecure, rather than as an effective witness to Christ. More radical still were the "death of God theologians" who after World War II, influenced by Hegel and Nietzche's notion that "modern man has come of age", argued that in our times religion must be replaced by a radical secularization.[92] With such talk liberal Protestantism seems to be reviving, although both among Protestants and Catholics there are evident in this last quarter of the century conservative movements (often of great popular appeal) calling themselves "evangelical" (or even "fundamentalist") in vigorous reaction to this lieberalization.[93]

More theologically significant are those who are intellectually progressive yet reject the liberal conformity to Humanism. To oppose this cultural conformism and to take a prophetic stance, they have found the eschatological, revolutionary, and communitarian ideas of the other secular religion of our times, Marxism, more theologically suggestive. Thus among Protestants Jürgen Moltmann has thematized a "theology of hope" in which the traditional Christian doctrines are rethought in terms of God as the goal of history bringing about historical change, rather than a God whose existence guarantees the legitimacy of the political status quo.[94] Among Catholics the most powerful theological movement since the end of Vatican II is that of the "political theology" of Johannes Metz,[95] a pupil of Karl Rahner, or the "theology of liberation" developed by Latin American theologians.[96] Catholics have begun to recognize that not only is theology historically conditioned, but that it is conditioned by class interests, so that it sometimes functions as an ideology to justify the existing order in state and church, and sometimes as an instrument of radical social change. The question is asked whether the theologian can be true to the Gospel unless he identifies with the oppressed whom Jesus came to liberate. Consequently, might not Marx's "materialism" with its concern for the real needs of the oppressed be a better instrument for theology, than "spiritualistic" philosophies which ignore or even conceal these needs? In such theologies the communitarian "Kingdom of God" which Jesus Himself preached tends to replace the Pauline language of salvation

234

by faith which seems to reduce the Gospel to a purely "spiritual" i.e. private, individualistic message.

Thus for both Catholics and Protestants the second half of this century is marked by an effort to overcome the long trend toward a theology centering in subjectivity, which I have traced from the late Middle Ages, by a stress on the eschatological character of the Gospel in the sense of a transforming power within history. History itself is less understood as spiritual autobiography or "salvation history" and more in the sense of a call to an active role for the Christian in "making history."

In spite of this general trend, however, we must note that the thinker who has most deeply influenced both Protestant and Catholic theology in this century has probably been the enigmatic philosopher Martin Heidegger (d.1976), although he was for a time associated with the most obviously unchristian movement of our dark times, German National Socialism.[97] Heidegger's thought derived from the phenomenology of Edmund Husserl (d.1938) and is in the line of the Idealism of Descartes and Kant, but Heidegger attempted to overcome the subject-dualism of Kant and the apocalyptic nihilism of Nietzche whose works he thought to be especially significant.

In his *Being and Time* Heidegger reflected deeply on human historicity and concluded that human "existence" (i.e. our subjectivity) becomes "authentic" only when we honestly face the inevitability of death, the perishing both of our individual egos and our "worlds" i.e. the historic culture in which each of us has been fated to live with its heritage of truths.[98] Heidegger rejected Idealism (to which Husserl seems to have succumbed) because the subject and the object, the human existent and the world through which he exists and which exists through him, are inseparably correlative. They come to be and perish together. What especially struck Heidegger was the prevalence of the illusion in contemporary culture that we are "making history" when in fact it is rather that history has made us. Thus for Heidegger the term "Being" no longer has the metaphysical sense of a noumenal reality behind the phenomena, but is the totality of humanity "thrown into" the stream of history and fated to perish in it, yet that history is not a reality outside humanity but is historical experience itself.

The crucial fact, however, for Heidegger is that "western man", since the time of the rise of Greek philosophy, has more and more "forgotten Being" because he has more and more been obsessed by the idea that he can *control* and master Being.[99] This progressive forgetfulness is itself the chief feature of western history which we have been doomed to work out to its final conclusion. Philosophy and science (and Christian theology insofar as it has deserted a purely religious attitude by becoming philosophical) have been the chief instruments of this forgetfulness of Being in its totality and the turning toward "beings" which can first be controlled in thought so that they can be controlled in technology.

For Heidegger this steady progress of which Humanism is so proud is ending in our century in a kind of apocalypse which Nietzsche, the ultimate figure in the history of philosophy, announced by his nihilism. It remains for us today to turn from philosophy to another mode of "thinking" which instead of trying to control the world by reducing it to "beings", patiently waits for Being to reveal itself in the next turn of history. Whether this new revelation of Being will take place when the West begins to learn from the contemplative attitude of the East, or through a revival of mystical religion (Heidegger himself requested and received Catholic burial) or through something quite different, Heidegger did not presume to prophesy, but only to call us back to a remembrance of Being.[100]

It is certainly astonishing that this cryptic and increasingly oracular kind of philosophy which purported to announce the overcoming of philosophy itself should have become so influential in recent theology. Yet the most influential of post-Barthian Protestant exegetes Rudolf Bultmann (d.1976) found in Heidegger a new hermeneutic which he believed would make it possible to make the Gospel relevant to modern man by finding its significance not in some historical account of past events, but as a challenge to our present situation in which we find ourselves living inauthentically.[101] In a very different way, certainly the most influential of Catholic theologians in the Vatican II period, Karl Rahner, has combined his "transcendental" Thomism with Heidegger's hermeneutical method in order to translate the content of Catholic tradition into the language of contemporary experience.[102] Again, philosophers strongly influenced by Heidegger such as Paul Ricoeur and Hans-Georg Gadamer[103] have developed a hermeneutic philosophy which promises to be of considerable help not only in biblical exegesis but in dealing with the whole problem of cultural and religious pluralism.

What is obvious is that it is this problem of pluralism which troubles all of current theology. Ecumenism demands that we try to interpret one Christian religous tradition to another, and this extends beyond Christianity to all the world religions, as I emphasized in Chapter 1 of this book.

In Great Britain and the United States where Phenomenology has not been as popular among philosophers as Analytic Philosophy, and recently Structuralism and Semiotics, theologians are more inclined to approach this problem of pluralism from the side of language than of phenomenological analysis.[104] Thought has expression only in language, and language, because it is social, reflects not just the subjective but also the objective world of common experience. Consequently by clarifying the language with which we communicate, untangling its confusions, we are also both recognizing and reconciling the plurality of our individual and group experiences. For theology, which uses its own peculiar "God language" so little understood in a world dominated by Humanism, this requires on the part of theologians an effort to find the basis of the Gospel in common human experience.

Thus we are brought back to the problem of the body. It is possible that Heidegger is right in saying that Western culture with its scientific obsession is coming to an end and Being is preparing a new revelation of itself. Nevertheless, as we await the future we find that our world is actually dominated by the language of science. This is now true of the East as well as the West. Even our efforts to revive the mystical, contemplative or esthetic traditions of the past, like our efforts to exegete the Bible for modern man by demythologizing it, or our efforts to make science humanly meaningful by mythologizing it, always involve us in the language of psychology, rather than of metaphysics, because psychology is "scientific." We cannot escape what science has discovered about the physical world in which we live and about our own selves as bodily beings.

For these reasons some regard Teilhard de Chardin, S.J., (d.1955) as the most significant twentieth century Christian thinker, not precisely as a philosopher or a theologian, but as a prophetic visionary.[105] Teilhard, a distinguished paleontologist, was strongly influenced by the philosopher of "creative evolution" Henri Bergson (d.1941) in developing a philosophy of history as an evolutionary process involving the whole cosmos. He gave this a Christian interpretation as the evolution of matter toward spirit manifesting itself in humanity and culminating in Christ as the Head of the Mystical Body, which will eventually encompass all of humankind and the whole universe. Although Teilhard left many gaps in his system, we must be grateful for his breadth of vision and his courageous conviction that Christian faith does not need merely an apologetic in the face of modern science but much more the wisdom to integrate what God is revealing about Himself to us through modern science into a "new theology."

Teilhard did not conceive this new theology as overthrowing any of the elements of Christian orthodoxy, but Pius XII in *Humani Generis* (1950) expressed alarm at the consequences evolutionary thought seemed to be having in this new theology, although the pope himself was fascinated by the possibilities of the dialogue between science and faith and expressed this hope in numerous allocutions.[106] Undoubtedly Teilhard's influence was felt at Vatican II and especially in the pastoral constitution *The Church in the Modern World*,[107] in which for the first time in an authoritative way the Catholic Church affirmed that the Holy

Spirit is at work in modern science. Unfortunately the conciliar documents, however, only vaguely reflect the lessons of modern science. Probably this was because of the reluctance of most theologians today (as I have indicated in this chapter) to deal directly with the data of objective science and their tendency to turn to a more subjective type of philosophy.

Bergson's influence, moreover, has also been reaching Protestant theology through the "Process Philosophy" of Alfred North Whitehead (d.1947) and his follower Charles Hartshorne, a system which I will discuss more fully in the next chapter.

3. The Lessons of History

We saw in Chapter 4 how the long rivalry of Platonism and Aristotelianism as instruments of Christian theology lead to the conclusion that a Christian anthropology must retain the Platonic defense of human spirituality and interior life, but it ought also to support Aristotle's criticism of Platonic dualism and accept the Aristotelian concept of the soul as the vital principle of the body and the body as the necessary instrument of the soul in thinking and willing.

In Chapter 5, we saw how from the time of the late Middle Ages philosophy and theology shifted from an ontological study of human nature to an understanding of human life in terms of relations of power, and in this context developed the notion of human spirituality as subjectivity or interiority. At the same time the study of the objective, material world in which the human subject is inserted by reason of its body was handed over to the rapidly developing natural sciences, whose method was merely to see the phenomena without pretending to any ontological or noumenal insight. In such a philosophy the human body and its physical environment are understood only as mechanisms. At the same time the human condition was seen in more and more dramatic and historical terms, sometimes optimistically as a share in the Divine creativity and sometimes pessimistically as a state of depravity, absolutely helpless before the Sovereign God.

In Chapter 6, we saw that Christian theology confronted by the new theology of non-Christian Humanism sought some common ground and found it within the romantic wing of Humanism described in Chapter 4, while regarding the scientific wing (also described in Chapter 4) as the enemy. This had the positive result that Christian theology was able to make fruitful use of the romantic exploration of human subjectivity and historicity, but with the negative effect of an Idealism incompatible with the Christian understanding of the relations of faith and reason affirmed at Vatican I and reaffirmed at Vatican II. At the end of the century a "return to sources" led Catholics to seek a firm basis for dealing with modern thought in the metaphysics of St. Thomas and Protestants in the neo-orthodox movement. Finally, we saw that while there is now a growing scepticism, expressed in the works of Martin Heidegger, concerning the ultimate value of modern science and its technological power on which Humanism has built its hopes, a Christian theology of today must take full account of the scientific world-picture outlined in Chapter 2. We must take a more positive approach to the material world of our experience, the world of which we are a part through our bodies. We must also take positive account of the advance of our technological control over nature and what this can contribute to the advance of the Reign of God in human society. At this point also a convergence with the other great secular religion of Marxism through a theology of liberation seems a possibiliy, as well as with the other world religions which have for centuries guided the human community whose flesh and blood we all share. In the following chapters I will propose in broad outlines what such an ecumenical, Christian and Catholic "theology of the body" might look like.

Notes

1. See *Cursus philosophicus* (ed. B. Reiser, Turin, 1933), 2, 844 ff. (This work was first published at Rome 1637-8). In Tractatus I, c.2, p. 847, he accepts the Ptolemaic system without question; in Tractatus II, cap. 1 he struggles to defend Aristotle's view of the Milky Way although he recognizes that experts reject it. In cap. 3, he mentions the novae of 1572 and 1603 but decides they were miraculous! For Dominic Soto's anticipation of Galileo see Alexandre Koyré's discussion in René Taton, ed., *History of Science: The Beginnings of Modern Science from 1450 to 1800* (New York: Basic Books, 1964), pp. 94-195.

2. See William A. Wallace, *Galileo's Early Notebooks: The Physical Questions* (Notre Dame, Indiana: University of Notre Dame Press, 1977); R.E. Butts and Joseph C. Pitts, eds., *New Perspectives on Galileo* (Dordrecht-Boston: D. Reidel, 1978). On the renewed interest in mathematics in the Middle Ages, see J.E. Murdoch, *"Mathesis in philosophiam scolasticam introducta"* in *Arts Liberaux et Philosophie au Moyen Age* (Montreal: Institut d'études mediévales), pp. 215-54.

3. On the Aristotelian conception of the relation of mathematics and physics see B.L. Mullahy, "Subalternation and Mathematical Physics", *Laval Theologique et Philosophique*, 2 (1946): 89-107.

4. See William A. Wallace, O.P., *Causality and Scientifc Explanation* (Ann Arbor: University of Michigan Press, 1972), 1, 117-183, and "Galileo and the Thomists" in *St. Thomas Aquinas 1274-1974: Commemorative Studies* (Toronto: Pontifical Institute of Medieval Studies, 1974), 2, 293-330; and A.C. Crombie, "Sources of Galileo's Early Natural Philosphy," in M.L. Righini Bonelli and R.W. Shea, eds, *Reason, Experiment and Mysticism in the Scientific Revolution* (New York: Science History Publications, 1975), pp. 160-3.

5. The unanswerable empirical objection against the heliocentric theory was not empirically answered until 1838 when Friederich Bessel first detected the annual parallax of the star 61 Cygni. Until then the acceptance of the theory of Copernicus rested simply on its mathematical elegance. See Colin A. Ryan, *The Ages of Science* (London: George G. Harrap and Co., 1966), pp. 221 f.

239

6. It was a genuine issue because (a) Galileo claimed greater certainty for the Copernican theory than the then known empirical evidence warranted, and supported it by his theory of the tides which has proved fallacious; (b) the theologians claimed a certitude for their interpreetation of the Bible which even the hermeneutics of their day did not warrant. Catholics would still maintain that if the theologians had correctly interpreted the Scriptures when they asserted that the sun does not move, then it would also be certain that a scientific theory contradicting the authentic teaching of Scripture would never be empirically confirmed. In spite of some assertions by theologians to the contrary, Vatican II still maintains the inerrancy of the Scriptures, "Since, therefore, all that the inspired authors, or sacred writers, affirm should be regarded as affirmed by the Holy Spirit, we must acknowledge that the books of Scripture, firmly, faithfully, and without error teach that truth which God, for the sake of our salvation, wished to see confided to the Sacred Scriptures." *Dogmatic Constitution on Divine Revelation,* chapter 3, no. 11, in Austin Flannery ed., *Vatican Council II: The Conciliar and Post Documents* (Collegeville, Minn.: Liturgical Press, 1981 ed.), p. 757. Note that the phrase "for the sake of our salvation" establishes the fundamental hermeneutic rule that the Biblical authors are to be understood to be concerned to teach matters of cosmology, anthropology and history only insofar as these relate directly to religious truth. It is this rule that the theologians of Galileo's day incorrectly applied, although both they and Galileo (as is evident from his defense) accepted it.

7. See Eduard Jan Dijksterhuis, *The Mechanization of the World Picture* (Oxford: Clarendon Press, 1961).

8. On Mersenne and Gassendi see references in Chapter 3, note 14.

9. See Gaston Sortais, "Le Cartésianisme chez les jésuites francais au XIIe et XVIIIe siècle", *Archives de Philosophie* 6, 3 (1929), pp. 1-93; Étienne Gilson, *Études sur le rôle de la pensée mediévale dans la formation du système cartésienne,* 3rd ed. (Paris: J. Vrin, 1967).

10. "While in arithemetic the only exact roots obtainable are those of perfect powers, in geometry a length can be found which will represent exactly the square root of a given line, even though this line be not commensurable with unity." Note 3 in David Eugene Smith and Marcia L. Latham, translators and editors of *The Geometry of René Descartes* (New York: Dover, 1954), p. 5. Or see John Burnet, *Early Greek Philosophy* (London: A. & C. Black, 1948), pp. 105 and Robert S. Brumbaugh, *Plato's Mathematical Imagination* (Bloomington: Indiana University Press, 1954), pp. 38-46.

11. Étienne Gilson, *Études sur le rôle de la pensée mediéval,* (note 9 above), pp. 27-50, 190-201.

12. On Montaigne's Christianity and the widespread fideism of this period see Maturin Dréano, *La pensée religieuse de Montaigne* (Paris: Beauchesne, 1936), especially pp. 471-473, and H.J.J. Janssen C.S.S.R., *Montaigne Fidéiste* (Nijmegen: N.V. Dekker, 1930).

13. *De Libero Arbitrio,* III, 7, see Chapter IV, note 80.

14. The famous *Discourse on Method* is a kind of intellectual *Confessions* in the Augustinian manner. Daniel J. Bronstein, in his introduction to *Essential Works of Descartes* (New York: Bantam Books, 1961), p. vii, writes: "Philosopher, scientist, mathematician, Descartes is none the less an engaging writer. Without the usual quotations from authorities, and with a minimum of technical jargon, his *Discourse on Method* tells about his schooling, his doubts, his travels and discoveries, and manages to convey the excitement and deep satisfaction which he himself experienced in applying a newly discovered method to problems in the sciences." The standard biography is Charles Adam, *Descartes, sa vie, son oeuvre* (Paris: J. Vrin 1937), vol. 12 of Charles Adam and Paul Tannery, *Oeuvres de Descartes.*

15. See Paul Mouy, *Le développement de la physique cartésienne 1646-1712* (Paris: J. Vrin, 1934).

16. It must be remembered that the scholastics never aimed at producing *literary* works, but wrote in a purely technical style. The Renaissance humanists, on the other hand, had given the cultivated public a taste for literary style which made scholastic works, whatever their intellectual merit, unreadable. This factor probably had more to do with the long oblivion into which such works fell, than did their often alleged dogmatism, deductivism, and logic-chopping.

17. See Francisque Bouillier, *Histoire de la Philosophie Cartésienne,* 3rd ed. (Paris, 1868), 2 vols., Reprint, Culture et Civilisation, (Brussels, 1969), vol. 2, p. 287 and 466-69.

18. On the contrast between the mentality of Descartes and that of Pascal, see Hans Küng, *Does God Exist?* (Garden City, N.Y.: Doubleday, 1980), p. 3-80. On the relation of Cartesianism to atheism see Cornelio Fabro, *God in Exile: Modern Atheism: A Study of the Internal Dynamic of Modern Atheism from its Roots in the Cartesian Cogito to the Present Day* (Westminister, Md.: Newman Press, 1986).

19. See André Malet, *Le Traité Théologico-Politique de Spinoza et la pensée biblique* (Paris: Société Les Belles Lettres, 1966).

20. See W.H. Barber, *Leibnitz in France* (Oxford: Clarendon Press; 2nd ed. 1955), and Lewis W. Beck, *Early German Philosophy* (Cambridge, Mass: Harvard University Press, 1969), pp. 196-242.

21. *Ibid.,* pp. 115-138, and Norman E. Fenton, *A New Interpretation of Leibnitz's Philosophy with Emphasis on His Theory of Space* (Dallas: Paon Press, 1973).

22. Barber, *Leibnitz in France* (note 20 above), pp. 123-140, and Beck, Early German Philosophy (note 20 above), pp. 275-305. Also John V. Burns, *Dynamism in the Cosmology of Christian Wolff* (New York: Exposition Press, 1965) discusses Wolff's natural philosophy and John E. Gurr, S.J., *The Principle of Sufficient Reason in Some Scholastic Systems,* 1750-1900 (Milwaukee: Marquette University Press, 1959), is informative on Wolff's influence on the textbook tradition of philosophy.

23. See Craig B. Brush, *Montaigne and Bayle* (The Hague: Martinus Nijhoff, 1966), and Lucien Febvre, *The Problem of Unbelief in the Sixteenth Century: The Religion of Rabelais* (Cambridge: Harvard University press, 1982). On the sceptical and rationalistic currents as they affected English thought in particular see, Basil Willey, *The Seventeenth Century Background* (New York: Doubleday Anchor Books, 1953) (1935); Margaret L. Wiley, *The Subtle Knot: Creative Scepticism in Seventeenth Century England* (Cambridge, Mass.: Harvard University Press, 1952); Henry G. Van Leeuwen, *The Problem of Certainty in English Thought,* 1630-1690 (The Hague: Martinus Nijhoff, 1963), and John Redwood, *Reason, Ridicule and Religion: The Age of Enlightenment in England, 1660-1750* (London: Thames and Hudson, 1976).

24. See Chapter III, note 11 for references.

25. F.C. Hood, *The Divine Politics of Thomas Hobbes: An Interpretation of Leviathan* (Oxford: Clarendon Press, 1964), pp. 1-13, discusses Hobbes' religious background and defends the sincerity of his admittedly unorthodox "Christianity." It can hardly be an accident that the term "covenant," with Biblical allusions, plays so large a part in Hobbes' ethical theory.

26. Locke could see no reason that God could not give matter the power to think; see *An Essay Concerning Human Understanding,* IV, iii, 6, and Richard I. Aaron, *John Locke,* 2nd ed. (Oxford: Clarendon Press, 1971), pp. 144-148.

27. See Ernst Cassirer, *The Platonic Renaissance in England* (Austin: University of Texas Press, 1953), pp. 36-41 and 157-202, and Henry R. McAdoo *The Spirit of Anglicanism: A Survey of Anglican Theological Method in the Seventeenth Century* (New York: Charles Scribner's Sons, 1965), pp. 81-239.

28. On Newton see William A. Wallace, *Causality,* note 4 above, 1, 194-210. On Boyle see the introduction by Marie Boas Hall, *Robert Boyle on Natural Philosophy, An Essay with a selection from his writings* (Bloomington, Ind.: Indiana University Press, 1965).

29. See Earl M. Wilbur, *A History of Unitarianism in Transylvania, England and America* (Boston: Beacon Press, 1952), pp. 166 ff.

30. "The propagandists of the Enlightenment were French, but its patron saints and pioneers were British." Peter Gay, *The Enlightenment* (New York: Alfred A. Knopf, 1967), 1, 11. While Leibnitz (d.1716) is sometimes included as the beginning of the German Enlightenment, Frederick Copleston, S.J., *A History of Philosophy* (Garden City, N.Y.: Doubleday Image Books, 1960), 6, pt. 1, 121, is surely right in beginning with Christian Thomasius (d.1728) who "emphasized the superiority of the French to the Germans in philosophy."

31. See J.A. Dorner, *A History of Protestant Theology* (Edinburgh: T. and T. Clark, 1871), vol. 2, pp. 252-348.

32. Alexander L. Drummond, *German Protestantism Since Luther* (London: Epworth Press, 1951), pp. 80-103.

33. See J. Diebolt, *La théologie moral catholique en Allemagne au temps du philosophisme et de la restauration (1750-1850)* (Strasbourg: F.X. Le Roux, 1926); Paul Hazard, *The European Mind: 1680-1715* (Cleveland: Meridian Books-World Publishing Co. 1964), and *European Thought in the Eighteenth Century* (same publisher, 1963); Albert Monod, *De Paschal à Catheaubriand: Les défenseurs francqis du Christianisme de 1670 a 1802* (Geneva: Slatkine Reprints, 1970) (1916), and Jean DeLumeau, *Catholicism between Luther and Voltaire: A New View of the Counter Reformation* (London: Burns and Oates, 1977).

34. F. Ernest Shoeffler, *German Pietism During the Eighteenth Century* (Leiden: E.J. Brill, 1973), and H. Daniel-Rops, *The Church and the Eighteenth Century* (New York: Dutton, 1964).

35. For an analysis of Kant's theory of teleology see the introduction by James Creed Meredith to his edition of *Kant's Critique of Teleological Judgment* (Oxford: Clarendon Press, 1928), pp. xi-xcvii.

36. See A.E. Teale, *Kantian Ethics* (Westport, Conn: Greenwood Press, 1975), and Clement C.J. Webb, *Kant's Philosophy of Religion* (Oxford: Clarendon Press, 1926) (Kraus Reprint, New York, 1970). See also Ernst Cassirer, *The Philosophy of the Enlightenment* (Boston: Beacon Press, 1951), pp. 104 ff., for the way that Enlightenment rationalism shifted from pure rationalism to an emphasis on human *affectivity* as a determinant in judgment, thus preparing the way for Kant's ethical theory.

37. W.H. Werkmeister, *Kant: The Architectonic and Development of His Philosophy* (Lasalle, Il.: Open Court, 1980), shows that Kant most admired Newton and Rousseau and said, "After Newton and Rousseau, the ways of God are justified" (*Fragmente* in vol. 8 of the G. Hartenstein ed. of Kant's works, Leipzig: L. Voss, 1869), p. 630.

38. Ralph M. Wiltgen, S.V.D., *The Rhine Flows into the Tiber* (New York: Hawthorn Books, 1966) gives a detailed account of the history of Vatican II based on the thesis that its work was dominated by the Rhineland episcopates of Germany, Austria, Switzerland, France, and the Netherlands, with the German theological tradition especially influential.

39. See Thomas F. O'Meara, O.P., *Romantic Idealism and Roman Catholicism: Schelling and the Theologians* (Notre Dame, Ind.: University of Notre Dame Press, 1982). See also Louis Folcer, *La philosophie catholique en France aux xix siècle avant la renaissance thomiste et dans son rapport avec elle (1800-1880)* (Paris: J. Vrin, 1955), for a discussion of traditionalism and ontologism in that country, and E. Hocedez, *Histoire de la théologie au XIXème siècle* (Paris: Desclée de Brouwer 1951); T.M. Schoof, *A Survey of Catholic Theology, 1800-1970* (Glen Rock, N.J.: Paulist Newman, 1970) for various interpretations of this period. Also Gerald McCool, *Catholic Theology in the Nineteenth Century,* (New York: Seabury, 1977). On Protestant trends see Karl Barth, *Protestant Theology in the Nineteenth Century* (London: SCM Press, 1972), and Claude Welch, *Protestant Thought in the Nineteenth Century* (New Haven: Yale University Press), vol. 1, 1972-.

40. See Joseph L. Esposito, *Schelling's Idealism and Philosophy of Nature* (Lewisburg, Penn.: Bucknell University Press, 1977), especially pp. 125-159, who shows that Schelling, in his search for a synthesizing view of science, prepared the way for the evolutionary view of the cosmos and of life.

41. For the *objective* character of Platonic idealism see I.M. Crombie, *An Examination of Plato's Doctrines* (London: Routledge and Kegan Paul, 1963), 2, 247-472. Crombie concludes that for Plato Reason or Mind is prior to the material world and the objects of Mind are the Forms or Ideas (p. 471-72).

42. Aquinas correctly interprets Aristotle's notion of the way an essential rather than a merely phenomenal science of nature is possible when he writes, "Since essential forms are not known to us *per se,* it is necessary that they should be manifested through some accidents, which are signs of that form, as is evident in Metaphysics VIII. It is not necessary, however, to take accidents proper to that species, since it is necessary to demonstrate these through the definition of the species; but it is necessary that the form of the species should be known through some common accidents; and thus the assumed differences are said to be in a manner "substantial," in as much as they are used to declare the essential form; they are, however, more common than the species, in as much as they are taken from certain signs which follow on the superior genera." *Commentary on the Posterior Analytics of*

242

Aristotle II, lect. 13, n. 7. See my articles, "Are Thomists Selling Science Short?," The 1960 Lecture Series in the Philosophy of Science (Mt. St. Mary's Seminary of the West, Cincinnati, Ohio, 1960), pp. 13-33, and "Does Natural Science Attain Nature or Only the Phenomena?" in Vincent E. Smith, *The Philosophy of Physics* (St. John's University, Jamaica, N.Y., 1961) pp. 63-82.

43. See C.D. Broad, *Kant: An Introduction* (Cambridge: Cambridge University Press, 1978), Chapter 2, pp. 16-71. For an exposition of the Transcendental Aesthetic and for a detailed criticism H.W. Cassirer, *Kant's First Critique*, (London: George Allen and Unwin, 1954), Chapter 1, "Kant's Theory of Sensible Intuition", p. 1-51.

44. Nicola Abbagnano in his article "Positivism" in Paul Edwards, ed., *The Encyclopedia of Philosophy*, (New York: Macmillan and the Free Press, 1972), 5., 414-419, very usefully distinguishes between "Social Positivism" originating with Saint-Simon and Comte, "Evolutionary Positivism" of which Herbert Spencer is representative, and "Critical Positivism" of Mach out of which "Logical Positivism" arose. Critical Positivism definitely has Kantian roots although it has been in many respects anti-Kantian.

45. See Frederick Copleston, S.J., *A History of Philosophy* (note 30 above), 7(1), 50-182 for a good account of the transition from Kant to Hegel.

46. For a recent exposition of Fichte's concept of humanity as "the active being" see Tom Rockmore, *Fichte, Marx and the German Philosophical Tradition* (Carbondale and Edwardsville, Ill: Southern Illinois University Press, 1980), "Conclusion," pp. 162-164.

47. On the influence of Spinoza on Fichte and Schelling see Esposito, *Schelling's Idealism* (note 40 above), pp. 31-46.

48. For a Thomistic criticism of Hegel's notion of a philosophy of history see Jacques Maritain, *On the Philosophy of History* (New York: Charles Scribner's Sons, 1957), pp. 19ff. On the sense of the famous Hegelian dictum that "The ideal is real, and the real the ideal," see J.N. Findlay, *Hegel: A Re-Examination* (George Allen and Unwin, 1958), pp. 34-57, who stresses the *theological* character of Hegel's view of the Real. It must be noted, however, with Karl Lowith, *From Hegel to Nietzsche* (New York: Holt Rinehart and Winston, 1964), that Hegel still "excluded mere transitory 'accidental' existence from the interest of philosphy considered as a knowledge of 'reality'." p. 46.

49. See Benedetto Croce, *The Philosophy of Giambattista Vico* (London: H. Latimer, 1913); A.R. Caponigri, *Time and Idea—The Theory of History in Giambattista Vico* (Chicago: Regnery, 1953). For a brief statement see John Edward Sullivan, *Prophets of the West* (New York: Holt Rinehart and Winston, 1970), pp. 3-20.

50. See R.T. Clark, *Herder: His LIfe and Thought* (Berkley and Los Angeles: University of California Press, 1955). On Schlegel see René Wellek, *A History of Modern Criticism*, (New Haven: Yale University Press) 2:5-35.

51. *Hegel's Philosophy of Nature* (Part 2 of the *Encyclopedia of the Philosophical Sciences*, 1831): ed. by A.V. Miller (Oxford: Clarendon Press, 1970), #249, pp. 20-22, 26, and #339, p. 284. Hegel rejected the origin of species both by way of evolution (from imperfect to perfect) and by way of emanation (from perfect to imperfect) and held that they were simultaneous, or perhaps successive in the manner of *Genesis* literally taken. He did admit, however, that the earth may have existed without life for many ages.

52. "Almost a century before Nietzsche's announcement: 'God is dead! God remains dead! And we have killed him.' [*The Gay Science* (New York: Random House, Vintage Books, 1974), p. 181]. Hegel, in his 'Faith and Knowledge', [*Faith and Knowledge or The Reflective Philosophy of Subjectivity* (New York: State University of New York Press, 1977) p. 190] gives the title of 'Death of God' to the history of modern times, at the same time invoking Pascal." See Hans Küng, *Does God Exist?* (note 18 above) p. 138 f.

53. J.N. Findlay, *Hegel: A Re-Examination* (note 48 above), pp. 34-57, stresses the teleological character of Hegel's view of the Real. Walter Kaufmann, *Hegel: A Re-Interpretation* (New York: Doubleday Anchor Books, 1966), gives a secular humanist interpretation of Hegel, speaking from the view that the work of philosophy is essentially a negative task of destroying illusions (cf. p. 67). A very different reading by a Jew who has deep insight into Christian theology is Emil L. Fackenheim, *The Religious Dimension in Hegel's Thought* (Beacon Press, Boston, 1967). A less favorable view of Hegel's philosphy of religion is given by

James Collins, *God in Modern Philosophy* (Chicago: Regnery, 1959), pp. 201-237, and by André Leonard, *La foi chez Hegel* (Paris: Desclée, 1970), who argues that the world and its "God" are for Hegel eternally "immature" and hence such a God is not really the Absolute Spirit. See also Hans Küng, *Menschwerdung Gottes: Eine Einführung in Hegels theologisches Denken als Prolegomenon zu einer künftigen Christologie* (Freiburg-Basel-Vienna: 1970), the results of which are summarized in his *Does God Exist* (see note 18 above) p. 129-169.

54. See Karl Barth, *Protestant Theology* (note 39 above), pp. 425-473 on Schleiermacher's crucial importance for this whole period.

55. See David McLellan, *The Young Hegelians and Karl Marx* (London: MacMillan 1969).

56. On the theological consequences of Kantianism see Otto Pfleiderer, *The Development of Theology since Kant and its Progress in Great Britain* (London: George Allen and Unwin, 1909).

57. See W. Lowrie, *Kierkegaard* (London: Oxford University Press, 1938) and James Collins, *The Mind of Kierkegaard* (Chicago: University of Chicago Press, 1953).

58. On Catholic Traditionalism see McCool, *Catholic Theology in the Nineteenth Century* (note 39 above), pp. 37-58, and Louis Foucher, *La philosophie catholique en France au xixe siècle avant la renaissance thomiste et dans son rapport avec elle (1800-1880)* (Paris: J. Vrin, 1955), pp. 11-29 and 72-98.

59. On Ontologism see A. Fonck, "Ontologisme" in *Dictionnaire Théologique Catholique* t.11 (1): 1000-61; Michele Frederico Sciacca, *Interpretazioni Rosminiana* (Milan: Marzorati, 1958;, A. Robert Caponigri, *A History of Western Philosophy* (Notre Dame, Ind.: Notre Dame University Press, 1971), 4, pp. 26-34; and McCool, *Catholic Theology in The Nineteenth Century* (note 39 above), pp. 113-128.

60. See Frederick Copleston, S.J. *History of Philosophy* (note 30 above), 9, pp. 37-68, on French Idealism initiated by Maine de Biran but greatly influenced by Hegel.

61. On the Catholic faculty of Tübingen see O'Meara, *Romantic Idealism* (note 39 above), pp. 94-108 (on Johann Sebastian Drey, 1777-1853, its founder) and pp. 138-168 (on R.A. Staudenmaier, J.A. Möhler and J.E. Kuhn).

62. McCool, *Catholic Theology in the Nineteenth Century* (note 39 above), pp. 59-87.

63. *Ibid.*, pp. 88-112.

64. The great controversy over papal infallibility has obscured the importance of Vatican I for the doctrine on the relation between faith and reason, a topic of much more importance for Christian anthropology.

65. Msgr. Antonio Piolanti is editor *Studi e Richerche sulla Rinascita del Tomismo* (Pontifica Accadema Teologica Romana: Liberaria Editorice Vaticana), some 7 monographs by various authors, 1965 — and *Biblioteca per la Storia del Tomismo,* (same publisher), some 9 volumes by various authors, 1972 —. The best summary of research on this subject is Thomas J.A. Hartley, *The Thomist Revival and the Modernist Era* (Toronto: University of Toronto, St. Michael's College, 1971).

66. On Kleutgen see McCool, *Catholic Theology in the Nineteenth Century* (note 39 above), pp. 167-215.

67. It is noteworthy that among modern authorities to whom reference is commonly made for interpretation of the thought of Aquinas, Étienne Gilson drew his conception of Thomism chiefly from the *Summa Theologiae* to the neglect of the commentaries on Aristotle, and Jacques Maritain depended strongly on John of St. Thomas. Others have read Aquinas largely through the eyes of Cajetan. A truly historical reading of Aquinas still remains to be done, although James A. Weisheipl, *Friar Thomas D'Aquino* (Garden City, N.Y.: Doubleday, 1974), within the limits of a biography has outlined such a study and M.D. Chenu, *Toward an Understanding of St. Thomas* (Chicago: Regnery, 1964) provided some of the indispensable tools for it.

68. Ample bibliography on the Modernist heresy can be found in Thomas M. Loome, *Liberal Catholicism, Reform Catholicism, Modernism: A Contribution to a New Orientation in Modernist Research* (Mainz: Matthais Grünewald Verlag, 1970). Under Pius X controversy over just what was the Thomism approved by the Holy See led to the *Motu Proprio* of 29 June, 1914 establishing the so-called "Twenty-Four Theses" as characterizing authentic Thomism (see Édouard Hugon, *Les Vingt-Quatre Thèses Thomistes* (Paris: P. Tequi, 1926) and this was confirmed by Benedict XV in 1916.

69. *An Essay on the Development of Christian Doctrine* edited with an introduction by J.M. Cameron (Baltimore, Md.: Penguin Books, 1974).

70. See Jan H. Walgrave, *Unfolding Revelation: The Nature of Doctrinal Development,* Theological Resources (Philadelphia: Westminister of Philadelphia, 1972), on Newman's theory of development, pp. 293-313.

71. *Ibid., passim.*

72. *Ibid.,* Chapter 8, "Origin and Development of the Transformic Theory", especially the section on Modernism, pp. 179-277.

73. See Richard T.A. Murphy, O.P. ed., *Lagrange and Biblical Renewal* (Chicago: Priory Press, 1966).

74. Romanus Cessario, *Christian Satisfaction in Aquinas* (Washington, D.C.: University Press of America, 1982), Chapter 1, note 30, p. 275-276 shows that, because Aquinas believes theology is a science, in the *Summa* he moves from what is most necessary to what is contingent and historical. That is why the Incarnation is treated only in the Third Part *after* the treatment in the second part of God whose existence is absolutely necessary and of Man whose nature is hypothetically necessary, because God could have saved us in some other way than by the Incarnation.

75. Helen James John, S.N.D., *The Thomist Spectrum* (New York: Fordham Press: 1966), provides a useful survey, unsympathetic to Thomism "of the strict observance". T.M. Schoof, *A Survey of Catholic Theology* (note 39 above), p. 146-200, is also little sympathetic to neo-scholasticism. Another point of view is provided by J.A. Weisheipl, "Contemporary Scholasticism" in article "Scholasticism" in *New Catholic Encyclopedia,* 12, 1165-1170. On Transcendental Thomism see Otto Muck, S.J., *The Transcendental Method* (New York: Herder and Herder, 1968); James B. Reichmann, "The Transcendental Method and the Psychogenesis of Being", *The Thomist* 32 (1968): 448-508, and the severe critique of Robert J. Henle, S.J., "Transcendental Thomism: A Critical Assessment" in Victor B. Brezik, C.S.B., ed., *One Hundred Years of Thomism: Aeternis Patris and Afterwards* (Houston: University of St. Thomas, 1981), pp. 90-116. For recent reflections on the historic significance of *Aeterni Patris,* see the many essays in *Atti dell' VIII Congresso Tomistica Internazionale Pontificia Academia di San Tommaso,* 3 vols. (Rome, Liberia Editrice Vaticana, 1981).

76. See note 39 above for reference.

77. Some see in Vatican II the triumph of Modernism when in fact the Council faced some very real problems raised by the modernists and provided fundamental principles for their solution. While the severe measures taken by Pius X to suppress the heresy are understandable in view of the lack of preparedness of Catholic theologians to meet these problems at the time, these measures only postponed the necessary confrontation at Vatican II.

78. See Chapter 5 note 13 for references.

79. Karl Rahner, *Spirit in the World* (New York: Herder and Herder, 1968, original German, 1957). In his introduction to the English edition Francis Fiorenza points out that Rahner agrees with Kant (and, it should be added, with Aristotle) that all human knowledge is related to sense intuitions, but "he rejects those philosophical positions which maintain that a metaphysics of transcendence is possible because of a special innate idea or because of a specific and immediate intuition of a metaphysical object, be it an eternal truth or an objectively conceived absolute being. He denies explicitly that the absolute is known as some object or that the human mind could form an adequate objective concept of God. Instead he proposes a transcendental understanding of God, who is not known by man as an object of reality, but as the principle of human knowledge and of reality. This fundamentally non-objective transcendental knowledge of God as the principle of knowledge and reality is central to Rahner's whole theology." (pp. xliii-xliv).

This is Rahner's way of meeting Kant's arguments against the possibility of a transcendental metaphysics. Fiorenza finds "a basic difference between Rahner and other students of Maréchal" in this, that "Whereas most of Maréchal's followers have carried on their dialogue within the discipline of philosophy, Rahner has seen that a philosophical and existential theology is the only adequate horizon for a dialogue with modern philosophies and their emphasis on the dimension of history. Whereas Lonergan rejects phenomenology and existentialism as merely descriptive and as a purified empiricism, Rahner stresses . . . the ontological and theological relevance of the problems of historicity and 'facticity.'" (pp. xliv-xlv). Fiorenza considers (note 39, p. xlii) that Rahner's "rejection of a metaphysical intuition" is "his major correction of the Thomistic or neo-scholastic interpretations." I would also reject "a metaphysical intuition" of the type proposed by Gilson or Maritain (whom Fiorenza seems to have in mind); see note 82 below.

What must be emphasized, however, is that Rahner remains within the Maréchalian and Kantian camp in seeking an *a priori* element in knowledge and, therefore, although he "is aware that all human knowledge is related to sense intuitions", he stands over against the Aristotelian epistemology that rejects any such *a priori* and thus he remains within Platonic dualism. Rahner himself puts this explicitly in his great work *Foundations of Christian Faith: An Introduction to the Idea of Christianity* (New York: Seabury, 1978) (Note the term "Idea" (*Begriff*) with its overtones of German Idealism). "It is not the case that this co-known, unthematic self-presence of the subject and its self-knowledge is merely an accompanying phenomenon in every act of knowledge which grasps an object, so that knowledge of this object in its structure and content would be completely independent of the structure of the subjective self-presence. Rather the structure of the subject is itself an a priori, that is, it forms an antecedent law governing what and how something can become manifest to the knowing subject." (p. 19). This point of departure which grounds metaphysics in self-knowledge as subjective, rather than in the objective knowledge of the sensible world and of the human person as part of that sensible world is the Platonic source of the dualism against which the present books argues.

80. Rahner, *Foundations* (previous note), "If man really is a subject, that is, a transcendent, responsible and free being who as subject is both entrusted into his own hands and always in the hands of what is beyond his control, then basically this has already said that man is a being oriented towards God. His orientation towards the absolute mystery alway continues to be offered to him by this mystery as the ground and content of his being. To understand man in this way, of course, does not mean that when we use the term 'God' in such a statement, we know what this term means from any other source except through this orientation to mystery. At this point theology and anthropology necessarily become one." (p. 44) It is evident from this that for Rahner what science tells us about the cosmos and about ourselves plays as little a part in his theology as it did for St. Augustine when he said that nothing is worth knowing but God and self. How different from the viewpoint of Aquinas!

81. See McCool, *Catholic Theology in the Nineteenth Century* (note 39 above) pp. 255-267.

82. See William H. Kane, O.P., "The Subject of Metaphysics" *Thomist,* Oct. 1955, pp. 503-531 and Melvin A. Glutz, C.P., "The Formal Subject of Metaphysics", *ibid.* Jan. 1956: pp. 59-74. The latter contains references to the articles of Geiger, Robert, Smith and others.

83. See James A. Weisheipl, O.P., ed., *The Dignity of Science,* Studies in the Philosophy of Science presented to William H. Kane, O.P. (Washington, D.C.: The Thomist Press, 1961). Besides Kane and Weisheipl, other writers associated with this type of Thomism include Albert Moraczewski, O.P., Raymond J. Nogar, O.P., Ralph A. Powell, O.P., Vincent E. Smith, William A. Wallace, O.P., Denis Zusy, O.P. and myself.

84. See Bernard M.C. Reardon, ed., *Liberal Protestantism,* (A Library of Modern Religious Thought, Henry Chadwick, ed., (London: A. and C. Black, 1968) introduction; and Robert T. Handy, ed., *The Social Gospel In America 1870-1920* (New York: Oxford University Press, 1969).

85. See *Natural Theology: Comprising "Nature and Grace" by Prof. Dr. Emil Brunner and the Reply "No!" by Dr. Karl Barth* (London: Geoffrey Bles, 1946).

86. See Alan M. Fairweather, *A Critical Examination of the Christian Doctrine of Revelation in the Writings of Thomas Aquinas and Karl Barth* (London: Lutterworth Press, 1944). Jerome Hamer, *Karl Barth* (Westminster, Maryland: Newman Press, 1962), and Hans Urs von Balthasar, *The Theology of Karl Barth,* (New York: Holt, Rinehart and Co., 1971) for Catholic evaluations of the work of Barth.

87. See *The Humanity of God* (Richmond, Virginia: John Knox Press, 1960) and Eberhard Busch, *Karl Barth: His LIfe from Letters and Autobiographical Texts* (Philadelphia: Fortress Press, 1975), pp. 423-430.

88. *Church Dogmatics* (Edinburgh: T. and T. Clark, 1936-1960), 4 vols.

89 Paul Tillich (1886-1965), *Systematic Theology* (Chicago: University of Chicago Press, 1951), 3 vols. in one, 1967.

90. Note Tillich's brief treatment of Eckhart in his *A History of Christian Thought,* edited by Carl E. Braaten (New York: Simon and Schuster, 1967), pp. 201-203. Braaten in his introduction pp. xxii-xxv, on the basis of his first hand experience of Tillich as a teacher, shows how Tillich rehabilitated the Middle Ages for Protestants, especially as regards mysticism. He also shows that Tillich favors Augustinianism and traces the rise of modern rationalism to Aquinas. "In 'The Two Types of Philosophy of Religion', he traces the roots of the modern split between faith and knowledge back to the Thomistic denial of the Augustinian belief in the immediate presence of God in the act of knowing. For Thomas, God is first in the order of beings, but last in the order of knowledge. The knowledge of God is the end result of a line of reasoning, not the presupposition of all our knowing. Where reason leaves off, faith takes over. The act of faith, however, becomes the movement of the will to accept truth on authority. Tillich's verdict is clear: [quoting *Theology of Culture,* 1919] 'This is the final outcome of the Thomistic dissolution of the Augustinian *solution.'"* See also John R. Dourley, *Paul Tillich and Bonaventure: An Evaluation of Tillich's Claim to stand in the Augustinian Tradition* (Leiden: Brill, 1975) which concludes that Tillich's claim was largely justified.

91. See André Dumas, *Dietrich Bonhoeffer Theologian of Reality* (London: SCM Press, 1959) for an exposition of Bonhoeffer's basic ideas and the very wide range of different interpretations that have been given to them.

92. *Ibid.,* pp. 162-214.

93. See John C. King, *The Evangelicals* (London: Hodder and Staughton, 1969); Ernest R. Sandeen, *The Roots of Fundamentalism* (Chicago: University of Chicago Press, 1970); Donald Bloesch, *The Evangelical Renaissance* (Grand Rapids: Eerdmans, 1973); and George M. Marsden, *Fundamentalism and American Culture* (New York: Oxford, 1980). Fundamentalism by reason of its millenarianism and its total rejection of historical critical methods in the study of the Bible must be distinguished from Evangelicalism.

94. Jürgen Moltmann, *Theology of Hope* (New York: Harper and Row, 1967).

95. See Johannes Metz, *Poverty of Spirit* (Glen Rock, N.J.: Newman Press, 1968); *Theology of the Word* (New York: Herder and Herder, 1969); *Faith in History and Society,* (New York: Seabury, 1980) and *The Emergent Church* (New York: Crossroads, 1981).

96. A representative selection of essays by theologians of this school can be found in Rosino Gibellini, ed. *Frontiers of Theology in Latin America,* (Maryknoll, N.Y.: Orbis Books, 1979).

97. On the significance of Heidegger's temporary collaboration with National Socialism see Hannah Arendt "Martin Heidegger at Eighty" and Karsten Harries, "Heidegger as Political Thinker" in Michael Murray ed., *Heidegger and Modern Philosophy* (New Haven: Yale University Press, 1978), pp. 293-328. Arendt excuses Heidegger as an academic, remote from political reality, who had never even read *Mein Kampf.* Harries, however, shows that Heidegger believed that European civilization has come to an end from which it can be saved only by creative leadership, which he hoped to provide in the universities and imagined that Hitler might provide in the state. When (like Plato) he was quickly disillusioned his only recourse was to withdraw completely from public life and await for Being to reveal itself in a new historic age.

98. The following are helpful in understanding this very difficult author: Thomas Langan, *The Meaning of Heidegger,* (New York: Columbia University Press, 1961); Vincent Vycinas, *Earth and Gods* (The Hague: Martinus Nijhoff, 1961); George J. Seidel, *Martin Heidegger and the Pre-Socratics* (Lincoln: University of Nebraska Press, 1964); John Macquarrie, *An Existentialist Theology,* (New York: Harper and Row, 1965); Stephen A. Erikson, *Language and Being: An Analytic Phenomenology* (New Haven: Yale University Press, 1970). Bernd Magnus, *Heidegger's Metahistory of Philosophy: Amor Fati, Being and Truth* (The Hague: Martinus Nijhoff, 1970); William Richardson, *Heidegger: Through Phenomenology to Thought* (The Hague: Martinus Nijhoff, 1967); Stephen A. Erikson, *Language and Being: An Analytic Phenomenology* (New Haven, Yale University Press, 1970); John N. Deely,

The Tradition via Heidegger (The Hague: Martinus Nijhoff, 1971); John D. Caputo, *The Mystical Element in Heidegger's Thought* (Athens: Ohio University Press, 1978) and *Heidegger and Aquinas* (New York: Fordham University Press, 1982.

99. By this "forgetfulness of being" Heidegger seems to mean that western thought from the time of Plato by assuming an analytic attitude in the attempt to gain control over "beings", a control that has produced scientific technology, has lost sight of the more fundamental question, "Why is there anything at all?" Yet this question is the one which really makes us human, since it is the question of the ultimate meaning of the world and of our existence in the world. Heidegger did not pretend to give an answer to this question but only to awaken modern man to it. Apparently he believed that the answer cannot be found by philosophical or scientific methods, because it is precisely these that have turned our attention from this question, but only by a kind of "thinking" that permits Being to disclose itself to us through the fated unfolding of the drama of history which is beyond our control but which expresses itself in the development of human culture, which itself is embodied in language. It is probable that for the late Heidegger this "waiting on Being" was open to the God of faith, although rejecting the God of metaphysics as "not divine enough". This question is thoroughly treated in the as yet unpublished thesis of Vincent A. Guagliardo, O.P., *The Future of an Origin: Being and God in the Philosophy of Martin Heidegger* (Graduate Theological Union, Berkley, Calif., 1981).

100. See John D. Caputo, *The Mystical Element in Heidegger* (note 98 above), pp. 203-217. Heidegger's *Letter to Albert Borgnmann* is printed in *Philosophy East and West* 20 (July 1970): 221, which entire issue is devoted to a discussion of the relation of Heidegger and eastern thought.

101. On hermeneutic philosophy see the works of R.E. Palmer, Gadamer, and A.D. Thiselton referred to in Chapter 1, notes 27 and 30.

102. See Louis Roberts, *The Achievement of Karl Rahner,* (Herder and Herder, 1967); Anne Carr, *The Theological Method of Karl Rahner* (diss.) (Missoula: Montana: Scholars Press, 1977); and Leo J. O'Donovan, ed. *A World of Grace* (New York: Seabury, 1980). The original translator of *Theological Investigations* (Baltimore: Helicon Press, 1961), vol. 1, p. xiii, note 1, Fr. Cornelius Ernst, O.P. raised the fundamental question about the philosophical idealism at the basis of Rahner's theology.

103. See Don Ihde, *Hermeneutic Phenomenology: The Philosophy of Paul Ricoeuer* (Evanston, Ill.: Northwestern University, 1971); John B. Thompson, *Critical Hermeneutics: A Study in the Thought of Paul Ricoeuer and Jürgen Habermas* (Cambridge: Cambridge University Press, 1981). On Gadamer see Richard E. Palmer, *Hermeneutics: Interpretation Theory in Schleiermacher, Dilthey, Heidegger and Gadamer* (Evanston, Ill: Northwestern University Press, 1969).

104. "One may well say that the most lively form of contemporary philosophical theology has been based not at all on speculative metaphysical system, but on the methods and practices of ordinary language philosophy", Langdon B. Gilkey, *Naming the Whirlwind: The Renewal of God Language* (Indianapolis: Bobbs-Merril, 1969) p. 235.

105. For Teilhard's life see Mary and Ellen Lukas, *Teilhard,* (New York: McGraw-Hill, new ed. 1981). A handy up-to-date bibliography will be found in Ursula King, *Towards a New Mysticism: Teilhard de Chardin and Eastern Religions* (New York: Seabury, 1981), pp. 293-306. On the scientific value of his thought see Paul Chauchard, *La Pensée Scientifique de Teilhard* (Paris: Editions Universitaires, 1965). *The Phenomenon of Man* (New York: Harper Torch Books, 1959) with a preface by Julian Huxley received a very critical review by P.B. Medawar, *Mind,* 70 (Jan. 1961): 99-106 and by George G. Simpson, *Scientific American,* 202 (April, 1960) 201-207; cf. Bernard Towers, "Scientific Master Versus Pioneer: Medaward and Teilhard" in his *Concerning Teilhard and Other Writings on Science and Religion,* (London: Collins, 1969), pp. 89-99. In understanding his religious thought I have found especially useful Robert L. Faricy, S.J., *Teilhard de Chardin's Theology of the Christian in the World* (New York: Sheed and Ward, 1967) and Christopher F. Mooney, S.J., *Teilhard de Chardin and the Mystery of Christ* (New York: Desclée, 1967); Henri de Lubac, *The Religion of Teilhard de Chardin* (New York: Desclée, 1967) and for the particular problem with which

this book is concerned Robert North, S.J. *Teilhard and the Creation of the Soul* (Bruce: Milwaukee, 1966).

For severe criticism see Phillipe de Trinité, O.C.D., *Teilhard: Étude Critique,* 2 vols. (Paris: La Table Rond, 1968) and Jacques Maritain, *The Peasant of the Garonne* (New York: Holt, Rinehart and Winston, 1968), pp. 116-126 and appendices. Maritain accuses Teilhard of "sins against the intellect." "From the start one mixes everything together, since, scanty as it may be in such and such an age, philosophy of nature, metaphysics, theology, natural mystique, even touches of supernatural mystique, all of which are made to contaminate and corrupt one another in a powerful high-soaring lyrical flight — unnatural and deceptive because it is pseudo-angelic," p. 115. This reflects Maritain's earlier opposition to Bergson, *Bergsonian Philosophy and Thomism* (New York: Greenwood Press, 1968). As to the classification of Teilhard's type of thought see John O'Manique, *Energy in Evolution: Teilhard's 'Physics of the Future',* The Teilhard Study Library (London: Garnstone Press, 1969). "In the *Phenomenon of Man* Teilhard also states that his work is not metaphysics. As present, however, this statement is somewhat misleading, for it could be interpreted to mean that, recognizing the nature and value of metaphysics, Teilhard realizes that his phenomenological approach is inherently limited and cannot encroach upon the domain of the metaphysician. A reading of some of the author's correspondence, however, would lead one to believe that Teilhard actually rejects metaphysics, on the grounds that it is rationalistic and static, and he sees his hyperphysics as a replacement for metaphysics." O'Manique refers to Claude Cuénot, *Teilhard de Chardin* (Baltimore: Helicon, 1965), pp. 213 and 233.

106. *Humani Generis,* Aug. 12, 1950, text in *Acta Apostolicae Sedis,* 1950 (561-578) translation in Claudia Carlen, I.H.M. ed., *The Papal Encyclicals* 1939-1958. (New York: McGrath, 1981), pp. 175-184. On its historical context see J.M. Connolly, *Voices of France,* (New York: Macmillan, 1961).

107. *Gaudium et Spes,* Austin Flannery, O.P. ed., *Vatican Council II: The Conciliar and Post-Conciliar Documents* (Collegeville, Minn: Liturgical Press, 1975), 1:903-1001.

PART III

A RADICAL PROCESS INTERPRETATION OF SCIENCE

Chapter 7

Primary Units In Process

I: The Meaning of the Evolutionary Process

1. The Variety of Evolutionary Processes

The foregoing historical sketch has shown that Christian theology to date has failed to rethink the Christian world-view in terms of God's revelation of Himself through modern science, while Humanism has used modern science as an effective apologetic for its claim that we human beings must ourselves give meaning to a cosmos which is in itself meaningless. Now I must try to meet my own challenge by proposing, at least in outline, some way in which Christian theology might advance in accomplishing this long neglected task. Such a project first demands a philosophical interpretation of the scientific world-picture described in Chapter 2.[1] Since, as I tried to make clear in that chapter, science shows us a world and a humanity in *process,* moreover in an *evolutionary,* not a merely cyclical process, it is on evolution that we must first focus our attention. It was the great merit of Teilhard de Chardin as a prophet of our times to have done just that.

Modern science, however, has not yet uncovered a single unified, evolutionary process in the world, as Teilhard thought.[2] In fact the most obvious feature of our cosmos is that it is running-down, inevitably sinking deeper and deeper into the chaos of increasing entropy. Apparently the most energetic, negentropic (i.e. negative entropic) state of the world was that primal moment before the Big Bang. Since that time matter and energy have been scattering in an expanding universe, which will finally arrive at a condition in which matter and energy are so dissipated and reduced to so low grade an intensity that all creative process will cease. Evolution conceived as an increase in complexity of organization resulting in the higher elements, complex compounds, increasingly elaborate organisms, and finally in big-brained, intelligent humans, is a purely *local* event in a vast universe, which is eventually doomed to be completely undone by the general and inevitable encroachment of entropy.

Moreover, within those local patches in which evolution to more complex entities is able to advance in a negentropic back-eddy moving against the current of entropy, that evolution is not a single, unified process. When biological evolution was first discovered in the nineteenth century, there was a temptation to talk as if nuclear and chemical and psycho-social "evolution" were phases of one process, but these three types of evolution are at best merely analogically the same. The recent sociobiology controversy has once more made clear that explanations of cultural development cannot be simply reduced to the principles of genetic mutation and natural selection used to explain biological evolution.[3]

Human culture develops not so much by adaptation to the environment as by our increasing power to transform our environment. Nor can we say that the development of the various forms of non-living matter in the course of the formation of the stars and planetary systems is more than vaguely analogous to the process of biological evolution. Finally, the rise of living things from non-living, even if it has taken place many times and in many parts of the universe — a question we at present have no way of settling[4] — was not the result of some inevitable development of inanimate matter endowed with an intrinsic tendency to become alive, but the outcome of certain unusual initial conditions on our planet earth, which as far as natural laws are concerned, need never have been actually realized in a universe where matter is scattered about so wildly.

Although some biologists believe that in biological evolution, the process of natural selection insures that given enough time and the proper environmental conditions, it is inevitable that more complex and highly organized life forms will arise, current evolutionary theory makes no claim that it was inevitable or even determinately probable that any particular species, including our own human species, should have emerged. Not only may insects prove better fitted than we to survive,[5] but there is no natural law or combination of laws demanding that either insect or human life should ever have actually arisen. At every branching of the "evolutionary tree" any off-shoot might have been terminal, and many were and all will be. Thus modern science does not show us a universe like the Orphic egg in which the whole subsequent series of forms was pre-programmed. Instead we find ourselves in a universe which has had a *history* in the strict sense of the word, a sequence of events which if it could have been predicted at all was predictable only in a very vague, general manner, whose details were predictable only in the short-range.

Considering the enormous (although finite) size of the universe, the vast time it has been in process, and the purely statistical character of natural laws, this historicity of the universe is the only thing about it which is inevitable. We must even admit the possibility (on present evidence highly unlikely) of a reversal of the increase of entropy which would recurve history into a cyclical pattern.[6] Thus, while science shows us a universe whose reality is that of evolutionary processes, these processes lack any single unified development, but constitute an interlacing of many very different processes all of which will probably finally run down, but maybe not.

What if, in fact, the present expansion of the universe does reverse itself? Some speculate that if we should discover that the present average density of matter in the universe is greater than a certain critical figure, then the process now producing expansion will produce a contraction of matter and a return to a state of extreme condensation of matter and energy such as that which existed just before the Big Bang. Presumably in this case a new Big Bang will someday occur and the universe may thus go on forever in a cycle of expansion and contraction. Others have guessed that perhaps there is a continuous creation of matter and energy in amounts just sufficient to off-set expansion and the increase of entropy, so that our universe is in a "steady-state." This latter hypothesis has not been confirmed and has the great disadvantage of introducing into scientific theory a continuous "creation" of matter for which there is no causal explanation.[7]

As for the cyclical theory, not only is it not in agreement with presently known facts[8], but it would not give us infinite time, any more than we have infinite space. During the phase of contraction there would be a reversal of evolutionary developments in which no further advance in the level of complexity would occur, but a gradual regression. In the last phases of this contraction all traces of what occurred in our present cycle would be obliterated, so that if in the next cycle intelligent life should develop (and, as we have seen, this would not be inevitable), those intelligent beings could never discover anything about the history of our cycle. There would be no fossils or ruins left from which they might reconstruct the events of that cycle. Thus *our* universe is the only one of which we can have knowledge and its life-span is probably finite. It is our history and our evolution (including of course all parts of our present cosmos) which is pertinent to our inquiry. This history exhibits certain local patches of temporary negentropy, but it is not a single unified development toward any predetermined goal.

2. Evolution and Causality

Does this mean that evolutionary processes in the sense of local negentropic advances in organization and complexity such as we see have actually occurred on our earth are a matter of *chance*? It is tempting to counter this possibility by arguing that at least the evolution of living things is a kind of unidirectional (orthogenetic) unfolding of active tendencies in organic matter on the analogy of the way a single fertilized ovum is pre-programmed by its genetic structure to develop regularly into a specific type of plant or animal. But the course of evolution is not included in the genetic code, but consists in mutations, i.e. accidents which cause reshufflings or omissions in the code.[9] Thus biological evolution involves real novelties. From these new genetic programs a few are selected by interaction with the changing environment as advantageous for survival in that environment.

This process of natural selection by the environment does not imply that evolving organism are merely passive factors in their own evolution. Living things have an astonishing capacity to adapt themselves to their environment, and without this life struggle on their part, environmental factors would not promote evolution. Nevertheless, this capacity of the members of any existing species to survive in a particular environment always has limits, and it is precisely at these limits that changes in environment determine which new variations in the species will survive. Moreover, these variations resulting from genetic mutations themselves ultimately have environmental causes, since organisms tend to reproduce their kind and to this end are sometimes able to repair their own genetic defects or at least eliminate defective germ cells from the reproductive process.[10] Thus the primary determinants of biological evolution are not to be found in the organisms themselves but in the external environment.

If living organisms had always existed in a perfectly stable environment there would have been no evolution. Therefore, to explain biological evolution we must not simply confine our search to the living realm itself but we must also seek its ultimate causes in the evolution of the environment. Of course once life appeared on earth this environment always included other organisms which were both competitive and symbiotic, but this living environment has always been dependent on changes in the non-living, geological environment. The more we learn about the history of our terrestrial globe, the more we realize how extraordinarily it has been fine-tuned to permit the development of life, and especially the actual variety and range of terrestrial life forms including our own thinking selves.[11]

Thus the common objection of conservative philosophers to the theory of biological evolution, namely, that it defies the principle of causality by positing that "the higher comes from the lower," can be satisfactorily answered by the present form of the theory. Since this theory, as we have just seen, does not claim that life has evolved from simpler forms of matter, nor more complex and highly organized forms of life from lower forms of life because of any guiding agency within the lower forms themselves, but from the interaction of each stage of living things with their environment, it does not stand in contradiction to the principle of causality if that environment is, in some effective sense, "higher", i.e. a more complex and more ordered system than the living things themselves.

To find an analogy, we can think of what happens in a chemical laboratory when a complex compound is synthesized from simpler substances. If the simpler substances were left to themselves they would never develop into anything more complex, but if they are mixed under precisely the right sort of conditions of temperature, pressure, etc. in an ordered sequence, a process of "chemical evolution" will take place which can result in new compounds of extraordinary complexity. The raw materials must be in contact with each other and must be supplied with just the right amounts of energy at the right times, and any by-products which might interfere with the on-going process must also be removed from the environment at the right time. Biological evolution in its earth laboratory has required, analogously, a similar exact, sequential ordering of environmental conditions and events by which natural selection has so favored certain random variations and rejected others, that gradually more and more complex organisms have arisen.[12]

Perhaps someday we will be able to place the basic raw materials of living organisms in a laboratory environment whose sequential changes are regulated by a pre-programmed computer and see life arise and evolve just as it has done in this terrestrial laboratory. Of course there is the problem of the immense time historical evolution has required, but perhaps once the crucial steps are known, we might be able to short-circuit the wandering course of history and telescope the time.

The evolution of life, therefore, has been in one sense a chance process and in another not. It has in no way violated the principle of causality, since we can trace according to exact causality every step that has been taken, and we can show that only a certain sequence of events could ever have produced the actual life forms which now exist (although as I have just suggested, perhaps this sequence could have been shorter and more direct). On the other hand, biological evolution could have taken an entirely different course. It could never have taken place at all, or when it did actually begin it might have stopped short at any point, or gone off in various very different directions than those actually taken. Ultimately what determined it to have taken its historic course was nothing in the original organisms themselves, but in the history of their environment.

Now what of the history of our globe? Again we can explain the changes which have occurred in that environment by tracing its present state back geologically epoch by epoch; but to do so we have to introduce factors from a wider environment, namely, the solar system. We have to hypothesize changes in solar energy, the fiery rain of meteors, the influence of other planetary bodies on the tides, etc.[13] Given certain initial conditions we can always explain *over a short-range* why certain events rather than others took place, for example, why the dinosaurs disappeared, but these initial conditions are themselves not the result of any single natural law or governing agency, but must be explained by going outside the solar system to a still wider area of the universe.

Therefore, we cannot look to science to explain our world simply in terms of natural laws. Natural laws formulate *regular* occurrences. Hence no single law can explain novelty, and evolution is by definition the rise of the new. In order to give a causal explanation of evolution we must have recourse to many natural laws coming into play in an orderly sequence so as to produce "the higher from the lower", and such a sequence is itself not reducible to any single law, but only to *history,* that is, to a unique sequence of occurrences. Many conditions and forces at any point in this sequence must meet in conjunctions which may never again occur, or if they do reoccur, could never be predicted. To explain any historical event, for example, the assassination of President John F. Kennedy or what happened when Mount Helen exploded, we could call upon all the different physical or psychological laws which may have operated in these events, but ultimately we would not be able to explain why all these conditions and forces concurred at that time and place except by taking for granted the initial conditions, that is, the historical situation immediately preceding that event. That supposition would then force us back to the various antecedent situations and the narrative of such a sequence of situations is not a question of some universal natural law, but simply their history. Furthermore when we recall that natural laws themselves are only statistical, as contemporary physics insists, we can only conclude that the ultimate explanation of our evolving world must be in terms not of nature but of history.

To summarize: the world-picture revealed by modern science is most plausibly interpreted as that of a somewhat loosely organized universe whose history from the Big-Bang to its entropic death we can broadly trace. In the midst of this vast but finite time and space our little earth occupies a remarkable negentropic counter-eddy in which we can track the evolution of life as an intelligible but not predictable history. This history exhibits the operation at every point of a few fundamental natural forces working with a beautiful, but not invariable, regularity, which have finally produced the enormously complex little organism we call a human being through a remarkable sequence of historical events, each of which might have taken a very different course in any other, even slightly, different environment subject to any other interferences than those which have actually occurred through various chance encounters within the solar system and from beyond.

Thus the development of modern science has not only exploded the ancient Greek conception which Christian theology inherited from Plato and Aristotle of a neatly organized, cyclical system in which chance was reduced to a minimum by the action of the inalterable, perpetually rotating heavenly spheres, but it has also exploded the conception of the early Enlightenment that our universe is a clockwork machine obeying Newton's laws in which, as Laplace supposed, the whole future is theoretically predictable from its initial state.[14] Modern science has revealed a universe in which there exists a range of beings of greater or less complexity, but in which even the simplest, such as the helium atom, are of marvelous complexity. It has shown that this complexity has emerged out of an enormous but spatio-temporally finite mass of matter and energy existing at every stage of complexification, but remaining for the most part at a relatively low level of organization, forming planets, stars, clusters of stars, galaxies, clusters of galaxies, and super-clusters of galaxies without any over-all plan or systematic unity.

It has often been suggested that this universe is a set of Chinese boxes fitted together in such a way that no matter how small or great the unit, it is equally organized at every level. This theory does not seem to be confirmed by what we actually observe.[15] In fact the macrocosm of galaxies and super-galaxies seems to be minimally organized, while the microcosm of the atom, although in some measure indeterministic, is wonderfully organized and unified. On the whole our universe is full of order, but it is an order which here and there emerges out of a rather chaotic background and then disintegrates again like beautiful images reflected on troubled waters. This universe unveiled by modern science is very different from ancient models of the cosmos, yet, as some have recently observed,[16] it has less in common with classical Greek cosmologies than with the oriental view that the physical world is like a transitory dream and perhaps even more agrees with the Hebraic notion that the stability of the earth in the midst of the primeval chaotic ocean of the cosmos is its greatest marvel.[17]

II: False Clues

1. Reductionism

In all efforts to give a satisfactory interpretation to the data of modern science there is an ever present tension which never seems to be satisfactorily resolved centering around the notion of "reduction."[18] It seems clear that science has been successful in uncovering more and more of the orderliness in our world by a method of reduction, that is, by analyzing whatever appears complex into its simplest components. Recently this method is achieving some of its greatest triumphs in the field of molecular biology.[19]

From the time of Newton and Boyle it seemed a well-founded hope that sooner or later it should be possible to explain all physical phenomena by reducing matter to some type of elementary particles acting on each other by a few simple forces. Maxwell's unification of electric, magnetic, and optical forces seemed to confirm this hope, and, although it has not yet proved possible to realize Einstein's dream of a completely unified field theory, we seem to come ever closer.[20] While the phenomena of life, especially psychic life, for a long time seemed to demand an irreducibly different type of explanation, it now seems possible to reduce organic life to purely chemical terms through the discovery of DNA whose complex molecules are able to encode all the information required (along with environmental factors) to explain embryology, physiology, and evolution and to reduce psychic life to electro-chemical messages passing through the wonderfully complex circuitry of the brain.[21]

Thus to the degree that we can explain the world by natural laws, it seems clear enough that we should be able in time to reduce all natural objects and their characteristic activities to a few fundamental, relatively very simple units possessed of a very few fundamental properties expressed by a very few fundamental laws. The intellectual beauty of such a system is a great consolation to thinkers who feel frustrated by their realization of the indeterministic, unpredictable, historical character of reality described in the previous section of this chapter.

Yet over against this satisfaction with the reductionist ideal of science is a constantly recurrent opposition against this ideal within the scientific community itself. This opposition is a loyal one. It does not question either the goals or the methods of modern science, but it is again and again compelled to criticize what it regards as a tendency to be satisfied with simplistic theories which neglect the full richness of the data. These protests are often made in the name of "holism" or of what is sometimes called "emergent evolution", or more recently what is called "general system theory."[22] The point of all these objections is sometimes expressed by saying "the whole is *not* equal to the sum of its parts, but is greater."

Certainly it is obvious enough that there is a great deal of difference between an assembled, operating machine, and its disconnected parts lying in a heap. Similarly we may analyze a living organism into its organs, these into tissues, these into cells, these into organelles, these into macromolecules, these into simpler molecules, these into atoms, these into subatomic particles, and perhaps these into quarks. Yet no one would admit that the ashes and gases produced by incinerating a living thing are *equivalent* to an organism in which all these particles are organized and acting as an inter-related dynamic system. What remains after a nuclear explosion? The same sum of matter and energy, yes. The same realities, no! The general system theorists have shown that a system has properties not possessed by its separate components.[23] As we study more complex systems ranging from the atom to the human organism we discover many levels of organization each with new properties which could not have been confidently predicted by an examination of entities having lower levels of organization.

Is there anything mysterious about this? Not really, since it is obvious that the difference between the components of a system and the system itself is the *organization* of the components into a system. What do we mean by "organization"? We mean that the components have been given certain new *relations* to one another, e.g. the parts of a clock have been fitted together in a definite way and subjected to the action of the watch-spring which itself has been wound into tension, so that the clock begins to run as a unified system.

The essential point, neglected by reductionist habits of thinking, is that these relations are not mental fictions, projections or interpretations but are objectively *real,* just as real as the parts which they relate.[24] These real relations consist first of all of static arrangements of the material parts in space and time, and second of the interactions of cause and effect between these parts which cause the system to undergo a series of transformations involving the in-put and out-put of energy. Reductionism is the habit of talking as if the system were "nothing but" the component parts, while treating the static and dynamic relations by which these parts form a whole as if these relations were somehow less real than the parts they relate, or even as if they were merely mental fictions. To say that a human being is "nothing but" a collection of (or that an atom is "nothing but" a collection of) sub-atomic particles is reductionist. Underlying the reductionist habit of mind is the unconscious conviction that "matter" is the really real or at least more real than the relations and actions of matter, that is, than "energy."

2. Process Philosophies

There is, however, a kind of counter-reductionism which ends in what has sometimes been called "dynamism," exemplified in the philosophy of Leibnitz and Boscovitch[25] and which has recently re-emerged in the so-called "process philosophies." In opposition to the reduction of all physical reality to inert matter, dynamists argue that matter is "nothing but" energy. They wish to reduce physical reality to pure process, to "events." This tendency which goes back to Heraclitus and which underlay the Stoic notion of the world-soul as energy, also was evident in the neo-Platonic effort to picture all of reality as light and in the German idealists' effort to understand reality as a spiritual "dialectic". In our century it has emerged again in Bergson and Whitehead as "process philosophy." Bergson conceived the universe as an evolutionary process driven by the *élan vital* or life impulse.[26] Whitehead, who grappled seriously with relativity and quantum physics, attempted to reduce reality to "actual occasions," minute, sub-atomic quanta of energy which come into existence and perish in a quantum of time — a strange synthesis of dynamism and atomism.[27]

The difficulty with dynamistic or process philosophies is that they tend to idealism, to an over-spiritualization of our tangible material world. What would a physical "process" be that did not move or transform a *body* in space and time? But if we use the categories of space and time we are speaking of *extension,* of parts outside parts, or a "here" and a "there."[28] Either this extension is considered to be that of a material body or it is considered to be a property of empty space. Certainly we do not need to identify "body" with "extension" as Descartes did,[29] but we can hardly deny that all the bodies of our sensible experience are extended, and if they were not we could not observe them.

On the other hand I will argue in the next section that the notion of "empty space" is self-contradictory. Einstein replaced "empty space" with the concept of "field" and showed that a "field" always implies the existence of a body in that field.[30] Even if we think of an empty space free of matter, we must still think of it as transversed by waves of energy, of light, and gravitation. In modern physics such energy is quantified, that is, it can be understood as massless particles (photons, neutrinos, gravitons) but is convertible into matter having mass.[31] Thus, processes do not go

262

on in empty space but in bodies having spatial relations, undergoing movement or transformation in time.

Nor is it helpful to attempt, as did Russell and Whitehead[32] to reduce material objects to "events." What is an "event" if it is not conceived as an encounter of bodies or the transformation of bodies by mutual interaction? It must be admitted that physical reality always exhibits a certain duality; not the dualism of Descartes' mind and matter, but the polarity of matter and energy, of that which undergoes change and of the process of change itself. We never observe changes or processes that are not changes or processes of something *relatively* inert and passive. We never observe events which are not interactions between bodies which pre-exist the event and are transformed by it; or, if they perish in the event, somehow provide the substrate for the formation of new bodies.

Whatever interpretation we give of the scientific world-picture we must respect this duality of physical reality by using at least two basic terms corresponding to "matter" and "energy", which cannot simply be reduced to each other. Even when physics says that "matter" and "energy" are convertible, it is affirming, not denying, that physical reality always has two faces. It is static *and* dynamic. This is why any reductionist interpretation which attempts to say that the physical world is "nothing but" elementary particles or "nothing but" energetic processes is inadequate to the observed data.[33]

The most serious result of both atomistic and dynamistic reductionism is that they make the scientific process itself absurd and reduce the scientist — the very type of intelligent humanity — to absurdity. If the scientist is nothing but a collection of sub-atomic particles, how can it be possible for him to observe and think about the world of sub-atomic particles? It is only because in this scientist as a thinking organism particulate matter has been highly organized so as to form a single unified system, more highly organized than the most intricate computer scientists have yet contrived, that this scientist can be a thinking being. The network of real relations and dynamic causal interactions which constitute this thinking organism which is a scientist is just as real as are the particles of matter which this network dynamically organizes.

This real unification of the components of the scientist as organism is so intense that it enables the scientist to become *self-conscious.* He or she can not only receive information from the environment, but can process it and react to the environment in accordance with this processed information, and still more importantly can be conscious that he or she is doing all this, that he or she is a thinking organism performing the scientific process of gathering data and theoretically analyzing it. The computers which scientists have invented as their servants can perform many phases of this scientific process, but they cannot (at least yet) as I will show in the next chapter, perform the essential work of self-consciousness, because they lack this kind of unification or integration. Without that power of self-reflection which the scientist possesses and which the computer lacks there can be no science.

On the other hand a dynamistic reductionism also makes science absurd and the human scientist an angel. If matter is nothing but energy, process, a stream of events, and scientists themselves nothing but streams of consciousness, what has become of the material universe and of the scientists as human beings, bodily parts of that universe? Dynamism thus tends to evaporate into idealism and the scientist into an angel, a being of pure thought. Surely, any interpretation of modern science which ends in reducing it to a play of ideas in an angelic, disembodied mind, goes counter to the very first truth on which science is built, namely that a scientist is part of a material world to be explored by our bodily senses, to be known *empirically.*

I must, however, pay tribute to the process philosophers at least for this achievement, that they have focused modern discussion of the scientific world-picture on the *dynamic* aspect of physical reality, since undoubtedly it is this aspect which must be at the center of our attention. Granted that the world is material-being-in-process, yet it is only through the processes which matter undergoes that we can understand matter itself. Hence my intention in this chapter is to sketch a process philosophy of science, but a process philosophy which does not neglect matter, nor the human bodies of the scientists who strive to understand themselves and the material-world-in-process of which they are primary units.

3. Clearing Away the Nonsense

As I attempt to sketch what a process philosophy interpretation of the scientific world-picture which would respect the reality of matter and the human body might be, I must first point out that certain terms which often enter into discussions of modern science must be eschewed from the outset because they have been shown by the progress of science to be meaningless, since no empirical basis for them can be found. It might seem that if these notions have been rendered obsolete by science itself then this purification of our vocabulary is unnecessary, but the annoying fact is that although these notions play no useful role in science itself, they continue to wreak havoc in the philosophical interpretations of science.

The first of these obsolete notions, one which played a decisive role in Kantian philosophy and throughout the nineteenth century and which still survives in some writers today is the idea that scientific explanation is *prediction.*[34] Of course there is some sense in saying that one of the criteria of a good scientific hypothesis is that it "predicts" some previously unnoticed phenomena which can then be verified by observation or experiment, but it would be much more accurate to say that it "suggests" rather than it "predicts." Today we know that scientific laws need not be absolute, but only probabilistic. Because physical reality is historical, the result of multiple coincident causes in which some degree of chance (entropy) is always present, we cannot predict the future, not just because we lack complete information about the present, but because in actual fact the future is not wholly determined by the present state of affairs.

Scientific explanation, therefore, is not prediction but postdiction. It is an explanation of what has already actually taken place, just as is human history. I can give a reasonable explanation of why President Kennedy was killed, but I could not have predicted the event with certitude. In the same way I can give a reasonable explanation of the explosion of Mount Helen, but I could not (at least accurately and with any confidence) have predicted that event in concrete detail either. We can explain to a degree how biological evolution has taken place, and even cosmological evolution, but we cannot predict the future except by conjectures which may prove very wide of the mark. It is true that we have learned to predict eclipses and the positions of the planets with remarkable accuracy. We have also predicted the discovery of previously unknown planetary bodies, of previously unknown elements and elementary particles. But these successes, beautiful as they are, are not typical of most scientific efforts and cannot provide us with the paradigm of scientific explanation.[35]

A second set of useless notions (already discussed in a previous section) is that of *absolute space and time* independent of bodies and their processes. A space independent of any body is a contradictory notion because it would be a true "void," i.e. absolute non-being and yet it would have structure, i.e. extension or parts outside each other. How could a true void be curved, as Einstein postulates? It could not even be Euclidean, since Euclidean space has a uniform structure and is in fact only a fiction of the mathematical imagination. As for a time independent of bodies in process, that too is a contradictory notion, since time is only a relation between on-going processes by which one is measured by the other. Einstein has revived in a new form the Aristotelian view that space and time are inseparable from each other and from the actual motion of bodies. The source of Newton's absolute time and space was his identification of imaginary mathematical space with real space, and of the mathematical representations of motion and time as lines in space with the physical realities of time and process.[36] We ought not substitute a static diagram for a dynamic on-going sequence of events.

A third useless notion which often recurs in cosomological theories is that of *infinite structural complexity,* i.e. that no matter how far we explore macro-space we will always find greater and greater worlds and no matter how far we explore the microworld we will always find smaller and smaller units of matter and energy. I have referred in a previous section to the hypothesis often revived which is supposed to answer the famous paradox attributed to Olbers[37] according to which if the universe were spatially infinite the night sky would be infinitely bright. This hypothesis is that matter is so arranged that the universe is like an infinite set of Chinese boxes, worlds within worlds, so that every particle is a super-galaxy of particles each of which is a super-galaxy *ad infinitum,* while every super-galaxy of stars is but a star in super-galaxy of a higher order and so *ad infinitum.* This notion has a certain plausibility because we have broken down the supposedly indivisible atoms into sub-atomic particles and even these into quarks and we cannot foresee how far this divisibility of matter may go. Similarly we have greatly expanded our notion of the size of the universe. Nevertheless, this hypothesis by its very terms is incapable of empirical verification because we cannot extrapolate what we know about finite structures to infinite ones. Moreover, it plunges us into the same type of paradoxes as that of Olbers, paradoxes that always arise when we try to talk about an actual physical infinite.

Hence we must also rid ourselves of the notions of infinite space or time. Koyré has shown that the introduction by Nicholas of Cusa of the notion that the universe is spatially infinite had a profound effect on the Copernican paradigm shift.[38] Cusa espoused the notion for mystical reasons, because he felt that such a universe would be a better mirror of God whose infinity and ineffability were a central theme of his neo-Platonic, Eckhardian mysticism. Modern science, however, has found no use for these notions, since the universe of Einstein although "unbounded" is finite in its dimensions.[39] The Big Bang hypothesis also implies that the universe as a dynamic unity is also finite in time. Yet even if this hypothesis fails and we discover that the universe has a cyclical or steady-state structure, still in this case we could have no way to verify empirically any events except those which take place in our finite epoch.[40]

A fourth useless notion, which Alfred North Whitehead has eloquently denounced,[41] is that physical reality is constituted by *elementary substances which are inert or vacuous.* Democritus first proposed the fascinating idea that physical reality consists of inalterable atoms which have no qualities except shape, weight, and indivisibility and which are related to each other only by spatial position and surface contact. Science

266

has never discovered anything like Democritus' atoms, yet they haunt philosophical interpretations of science. Every material object we know of undergoes *internal* transformations and manifests its own inner dynamism by acting on other bodies in diverse ways, which in turn react on it. While it is true that these interactions seem reducible ultimately to a few fundamental "forces" (gravitation, electromagnetic attraction and repulsion, and the strong and weak nuclear forces); yet these forces are combined in such a vast variety of ways that any body is characterized by many dynamic qualities.

A fifth useless notion is the oft-revived theory that the universe of our experience is an *organism,* i.e. that it is has that kind of substantial unity found in living things. Whitehead speaks of his cosmology as a "philosophy of organism",[42] a term which is self-refuting unless we understand it merely as an analogy and a risky one at that. This tendency to monism has resurfaced again and again in the history of science and is based on the ancient myth of the Macroscopic Man and the Stoic notion of the "World Soul".[43] We still encounter the notion that the organisms of our experience are simply cells within some greater Cosmic Organism. This notion, however, has never proved scientifically fruitful and it stands in contradiction to all we know of the universe. In fact the more we explore the cosmos, the less plausible any monistic interpretation becomes and the more obvious becomes the *pluralism* of the physical world.

For Greek astronomers the cosmos was a great machine rather than an organism, but it was a very neatly integrated machine. For modern astronomers, on the contrary, it is a vast collection of highly dispersed knots of matter floating in a very thin soup, held together by very weak gravitational forces. Our solar system is more than 4-1/2 light years away from the nearest star and from that star we receive only a tiny pulse of energy in the form of a faint ray of light and an even weaker gravitational pull.

Moreover, even on the surface of our earth the fish that swim in the ocean are not organically related to the water, nor are the animals that run through the forest so related to each other or the rocks or the trees. We can speak of an ecological system uniting the swimmers in the stream and the creatures of the forest, but this kind of system has a unity of an entirely different and lesser order than the unity of an organism such as a trout, a deer, or even a fir tree. I have already pointed out that the term "system" meaning an "organized whole of several parts" is an analogous term which can be applied equally to the universe or an organism or an atom, but it is a crude error to take it so univocally that we cannot sharply distinguish the kind of tight unity possessed by a stable atom, a crystal, or an animal from that of a loose collections of objects such as a forest or a solar system.

Finally, I have to raise a question about the sense of that notion which Whitehead made central to his whole process cosmology — the notion of "creativity."[44] For Whitehead creativity is the ultimate principle which is embodied in every existent and which brings it into existence. Something in this notion certainly should be retained, namely, that all process or change involves genuine novelty. It is astonishing how this fact of novelty has been denied or ignored in so many interpretations of science.

This blindness to novelty is another of the less fortunate effects of mathematization on the natural sciences. Modern science has progressed largely through using mathematics as its instrument of formulation, for its wonderfully clear and versatile language. Mathematical objects, however, are ideal objects existing only in the constructive human imagination and abstracted from all change. When a mathematician wishes to represent the measurement of a change or process, he reduces it to a *function,* i.e. to a relation between "variables", but these "variables" do not in fact "vary", but are simply a series of entirely static numbers. Consequently, the mathematical representation of the changing world reduces the past and future to segments on a line and the present to a point that marks their juncture. To do this is to eliminate our fundamental experience of time and of change through time as moving in a single irreversible direction into a future which is not yet real and out of past which has ceased to be real, but which was irretrievably real and whose effects perdure in the present.[45]

Thus if there is to be process in time, if there is to be real change, if the future is not identical with the past, then there must be real newness in every moment of change. That is why it is meaningless to think of science as prediction and natural law as absolute determinism. The future can never be totally predictable in a world where change and novelty penetrate every reality. With Whitehead I want to make this notion of creativity as genuine novelty central to my radical process interpretation of the scientific-world picture — which is a *moving* picture. Process *is* creativity, and every entity in the world of our experience is in process.

III: Primary Units

1. Empirical Being

After eliminating certain allegedly explanatory concepts which in fact are useless for our purpose, we must ask ourselves where to begin. Sometimes it is said that ancient and medieval thought was geocentric and therefore anthropocentric, while modern thought sees humanity only as a speck of dust in an incomparably greater universe. The very opposite is the historic truth. For ancient and medieval thought human beings were the center of the universe only in the sense that humanity is at the border-line between the spiritual world of the angelic intelligences dwelling in the inalterable heavenly bodies which move in perpetual cycles and the irrational world of things subject to change and chance, but we are the *least* of intelligent beings who dwell, literally, at the *bottom* of the universe.[46]

Both Plato and Aristotle (and the medievals to the degree they followed these mentors) strove to form a conception of the cosmos as a whole in which they assigned to humans a very lowly part. Plato tried to grasp this whole by transcending the human mode of knowing so as to share intuitively in the Eternal Mind and its comprehensive vision of the whole. Aristotle tried to grasp the whole by developing an astronomy in which the whole was understood, at least in its basic features, through an observation of the perpetual movements of the planets and stars.[47] Modern thought, however, at least by the time of Nicholas of Cusa, abandoned this hope, since if it is possible that the world is infinite or at least incomprehensibly vast, then it is also unknowable as a whole. What we *can* know is not the cosmos but our local region.

Thus it was not the ancients, but ourselves who are anthropocentric. From Descartes on, moderns have come to resign themselves to the fact that we cannot grasp the world as a whole. We cannot stand outside the cosmos, as we can now look down on planet earth, but must begin from within the cosmos, from the ground on which we stand, from our own experience of ourselves in a world that stretches beyond our horizon in every direction. Our times have produced the Einsteinian cosmology in which for the first time this relativity to the observer of all our knowledge about the universe is thematized.[48] This insistence on the relativity of all knowledge to the knower, to the subject, is (as we saw in Chapter 6) what marks all modern thought as modern. I too accept it, not because it is "modern", but because I think it has been a major advance in realism over the ancient and medieval view.[49] This anthropocentrism is in fact the logical consequence of the Aristotelian rule that we must always begin in thought from what we know best to explore what we know less well, and the other rule, that we must remember that all valid knowledge must begin in sense experience.

Nevertheless, we must not succumb to the Cartesian understanding of this anthropocentric point of departure. Influenced by the Platonism of Augustine, Descartes understood this beginning with the subject not as beginning our search for truth with our experience of ourselves as *bodies* (for him that was a discredited Aristotelian option), but with ourselves as minds, *cogito ergo sum.* To begin this way is to treat the body and the material world of which it is a part as pure *objects* set over against the mind in dualistic opposition, to be understood as knowable only extrinsically and superficially in a merely mechanical way, describable in purely mathematical terms. By thus beginning with his own disembodied mind, Descartes also began solipsistically with his own individual mind and not with human bodily beings as *social* animals. No doubt his mathematicism contributed to this solipsism, since it lead him to think that thought is true only when it has that kind of clarity which can best be expressed in mathematical symbols, and thus blinded him to the fact that human language invented to communicate between human beings in a community is never clear, but always required a dialogue and an hermeneutic. The warranty of human thinking does not come from its subjective clarity but from its verification through social experience and critical dialogue. Science has progressed effectively because it has become institutionalized in a community which has developed criteria of experimental and critical testing. Consequently, human thought is not only a social and historical process, it is also pluralistic, a competition and cooperation between apparently conflicting views of the world which can refute or complement each other.

If we grant that any valid understanding of our world must begin from our experience as human beings in community, not as isolated individuals, and as a community which daily has to cope with an environment partly natural and partly of its own manufacture, i.e. a culture; then we must also admit that this primordial knowledge of reality is not purely "objective" but is natively "subjective", that is, "connatural", resonant to our self-consciousness. Indeed, in the beginning of our awareness of reality this distinction between subject and object is not thematized. The tension is not so much between the self and the other as it is between what is *pleasant* (or at least comfortable, convenient) and *unpleasant,* between what I desire or at least can live with and what I reject and what my community with which I identify desires or approves and rejects.

The pleasant and the unpleasant themselves imply consciousness, although (as Freud forced us to recognize) this consciousness may be a sub-consciousness very obscure and difficult to verbalize, or as some would say "non-conceptual." Hence our primordial knowledge is not "objective" in the sense of value-free, but is permeated with value, with "subjectivity" both in the sense of consciousness and in the sense of *evaluative* consciousness. It is even permeated by an evaluation of the reality of the real in the sense that waking consciousness has a certain kind of pleasantness that dreaming or hallucinatory consciousness lacks. There is a certain feeling of being really there which I have when fully awake and an insecurity of being afloat when I am between unconsciousness and wide-awake awareness. For this reason also, an important aspect of our primordial knowledge is the tension of curiosity. Freud believed that curiosity, even the curiosity of the scientist, originates in sexual curiosity.[50] It might be truer to say that it originates in libidinal curiosity, that is, in the desire to become fully conscious of pleasurable states along with the opposite tendency to escape and forget unpleasant states of things. Yet we soon learn that we need to know even what is unpleasant (e.g., children's fascination with tales of horror), in order to be able to avoid or overcome it. Thus at the beginning our knowledge is indistinguishably both theoretical (awareness of what is) and practical (evaluative, tinged with what should or should not be).

Since the Greeks, philosophers have talked much about "Being" in a rather mystifying way. Eastern philosophers tend to talk also a great deal about "Non-Being" and to assert that ultimate reality is beyond Being and Non-Being. Heidegger at first claimed that the whole history of western thought is a story of the "forgetfulness of Being", and ended by appealing for an opening of western to eastern thought. What is this "Being" which comes first in human knowledge and which we moderns have forgotten? As John N. Deely has pointed out,[51] the sense of this question for Heidegger is by no means the same as for Aristotle (for whose epistemology as against Plato's I have already opted),[52] but we cannot understand or evaluate Heidegger's position without first understanding Aristotle's position in its own light, rather than in the perspective of the idealist tradition through which Heidegger viewed it.[53]

For Aristotle (and for Aquinas, who as we saw in Chapter 4 is the Catholic thinker who most strongly advocated the Aristotelian epistemology), Plato was mistaken in supposing that the Being directly known to us is some transcendent reality accessible by an innate and private intuition. The only Being directly knowable by us is the world of ordinary human life of which we are part by our bodily existence.[54] Aristotle's epistemology (unlike Kant's), does not exclude the possibility that there exist entities outside our direct experience, but it maintains that if we are ever to know anything of such entities it must be through interference from the beings of this material world which we directly experience. Aquinas calls this first known Being *ens commune,* the common reality of sensible, material things whose observed existence may or may not entail the existence of other unobservable things — that question is left open.[55]

What is most evident about this Being which is first known by us is that it is changing, in process. It is not a Being which is fixed, but it

271

appears and slips away, so that it is easily confused with Non-Being. No doubt this is why for Pythagoras, Parmenides, Plato and for many oriental thinkers, empirical being more deserves to be called Non-Being than Being. Yet surely Aristotle was right in saying that the being of our everyday world is, however imperfectly, truly Being. If it is not real, then we know of nothing that is real. Aristotle also is certainly right in making what is after all a very modest claim that our human intelligence, dull as it may be, is capable (at least sometimes and in some cases) of distinguishing Being from Non-Being, i.e. what here and now exists from what has ceased to exist or has not yet begun to exist.

This intelligent discrimination within our confused experience of what is from what is not makes possible our most basic human judgment: "What exists cannot *not* exist at the same time and in the same respect." This principle of non-contradiction is not merely a logical rule but our first rational judgment about the world of our experience. It is a world in which a fact is a fact at the moment that it is a fact. It is the starting point of all our thinking, long antecedent to Descartes' *cogito ergo sum* which presupposes it, because it is the least possible yet necessary assertion about our experience. We cannot even ask a meaningful question (in other words, we cannot really *doubt* anything) without presupposing it, because a contradictory world would be an absurd, meaningless world in which doubt could not be distinguished from certitude. It is rooted in our experience of process, but a process of something, namely what is not nothing. It constitutes the first stable point in my struggle to understand the world, the bedrock on which I can escape the original vertigo of being confronted by a puzzling, pleasurable-painful life.

The principle of non-contradiction also implies (not thematically but *in actu exercitu,* in the very process of knowing) a distinction between sense knowledge and intellectual knowledge, since in our sense experience Being and Non-Being are confused in the process of change, in the coming-and-going of what we see, hear, touch; and it must be some other knowing capacity, which we label "intelligence," which discriminates the positive from the negative in this experience, which recognizes Being as such. Non-Being as a concept is not derived from reality itself, but is a concept created by the intelligence as a being of the mind (*ens rationis*) to which nothing corresponds in reality except the negation of reality (*ens reale*) by the process of change.

This discrimination also has a libidinal aspect since Being seems preferable to Non-Being, and the discrimination of Being from Non-Being seems preferable to their confusion. In my dim awareness of Being I struggle to grasp and possess Being and to avoid slipping away into Non-Being. "To be or not to be, that is the question," but normally we humans opt for being. There is even here a certain sexual element, because the knower, as empty of Being and seeking to possess Being more clearly experiences her or his femininity (for femininity is half of every human person, female or male) in the desire to be impregnated and filled with new life. Yet the human knower, in also desiring to penetrate, control, and transform Being experiences his or her masculinity. In biblical language "to know" can mean to know sexually, "Adam knew his wife Eve and she conceived and gave birth" (Genesis 4:1). The knowledge of Being is the beginning of our human affirmation of life, and its social transmission.

After we have understood that Being-in-process (for that is the only kind of Being I experience) can be distinguished from Non-Being, we begin to see the distinction between the One and the Many, the Whole and the Part, the Same and Other. We recognize that the Being of our experience is not simply one but many, and many in many ways, and that within each single being there are parts, since whatever undergoes change has parts,[56] nor would there be any change unless there were many beings acting on each other, or at least one part within a being acting on another part. Thus the ones which we experience are not simply one but in another way many. They are wholes having parts. They are "systems."

With this discrimination of the One from the Many arises also the concept of Relation and its correlate the Absolute, meaning by "absolute" not some first being, but simply those beings which are the *terms* of relations. Thus there are real relations between the parts and the whole in every being, and between beings as they are same or different. We begin now to see that the reason Being-in-process appeared to us first so confusedly was that it is not one thing, but many things held together in a whole not by an absolute unity, but pluralistically by a network of real relations, not in some fixed, static pattern but in the ever shifting relations of process.

Only after we have achieved this kind of intellectual analysis of Being-in-process which every growing child learns to perform in the very process of learning to speak intelligently,[57] but which philosophers must retrace critically, do we finally come to what Descartes mistook for the *starting* point of certainly valid thought, namely, the discrimination between the known and the knower. Descartes spoke as if what is clearest to us is our thought from which we infer our own existence. But is this true to our experience? Does the child know he is thinking? Or rather does he know his parents, his dog, and his breakfast? We have gone many steps in learning to think before we realize that the *relation* between what we know and ourselves as knowers is what is called "thought." Moreover (and this fact must never be lost sight of in our theory of knowledge), we come to know ourselves as knowers, as a special kind of being among

273

the other beings of the world, not by first exploring ourselves as objects, but by our interest in things other than ourselves: food, people, playthings. Then we discover the relations between these others and our own bodies — pleasurable or painful relations — and thus we discover ourselves not in splendid isolation like Rodin's "Thinker" but always in dynamic relation to other things. Even in exploring our bodies we first know ourselves as toes, hands, torso, head — as parts and then as a whole. No doubt prior to all these discoveries some kind of awareness of our bodies, of feelings of pleasure and pain, and of images of things not actually present are part of our experience of Being-in-process. We are already, in Heidegger's meaning of the term[58] in a "world", but that world which includes ourselves is the confused vision of Being-in-process and it is only by several steps of discrimination that we arrive at explicit self-consciousness.

This discrimination of the knower from the known, of subject from object is not possible until it has become clear that there is a difference in our experience between real objects and purely mental objects, between the real relations given directly in experience and the mental relations which are constructed by the process of knowing.[59] What is the criterion by which we are able to make this all important discrimination between the really real and illusion? I can thus discriminate because mental beings can be self-contradictory (for example, I have already formed the notion of Non-Being as the opposite of Being, yet the notion of Non-Being is contradictory), while real Being must conform to the principle of non-contradiction.[60]

Once we can discriminate between what we experience and what we think about it, between the real and the mental, the concepts of Truth and Falsity appear, since it becomes possible to discriminate between knowing that is verified in real experience and that which can be falsified. Today some teach that while we can falsify propositions we can never verify them with certitude. While this rule has some heuristic value in the sciences, it cannot be taken literally, since we could never falsify a proposition except by showing that it is counter-factual, i.e. contradictory to some certainly true proposition, at least the principle of non-contradiction which, as we have seen, is more than a logical rule, since it is a true proposition about our experience.

The idealism to which Descartes committed himself and which conditioned Kantianism and the mainline of German philosophy right down to the phenomenologists including Heidegger, in spite of valiant efforts by the phenomenologists to free themselves from it, rests on this wrong turn of taking the knowing subject as the departure point of knowledge as if this were the first vision of Being (the *primum cognitum*), instead of starting with empirical Being-in-process *within* which we come step-

274

wise to discriminate subject and object. In a measure the phenomenologists have become aware of this wrong turn in their insistence that we must begin with man-in-his-world, but their notion of "world" remains ambiguous because they are held back by their phenomenological method from pushing all the way back to Being-in-process. As John N. Deely has shown,[62] Heidegger deliberately confines his notion of "Being" to Being in its *intentional* existence (*esse intentionale*) i.e. as purely phenomenological.

In the Chapter 9 on human historicity, I will discuss the importance and fruitfulness of Heidegger's problematic and show that in a sense he is correct in saying that western thought has suffered from a "forgetfulness of Being" in his sense of phenomenological Being; but it is also true that since Descartes western thought has suffered even more severely from a forgetfulness of Being-in-process as experienced anterior to the so-called "critical point of view", i.e. before any choice between an idealist and a realist epistemology is even possible. If we do not push all the way back to this Common Being we cannot intelligently make a choice between idealism and realism and will be uncritically imprisoned, as the phenomenologists seemed to be, in the Cartesian perspective.

On the other hand, if we do push back all the way to Being-in-Process as Aristotle was forced to do by the paradoxes which he recognized in his master Plato's system[63], we will be persuaded that the choice of realism is fully justified. If we can (at least in some favorable cases) discriminate between things as they are and our thought about them, between the existing reality (which includes ourselves as a certain kind of thing — a knowing thing) and our thinking about them which is sometimes true and sometimes false, and if (at least in some favorable cases) we can verify some of our judgments about this reality, then there is no reason to choose idealism and every reason to choose realism as our philosophical tradition and to correct it critically by reference to empirical verification. The idealism of Plato which supposed that the proper object of our intellection is innate ideas which somehow have a transcendent existence, or modern idealism which supposes that the proper object of our intellection is our own thought or the world as it appears in our thought but not as it is independent of our thought, are thus exposed as self-contradictory, since these suppositions presuppose that we know our thinking before we know the Being-in-process of which our thinking is only a dimension, a dimension which can be discriminated from other dimensions, but only in relation to them.

In our efforts to understand our originally confused view of Being, we cannot rest simply with intuitions or insights which arise by this process of discrimination and distinction and by forming judgments that

Being is not Non-Being, that the whole is greater than the part, etc. We discover that we desire to understand our world and ourselves within the world more fully and clearly, because puzzlement is painful, and also in order to control, use and transform the world which we evaluate as pleasurable or painful, good or bad. At first our theoretical and practical knowledges are confused, but little by little we come to see that they have different but interrelated purposes. The Marxist insistence on the unity of theory and praxis, or the various pragmatic theories of truth congenial to Humanism are certainly correct in insisting that human knowledge arises from practical life and is tested in it; but it is equally true that practical action always presupposes an effort to obtain a true, theoretical analysis of the world. Moreover, ultimately our purpose in trying to control the world is to contribute to our own human satisfaction, a satisfaction which is completed only in reflection on and contemplation of what we have accomplished in the context of a wider cosmic wisdom.

These fundamental discriminations in Being-in-Process which yield the basic vocabulary of all thought (Being and Nothing, One and Many, Whole and Part, Real and Ideal, Truth and Falsity, Good and Evil) are not peculiar to the philosopher, but belong to all human knowing and all language. They are very vague, broad, crude, analogous, not univocal, naive, not thematic or critical. They are not "categories" or classifications of beings but are clarifications and discriminations of the general experience of our being in the world, of Being-in-process. They supply us, however, with a basic set of non-arbitrary notions and principles with which to build a process philosophy, since without them the very notion of process remains empty and meaningless.

It should be noted that these basic terms are not "metaphysical" terms, whether that abused term be taken in its current pejorative sense as a meaningless word, or its Kantian sense of a word illegitimately extended beyond its empirical usage, or finally in the ancient sense of a term proper to that first philosophy whose object is "Being as such" i.e., Being as it includes both material and immaterial realities. These basic terms in the sense so far used relate only to Being-in-process and are proper to a philosophical study of empirical reality.[64]

2. The Natural Unit

In order to advance from this general description of Being-in-process we must next raise the analytical question of the One and the Many, the Whole and the Parts of the reality we experience. One approach would be to begin from the whole, the universe or cosmos as if it were absolutely one, the primary reality. In ancient times this was the Stoic approach and it still seems to be that of Marxism which asserts that the really real is a vast mass of matter in motion.[65] As we have seen, however, nothing in modern science makes this way of preceding very promising, since our expanding universe is so loosely organized. Even our solar system and our terrestrial environment, although they are systems, are held together only by a network of relations rather than by a single unified order.

It is of fundamental importance in any interpretation of the scientific world-picture to determine what are the primary natural units. We have already argued that nothing is gained by supposing, as did Whitehead,[66] that these units are "actual occasions" or "events", because there is little sense in talking about an "event" until we have decided what are the beings to which events happen. Nor can we be content to say that the selection of such units is merely a matter of observational, experimental, or mathematical convenience like measuring off a space in yards or meters. What we need to determine is what units are beings in the absolute sense, that is, the natural terms of the network of relations which form the universe.

The reductionist bias of so much modern science has led many theorists (including Whitehead who is an opponent of most types of reductionism) without sufficient reason to suppose that these natural units must not be macroscopic objects of direct experience but hypothetical particles of microscopic (or less!) size. Back of this supposition is the long tradition of Democritus who posited his indivisible, structureless, inert atoms as the primary natural units because he thought that these units must be absolutely simple. There is no logical reason, however, why a primary unit of Being-in-process should be absolutely simple; or why it should not have structural parts, provided these parts have no absolute existence but only existence relative to the whole formed from them. Moreover, since these primary units are units of Being-in-process, they need not be inert, indivisible, or unalterable but rather should be conceived as dynamic units, coming to be, maintaining themselves for a time, mutually interacting, and inevitably perishing sooner or later as a result of such interactions.

What then are the primary natural units which we know empirically? The answer is so obvious that it is usually ignored. The natural unit most empirically evident to a scientist is himself or herself, not as a mind, but as a thinking body presupposed to every act of observation or experimentaion. Next are the other human persons who are his co-workers or his fellow-scientists for whom he publishes the reports of his research, or, if he works in the life sciences, other members of the species *homo sapiens,* and other animals and plants of various spcies. Of course these natural units are highly complex and they have parts and these also have sub-parts all the way down to the atoms which exist as dependent constituents of the unit; but the complexity of these units need in no way raise doubts about the fact that these are empirically determined natural units clearly distinct from their environment, although related to it in many ways.

It is one of the strangest paradoxes of modern thought that at the very time when with Descartes scientists became conscious that the self-conscious subject is at the center of our exploration of the universe, they also began to doubt with Locke and Hume that we can know the human self, and to suppose that what we know is not the self but merely our stream of consciousness. Although Kant attempted to overcome the scepticism of Hume, he too felt compelled to deny that the noumenal self can be known to us.[67] We have seen that the cause of this dilemma was Descartes' assumption that what we are first conscious of is our thought, rather than of ourselves as body-persons in a world of bodies. Human thought is certainly not just sensation, but it always arises from and includes sensation — the awareness of myself as a body in contact with bodies. For human knowing the Being which is first known (and indeed the only Being we know directly) is Being-in-process, the world of bodies of which my body is one. Of course, I also am aware that I am a knowing body, a body-person and as such I am a primary natural unit because I know myself as an absolute unity, the term of relations to other bodies. I am simply *one*, not as a collection of smaller units that are *more* one than I am, but I am at least as absolutely one as anything in my experience, the very paradigm of unity to which any other kind of unity has to be compared. The fact that I have many organs in no way negates my unity, nor that each of these organs is in turn made up of tissues, molecules, and atoms, but rather evidences by its very complexity the very high order of unification that constitutes me one living organism, more truly a unit than my social community, my terrestrial environment, the solar system of which the earth is a part, the galaxy through which that system moves, or the vast reaches of the expanding universe. All the findings of modern science are congruent with this basic fact of direct experience.

So similar is my body to that of the higher animals that I cannot seriously doubt that they too are natural units, although I cannot confirm this by talking to them as I can to other humans. Less clear is it that certain lower animals or plants are units, that is, I am not always sure where one unit leaves off and another begins. Yet the experience of a single seed growing up to produce a plant, and the observation of many plants of the same species (e.g. a field of wheat) each growing separately from the others, leaves no doubt that these too are primary units, although I note they exhibit a less organized unity than do most animals. For both animals and plants among lower forms there are ambiguous cases, but these ambiguities in no way contradict the clear evidence of the other forms.[68]

278

Among inanimate things the designation of natural units at first sight seems very difficult. Is a rock a natural unit? A raindrop or the ocean? In liquids or gases it appears that a body can be divided into parts each of which without any further change becomes a natural unit; for example, a glass of water dipped from a lake. In the case of solids once this division is made it is usually difficult to get the parts to reunite without first melting them. Moreover, when we examine solids such as rocks we note that they seem to be aggregates of heterogeneous materials without any unifying structure. Nevertheless, it is precisely such puzzles which baffled ancient science that modern scientific investigation helps us to resolve, and not necessarily in a reductionist sense.

We are now able to explore the crystalline structure of homogeneous substances and to show that some gases and liquids are also homogeneous. These homogeneous substances have been shown to be composed of molecules. Therefore, the most reasonable interpretation seems to be that while in gases these molecules are free primary natural units, in solids these units coalesce to form larger *primary* units in which the molecules remain only as parts, and in liquids they exist in a transition state. The fact that homogeneous solids normally exists in crystalline form (amorphous forms can be considered transition states) in which the molecular lattice is repeated over and over indefinitely seems to show that the kind of unity proper to solids does not determine their dimensions, unlike the unity of a living organism which does determine its mature growth at least within a range of size. It should not be surprising that the unity of simpler, inanimate units should be of a lower order than that of living units.

It also seems clear that while molecules are divisible into atomic parts which sometimes exist in their own right as natural units, in the molecule these atoms undergo a transformation which makes them genuine parts of the molecule as a primary unit.[69] As for the parts of the atom, the subatomic particles, the present evidence seems to favor the conclusion that they are *not* primary units in their own right, since they have independent existences only as transitional fragments which continue to exist only by uniting with other particles or being absorbed by existing atoms or molecules.[70]

Thus, in spite of many not unexpected obscurities, it seems safe to say that we can correctly interpret the findings of modern science in such a way as to conclude that the primary natural units of the part of the universe which we have explored are animal and plant organisms, molecules and atoms, as well as inanimate homogeneous solids of indefinite dimensions.

The universe also contains a great deal of matter which is in transition from relatively stable existence in the form of primary units to existence in some other unit. Thus among living things we have corpses and even temporarily living cells that have survived the death of an organism and have not yet returned to the normal condition of inanimate substances or await consumption and re-incorporation in some living organism. We have germ cells after they have separated from the body of the parent and before they unite to form a new organism. We have liquified or amorphous (congealed but not crystallized) substances. We have chemical "radicals" or fragmentary molecules in a solution. We have subatomic particles in transit. We also need to include among the transitional entities the "vacuum" between subatomic particles, in the interstices of crystals, liquids, and gases, and the interstellar "vacuum." Formerly this "empty space" was called "aether" and was conceived as a mechanical entity, a kind of fluid, itself made up of some kind of particles. Today it is thought of simply as a "field" which is without mass or inertia but which is filled with photons and gravitons or (according to another model) is said to be transversed by electromagnetic and gravitational waves.[71] As I have already argued this notion of "empty space" is empirically meaningless if we understand it as Non-Being having an extended structure. Rather it should be understood as material, since it has extension, "curvature", and undergoes various processes by which it transmits energy in the manner of a material medium; but it is matter without mass.[72]

Modern physics readily admits the existence of material particles which have no mass (the photon, neutrino and graviton), so there is no paradox in considering so-called "empty space" as material. Probably the best way to conceive of this vacuum which occupies most of our universe (an increasing part as our universe expands) and which seems to fill the interstices of ordinary matter, not as a natural unit in itself, nor as made up of natural units, but rather as made up of transitional entities, since photons and gravitons seem to have this transitional character. In fact this is what modern physics often means by "energy", that is, matter-in-process, and when this matter is in transition from one stable form to another what we observe is a kind of chaos of transitional entities which make up the "space" between more stable bodies which alone have the character of natural units.

The result of the foregoing interpretation is that our universe can be viewed as a vast collection of natural units swimming as it were in a much vaster chaotic ocean of matter-energy in transitional states. Among these units the most complex and highly unified which we know are ourselves.

3. Units in Process

The units we have now identified as Being-in-process are the terms of those static and dynamic relations which link these terms together to form our not so highly unified universe and result in constant transformation or *change*. The word "change" etymologically comes from a word for a commercial transaction (as in "exchange" or "change" for petty cash). It has a somewhat broader usage than "process," since the latter seems to imply a movement toward greater organization and unity (pro-cess like pro-gress); and also broader than "motion", formerly the common philosophical term, now usually restricted in English to refer to change of place. When we speak of "natural processes" we have in mind primarily those kinds of change which reoccur regularly and which produce the structured primary units which are observed in our world and promote their full development. It is these regularly reoccurring processes that scientists attempt to isolate from the chaos of random events to which entropy dooms all material things and thus to formulate "natural laws" of uniformity.

As we have seen, the actual universe-in-process has an historical character which is explained by these laws as regards what is uniformly repetitive but not as to what is novel and evolutionary. This evolutionary aspect of the world cannot be reduced to one law or force, but only to the concurrence of many independent laws and forces, which concurrence appears to be irreducibly a matter of chance. Yet in fact the result has been the production of our solar system and our earth as suitable environments for life and for ourselves as thinking bodies who have arrived at a knowledge of scientific laws by which we can to a degree control natural processes.

Hume attempted to resolve the causal laws through which all this change and process is made partially intelligible by pointing out that all that we actually observe is the succession of one event after another which we ascribe to causality because we have become habituated to expect regular recurrence of certain patterns of events. [73] Yet, since he also resolved the "I" into a stream of conscious events, he never explained how I can anticipate regular recurrences nor recognize them when they occur.

The truth in Hume's position is that the fact that we observe a regular succession of events is not of itself sufficient evidence to convince us that these events are joined by a relation of cause and effect. But is this relation one of succession? Causality means that the existence of one state of affairs depends necessarily on another. Strictly speaking this implies that the cause and the effect exist *simultaneously,* not successively. Otherwise the effect could exist without the cause when the cause ceases to be and the effect which succeeds it remains. The paradigm of causality is not an act of striking a golf ball followed by the flight of the golf ball through the air, but the event in which as the club hits the ball the ball ceases to be at rest. What follows subsequently is a series of new events which involve other types of causality. A better paradigm of causality in the strict sense is a room which remains illuminated only while the lamp is shining. [74]

How then are we able to discriminate between events that are not causally related and those that are? Certainly regular recurrence of a sequence of events is a very useful but not sufficient sign of causal relationships, and so also is the simultaneous correlation of events, but ultimately we are convinced of such relationships by an intellectual analysis of different lines of empirical evidence. The whole work of science is to devise observations and experiments which eliminate false leads and provide useful leads in making this discrimination. The successful advance of science shows that such discrimination is possible. Who today would doubt that the cause of the rising of the sun is the rotation of the earth and not the orbiting of the sun? Is not Galileo supposed to have whispered under his breath in the chambers of the Inquisition, "Yet the earth *does* move!"? Kant tried to escape Hume's scepticism by saying that these causal laws are imposed on reality by the innate categories of our intelligence; but this is a gratuitous explanation. Our intelligence serves us well in dealing with the world both theoretically and practically because it has the capability of discriminating between unrelated events and related ones and is thus able to make progress in discovery and control. Only if the causal relation is in the things which change, and not merely attributed to them by our habits of thought (Hume) or by the inescapable pigeonholes of our minds (Kant), would such discrimination be possible and useful.

We can, therefore, justifiably say that we "observe" causal relations empirically, if by "observe" we do not suppose that a human observer is merely a camera or a tape-recorder but an intelligent person who observes with senses working under intelligent control. Hume's causality fallacy arose because (although he was attacking Descartes' rationalism), he assumed with Descartes that we are "minds" rather than thinking bodies in whom sense and thought are correlated activities of a single primary natural unit. [75]

Nor is it true, as some modern philosophers of science have tried to show, that modern science deals only with statistical correlations and not with causal laws. [76] While statistical correlations are important evidence of causal dependence (and otherwise would be of no interest), no scientific theory rests merely on such a correlation but always aims to provide some causal model to explain such correlations, or to reject them as merely coincidental.

In our effort to understand natural change and process in a causal way we need to recognize that some kinds of process are much more radical than others. Those processes which produce new natural units and the counter changes which destroy such units can be called *radical* in distinction to those changes which modify a natural unit or its relations to other units without, however, destroying it or producing a new unit, or which can be called *superficial* changes, although even such superficial changes can bring about very significant modifications of an existing unit. The most evident of radical processes are found in the living world in the birth and death of organisms. Most radical of all are those which in the evolutionary processes produce not merely a new individual organism but a new species of organism.

In the inanimate world such radical transformations are most clearly evident in the formation of chemical compounds, new kinds of molecules, from simpler substances; or when such a compound is decomposed into simpler substances. Radical transformations also seem to take place in the formation or disintegration of atoms of elemental substances. Thus radium decays radioactively into lead, and in certain extreme situations hydrogen can be fused into the higher element helium. No natural unit is immortal; every kind, even the most elemental can be destroyed and produced anew by natural changes and processes.

Thus we must opt for one of two possible interpretations of radical change: either new primary units are formed out of nothing, or they are formed by the destruction of previously existing natural units. To produce a new unit *ex nihilo* by "creation" in the strict sense of the term would seem to require an infinitely powerful agent, since the difference between absolute being and absolute nothing is infinite. On the other hand, to form a new unit out of pre-existing units would seem to require only a finite power, an agent merely capable of destroying what already exists as a unit and recasting it in some new form as a new unit — a transition from being to being, not from not-being to being.

What thus *seems* possible is what we actually observe in many cases. Living organisms do not suddenly appear but are observably produced from materials taken from other organisms. Molecules do not suddenly appear, but are compounded from pre-existing elements. This alternative to creation does not exclude the possibility that genuine creation may sometimes take place in our universe and, as previously mentioned, some physicists have actually proposed theories based on the hypothesis of the continuous creation of new matter in the universe, but it provides us with a valid interpretation of the ordinary instances of radical change.

It would be a serious error, however, to attempt to reduce radical changes of this type to a more intense degree of superficial change, a mere aggregation or linking of previously existing units into a new system. If such changes are merely superficial, then the primary units of our universe are eternal, changing only superficially and perduring through vast eons of change. Such were the atoms of Democritus and of nineteenth century physics. Such eternal entities have no place in science today. Even at the sub-atomic level all known entities come to be and perish. It is all too evident, moreover, that we human beings, the most evident of primary units, die, and it is for some, at least, good fortune to have been conceived.

This fundamental fact of experience that radical change takes place in our universe but cannot reasonably be attributed in most cases to *creatio ex nihilo* (what evolutionist would ever admit that?) demands that we accept the uncomfortable but very fruitful conclusion, drawn long ago by Aristotle,[77] that in a radical change, the substrate which is common to the unit (or units) destroyed and the unit (or units) newly produced cannot be any *actual* thing (i.e. any natural unit or units) but a pure *potentiality* or primary "matter" out of which all primary units are made and which exists only as a capacity within natural units to be transformed into other units.

Modern physics, after being long dominated by the Democritean notion of matter as some kind of elementary particles which are "building-blocks" of all macroscopic objects and which, therefore, would be the primary natural units, has been forced by the facts back to the notion of such a purely potential substrate, as one of its greatest exponents, Werner Heisenberg, has argued.[78] Since we have never discovered anything like the Democritean atoms, but only particles which undergo radical transformation, and since even these particles are so dynamic that they are conceived as wave-like disturbances in a universal, pervading field, it becomes apparent that we must either accept this field as the single fundamental unity, or we must consider it not a unit but a capacity in every natural unit to be radically transformed. Thus it is this primary "matter" or capacity for radical change which gives us a right to call our universe, pluralistic and loosely unified as it is, a uni-verse. It is made up of many natural units, but all these units have in common a substrate which is not some actual, more primordial substance, but simply the potentiality that all these units have to be transformed into each other. No other interpretation has been suggested which will account for two fundamental facts of our experience, facts which are more empirically certain than most of the accepted conclusions of physics: (1) all the primary units of our universe undergo radical change; (2) the primary units of our universe include directly observable macroscopic objects such as living organisms.

It is significant that Alfred North Whitehead, thoroughly acquainted as he was with the development of modern physics, felt it necessary to posit as the basic principle of his process, organismic cosmology, an equivalent to primary matter which he called "creativity." He writes:[79]

"Creativity" is the universal of universals characterizing ultimate matter of fact . . . It is that ultimate principle by which the many, which are the universe disjunctively, become the one actual occasion which is the universe conjunctively. It lies in the nature of things that many enter into a complex unity.

Aristotle's primary matter and Whitehead's creativity have in common that they are the universal principle embodied in every actual existent. They differ essentially in this, that while primary matter is a purely potential and therefore entirely *passive* principle, Whitehead's creativity is an active and therefore somehow actual principle. Whitehead's conception thus has more in common with the hylozoic and Stoic conceptions of matter (still present in Marxist cosmology and in many other process thinkers such as Bergson and even Teilhard de Chardin),[80] according to which matter is conceived after the analogy of an egg which develops from within. While we must grant that in living things we do seem to see a kind of self-development and that even in such inanimate units as the crystal, molecule, or atom there is a vague analogy to growth and self-development, it is an unwarranted leap to attribute this vital capacity for self-development to all primary units. No such leap is required to admit that all primary units have at least the passive capacity for change, which is all that is implied in the concept of a primary matter.

The great advantage of considering that this primary matter is purely passive is that this interpretation is consistent with the principle that "nothing changes itself," a principle which we find verified in all cases of process which we examine carefully. What scientist is content to explain a process by saying, "It causes itself"? Experience has taught scientists that all processes when studied will be found to have *agents* distinct from the primary unit which is undergoing change, if a search for such agents is pursued. Even in the development of an embryo, which would seem to be the most convincing example of something that "causes itself", research shows that this self-development is the result first of all of one part of the embryo acting on another part, and then of the in-put of energy which the embryo receives from its environment without which it would die in a few minutes. In other words, no primary unit is an absolutely closed system, but all act only when there is in-put, that is, an outside agent. Therefore, the primary matter of radical change out of which new units come into existence must be absolutely passive, hence purely potential, having no actuality of its own.

Thus we see that in every natural change of a radical kind, we need to distinguish three aspects: (1) the purely passive potentiality or primary matter which is acted upon; (2) the agent which acts upon it; (3) the organization or unity which the agent produces in the matter as the actualization of a new primary unit. Once this unit has come into existence it becomes itself an agent of change acting on other units, and also a subject of further *superficial* change by their reactions on it.

In living things, as I have already noted, these superficial changes (changes which are not radical enough to kill the organism) originate ultimately from external agents, but also more immediately from one part of the organism acting on another part so as to make of the organism a self-developing, self-maintaining, self-restoring, self-reproducing homeostatic system, ultimately dependent, however, on a constant in-put of energy from without. The cumulative effect of superficial changes in the unit produced from its interaction without outside agents eventually results in the destruction of its unified organization and unity and its transformation into new primary units. Thus living things die and their corpses are converted into new inanimate or (when eaten and digested) new animate units, while inanimate units such as molecules decompose and are converted in atoms or restructured as parts of more complex molecules.

The fact that all these processes, both radical and superficial, tend to run a regular course (i.e. they are *natural* processes) and that those which are not destructive but constructive (*processes* in the strict sense) result in the formation of new primary units having a definite organization which permits us to classify each of them as a member of a definite chemical or biological species, means that such processes are *directional* or teleological.[81] "Teleology" is often treated by scientists today as an unscientific notion, but this is because of its misuse to imply either that (1) a process requires some other occult force to bring it about than the ordinary agents recognized by science; or (2) natural agents have a certain psychic awareness by which they act purposefully. Actually, however, teleology implies neither of these propositions, but only that every natural agent which constructs a natural unit or contributes to its full development must be such as to tend to produce a *determinate* effect. For example, the human stomach would not contribute to the life of the organism unless its digestive action (and the structure which makes this action possible) was such that it regularly transforms food in a definite way; nor would the lungs so contribute unless they regularly supplied oxygen to the blood. Or to take examples from the inanimate world, the sun would not hold the planets together into a relatively stable system unless it exerted on them a regular gravitational pull.

If the notion of teleology does not introduce any new agent into a scientific explanation how is it useful? Its function is to help us characterize every natural process not only by the agent which causes the process, but also by the end product which is regularly produced. It will be observed that this end product is just as empirically observable as is the agent of change. In fact it is usually *more* evident, so that commonly we begin from observations of certain natural products and then search for the agents which produced them.

Thus to reject the concept of teleology amounts to rejecting the principle of causality or the cause-and-effect relationship, since "effect" implies teleology whenever the change in question is a process in the strict sense, i.e. a radical change which produces a new primary unit, or that kind of superficial change which is constructive rather than destructive of such units. Yet not every effect of change is teleological, since, as we have seen, the universe is full of chance effects. A teleological effect is empirically distinguishable from a chance effect because (1) a teleological effect occurs regularly and thus can be expressed by a natural law; (2) it produces a primary natural unit or some normal development of that unit; while chance effects are irregular and tend to impede rather than promote the normal development of a natural unit. For example, the normal action of the human heart is teleological because it follows a regular pattern and is functional for the health of the organism; while the effect of a coronary thrombosis is to impede this function, and is considered an "accident," although its causes can often be discovered.

To summarize: natural changes or processes can be adequately understood only if we attempt to determine empirically in each case (1) the subject in which this process takes place (primary matter if the change is a radical production of new primary units, some existing unit if the change is superficial); (2) the agent which produces the change, which sometimes proximately is internal to the unit, one part acting on the other, but which ultimately is always some other unit than the unit undergoing the process; (3) the new organization which the process tends to produce in the subject, in the case of radical change the new unity which constitutes the new unit, in the case of superficial change any modification of the existing subject unit; (4) if the change is a process, in the strict sense, the teleology of the process, i.e. the final product whether this be a new unit, or some *constructive* modification of the existing unit.

The great importance of maintaining the distinction of these four aspects of change and process is evident from the problems that have arisen in the history of science as the result of their confusion.[82] In the Middle Ages the influence of Platonism resulted in a tendency to assign a certain *actuality* to primary matter, instead of maintaining its pure potentiality, thus treating it as the lowest kind of unifying form, rather than as correlative to unity. This tendency is especially clear in the cosmology of Duns Scotus and survived in that of Suarez through whom it resulted in a considerable distortion of Thomism.[83] Galileo, in his eagerness to mathematize physics, reduced the dynamic unity of a primary unit to nothing more than static geometrical form and eliminated the teleological aspect, leaving nothing but the agent and the matter.[84] Subsequently Cartesian mechanists began to speak of the agent simply in terms of disembodied "forces." Moreover, mechanists, especially after Boyle,[85] adopted a Democritean notion of matter as atomic. Thus in nineteenth century physics, change and process had to be interpreted in terms of *matter* (conceived atomically) and *energy* (conceived as a force). Yet surreptitiously all four aspects of process were considered under different and ill-defined terms. If today we wish to interpret the scientific world-picture in a way that does full justice to process and novelty, it is essential to kept all four aspects of change clearly distinct and correlative to each other.

2. Kinds of Process

With this analysis of what we mean when we speak of a "process," the next question is how we are to differentiate different kinds of processes, since it is empirically evident that our pluralistic world, made up as it is of many different kinds of natural units, also undergoes a variety of processes, not just one uniform development. First of all it is necessary to classify all the kinds of natural units as accurately as possible. Modern science has accomplished this in a remarkably successful way beyond the dreams of ancient science, and is continuing to complete and perfect this taxonomy. Some have supposed that the theory of biological evolution makes the notion of "species" arbitrary, since new species are evolving gradually out of old. But Ernst Mayer has shown that this is by no means the case.[86] Although species merge into each other diachronically, at any synchronic cross-section of evolutionary history there exists a range of distinct species which usually survive little changed for long periods of time. Species are kept sharply distinct by the fact that they cannot interbreed. In the inanimate realm chemical species are also clearly distinguished by the fact that each of the elements has an exactly determined nuclear structure, and each molecule an exact atomic composition and molecular structure, which features are precisely expressed by the symbolism of modern physics and chemistry and the periodic table of the elements. As a result it is possible today to set up beautiful taxomomies of all the known natural units in the universe not only at present but throughout evolutionary history, and we have good reason to believe that as new species are discovered or evolve they can be fitted into these classifications without radical revision. In Aristotelian terms such a classification of the kinds of natural units is called the "Category of Substances."

Often Aristotle is accused of proposing a philosophy of "substance" as against a philosophy of "process", or a philosophy of "being" as against one of "becoming," or a "static" as against a "dynamic" philosophy. This is an anachronistic retrojection of the Cartesian notion of substance on Aristotle whose notion was essentially different. In any case, in my use of the term "substance" to apply to the classification of primary units, I am not speaking metaphysically (any more than was Aristotle), since this discussion is at the level of an interpretation of the natural sciences as empirical disciplines, and I have made no attempt to establish that there is such a valid discipline as "metaphysics" distinct from the philosophical analysis of the empirical results of scientific observation and experiment. Primary units are, if my analysis in this chapter has been correct, Beings-in-process accessible to empirical study. They are "being" but not as opposed to "becoming", *since they are beings that exist only in becoming, in process.* Although they have a static or inertial aspect, because they are acted upon as well as acting, passive as well as active; yet they are the causes of process and change, and their passivities are what makes them capable of entering into process by changing.

In the following chapter I will continue this analysis by considering the kinds of process which produce and destroy primary units.

Notes

1. In this chapter I have not attempted to document my philosophical argument with extensive references to current controversies in the philosophy of science which would distract from my principally theological purpose. An excellent bibliography on these topics can be found in Richard J. Blackwell, *A Bibliography of the Philosophy of Science: 1945-1981* (Westport, Conn.: Greenwood Press, 1983).

2. "Reduced to its ultimate essence, the substance of these long pages can be summed up in this simple affirmation: that if the universe, considered siderally, is in the process of spatial expansion (from the infinitely small to the immense), in the same way and still more clearly it presents itself to us, physico-chemically, as in process of organic *involution* upon itself (from the extremely simple to the extremely complex) — and moreover this particular involution 'of complexity' is experimentally bound up with a correlative increase in interiorisation, that is to say in the psyche or consciousness." Pierre Teilhard de Chardin, *The Phenomenon of Man,* (New York: Harper and Row, 1959), p. 300.

3. See Edward O. Wilson, *Sociobiology: The New Synthesis* (Cambridge, Mass.: Harvard University Press, 1975) and *On Human Nature, ibid.* 1978; M. Sahlins, *The Use and Abuse of Biology* (Ann Arbor: University of Michigan Press, 1976); Arthur L. Caplan, *The Sociobiology Debate* (New York: Harper Colophon Books, 1978); Michael S. Gregory, Anita Silvers and Diane Sutch, eds. *Sociobiology and Human Nature* (San Francisco: Jossey-Boss, 1978) and Alexander Rosenberg, *Sociobiology and the Preemption of Social Science* (Baltimore: John Hopkins University Press, 1980).

4. See Chapter 2 note 49 for references.

5. "In number of species, insects far surpass all the other kinds of living things and the total number of individual insects is beyond estimation. To date there are taxonomic descriptions of about a million species, and the tally is estimated to be in the neighborhood of three million. Without insects many plants and animals would cease to exist, and the world would be a different place to live.

Insects have a phenomenal ability to preempt every available space, to take possession of almost every square yard of our planet except the oceans, and to live and multiply in these places. In their billions and trillions these often fragile beings survive the thousand-fold dangers of an inimical environment. Their reproductive capacity is fabulous . . . Furthermore, many insects have resting stages capable of surviving seasonal extremes of weather." Walter Linsenmaier, *Insects of the World* (New York: McGraw-Hill, 1972), p. 10.

6. See Chapter 2, pp. 38-40, especially note 51.

7. *Ibid.* pp. 35 and notes 38 and 45.

8. *Ibid.* pp. 39 and note 51.

9. *Ibid.* pp. 29.

10. See Monroe W. Strickberger, *Genetics,* 2nd ed. (New York-London: Macmillan-Collier, 1976), "Inbreeding and Heterosis," pp. 783-801.

11. See Chapter 2.

12. *Ibid.,* pp. 29.

13. *Ibid.,* pp. 37-38; cf. 26-27.

14. *Oeuvres Complètes de Laplace,* (Paris, 1882), pp. 8, 144-145, as quoted by Roger Hahn, *Laplace as a Newtonian Scientist.* (William Andrews Clark Memorial Library, April 8, 1967, University of California at Los Angeles, 1967).

15. See Chapter 2, pp. 36-37.

16. See Fritjof Capra, *The Tao of Physics: An Exploration of the Parallels between Modern Physics and Eastern Mysticism,* (Boulder, CO: Shambhala, 1975). In his rather farfetched search for parallels Capra argues that the scientific is the rational or *yang* approach and mysticism the intuitive, *ying* approach to reality, but both have come to essentially the same result as modern atomic physics. Both end with a view which cannot be expressed in ordinary language, yet both are rooted in experiment. Modern physics has shown that the "organic" view of reality according to which "all phenomena in the universe are integral parts of an inseparable harmonious whole" (p. 304) is more fundamental than the mechanistic view, which is more for practical use than for understanding.

17. "He stretches the firmament over empty space, and suspends the earth over nothing at all." (Job 26:7); "Where were you when I founded the earth . . . Into what were its pedestals sunk and who laid the cornerstone?" (38:4,6) Also Psalm 104:5 and Psalm 119:89-91.

18. The term "reduction" is used in many different senses in contemporary philosophy. I use it here in the sense of "reduction of a whole to its parts." It is also used to mean the reduction of *theoretical* terms to the empirical terms by which they are verifiable or falsifiable. Again it is used by phenomenologists to mean the reduction of phenomena to their intuited essences, bracketing their existence or non-existence. See Mary W. Wartofsky, *Conceptual Foundations of Scientific Thought* (New York: Macmillan, 1968, pp. 344-369). Also Ernest Nagel, *The Structure of Science,* New York: Harcourt, Brace and World, 1961, c.11 "The Reduction of Theories", pp. 336-397; G. Radnitsky, *Contemporary Schools of Metascience,* 2nd ed. (New York: Humanities Press), vol. 1, pp. 72-92; Rom Harré, *The Principles of Scientific Thinking* (Chicago: University of Chicago Press, 1970), pp. 61-62; and Karel Lambert and Gordan G. Brittan Jr., *An Introduction to the Philosophy of Science,* 2nd ed. (Reseda, Calif.: Ridgeview Pub. Co., 1979). For the phenomenological sense, see William A. Luijpen, *Phenomenology and Metaphysics* (Pittsburgh: Duquesne University Press, 1965), p. 63 ff. For a critical symposium see Arthur Koestler and J.R. Smythies, eds., *Beyond Reductionism* (Boston: Beacon Press, 1969).

19. See Chapter 2, pp. 30f. For a vigorous statement of the reductionist point of view see Francis Crick, *Of Molecules and Men* (Seattle and London: University of Washington Press, 1966), but see also Kenneth F. Schaffner, "The Watson-Crick Model and Reductionism," *British Journal of the Philosophy of Science,* 20 (1969): 325-348. Also Charles De Koninck, *The Hollow Universe* (London: Oxford University Press, 1960), Chapter 3, "The Lifeless World of Biology," pp. 79-114, and Eugene P. Wigner, *Symmetries and Reflections* (Bloomington: Indiana University Press, 1967), "The Problem of the Existence of a Self-Reproducing Unit," pp. 200-210.

See the essay of Paul Weiss, "The Living System: Determinism Stratified" in Koestler and J.R. Smythies, *Beyond Reductionism,* (note 18 above), pp. 3-35. "There is no phenomenon in a living system that is *not* molecular, but there is none that is *only* molecular, either" (p. 10). "The 'more' (than the sum of its parts) in the above tenet ['the whole is greater than the sum of its parts'] does not at all refer to any measurable quantity in the observed systems themselves; it refers solely to the necessity for the observer to supplement the sum of the statements that can be made about the separate parts by any such additional statements as will be needed to describe the *collective behavior* of the parts, when in an organized group. In carrying out this upgrading process, he is in effect doing no more than *restoring information content* that has been lost on the way down in the progressive analysis of the unitary universe into abstracted elements." (p. 11).

20. Chapter 2, pp. p. 33. See also D. Paul Snyder, *Toward One Science* (New York: St. Martin's Press, 1978).

21. See Chapter 2, pp. 22-23.

22. See Ludwig von Bertalanffy, *General Systems Theory: Foundations, Development, Applications* (New York: Dover, 1968), Introduction pp. 3-29, and C.W. Churchman, *The Systems Approach* (New York: Dell Delta Books, 1968), who emphasizes as "fundamental in the general theory of systems" the concept of *hierarchic order*. We presently "see" the universe as a tremendous hierarchy, from elementary particles to atomic nuclei, to atoms, molecules, high-molecular compounds to the wealth of structures (electron and light-microscopic) between molecules and cells. . . ,to cells, organisms and beyond to supra-individual organizations . . . A similar hierarchy is found both in 'structures' and in 'functions'", p. 27. Essential to this hierarchical organization is what Bertalanffy calls "the principle of centralization", p. 71 ff. which recalls Teilhard de Chardin's "centrification" and goes back to Aristotle's concept of a central part which is the "unmoved mover" in each and every system.

23. Bertalanffy, *General Systems Theory* (note 22 above), pp. 54-88.

24. Modern discussions of relations do little more than discuss the problem of "internal" and "external relations." The disproof by medieval Nominalism of the existence of real relations is often taken as valid. More attention should be given to the classical medieval distinction between "mental relations" which result from a merely mental comparison of objects, and "real relations" founded in some extra-mental connection between the relata. These latter are either "transcendental" when this connection is essential to the relata, or "predicamental" when it is accidental. In this latter case the connection may be of three types: (1) causal relations; (2) similarity-dissimilarity; (3) measure-measured. See Jean Poinsot (John of St. Thomas), O.P., *Cursus Philosophicus*, Log. II, q. xvii, Reiser ed. vol. I, pp. 573-590, and Ralph Austin Powell, O.P., *Freely Chosen Reality*, (Washington, D.C.: University Press of America, 1983) pp. 11-96.

25. See Leroy E. Loemker, *Struggle for Synthesis: The Seventeenth Century Background of Leibnitz's Synthesis of Order and Freedom*, (Cambridge, Mass.: Harvard University Press, 1972), pp. 86-113 on some of the factors leading to dynamism. Also Charles C. Gillespie, "Elements of Physical Idealism" in *Mélanges Alexandre Koyré*, 2 vols. (Paris: Hermann, 1964), vol. 2, pp. 206-224 on how Leibnitz sought to base physics on *energetics* and time instead on Newtonian motion. For Boscovich see the brief life printed in Joseph R. Boscovich, S.J., *A Theory of Natural Philosophy* (reprint of 1763 edition, Cambridge, Mass.: MIT Press, 1966).

26. Henri Bergson, *Introduction to Metaphysics* (New York: Philosophical Library, 1961); *Matter and Memory* (London: George Allen and Co., 1913); *Creative Evolution* (New York: MacMillan and Co., 1938). For a critique see Jacques Maritain, *Bergsonian Philosophy and Thomism* (New York: Greenwood Press, 1968). On the antecedents of Bergson see Newton P. Stallknecht, *Studies in the Philosophy of Creation, with special reference to Bergson and Whitehead* (Princeton, N.J.: Princeton University Press, 1934) and Ben-Ami Scharfstein, *Roots of Bergson's Philosophy* (New York: Columbia University Press, 1943). On the influence of Bergson on Whitehead and their differences see Victor Lowe, *Understanding Whitehead* (Baltimore, John Hopkins, 1962) pp. 193 ff. and 257 ff. Also Robert M. Palter, "Bergson's Influence on Whitehead," *The Personalist*, 35 (1955): 250-257.

27. Alfred North Whitehead, *The Concept of Nature*, (Cambridge: Cambridge University Press, 1920); *An Enquiry Concerning the Principles of Natural Knowledge* (Cambridge: Cambridge University Press, 2nd ed. 1925); *Science and the Modern World*, (1925, Mentor:

New American Library, 1959); *Process and Reality: An Essay in Cosmology* (1929, corrected ed., New York: MacMillan-Free Press, 1978). For commentary see, William W. Hammerschmidt, *Whitehead's Philosophy of Time* (New York: King's Crown Press, 1947); Ivor Leclerc, *Whitehead's Metaphysics* (London: Allen and Unwin, 1958); William A. Christian, *An Interpretation of Whitehead's Metaphysics* (New Haven: Yale University Press, 1959); Robert M. Palter, *Whitehead's Philosophy of Science* (Chicago: University of Chicago Press, 1960); Lowe, *Understanding W.,* note 26 above; Donald W. Sherburne, ed., *A Key to Whitehead's Process and Reality* (New York: MacMillan, 1966); Dorothy M. Emmet, article "Whitehead" in Paul Edwards, ed. *The Encyclopedia of Philosophy* (New York, MacMillan) 8, pp. 290-296; R.M. Martin, *Whitehead's Categoreal Scheme and Other Papers* (The Hague: Martinus Nijhoff, 1975); Ann L. Plamondon, *Whitehead's Organic Philosophy of Science* (New York: University of New York Press, 1979).

On the history of process philosophy and theology see Gene Reeves and Delwin Brown, "The Development of Process Theology" in a book edited by them *Process Philosophy and Christian Thought* (Bobbs-Merrill, Indianapolis, 1971, pp. 21-64). I have not discussed Charles Hartshorne at any length here, because it seems to me that his contribution has not been so much to the idea of process itself as to the problem of what he calls "the divine relativity", a matter I will discuss below in Chapter 11. He attempts to show that relativity in the world requires an Absolute Being (ontological argument), but that this Absolute Being must also be relative to the world which He knows, and hence is di-polar. From this it follows that the process in the world implies process in God. See his *The Divine Relativity, A Social Conception of God* (Yale University Press, New Haven, 1964 edition), pp. 6-18.

28. The traditional definition of quantity as "a whole having part outside part," based on Aristotle, *Categories,* c.6, 4b 20, *Physics* VI, 1-230 a21-233b 31, seems to suffer from the circularity of the term "outside". However, its meaning in Aristotle is that the parts in continuous quantity may coincide at a point of contact but not as wholes, and in discrete quantity they do not at all coincide.

29. *Principles of Philosophy* I, 53, (C. Adam and P. Tannery eds.), *Oeuvres de Descartes* (Paris, 1897-1913), 8, 25 ff.

30. "From the standpoint of epistemology it is more satisfying to have the mechanical properties of space completely determined by matter, and this is the case only in a closed universe." Albert Einstein, *The Meaning of Relativity,* 5th ed. (Princeton, N.J.: Princeton University Press (1922, 1974), p. 108.

31. "According to modern theory, a vacuum is not exactly nothing but is teeming with quantum particles that fluctuate between being and nothingness. These tiny particles can come into existence for a fraction of a second before they annihilate each other, leaving nothing behind. A vacuum in that sense is like the surface of an ocean. Up close it is churning with waves, but from a longer distance such as from a jet plane, it appears smooth and placid. Similarly, any vacuum examined close up with the proper instruments is seen to be churning with tiny quantum particles." Heinz R. Pagels, "Before the Big Bang," *Natural History,* 92 (April, 1983): 22-27, p. 26.

32. Whitehead: "I will also use the term 'actual occasion' in the place of the term 'actual entity.' Thus the actual world is built up of actual occasions; and by the ontological principle whatever things there are in any sense of 'existence,' are derived by abstraction from actual occasions. I shall use the term 'event' in the more general sense of a nexus of actual occasions, inter-related in some determinate fashion in one extensive quantum. An actual occasion is a limiting type of an event with only one member." *Process and Reality,* (note 27 above) 2, 3, p. 73. For Whitehead an actual occasion begins to perish at the very moment (quantum of time) that it comes into existence, "In the organic philosophy an actual entity has 'perished' when it is complete." *Ibid.* p. 81 f.

33. "We have here actually the final proof for the unity of matter. All the elementary particles are made of the same substance, which we may call energy or universal matter; they are just different forms in which matter can appear.

If we compare this situation with the Aristotelian concepts of matter and form, we can say that the matter of Aristotle, which is mere 'potentia', should be compared to our concept of energy which gets into "actuality" by means of the form, when the elementary particles are created.

Modern physics is of course not satisfied with only qualitative description of the fundamental structures of matter; it must try on the basis of careful experimental investigations to get a mathematical formulation of those natural laws that determine the 'forms' of matter, the elementary particles and their forces." Werner Heisenberg, *Physics and Philosophy: The Revolution in Modern Science*, (New York: Harper and Bros., 1958) p. 160.

34. The emphasis on the *predictive* value of the scientific method is attributed to C.S. Pierce (1877-8) by R.B. Braithwaite, *Scientific Explanation*, (Cambridge: Cambridge University Press, paperback ed., 1968), pp. 264 ff. The emphasis on the pragmatic or *controlling* value of science goes back to Francis Bacon.

35. The tendency of current philosophy of science seems to avoid both the notion of "causation" and "prediction" and to speak simply of "explanation." Thus Karl R. Popper, *Objective Knowledge* (Oxford: Clarendon Press, 1973), p. 190, writes: "I suggest that it is the aim of science to find *satisfactory explanations* of whatever strikes us as being in need of explanation." He also refers to explanation as "causal", but then goes on to show that it amounts to saying that an explanation is "satisfactory" if it is "in terms of testable and falsifiable universal laws and initial conditions" (p. 193), but many current philosophers of science prefer to discuss "natural laws" without introducing the concept of "causation".

36. See Chapter 2, pp. 35-40, and notes 48 and 49 below.

37. F.B. Dickson, *The Bowl of Night: The Physical Universe and Scientific Thought* (Cambridge, Mass.: M.I.T. Press, 1968) pp. 50-60. Stanley L. Jaki, *The Paradox of Olber* (New York: Herder and Herder, 1969) gives an excellent history of how the infinity of the world became an unquestioned scientific dogma between the seventeenth and nineteenth centuries, leading scientists either to ignore Olber's paradox or offer superficial solutions, until the rise of Einstein's argument for finitude. Dickson shows the paradox was first stated not by Olbers in 1823, but by Edmond Halley in 1720 and then Jean-Phillipe Löys de Chéseaux in 1744. Oddly Jaki, who berates others for their prejudices in ignoring this paradox, himself fails to accredit its first statement and solution to Aristotle. He notes that Aristotle "worked hard on a long list of arguments against the infinity of the universe but failed to seize on what would certainly have forged into a powerful proof in support of his cosmology. His mind was too committed to his own world view to exploit in full in each case the opposite opinion to his own advantage" (p. 10). He then launches into a discussion of Aristotle's mistaken views on the nature of the Milky Way and his rejection of Democritus' sensible opinion that it was composed of stars. This was due, says Jaki, to "Aristotle's customary readiness to rush to conclusions and fit things into his scheme" (p. 11). Yet if Jaki had more carefully considered Aristotle's arguments against a spatially infinite universe (*Physics* III, cc. 4-8, 202b 30 — 208a 26); *De Caelo*, I, cc. 5-6 (to which alone Jaki refers) 271 b 1 — 276a 18; *Metaphysics* XI (K) c. 10, 1066a 35 — 1067b 1 (repeating the *Physics*), he would have seen that one of them is based on the proposition that an infinity of any one of the elements is impossible since "the infinite body will obviously prevail over and annihilate the finite body" (Physics 204b 18), i.e. the light and heat from an infinite number of stars would destroy the other elements on the earth.

38. See Alexandre Koyré, *From Closed to Infinite Universe*, (Baltimore: John Hopkins Press, 1957).

39. Einstein, *The Meaning of Relativity* (note 30 above), p. 99-108 and Chapter 2, pp. 39-40.

40. See Chapter 2, pp. 39.

41. "The four categories of explanation . . . constitute the repudiation of the notion of vacuous actuality, which haunts realistic philosophy. The term 'vacuous actuality' here means the notion of a *res vera* devoid of subjective immediacy. This repudiation is fundamental for the organic philosophy . . . The notion of 'vacuous actuality' is very closely allied to the notion of the inherence of quality in substance. Both notions — in their misapplication as fundamental metaphysical categories — find their chief support in a misunderstanding of the true analysis of 'presentational immediacy.' " Whitehead, *Process and Reality*, (note 27 above) pp. 28-29.

42. "In the philosophy of organism it is held that the notion of 'organism' has two meanings, interconnected but intellectually separable, namely, the microscopic and the macroscopic meaning. The microscopic meaning is concerned with the formal constitution of an actual occasion, considered as a process of realizing an individual unity of experience. The macroscopic meaning is concerned with the giveness of the actual world, considered as the stubborn fact which at once limits and provides opportunity for the actual occasion.", *ibid.*, p. 128-129.

43. See Conrad Bonifazi, *The Soul of the World* (Washington, D.C., University Press of America, 1978) for a history of this concept.

44. " 'Creativity' is the principle of *novelty*. An actual occasion is a novel entity diverse from any entity in the 'many' which it unifies. Thus 'creativity' introduces novelty into the content of the many, which are the universe disjunctively. The 'creative advance' is the application of this ultimate principle of creativity to each novel situation which it originates." Whitehead, *Process and Reality*, (note 27 above), p. 21. See Christian, *Whitehead's Metaphysics* (note 27 above), Chapter 3, "Novelty and Exclusiveness," p. 48 ff. for an exhaustive discussion. Christian shows that for Whitehead each actual occasion is novel because (a) it has a different set of data from any other; the data are the causal or deterministic element, yet they are not exactly identical in any two entities; (b) it has a unique "subjective aim" by which these data are unified; (c) it has a unique "subjective form" or feeling as the outcome of this unification of the data (note, however, that this aim is derived from the "pure potentials" or eternal objects in the primordial nature of God); (d) it has a unique "satisfaction".

45. Chapter 2, pp. 35.

46. Benedict M. Ashley, *Aristotle's Sluggish Earth: The Problematics of the De Caelo* (Albertus Magnus Lyceum, River Forest, IL, 1958) p. 31 ff., on the considerations which led Aristotle to his theory of the superiority of the celestial bodies to man.

47. Going from whole to part, *ibid.* p. 14-18.

48. See Adolf Grünbaum "Relativity Theory, Philosophical Significance of", in Paul Edwards, ed. *The Encyclopedia of Philosophy* (New York-London: Macmillan-Collier, 1967), 7, 133-140. Grünbaum concludes "Thus, there is an important sense in which GTR has not repudiated the concept of absolute space" (p. 140), because (contrary to Mach's hypothesis that the inertia of any mass is a function of the total mass of the universe), in the General Theory of Relativity the gravitational acceleration of the earth toward the sun is independent of the amount of distant matter uniformally distributed about the sun. Since, however, the measurement of this total mass of the universe is not knowable to us in any *directly* empirical way but only on the basis of some theory like that of Einstein's, it seems this caveat is not very significant.

49. Einsteinian relativity is more "realistic" than Newtonianism in the sense that for Einstein the notions of "space" and "time" (as for Aristotle) are defined in empirical not ideal terms; while Newton defines them in ideal mathematical terms.

50. "Historians of civilization appear to be at one in assuming that powerful components are acquired for every kind of cultural achievement by this diversion of sexual instinctual forces from sexual aims and their direction to new ones — a process which deserves the name of 'sublimation'. To this we would add, accordingly, that same process plays a part in the development of the individual and we would place its beginning in the period of sexual latency in childhood." *Standard Edition of Complete Works* (London: Hogarth, 1953-74), "Three Essays on Sexuality", vol. 7, 178. Freud's scientism, however, was unqualified; see Ernest Jones, M.D., *The Life and Work of Sigmund Freud,* (New York: Basic Books, 1957) 3: 357-360.

51. John N. Deely, *The Tradition Via Heidegger* (The Hague: Martinus Nijhoff, 1971) and Ralph A. Powell, O.P., "The Late Heidegger's omission of the Onti-Ontological Structure of Dasein," in John Sallis, ed., *Heidegger and the Path of Thinking* (Pittsburgh: Duquesne University Press, 1970), pp. 116-137.

52. See Chapter 6, pp. 238 f.

53. The idealist character of Heidegger's thought is well argued by Deely, *The Tradition* (note 51 above) pp. 171-177.

54. Aristotle, *De Anima,* III, 8. 432a, 1-8 and St. Thomas Aquinas, *In De Anima* III, lect. 13, n. 791, cf. lect. 9, n. 722; lect. 12, n. 770.

55. William H. Kane, O.P., "Thomistic Critique of Knowledge" *New Scholasticism,* 30, 2 (April, 1956) and "Details of a Thomistic Critique of Knowledge" *ibid.,* 35 (Oct. 1961) 445-477.

56. Aristotle, *Physics VI,* 10, 240b, 8 sq. see commentary of Vincent E. Smith, *The General Science of Nature* (Milwaukee: Bruce, 1958), pp. 347-352.

57. On current views as to how children learn to speak, see Hermina Sinclair-de-Zwart, "Developmental Psycholinguistics" in David Elkund and John H. Flavell, eds., *Studies in Cognitive Development: Essays in Honor of Jean Piaget,* (Oxford University Press, New York, 1969) pp. 315-336. Piaget and his followers insist that a child first learns to think and then to talk. He learns to think by exploring and interacting with the environment, first learning to recognize the continuity and permanence of objects, and then to "know that he knows" these objects. Only then is he ready to name them and talk about them. Also Heinz Werner and Bernard Kaplan, *Symbol Formation* (Clark University: John Wiley and Sons, N.Y., 1963), which emphasizes that language does not originate in merely pragmatic activities of the child but in cognitive ones (see especially the analysis of Helen Keller's experience on p. 110 ff.), and requires the social experience of "sharing" a common interest in objects, pp. 110 ff.

58. On Heidegger's use of the term "world" see *Being and Time,* translated by John Macquarrie and Edward Robinson (New York: Harper and Row, 1962) 12, 24, pp. 78-148, with the commentary of Michael Gleven, *A Commentary on Heidegger's 'Being and Time'* (New York: Harper Torchbooks, pp. 52-64). For Heidegger "world" is an *a priori* ontological existential i.e., a necessary way in which man *(Dasein)* encounters Being.

59. See note 24 above with references.

60. Aristotle, *Metaphysics* IV (Gamma) is devoted to a discussion of the principle of non-contradiction as the first axiom in understanding Being, see also Aquinas, *In XII Libros Metaphysicorum, IV.*

61. Karl R. Popper, *The Logic of Scientific Discovery,* (London: Hutchinson, 1959) and *Postscript* (Totowa, NJ.: Rowan and Littlefield, 1983), 3 vols., is the great proponent and defender of this theory of falsification which he couples with a vigorous opposition to idealism.

62. See note 51 above.

63. Aristotle, *Physics* I, c. 9, 191b 35 sq.

64. In an Aristotelian epistemology "physical terms" are reducible to sensible observations. "Metaphysical terms" apply to all beings, sensible and non-sensible, whether real or merely mental. If there are real beings which are not sensible, then this must be demonstrated by causal inference from their sensible effects (or known by revelation).

65. See Georges M-M. Cottier, *L'athéisme du jeune Marx: ses origenes Hégéliennes,* 2nd ed. (Paris, J. Vrin, 1969) for the influences which led Marx to materialism and atheism, and David-Hillel Ruben, *Marxism and Materialism,* (Atlantic Highlands, N.J.: Humanities Press, 1977) for a defense of Marx which tends to identify "materialism" with "realism" i.e., anti-idealism.

66. See note 32 above.

67. *Critique of Pure Reason,* 2nd ed. translated by Norman Kemp Smith (London: MacMillan, 1963), I, Second Part, First Div., Chapter 2, Sec. 2 (B, 157), pp. 168-169. It was important to Kant to remove the noumenal ego from the empirical realm, because he believed that Newton had established that the empirical realm is determined by necessary physical laws which would make human freedom impossible if it pertained to that realm.

68. Aristotle, *De Partibus Animalium,* Book II, discusses at length the practical problems of biological taxonomy, see my article, "Does Natural Science Attain Nature or Only the Phenomena?" in Vincent E. Smith, *The Philosophy of Physics* (St. John's University Studies, Jamaica, N.Y., 1961), pp. 63-82.

69. See Chapter 2, pp. 34.

70. *Ibid.,* pp. 33.

71. *Ibid.*, pp. 32f. For the history of this problem see Edmund Taylor Whittaker, *A History of Theories of Aether and Electricity* (London: T. Nelson and Sons, 1951).

72. "Mass" in modern physics is not "matter" as such, but matter having the properties of *inertial* resistance (i.e., energy must be applied to set it in motion) and/or gravitational attraction. Its measure is increased in bodies accelerated to velocities approaching that of light, and it can be "converted" into an equivalent amount of energy according to equivalency m/c^2 where c is a velocity of light in a "vacuum".

73. Hume, *A Treatise of Human Nature,* I, 3. An excellent analysis of Hume's position is given by Frederick Copleston, S.J., *A History of Philosophy* (Garden City, N.Y.: Doubleday Image Books, 1964), vol. 5, Part 2, pp. 82-96.

74. Hume (*Enquiry Concerning Human Understanding* 7, 2, 60) says "We may define a cause to be an object, followed by another, and where all the objects similar to the first are followed by objects similar to the second. Or, in other words, where, if the first object had not been, the second never had existed." Note that the second formula admits the simultaneity of causes, the first makes the cause precede the effect. For Hume the difference of the two definitions seems of little importance.

75. See Norman Kemp Smith, *The Philosophy of David Hume* (MacMillan, London, 1964), "Concluding Comments", pp. 543 ff. for a penetrating analysis of the source of Hume's views on causality. Hume was not a sceptic about causality as objectively real, but was sceptical about the power of "reason" to attain this reality, which is manifest to us rather through "feeling" and "imagination" than through reason. Hume, Rousseau, and Kant are all influenced by that line of thought (first clearly stated by Francis Hutcheson) which seeks to supply by affectivity what is lacking to human rationality.

76. In spite of the tendency in current philosophy of science to avoid the notion of causality, distinction is made between statistical correlations that indicate causal connection and "spurious correlations" that do not. See Herbert A. Simon, "Spurious Correlation: A Causal Interpretation" in *Models of Discovery,* Boston Studies in the Philosophy of Science, vol. 54 (Dordrecht, Holland: D. Reidl, 1981), pp. 93-106, on empirical criteria for distinguishing the genuine from the spurious.

77. See my article, "Change and Process" in John N. Deely and R.J. Nogar, O.P., *The Problem of Evolution* (New York: Appleton Crofts, 1973), pp. 267-284, for a discussion of the Aristotelian conception of matter.

78. See 33 above.

79. *Process and Reality* (note 27 above), pp. 21 f.

80. "We are logically forced to assume the existence in rudimentary form (in a microscopic, i.e., and infinitely diffuse state) of some sort of psyche in every corpuscle, even in those (the mega-molecules and below) whose complexity is of such low or modest order as to render it (the psyche) imperceptible — just as the physicist assumed and can calculate those changes of mass (utterly imperceptible to direct observation) occasioned by slow movement." Teilhard de Chardin, *The Phenomenon of Man* (note 1 above) p. 301.

81. See my articles "Final Causality" (vol. 5, pp. 915-919) and "Teleology" (vol. 13, pp. 979-981), *New Catholic Encyclopedia* (New York: McGraw-Hill, 1967), and "Change and Process" (note 77 above). For an attempt to deal with the problem by the methods of analytic philosophy (not altogether successful) see Andrew Woodfield, *Teleology* (Cambridge: Cambridge University Press, 1976).

82. See William A. Wallace, O.P., *Causality and Scientific Explanation,* 2 vols. (Ann Arbor, Mich., University of Michigan Press, 1972). For recent discussion see Myles Brand ed., *The Nature of Causation* (Urbana, IL.: University of Illinois, 1976).

83. Chapter 5, pp. 158 and 183.

84. See Marie Boas Hall, "Matter in Seventeenth Century Science" in Ernan MacMullan, ed., *The Concept of Matter in Modern Philosophy* (Notre Dame, Ind.: University of Notre Dame Press, 1965), pp. 76-103.

85. *Ibid.,* with the appended comment by James A. Weisheipl, O.P., on the nature of seventeenth century "mechanism".

86. Ernst Mayr, *Animal Species and Evolution* (Cambridge, Mass.: Harvard University Press, 1963).

Chapter 8

Process And Creativity

I: Natural Processes

1. Categories and Causes

A theology of the body, we have seen, is a theology of a certain kind of primary natural unit and such units are beings-in-process. They originate through evolutionary processes and develop, act and are acted-upon by other units of various kinds. In this chapter it is necessary to consider in more detail the types of natural processes and how they produce and develop the primary units through their mutual interaction. Through these interactions these units exist not in isolation but in a complex network of relations by which they form our universe. Thus any account of our wondrous natural world must be in terms of at least three kinds or categories of being: *substances* (in the non-metaphysical sense of primary natural units, as explained in the last chapter), *actions*, and *relations*. Since for every action of one agent unit on another there must be a correlative "reception" of that action by the patient unit, we must also add a fourth category of *"receptions"* or "patiencies". For example, when one unit "heats" another, that other is "heated."

The reader may wonder why it is necessary to revive these classical categories of pre-Kantian, pre-critical philosophy. Doesn't the very difficulty in translating some of these Latin or Greek terms into current philosophical English prove they are hopelessly obsolete? Moreover, it is commonly asserted that these classical categories were based not on universal human experience but simply on the grammar of the Greek language.[1] Wittgenstein, however, has taught us that grammar can be an important clue to the structure of human experience; and Heidegger that philosophical terminology may both reveal and conceal fundamental cultural attitudes toward Being.[2] In fact the classical categories were not proposed by Aristotle merely on the basis of his analysis of language, but of his analysis of change and process.

Historically the chief alternative to the classical categories has been those of Kant. It was his "Copernican revolution" in epistemology with its advocacy of a consistency theory of truth which accounts for the general abandonment of the older categorial system and the difficulty we experience even to express that older system in current English. Does this mean that modern *experience* has demanded the newer categories? Or does it mean that this shift is an ideological function of the Humanist world-view of which Kant was a principal architect and which, because of its political and cultural dominance today, shapes our attitudes toward reality and the language we use to describe it?

I have already given my reasons for opting for Aristotle's epistemology in preference to that of the idealist tradition whether in its ancient Platonic form or in its modern "critical" Kantian form. Kant's categories (quantity, quality, relation, modality), although the first three terms are identical with three of Aristotle's, have an entirely different character. Aristotle's are derived from his analysis of Being-in-process. Kant's, in accordance with his idealist epistemology, are derived from logic (curiously enough the Aristotelian logic as transmitted very inadequately in the textbooks of Christian Wolff and his school[3]), and are supposed to represent the necessary structure of human thought, not the contingent but real structure of the empirical world.

In view of my epistemological option, it is necessary for me to show what system of categorial terms in addition to that of "primary unit" (substance) already discussed in the last chapter, can be empirically justified in view not of ancient but of modern science. If we are accurately to describe and analyze Being-in-process as it is empirically observable, we must carefully avoid on the one hand the fallacies of reductionism which can arise from too restricted a set of descriptive and analytic categories, and on the other hand the opposite fallacies which can result from multiplying terms which have not been shown to be securely grounded empirically in our experience of a world in process.

Certainly it is not possible to dispense with the category of *action* and its correlative *reception* in the analysis of Being-in-process, since obviously if one unit did not act on another, or one part of a unit act on another part, there would be no process. Within this category we can distinguish at least three major kinds of actions: (1) actions which produce radical change, i.e., which produce or destroy natural units (for example, a chicken lays an egg or is killed by a cook); (2) actions which produce only superficial changes but which nevertheless do modify a primary unit *internally* (for example, a plant grows and flowers, water is heated); (3) actions which produce only a superficial change which does not even change a primary unit internally but only in its *external* relations (for example, the free fall of a body).

Motion, in the ordinary sense of a change in the distance of one body from another through some intervening material medium, is the prime example of this third type of change, since a body can be moved without any apparent internal modification. It is, therefore, the most superficial kind of change, yet it is basic to all other kinds of change, because one unit cannot act on another unless it moves into contact with it, or at least acts on it through a medium of waves or particles that move into such contact. Newton's tentative support of the notion of *actio in distans* has never received any support from scientific investigation, since, as I showed in the last chapter, the notion of an absolute vacuum or empty space is a mathematical fiction which is impossible to verify empirically. Because motion is the basic type of change, we can concede to the mechanists that without the motion of matter there could be no change at all, but this is no reason to concede to them that all change is reducible to motion. To do so would be to deny the evident fact of radical process by which primary units are both produced and destroyed, not merely reassorted into new combinations as the mechanists claim.

Because primary units are in motion, we must recognize the category of *places* (in the classical theory *positio* and *habitus* were also distinguished as categories distinct from that of *ubi*, place, yet closely related to it; but I will omit discussion of this minor topic[4]). By definition motion is a change of place, but the "place" in question is not location in abstract mathematical space, nor in a fictional "empty space", but in a real, physical space, i.e. location by contact with surrounding bodies which are either stable primary units or a material media existing in those transitional states which I discussed in the last chapter.[5] Real physical space, therefore, is not the homogeneous space of mathematics, but consists of a continuous plenum of various kinds of bodies contiguous to each other.

Thus for human beings earth provides our natural place — "natural" in that it is appropriately located in the gravitational field of the sun where it can receive just the right amount of solar energy and that it is surrounded by just the right atmosphere and water supply, conditions not duplicated on any other planet and which we must imitate in our space ships to survive "out there." This little earth is truly our home. Birds fly to their nests, fish swim in the water, crystals "grow" in certain kinds of solutions, not in others, electrons behave differently within and without the atom, etc. Thus the category of place is a classification of the normal environments in which primary natural units can exist in stable form and without which they perish.[6] Geography, ecology and what today is commonly called "cosmology" are concerned especially with this category of places.

Motion from place to place is sometimes relatively uniform like the steady rotation of the earth, but it is sometimes highly irregular like the Brownian movements of molecules knocked every-which-way in a gas. As entropy increases, this irregularity and randomness of motion increases. The present order of the universe, however, and the existence of the primary units of which it is constituted is made possible only by the fact that in certain regions relatively regular systems of motions are maintained. The most notable example is our own earth, where the environment without which life would be impossible depends on the earth's regular rotation and orbital motion around the sun which guarantee organisms a regular rhythm of energy in-put. We all are only too well aware of the randomness of our weather, yet this randomness is effectively limited by the larger patterns of the seasons. It is this physical regulation of less regular processes by the more regular ones which forms the systems which require us to distinguish a special category of *times*.[7]

Einstein re-established the doctrine of the correlative character of motion in place and real time by showing that time is always relative to the observer and the observer's means of communication by which events are correlated.[8] The simultaneity of events (i.e. actions) is determined by the contact between natural units and the observer is such a unit. Thus the category or classification of natural times is not determined by coordinates on a homogeneous mathematical graph, but by different ways in which the more uniform motions in a given locality of the universe regulate the less uniform. Natural processes have natural times. An organism takes a certain number of months or years regulated by the rotation and orbiting of the earth and the resulting variations of sunlight to reach maturity. The heart has a normal rhythm and pace on which the physiological processes of the body depend. Chemical reactions take place faster or slower and with different results according to local conditions of temperature, pressure, etc. Radioactive elements decay at a rate that is amazingly uniform, but which is a statistical averaging of very irregular occurrences. Thus natural time is an essential feature of all processes which no scientific analysis of these processes can ever neglect.

While motion is only the most superficial kind of change, a change in external relationships, it sometimes results in internal changes in a primary unit. The second kind of change listed above, *alteration*, is an internal, although still not radical modification of an existing primary unit. These interior changes are of two types: *quantative* or *qualitative* which demand the recognition of two more categories. Obviously living things grow in size, i.e. change in quantity, the number of their observable parts. What a change from an elephant fetus to a mature elephant! In living things, however, growth is not indefinite. Mature elephants have a range but a very limited range of sizes. Nor are these normal sizes for each species of living primary units insignificant. Biologists know there is an intimate relation between body-size and the structure and activities of organisms.[9] The mouse and the elephant are adapted to very different environmental niches. It is also significant that the less complex, inanimate primary units, although they have minimal dimensions without which they cannot exist, do not seem to have any upper range of size. Crystals can "grow" by aggregation more or less indefinitely, and perhaps the oceans of the earth can be considered one huge primary unit in its liquid state. Each species of atom when "excited" can expand enormously[10] and perhaps the tenuous medium between the galaxies consists of such expanded atoms.

Since the quantity of a primary unit can vary without the unit being destroyed, it is obvious that quantity is a different category of Being-in-process than is "substance". Nevertheless, it is also true that no primary unit of Being-in-process can exist without having some actual quantity, i.e. without begin extended, a body, a whole having parts outside each other. If such a unit did not have such parts, it could not be in process, since, as we have seen, every being-in-process must at least be capable of motion, and to be in motion a unit must have one part entering a new place and another part which has not yet entered. Otherwise it would already be where it is going! As no unit can be in process without having extension or quantity, so conversely whatever is extended must be a being-in-process, because, since extension implies divisibility, it must be a entity which has potentiality and, as we saw in the last chapter, to exist really but potentially is characteristic of Being-in-process. Thus every body is extended and subject to process, and every process (in a univocal sense of that term derived from our direct experience) is first of all based on quantitative change as its fundamental condition.[11]

Thus by his Pythagorean mathematization of physics Galileo did science the great service of showing (what Aristotle had failed fully to appreciate) that we cannot make much progress in understanding the details of nature without measurement, because quantity is the primary condition of all process. Yet it was a fatal error on Descartes' part to make the further step of identifying "substance" (i.e. the category of primary natural units) with quantity or extension, thus committing the fallacy of identifying physics with mathematics, Being-in-process with the abstract, unchanging mathematical models used to give a partial description of nature — strangely forgetful of the elementary but fundamental fact that physical reality can be only *approximated* by a mathematical description. Mathematical models provide science with an indispensable instrument of scientific research, but they always require to be given a physical interpretation. We can never conclude that because something is mathematically true it is therefore physically true, although we can argue conversely: what is mathematically impossible is certainly also physically impossible. [12]

Why cannot mathematics be exactly applied to physical reality? Because primary units not only have quantity but also *qualities* and these qualities cannot be simply reduced to quantities. Thus qualities form a specific category of Being-in-process. Galileo proposed the thesis, which Locke popularized [13] that although the quantities of primary natural units are objectively real, their qualities are merely subjective impressions. Today many philosophers still think that science has proved that objects are not really red or warm, but only appear so to us, while in reality they are extended bodies which reflect light of a certain wave length and are made up of molecules that vibrate at a certain frequency.

In fact, however, all that science has established are the quantitative conditions necessary for a body to have certain qualities (since, as we have seen, quantity is the primary condition of all process), but it has in no way overthrown the evidence of our senses that bodies can be really red or warm. To deny the validity of what our senses observe is to undermine science itself, because if the redness and warmth we immediately perceive are not really there in the external world, how can we be sure that these objects have quantities either, since although quantity is the physical basis of quality, it is by qualities that we are able to sense the quantities of things? I can measure a red square only if I can first perceive redness. Sir James Jeans argued that all science can be reduced to "pointer readings", but that means that the scientist *really* perceives the qualitative difference between the black hand and the white dial of the measuring instrument. [14]

Obviously what we mean by saying that something is red is not the same as saying it is large, since we can vary the color without changing the size, and the size without changing the color. How then does the category of qualities differ from that of quantities? Quantity as such is a static aspect of Being-in-process; that is why it can be dealt with in abstract mathematics. Quality, however, is essentially dynamic, a characteristic of a primary unity by which it is adapted to act on other bodies in a specific way. Color is a quality because it is the characteristic of a body which is able (when activated by light) to act upon our light-sensitive organs, our eyes, in a certain specific way. Thus we are able to sense the quantity of things only by their qualities through which they are able to act on our sense organs. Histories of science often cite as an example of the pseudo-explanations supposedly offered by pre-Galilean philosophers the statement, "Opium puts people to sleep because of its somniferous quality." Is it not, then, a pseudo-explanation to say that an apple looks red because it has the quality of redness? No, because such statements are not really *explanations* at all but poorly formulated descriptions. We directly observe that the apple is red and we infer that opium is somniferous and we classify these descriptive terms in the category of qualities, just as we might say that the apple is three inches in diameter or infer from the size of the opium pill that it weighs less than two grams and classify this descriptive data in the category of quantity.

To *explain* such data is to take the further step of discovering the causes or pre-conditions of the processes by which the apple became red or the opium is able to cause sleep. Qualities always have quantitative pre-conditions, as I have shown, consequently it is altogether proper for scientists not to be content with saying that the apple is red or opium somniferous, but to seek to explain this data (not, however, to deny its validity and meaningfulness) by looking for the quantitative conditions which, in part, "explain" the data, e.g. to determine what in the molecular structure of the apple's skin is the pre-condition of its reflecting light of a certain wave-length, and why such light is the pre-condition of it having the actualized, dynamic quality of redness by which it is able to produce in us the awareness of its redness.[15]

The classification of quantities is familiar enough to us from elementary mathematics. Real physical quantities are those which are studied by physics, but which in abstract, idealized form are studied in geometry as continous quantities having the three Euclidean dimensions. Einstein is often misinterpreted as maintaining that physical space has more than these three dimensions, when in fact he merely used non-Euclidean mathematical models as convenient ways to express the physical laws of motion. There is no contradiction between our immediate experience that bodies have only three dimensions and the assertion that free moving bodies follow not straight but curved paths, because physical extension, even if it is three dimensional is not perfectly homogeneous as is ideal Euclidean *mathematical* space. Moreover, it must be noted that according to Gödel, non-Euclidean mathematics cannot be proved self-consistent except by reference to Euclidean geometry, and that in turn by its approximate validity in ordinary experience.[16]

The other type of quantity is discrete quantity or number. Because bodies have extension or continuous quantity they have parts outside each other which can be numbered. More fundamentally, because our extended universe is made up of primary units these also can be numbered, providing us with the real integers which, considered abstractly, are the object of arithmetic, algebra and number theory.

The classification or category of *qualities*, on the other hand, has never been thoroughly worked out in modern physics, no doubt because of physicists' post-Galilean bias for mathematization. Nevertheless, as I showed in Chapter 2, modern physics has uncovered four fundamental "forces": gravitation, electromagnetism, the weak and the strong nuclear forces. Although these forces are measurable quantitatively, as physical realities they are not mere dimensional numbers but capacities for action and reception, i.e. qualities. These fundamental forces operate in many specifically different ways within different systems of natural units and between their parts. Thus color results from light, the electromagnetic force, but each species of atom or molecule is able to emit a specific spectrum of colors. Even a simple visual inspection sometimes is able to distinguish one element from another by its color, e.g. pure gold from pure silver.

By abandoning the reduction of the qualities of Being-in-process to their quantitative pre-conditions, a way lies open to overcoming the dualism between the physical and the psychic and between "facts" and "values", because the language of descriptive psychology, of esthetics, and of ethics historically has proved to be irreducibly *qualitative* and the efforts to quantify these disciplines has never proved very fruitful. At the same time it would be a mistake to introduce a new dualism between quality and quantity by failing to maintain that quantity is the more fundamental aspect of bodies which is always the pre-condition of their qualities, as in anatomy function is always conditioned by structure.[17]

Thus any interpretation of modern science as an effort to understand Being-in-process without falling into mechanistic or dynamistic reductionism requires at least the following eight categories, irreducible to each other but existentially interdependent:

1. *Substances*, since the universe is made up of natural units of many different species, non-living and living.

2. *Quantities*, since in order to undergo change of any sort a unit must at least move, and motion implies that the unit is extended i.e. has parts external to each other.

3. *Qualities*, since to be in process a unit must be subject to internal modification by other units and must be able to act on them.

4. *Actions*, since the qualities of a unit are its *capabilities* to act and these must be actualized for it to become an actual agent of change.

5. *Receptions*, since the qualities or modifications of a unit which result from the action of another unit must be actually received by the unit which is patient.

6, *Places*, since one unit cannot act on another or be acted upon without immediate or mediate contact with other units which by their juxtaposition form a system of places in which motion from one place to another can occur.

7. *Times*, since some natural motions are more regular than others so that the motions of units form a system in which irregular motions are regulated by the more regular.

8. *Relations*, since these interactions of units-in-process set up a network of various kinds of relations by which they form various kinds of dynamic systems, including the universe.

2. Too Many Categories?

The category of relations raises an obvious question: Why is it not possible to reduce categories 2 to 7 to the category of relations, since all of them relate natural units to each other in process? To answer this we must first note that quantity and quality, although they are the conditions of units entering into relation with each other, are not *as such* relational. Rather they are manifestations of the inner structure of each unit. Quantity manifests its static structure as a complex whole having parts. Quality manifests its dynamic structure as capable of changing other units or being changed by them. In biology, for example, an organism must be described in terms of its structure and its function, its static anatomy and its dynamic capacities.

Categories 4 to 7 are indeed all relational because they necessarily involve relations, but they are not relations as such.[18] No aspect of primary units-in-process is more tenuous than pure relations. This is no doubt why reductionists have spoken of them as if they were mere mental fictions, an error that goes all the way back to William of Ockham.[19] Yet they are as real as their bases in the other categories and as the universe which they bind together. They are of four kinds: (1) those based on the primary units as such: the relations of *identity* and *diversity*; (2) those founded on the qualities of these units: *similarity* and *dissimilarity*; (3) those founded on the quantities of these units: *equality* and *inequality*. (4) those founded on actions and receptions: relations of *cause* and *effect*. Since the relations of places and times can be reduced to relations of quantity they do not constitute a distinct species of relation.

The relations of cause and effect are clearly the most important for science, since scientific explanation consists in discovering the proper causes of observed effects.[20] In Chapter 7 I have already showed that the notion of "cause" should not be limited to agent or efficient causes as is characteristic of mechanist interpretations of science, but must include matter as the potentiality for change and form as the actual organization and unity of this matter in any existing natural unit, as well as final causality or teleology as the tendency of the agent to produce regular or uniform effects.[21] Only when all four aspects of causality in a Being-in-process are taken account of do we have an understanding of its dynamism.

To sum up the discussion so far, we have seen that a radical philosophy of process enables us to understand the world as profoundly and intricately interrelated, yet by no means all of one piece. It is a pluralistic universe, since without a multitude of *relata* there would be no relations. Yet each of these relata or primary units is formed out of the primary matter common to the whole universe by the action of existing units on each other, an action which even as it destroys some units produces others. These units are not permanent, therefore, but are all produced and destroyed. Yet they perdure in relative stability for a time during which each undergoes various transformations by interaction with others, and by the internal interaction of their parts. Some of these transformations affect their size or quantitative extension and structure of parts, others modify their qualities or capacities to act on each other or to be acted upon. Through their contact with each other, contacts necessary for any action to occur, they form systems of places which can be mathematically described as "space." Through the interrelation of these processes with their different degrees of uniformity and regularity they form the dynamic system which can be mathematically described as "time." Many of their changes are random and this entropy tends to increase, yet we know that in our local region of the universe many changes are processes in the strict sense of being regular and productive of new units.

Although the universe as we now understand it is a vast, very loosely organized whole made up of units of relative simplicity and of many transitory forms of matter, yet on our earth natural processes have been able to produce complex chemical compounds and then organic life. This organic life has not only been subject to those accidents which by causing mutations in its genetic material tend to destroy its capacity to maintain its species, but the environment has been such as to select from these apparently injured forms of life some with survival advantages in the changing environment, and so to produce biological evolution resulting in the appearance of us human beings, capable by our intelligence of understanding the forces which have produced us and even of controlling to a degree these same forces, so as to re-create ourselves and our world in ways we freely choose.

II: Transcendent Process: Body and Mind

1. Interiority and Transcendence

In our attempt to understand the scientific world-picture as Being-in-Process, we have noted repeatedly that process implies real *novelty*, because all change is a movement from past to future, and because the future is not absolutely predictable, it always involves an element of novelty. This novelty is not merely the result of our ignorance, but is a real feature of the universe, inseparable from Becoming-Being. Moreover, in the universe just described, change is not merely cyclical but it is sometimes pro-cess in the strict sense. It involves a certain evolutionary progress toward more complexly organized natural units and systems of units, although this evolution is a local, rather than a universal feature of our world. Hence we must look more closely at the notion of "creativity," understanding this term not as creation out of nothing, since this is not an evident feature of our world, but as that aspect of change which produces the authentically novel.

For Whitehead, "creativity" is not only in some respects like what I have called "primary matter", but it is also the active tendency to produce unity out of multiplicity, to cause the "concrescence" of a number of forms which exist as possibilities in the Divine Mind, but which can take on actual existence only when they are "prehended" in this formation of an "actual occasion". These actual occasions are the only real existents and correspond to the primary natural units of the forgoing exposition, but differ from them in being sub-atomic quanta which actually exist only for a quantum of time.[22] Whitehead especially emphasized that

307

such actual occasions are not "vacuous entities," because each has an *interior* unity by which it has a certain subjectivity, "feeling," or degree of consciousness and a teleological "subjective aim" as a result of its participation in the Divine Mind through its prehension of the possibilities or "eternal objects" existing in that Mind. Nevertheless, this interiority or psychic life of every occasion does not isolate it (unlike the "windowless monads" of Leibnitz, whose thought greatly influenced Whitehead but from whom on this point he differs, [23]) but consists precisely in "prehensions" of other actual occasions which had previously embodied these same eternal objects in other combinations and which constitute its own past history.

Thus an actual occasion is a kind of remembering of its own past made up of countless events, of a plurality of forms which it now gathers into a unique synthesis or concrescence, thus constituting this *new* actual occasion, this instance of creativity. Yet at the very moment, in the very same quantum of time when this new occasion creates itself, it perishes like a spark, leaving only its memory in newly forming occasions and in the supreme, and only permanent actual occasion, the Divine Mind where it achieves "objective immortality" as God's memory of all that has been.

My previous analysis makes it unnecessary to adopt Whitehead's panpsychism in order to agree with him that a primary natural unit is a dynamic concrescence of elements from the past and that it can be understood as a synthesis of relations to other units. According to my interpretation of the scientific facts each primary unit comes into being because it has been produced from the matter of previously existing units and hence synthesizes many of the forms already present in them. For example, a molecule of water produced by supplying sufficient energy to a mixture of hydrogen and oxygen atoms is a new natural unit which synthesizes many of the characteristics of hydrogen and oxygen in a unity which also has new properties specifically its own. When such a unit comes into being (not through itself but through the agency of other units), it at once begins to interact with all the surrounding units and thus enters into that network of dynamic relations which constitute our universe and on which its own existence as a unit and ultimately its destruction depend.

Consequently, as we go up the scale of more and more complex natural units, the number and interrelations of the parts increases, that is, there is an increase in the *interiority* of its activities. Even in the atom there are interior activities, interactions which go on within the nucleus, as well as the changes in the positions and velocities of the orbital electrons as the atom receives energy from without or interacts chemically with other atoms. In the molecule, such interior interactions grow more complex. Finally in living organism there is that intense level of inner activity which we call "life." The difference between the living and non-living is not that the non-living lack all interiority, but that in the living interior activity begins to *dominate* exterior activity. Even in a plant which seems entirely at rest, all sorts of things are happening below the surface of its leaves which will eventually manifest themselves exteriorily when it puts forth its flowers and then the seeds that will transmit and multiply its life in new units.

When we reach the level of animal life, this interiority reaches a new phase which we call psychic life. Psychic interiority is truly novel, and we should not attempt to reduce it, as do panpsychists, to the kind of interiority possessed by inanimate units or even by plants, except by way of a very remote analogy. In psychic life, the animal produces within itself a *representation* (how literal a representation is another question) of the outer world through which it can respond to exterior reality. This makes possible a response which is to some degree freed from the limitations of an immediate reaction. For example, a cat can wait before pouncing on the mouse just at the right moment. But such internal representations are not mechanical copies of the reality like a shadow or a photograph.

A psychic image is a dynamic reality produced within the organism by the organism itself in response to the action on its sense organs of some stimulus from outside. This *information* (in the etymological sense of the term) received from outside actualizes the organism so as to give it a new capacity to act. A cat seeing a mouse *becomes* the mouse in a dramatic way so that the mouse's presence and movements are incorporated in the cat's own activity. When the mouse runs east, so runs the cat. When the mouse runs west, the cat follows. Similarly, within the internal system of the organism, the living thing receives information from its own body; the condition of the parts is communicated to the whole and enters into the way in which the whole performs. Thus an animal is aware both of its environment and of its own body in relation to the environment and to its own internal condition. It is only at this animal level of organization that we can really speak of what Whitehead calls "feeling" or "subjectivity", because it is only in animal organisms that there are unambiguous behavioral signs that the organism is responding on the basis of internalized representations. The tropisms of plants, such as the turning of plants toward the light or the folding of their leaves at a touch are not unambiguous because they can be explained by a simple triggering of physiological changes by the in-put of energy. Such tropisms provide an anticipatory analogy, but only an analogy to the responses of animals to sensation.[24]

At the level of the human person, interiority and subjectivity or consciousness come to still a new level, that of *self-consciousness* in the strict sense of the term, namely the awareness not only of the external world as impinging on my body and of the parts of my body on one another, but of *myself* as uniquely one and other than the world about me. The fullness of this notion of self is achieved when I confront another self or person and by our dialogue recognize that although we are both human selves, yet I am not You. Such self-understanding is the ultimate interiority and transparency in our bodily universe, yet it is not perfect, since I by no means know every aspect of my body-person. I cannot see the back of my own head, and I cannot feel the convolutions of my brain. I cannot easily recall all that is stored in my memory. Furthermore, I know myself and my own body, not directly, but by interacting with other bodies in my world and by comparing myself to other body-selves and to other bodies that are not selves at all. Of course I also know myself by "introspection", but when I look within my own mind what I find there is not innate information about myself, but images derived from my interaction with other things and with the parts of my body, from which images I have to draw all the information I have about myself for my reflections on my own identity. This is the fundamental insight of Aristotle by which he corrected Plato's mistaken notion (the notion which through Descartes has given all modern thought its idealistic turn), that the way to know ourselves is to turn away from the world and look within. Hegel was closer to Aristotle when he pointed out that it is only through going out to the Other that we can return to discover the Self.[25]

At this psychic level of self-consciousness, "process" begins to take on a new meaning, only analogous to that of physical change. In order to think humanly, we must in some way transcend the conditions of physical processes or we would not be able to escape their control by controlling them. We must somehow be able to transcend space and time. We must be able to distinguish between the accidental aspects of things and their essential characteristics. We must be able to distinguish between natural history and natural law. We must come to an insight into the principles of Being-in-process by which that Being not only is, but is *understood*. Plato was in error about how we know ourselves, but it was his great discovery that there is a fundamental difference between physical and psychic processes. Thought is indeed a process, but not one which consists in the combination of physical elements, but of *symbols* which stand for these elements.[26] It is because we human animals can guide our activities in the world by the manipulation of symbols which provide us with a wealth of information about the world, that we can control the world and ourselves.

310

Although Plato was profoundly right in pointing out the transcendence of human thought, since he was just as profoundly wrong in explaining that transcendence by innate knowledge, he was unable to explain how our transcendent thought is human, that is, the thinking of a human who is a primary natural unit, a body among bodies. It was Aristotle who confirmed Plato's great discovery by showing how human thought, although not *reducible* to physical process, is nevertheless *rooted* in physical process.

What characterizes human thinking as distinct from animal cognition is not merely our human capacity to form symbols, since animals also have a psychic life and thus make some use of symbols,[27] but that we use symbols of the *universal* aspects of things, so as to transcend particular spatial, temporal, and other superficial differences and similarities, and to signify the *essential* features of the primary units we encounter. For example, we use symbols (different in different languages but mutually translatable) to name not merely Joe Smith or Mary Brown, nor merely to name a kind of composite photo by which a human shape can be recognized among animal shapes, but to name "human-ness", that is, those features which make a whole group of natural units members of the human species in spite of geographical, historical, racial, or sexual differences. The point is sometimes missed that what is important about our human capacity to think abstractly or universally is not abstraction or universality as such (in fact these are of themselves defects in our thinking since reality is concrete and singular), but that such concepts permit us to grasp the *essential* features of things so as not to be misled by their superficial similarities or differences.

This human capacity to grasp the essential features of primary units by the use of universal symbols does not imply, as Plato thought, that these universals are known by some interior recollection of forms innate to our minds, nor does it imply, as some scholastic writers seem to have thought, that it is a capacity to arrive at an exhaustive understanding of the essences of primary units by a simple empirical inspection. We are able to arrive at universal notions by comparing the things we know by sense experience and thus eliminating their superficial or accidental features. Moreover, our grasp of their essential natures is a long, careful, observationally and experimentally strategic campaign in the face of many false clues, marching from a map of the very broadest features to the conquest of the specific. Such conquests are never final. We never arrive at an exhaustive understanding of any natural species, not even our own, but we are able to form a taxonomy of natural units without which ordinary language would be impossible. It has been recently shown that primitive people are able to recognize the natural species of animals with an accuracy (90%) comparable to that of trained zoologists and this accurate discrimination is reflected in their languages.[28]

Our study of natural species, however, can go beyond taxonomy and achieve an understanding of the functional *unity* of these features. Thus we not only know that a cat is essentially living, an animal, a mammal, a carnivore more like a lion or tiger than a dog, yet different in several features from lions and tigers; but we also are able to explain why in such a carnivore there is a functional relation between the kind of teeth and claws it has, its anatomy designed for quiet slinking and swift pouncing on small prey, its eyes adapted for nocturnal searches, its digestive system for flesh rather than fodder, etc. To the scientist who has studied cats anatomically, behaviorly, ecologically and in their evolutionary history, a cat is not just a bundle of traits, but a highly intelligible system, a whole in which every part is functional in relation to the way a cat has to live to survive and reproduce. To know the "essence" of a cat is a matter of extensive research that can be indefinitely pursued.

Hence it was that Aristotle as an experienced biologist, and Aquinas, who had been taught by Albert the Great who was also a biologist,[29] said that although the essences of material things are the proper object to know which the human intelligence is adapted, yet these essences remain hidden from us, that is, they are not inaccessible, yet can be known only by great effort and then only incompletely.[30] What is true of the species of living things is also true for chemical species, although it would seem that (1) chemical species because of their relatively simple organization can be more accurately classified and their characteristics more precisely expressed in chemical formulae than can organic species; (2) on the other hand, the functional unity of such natural units is more difficult to ascertain than in organisms, because the teleological aspect is less evident.[31]

2. Human Creativity

It is this remarkable human capacity to think in such a way as to arrive at an essential understanding of the physical world, of Being-in-process, which makes it possible for us to be "creative", not merely in the way that all the sub-human world is blindly creative through its evolutionary struggle, but in the sense that we can become self-conscious of our creativity, make it our own and learn through it to control our environment and our own body-selves in freedom. We can creatively discover the hidden structures of the world and perhaps even its significance for some wider world beyond the horizons of our direct experience.

312

To take full account of this human creativity we need a theory more satisfactory that those usually put forth today when the study of this human creative capacity has taken on great importance in our efforts to free and enhance it not only in a few geniuses but in all human beings.[32] Research on the ways in which creative people, inventors, scientists, artists, actually work has showed that creation requires moments of "creative insight" in which new synthetic, formative ideas appear; but that such insights come only after long periods of analytic study of a problem and the laying out and testing of various alternative solutions. Moreover, such insights have little value until they have been elaborated and verified by further critical analysis, observation and experimental testing. It has been shown that creative people are characterized by their ability to notice and take seriously alternative solutions to a problem which more conventionally minded people would have rejected *prima facie*.[33]

How, then, can these creative "insights" be explained? Some studies of the history of inventions seem to show that many people make the same inventions at almost the same time.[34] This, however, only shows that many people in a common cultural milieu are sharing in the same process of research. One of the chief theories of creativity is that its source is the unconscious mind of certain individuals in whom the activity of the unconscious (Freud) or the collective unconscious (Jung) is especially intense.[35] Ernst Kris has suggested that creativity is "regression to the unconscious in the service of the conscious," by which he means that creative people seem to have the ability to suspend ordinary rational processes in a controlled way so as to gain access to the creative unconscious.[36]

The weakness of such theories, in spite of their element of truth, is that they attribute what is highest, freest, and most transcendent in human thought to the part of the psyche which is presumably the most primitive, irrational, and deterministic.[37] The unconscious mind in Freudian theory is the Id which comprises the basic biological drives which we share with other animals. How can it be the source of precisely those human abilities which distinguish us from other animals? The collective unconscious in Jungian theory, on the other hand, is a kind of ancestral memory. Are we then to suppose that human creativity is simply a recollection of the past? If so, where is the *novelty*?

The Platonic theory of creative genius attributed it to "inspiration" that is, to guidance from a higher intelligence.[38] Aristotle did not rule out the possibility of such inspiration, but was inclined first of all to explain human art as an "imitation of nature", a theory which also fails to deal with the novelty of such works.[39] A much more adequate theory in keeping with Aristotle's own psychology and epistemology has been

313

developed by Jacques Maritain.[40] For Maritain human creativity is a function of the human intellect not primarily under its rational but its intuitive aspect. Aristotle showed on logical grounds that human reason necessarily presupposes that we also have an intuitive capacity, since all reasoning rests ultimately on indemonstrable principles, which therefore must be "self-evident" not from innate ideas, but from our direct intuition of essences derived by abstraction from sense experience. I have already explained that this intuition, although logically immediate,[41] requires to be mediated by sense experience, and ordinarily requires a long process of investigation before it achieves any richness of content. A scientist, therefore, must come to his creative insights only after a great deal of laborious work. For the artist this long work is not scientific research, but it is disciplined observation and experiment guided by his or her sensitivity to form. Aquinas believed that the rational and intuitive capacities of the human intelligence are not distinct faculties, but two necessarily related functions of a single faculty, and argued that the Platonic separation of these into what St. Augustine called the *ratio superior* and *inferior*, the higher and lower reason, was precisely the fatal dichotomy that forced Plato into the theory of innate ideas.[42]

Maritain, moreover, shows that the intense operation of this human intuitive capacity is enhanced by a kind of "connaturality" or *empathy* by which the creative knower comes to understand what she or he loves not only by rational study but by being sympathetically *attuned* to it. For example the artist in writing about a certain character, or the clinical psychologist in trying to understand the psychological processes of a certain subject, are assisted in this understanding by their own share in the subject's feelings which gives them an insight not easily accessible to the purely objective observer. Even the physicist or chemist is sometimes helped in his or her researches by "hunches" or "gut-feelings" arising from an imagination of "what it feels like to be an atom" or from some "taste" for elegant simplicity or symmetry in the mathematical models among which she or he is trying to choose.[43]

Such intuitions, it seems, often arise from a kind of dialectical thinking in which there is a *coincidentia oppositorum* or reconciliation of apparently contradictory ideas. The creative imagination is not afraid of paradoxes while the routine reason is. Arthur Koestler put forward what he called a theory of "bisociation" according to which new intuitions are most likely to arise by the confrontation of two very different views of the same reality, a confrontation which shocks the mind into a new way of perceiving that reality which somehow combines both views in a new synthesis.[44] He points out that a joke is the paradigm of such an intuition when "we get the point"! No wonder the ancients considered

314

that human beings can be distinguished from the brutes by the property of "risibility". We are "laughing animals."

These theories all offer valuable elements for a more adequate theory of creativity. Such a theory of creativity as it is a specifically human *process* must consider all four aspects which, as we have seen, characterize any process. It must explain how that primary unit which is the human being can be a creative *agent*. Also it must consider what kind of *matter* is susceptible of being organized in a novel manner; and, correlative to that matter, what kind of organizing, unifying *form* can make that matter into something new. Finally, it must discover what is the purpose served by this act of creation, whether it is some practical purpose beyond the completion of the work itself, or simply the work as an object to be contemplated as such. The theories already discussed focus on the psychology of the *agent*, or on the *matter* (i.e. the images to be selected and combined in a new way) often stored up in the unconscious to which creative people seem to have a freer access than most of us. As for the purpose of creativity, it is obvious that it can be put to many uses and, especially in the case of the fine arts, transcends the useful by producing the beautiful.[45]

The hard problem to be solved by a theory of creativity, however, is that of the *form* by which the agent achieves some purpose by combining already given elements into something truly new. Is this puzzle not analogous to the problem we noted in theories of evolution which deal so well with the agency of natural selection and the matter of genetic mutations, but have so little to say about the unifying form which makes the new species new?[46]

What is the nature and source of this new form or what Maritain calls the "formative idea" which guides an artist in producing a work, which suggests to the scientist a fruitful new hypothesis, or to a practical person a new plan of action? Whitehead supposes that novelty arises in the world by the action of the Divine Mind which initiates the concrescence of a new actual occasion by giving it a "subjective aim" as its teleological goal. But whence this new subjective aim? Its source is found in the infinite number of still unrealized *possibilities* in the "antecedent nature" of the Divine Mind, just as its "consequent nature" consists of the memory or "objective immortality" of all those possibilities which have already been realized in the history of the universe.[47]

The problem with this explanation is that Whitehead does not tell us how the human mind "prehends" some as yet unactualized possibility in the Divine Mind which the human mind has never experienced in the past. We seem driven back to some kind of a Platonic illumination or inspiration. Yet Whitehead's theory does us the service of underlining the fact that "novelty" means an as yet unrealized *possibility*, since the impossible can never be realized. How, then, does the human mind think of new possibilities, that is, possible realities which it has never yet experienced? If Aristotle was right, as I have been arguing he was right, in saying that all human knowledge is derived from intelligent reflection on sense experience of *actualities*, how are we able to know what is really possible but not yet actual?

Aristotle tried to answer this by saying that artistic production is *mimesis*, an imitation of nature, that is, of possibilities already realized in nature and experienced by us. This of course, as I have already remarked, does not tell us whence new possibilities arise in our minds. But Aristotle also notes that "poetry is more philosophical than history"[48] because history deals only with what is or has been, while poetry deals with what could be. In fact in poetry, he says, what is "likely" is better than what is "unlikely" even if what is unlikely is factually true. The old saw "truth is stranger than fiction" is another way of saying the same thing. Artistic *mimesis* is more convincing than history because history presents us with possibilities which have been actualized not merely in their essential features but with all the confusion of the accidents (the entropy) of history. The human mind boggles at the absurdity of the assassination of a great man, of a plane crash, or the bombing of Pearl Harbor, and quickly devises some "conspiracy theory" to reduce the craziness of chance to meaning and purpose. The poet, unlike the historian, is free to give such meaning to every detail of his narrative.

Thus, Aristotle pointed out that even children love to imitate, but he had in mind something different from the kind of copying we observe in monkeys (although no doubt this is an evolutionary anticipation of human imitation); he was thinking of our intelligent capacity to abstract the essential from the accidental (as even children do in playing make-believe), to cut through the randomness of historical experience and to apprehend in this experience the essential possibilities which are there actualized, but imperfectly so. The most obvious example of this capacity is to be found in the mathematical imagination by which we are able to construct perfect circles, numbers larger than any collection we have actually counted, and spaces of more than three dimensions.[49] Undoubtedly, this is why mathematics plays so large a part in scientific discovery and techological invention, but it is also an essential factor in artistic creation.[50]

Is what is novel in human creations, therefore, nothing more than the "idealization" of possibilities already experienced in nature in actual but imperfect and confused form? That would seem to be as far as Aristotle went, but it is easy to extend his theory a step further by considering another aspect of human intellection which plays a large part in Aristotle's

316

cognitive theory and especially in that of Aquinas, who developed certain Platonic notions familiar to Aristotle but which he failed to exploit thoroughly, namely, the fact that human thinking and language, although they seek univocity, do so by a skillful use of *analogy*.[51]

As we struggle to formulate our experiences, our knowledge is most perfect when we can find the precise concept and name for every given reality. Descartes was right in hungering for "clear and distinct ideas." Nevertheless, because of our initial ignorance and the laborious path we must take to arrive at insights in a very confused world, the strategy which we have to employ is to use a few concepts and an impoverished language to make do in thinking about an extremely varied reality. That requires us to *stretch* our concepts and words so that each covers many really different realities.[52] We are justified in doing this when those realities, even if they are essentially different, nevertheless in some ways more or less accidental are similar. This process of stretching concepts and names is analogical thinking.

Analogy can be *metaphor*, based on a comparison between two things whose resemblance may be purely superficial. Aristotle pointed out that the basic gift of every poet is to notice similarities others may miss and thus to use language metaphorically.[53] Again, analogy may be based on a comparison which is more than metaphorical, namely, on a similarity which does not amount to univocity but which consists in a similarity of *relations* between four terms. For example we notice that the relation between the nucleus of an atom and its peripheral electrons is similar to the relation between the nucleus of a living cell and its cytoplasm, and hence we use the term "nucleus" of these essentially different entities. This is called an analogy of *proportionality*. It is this type of analogy which is especially fruitful in imagining new *possibilities* never actually experienced.

It is essential to note, however, that when we have imagined such a "possibility" we are not really sure that it is really possible, since it might contain some hidden contradiction. A square-circle is clearly not a real possibility, but it is not so evident that it is impossible to "square the circle" i.e. to reduce pi to an even fraction. In order to establish the real possibility of a concept we have to point to an empirical instance, or lacking this we have to prove its existence *a posteriori*, from its empirical effects. When we argue from the existence of an effect to the existence of its cause, we can also conclude that there is at least a similarity between the two, since the cause must itself possess what it contributes to its effect. Analogy based on this similarity of cause and effect is called an analogy of *attribution*, because the characteristics of the effect are attributed at least analogously to the cause. By the use of a combination of analogy of proportionality and of attribution we can establish the real existence of entities which we have never actually experienced, although of course our knowledge of these entities is imperfect and requires to be further tested by empirical experience either of the entity after it is discovered or constructed, or at least by more experience of its effects. [54]

For example, the discovery of radium by Madame Curie was achieved by observing an effect for which a cause was to be sought (attribution); but, to imagine what this cause might be like, Madame Curie made use of an analogy of proportionality in imagining that radium might be similar to the already known uranium and by this analogy developed a model of what radium might be like in its various properties, making use of the periodic table of elements which is a schema of such proportionalities. Then she applied various techniques based on these similarities to isolate and verify the existence of radium. [55]

Similarly Beethoven was able to write his last quartets by constructing new sound patterns by analogy to music he had once heard. The novelty of these new compositions is revolutionary, yet musicologists are able to trace in his earlier music the steps by which he was prepared to attempt these new constructions which to other composers would never have occurred as possibilities. No doubt Beethoven was also able to test these novelties in his imagination, but until they were actually played no one really knew if this music would "work."

I would add that in the foregoing I have used the term "imagination" in the broad way common today in literary criticism. To be more accurate the creative power of the human mind is intelligence and not imagination, if imagination is understood, as it should be, to mean our power to form, recall, and combine images derived from sensations. Imagination in this sense is not characteristically human and is common to us and the animals. In humans, however, because our intelligence depends on our bodily senses, we cannot think without images of some sort [56] and our intelligence in turn habituates the imagination to serve its purposes. Thus the mathematician acquires the habit of imagining mathematically, and the poet of imagining verbally, the composer of imagining auditorily. The disciplined imagination is thus indispensable to our creative intelligence in readily suggesting those images from which we can most easily arrive at the concepts and intellectual intuitions which constitute the creative process itself. No doubt an important psychological predisposition to become a creative person is a lively imagination, but many people with lively imaginations are not very creative. On the other hand persons who are too rigidly "logical" are not likely to be very creative either if they permit this habit to exclude from their thinking those unusual images that could be the fruitful source for analogical development of new concepts and insights.

318

A final point must be added to this theory. Not every new actualized possibility is worth actualizing. In fact most possibilities are of such a low level of organization that they are without value. In the works of modern art of the minimalist style some hardly reward our contemplation. The famous monkey who was given a typewriter and as much time as he needed to produce a play as good as *Hamlet* must have used up many reams of paper before he even came up with *Charlie's Aunt.* When we speak of "novelty" we have in mind the contribution of something of *value,* something beautiful, true, useful. This is the *teleological* aspect of creativity which must serve some human purpose and thus be related to the ultimate and necessary goals of human life, and which raises questions about human freedom which I must postpone until the next chapter.

Thus we can conclude that we humans can be creative because we are primary units who as agents of a unique kind have the capacity to use our intelligences not only rationally but also intuitively. Morever, we have imaginations and memories which supply us with a flow of images and combinations of images, often from the depths of our unconscious, along with empathetic feelings which make us sensitive to unnoticed aspects of our experience, thus providing us with new materials of creations. Furthermore, we have the capacity to construct new possibilities, never actually experienced before, by a process of analogical thinking based on actual experiences, and then to choose from these possibilities a form to unify these materials into something genuinely new and also worthwhile because it fulfills worthy human purposes.

3. The Mind-Body Problem

According to the foregoing interpretation the whole universe is the result of continuously creative processes which constantly produce primary natural units which are specifically or at least individually new. Moreover, in those primary units which are human persons, this creativity becomes self-conscious and self-controlled. This raises the difficult question: how can human creativity *transcend* the evolutionary process from which it has emerged? Or in other terms, how can creative self-consciousness arise from a creativity which is unconscious? How can *purpose* arise from simple teleology?

For current thought this question is usually posed rather confusedly in terms of the so-called "Mind-Body Problem." The two extremes of the debate are the dualism of Descartes and the Identity Theory of Hobbes. Descartes, it will be recalled, maintained that Mind and Matter are the two absolutely distinct kinds of substance, the former consisting of "consciousness" and the latter of "extension." He denied any essential

319

distinction between living and non-living bodies and thought that animals are automatons without consciousness. He even suggested that in the human being mind and body are united at a single point, namely the pineal gland in the center of the brain. How these absolutely different substances could interact Descartes never explained, apparently attributing this to the special action of God. [57]

Hobbes, a contemporary of Descartes, proposed a form of what is now called the Identity Theory according to which thought is simply identical with the physiological processes going on in the central organ of the human body. Hobbes (following Aristotle) still supposed this central organ to be the heart, while, of course, identity theorists today (following Plato) suppose it to be the brain. [58] Hobbes conceived these physiological processes in dynamistic, Stoic manner as material "spirits" working in a kind of pneumatic machine to move the human body. Today the brain is thought to operate as an extremely complex system of integrated neural circuits transmitting electro-chemical impulses. This Identity Theory seems to be taken more or less for granted in popular literature and in medicine and is defended by many philosophers of the analytic school. [59]

Because both of these extreme positions seem simplistic, there have been various attempts to mediate between them. Marx and Engels and many others have favored *epiphenomenalism*, the theory that although thought is a material brain-process it is an evolutionary emergent distinct from lower forms of material process. [60] Recently Popper and Eccles, a distinguished philosopher and an equally distinguished neurologist, have collaborated in a defense of *parallelism* according to which brain-events and mind-events are regarded as existing at different levels of reality yet with a more intimate relation than in the Cartesian dualism. [61] Some link this parallelism with the theory of complementarity so successful in physics according to which the same reality can be viewed as a wave or a particle (seemingly incompatible models) by a translation of one model into another according to consistent rules, but cannot be represented by a single model. I have already frequently mentioned the *panpsychism* favored by Whitehead, Teilhard de Chardin, Julian Huxley and others, according to which human mentality is an intensification of the rudimentary mentality present in every physical entity. [62]

One of the sources of confusion in the discussion has been the failure to distinguish three different questions: (1) Can psychic activities involving some kind of consciousness be reduced to mechanical processes (Hobbes' position as against Descartes who believed that no material system could be conscious and therefore that animals are unconscious)? (2) Can human consciousness be related to the body in the same way as

320

animal consciousness? (3) Is the relation of human consciousness to brain-processes one of identity or transcendence? Since I have already rejected Platonic dualism according to which the human soul is only imprisoned in the body, and Cartesian dualism as both gratuitious and inconsistent with my option for an Aristotelian realism in epistemology, I will not discuss them further, except to inquire whether the solution I will present sufficiently saves the valid elements in the Platonic tradition which has contributed so much to Christian theology.

Can psychic activities involving some kind of consciousness be reduced to mechanical processes? In a famous book *The Concept of Mind* Gilbert Ryle attempted to refute Cartesian dualism by showing that we are not really "conscious" of thought, but only of our activities in relation to the outside world or to our own bodies. [63] Critics have pointed out that Ryle never explained why *awareness* of our physical activities is considered by most of us to be very different from these physical activities themselves. [64] It seems to me that we must distinguish: (1) the object of which a subject seems aware; (2) the subject's awareness of the object; (3) the subject's awareness of itself as aware of the object, i.e. of a relation of identity-in-difference between the subject and object. It is this third kind of awareness which Descartes clearly had in mind in his *cogito ergo sum* and which is properly called "self-consciousness."

Ryle rightly raises a question about this Cartesian "self-consciousness" by making the perfectly Aristotelian point that when we think, what we *directly* think about is not our thoughts but the outside world or our own bodily self. Ryle's critics are correct, however, in pointing out that Descartes' mistake was not to assert that we have "self-consciousness", (it is obvious that we do and Ryle goes too far in trying to explain away that obvious fact) but in failing correctly to relate the implicit or *indirect* self-consciousness, which we normally have when we are *directly* conscious of the outer world or our own bodies, to the direct self-consciousness we have when we *reflect* on our own thinking and become explicitly aware of ourselves as thinking. Descartes made this reflective consciousness primary, when it can only be secondary. We cannot be self-conscious before we are aware of the bodily world including our own selves as bodies.

Granted these distinctions, the first question just raised can be reformulated as, "Can a machine have "awareness", at least in the weaker sense, of objects other than itself or of itself as a material object?" Certainly we human beings cannot doubt that we are aware of the objects of the outer world by our eyes, ears, touch; and we can explore our own bodies in the same way. Moreover, we are aware of interior states; we feel our limbs move, and we feel pleasure or pain as conditions of our bodies. Finally, we are also aware of *images*, representations which we recognize to have been derived from past sensations, which we know can be forgotten or recalled from somewhere within us, and which can be combined spontaneously or deliberately as new compound images. We also have good reason to suppose (as against Descartes) that animals are sufficiently like us in anatomy and behavior that they too have a similar kind of awareness; while we have no sufficient reason to think that plants, let alone less complex primary natural units have such awareness.[65] To doubt these facts of experience is useless, since reasonable doubts must be based on something more certain than what is questioned. Nothing is more certain to us than that we animals have this minimal kind of consciousness, and that plants and inanimate units do not.

Today, however, it is not so obvious that some machines do not also have this kind of minimal awareness. We are familiar with automatons which are equipped with sensors which supply them with information about their environment and their own internal condition just as we animals see, hear, feel ill. Moreover, these machines equipped with computers can store this information, recall it selectively, and combine it in new ways just as we animals remember and imagine. Some even argue that such machines are creative, since they can be used to compose music, make scientific discoveries, write poems, and compete in chess and other games.[66]

To answer this we must ask phenomenologically: What is it for a subject to be aware of an object?[67] First it means that the object is present to the subject, i.e., united to it or identified with it. Second, in this union the object is not absorbed by the subject or made part of it, but remains itself. Therefore, this union is not a *material* union such as occurs when an organism assimilates food and incorporates its matter into its own formal unity. Third, since it is not a material union, it must be a *formal* union in which the formal unity of the subject is enriched in its complexity so that as a living primary unit it is enabled to carry on its life functions more successfully. Instead of acting blindly, it can now act in a manner that takes better account of its environment. Fourth, this enhancement of the formal unity is neither a radical change of the unit (since in that case the subject would be destroyed, and a unit of a new species produced as in evolution), nor is merely a superficial change by which the matter of the subject is modified (since it is an enhancement of the form of the unit), but is a special kind of qualitative enhancement consisting in *intentionality*, namely, a relation to the object of the kind just described, and this we call "awareness" of the object; for example, seeing, hearing, touching.[68]

From this phenomenological description of what our human awareness of an object consists in and which we reasonably attribute to animals also, it becomes evident that we cannot contribute awareness to machines, no matter how closely they are made to imitate the behavior of an animal, because the machine as such is not a primary unit and so it cannot be alive. As a non-living thing it receives information not in an intentional way, but only as a replication of a pattern existing materially. Such a replication does take place in every act of sensation, e.g. the chemical pattern on the photo-sensitive retina, but this pattern as such is a superficial change modifying the eye materially, not psychically, and is not different from what happens to a photographic plate. This physical modification becomes psychic only because the retina is not merely a photo-sensitive surface but an organ of a living organism which receives this physical impression not in a merely passive manner but as an enhancement of its own power to function as a living unit in its environment. If someday we can produce a machine which can do this, rather than merely imitate it analogically, we will be producing a living animal unit which is aware of what it sees, hears, touches; and of course then it will no longer be a machine but an organism.

Some may object that such a distinction between the animal seeing and moving toward food and an automaton which seems to do exactly the same thing but to which "seeing" in the sense of awareness is denied, can have no place in empirical science. This has always been the objection of the behaviorist school of psychology which wishes to observe animal and human behavior without introducing the notion of consciousness or awareness. [69] Critics have often pointed out, however, that behaviorism involves a contradiction. If the scientist is not conscious, how can he or she make an observation? The behaviorists claim a special privilege of consciousness as scientific, objective observers which they deny to the subjects they study. Moreover, there is no reason to deny that our human experience of awareness is any less empirical or any more private or unverifiable then any other kind of data admitted in science, since any observation ultimately must be tested by individual observers who have to be aware of what they observe whether this be an animal or a pointer-reading on an instrument. [70]

The Platonists taught that sensation is an act of the soul for which the body is only a channel. For example, the soul sees *through* the eyes not with the eyes. Descartes went even further in maintaining that the soul knows what is going on in the eyes only through messages conveyed from the eyes to the pineal gland. On the contrary, it is necessary to admit that seeing is an act of the eye itself, not in isolation, but as the organ of the unified animal body whose function is to see just as the function of the stomach is to digest. It is common to argue that we see, hear, or feel not in the eye, ear, or skin but in the brain, because if the nerves connecting these sense organs are cut there is no awareness of the object. This argument, however, is not conclusive, any more than we could argue that we do not walk with our feet but our brain, because if the nerves connecting foot to brain are severed the foot will not move. Because of the unity of the body of which the brain is the central organ, it is not surprising that the sense organs at the periphery cannot operate without being maintained in dependence on the brain in their own homeo-static condition necessary for their proper function, but this does not mean that when the eye functions it is not the eye that is the seat of seeing, of visual awareness, etc.[71] Consequently, it is not the brain that sees, hears, feels, but the animal who sees in its eyes, hears in its ears, feels in its skin.

So far the discussion has concerned *sensation* as the awareness of actually existing objects, whether these objects be external to the body or parts of the body, e.g. feeling one hand with the other, or simply feeling muscular tensions. But we also experience awareness of objects which are not actually present or existent. We have memories, images which we recognize as representing things sensed in the past but no longer present. We also have fantasies, images abstracted from any definite location in time or space, and often spontaneously or deliberately combining many images in representations of objects never actually sensed in such a combination. We also interpret the data of sensation so as to construct from it a more complete representation of reality than is actually given; e.g. from the two-dimensional picture given in vision we can construct the third dimension in perspective; and from a jumble of noise and patterned speech sounds we can separate the pattern from the background noise, etc.

Moreover, we can not only interpret sense data, we can *evaluate* them relative to the biological needs of our organism. Thus, we can recognize the differences between the data of the different senses and integrate them in a unified picture. We can usually know that we are awake and not asleep (when we are asleep we are not sure of the difference). We can distinguish between a sense object and the image of such an object. We can compare and estimate sizes, and above all we can experience some sensations or images as pleasant or painful. All of these activities can be lumped together as sense *perception* in distinction from sensation. They are an awareness of objects not merely in themselves but in relation to the organism, i.e. they have a certain subjectivity.

Of all the kinds of sensation, that of touch, the most fundamental of our senses whose organ is the whole surface of the body and even the muscles and some internal organs, is the one in which this subjectivity orginates even before perception begins. The information touch gives us is very vague, but it is the most existential. It makes us aware of objects in immediate contact with our bodies and in doing so makes us aware of the condition of those bodies. Hence, the perception of pain and pleasure are evaluative of the condition not of the object only but rather of the condition which an object produces in our bodies, a condition that is in accordance with biological needs or destructive of them.[72]

Epistemologically considered, sensation as such never deceives us because our sense organs cannot act unless stimulated by an object really present and therefore really existent. Of course our sensations do not furnish awareness of *all* the features of the object. Our eyes for example distinguish only a few colors in the continuous range of the visible spectrum and are blind to shorter or longer waves; and we cannot resolve minute patterns, as can a microscope, or detect very faint light as can a telescope. Such limits, however, do not imply that the senses deceive us as to what they positively present. At the level of perception, on the other hand, the possibility of errors are great and explain the various sensory illusions (the bending of a stick when seen partly through water, the "phantom limb," etc.) which are often cited by sceptics. Such errors are possible because perception results from a processing of images derived from sensation. Images are compared, combined, selected, recalled, stored, etc. Since such processing is a complex operation, it can go wrong at many different points of the sequence.[73]

In this processing of images there is an anticipation of human creativity, since the imagination creates out of a relatively small amount of data a rather complete representation of the environment. Moreover, it reconstructs the past and projects the future, and in dreams and fantasies it constructs a never-never land.

Such an ability in an organism also gives rise to *communication* between members of a species. The songs of birds warn other birds not to intrude on their food gathering territory, of the presence of predators, or signal a mate or their young. Bees have a "dance language" by which they inform other bees as to which direction to fly in order to find honey. There can also be communication by imitation, and it is now known that animals really teach their young and other members of their species modes of behavior which they themselves have learned, or even originated. Recently painstaking efforts have been made to teach primates a symbolic language similar to human language with some measure of success.[74] These facts make clear that animals have a very extensive and remarkable capacity for a kind of imaginative behavior. We have already seen that in our human thought this role of imagination is very large and susceptible of training. Why then should we be surprised that other animals also exhibit it? Yet I have also argued that this "creativity" is only the *material* aspect of what we call human creativity. It goes no further than the permutation and combination of images. The animal behavior so far observed does not require us to suppose that they have intelligence in the strict sense of a capacity to abstract the universal from the singular and so to achieve a recognition of the difference between the essential and the accidental, the really real and the merely intentional.

An obvious objection to this last statement is that animal behavior sometimes does seem to suggest that the animal acts with conscious purpose and selects from its perceptions those combinations of images which serve this purpose. Since it is this teleological aspect which, as we have seen, is so essential in defining genuine human creativity, it would seem that animals like Köhler's ape who fitted two sticks together to knock down a banana out of reach, or those other apes who have learned a language by which they can ask for things they want, are exhibiting something more than perception.[75] In answer to this difficulty, I would not deny that animals act teleologically, but the "purposes" which they have are innate biological drives which no more imply an intelligent understanding on their part than does the teleology found in a crystal. Because the animal, unlike the crystal, can have an image of the object which stimulates its instinctive drives even when the real object is not present, an animal has a great repertoire, some of it learned, of behaviors which may achieve this instinctive teleology. Consequently the processing which takes place in its imagination does not consist in merely random combinations and permutations but in pre-programmed selections of those combinations which serve its biological drives. The analogy with artificial "brains" is significant. Whatever computers can do animals may be able to do or learn to do without requiring us to attribute intelligence to either computer or animal, although we must attribute to the animal an awareness of the data it processes which there is no reason to attribute to the computer.

The organ of all these perceptive activities, of the processing of data received from the senses, is undoubtedly the brain.[76] Everything we now know about the brain (and of course there is much we do not know) indicates that the brain receives data from the senses and subjects it to an elaborate process of selection, storage, comparison, and combination so as to provide the organism with a relatively complete picture of the past, the present, and of the various possibilities which may be met in the future, as well as alternative ways of responding to these objects. Not all brain processes are mere shufflings of material markers as in a computer, but some are psychic processes in the sense explained, that is, acts of awareness of objects real and imaginary. As such they are actions of the brain, of a material system, but a material system operating in a psychic manner by the enhancement not primarily of the matter of the organism but of its qualities and its primary form as a living unit. Descartes' error (originating in his identification of matter with quantity), still so prevalent in current thinking, is that such psychic activity involving awareness is impossible for a material organ, but only for a disembodied mind. I have tried to show that both ourselves and animals sense with our sense organs and perceive with our brains and yet that neither activity is simply identical with the physiological functioning of these organs but is also a psychic process involving awareness of objects.

My third question was: is the relation of human consciousness to brain-processes one of identity or transcendence? In other terms, is there something unique about human thought compared with what we share with other animals which demands that we reject the Identity Theory without accepting a Platonic or Cartesian dualism?

This question can be answered in terms of my previous analysis of human intellection. If the human intelligence is able not only to use data derived from the senses and combined in various ways by perception in the brain, but also to derive from this data universal concepts and to make judgments about what is accidental and what is essential (thus coming to a step-by-step understanding of the natures of primary natural units, especially the human unit, and taking control of its world and the evolutionary processes by which this world and the human being has come into existence), then it is clear that this ability essentially exceeds the abilities of the animals lower in the evolutionary scale and constitutes a unique human transcendence of material processes, and even of the human brain which differs only in degree of complexity from animal brains. On the other hand, this human transcendence of material processes does not imply that the human self is a mind which merely dwells in the body, because the human self is the primary unit which transcends its world only by being a bodily part of that world, deriving its knowledge of itself from interaction with that world.

It is in the human being that awareness common to all animals also become consciousness in the full sense of consciousness, namely *self-consciousness*. The animal is aware of objects in the world. It is aware of the parts of its own body. A chimpanzee has been observed seemingly to recognize its own visage in a mirror. And of course animals recognize their mates, their offspring, and their human masters. Yet none of this awareness is unambiguously what we humans experience as self-consciousness and the consciousness of other human selves whom we know not merely as objects but as subjects like ourselves. Self-consciousness is not only awareness of an object, but an awareness of the knower as knowing the object. Since knowing is a kind of identity-in-

327

difference, an identity between the knower and the known, self-consciousness is a knowledge precisely of this identity-in-difference as such. It implies that the knower is able to *reflect* on the knowing of the object as the action of the knower and thus on him or herself as a primary unit and not merely as a stream of awareness.

Because of this self-awareness which involves reflection on the knower knowing, human beings are able to be creative in the strict sense of finding or making new objects in view of a purpose. Because we know ourselves, we not only have teleological tendencies as do all primary units, and biological drives as do animals, but *purposes* in the strict sense. In knowing ourselves, we know what we want and we freely choose ways to get it. It is not merely that we want something and we pursue it when it is presented or imagined, as do animals; but we understand it as the *reason*, the principle of our actions. Consequently, if we do not find at hand the means to achieve our goals, we invent them. Because we know ourselves and recognize other human beings as selves, and there is not at hand any adequate natural medium of communication, since gestures and grunts and cries prove insufficient, we modify these into languages. Whatever the debate about animal languages, it is clear that animals do not spontaneously invent anything like human language which is everywhere common to our species in thousands of different invented varieties.

Genuine language not only uses names but relates them syntactically, and syntax is not merely the combination of names but depends on the perception of grammatical and logical relations which are not real relations but mental relations. No doubt animals can learn to combine and select symbols according to pre-programmed rules, just as computers can do. That is not the same as to be aware of such relations, and therefore, to be able to use these relations purposefully in order to extend a language by analogy, or to invent a new one. Yet fairly early in life children, having learned a language by imitation, begin to play with it, to expand it, and even to make linguistic inventions.

Most important of all is the fact that this self-consciousness which gives the knower mastery over what he or she knows, brings us to the *ontological* level of awareness, that is, to the ability to distinguish between the really real and our knowing of the real. I have elaborated this point earlier in Chapter 7 by showing that as we explore Being-in-process we come to the distinction of the subject and the object by the fact that we can mentally conceive of contradictions and other mental objects which cannot possibly be, which are Non-Being, and therefore among the objects of our consciousness we can sort out beings from non-beings, reality from illusion.

328

Is this knowledge of the real as such an act of our intelligence only, or is the body involved? I have indicated that even at the level of sensation the sense of touch puts us in contact with the existential reality of our bodies in relation to the world, thus permitting us to distinguish between sleeping and waking, sensation and fantasy. No doubt there is also within perception a certain distinction between data which actually originates in the senses and our constructive interpretation of such data, e.g. between the actual perception of a two-dimensional visual pattern and our three-dimensional interpretation of it. Nevertheless, these qualitative differences in sensation and perception are only *signs* of the distinction between the real and the mental. We have no reason to suppose that animals can distinguish between the real and the unreal. Probably their awareness of the world remains dream-like, a stream of sensations and images differing only in vividness or faintness, as Locke and the British empiricists claimed to be the case in human consciousness.[77]

To distinguish the real from the unreal, the human being must reflect on itself as a primary unit, a Being-in-process, and distinguish itself as a real being among other real beings from its own mental representations, whether images, perceptions, or concepts, which derive from real being but which are produced by the mind and exist only in the mind. If the human person did not know itself as a real body through at least its sense of touch, it would lack the necessary data to make this discriminating judgment; but unless it was more than just a sensing, perceiving body, unless it was also an intelligent, reflective person able to tell the essential from the accidental, it could not correctly interpret this data and thus make an existential judgment about its own existence and the existence of the world and of other persons.

Thus, the human knowledge of the really real requires that we not only have sense and perception but a power of intelligence distinct from sense and perception. Yet even our intelligence does not achieve the real merely at the level of concepts (which may be of both the real and the unreal) but only in the act of judgment.[78] How is such a *reflective* act possible? Reflection is, of course, a metaphor, but it is derived from the fact that all awareness implies an identity of presence or contact between the knower and the known, as we have already shown. That is why to know their environment animals have to move about in order to contact by touch or through some medium in proper perspective all aspects of the environment. But in order to know itself, an animal needs to touch one part by another, as when we feel our own body with our hands. Undoubtedly this is why the brain is a mass of interconnecting pathways and neurons touching each other at synapses, and why in imitation of such a communications system any computer is a network of circuits. Each of these neurons or wires puts one spot of the brain or of the computer in contact with another. Only in this way can one part be brought into contact with another in a kind of inadequate *reflection* by which one hand knows what the other is doing, one part is aware of another part, and a certain psychic integration or unification takes place.

Adequate intellectual reflection, however, has to transcend this mode of contact, since no matter how intricate the circuitry, it can never put all parts of the system in contact with all parts simultaneously; it can only relate part to part, not the whole to itself. Yet human self-consciousness is precisely that: the *presence of the whole of the primary unit to the whole;* not in the sense that in self-consciousness we know all parts of ourselves in detail (since we are never totally transparent to ourselves and parts of the self lie in the unconscious), but in the sense that we know ourselves to be this really existing unit. Such total reflection, however, cannot be a physical process carried on simply by the brain because, all material processes, as we have seen, are extended, since although matter is not quantity it must have quantity or extension in order to enter into process. [79] Sensation and perception, which are forms of awareness but do not achieve self-consciousness, can and must be the psychic activity of the bodily sense organs and the brain. Human intelligence, although an act of the primary unit of which the human body provides an essential constituent, is a psychic activity which uses the sense organs and the brain to acquire its data but not to perform its own proper act of self-consciousness by which this data is given its realistic interpretation.

This transcendence by human intelligence of the human body and its bodily activities of sensation and perception explains why such intelligence can create computers that serve it in the processing of data. Turing has shown that in theory at least, every possible operation of formal logic can be performed by a machine. [80] On the other hand, Gödel has shown in the famous theorem bearing his name [81] that no purely formal system of logic complicated enough to deal at least with elementary arithmetic can ever be demonstrated to be both self-consistent and complete, which implies that it is always possible for human intelligence to invent problems in the terms of that system which cannot be solved without enlarging its set of axioms. Since computers operate by programmed rules that amount to a formal system of logic, this means that the human inventor can always propose a problem which his computer cannot solve; although, of course, he can then invent a new machine to solve it. He can even invent machines that will systematically or at random propose new problems and enlarge their set of axioms so as to have some chance of finding their solutions. But Gödel's theorem means that unless human intelligence is itself limited by a programmed set of axioms it will always be more creative than the machines it creates. [82]

Why is a formal system of logic and the computers based on it limited in this way? Modern symbolic logic is mathematical in character, which means that it deals with collections or sets of items that are essentially *quantitative.* I have shown that quantity is extension i.e. a whole whose parts are external to each other, part outside part. This is why computers which process such quantitative items by a mechanical process are possible. If intelligence however is, as I have tried to show, non-extended then it is not bound by the limits of Gödel's theorem. In fact that theorem is simply a way of saying that human problem solving transcends a logic that can be mathematized and computerized.[83]

To summarize: human knowledge at the level of sensation and perception, since it is awareness, cannot be the operation of a non-living machine, but is the psychic operation of the human sense organs and brain and is dependent on their vital physiological activities. Human knowing at the level of intelligence and self-consciousness depends on sensation and perception and therefore on the sense organs and brain for its data, but it transcends any physical operation by the fact that it involves total self-reflection, a process which cannot be achieved by a material, and therefore extended organ. Thus it is an activity of a primary natural unit which is material but by reason of its own special type of unifying form is not limited to material processes.

The chief difficulty against this conclusion that human intelligence is the unifying and transcending form of the human body is that raised by Teilhard de Chardin.[84] How can we conceive of this transcendent form, which can rightly be called *spirit,* emerging from matter by an evolutionary process? Yet if we do not admit this origin of human spirituality from matter are we not falling once more into the Platonic dualism of a soul inserted into a body from outside the evolutionary process? It seems to me that this objection (which I will treat more fully in Chapter 11) fails to allow enough to creativity as characteristic of our universe. Once we admit that the universe is Being-in-process which is continually producing genuine novelty, we must also admit that at every level where novelty emerges the mode of "emergence" itself has novelty, i.e. it differs from previous types of "emergence."

Thus when life emerged from non-living matter, the way in which this took place was different from the way in which the molecular level of organization arose (and continues to arise) from the elemental level of organization, and similarly for the emergence of psychic animal life from plant life. Why then should we suppose that spiritual self-consciousness emerges in the *same way* as previous emergents? What is common to these previous critical emergences is that they all take place by physical processes, but why then should not spirit emerge in a novel way by a spiritual process, having an analogy to physical processes, but transcending them? The consideration of what such a spiritual process might be I will postpone until I have discussed in Chapter 11 the ultimate creative sources of the whole cosmos, but it would be contrary to the very notion of an evolutionary universe to exclude the possibility of a creativity whose mode is not physical but spiritual.

4. Bodily Beauty

The unity between transcendent human intelligence and the body to which it gives unity and existence is manifest to us in an esthetic way which should not be overlooked. All works of fine art refer ultimately to the human body.[85] Thus the plastic arts would not have meaning or emotional content if the shapes, volumes, colors, and textures of these objects, even in the case of abstract or non-objective art, were not referable to our perception of the human body which furnishes the fundamental vocabulary for the plastic arts. Similarly music is rooted in the rhythms of dance and the natural expressiveness of the human voice which reflects the muscular tensions of the human body. Both poetry and prose receive their tones and rhythms from the voice, while the imagery and symbols which form their content ultimately refer to the human body. Finally, it seems likely that our perception of beauty in the physical world arises from the fundamental sexual attraction of the healthy human body.

This does not mean that the beauty of the world is merely a subjective interpretation read into the world as a projection of human drives, but that we are awakened to the beauty of our world by seeing the same esthetic principles of symmetry, splendor, and proportion which we first discover in the human body and then find manifest also in animals, plants, and inorganic forms. Thus the beauty of sunlight, that splendor which everyone recognizes makes a fine day, is first of all known to us because it illumines and warms our human life, providing the daylight in which we see each other at the family meal.

The beauty of the body throughout the human species, unlike the situation in many animal species, seems to be primarily attributed to the female. The male body also has its beauty and some artists like Michelangelo have considered it more beautiful than the female, yet it would seem that men are more attracted by the appearance of the woman, while for women attractiveness is found more in the man's strength and tenderness.[86] The beauty of the body seems primarily to express sexual attractiveness, yet the sexuality expressed is not animal but human sexuality, a sexuality which itself manifests the spiritual relationship of self-conscious persons. In this self-consciousness there is inherent a certain modesty (not shame which implies guilt and defect), precisely because the self is revealed through the body to another self, and this intimacy cannot be something merely casual but has deep significance. This spirituality of the nude body is evident from the fact that we commonly make a distinction between the appearance of the woman (or the man for that matter) who invites sexual desire in a merely animal sense who is by no means always beautiful, from one whose beauty implies a more human, romantic, or truly intimate relationship.

Thus the beauty of the body, as Plato rightly said,[87] is a shadow of the beauty of the spirit, and when we discover in a particular case that this is an illusion, we are deeply disappointed. This expressiveness is most clearly seen when physical beauty is not merely in static appearance, but comes from the dynamism of the figure. In the great dancer, singer, or even the skillful musician using an instrument as the auxiliary of his own body, or in the great athlete, or in the face and hands of a great actor or actress, we see the human intelligence become manifest. In the dancer or athlete the whole body seems transformed into spirit, the limbs become intelligent.

In such epiphanies we seem to have a direct, almost intuitive understanding of the unity of spirit and matter, because we see matter in process which is orderly in the highest degree, a unified process of many processes in rhythm and harmony. Yet none of this would be possible without the guidance of the intelligence of the dancer, surpassing the instinctive movements of the graceful animal. This intelligence is not merely "objective," but it also expresses subjectivity, the self-consciousness of the human person in its freedom, love, striving, and suffering. How strange that Plato, so sensitive to beauty, should have thought of the body as a prison or a beautiful tomb of the spirit, or even as its shining but concealing garment! Rather the body is the expression, the voice, the glory of the spirit.

5. A Radical Process Philosophy

I have attempted to sketch a philosophy of radical process which is an interpretation of the scientific world-picture sketched in Chapter 2, a picture of matter and energy in development and evolution. I attempted to show that if we do not make the mistake of trying to reduce the reality of process to a mere reshuffling of unchanging particles, or the opposite mistake of evaporating the world into bodiless energy in the void, but strive to interpret it in terms of matter as the capacity to undergo process and of energy as the dynamic actualization of primary units of matter interacting with each other to produce new units out of the old, we can understand this scientific world-picture in a way that does not contradict our ordinary, common-sense view of the world as made up of minerals, plants, animals, and humans, but deepens and clarifies it.

I have also argued that this interpretation of the scientific data requires us to accept the universal fact of genuine novelty in change, so that the emergence of life from non-life, and of intelligent free life from animal life is congruent with the whole character of our universe in process, unexpected as this emergence might be. Finally, I have tried to show that in these terms we can come to a deeper understanding of ourselves as bodies that are also transcendentally spiritual. If it were not so we would have been unable to come to know the world and ourselves scientifically, but because it *is* so, we can give to these scientific discoveries an interpretation which agrees with our own inner sense of ourselves as bodies who can think scientifically, who recognize the novelty in the world, and who can ourselves contribute freely and creatively to that novelty.

Notes

1. Sir David Ross, *Aristotle*, (London: Methuen and Co., 1949), p. 22, traces this opinion to Trendelenburg. Ross (p. 21-25) modifies this opinion but does not connect the *Categories* with the *Physics*; but see Marjorie Grene, *A Portrait of Aristotle*, (Chicago: University of Chicago, 1963), pp. 20-80 and 117-121. See Hippocrates G. Apostle's translation with notes, *Aristotle's Categories and Propositions (De Interpretatione)*, Grinnell, Iowa (The Peripatetic Press, 1980). The categories are also listed in *Topics*, I, 9, 103b 20 sq.

2. See James Bogen, *Wittgenstein's Philosophy of Language*, (London: Routledge and Kegan Paul, 1972), p. 14-101, on the ontological implications of the *Tractatus*, which Wittgenstein ultimately abandoned. On Heidegger's conception of the role of language in philosophy see John D. Caputo, *Heidegger and Aquinas* (New York: Fordham University Press, 1982), p. 158-167.

3. See L. W. Beck, *Early German Philosophy: Kant and his Predecessors* (Cambridge, Mass.; Harvard, 1969) on Kant's background.

4. *Place (ubi,* Greek *tópos)* is defined by Aristotle as "The first inner motionless boundary of a containing body" *Physics* 212a 20-1, but *position (positio,* Greek *thésis)* is the relative orientation of body within its place, *Categories* 6b 2-14. Thus a cylinder rotated within a block changes its position but not its place. *Vestition (habitus,* Greek *échein)* is explained by Aristotle as the state of being clothed or naked — a very puzzling kind of entity. Later scholastics explained it as differing from place as a *loose* container differs from a tight container ("first inner boundary"). Thus it would apply not only to such an artificial thing as clothing but to many situations in nature e.g. the relation of peas to the pod which contains them, or of the "loose" electrons within a conductor. See John of St. Thomas (Jean Poinsot), *Cursus Philosophicus*, Reiser, ed., (Turin: Marietti, 1930), Vol. I, *Logica,* part 2, q. 18-19, pp. 630-638.

5. See Chapter 7, pp.

6. Aristotle, *Physics* IV, 1, 208a 25 sq.; see also Jean Poinsot, references in previous note 4; and Yves Simon, *The Great Dialogue of Nature and Space* (Albany, N.Y.: Magi Books, 1970), especially 113-128.

7. Aristotle, *Physics* IV, 10, 217b 29 sq. "Time (*quando*, Greek *chronos* is the number (measure) of motion with respect to before and after." Jean Poinsot, *Cursus* (note 4 above), and Simon, *The Great Dialogue* (note 6 above) pp. 129-138.

8. J. R. Lucas, *A Treatise on Time and Space* (New York: Barnes and Noble, 1973) gives an excellent analysis of the current problems connected with these topics.

9. See F. W. Went, "The Size of Man", *American Scientist* 56 (1968): 400-413 who argues that the size of the human body is essential to our development of technology which is possible only at a mid-point between micro- and macrocosmic physics. Went compares the possibilities of the human body to that of insects and shows how size is an essential factor.

10. See Daniel Klepner, M. G. Littman and M. L. Zimmerman, "Highly Excited Atoms", *Scientific American*, 244 (May, 1981) 130-149, which reports that the diameters of highly excited atoms can expand to 100,000 times their measurements at low energy states.

11. St. Thomas Aquinas, *In Libros Physicorum Aristotelis* (Naples, 1953, A. M. Pirotta, O.P. ed.), Bk 3, lect. 4, 585. See Vincent E. Smith, *The General Science of Nature* (Milwaukee: Bruce, 1958), Chapter 13, pp. 255-273 for a detailed exposition.

12. See my *Aristotle's Sluggish Earth: The Problematics of the De Caelo* (River Forest, Illinois: Albertus Magnus Lyceum, 1958), *(The New Scholasticism,* 32, April, 1958: 202-234), pp. 31-36.

13. See E. A. Burtt, *The Metaphysical Foundations of Modern Physical Science*, rev. ed. (Garden City, N.Y.: Doubleday Anchor Books, 1954) on the distinction between primary and secondary qualities in Galileo pp. 83-87, Kepler, pp. 67-68, Descartes pp. 115-121; Hobbes, pp. 130-132; Henry More, pp. 135-137; Cudworth, pp. 148-150; Boyle, pp. 180-184 (who is unclear on this subject); and Newton, pp. 235-237. Locke discusses it in the *Essay Concerning Human Understanding*, Bk. II, c. 8, 9-26 (pp. 104-111) in Everyman's Library, ed. J. W. Yolton (New York: Dutton, 1965).

14. "To sum up, physics tries to discover the pattern of events which controls the phenomena we observe. But we can never know what this pattern means or how it originates and even if some superior intelligence were to tell us, we would find the explanation unintelligible. Our studies can never put us into contact with reality and its true meaning and nature must be forever hidden from us" — Sir James Jeans, *Physics and Philosophy* (Cambridge: Cambridge University Press) 1944, p. 16. Against this attitude see Ernan MacMullan, "Realism in Modern Cosmology" in Alden L. Fischer and George B. Murray, S.J., eds. *Philosophy and Science as Modes of Knowing* (New York: Appleton-Century-Crofts, 1969), pp. 116-130.

15. This is not to assert that our senses are sensitive to all differences of quality. Thus our eyes detect only six colors in the spectrum and are blind to infra-red and ultra-violet light, although the spectrum is a continuous qualitative variation of color. So-called "color blind" persons are not able to discriminate even as many as six colors. This lack of ability to discriminate is not *positive* error (i.e. we do not see colors that are not there), but purely negative (i.e., we lump some colors that are really there together). Similarly the tone-deaf persons hear some differences between tones, but is insensitive to most of them. Many of the arguments of scepticism are based on a failure to distinguish between positive error and negative "error" which is not really error at all, but merely a lack of perfect knowledge.

16. Gödel's theorem (for references see note 81 below) eliminated the possibility of proving that a formal system as complex as arithmetic (such as Non-Euclidean and Euclidean geometry) can be proved to be self-consistent from within its own assumptions (axioms). In order to show that it is self-consistent, therefore, it must be reduced to another system which is known to be self-consistent. Thus we can be assured of the consistency (and, therefore, the truth) of the Non-Euclidean Geometries only by showing that they are true *if* Euclidean geometry is true (self-consistent). But to show Euclidean geometry is true we have to go outside the ideal mathematical realm and verify its truth by reference to empirical experience. If experience of the real world is inconsistent, then the principle of non-contradiction would be violated. On the ways in which constructive definitions are used in this reduction to the axioms of arithmetic and these to the natural integers which can be verified in empirical experience by counting, see Stephen F. Barker, *Philosophy of Mathematics*, (New York: Prentice-Hall) pp. 74-101.

335

17. See Norman K. Wessels, ed., *Vertebrate Structure and Function; Readings from Scientific American* (San Francisco: W. H. Freeman, 1974), for ample illustrations of these principles from contemporary evolutionary biology which explains them chiefly by adaptation.

18. See Jean Poinsot (John of St. Thomas), *Cursus* (note 4 above), Vol. 1 *Logica,* Part 2, q. 19, pp. 621-638.

19. More accurately, Ockham's position is stated as follows by Philotheus Boehner, O.F.M. ed. and translator, *Ockham: Philosophical Writings,* (New York: Nelson, 1957), Introduction pp. xlvii, "Every reality found in creatures can be reduced ultimately to two classes or categories: substances and qualities. Only these two categories signify or denote distinct entities; all the other categories mentioned by Aristotle are connotative terms, which denote either a substance or a quality and connote something else. Thus 'quantity' for instance is a term denoting either substance or a quality; what it denotes is that the substance or quality in question has parts distinct from each other. 'Relation' denotes two entities (substances or qualities), and connotes that one of these entities is being compared with the other, and so on. So too, for all other categories." Gordon Leff, *William of Ockham* (Manchester: Manchester University Press, 1975), pp. 213-237, discusses Ockham's view on real and mental relations at great length and concludes, "Relation for Ockham expresses a real relation between real things but not something real in its own right. It is not therefore a thing, nor is it merely a name: it is a sign for an actual way in which real substances and qualities are ordered in respect of one another."

20. On the Aristotelian notion of explanation see Melvin A. Glutz, C.P., *The Manner of Demonstrating in Natural Philosophy* (Aquinas Institute thesis: River Forest, IL., 1956) and William A. Wallace, O.P., *Causality and Scientific Explanation,* 2 vols. (Ann Arbor, Michigan: University of Michigan Press, 1972.)

21. Chapter 7, pp. 286-287. For a recent discussion of the problem of teleology see Ernst Nagel, "Teleology Revisited" in his *Teleology Revisited and Other Essays,* (New York, Columbia University Press, 1979) pp. 275-316.

22. "'Actual entities' — also termed 'actual occasions' — are the final real things of which the world is made up. There is no going behind actual entities to find anything more real. They differ among themselves: God is an actual entity, and so is the most trivial puff of existence in far-off empty space. But, though there are gradations of importance, and diversities of function, yet in the principles which actuality exemplifies, all are on the same level. The final facts are all alike, actual entities; and these actual entities are drops of experience, complex and interdependent." *Process and Reality* (corrected edition, New York: MacMillan-Free Press, 1978) 2, 1, p. 18. But, although Whitehead regards God as an actual occasion, he insists that God differs from all other occasions in that He endures forever, while all others on coming into existence immediately perish, to remain only in their causal influence on new occasions and in their "objective immortality" in the memory of God," *ibid.,* closing paragraph, p. 351. "In the actual world we discern four grades of actual occasions, grades which are not to be sharply distinguished from each other. First, and lowest, there are actual occasions in so-called 'empty-space'; secondly, there are the actual occasions which are moments in the life-histories of enduring non-living objects, such as electrons or other primitive organisms; thirdly, there are the actual occasions which are moments in the life-histories of enduring living things; fourthly, there are the actual occasions which are moments in the histories of enduring objects with conscious knowledge.", *ibid.* Pt. 2, c.8, sec. 3, p. 177. This conscious knowledge is characterized by Whitehead as follows, "The final percipient route of occasions is perhaps some thread of happenings wandering in 'empty' space amid the interstices of the brain . . . In its turn, this culmination of bodily life transmits itself throughout the avenues of the body. Its sole use to the body is its vivid originality; it is the organ of novelty.", *ibid.* Pt. 5, c.1, sec. 3, p. 39. Thus it is evident that, except for God, the actual occasions are sub-atomic (even smaller than the electrons which they constitute) and exist only for an instant, or quantum of time.

23. Gottfried Wilhelm von Leibniz, *Monadology and Other Philosophical Essays,* Library of Liberal Arts (Indianapolis: Bobbs-Merrill, 1965).

336

24. See unsigned article "Stereotyped Responses", *Encyclopedia Britannica* (1980) vol. 17. 671-676, and Robert M. Devilin, *Plant Physiology*, 2nd ed. (New York, Van Nostrand Reinhold, 1969), pp. 411-428.

25. "If God is all sufficient and lacks nothing, how does He come to release Himself into something so clearly unequal to Him? The divine Idea is just this self-release, the expulsion of this other out of itself, and the acceptance of it again, in order to constitute subjectivity and spirit. The philosophy of nature itself belongs to this pathway of return, for it is the philosophy of nature which overcomes the division of nature and spirit and renders the recognition of its essence in nature." *Hegel's Philosophy of Nature*, (Second Part of the *Encyclopedia of the Philosophical Sciences*) 3 vols. M. J. Petry, ed. and translator (London: George Allen and Unwin, 1970), 1, p. 205.

26. For current discussion of the nature of symbols, including the theories of Cassirer, Langer, Ricoeur, and Merleau-Ponty see William A. Van Roo, *Man the Symbolizer* (Rome: Gregorian University Press, 1981).

27. On the question of animal "language" see Thomas A. Sebeok, *Perspective in Zoosemiotics* (The Hague: Mouton, 1972).

28. See J. M. Diamond, "Zoological Classification System of a Primitive People", *Science*, 151 (3714), March 4, 1966, pp. 1102-04.

29. On Aristotle as a biologist see Marjorie Grene, *A Portrait of Aristotle* (Chicago: University of Chicago Press, 1963) pp. 122-174, and the bibliography in Jonathan Barnes, M. Scofield and R. Sorabji, eds. *Articles on Aristotle,* (London: Duckworth, 1975), 1, 203-205. On Albert the Great as a biologist see the essays in James A. Weisheipl, O.P., ed., *Albertus Magnus and the Sciences: Commemorative Essays* (Toronto: Pontifical Institute of Medieval Studies, 1980) pp. 321-479.

30. Thus Aquinas says "Since essential forms are not known to us *per se*, it is necessary that they should be manifested through some accidents, which are signs of such a form, as is evident in Metaphysics VIII". *In Libros Posteriorum Analyticorum* II, 13, n.7. In book VIII of the Metaphysics (Eta) Aristotle deals with the many difficulties of defining substances, attacking those philosophers, especially the Platonists, who supposed that this was a simple matter of direct intuition. See my article, "Does Science Attain Nature or Only the Phenomena", in Vincent E. Smith, ed., *The Philosophy of Physics* (Jamaica, N.Y.: St. John's University, 1961), pp. 63-82.

31. *Meteorologica* IV, 12, 389b 22 to 390b 21. See also "The Ascidians differ but slightly from plants, and yet have more of an animal nature than the sponges which are virtually plants and nothing more. For nature passes from lifeless objects to animals in such unbroken sequence, interposing between them beings which live and yet are not animals, that scarcely any difference seems to exist between two neighbouring groups owing to their close proximity". *De Partibus Animalium* IV, 5, 681a 10-15 (Oxford translation works edited by W. D. Ross, Oxford University Press, 1931).

32. An annotated bibliography of the vast literature is to be found in vol. 1 of A. Reza Arasteh, *Creativity in the Life Cycle* (Leiden: Brill, 1968). I have found the following works helpful: C. Spearman, *Creative Mind* (Cambridge: Cambridge University Press, 1930); Newton P. Stallknecht, *Studies in the Philosophy of Creation* (Princeton, N.J.: Princeton University Press, 1934); Jacques Maritain, *Creative Intuition in Art and Poetry* (New York: Pantheon Books, 1953); Nicholas Berdyaev, *The Meaning of the Creative Act* (New York: Harpers, 1954); W. I. B. Beveridge, *The Art of Scientific Investigation* (New York: Vintage Books, 3rd ed., 1957); Charles H. Clark, *Brainstorming* (Garden City, N.Y.: Doubleday, 1958); Nathaniel S. Lehrman, "Creativity, Consciousness, and Revelation," in *Diseases of the Nervous System*, 21, nos. 8-9 (Aug.-Sept.), 1960; William J. Gordon, *Synectics: The Development of Creative Capacity* (New York: Harper and Row, 1961); Jacob W. Getzels and Philip W. Jackson, *Creativity and Intelligence* (London-New York: Wiley, 1962); Myron A. Coler and Paul A. McGhee, eds., *Essays on Creativity in the Sciences* (New York: New York University Press, 1963); Harold Rugg, *Imagination* (New York: Harper and Row, 1963); Arthur Koestler, *The Act of Creation* (New York: Macmillan, 1964); (Ernest R. Hilgard's review of Koestler, "Creativity: The Juxtaposition and Integration of Disparate Categories" *Science*, 147

337

(Jan., 1965), 37-38; Hendrik M. Ruitenbeck, *The Creative Imagination: Psychoanalysis and the genius of inspiration* (Chicago: Quadrangle Books, 1965); George F. Kneller, *The Art and Science of Creativity* (New York: Holt, Rinehart and Winston, 1965); A. M. Taylor, *Imagination and the Growth of Science* (London: John Murray, 1966); Jost A. M. Meerloo, *Creativity and Eternization* (Assen: The Netherlands, 1967); Ray S. Wilbert, *The Experimental psychology of original thinking* (New York: Macmillan Co., 1967); H. Edward Tryk, "Assessment in the study of creativity," in Paul McReynolds, ed., *Advances in Psychological Assessment* (Palo Alto, California: Science and Behavior Books, 1968), 1, 34-54; Ruben A. Alves, *Tomorrow's Child: Imagination, Creativity, and the Rebirth of Culture* (New York: Harper and Row, 1972); Carl R. Hausman, *A Discourse on Novelty and Creation* (The Hague: Martinus Nijhoff, 1975); Silvano Arieti, *Creativity: The Magic Synthesis* (New York: Basic Books, 1976); Albert Rothenberg and Carl R. Hausman, eds., *The Creativity Question* (Durham, N. C., Duke University Press, 1976); Erich Neumann, *Art and the Creative Unconscious* (New York: Pantheon, 1959) and *Creative Man* (Princeton: Princeton University Press, 1979); Irving A. Taylor "A Retrospective View of Creativity Investigation", pp. 1-36 and J. P. Guilford, "Creativity: A Quarter Century of Progress" pp. 37-59 in Irving A. Taylor and J. W. Getzels, *Perspectives in Creativity* (Chicago: Aldine Press, 1979); Albert Rothberg, *The Emerging Goddess: The Creative Process in Art, Science and Other Fields* (Chicago: University of Chicago Press, 1979); and Herbert A. Simon, *Models of Discovery and Other Topics in the Methods of Science* (Dordrecht: D. Reidel, 1978), with the critical review by L. J. Cohen, *British Journal of the Philosophy of Science*, 30 (1979): 293-297.

33. On the psychological processes involved see Silvano Arieti, *The Intra-psychic Self: Creativity and its Cultivation* (New York, Basic Books, 1967).

34. "From the uncanny way in which the invention of the steamship, telephone, airplane, photography, automobile, and hundreds of other patentable devices have been subject to conflicting claims of priority by independent individuals and laboratories (cf. Kroeber, 1948), the conclusion seems inescapable that when the infrastructural conditions are ripe, the appropriate thoughts will occur, not once, but again and again". Marvin Harris, *Cultural Materialism: The Struggle for a Science of Culture* (New York: Random House, 1979), p. 59. The reference is to A. L. Kroeber *Anthropology* (New York: Harcourt Brace and World, 1948) pp. 341-343 and 352-367.

35. See Daniel E. Schneider, *The Psychoanalyst and the Artist* (New York: New America Library: Mentor Books, 1950) (Freudian), and Carl G. Jung, "Psychology and Literature" from *Modern Man in Search of a Soul* (London: Routledge and Kegan Paul, 1933), reprinted in Brewster Ghiselin, ed. *The Creative Process* (New York: New American Library: Mentor Books, 1952), pp. 208-223.

36. Ernst Kris, *Psychoanalytic Explorations in Art*, (New York: Schocken, 1964).

37. See my article, "A Psychological Model with a Spiritual Dimension," *Pastoral Psychology* 23: 224 (May, 1972): 31-41, and the works of Maritain (note 32 above) and Kris (note 36 above).

38. *Phaedrus* 245, and *Ion*.

39. *Poetics*, 1-3, 1447a 7 sq. See Gerald F. Else, *Aristotle's Poetics: The Arguments* (Cambridge, Mass.: Harvard University Press, 1957) and the edition of the *Poetics*, translated by Leon Golden and commentary by O. B. Hardison, Jr. (Tallahasse: University Presses of Florida, 1981) for discussion.

40. *Creative Intuition* (note 32 above).

41. See Chapter 7, pp. 272 ff. Since the validity of all reasoned arguments rests on premises, and this reduction cannot go on *ad infinitum*, we must have the capacity to know some ultimate premises immediately by an intuition which either rests on innate ideas (Plato) or on abstraction from experience (Aristotle). Modern philosophy seeks to escape this dilemma by attributing the certitude of reasoning not to the object but to the thinking subject (Kant), thus ending in idealism, or to contending that all human knowledge is merely probable (positivists), or that knowledge is impossible (sceptics). The positivistic theory, which generally prevails in scientific circles runs into the dilemma that if all knowledge is only probable then every premise rests on an infinite series of probabilities and thus

338

approaches zero probability. The Aristotelian option is preferable if understood moderately, as Aristotle did, to mean that we have certainty about only relatively few matters of which we have extensive experience on which we have reflected critically.

42. See St. Thomas Aquinas, *Summa Theologiae* I, q. 79, a. 9 and I-II; II *Sententia*, dist. 24, q. 2, a. 2; *De Veritate* q. 15, a. 2; See St. Augustine, *De Trinitate* XII, c. 4; PL 42, 1000; cf. Julien Peghaire, *Intellectus et ratio selon S. Thomas d'Aquin* (Ottawa: Institute d'études médiévales, 1936), pp. 31-38.

43. Jacques S. Hadamard, *An essay on the psychology of invention in the mathematical field* (Princeton, N.J., Princeton University Press, 1949); Mary Hesse, *Models and Analogies in Science*. Henri Poincare in an essay "Mathematical Creation" from his *Foundations of Science* (Notre Dame, Ind: University of Notre Dame Press, 1966); (in Ghiselin, note 35 above, 33-42) compares the "dance" of ideas in his mind to the dance of atoms in a gas and these to a swarm of gnats buzzing around his head.

44. *The Act of Creation* (note 32 above).

45. On the way the useful shades into the beautiful see John Dewey, *Art as Experience* (New York: Minton, Balch and Co., 1934).

46. Chapter 7, pp. 256-258.

47. *Process and Reality* I, 3, sec. 1, pp. 31-34 (note 22 above).

48. "Hence poetry is something more philosophic and of graver import than history, since its statements are of the nature rather of universals, whereas those of history are singular. By a universal statement I mean one as what such or such a kind of man will probably or necessarily say or do — which is the aim of poetry, though it affixes proper names to the characters; by a singular statement, one as to what, say, Alcibiades did or has done to him . . . And if he [the poet] should come to take a subject from actual history, he is none the less a poet for that; since some historic occurences may very well be in the probable and possible order of things; and it is in that aspect of them that he is their poet." *Poetics* 9, 1451b 5-11, 29-32 (Oxford translation).

49. See Robert S. Brumbaugh, *Plato's Mathematical Imagination*, (New York: Kraus Reprints, 1968; originally Indiana University, 1954).

50. See Joseph Schillinger, *The Mathematical Basis of the Arts*. (New York, Philosophical Library, 1945); Hermann Weyl, *Symmetry* (Princeton, N.J.: Princeton University Press, 1952); William M. Ivins Jr., *Art and Geometry* (New York: Dover, 1964 [1945]); and Gyorgy Kepes, ed., *Structure in Art and Science*, (New York, George Braziller, 1965) and *Module, Proportion, Symmetry, Rhythm*, (same publishers, 1966).

51. See M. Pénido, *Le rôle de l'analogie en théologie dogmatique* (Paris: Bibliotheque Thomiste, 1931); Hampus Lyttkens, *The Analogy between God and the World* (Uppsala: Almquist and Wiksells, 1952); George Klubertanz, S.J., *St. Thomas on Analogy* (Chicago: Loyola University Press, 1960); Ralph M. McInerny, *Studies in Analogy* (The Hague, Martin Nijhoff, 1960) and *The Logic of Analogy* (same publisher, 1961); Battista Mondin, *The Principle of Analogy in Protestant and Catholic Theology* (The Hague: Martinus Nijhoff, 1963); Eric L. Mascall, *He Who Is* (London: Longmans Green, 1964) pp. 95-112; and *Existence and Analogy*, (London: Darton, Longman and Todd, 1966); Henry Chavvanes, *L'analogie entre Dieu et le monde selon saint Thomas d'Aquin et selon Karl Barth* (Paris: Ed. du Cerf, 1969); Santiago Maria Ramírez, *De Analogia*, Ed. by Victorino Rodriquez (Madrid: Instituto de Filosofia "Luis Vives"; 4 vols.) (*Opera Omnia*, tom. 2), 1972; David B. Burrell, *Analogy and Philosophical Language* (New Haven: Yale University, 1973).

52. See Humphrey Palmer, *Analogy: A Study in Qualification and Argument in Theology* (New York: Martin's Press, 1973) and John Stephen Morreall, *Analogy and Talking About God: A Critique of the Thomistic Approach* (Washington, D.C.: University Press of America, 1978) for arguments against the validity of analogical reasoning, which, however, founder on the fact that without such reasoning human creativity would be impossible.

53. *Poetica*, 22, 1459a 3. "It is a great thing, indeed, to make a proper use of these poetical forms, as also of compounds and strange words. But the greatest thing by far is to be master of metaphor. It is the one thing that cannot be learnt from others; and it is also a sign of genius, since a good metaphor implies an intuitive perception of the similarity in dissimilars." (Oxford translation).

54. For a recent defense of analogical language see James F. Ross, *Portraying Analogy* (Cambridge: Cambridge University Press, 1981).

55. See Eve Curie, *Madame Curie: A Biography* (Garden City, N.Y.: Doubleday Doran, 1937) for an excellent account of the process of discovery in which sudden insight is combined with immense and prolonged toil, both mental and physical.

56. See Alan Richardson, *Mental Imagery* (New York: Springer, 1969) and Alasdair Hanney, *Mental Images: A Defense* (London: Allen and Unwin, 1971) on the role of images in thought.

57. This was the conclusion which Descartes' follower Geulinckx actually drew and which is called "occasionalism", but Descartes himself seems to have hesitated to go so far. See the discussion in Frederick Copleston, S.J., *A History of Philosophy*, (Garden City, N.J. Doubleday Image Books, 1963), vol. 4, pp. 129-132.

58. Plato, *Timaeus*, 70 and 74-76. Thomas Hobbes, *Elements of Philosophy, Works* (ed. by Sir William Molesworth) (London, 1839, reprinted by Scientia Verlag, Aalen, 1966), vol. 1, Part IV, c. 25, pp. 387-410. Although Hobbes attributes human thinking to the activity of the brain, he compromises with the Aristotelian view by saying that it acts "reciprocally" with the heart, which is "the fountain of all sense".

59. Much of this discussion was started by Gilbert Ryle, who in his *The Concept of Mind* (New York: Barnes and Noble, 1949) attempted to eliminate the notion of "consciousness". Some of the subsequent literature is as follows: Herbert Feigl and Willard Sellars, eds. *Readings in Philosophical Analysis* (New York: Appleton-Century Crofts, 1949) "Mind-Body Problem" pp. 351-458 including the essay of M. Schlick, "On the Relation between Psychological and Physical Concepts" pp. 383-407; same editors, with Keith Lehrer, *New Reading in Philosophical Analysis* (same publishers, 1972), pp. 371-460, especially Feigl's "Mind-Body, Not a Pseudo-problem, pp. 371-377; James W. Corman, "The Identity of Mind and Body" pp. 386-390, and Richard Rorty, "Mind-Body Identity, Privacy and Categories", pp. 391-407; Herbert Feigl, "Mind-Body Identity, Privacy and Categories", pp. 391-407; Herbert Feigl, "The Mind-Body Problem in the Development of Empiricism" in H. Feigl and May Brodbeck, eds., *Readings in the Philosophy of Science* (New York, Appleton-Crofts, 1953), pp. 612-626, and "The 'Mental' and the 'Physical' " in H. Feigl, Michael Scriven and Grover Maxwell, *Minnesota Studies in the Philosophy of Science* (Minneapolis: University of Minnesota Press) 2, pp. 370-497 with an important bibliography, pp. 483-497 published separately with a "Postscript" (Minneapolis: University of Minnesota Press, 1969); G.N.A. Vesy, *Body and Mind: Readings in Philosophy* (London: George Allen and Unwin, 1964) (many of the best papers up to that time); Jerome Schaffer, "Recent Work on the Mind-Body Problem", *American Philosophical Quarterly* 2 (1965), pp. 86-104. (Schaffer was critical of the theory); J. R. Smythies, ed. *Brain and Mind* (London: Routledge and Kegan Paul, 1965), (Smythies also sponsored criticism of the theory); Jaegwon Kim, "On the Psychophysical Identity Theory" *American Philosophical Quarterly*, 2 (1966) pp. 227-235 (critical); C. A. Peursen, *Body, Soul, Spirit: a Survey of the Body-Mind Problem* (London: Oxford University Press, 1966); Paul K. Feyerabend and Grover Maxwell, eds. *Mind, Matter and Method: Essays in honor of Herbert Feigl* (Minneapolis, University of Minnesota Press, 1966); Jordan M. Scher, ed. *Theories of Mind* (New York: Free Press, 1966); D. M. Armstrong, *A Materialist Theory of Mind* (London: Routledge, and Kegan Paul, 1968); Jerome A. Shaffer, *Philosophy of Mind* (Englewood Cliffs, N. J.: Prentice-Hall 1968) (critical); H. D. Lewis, *The Elusive Mind* (London: George Allen and Unwin, 1969) (critique of Ryle); David M. ed., *Materialism and the Mind-Body Problem* (Englewood Cliffs, N. J.: Prentice-Hall, 1971); C. H. Whiteley, *Mind in Action: An Essay in Philosophical Psychology* (New York: Oxford University Press, 1973); Fanny L. Epstein, "The Metaphysics of Mind-Body Identity Theories", *American Philosophical Quarterly* 10 (April, 1973): 11-121; Kendrick V. Walker, "Armstrong's Analysis of Self-Awareness", *Personalist* 57 (Autumn, 1976): 395-402; R. J. Nelson, "Mechanism, Functionalism, and the Identity Theory", *Journal of Philosophy* 73 (July 15, 1976): 365-385; Mark L. Conkling, "Ryle's Mistake About Consciousness", *Philosophy Today* 21 (Winter 1977): 376-388; Wallace I. Matson, *Sentience* (Berkeley: University of California Press, 1977); Roland

Puccetti, "Unraveling the World Knot: Scientists and Philosophers in the Mind-Brain Controversy", *British Journal of the Philosophy of Science*, 29 (1978), 61-68; Robert A. Jaeger, "Notes on the Logic of Physicalism", *Mind* 88 (351) (July, 1979): 424-429; Michael E. Levin, *Metaphysics and the Mind-Body Problem* (Oxford: Clarendon Press, 1979) ("Topic neutralism"); K. V. Wilkes, *Physicalism*, (Atlantic Highlands, N.J., Humanities Press, 1978); Edgar Wilson, *The Mental as Physical*, (London: Routledge and Kegan Paul, 1979). It is noteworthy how much of this debate is formulated in Cartesian terms, see Rebecca R. De Boer, "Cartesian Categories in Mind-Body Identity Theories", *Philosophical Forum*, N.S. (Fall, 1975), 139-158. The Aristotelian alternative is hardly considered (but see Edwin Hartmann, *Substance, Body and Soul* (Princeton, N.J., Princeton University Press, 1977), evidently because the matter-form analysis is supposed to be excluded by modern science. See also Jerry A. Fodor, *The Mind-Body Problem, Scientific American* 244 (Jan., 1981): 114-123, who proposes "Functionalism" as a solution, admitting that mind is a function of the brain but also might be a function of some entirely different structure.

60. "This is materialism: matter acting upon our sense-organs produces sensation. Sensation depends on the brain, nerves, retina, etc., i.e., on matter organized in a definite way. The existence of matter does not depend on sensation, Matter is primary. Sensation, thought, consciousness are the supreme product of matter organized in a particular way. Such are the views of materialism in general, and of Marx and Engels in particular." Vladimir Ilich Lenin, *Materialism and Empirio-criticism* (New York: International Press, 1970), Chapter 1, p. 48. It is noteworthy that this formulation as far as it goes is susceptible of an Aristotelian interpretation. Lenin seems to think that the only alternative to the idealism he is attacking is materialism.

61. Karl R. Popper and John C. Eccles, *The Self and Its Brain* (New York/Berlin: Springer International, 1977). In his Gifford Lectures, *The Human Mystery*, (New York-Heidelberg-Berlin: Springer International, 1979), pp. 94-95, Sir John speaks in a less dualistic and more Aristotelian way of the mind-body relation.

62. See Giovanni Blandino, S.J., *Theories on the Nature of Life* (New York: Philosophical Library, 1969), on various forms of panpsychism, pp. 211-221, with Blandino's criticism pp. 287-288.

63. See note 59 above.

64. See H. D. Lewis, *The Elusive Mind* (note 59 above).

65. See note 24 above.

66. See Michael A. Arbib, *Brains, Machines and Mathematics* (New York: McGraw-Hill, 1964); Kenneth M. Sayre, *Recognition: A Study in the Philosophy of Artificial Intelligence* (Notre Dame, Indiana: University of Notre Dame, 1965) and *Cybernetics and the Philosophy of Mind* (London: Routledge and Kegan Paul, 1976) and Leon Brillouin, *Science and Information Theory*, 2nd ed. (New York: Academic Press, 1962).

67. Ryle (see note 59 above) attempted to show that what we are conscious of is not some interior state but our dealings with external objects. This is essentially an Aristotelian position and refutes the Cartesian *cogito* (which was Ryle's chief aim); but Ryle fails to deal adequately with the fact that we not only know the object but we know that we know the object. It is this reflex knowledge that is "consciousness" in the strong sense of "self-consciousness" which cannot be reduced to statements about physical objects. On the relation of consciousness to brain state see the articles in G. G. Globus, G. Maxwell and I. Savodnik, eds. *Consciousness and the Brain* (New York: Plenum Press, 1976).

68. Modern phenomenology derived the term "intentionality" from scholasticism through Clemens Brentano, but it is used in different senses by Aquinas, Husserl, and Heidegger. See John N. Deely, *The Tradition via Heidegger* (The Hague: Martinus Nijhoff, 1971), pp. 62-110, for a detailed discussion of the difference between Husserl and Aquinas in this usage.

69. B. F. Skinner, *Beyond Freedom and Dignity* (New York: Knopf, 1971) and *About Behaviorism* (New York: Random House, 1974) with the severe criticism of Noam Chomsky, "The Case Against B. F. Skinner," *The New York Review of Books*, 17 (Dec. 30, 1971) 18-24.

70. Nor is it true that when I or you report what we have experienced introspectively and what we experience by external sensation that the former always is "private", while the latter is "public". In both cases the experience becomes public only when we report our experiences to each other and conclude that they must refer to numerically or at least specifically the same object. Thus we can agree that giraffes exist because we have seen them and that people get hungry because we have both felt hungry. It is true of course that each of us can have private experiences, both introspective and of external sensation, which others cannot verify. On the notion of "inner life" see Louis Dupré, *Transcendent Selfhood:* The Loss and Rediscovery of the Inner Life (New York: Seabury, 1976).

71. On the problem of the "phantom limb" phenomena, the amputee's impression that an absent limb is still painful see I. P. Howard, "Orientation and Motion in Space" in E. C. Carterette and Morton P. Friedman, *Handbook of Perception* (New York: Academic Press, 1974), 3, 291-315, "The Body Schema", p. 295 and references.

72. See Thomas S. Szasz, *Pain and Pleasure: A Study of Bodily Feelings*, rev. ed. (New York: Basic Books, 1975); also Frederik J. J. Buytendijk, *Pain: Its Modes and Functions* (Chicago: University of Chicago Press, 1962).

73. See Matthew Luckiesh, *Visual Illusions* (New York: Dover, 1965); Peter From, *Science, Art and Visual Illusion* (New York: Simon and Schuster, 1971); and Richard L. Gregory and E. H. Gombrich, eds. *Illusion in Nature and Art* (New York: Scribner's, 1973). Also Paul A. Kolers, "Illusion of Movement", *Scientific American*, 233 (Oct., 1964), 98-106 and Diana Deutsch, "Musical Illusions", *ibid.*, 244 (Oct., 1975) 92-104.

74. See references Chapter 2, note 4.

75. On tool using by animals see Chapter 2, note 3.

76. It seems odd to us that this was ever doubted by the ancient physicians but see Aristotle, *De Partibus Animalium*; Edwin Clarke and C. D. O'Malley, *The Human Brain and Spinal Cord* (Berkeley: University of California Press, 1968). Bk. 2, c. 7, 652a 24 sq. and Bk. 3, c. 4, 665a 28 sq. Aristotle's difficulty was that he believed the organ of sensation, since it had the highest coordinating function in the animal organism, must also have the highest energy-state (we would say "rate of metabolism"). This he identified mistakenly (based on his still primitive chemistry) with the organ which was *hottest*, which was obviously the heart, not the brain. Modern data reverses the argument.

77. See James Gibson, *Locke's Theory of Knowledge and Its Historical Relations* (Cambridge: Cambridge University Press, 1964), pp. 52-70 and 321-326. Locke did admit Ideas of Reflection by which the mind knows its own acts, but his General Ideas are only collections of ideas derived directly from sense impressions.

78. See Bernard J. Muller-Thym, "The 'To Be' which Signifies the Truth of Propositions", *Proceedings of the Sixteenth Annual Meeting of the American Catholic Philosophical Association*, 16 (1940); Peter Honen, S.J., *Reality and Judgment According to St. Thomas Aquinas* (Chicago: Regnery, 1952) and Frederick D. Wilhelmsen, *Man's Knowledge of Reality*, (Englewood Cliffs, N.J., Prentice Hall, 1956), pp. 122-184.

79. Aristotle does not assume that every entity subject to change is a body i.e., extended or quantitative, but demonstrates this to be so: *Physics* VI, c. 4-5, 227b 3 sq.

80. A.M. Türing, "Computing Machinery and Intelligence" in E. A. Feigenbaum and J. Feldman, eds. *Computers and Thought*, (McGraw-Hill, N.Y., 1963). Cf. Discussion in Mortimer J. Adler, *The Difference of Man and the Difference it Makes*, (Holt, Rhinehart, Winston, N.Y., 1967), p. 244 ff., and for further references, the bibliography noted on pp. 351-354. Also James H. Moor "An Analysis of the Türing Test," *Philosophical Studies* (USA, University of Arizona) 30 (Oct., 1976), 249 ff., and J. Bronowski, "Logic of Mind" in *Science in Progress*, 16th Series (New Haven: Yale University, 1967) pp. 217-237., which also discusses the relation of Türing's argument with those of Gödel, Church, and Tarski.

81. Ernest Nagel and James R. Newman, *Gödel's Proof*, (New York: New York University Press, 1958). Gödel's theses were that (1) for any formal logic L, satisfying certain conditions (i.e., adequate at least to apply to simple arithmetic), there are undecidable propositions in L (i.e., propositions F such that neither F nor not-F is probable); (2) for any suitable L the simply consistency of L cannot be proved in L.

82. For collections of essays on the question of whether Gödel's theory is relevant to the mind-body problem see Sidney Hook, ed. Dimensions of Mind (New York: New York University Press, 1960); J.J.C. Smart, *Philosophy and Scientific Realism* (London: Routledge and Kegan Paul, 1963); Kenneth M. Sayre and Frederick J. Crosson, ed., *The Modelling of Mind: Computers and Intelligence*. J. P. Lucas has carried on the most thorough defense of the affirmative position with which I agree; see his "Minds, Machines and Gödel", *Philosophy* 26, 1961, pp. 112-127 with replies by David Lewis, "Lucas against Mechanism", *ibid*. 44, 1969, pp. 231-233 and David Coder, "Gödel's Theorem and Mechanisms", pp. 234-236. Also I. J. Good "Human and Machine Logic", *British Journal of the Philosophy of Science*, 18, 1967, pp. 145-6; P. Benaceraff, "God, the Devil, and Gödel", *Monist*, 51, 1, pp. 9-33 and Judson Webb, "Metamathematics and the philosophy of mind", *Philosophy of Science*, 35 (June, 1968), pp. 156-178. Webb argues that the force of Gödel's researches is only to show the limitations of any logical argument so that it is wrong to use his theorem as the basis of metaphysical conclusions. Lucas replied to some criticisms in "Human and Machine Logic", *British Journal of the Philosophy of Science*, 19, 1968, pp. 156. Later Anthony Hutton returned to the attack in "This Gödel is Killing Me," *Philosophia* 6 (March, 1976): 135-144, and Lucas answered *ibid*. pp. 145-148. Other articles and books of interest on the subject are Michael Scriven, "The Compleat Robot: a Prologomena to Androidology", in Hook, pp. 118-142; and "The Mechanical Concept of Mind" in Sayre and Crosson, pp. 243-254; Donald McKay, "Mind-like Behaviour in Artifacts", *ibid*. 225-242; Hilary Putnam, "Minds and Machines" in Hook, pp. 149-179, J.J.C. Smart, "The Argument against Gödel in Smart," pp. 116-118 and "Gödel's Theorem, Church's Theorem and Mechanism", *Synthese*, 13, 1961, pp. 105-10; and F. H. George, *The Brain as Computer* (London: Pergamon Press, 1961) and *Cognition* (London, Methuen, 1962), p. 209.

83. Herbert Simon, "Is Thinking Uniquely Human?" *University of Chicago Magazine*, Fall, 1981, pp. 12-21, has recently argued "We give people tasks; on the basis of performance in a task we consider that some thought has taken place in reaching a solution to a problem. Similarly, we can give computers the same task; then, it would seem to me, that it is only some kind of vulgar prejudice if we refuse the accolade of intelligence to the computer" (p. 14). For Simon, intelligence is "pattern manipulation" involving (1) means-ends analysis and (2) pattern recognition as in chess. He states as an "hypothesis" "that a necessary and sufficient condition for a system to exhibit intelligence is that it be a symbol system, that it have symbol manipulating capabilities" (p. 19). Since for many tasks that humans perform we have been able to devise computers to perform the same, and since theoretically we know of no reasons why we eventually may not be able to do the same for any task we can exactly formulate, it seems probable that computers are intelligent. This misses the point of the Gödel argument (as in my opinion do all the various attempts to refute it) that human intelligence can invent a machine to do anything mechanically which we can do mentally, except that intelligence can always pose a new problem which the old machine cannot solve, but which our intelligence can invent a new machine to solve. The problem-solving capacity of a *given* machine is only a reflection of a certain level of human thought which we can always transcend. This is not to claim that human intelligence has no limits, but only that it is unlimited in inventing the kinds of problems that can be reduced to computers.

84. See Robert North, S.J., *Teilhard and the Creation of the Soul* (Milwaukee: Bruce, 1966) for a spirited defense of Teilhard on this issue. According to North the original state of matter must have contained the active potentiality to develop itself into the present universe just as a fertilized ovum does to develop into a mature animal. "The Alpha Point must have exhibited a millionfold more intensely that 'complexity latent in simplicity' which every ovum exhibits. Moreover, this Alpha Point must, like Omega, possess some special identifiability with Christ or God." (p. 116). I showed in Chapter 7 why this conception of evolution is not a very plausible interpretation of the present scientific account of evolution.

85. See my article, "Significance in Non-Objective Art", *Proceedings of the American Catholic Philosophical Association,* 1965, pp. 156-165.

86. See "Characteristics of Male and Female Responses" in *Sexuality and Man,* SIECUS (Sex Information and Education Council of the United States) (New York: Scribner's, 1970) pp. 24-36.

343

87. *Symposium* 209 sq. Note that although Plato in this dialogue has repeatedly referred to homoeroticism, in this central part of the dialogue he speaks of sexual love as essentially related to reproduction.

Chapter 9

Historicity and the Human Body

I: Two Conceptions Of Human Historicity

1. Humanism and Historicity

A radical process philosophy has given us an interpretation of the modern scientific world-picture, in which the uniform laws of nature as Galileo and Newton conceived them appear secondary to the historical evolution of the universe out of which humanity in its intelligent freedom has emerged. Nature is infused with creativity, with the constant birth of the new and surprising, not merely the majestic repetition of orderly cycles. This creativity becomes self-conscious historically in humanity and supremely in the scientific intelligence, which has been able to uncover its own evolutionary history and seems now able to take control of it. But this human self-understanding also makes clear to us that our creative freedom can only be realized by using the determinisms of the laws of nature which have their own teleologies which we must respect. We ourselves because we are bodily beings, part of nature, are subject to the laws of natural process and its temporal rhythms. This is why we are through and through historical beings, exercising our creative freedom within history and helping to shape it.

Therefore, to complete this radical process philosophy it is necessary to examine more carefully what it means to say that we humans are historical beings, how this relates to our bodily existence, and what significance this all has for the original question raised in Chapter 1: "Can we create ourselves?" To pursue this more detailed examination of human historicity in this chapter, I will deal with three questions.

First, I will ask how Humanism and Christianity differ on the nature of history and its relation to nature. Second, I will focus on the principal dilemma which both theologies have to face, namely, does human historicity imply that human values are purely man-made, entirely relative to a given culture, or does it mean that through its historical experience humanity is coming to understand itself as a universal community sharing common goals? Third, I will turn to the Biblical revelation to ask whether it provides us with any clues to solve this dilemma. This third section of the chapter constitutes a transition from the primarily philosophical questions with which this Part Three of my book has been concerned, to the purely theological questions with which I will be dealing in Part Four.

To initiate the comparison of Humanist and Christian views on the nature of history and its relation to human values, it is convenient first to list what seem to me to be the chief trends of Humanist thinking on this topic, some of which have been discussed earlier but require summary here. First, it seems a basic conviction of most Humanists that we have arrived at that point of history when we can and ought to remake our world and ourselves. Our human world, precisely because we are human, has always been a human product, but we have produced it without knowing what we were doing, the way that a primitive artist works. Now we must do so knowingly, deliberately. The notion of "human nature" as something given, universal, and permanent must be abandoned. Even the notion of the noumenal self must be replaced by the view that each individual is created by his or her personal experience and education within an historically developed culture. All ethical norms are relative to a cultural epoch.[1] Finally truth, even scientific truth, must be evaluated as relative to a time and a culture.

Second, Humanists seem now to be freeing themselves from what Karl Popper has criticized as "historicism", by which he means the fallacy of trying to make history into a predictive science.[2] In Chapter 7 I showed that modern science itself no longer claims to predict the future of natural events. All the more is it impossible to predict the course of history even in the broad, dialectical way that Marxists suppose. Consequently, to study ourselves under the aspect of historical existence, we must be content as was Hegel to reflect on what has already occurred. Today, however,

346

Humanists no longer even have Hegel's confidence that history is rational when viewed *a parte post*. The question this raises, of course, is how Humanism can provide sufficient motivation for its disciples so as to function as an equivalent to religion without the eschatological assurance of human progress.

Third, in this latter part of the twentieth century, Humanists tend more and more to understand history in terms of *tragedy*. This does not mean that the ideals and the hopes for social reform have been erased from the Humanist creed, but that Humanists frankly admit that their hopes rest on questionable evidence. There is nothing in the world-picture provided by science in its self-interpretation, nor in the lessons of history, as Humanists now write history, that can guarantee that the control of nature provided by technology or that the social organization provided by human autonomy will lead to human freedom rather than end in nuclear globicide or a thousand years of tyranny. The very fact that Marxism, the chief rival of Humanism, seems to inculcate a greater faith in the future than the latter can inspire only makes Humanists the more sceptical, since the future which Marxism seems to be actually building appears to be just such a tyranny.[3]

Fourth, this scepticism about its own hopes does not prove to Humanists (nor do I draw this conclusion) that Humanism is in decline as a world religion. Humanists are convinced that this sceptical realism simply marks its coming to maturity. In the twentieth century the world-wide influence of Humanism has become ever more dominant with the spread of scientific technological culture. Christianity and the other older world religions seem to be in retreat before it everywhere.[4] In my own opinion Humanism may now possibly be renewed and strengthened by this greater realism, especially if Humanists find ways to heal Humanism's inner contradiction between Scientism and Romanticism, as they may do by developing a more warm-hearted view of what it is to be human while retaining a hard-headed scientific objectivity about human limitations.

For both Humanism and Marxism the contrast between their hopes and the actual history of the twentieth century has raised the hard question of how these world-views can explain the *evil* in the world, evil that neither progress nor revolution seems to overcome. In their earlier phases, both religions confidently explained evil simply as the burden of the past, or of reactionary resistance to enlightenment and reform. The evils of human life, whether caused by ignorance or powerlessness, it was claimed, would be inevitably overcome by the advance of science and technology. Once this confidence in progress has been abandoned, we are all faced with the existential question: Why is the world, of which we are so much a part, so evil? The agnosticism or atheism of these religions puts the full burden of responsibility on our own shoulders, or simply is forced to concede (contrary to the Enlightenment theory that gave it birth), that the world is absurd.

2. Christianity and Historicity

In co-opting the notion of human historicity, the Humanism of the Enlightenment was actually renewing an element of Christianity which the Middle Ages with its Neo-Platonic other-worldliness had tended to neglect. The religion of the Bible, as everyone recognizes, is profoundly historical in its conception of what it is to be human.[5] Generally speaking, the ancient religions other than Judaism accepted the view that the universe is an eternal emanation from God or the Absolute which undergoes an eternal cycle, the "eternal return", but the Bible (with the exception of certain isolated passages and the somewhat hellenized book of Qoheleth[6]), clearly proclaims that time is linear, beginning with the free creation of the world *ex nihilo* and moving toward the final *eschaton* when the Reign of God will be perfectly realized. The Biblical view of history rests on faith in God's creation and providence over the world; while for Humanists it is not God's providence that overcomes evil, but only human effort.

It was the singular achievement of St. Augustine, neo-Platonic as he was, to give full emphasis to this biblical historicity in the context of a theology whose philosophical categories were generally acceptable to the Greeks.[7] This aspect of Augustinianism, however, was not much stressed in medieval theology except in the heretical movement stemming from the prophecies of Joachim of Flora predicting the coming age of the Holy Spirit.[8] St. Thomas Aquinas, arguing that theology is a science, raised to himself the obvious objection that according to Aristotle a science can only be about universals, not singulars, while theology is a commentary on the Bible which is in large part a narrative dealing with singular persons and events. Aquinas was forced to give to his own question a reply which for him was unusually evasive:

> Singulars are treated in sacred doctrine not because its principal task is to deal with them, but they are introduced either as examples for living, as in the moral sciences, or to establish the authority of the men through whom divine revelation has been transmitted to us, on which authority Sacred Scripture or doctrine rests.[9]

Yet in principle the philosophy of Aristotle was in some respects more open to history than that of Plato. It is true that Aristotle did not believe that a science (in his sense of the term) of history was possible. Moreover, he believed in a steady-state universe whose celestial spheres rotated in a perpetual cycle with the result that on the earth (where change is ultimately due to energy from the sun and other celestial bodies), there could be no evolution of species. Nevertheless, he insisted that these cycles only repeat themselves in a universal way, the individuals never are twice the same;[10] and this cyclical repetition of the general state of things is not due to the nature of terrestrial realities themselves, but to the governance of the heavenly bodies. What closed Aristotle off from allowing for any real evolution in the world was not his analysis of Being-in-process, but his notion that the heavens were radically different from earthly entities, a conviction based partly on the Greek religious veneration of the celestial bodies and ignorance of the notion of creation from nothing, and partly on the claim of the Babylonian astronomers that their observations proved that the heavens never change.[11] Aquinas explicitly recognized that these Greek astronomical systems were only hypothetical, but he saw no empirical reasons to discard them, and was not inclined to use the Bible to refute the accepted views of scientists.[12]

The rise of critical history in the sixteenth and seventeenth century was partly the result of the interest of the Renaissance humanists in classical literature and antiquities, but it was powerfully motivated by the religious controversies of the Reformation and Counter-Reformation, culminating in the attack on the historical credibility of the Bible by the Jewish rationalist Spinoza (d.1677) and the Huguenot apologist Pierre Bayle (d.1706). The wide-spread scepticism of this period, of which Bayle is the typical figure, was not itself Humanist, but it prepared the way for Humanism by its desperate fideism which sought to drive people to blind faith by discrediting every philosophical and historical argument for the reliability of the tradition of the Christian Church.[13] This critical, and often hyper-critical history, played a large part in dismantling the medieval world-view; but Christian theologians soon came to see that it also had positive value for a better understanding of the Church's own foundations. An important factor in this widening and deepening historical understanding was the discovery and exploration of the non-European world which was going on during this same time, and which revealed to Europeans the variety of world cultures which were equal to or in some respects superior to that of Christian Europe, thus raising the problem of cultural relativism.

In the eighteenth century, as we have seen, Christianity although under attack by Humanism received some help from the romantic wing of Humanism, especially as humanist historians abandoned the anti-medieval prejudices of the early Enlightenment for a more tolerant cultural relativism. The Catholic Vico (d.1744) and the Protestants Herder (d.1803), Schelling (d.1854) and Hegel (d.1831) developed the truly novel notion of a "philosophy of history", which would have been unthinkable to Aristotle or Aquinas, in which all history becomes a manifestation of God to human reason while Christian revelation, is seen only as a culminating phase of this rationally accessible revelation.[14] The Bible, although it had largely lost its reputation as reliable history, was thus able to regain standing as a religious interpretation of history, to be understood not literally but poetically.

In Chapter 6 we saw that this Christian attempt to reconcile faith with critical history in the nineteenth century led to the development of Liberal Protestantism, and in Catholicism led first to Traditionalism and then to Modernism, with the result that the essentially historical character of the Christian faith was seriously imperiled. That this crisis is still by no means completely resolved is evident from the so-called "Quest for the Historical Jesus," the existentialist "demythologizing" of Rudolf Bultmann with his distinction between the inaccessible "Jesus of history" and "the Christ of faith" and the controversies over the recent Christological efforts of Hans Küng and Edward Schillebeeckx.

Nevertheless, beginning with the theory of the "development of doctrine" of John Henry Newman (d.1890), the critical biblical studies of M. J. Lagrange, O.P. (d.1935), and the theologians of Vatican II such as Karl Rahner, S.J., and Yves Congar, O.P., Catholic theology has thoroughly accepted the view that if Christian theology is to be consistent with the historical character of biblical revelation and the development of doctrine in Christian tradition it must be not only philosophical, as was scholastic theology, but also historical in its method. Although the Orthodox Churches as yet seem to be only on the verge of this development, and the Protestant Churches have been deeply divided by it into liberals and conservatives, it can hardly be doubted that it is this method which is enabling the progress of ecumenism and the reunion of the churches who have been divided, not so much by formal heresy as by failures in communication resulting from historico-cultural factors.[15]

The central problem which faces modern Christianity is not the truth of its historical claims which have withstood the test of modern critical methods, whose own reliability is, after all, not free from question; but the same problem which we have seen confronts Humanism and Marxism: How do you explain in your theology the continued existence of *evil* in the world, evil so vividly manifested in history and in particular in the history of your own religion, with its claims to save the world from evil? Humanism arose from the disillusionment with Christianity produced by the historical tragedy of the wars of religion. Marxism arose from the disillusionment with Humanism produced by the historical tragedy of class warfare and imperialist wars. To be credible today, Christian theology must deal honestly with these historical scandals.[16]

For Christians, evil can have its origin only in the free will of created persons, and comes about because these creatures prefer an absolute, god-like autonomy in preference to entering into that community of love which centers in God, or, better said, is included in the community of the Trinity. This rebellion against God has taken place on a cosmic scale in the fall of superhuman intelligences, who cannot be reduced to mere myths, but help to explain the cosmic scope of evil which cannot be attributed to God and which seems to exceed human influence.[17] "Our battle is not against human forces, but against the principalities and powers, the rulers of this world of darkness, the evil spirits in regions above" (Eph 6:12). In human history evil is rooted in original sin, which modern theology conceives not merely as a single act at the beginning of human history but also as the accumulation of evils produced by sins committed throughout the ages in complicity with the first sin of the human community, which has distorted our cultural world, our earthly environment, and even our individual persons from the moment of our conception as bodies living spiritually in a state of alienation. We are "thrown into" a world, as Heidegger would say, which was created with an orientation toward God, but which is now an obstacle to our finding Him again until it is itself redeemed.

Yet the Good News of the Christian Gospel is that from the beginning of human history, from the very moment when our race turned from God, God in his mercy out of the same wisdom and love by which he created us has begun the work of redeeming the world in view of his determination to restore it in the New Man, who is also his divine Son become a member of our race. To prepare the coming of this New Man in whom the Reign of the Father will be restored throughout the cosmos, the Father has sent his Spirit who is everywhere at work in human history among every people and in every time, reaching to every human person in order to turn us back to him in faith, hope and love and to make this possible by healing the wounds of our sins and restoring us to friendship with God, by renewing in us that share in the Divine life which we call "grace" and with which God originally endowed us beyond the hopes of our own creaturehood.

Human history, therefore, in this Christian view, is to be interpreted as a drama or warfare, both visible and hidden, in which the spirit of God is in constant conflict with the spirits of evil, whether these be the unredeemable cosmic spirits or the redeemable human ones, redeemable precisely because our mortal human bodily ways of knowing and willing

351

are so limited and variable that our rebellion against God is not definitive before death. This means that human institutions remain ambiguous. On the one hand they are unjust and oppressive inventions of human sinners, sometimes inspired by evil cosmic intelligences worshipped by us as false gods. On the other they are efforts of persons endowed by God with creative intelligence and inspired by his Holy Spirit to carry on his ministries of healing the distortions of the world wrought by sin.

The Christian Church, as Vatican II has made clear,[18] has the responsibility to acknowledge the work of God's Spirit in all that is good in human works, of science, art, technology, philosophy, scholarship, social organization and government, even in the other world religions that seem to be its rivals. Consequently, human history can be understood as a genuine *progress*, not only in the Church but throughout the world. Yet this progress is confused by the constant opposition of evil forces which strive to co-opt whatever is good for evil purposes and to deceive even those with the best intentions, so that even the work of the Christian Church is distorted and divided and thus often bears the appearance of the very tyranny and inhumanity from which it was sent to liberate all the forces of good in the world. It is only the hidden Spirit which perserves the Church essentially faithful to its witness to God's will to redeem the whole world and to restore it as his own Reign in his incarnate and resurrected Son, Jesus Christ.

From this perspective, therefore, history is seen with a fundamental optimism as to the eventual outcome of history, but with an evenly balanced optimism and pessimism about the course of history. To interpret historical movements rightly and realistically, "to read the signs of the times" (Mt 16:3) is a theological task, but a task which cannot escape the dangers of demonic deception which puts the good at war with the good (as in the divisions within the Church) or the blindness that led the Pharisees to attribute the work of the Spirit to Beelzebul (Mk 3:20-30).

This "reading of the signs of the times", however, implies one very fundamental difference between a Christian reading of history and what Humanism understands by "critical history", a difference which becomes very important in biblical hermeneutics. Humanism has no faith in Divine Providence; history in no way favors humanity except to the extent that we learn to control history and to shape it to our purposes. Today few humanists are so optimistic as to believe human control over history will ever become so complete as to avoid the ultimate destruction of the human species by the inevitable tide of entropy. Consequently, history has no meaning except the meaning we have given it to the degree that we control it. Ultimately it has no meaning at all. This implies that critical history attempts to establish the facts of the past in their sheer facticity,

352

and then to explain them causally but non-teleologically, except to the degree they manifest purely human purpose.

For the Christian, however, history does indeed have an element of the absurd as a result on the one hand of the existence of real chance in the universe[19] and on the other as a result of human folly; but all of history is under the control of God's providence by which even chance and sin are used by him for the ultimate good of persons. It is not the best of all possible worlds as Leibnitz argued, thus provoking Voltaire's scandalized mockery,[20] but it is in its ultimate outcome *good* with a total goodness that justifies its partial evils, just as an excellent symphony in its totality justifies its transitory discords. This implies, therefore, that history is a *revelation* of God's purposes, a *communication* or dialogue between God and his creatures in which they can raise questions and to which He ultimately gives answers.

It seems to me that this is the reason that the historico-critical method of reading the Scriptures (which Humanism fostered to destroy faith in the Bible but which Christian exegetes with great profit now employ in its defense) cannot be the complete hermeneutical method, as some exegetes seem to think. Undoubtedly this method is useful and necessary today in order to reconstruct the sources and stages of composition of the literary text, to place it in its historical context, and to determine its relation to other historical documents, as well as its literary form.[21] Without this help an exegete is in danger of serious errors of interpretation. Yet when all this work has been done, the true sense of the biblical text is not yet attained. The formalist and textualist literary critics have made us aware that a text must ultimately be understood in its own terms and not in terms of the history of its composition.[22] In the case of the Bible this also means that individual books must also be interpreted as parts of the whole canon, since for the Christian it is the Bible as a whole which is inspired and its parts have their inspiration as parts of that whole whose principal author is God.[23] Finally, as Aquinas pointed out,[24] the Bible is unique in that it relates the history of the world as that history itself is, as it were, a cosmic book written by Divine Providence in which the actual events speak to us.

The language in which these events speak to us cannot be interpreted in a merely literal way but must also be read *symbolically*. For example, the Exodus event of the liberation of Israel from Egypt is full of meaning, but a meaning so polyvalent that efforts to spell it out are always inadequate. We can return to it again and again for meditation and discover new meanings in it. The whole Bible keeps reinterpreting this event as its historical significance comes more and more to light in the subsequent events of history which it has not only influenced, but to the meaning

353

of which it gives a clue in a kind of hermeneutic circle. Who would have known in advance that the Exodus would find its explanation in Christian baptism, and at the same time it would help to explain Christian baptism?

This is why patristic and medieval exegesis were so given to typology and allegory. Modern exegesis, after scorning this "Alexandrian method" in preference for the "Antiochene" literalism has come to realize that the former had a real affinity to the thought-forms of the world in which the biblical texts were composed.[25] Yet it must not be thought that the recognition of this symbolical character of biblical texts in general and of biblical history in particular means that the historical narrations of the Bible can simply be reduced to religious fiction having no necessary connection with "real history". The Jewish and Christian religions are so essentially historical that if their fundamental symbols are not derived from real historical events, their claims on reasonable human faith are invalid. These religions may make occasional use of myths but they are essentially anti-mythical.[26]

In reading what the Bible presents as history, therefore, it is necessary first to recognize that some books and some parts of the Bible (e.g. the *Book of Jonah* or *Judith*) may be historical fiction, but not its most fundamental historical narratives such as the Exodus, the Exile, the Gospels or *Acts of the Apostles*. The manner in which these genuine historical narratives are told may admit of a wide range of literary forms from accurate eye-witness accounts to popular and traditional histories transmitted in largely legendary form, yet the author's religious purpose is essentially bound up with the historic reality of those related events which are *revelatory* of God's saving purpose. The problem of exegesis is to uncover this historical foundation, not simply for the sake of gaining historical information, but in order to understand the religious message in its historic truth.

This means that the criteria by which a critical historian judges the historical reliability of an account are not identical with those of a Christian exegete, although the exegete must respect the validity of the critical method as far as it goes. What is peculiar to Christian exegesis is the conviction that because God is the author of the text and God is Lord of history, there must be a meaningful *coherence* between the text and the events of history. As in reading a defective manuscript, we use our general grasp of the meaning of the text as a whole as a criterion for judging the merits of a variant reading or for filling in a lacuna, so if we believe that God is communicating to us through the events of history and we are faced with the problem of whether the Bible correctly reports an event, or if it gives seemingly inconsistent accounts, we may be able to judge the truth of the matter by the context of the event in the whole of salvation history. For example we cannot critically verify the historicity of Abraham's call by God, yet the Christian historian will not doubt that behind the evidentially tenuous account which we have lies an historical event of great importance.

This means that the Christian in reading the Bible, after listening with respect to the results of historical criticism, may still have a rational conviction that certain events related in the Scriptures are in substance real history although they cannot be verified by critical methods and (although not contradictory to known fact) may even on the basis of the available evidence seem improbable. The point I am making is not that the Christian simply believes these on faith although they cannot be scientifically established, but that because of his faith in Divine Providence the Christian judges the scientific evidence differently than would the critical historian relying only on his own methodology. The critical historian judges history as would a textual critic who had no notion of the general import of the text or even any conviction that it made good sense, while Chrisians judge the evidence in terms of their faith that the text makes good sense consistent with the general plan of God whose outlines they already know by faith.

Obviously this Christian judgment on historical evidence is a very risky business. In the Middle Ages it led to that fondness for the legendary, out of which grew up the mass of hagiography and the pseudo-histories such as that of King Arthur in which all sorts of historical data are embedded in a tissue of fictions and allegories. We have only to think of the legend of the Cross to see how far the symbolic imagination could go in inventing history.[27] We have a good reason, therefore, to be grateful for the development of a critical historical approach to the Bible to which Humanism, for its own purposes, has so much contributed. Nevertheless, the critical historical method has its own limitations. It is deaf and blind to the meaning of history as revelatory of Providence, and hence it lacks an important set of criteria for determining what has really happened in history.

Thus Christians and Humanists not only interpret the facts of history differently, but they may come to different judgments about the evidence on which these "facts" are asserted, yet both may and should employ a critical and objective methodology.

II: Cultural Relativism

1. Nature, Nurture and History as a Science

For both the Humanist and the Christian view of human historicity there exists the problem of how nature and culture are related. The division within Humanistic thought between the sciences of nature and culture has led to the result that natural sciences seek to explain human behavior in biological terms. This division has culminated in the current theory of sociobiology which sees human conduct as traceable in great detail to events of genetic selection which took place millions of years ago, while the social and cultural sciences tend to deny genetic foundation even for such apparently "natural" features of human life as the difference of sex roles which they attribute to purely historico-cultural factors.[28] Therefore, if we believe the sociobiologists, our present life is severely limited in its creative options by archaic features of human nature; while if we believe the cultural scientists, our future is wide open to almost indefinite modification of traditional patterns of behavior. As I indicated in Chapter 1, some would like to resolve this contradiction by a genetic recreation of human nature to make it corresond to cultural changes.

Certainly what characterizes and specifies human nature is its intelligent freedom, its self-conscious participation in creativity, as I argued in Chapters 7 and 8. Hence human culture as a human creation predominates in explaining human behavior over the more or less fixed, genetic aspects of human nature. Nevertheless, it also cannot be denied that, because human beings are creatures who did not produce themselves but were produced by the processes of evolution, there are aspects of our human nature which remain unknown to us, which constitute the human *mystery* and which we cannot, at least as yet, control because we do not yet understand them.

Because we are not disembodied minds we are not transparent to ourselves, but come to know our psychological powers through the structures and functions of our bodies. This self-discovery takes a long time and laborious research that can never exhaust the human mystery. To understand ourselves is possible only through understanding our universe and our evolutionary history. Consequently, to treat human nature as a cultural artifact susceptible of indefinite modification may lead to our self-destruction. Christians believe that human nature is a masterpiece of God and therefore sacred. If Humanists are to treat anything as sacred it has to be scientific truth.[29] Scientific truth, however, is dependent upon human intelligence, and that is dependent in turn on a reverence for human nature which cannot be lightly tampered with without risking its destruction and the destruction of the natural environment through which human nature has evolved and to which it has been adapted.

Some would say that the nature-nurture controversy is a dead-end because human nature exists only in on-going historical cultures, and cultures are always conditioned by what is genetically given as human nature.[30] That is very true, yet it leaves us with the problem of determining how nature limits cultural change and deciding what cultural changes are responsible social experimentation and which are not. The rule which has to guide us in such experiments is still that human nature must be sacred to both Humanism and Christianity alike (although for somewhat different reasons), and hence must be treated with reverence, with *conservation* if not conservatism. Otherwise, our crude re-making of ourselves based on very little knowledge and very great ignorance of our own mystery may in the end destroy the very creativity by which we hope to improve ourselves. Human creativity thus is not *creatio ex nihilo* in the strict sense. Rather it is a reverent cooperation with the creativity already manifest in nature. The Christian knows that Adam was placed in the garden "to cultivate and guard it" (Gn 2:15). The Humanists knows that our brain and our earth are the only ones we have.

This conclusion, namely, that culture rests on nature and cannot replace it, is reinforced by the fact that history, on which our understanding of culture depends, cannot be a science in the same sense as natural science nor equal to it in its objectivity and certitude. In the final analysis, Aristotle was not mistaken in saying that "science is of the universal," and thus concluding that history is not science in the strict sense.[31] History relies on a principle of uniformity, i.e. it assumes that what happened in the past is the result of the same fundamental factors that are at work today and subject to our observation. But these factors are precisely what is meant by human nature and the natural laws of our physical environment.[32] To the extent that historians explain events by such universal factors, geographical, biological or psychological, they are reducing historical explanation to the natural sciences, not providing specifically historical principles of explanation. I have already argued that the laws of nature can only explain what is universal, regular and uniform, not those unique, novel events in which the course of history and of culture, as well as the course of natural evolution, differ from that static nature described by natural law and the principle of uniformity.[33]

Wilhelm Dilthey (d.1911) put forward the view that history and the cultural sciences (*Geisteswissenschaften*) can claim to be true sciences because they have a proper method which differs from the "objectivising" method of the natural sciences (*Naturwissenschaften*). This non-objectivising method is possible because we can enter empathetically

357

into the minds of other persons, including those of the past. The natural sciences deal with objects, the historical and cultural sciences with subjects as such.[34] I have already referred to Karl Popper's refutation of this claim (which he calls "historicism" using that term somewhat eccentrically, since it is commonly used to mean the reduction of all human knowledge to history as in the philosophy of R. G. Collingwood).[35] As Popper says, we can only enter into the minds of others through their behavior, so that the social sciences also require the objectivizing method of the natural sciences.[36] Thus history, insofar as it is scientific, rests on the more certain natural sciences.

Yet even if we cannot consider history a science in the same sense as the natural sciences, Hegel was right in defending its genuinely philosophical character as against the Aristotelian tradition which saw it merely as a conveniently arranged collection of facts.[37] Of course history is scientific in the sense that it uses a critical method in collecting and verifying its data; but this too reduces to a reliance on the principle of uniformity and consequently misses the unique and novel. Hume argued that miracles can never be historically verified because such unique events would always have to be judged less probable than the probability of mistake or deception, since the latter are more common in ordinary experience.[38] If this rule were applied consistently it would level historical narrative to a repetitious chronicle in which all unique acts of heroism, scientific discovery, invention, and artistic creation would be passed over as legends. The measure of history would become the world-view of the historian and we would be deprived of the greatest contribution which history can make to our understanding of ourselves, namely, that human nature is something much deeper and wider than its overt manifestation in ourselves. To understand human nature in its mystery we must communicate through history and the cultural disciplines with the total human community spread out in time and space in which alone the rich spectrum of human possibilities can be revealed.

If Dilthey was wrong in claiming that history is a science in any strict use of the term, he was certainly right, as against Hume, that the method proper to history and the cultural sciences is *hermeneutic*,[39] that is, history seeks to translate the self-understanding of other people in another age and culture into the language of our own contemporary self-understanding without losing its own uniqueness. This possibility of translation from one language to another would not be possible unless we were all human, members not only of one biological species but potentially at least members of a single global community. Consequently history would be impossible unless there was such a thing as human nature. The famous dictum of Jean Paul Sartre, "Man has no nature, but only a history"[40] should be revised to read, "Because man has a nature, he can have a history."

Once we have granted that history is not science because it does not rest on universal law, but on the communication of experience within the community grounded in a common human nature, we must also acknowledge that history rests on faith in the testimony of others. Not indeed that we should trust everyone, but the critical, scientific aspect of history does no more than assure us of the credibility of documents and ultimately of human witnesses. Once the credibility of witnesses has been established, we are obliged to believe their testimony even when they testify to unique events which we cannot test by the principle of uniformity. I must believe that it was Shakespeare who produced those unique plays on the word of his contemporaries. There will never again be a Shakespeare nor will anyone else duplicate his plays.

Consequently, all history demands of us that we have faith in the testimony of human beings of other times and cultures than our own. The men of the Enlightenment doubted the whole history of the Jews, of the early Christian Church, and of the Middle Ages, but generally they trusted the histories of classical times.[41] Why? Because they considered Jews and Christians to be superstitious barbarians and the Greeks and Romans men of reason. Similarly, we have often been unwilling to believe traditional accounts of primitive people on the grounds that they could not distinguish between reality and fantasy. Today both Humanists and Christians recognize that no culture can claim to be "enlightened" or consign other peoples and times to darkness. As far back as real history extends and in all existing human cultures with which we have been able to communicate in depth, human beings have had the same basic intellectual abilities, some measure of discernment of the real from the unreal, and some degree of freedom and creativity.[42] In every time and culture there are liars and deluded fools, but there are also honest and trustworthy men and women whose testimony should be believed by those who are themselves honest enough not to dismiss such testimony merely because it differs from their own limited experience.

Such openness to history in its full human breadth is not contradictory to the claim that some periods and places have been of special significance for the intellectual, political, and religious development of human culture in its totality. If history is not uniform but open to novelty, then it will include crises, turning points, origins of great movements, "world-historical figures", "classical" eras as well as stagnant backwaters and places and times where "nothing much happened." It would only be another misuse of the principle of uniformity to evaluate every historical event as of equal importance.

2. Morality and Cultural Relativism

The view that Christian morality, although based on the revealed Gospel, must also conform to "the natural moral law" accessible to human reason by reflection on our universal experience of our common human nature was generally accepted by the Church Fathers and the medieval scholastics, but today is widely doubted by theologians. After the great explorations of the New World, Africa, and Asia in the sixteenth century, this notion of natural law took on special significance as Europeans encountered non-Europeans and discovered they were moral beings. This led to that development (discussed in Chapter 5) of international law and inalienable human rights by theologians such as Francis of Vitoria, Francis Suarez, Robert Bellarmine and Hugo Grotius, which still survives in the United Nations *Universal Declaration of Human Rights* of 1948.

With the Reformation, however, Protestant theologians began to attempt to ground ethics exclusively on Biblical revelation. Although both Luther and Calvin admitted the existence of a natural law, they believed that the corruption of human nature by the Fall had so obscured it as to make it of little value for the construction of a Christian ethics.[43] On the other hand, the Enlightenment in its earliest phase greatly favored the notion of natural law as a substitute for the revealed biblical law; and liberal Protestantism retained from biblical morality only what could be easily rationalized. For Humanism, natural law seemed to establish universal human rights independent of any confessional commitment, and thus to provide a bulwark against the tyranny of priests and kings. Just as many Protestants claimed that the Bible was clear to all without any need of the Church to interpret it, and thus were able to use it as a revolutionary weapon against ecclesiastical resistance to reform; so the early Humanists maintained the natural law to be "self-evident" (in Thomas Jefferson's classic phrase), and thus could employ it against the traditional aristocracy.[44] In our century the notion of natural law has been revived against legal positivists and cultural relativists by Thomists such as Jacques Maritain (d.1973) and freed of the rationalistic notion of self-evidence given to it by Enlightenment thinkers. It still receives strong support in the documents of Vatican II and recent papal documents on questions both of private and public morality.[45]

Nevertheless Catholic theologians are today much less confident about the role of natural law than formerly, for the following reasons: (1) the desire to insure that Christian ethics is firmly rooted in biblical revelation, rather than in human philosophy; (2) the influence of existentialist philosophies, particularly that of Heidegger, which emphasize human historicity as against a fixed human nature; (3) doubts raised by anthropology and the cultural sciences about the existence of cultural universals; (4) disillusionment with the apparent inadequacy of natural law theories to deal with the realities of current moral dilemmas, e.g. the contraception controversy, the abortion controversy, and the problems of just war and of private property. A curious illustration of this last difficulty is the way in which some theologians deplore papal teaching on sexual morality because it is based on natural law arguments rather than on the Gospel, while at the same time applauding papal teaching on social matters because it is a defense of the inalienable rights established by the natural law independent of any religion![46]

In Protestant circles natural law had been given some support by liberal Protestantism but the neo-orthodoxy of Karl Barth (d. 1968) and his followers, in reaction to liberalism, rejected even the minimal support given to it by Luther and Calvin and treated it simply as another man-made idol that could only distort the authentic Word of God.[47] The decline of neo-orthodoxy has made some Protestant authors more sympathetic to the natural law tradition,[48] but generally it plays little part in the writings of Protestant moralists for much the same reasons as those given above for recent Catholic authors. Although at one time Catholic authors tended to use natural law arguments as a bridge between Catholic and Protestant positions, since this approach seemed to by-pass confessional differences about the Law-Gospel problem, such a strategy no longer seems to be very helpful.

The result of these difficulties is that some Catholic moralists, especially in Northern European faculties where the Kantian ethical tradition as well as the devotion to biblical and historical scholarship has always made theologians uncomfortable with the "objectivizing" method and the "metaphysical," non-historical orientation of Thomistic natural law theory, are searching for some other base for ethics, a search made practically urgent by the painful controversy over Paul VI's encyclical on contraception, *Humanae Vitae* (1968). Bernard Häring, C.SS.R., in a number of influential works, has tried to show how the legalism of the deontological system of ethics (of which the founder of his own religious congregation, St. Alphonsus Ligouri (d. 1787) was the representative most approved by the Church), might be softened by greater emphasis on the primacy of the Commandment of Love, an emphasis already present in the teleological ethics of Aquinas.[49] This effort of Häring to save both the deontologism of the Counter-Reformation moral theology and the teleologism of Thomism has seemed to many moralists to be very attractive but too eclectic to be convincing.

One theologian who paved the way for some new approaches (but who oddly enough proved to be a strong defender of *Humanae Vitae*) was Dietrich von Hildebrand (d.1977), who during the neo-Thomistic hegemony had championed Augustine as against Aquinas. In fact, however, Hildebrand's thought was based philosophically on the phenomenology of his teacher Edmund Husserl (d.1938), and especially on the most important application of that phenomenology to ethics by another pupil of Husserl's, Max Scheler (d.1928).[50] Recently in the wake of the *Humanae Vitae* controversy, this ethics of Scheler has again been taken up by Josef Fuchs, S.J., of the Gregorian University in Rome and his pupil Bruno Schüller, S.J., and has been influentially expounded in the United States by Richard A. McCormick, S.J.[51] In the writings of these Jesuit moralists it is evident that like Häring, they too are reacting to the Counter-Reformation deontological moral theology in the development of which the Society of Jesus played so important a role. This fact is important for an understanding of these developments, because in discussions of the issues involved it is often taken for granted that what is in question is the natural law theory of Aquinas, when in fact what is going on is really an effort to revise a voluntaristic deontological tradition stemming from the Augustinian tradition of late medieval theology in the line of Scotus, Ockham, and Suarez, rather than a revision of the quite different intellectualist teleological system of Aquinas. No wonder the debate is somewhat confused![52]

Aquinas himself may have contributed to this confusion by his usual harmonizing effort to assimilate a traditional deontological terminology to the Aristotelian teleological ethic which he had very deliberately preferred. In the Old Testament there is a sound basis for a teleological approach to ethics in the Wisdom Literature. Even the term Torah is better translated "instruction" than "law" and it is interpreted by the prophets as a way of life more than as a set of rules. Jesus is in the line of this prophetic understanding of the Torah in his protest against the scribes and Pharisees who had adopted a rigorously legal hermeneutic in which the "instruction" had become a detailed code of laws. In the Septuagint Greek, *torah* became *nomos* and in the Latin Vulgate *lex*, thus taking on the overtones of the Roman legal tradition which was itself influenced by Stoic philosophy.[53]

For the Roman Stoics, "law" was *logos, ratio*, "reason" but this was understood much as we today understand the term "natural law" in physics, i.e. it was the strictly determined order of the world which manifests itself in animals as *instinct* and in human beings as *moral law*. This moral law was the "natural" human way of behavior common to men and animals antecedent to human civil legislation which the Stoics believed was sometimes invalid because it contradicted these more basic human instincts. This conception of natural moral law as *instinctive* was in agreement with Aristotle's philosophy (from which it was derived) as regards sub-human nature, but in radical disagreement with his ethics, because human reason differs from the blind *ratio* of nature in that it is conscious and free. The Stoics who were materialists and determinists erased this distinction.[54] Hence, for them the moral law was "natural law" in the sense that it was simply a manifestation of the universal determined order of the cosmos against which it was vain for men to struggle.

Obviously such a "law of nature" is not the "law" of the Bible which consists in the commands of a personal God given to his free human creatures as a guide to their own free action, just as a father by light of his superior wisdom helps his children avoid mistakes and take the right path, but leaves the children as they grow in maturity to apply that guidance in the light of their own reason and experience. Aquinas, while retaining the biblical and Stoic legal terminology, sought to make use of Aristotle's more nuanced anthropology and teleological systematization of ethics.

Often Aquinas' treatise on law (*Summa Theologiae* I-II, q.90-108) is referred to as if it were his treatment of the fundamental principles of ethics. In fact this treatment of the nature of morality is found in the 89 preceding questions. In q.90 he tells us that after considering the *intrinsic* principles of morality, he is now turning to the question of how God helps us to fulfill our moral obligations. God's help is provided in two ways: first God gives us guidance both through the natural order known to our reason and through special revelation, and second by His grace which empowers us to understand this guidance and to apply it effectively in our lives. The treatise on law is an explanation of this guidance, and is then followed by the treatise on grace (q.109-114). It should be noted, moreover, that in q.108 Aquinas explains that while in the Old Testament humanity was in need of an external law to guide it because of the darkening of the human reason by sin, in the New Testament this exterior law is replaced by an interior law which is the indwelling Holy Spirit who guides the Christian by grace. This does not mean that the Old Law is no longer of any service, but it does mean that its role is secondary.

What then is Aquinas' account of the intrinsic nature of morality (I-II, q.1-89)?[55] Aquinas, following Aristotle, develops a strictly teleological ethics according to which true morality consists in a free commitment to live "according to reason." Reason shows us that we have a human nature common to all members of the human family which has certain innate needs, including our need to live in society and to worship God. Therefore, we must first of all commit ourselves to the ultimate goal (*telos*) of human life which consists in satisfying these basic needs and helping others to do the same for themselves in a consistent and unified manner; and second, we must choose each of our free acts in life in such a way that they do not divert ourselves or others from this goal, but lead us toward it. To act in this way is moral because it is in harmony with the wisdom of the Creator who formed our human nature with these basic needs, including our supreme need for union with him and through him with other persons, and thus is cooperative with his love for us all. On the other hand, to commit ourselves in life to seeking some particular good in such a way that this commitment prevents our commitment to our true end, or to choose to perform acts which frustrate ourselves or others in achieving this true end is immoral, and is said to be "sin" because it is a rejection of God's love for us and contempt of His wisdom.

In such a teleological ethics the ultimate norm of morality is the practical judgment (conscience) of the human agent. If the agent acts contrary to his or her best judgment, then the act is *subjectively* immoral; if in conformity with this personal judgment then the act is moral. We cannot, however, act according to our consciences unless we first try to inform our consciences *objectively*, since it would be contrary to human reason to attempt to achieve a goal without finding out whether a given action will really (objectively) lead to that goal or away from it. In moral matters it is often not easy to establish the objective morality of a given action, since we always act in concrete situations which have unique features. To acquire the ability to make such judgments effectively we need the virtue of *prudence* which is acquired by experience, but which can also be aided by *ethics* as systematic reflection on general human experience. When we act with a subjectively good conscience our act is morally good, but if we are mistaken about its objective morality we and others will still suffer the harmful consequences of that objectively wrong act.

How is this objective morality determined by ethics and prudence? It is determined by considering the alternative means to our true end and choosing the one which will most probably lead to that end. Obviously this judgment will be successful only to the degree that we really understand the basic needs of human nature and their integration as the unified goal of human life, and also to the degree that we have correctly understood the alternative means and their probable consequences. For Aquinas this demands that we consider two aspects of the act:[56] first we must consider the act in itself in relation to the true goal of life (this he calls the *moral object*). Some suggested means will prove on examination to be intrinsically contradictory to this goal because they are contradictory to one of the basic human needs from which that goal is integrated (e.g. suicide deprives the agent of his own life, an essential need of human nature without which the goal of human life is unattainable). Such acts which are *intrinsically immoral (malum per se)* can never be ethically approved or prudent even if they have some good consequences for the agent or for others, because their first effect (antecedent to any further consequences) is an injury to the agent himself or herself since it constitutes an act against reason and against his or her commitment to the true goal of human life.

On the other hand, if the act in question is not intrinsically evil, it still remains necessary to determine whether in the concrete circumstances in which this act is to be performed, it will serve to achieve the true goal of human life or hinder this achievement. Only in the former case is it ethically and prudentially justified. The role of law, whether it be the revealed law of God, the natural law as an ethics of the type just described, or the positive laws passed by legitimate human authorities, is to guide the human conscience in arriving at sound judgments of objective morality; but the authority of all valid laws is itself grounded in the objective relations of means to end, as these are in turn rooted in human nature as God created it or as He has elevated it by grace to a more intimate share in His divine life than it was destined to have by its creaturehood.

The fact that Aquinas stresses that human acts have an intrinsic morality prior to their consequences and apart from their circumstances does not mean, as some have mistakenly supposed,[57] that Aquinas is really a deontologist for whom the consequences of an act are morally irrelevant. For him the intrinsic morality of an act is always relative to the end toward which it is directed. An act is moral only if it is suited by its intrinsic character to bring the actor to his or her true goal;[58] immoral if by its intrinsic character it impedes this movement toward the true goal. The goal to which we are striving, however, is not merely the modification of our external environment (which is what is ordinarily meant by the "consequences" of an action). Such productions are the task of technology which is in the service of moral purposes but not determinative of them. Our goal as human is not just to make things but to become and to be fully human persons. Morally good actions make persons good and good persons usually act morally.

Moreover, our human goal is not just to become good individuals, to be self-fulfilled. It is to become good social beings, to enter into a community of love and mutual self-giving. Hence the morality of any human act does not depend merely on its external consequences but first of all and intrinsically on the way it accords with the nature of the agent as individual and as a member of society acting in common, on how it promotes growth in personal integrity and maturity in community.

Hence the ethics of Aquinas, although purely teleological, is not merely an ethics of act, but rather one of *virtue*.[59] By virtue he means a developed capacity to act freely and intelligently in accordance with all the inherent purposes of human nature in such a way as to realize harmoniously all these teleologies inherent in the human body-person. Such a capacity for moral action is contrary both to a rigid conformity to external law and to a blind submission to the instincts common to man and animal. It is a disciplined, rational capacity to find ways, even novel ways, to reconcile all the different and sometimes apparently conflicting needs of human nature with the opportunities of a given environment and situation.

The Christian character of this type of teleological ethics arises from the fact that the Scriptures reveal to us that from the beginning God willed to elevate humanity to this more intimate relation with Himself. Human sin, therefore, has not only set us on a course of action which will inevitably lead us away from the true goal of life to which our human nature tends, but first of all it is a rejection of God's invitation to a more intimate relation to Him than our human nature can claim. This relation could never have been restored by mere finite human power. Consequently, it is only in Christ who is God made man that it is now possible for human persons to set out once again on that road by which they will not only become once more good human beings, but also children of God in that higher sense of sharers in His own divine life.[60] Christian ethics, therefore, *includes* natural law ethics, because "grace perfects nature"[61], but it modifies it in the same way that actions taken to achieve an end may be modified if that end is itself viewed as a means to some further end.

To the foregoing exposition of Aquinas' ethics it is important to add that in his own use of the term "natural law", he by no means takes the view, as did the philosophers of the Enlightenment who used this term, that this "natural law" is clearly evident to human reason. I have already showed[62] that Aristotle and Aquinas believed that the nature of primary units is never known exhaustively and is always open to further research. This holds as well for the nature of human beings. Therefore, it is possible that in some times and cultures the knowledge of natural law is very rudimentary or distorted, especially as a result of human sin. That is why God's revelation of the Law was necessary to revive a true understanding of the principles of the natural law, although the progress of human science also contributes to a better understanding and application of these principles.[63]

I have sketched the chief features of Aquinas' ethical thought as I understand them, to make intelligible the revision of this system which is now being proposed under the name of Proportionalism by the followers of Max Scheler and Josef Fuchs. Max Scheler, even in his Catholic period, was not a Thomist, but a phenomenologist,[64] but he addressed himself to the question which has also preoccupied Thomists, namely, "How can we find a foundation of ethics which is trans-cultural?" His solution was to distinguish between human *values* which can be isolated in their pure universal and transcultural essences by the phenomenological method from their concrete realization in particular circumstances. Josef Fuchs, S.J., in an important article in 1971 proposed that this notion of value might be used to revise the traditional Principle of Double Effect which had been so widely used by moralists (especially by deontologists)[65] to solve problems which involve a conflict of obligations. According to that principle, when an act entails both good and bad consequences it is necessary to meet certain criteria before it can be judged ethical. The first of these criteria is that the act not be *intrinsically* immoral, and the second is that the good consequences be *proportionate* (greater or equal to the bad consequences).

Fuchs, without rejecting the principle, suggested that this first condition was unnecessary since in every case what makes an act *morally* evil is not the act considered in the abstract but whether in the concrete circumstances in which it is performed there is a proportion between the values and the disvalues which that act embodies. He argued that such a method is more objective than the traditional method since it takes into account *all* the values and disvalues involved, and he drew the conclusion (clearly Schelerian) that although there are "absolute" values expressed in general norms, in the concrete application of these norms there may always be (at least in theory) room for *exceptions*. This system has become widely influential and in practice means that its proponents defend the commonly accepted norms of Catholic morality, but contend that in difficult cases exception should be made to these norms when a careful weighing of the values and disvalues warrant it.

The critics of this theory[66] have raised many objections against it, to which its defenders have attempted replies. In particular the defenders deny that they believe that it is ever permissible to perform an evil (immoral act) to accomplish a good end. They adopt the distinction put forth by Louis Janssens[67] between an "ontic" or pre-moral evil and a moral evil. For example, *homicide* is a negative *ontic* value, which can become either a negative *moral* value (murder) or a positive *moral* value (capital punishment, or self-defense, or abortion to save the life of the mother) in view of the circumstances and intention of the agent. In judging the morality

367

of an action, the ontic values and disvalues to be weighed are not in themselves either moral or immoral. Morality enters only in the *proportion* of values to disvalues. Consequently, an act in which the values exceed the disvalues is by definition never an immoral or evil means even if it is contrary to some generally valid norm.[68]

The problem which remains, however, is how this proportion is to be estimated. The critics of Proportionalism argue that in practice it comes down to utilitarianism, and this is what accounts for the popularity of the theory in our pragmatic culture. All Catholic moralists, including the proportionalists, repudiate utilitarianism because in its *act* version (as exemplified by the Situationism of Joseph Fletcher)[69] it repudiates moral norms altogether (except perhaps some single norm such as "act lovingly" or "the greatest good of the greatest number"); while in its *rule* version (accepted by most utilitarians) it reduces all moral values to a quantitative calculus of good and bad consequences (Consequentialism), a calculus that seems inapplicable to the spiritual values given highest rank by the Gospel.

Proportionalists generally retort that the traditional Principle of Double Effect also requires that we estimate the proportion of ontic good and bad in an act. To this John Connery, S.J. has replied that in the Principle of Double Effect this proportion is only a secondary determinant of morality. Hence a rough estimate suffices to apply the principle in practice. For Proportionalism, however, this proportion is the primary and only determinant. Hence it requires an accurate and exhaustive calculation practically impossible to fulfill; thus making room for utilitarianism.[70]

My own reason for rejecting Proportionalism as a valid and practical system of moral evaluation is that it appears to me contradictory. It claims first to establish the proportion of "ontic" or pre-moral values, and then if the values outweigh the disvalues, to judge the act to be morally good because of this proportion, even if this requires making an exception to a generally valid moral norm. Proportionalists admit that if these values were not ontic but moral, then no exception could be made to the norm, because this would be doing evil to achieve the good. Yet if these values are not evaluated morally, i.e. in relation to the true end of human life in which relation, according to any teleological ethics, the essence of morality consists, then how can they be evaluated in any way relevant to ethics?[71]

I believe that this contradiction reveals that in fact Fuchs' theory is not really teleological at all because it does not rest on the principle that the essence of morality is in the relation between means and ends. What is its real basis? Max Scheler wrote his most famous book against the *formalism* of Kant's deontological ethics; but what he objected to was not its deontologism but its formalism.[72] In order to escape the emptiness of this formalism, Scheler thought he could give positive content to ethics by introducing an *intuitive* vision of absolute moral values attained by the phenomenological method. These intuited values are "pure values", that is, freed of every reference to the concrete human condition, a fact which betrays the essentially Idealist character of Scheler's system. It is really, it seems to me, on this phenomenological intuition of values that Fuchs is relying to determine their proportion. I wonder if his American followers realize that their "revision" of Thomism is really a capitulation to philosophical Idealism? In any case such an approach to ethics seems to me incompatible both with the realistic epistemology which I have adopted and with a teleological ethics which I am here defending.

If, leaving aside Thomism, we wish to explore a little further the outcome of Scheler's phenomenological ethics, we can turn to his fellow disciple of Husserl, Heidegger (whose influence on Fuchs also seems evident).[73] Heidegger did not concern himself with developing an ethical system, but in his anthropology of humanity as *Dasein*, the place where Being reveals itself, he raised the question as to the source of this phenomenology of values on which Scheler's ethics is based. Obviously these values are inherent in our vision of Being. Modern man is in confusion about the authentic values of life because he has "forgotten Being" in his efforts to gain technological pragmatic control over the world. In our times, this effort is exposing its own emptiness, and we must be recalled to a remembrance of Being. For Heidegger, what is this Being which both reveals and conceals itself in human life? It is not at all the Being-in-process of modern science with which I dealt in Chapters 7 and 8. It is *historical* Being, i.e. the way that life appears to human beings in a given time and culture, a world into which they are thrown by the accidents of birth and which will pass away with them when they die. This Being is "fated" to unfold in the passage of time, because for Heidegger Being and Time are in a sense identical, and to be human is to be the participant spectator of this drama, authentically conscious of it, or unauthentically oblivious to its passage.[74] Hence Scheler's ethical method in the hands of Heidegger becomes a frank cultural relativism.

Proportionalism seems to me to be only one of a number of current proposals for a revision of traditional Christian ethics (I have chosen it merely because of its current popularity in Catholic circles), which are based on the belief that modern science has rendered obsolete the notion of a human nature which can be empirically explored, and that the cultural sciences have made it necessary for us to accept cultural relativism. Another example is the concern of Charles A. Curran to revise traditional Catholic ethics in such a way as to eliminate from it what he calls "physicalism" or "biologism," which he identifies as the reason that the popes have been unable in their teachings on questions of sexual morality to provide credible answers to the very practical problems of the Christian laity. [75] The fallacy of "biologism" (with which Curran charges the Thomists, but not St. Thomas), is to attempt to determine the morality of a human act by the teleology of some part of the human body, e.g. masturbation is wrong because it is an unnatural use of the genitals. To Curran this is to accept the Stoic reduction of human morality to the teleology of animal instincts, and fails to do justice to Aquinas' distinction of the moral object from the physical object (i.e. the ontic or premoral object). Certainly Curran is correct, as I have already indicated, in rejecting the Stoic notion of natural law as instinctive. But what is the consequence of this separation of human morality from the teleology of the human body? Does it not return us to that Platonic dualism against which I have labored throughout this book?

In Chapters 7 and 8 I showed that modern science remains open to a radical process interpretation which permits us to retain and continue to explore the notion of "human nature" as common to all human persons everywhere and throughout the whole of human history, and therefore to break through cultural relativism in order to establish an empirically grounded ethics. This does not commit us, however, to "biologism" as Curran defines it, yet neither does it drive us in desperation to some kind of dualism or phenomenological intuitivism. The human person is an animal, but transcends the animal by intelligence and freedom. Consequently "life according to reason" is not determined by nature or natural instincts, but by our intelligent understanding of our own natures, including our own bodies, which permits us to control our own actions freely and creatively. Nevertheless, (and here is where I disagree with Curran), the teleological structures of our human body must be respected by us as sacred because they are integral to our human nature. To violate these teleological structures is destructive of human intelligence and freedom. In arguing that some acts are intrinsically wrong because they contradict fundamental human needs (the satisfaction of which, taken together, constitutes the true goal of human life), Aquinas understood that the teleological structures of our bodies are important signs of these basic needs. [76] Masturbation is intrinsically wrong, in such an ethics, not because it is just an unnatural use of the genitals, but because to perform it, human beings have to use their own bodies in a way which their reason tells them is not in accordance with their basic needs as sexual beings, which needs are manifested in their own personal bodily architecture as well as in their psychology.

370

But may it not still be objected that the historical and cultural sciences have exploded the notion of natural law as a universal code of ethics known and honored throughout all cultures and times? Certainly experts in these disciplines often make such a claim, and they have accumulated a vast amount of data to show the almost unlimited variety of customs and moral attitudes within human cultures past and present. In evaluating such assertions, however, it is necessary to keep several hermeneutical principles in mind.

First, this assertion is not so much a conclusion from the data as an *heuristic* rule for the guidance of research. Cultural scientists, just as they heuristically reject sociobiological reductions of human behavior to genetic factors because this might discourage the search for historical and sociological factors which are the proper concern of their own disciplines, so they are inclined to reject philosophical talk about human nature because it too might discourage their own line of research. The second principle is that historians and cultural scientists, when describing cultures, are very interested in details, especially in what is specific and unusual. Consequently, broad and bland generalizations such as that all human beings need to eat and reproduce are of little interest, compared with the description of the curious kinds of food some people eat but others would not touch, or the very odd sexual customs in certain little known tribes. Third, it is a fallacy in this kind of research to apply the logical rule that a counter-factual disproves a statement. Human behavior, because it is to some degree free and creative, is able to violate every rule. If it were not, the very notion of immorality would be impossible. Moreover, this freedom can lead to prejudice, an unwillingness to admit that black is black or white is white, and an ability to rationalize black as white and white as black. Such prejudices can become institutionalized, and thus blind innocent generations to the truth. If this were not the case, the doctrine of original sin would not make sense. Thus broad generalizations about human nature and the natural law can be valid in the face of widespread ignorance and of customary moral norms and attitudes contrary to natural law. If this were the not so the distinction between subjective and objective morality would not be necessary. Consequently, the facility with which cultural relativists can cite an exception to every supposed norm of the natural law by referring to the customs of some tribe of 2000 people on its way to extinction on some Pacific Island should not overawe us. [77]

If we make these allowances in reading the data so patiently collected and analyzed by historical and cultural scientists, we note that many of the most curious human customs are found just where we might suppose, in small groups in backwaters of history among peoples not notably successful in their struggle to survive. A comparison of the great cultures of the world, including their religions, does show an amazing variety, yet as the ecumenical movement is demonstrating (not to mention the less gentle spread of the technological cultures of Humanism and Marxism), these cultures have fundamental values in common.[78]

Cultural relativism, if applied practically and consistently, would lead to the conclusion that the human global community should remain divided into isolated and fixed cultures. The moment that cultures begin to communicate with each other the people who live in them are faced with moral alternatives and a debate necessarily begins as to which way of behaving is *really* right. To answer this problem simply by mutual toleration in cultural pluralism never proves an adequate answer. Undoubtedly such pluralistic toleration is itself ethically sound, because we have no right to judge our neighbor; but the need of human beings to live and work together in common action has to cut across such mere toleration, and tends toward consensus on certain basic values founded in transcultural human needs.

Once this debate begins, how is it to be conducted? Here I must refer to my argument given in Chapter 1 that an ecumenical, hermeneutic methodology is the only fruitful form of dialogue. In such a dialogue, however, it is inevitable that appeal be made to some concept of our common human nature, its innate needs, and the values by which these needs are satisfied. Such an appeal is an appeal to "natural law," in a broad sense that abstracts from particular philosophical formulations of this law. Here too, different notions of the nature of natural law and its precise contents will arise, but these differences also can only be resolved by an appeal not to the ideologies of the dialogue partners, but first of all to what the natural sciences can tell us about human nature, and second, to what history and the cultural sciences can tell us about how such views have arisen, why they differ, and what they have in common. Once again this is an appeal to "natural law", i.e. to basic human needs.

III: The Bible, Human Historicity and Ethics

1. Adam and Biblical Anthropology

Any ethics rests on an anthropology, some conception of what it is to be human. A Christian ethics rests on Christology, that is, not merely on an abstract ideal of what it is to be human, but on the historical Jesus Christ in whom alone humanity has been truly realized. But Jesus of Nazareth is known to us principally through the New Testament, and the New Testament rests on the Old Testament as its indispensable foundation. It is in the historical unfolding of the Scriptures from the

372

creation of Adam to the victory of Christ that what it is to be human is manifested as a story rather than as an abstract description, and this narrative anthropology provides the basis for a Christian ethics.

In this book there is no room for an extensive commentary on this whole history of what it is to be human, so I have chosen only to look at the beginning, the first Adam, and at the end, the New Adam, because the structure of the Bible itself invites this comparison and it is traditional in Christian thought. But is it not dangerous to compare Adam to Jesus, thus reducing Jesus to a myth as Adam is a myth? This difficulty vanishes, I believe, once we realize that the historical understanding of what it is to be human must itself have a history, so that the final author of *Genesis* could only write the history of the first human beings in the manner of traditional narratives, just as we can only write it in the manner of Ph.D. dissertations. Myths are stories that take place in a purely ideal, fictional time and must be interpreted as such, but the author of *Genesis*, although he may have used mythical materials, did not write a myth, but an account of "what really happened" at the beginning of *history*, of real time, in the only way possible to him, namely, by conjecture based on the actual human condition as he observed it in the Mesopotamia of his day, by filling this in with the traditions of his people and with well-known myths, but only after historicizing these myths by a guess as to what they reflected about human reality. This Biblical effort to reconstruct the past proceeds by very different methods than does modern history, but in principle both are attempts to recover the real past by conjectures based on the flotsam and jetsam of data that the flood of time has left behind, not the mere recounting of symbolic tales. To think historically is to try to read such narratives in their own historic mode.

I will begin, therefore, with a reading of the first chapters of *Genesis* aimed at uncovering what the author wanted to say about anthropology, or what it is to be human as an introduction to the *Torah* or moral instruction which is the foundation of Christian ethics, since Jesus in all his teachings presupposes, confirms, but completes that instruction. My interpretation, therefore, does not attempt to cover every detail of these three chapters (especially of Chapter 1). It is based on current commentaries,[79] but I have felt free to dwell more on the *symbolism* of these accounts than do more sober-minded exegetes, because in view of the findings of modern comparative mythology it seems to me possible to explain how the final author of *Genesis* "demythologized" his materials to convert them into a religious history.[80]

The first chapter of *Genesis* is a narrative so symmetrical and liturgical that, as Augustine saw long before Darwin,[81] it is better read as a description of the order of creation as God intended it to be than as a chronological account of the stages by which it was brought into being. The scheme of seven days which ends with God's sabbath rest is obviously proposed as a model for human beings to rest from their daily work and contemplate the work of the Creator so as to remember Him in praise and thanksgiving, therefore, it cannot be read simply as history. It does, however, furnish the fundamental principle of any Christian interpretation of history, namely, that the universe was made by God for a purpose which is totally good and that consequently all history is a manifestation of God's wisdom and care.

Furthermore, this first chapter of *Genesis* establishes the basic principle of a Christian anthropology. Humanity is the greatest work of the visible creation because we are made in "the image and likeness of God" (Gn 1:27), in the sense that alone of these visible creatures, we are endowed with intelligence, freedom, and co-creativity, so that as God has created the world and will bring it to its perfection, we are His stewards commissioned to work with Him in the governance of lower creatures and their perfecting. Just as evolutionists today acknowledge that it is in human beings that nature first comes under intelligent control, the author of *Genesis* sees this fact as the result of God's gift to humankind of a share in his own supreme control over all his creatures. Since human control over nature is not an absolute autonomy but something shared with God, it has its limits. This chapter also tells us that Humanity is divided into male and female, both equally in God's image (1:27b) but differentiated so as to share in that fertility common to all living things, a fertility which is itself a reflection of God's creativity, but which in human beings is joined to their human dignity in such a way that they can use it intelligently and freely in their control of nature. God says to Humanity not only "Be fruitful and multiply" as He has said to the animals (v.22), but also adds "fill the earth and subdue it. Have dominion over the fish of the sea, the birds of the air, the cattle and all the animals that crawl on the earth," and also over vegetative life (1:28-31). This view of Humanity as the *imago Dei* is the first principle of Christian ethics, but since our understanding of what God is like is so imperfect, it still raises for us the question: What is Humanity really like?

In the second and following chapters of *Genesis*, the final author of the work drew from other sources the materials with which to fill out this anthropology in more concrete detail and in a form which is unmistakeably historical, not indeed in the sense that he knew just how the first human beings came into being or just what occasioned the fatal decision by which they gave to human history its fundamental direction, but in the sense that he does intend to tell us it "really happened" that human beings made in God's image, superior to animals, appeared on earth, were

374

faced with a fundamental option and made that option sinfully, with endless historical consequences.

The medievals supposed that since the first man and woman were works of God, originally they must have been perfect in intelligence, freedom and in bodily beauty appropriate to that intelligence. Yet to be human, and therefore made in the image of God, is not to have a certain IQ or a particular physique, but to be intelligent and capable of free choice. To say otherwise would be to deny that human nature is essentially bodily, and therefore necessarily fully realized, not in this or that ideal human being, but in the whole range and variety of human beings that make up a human community. Adam and Eve were not defectives, but there is no need to say that they were "perfect" according to some abstract ideal. Nor, once we accept the view that evolution is the mode by which God creates living things, need we deny that evolution has continued since these first human beings to produce the racial and individual variety which we now see, and thus further to perfect humanity by producing human beings of better integrated brains and physiques better adapted to those brains than those of the first human beings.

More puzzling is the problem of how the first *two* human beings came into existence simultaneously. In current evolutionary theory, it is not individuals but populations which evolve from one species to another.[82] A species is a single population which exhibits a range of genetic variation, but a range not so great as to make interbreeding impossible. By interbreeding, this population maintains its identity and relative stability. If, however, this population becomes separated into two groups by environmental barriers (e.g. the separation of the islands in the Galapagos archipelago which led to the evolution of several species of Darwin's finches), the continued genetic variation through accidental mutations which goes on in both populations, no longer offset by the sharing of these new genes through interbreeding, may result in the two populations becoming so different from each other that even if the barriers are breached they cannot any longer interbreed and thus are recognized as new species. In this explanation, the ability to interbreed is the *defining* trait by which the species are identified or distinguished, yet this trait is not viewed as something absolute but relative (i.e some hybridization may remain possible, but of such low frequency as not to cause a refusion of the species), and it is produced by a combination of many traits rather than being unitary.

It is somewhat difficult to apply this theory of speciation to the origin of human species if we consider that truly human intelligence as it was defined in Chapter 8 must be a unitary, all or none trait, so that we can say this organism is human because it has intelligence or it is not because

375

it does not have that trait. Persumably that unitary trait would have first appeared ontogenetically by a mutation in the chromosomes of a subhuman individual and phylogentically in one of its offspring who would have been the first human being. Would such an individual still have been able to interbreed with other members of the population and thus produce offspring with this trait, who could then have interbred and produced an intelligent variety of the species, which only after isolation from the general population would have become so physically different that such interbreeding would no longer be possible? We do not know, apart from unverified rumors, that hybridization between the human species and other now existing primates is possible. In my opinion such speculations, although they show how according to evolutionary theory the human species probably originated, have to be interpreted in the light of the discussion in Chapter 7, in which I showed that evolutionary theory as a purely physical theory in terms of natural laws is necessarily incomplete. In order to complete it as a theory which is not merely an historical description but fully consistent with the principle of causality, it is necessary to refer to some superphysical creativity. This superphysical creativity is required in a special way to explain the origin of human intelligence (or, better say, intelligent human beings), because as shown in Chapter 8, intelligent, creative thinking cannot be reduced totally to a function of the brain. Consequently, the evolutionary origin of human beings, while it must have been entirely consistent with the natural processes of biological evolution, also was a unique, creative event which need not be, and indeed cannot be, reduced to a purely biological explanation.

To me it is theologically not incongrous to suppose that there may have been a considerable period of time in which the human species as a biologically distinct population may have been in the process of emergence during which time there were beings with human intelligence still interbreeding with primates who lacked such intelligence until a group of intelligent beings withdrew itself from this mixed population and by living in isolation and interbreeding, became a homogeneous human species. It was at this point that we can speak unequivocally about the "human race", and it is this historic race to which all human beings now belong and to whom the account of *Genesis* is directed and with which it deals. What then of those truly human beings who existed for a time and interbred with nonhumans? It would seem that they would not yet have recognized their humanity or arrived at that self-consciousness and actual communication through language which are necessary for human intelligence to be specifically actualized. Consequently, they would not yet have the actual use of human freedom, and thus would be radically human but not yet active in a truly human way, somewhat as seems to be the condition of "feral children",[83] who apparently have normal brains but cannot use their intelligence normally because they have not been raised in human social conditions based on communication through human language. But even if we admit that there were some humanoids who actually lived as intelligent and free but still not as constituting a *biologically* distinct species, we could still recognize them as members of our own species if we define that not biologically but in terms of the single trait of intelligence. The difficulty (as pointed out by Pius XII, in his encyclical *Humani Generis* (1950)) in such a theory of human origins for the doctrine of original sin, will be discussed later.[84]

On the other hand, there is nothing about evolutionary theory, except the laudable desire to avoid postulating unique events, to exclude the simple possibility that the origin of that final genetic trait responsible to produce a human brain capable of functioning at the human level depended on the mutation of one dominant gene that occurred in the germ-cells of a primate ancestor, which was not itself human but which then bred with another primate of its own kind to produce a male and a female child who were genotypically the first human beings having fully human brains, and who by interbreeding became the ancestors of the entire human race. Either this or the former explanation is consistent with the interpretation of *Genesis* which is not concerned with the exact way in which the human species came into existence and began as a single interbreeding and intercommunicating species to have a history determined by a primordial act of human choice. Whether this act of choice was the very first human act is not significant, but only that the human race as it now exists had its destiny determined in the days of its origins, whether we consider this period to have been very short or quite long.

What account then does *Genesis* give us of this primordial choice which has determined human history? I presuppose that the author had no certain way of knowing what then occurred by means of some continuous historical memory, since we now think that an immense time had intervened, but that he conjectured as we might do (although his was a divinely inspired conjecture), in view of the human condition as he experienced it. From what he could gather about history from the traditions he knew and the explanatory myths which he had heard, the first great events of human history must have been something like what he narrates.

To appreciate fully the profound anthropology contained in the story of Adam and Eve it is necessary to enter at some length into a reading of the two accounts of creation (Genesis 1 and 2-3) in terms of their

377

symbolic use of the notions of masculinity and femininity as ontological metaphors. I emphasize strongly that the anthropological significance of these categories will be seriously misunderstood if one attempts to draw *directly* from this symbolism some kind of practical ethical conclusions about the respective rights of men and women, because rights belong to real persons, not to symbolic persons.

Behind the creation accounts are myths in which God is conceived as masculine and the primeval chaos is pictured as a great feminine serpent or dragon, namely, the ocean waste. This ocean-serpent is feminine, as Carl Jung has shown from the study of comparative mythology,[85] because all things are born from her watery, maternal womb. She is a serpent because the ocean is filled with serpentine swimming, undulant creatures and was thought to circle the earth like the *uroboros* or snake that grasps its own tail in a circle and swallows itself, a kind of ancient "black-hole", an apt symbol of non-being. The male God in a violent struggle rapes and impregnates this dark mother with his "spirit" or fiery, yet hidden seed and thus out of her chaos produces the present orderly universe.

The final author of *Genesis* 1 and the sources on which he immediately drew have "demythologized" this account in order to elevate the Creator above all implications of sexuality, male or female. God creates by his Spirit but this Spirit is not clearly distinguished from His Word, although it is first mentioned as "hovering over the waters" like a bird hatching the primeval egg, and then as His creative words of command, "Let there be light" etc. Yet the masculine-feminine symbolism is not forgotten, but transferred to the creation (where it properly belongs and from which it was derived in the first place) as a metaphor through which to speak about the relation of creation to God.

It is clear that for Jesus God was experienced as Father, "Abba," as masculine. Moreover, frequently in the Old Testament the prophets speak of God as the Bridegroom in relation to His convenanted People, as His Bride (Hosea 1-3; Isaiah 1:21-26; Jeremiah 2:2; 3:1-12; Ezekiel 16 and 23, and according to the Jewish rabbis the whole of *The Song of Songs).*[86] The point of this comparison is not to stress the superiority of God or his dominative power, but his faithful love of his people whom he has freely bound to himself by covenant as if they were his equals.

What then of the opinion of some Christian feminists that the Holy Spirit is portrayed as feminine in Scripture because the Hebrew *ruah* is of that gender?[87] A much firmer ground for this notion is the explicit personification of the Divine Wisdom as feminine (Proverbs 8-9; Job 28; Baruch 3:9-4:4; Sirach 1:1-20; Wisdom 6-9). On the one hand this Wisdom is God Himself considered in His creative intelligence and power and on the other hand it is the *creation* as imaging or manifesting this wisdom and power. Thus the Spirit of God is the dynamic vitalizing principle or soul (the Logos of the Stoics or the World-Soul of the Platonists) *immanent* in the universe, and it can be considered as feminine, not in itself as identical with God in his fatherhood, but as embodied in the creation which is the Bride of the Creator. Thus it would be a confusion of symbols to speak of the Third Person of the Christian Trinity as the Bride of God the Father (or of the Son), but it is correct to speak of the *cosmos* living by the Spirit as that bride.[88]

This cosmos as mother brings forth at the command of God all the varied order of creatures by a process of differentiation and ornamentation, a process which culminates in the production of the Human Person made in the image of God (1:27a), and divided into male and female to whom God gives stewardship over the other creatures, as will be discussed at length in Chapter 10. In the second account of creation from another different but complementary source, the contrast between chaos and order is not between the ocean and the land, but between the desert and the oasis (2:4-6). Out of the dust of the barren, uncultivated desert turned into mud not by regular rainfall but by mist or oozing from the soil (the translation is uncertain), God fashioned Adam and "breathed into his nostrils the breath of life", and then prepared for him a garden oasis, fertile and well-watered (2:8-10), which Adam is to cultivate and guard and which he can enjoy as he pleases provided he does not eat of "the tree of the knowledge of good and evil." This permission to eat of all the other trees seems to include even "the tree of life" (1:9) a tree that would give immortality. What is the forbidden tree? Clearly it is the tree of death, as God warns Adam (2:17). Adam is god-like in his stewardship over the garden, yet as the garden itself is a limited domain surrounded by desert, so Adam's power is limited, not autonomous. Adam does not question this fact, nor is he yet conscious of being alone in the world, since he finds himself in God's presence and under his guidance.

God, however, knows that the Human Person he has created is incomplete, even within its appointed limits, in two ways. First, God sees that Adam needs a human partner with whom to share the enjoyment of the garden and its care (2:18). While Adam has heard God's voice he has not yet himself spoken to God, but as God brings the animals before him, Adam as God's steward recognizes their different natures and gives them names by which he can govern them. Simultaneously Adam comes to recognize his own uniqueness, less than God yet greater than all other creatures. Second, God by speaking of the possibility of death (2:17) to Adam has also implied to him that by his own nature taken from the dust he is liable to death, from which he can be saved only by the tree of life.

To meet these needs of Adam, who is only gradually coming to recognize his own needs under God's tutelage, God casts him into a deep sleep. Adam is certainly not tired, since he has not yet begun to work. Moreover, this deep sleep seems like an anticipation of death. It is in fact the sleep of vision[89] from which will come to Adam a revelation of the mystery of his own being, of its hidden depths, its origin from the earth out of which he has been created, but also of its promised heights, the eternal joy of immortal life to which God wishes to lead him. In that creative sleep which is like a return to the beginning and a foresight of the end, God forms Woman who is the hidden mystery of the Human Person, the total human self which cannot be named because it exceeds comparison with any other of the animals. This Woman is formed out of the very side of Man, out of his heart and as his equal partner.

God wakens Adam and brings Woman to him and he recognizes her as his *other* self, "bone of my bone and flesh of my flesh", i.e. she is his body taken from the dust, yet a living body "taken from man" (2:23), i.e. his *inner* self, and he names her "Woman." Does this naming of her imply that he rules her as he rules the garden and its animal inhabitants? In a sense yes, because as feminine she *is* the fruitful garden, she is his Paradise ("She is a garden enclosed, my sister, my promised bride; a garden enclosed, a sealed fountain. Your shoots form an orchard of pomegranate trees", *Songs of Songs* 4:1-13), his own body, and hence she is also the whole cosmos as that is the spouse of the Creator; but in another sense no, because as inner self, his very heart, given him by God, in vision she is his inspiration, the promise of immortal life, his spiritual self or intuitive creative intelligence and active, productive *will,* the Creative Divine Wisdom which is also feminine, and hence the superior that guides him. She is the tree of life from which he is permitted to eat and thus to become immortal. The symbol of the woman, therefore, manifests that in every human person surrounding the ego-mind which can be verbalized (Adam naming the animals) is a wider self, which on the one hand joins nature in its determinism (Woman as the earth, as body), and on the other joins the realm of free, creative, intuitive spirit (Woman as the tree of life, as vision, as Spirit).[90]

As he greets his companion and names her Woman, Adam is thus faced with a question: what will this woman be for him, will she draw him back to chaos, or will she raise him to immortal life? Is she his tree of life or his tree of death? The text does not suggest that they at once made love, rather, after Adam's first cry of recognition, for the moment they separate to think over the meaning of this encounter! The viewpoint of the story shifts now from Adam to the Woman whom we find alone near the forbidden tree. Woman must also discover herself as person, apart from Adam. She is not merely his helpmate but a person in her own right. By this separation is raised not merely the question of the equality of woman to man but also the more fundamental question of the relation of the individual to society. By going off to be alone, the Woman asserts the proper autonomy of the person. She is not just part of Adam or his instrument. She is Humanity in its subjectivity and inwardness as Adam is Humanity in its outward control over the world.

Woman alone, pondering her own identity, encounters the snake. As we have already seen, the serpent is the original feminine abyss out of which creation has been formed by God. This abyss, as pure potentiality for all forms, is ambiguously positive or negative. It is dual and so its embodiment in the serpent is ambiguous and dual, because the snake is the woman herself in her own ambiguity and potentiality. On the one hand the snake is an animal, naturally akin to the dust of the earth, but it is "more cunning than any beast of the field which the Lord God had made" (3:1). It is Wisdom ("You must be clever as snakes and innocent as doves", Jesus said), (Matt. 10:16), combining snake-wisdom and dove-wisdom, the cunning of the devil and of the Holy Spirit, because the ancients thought that the serpent with its bright eyes and its uncanny swiftness in escape was like the mind of a subtle, cunning thinker. Such wisdom is not that straightforward practical masculine reason that can be put into words, but the feminine intuition which can only be put into symbols and seems always to change its shape in protean fashion. We have only to think of Shakespeare's portrait of Cleopatra whom Anthony loves to call "my serpent of old Nile" and of whom Enobarbus says "Age cannot wither her, nor custom stale Her infinite variety" and who kills herself with asps that she compares to her children. This serpent image should not be understood as symbolizing women as persons but the ambiguity and duality of Humanity itself, in its polarity of flesh and spirit, or mortality and immortality.

In fact as Woman confronts herself she is, as Jung shows,[91] also discovering the other side of her own personality, her masculinity. The serpent is also phallic, and is the masculine Satan, the Don Juan symbol of restless sexual energy and will to power, the Faustian reason as controlling and dominating, of aggression and non-submission. As Adam discovered his *anima* or femininity in a dream, so Woman discovers her *animus* or masculinity in a vision in which an animal speaks. Because the Satan-Serpent is ambiguous, he is a "liar and the father of lies" — "who brought death to man from the beginning" (John 8:44), and he tempts the Woman by appealing to her curiosity which is an aspect of that which is highest in her, her intuitive wisdom, by suggesting that God has forbidden Humanity to eat of the tree of the knowledge of good and evil lest it obtain god-like power through knowledge, and thus become God's rival. "To know good and evil" for the Jews was to become adult, especially as regards the awareness of sexuality which they called "the evil impulse," not because sex as such was evil but because it tempted one to exceed the limits set by family relationships, thus disrupting the social order.[92] Woman, or rather the Human Person as it seeks to discover its own autonomy is tempted to grasp at an *absolute* autonomy rather than a relative autonomy which respects the relation to God and to other members of society. It is the temptation to become a law unto oneself, refusing to submit to the order of nature and social cooperation, and to usurp the divine and social power to determine what is "good and evil," not for the common good, but for one's own selfish interests. Because Woman symbolizes the *creative* in Humanity, it is to her that this temptation is directed. The serpent promises that her "eyes will be opened and you will be like God, knowing good and evil" i.e. "knowing and therefore controlling all things" (v.6)[93]

"Now the woman saw that the tree was good for food, pleasing to the eye, and desirable for the knowledge it would give. She took of its fruit and ate it, and also gave some to her husband and he ate. Then the eyes of both were opened, and they realized they were naked" (3:6-7). The tree, which symbolically is identical with the serpent, is poisonous, but it appears useful, beautiful, and true. It is the beginning of human illusion, false knowledge that appears true, because sin is not possible without *willful* ignorance,[94] and the Woman's ambiguity has now become self-deception. Instead of eating from the tree of life (which is her real self) she eats of the tree of death, her false self, and she uses the power she gains from her new knowledge to seduce her husband. The text does not describe Adam's temptation in detail because the Woman is his other self and his sin is only the final ratification of hers, the full and deliberate consent of the will by which sin becomes mortal.

If Woman and Man had not reached out to take the forbidden fruit by an act of rebellious autonomy, God would have given to them (indeed He had already given to them but they had not yet sought it out), the fruit of immortal life. They would have become gods with God, since God has no envy and fears no rival. They would have had what the serpent promised and could not give, but which God had promised and had already given. Since Woman is herself the tree of life, Adam would have received this perfect gift of life in all its abundance, and Woman would have achieved her own personhood as Adam's equal, having her own name as "the mother of all the living" (3:20), thus rooting Humanity in the soil as its mother and yet inspiring Humanity as its creative wisdom. Humanity would have had immortal life both in the sense that it would have multiplied and filled the earth so that death could not destroy it, and in the sense that it would have achieved that wisdom of faith and revelation which brings us into intimate relation with God in everlasting life.

The outcome of Adam and Eve's sin, however, was that they now discovered their sexuality, not as the complementary equality of their natures, but as a separation between them that could only be overcome by domination and subjugation, the "war of the sexes." And this outward war manifests the inward war of what St. Paul calls "the law of the inner self" (Romans 7:22-23) and the "law of the members" i.e. between reason and desire for immediate satisfaction, the loss of full freedom that comes from inner conflict. It is this sense of loss of integrity that makes them ashamed of their bodies and their sexuality. From God (to whom they have never yet spoken), who comes looking for them in the evening wind (the Spirit), to prepare them for that nuptial experience of their first holy union (the eating of the tree of life) they hide themselves, concealing the bodies of which they are ashamed with the great leaves of the fig-tree (presumably the leaves of the forbidden tree, since this concealment of shame is that of a refusal to face their guilt, a suppression of their new self-knowledge of their own limitations).

Their first words to God betray what liars they have become. Adam says "I heard you in the garden, and I was afraid because I was naked; and I hid". It is true he was afraid, but his implication is that God is an envious, spying guardian who wishes to curb the natural appetites he has given to humans. And then Adam puts the blame on the Woman and God for giving her to him; "The woman you placed at my side gave me fruit from the tree and I ate". If Adam could have forgiven the Woman, God could have forgiven him. And the woman blames the serpent; "The serpent deceived me and I ate" i.e. she blames her hidden Self and thus makes herself inferior to Adam, rather than admit it was her false ambition to autonomy which betrayed her.

God then passes judgment on the serpent, the Woman, and Adam, not by some added punishment, but by declaring to them the inevitable outcome of their own choices. The serpent (like the rainbow sign after the Flood (Gen. 9:16), will retain its place among the creeping things of the earth, but it will become forever a symbol to remind human beings of the origin of their present condition — a symbol of life and healing because of its ability to shed its skin, its mobility, its cunning, its phallicism; it will also be a symbol of death because of its poison and its dwelling in tombs, in the dust of death which will be its food. Because of this symbolism humans will ever seek to destroy serpents, i.e. to overcome death. Although the serpent in its life-giving aspect will also become a symbol of healing (the bronze serpent on the staff of Numbers 21:4-9 and the Greek association of the serpent with medicine), yet death will always be at hand.

In verse 15, God pronounces an enmity of death against life and life against death between the serpent and the woman, "between your seed and her seed," i.e. the remedy against the death of the human race lies in the woman "who is the mother of all the living." The people will somehow survive until God can bring ultimate redemption. For the woman, however, this means "distress in child-bearing", and "for you, your husband shall be your longing, though he have dominion over you," i.e. in seeking autonomy, Woman has let loose that lust for domination which will lead to her enslavement by her physically stronger husband, who will be her master rather than her partner. As for Adam, his lot will be the endless struggle to make a living, no longer in the garden but in the ungrateful and resistant desert, and in the end he will return to the lifeless earth from which he was made.

In a final scene, God expels the couple from the garden forever, saying ironically that if He were not to do so they might eat of the tree of life and live forever, which would in fact be their condemnation to an unending life of misery. Yet he clothes them with animal skins (3:21) to protect them from the harsh weather of the desert (but also to symbolize their brutalized condition) into which he sends them. The cherubim with flaming swords that bar their re-entrance to the garden indicate that they have no way to regain their original innocence by their own power. Yet God does not simply let them die, as would have been the natural consequence of their eating of the tree of death, but remains with them so that when they do unite sexually, Eve conceives Cain and cries out in joy, "I have given birth to a man-child with the help of the Lord" (4:1), thus at the same time acknowledging the continuing favor of God, triumphing over Adam (since she is no longer the one "taken from man" but the one from whom man is taken (I. Cor. 11:12), and yet she feels the consequence of her sin in the pains of child-birth and in her submission to sexism, since what she glories in is a *man*-child, and one who will prove a man of violence, and the father of Lamech, the first to introduce polygyny (4:19).

In what follows (chapters 4-11) we have a vivid account of the spread of the human race, its development of technology, its increasing moral corruption, a great natural disaster which almost destroys the race, but God's salvation of a remnant and his rainbow promise to be patient henceforth with his children whose evil ways he concedes to tolerate within limits, the renewed expansion of the human family, its false religion, its technological progress wasted on ostentatious display, its division into warring cultures. All these developments manifest the mixture of the creative use of God's gifts and their evil use for domination and destruction. Even the best falter. Does not Noah, the saintly man of faith, invent wine, become drunk, and commit incest with his daughters?

The teaching of the Bible on original sin has often been misunderstood and presented theologically in a manner that has made it seem incredible and anti-human.[95] How could it be that the sin of two people thousands or millions of years ago would condemn the whole human race to eternal damnation? How can this be reconciled with the clear teaching of the prophet Ezekiel who said, "If a man begets a son who, seeing all the sins his father commits, yet fears and does not imitate him . . . this one shall not die for the sins of his father, but shall surely live." (18:14, 17)? Yet there is a sense in which the sins of the fathers *are* visited on their children, not because the children are personally guilty in an ethical sense, but because children suffer from the natural consequences of their parents' evil actions, just as they benefit from the good consequences of their good actions. It is in this sense that the children of Adam have suffered from the orientation of human history which he gave it, and not only from his sin but from the whole accumulation of the sins of all those who have followed in Adam's way ever since. Every child who comes into this world comes not into a world oriented toward God as God intended it to be, making the living of lives truly worthy of human dignity easy and delightful, but into a world distorted by the wars, the tyrannies, the lying delusions of the centuries, the whole burden of history. This corruption affects us from the moment of our conception by the fact that our bodies have suffered from the effects of all the damage human beings have done to the environment and to their own health, and it enters into every phase of our psychological development from the defective relations within families to the false and brutalizing education received from sinful human institutions. What is more evident than such facts of the human condition?

The individual is profoundly affected from conception in his or her entire personality. This does not make the individual a sinner guilty before God, however. The infant is the *victim* of sin, deprived of its rightful inheritance of God's gifts by the sins of others, and weakened by the consequences of these sins in its own capacity for good. We become personally sinners only when we use the freedom that remains to us to become cooperators in this historic sinfulness of the race by conforming to evil ways that the light of reason which remains to us sometimes exposes to our conscience. Nor has God's grace of conversion into better ways ever been lacking to any human being anywhere in human history; since the Scriptures shows us that there have been just men even among those people whom God did not choose in Abraham for the mission of restoring His kingdom.[96]

2. Natural Law and Christology

We now turn to the other end of biblical history to look at the New Adam, chosen from the Chosen People to announce the fulfillment of God's promises to restore His kingdom over the world after the long centuries necessary to teach the human race its folly from experience, like the Prodigal Son (Lk 15:11-32), who would learn in no other way that his real happiness could be found only in his father's house. We find that the Old Testament Torah or instruction in morals based on the anthropology of *Genesis* is not destroyed, but deepened. A Christian ethics cannot be based on values known intuitively as abstract ideals, as Scheler thought, but only on the historical experience of the Christian community. The Christian community, however, receives its identity from the historical Jesus of Nazareth, in whom its ethical values were fully realized, and whose risen presence in which it believes remains the source of its life through which these same values can continue to be realized in its members.

This Jesus of Nazareth in whom the early Church believed "the fullness of deity resides in *bodily* form" (Col. 2:8) was an historical being, a living, thinking body whom other humans heard, saw and touched (I Jn 1:1). These first Christians were convinced that as Risen Lord he continued to be present with them sacramentally in his Eucharistic Body (I Cor. 11:17-34), on which they were fed and by whose life and death they were incorporated into his visible Body, the Church (Col. 1:18). Unless we are willing to accept a docetic or gnostic and dualistic Christology, we must look on Jesus as a human being who entered fully into our history and therefore into our natural order, subject to the Jewish law and to the conditions of the culture of his own times and to the laws of nature. St. Paul says of him, "When the designated time had come, God sent forth his Son born of a woman, born under the law" (Gal. 4:4), meaning by "law" of course the Jewish law with its historical and cultural particularism; but for St. Paul this always presupposes the more universal order of creation, the natural law.[97] The ethics of Jesus, therefore, and his historical exemplification of ethics cannot be separated from the natural law, but instead becomes for the Christian a way to recover the natural law in its authentic sense. To know what it is to be truly human, we must look to the historic Jesus who alone in history is the fully, undistorted realization of humanity. Ultimately, it is this conviction that makes Christian Humanism different from Enlightenment Humanism.

But how can we truly know what Jesus was really like? The problem of "the quest for the historical Jesus" is inescapable, because our

386

present day historico-critical approach to our only substantial record of the historical Jesus, the New Testament, seems to have reduced this record to shreds. At best these shreds of evidence themselves seem to represent nothing more than the *faith* of the early Christian community, not the historical Jesus himself.[98] This problem has been thoroughly dealt with in a manner far beyond my competence by recent Christologists, notably by Edward Schillebeeckx.[99] Although these works are certainly not the last word on the problem, they provide us with a sufficiently solid ground for supposing that an objective application of historico-critical methods can supply us a reliable if not detailed account of the chief acts and teachings of the historic Jesus.

Once we realize that all great religious leaders, precisely because they raise for their contemporaries the most disturbing and divisive questions of ultimate concern — "Do not suppose that I have come to bring peace to the earth; but a sword" (Mt 10:34), that they will always be known to us only through the eyes of the faithful or through the eyes of those who seek to kill them, then we will no longer be shocked by the difficulties which modern biblical scholarship raises about discovering the historical Jesus. When historians try to recover the historical Gautama the Buddha, or Mohammed, or even the historical Voltaire or Karl Marx, they find themselves faced by the problem of hagiography, the idealization of a revered leader. In the case of Jesus, of course, we have the additional difficulty that we do not have the contemporary records of his opponents; the rabbinical traditions are late and the secular records scanty.[100] Yet, again, this is also the case for the Buddha and Mohammed, if not for Voltaire or Marx.[101] The great religious leaders, just because they were so counter-cultural, often lived obscure lives. Recognized at first only by their disciples, their public importance only gradually became evident and then too late for outsiders to supply an independent witness. Yet even when our documentation is ample, as in the case of Luther, Calvin, Voltaire, or Marx, we are still forced to choose among the most varied interpretations.[102]

Must we, therefore, be reduced to historical scepticism? It seems to me that the historical methodology of some participants in the "quest for the historical Jesus" who attempt to separate sharply the "Jesus of history" from the "Christ of faith", is wrong in principle.[103] Historically, a leader can be known only through his or her historical *effects*. Augustus Caesar or Napoleon cannot be separated from the empires they built, nor Mozart or Beethoven from the music they composed, nor can a Jesus or a Mohammed be separated from the believing community which they inspired. No doubt such leaders are often partially misunderstood by their followers and their teachings and purposes distorted or modified, yet the vitality of the movements and institutions which they created cannot have come principally from those movements or institutions but only from their creators. It is this vitality, this dynamic form and inspiration which gives unity to the tradition which expresses the genius, the mind and heart of the leader.

Hence the object of historical inquiry cannot be Jesus of Nazareth abstracted from his disciples, but that same Jesus in whom they believed and of whom they are the only well-informed witnesses. Of course that point of departure implies that we also inquire as to the character of this community of faith. Was it a bond of fanatics, or mercenaries, or pitifully deluded people like the disciples of the Rev. Mr. Moon or the kool-aid drinkers at Jonesville?

A recovery of the real Jesus is possible just because the community of faith which has kept his remembrance is present today and subject to scrutiny. The biblical account of Jesus is a privileged record, but it is dependent for its own authentication only on the living faith of the Christian community as it has maintained its continuity through history. This fact seemed evident enough to Christians before the Reformation, but the Reformers, in their efforts to correct very real abuses, vigorously attacked the living Church from which they had themselves received their faith and went so far as to claim that it had become essentially corrupt, so that the Gospel was no longer to be found in the Church but only in the Bible. They were then forced to claim that the Bible authenticates itself by its power to generate faith and conversion,[104] a position which, as we see it surfacing in biblical fundamentalism on the one hand and Bultmannian existentialism on the other, renders vain the whole historical enterprise. If, as fundamentalists believe, the Bible is an historical record that requires no historical hermeneutic; or if Bultmann was right that this hermeneutic prepares a religious decision which is independent of historical fact, then in either case theology can ignore historical understanding.

The Catholic understanding seeks to save an historical interpretation of the Bible by reading it always in the light of the faith of the historic Christian community (including, not excluding the Protestant tradition). Jesus cannot be understood without his Church, nor can the Bible be read apart from the Church which wrote, preserved, and canonized it, just as there can be no insightful knowledge of Mohammed apart from Islam, nor an adequate reading of the Quran without the insight of that community of faith, nor of the Jewish Scriptures apart from the rabbinical tradition. This does not imply that such readings should not strive for scientific, critical objectivity, nor that outsiders to a tradition cannot make important contributions to its study, but it means that such objectivity is possible only if the objective scholar strives to be truly sympathetic to the understanding of believers within the tradition.

The reliability of the Church's witness to Jesus has been attacked by Humanists on the grounds that its "memories" are legends and mythical constructions having only a tenuous historical basis, and by Marxists on the grounds that these constructions reflect an ideology whose purpose is to conceal from the masses their own enslavement. Underlying such accusations are hermeneutical preunderstandings specific to the world-views of Humanism and Marxism. Humanists use the Humean principle of uniformity to devalue *a priori* any document which narrates the miraculous, or which pictures Jesus as the unique embodiment of humanistic values which Humanists consider incompatible with super-naturalism. Marxists use their dogma of materialism not only to disclaim the supernatural but to explain Jesus in political terms in face of the documents which represent him as transcending such partisan categories. Once these assumptions are bracketed, a different historical evaluation of the Church's witness is opened up.

The Christian community in its historic reality has to be judged as a reliable witness, not by some abstract standard of objectivity, but com-paratively in relation to the self-witness of other religious traditions, in-cluding those of Humanism and Marxism. When so compared, its historical fidelity to its origins, its universal transcultural endurance, its remarkable unity of testimony to the essentials of its teaching (in spite of its internal quarrels), and its "holiness", that is, its power to reproduce the essential features of Jesus' own life creatively in its saints as well as at least minimally in the average active member of the community, are historically impressive and show that its witness is to be taken seriously, not dismissed as fanaticism or delusion. The many obvious defects of the Christian church do not disprove this positive evidence of its credibility, since these defects are only those common to all human institutions, in-cluding those of Humanism and Marxism, while its postive traits make a unique claim on the open-minded observer. I state this as a fact, but of course it is a disputed fact and I will not here marshal my evidence for my assertion. I am here only concerned to make clear what is the Christian claim in the ecumenical dialogue with the advocates of other religions, because I am convinced that the first step of fruitful dialogue is clarity about our respective positions. We Christians are willing to put our claims to the test of historical experience.[105]

If this general reliability of the Church's memory is taken seriously, then we can turn with some confidence to the Bible in search of what it tells us about Jesus in his bodily existence, since it is this bodily ex-istence with which my book is concerned. Jesus can be viewed as a human being in his cultural context, a Jewish prophet whose ethical teaching must first of all be evaluated in terms of its adequacy to meet universal human needs, i.e. as an expression of natural law. Some of the Humanists of the Enlightenment, for example Locke and Thomas Jefferson,[106] appreciated him in this way, as does liberal Protestantism. I do not think they have been mistaken in this evaluation taken as the first step of a Christology, but only in failing to go further. The historic Jesus is first of all the Jewish prophet who in about 27 A.D. began to preach the imminent coming of the Reign of God, a reign which, whatever else it may be, is first of all an earthly reign of justice and peace, a renewal of the natural order established by God in creating us. This order was *included* in the plan of God's grace and can be restored only by grace, but it is an order intrinsic to human nature and the basis of natural law ethics.

Yet some exegetes deny that the historic Jesus had any real interest in ethics precisely because he saw his mission as the proclamation of the coming Reign of God, and he believed that the immediate apocalyptic arrival of this Reign would solve all the problems of human life.[107] Most exegetes today, however, would grant that Jesus in announcing the coming of the Reign of God also declared that certain moral conditions were required of those hoping to enter that kingdom. All were called to the banquet, but not all might be willing to make the sacrifices implied by the acceptance of that invitation (Mt 22:1-14). Thus these demands of Jesus which accompanied His universal invitation and which He explained in their full implications to those whom He called to be closely associated with Him in His mission (Mk 4:34), constitute the ethics of Jesus which He taught with authority (Mk 1:22), which He exemplified by His conduct (Mt 11:29), and which became the standard of His community as they awaited His return even when the conviction of its imminence had begun to fade (2 Pt: 3).[108]

Five fundamental themes characterized this ethics of Jesus: (1) He announced the coming Kingdom or Reign of God promised by the Old Testament prophets, a Reign which would be transcendent and everlasting, yet which would begin not in the remote future but in Jesus' historical presence, although it would be consummated only with the end of human history. (2) Jesus announced that all human beings, and especially the most neglected, alienated, and outcast are invited into this Kingdom and to messianic table fellowship with him. (3) This Reign of God was based on Jesus' new understanding of our human relation to God as a loving and merciful Father on whom even the lowliest of human beings can rely as God's sons and daughters. (4) Jesus understood his own relation to the Father as one of true Sonship which required his perfect obedience, a fidelity to the mission given him by the Father as the Suffering Servant mysteriously

390

proclaimed by the prophets. Hence he in turn required of his disciples who shared in His mission to lay aside all ambitions for power and to follow him as servants and stewards on his way of the Cross. (5) Jesus was convinced that he could fulfill this ministry because of the power and authority with which he was endowed by the Holy Spirit of God, formerly given to the prophets but to him in fullness, and he promised this same empowering Spirit to his disciples in their ministry after his ascension to his Father.[109] It is these themes which give a specific character to Christian ethics.

Does this mean, as some Protestant theologians have argued, that Christian ethics is purely a matter of attitude or motivation, of faith in God and love of neighbor, from which attitude of faith, right behavior proceeds spontaneously without the need of any ethics, that is, of any systematic set of moral norms or of a casuistic methodology of moral decision in difficult cases?[110] Does it mean that every individual must meet the existential situations of life *coram Deo* but without any moral code publicly taught and sanctioned?[111] Does it mean that Jesus gave us only one moral principle, "do the most loving thing", while rejecting all other moral rules as pharisaic?[112] These views seem to be the result of the predominate authority given by Luther to the chief writings of St. Paul in which Paul seems to declare that the Law is abolished by the grace of Christ. Hence if the revealed Law is annulled, so also is the force of any "natural law."[113] By this such Protestant ethicists do not mean that the Christian is free of moral demands, but that these demands can never be formulated as universal moral norms. The difficulty with this reading of the evidence about Jesus' own ethics is that it requires some theory of "a canon within a canon", by which the rest of the New Testament (including the Pauline Pastorals) are treated as reflections of an "Early Catholicism", a Christianity which had already lost its full authenticity through institutionalization and compromise with the world.[114]

The exegesis of Paul on this matter of the Law is of course difficult and here I can only state my understanding without argument, referring the reader to the notes.[115] I believe that once we allow for the polemic simplifications into which Paul was sometimes forced by his need to defend his Gentile ministry to the Judaizers, we find that what he means by the abolition of the Law are two related points: (1) Jesus has brought to completion the teaching of the prophets that salvation does not come through a mere external obedience of the Law but through the interior spirit of faith like that which saved the patriarchs before the Law was given through Moses, and this faith is possible to us only by the grace of the Spirit given through Christ. (2) In view of this true understanding of the purpose for which God gave the Jews the Torah, namely, to prepare them for the coming of Christ, it has now become obsolete for Christians insofar as it is a *particularistic* law for the Jewish people. What remains valid for Christians, therefore, are only those universal demands of God's will given in creation to all people alike and renewed in the teaching of Jesus. That this explanation is correct is evident from the fact that Paul never ceases to exhort the Christians not to rely on their own righteousness but on the power to live a life pleasing to God through faith in Christ, a faith that causes a total change of attitude, a new creation, an interior rebirth; yet at the same time he exhorts and commands them to carry out moral duties which are identical with the chief commandments of the Torah, freed of whatever was peculiar to the Jews as a particular nation.

If this interpretation of Paul is correct, Paul's teaching is not contradictory, but complementary to other New Testament writings which take a somewhat different approach to the question, for example, St. Matthew's version of the Sermon on the Mount in which Jesus rejects the notion that He has come to abolish the Law (Mt 5: 17-20), or the *Epistle of St. James* in which the necessity of good works in addition to faith is extolled. Moreover, it seems entirely in harmony with Jesus' own teaching and conduct. Jesus is never reported to have repudiated the Torah, but only *some* of the interpretations given it by the scribes and Pharisees. His own interpretation of the Law free it from particularistic elements and lead it back to God's original intention in creation (e.g. see Mk 10:2-12).[116]

Thus the notion of a natural moral law in the sense of a universal moral order given in the creation of human nature, and accessible to human reason as a reflection on human experience, is by no means foreign to the Bible but is implied whenever the biblical writers speak of God's original intentions for humanity in His creation and of His universal providence over all the children of Adam and Eve. The Torah itself embodies this natural law but in a particular, historical, culturally conditioned form in view of the needs of the Jewish people and their unique vocation, to which they are still called to be faithful. With the coming of Jesus, the hopes of the prophets are fulfilled by opening to all nations, whatever their history and culture, the special election given to the Jews for the sake of all the nations with the result that the Torah remains as a confirmation of the universal moral law. It is freed from its historical limitation so that it can serve to guide all people, without subjugating them to the special culture of one nation.

When in Jesus Himself as the New Adam (Rom 5:15) the original order of creation is restored, the natural law which had been partially obscured by sin (Rom 1:21) is clarified; but the new order surpasses the original order. "Scripture has it that Adam, the first man, became a living soul; the last Adam has become a life-giving spirit . . . The first man was of earth, formed from dust, the second is from heaven . . . Just as we resemble the man from earth, so shall we bear the likeness of the man from heaven" (I Cor 15:45,47,49). Paul does not distinguish in the first Adam, as Aquinas does, between what was of nature and what was of grace, yet he does here indicate that human nature, even apart from sin, has been elevated in Christ to a level that transcends the earthly order of nature, the order of the "spirit" in the sense of a share in the very life of God. Thus the goal of human life in Christ transcends the goal of life as known from the natural law. In the Old Testament Psalm 115:16 we read, "Heaven belongs to Yahweh, earth he bestows on man," but Paul sees mankind as called in Christ to share the life of heaven. Such an expectation exceeds what belongs to human creatures as human, but is possible to them in Christ who is truly human, the New Adam, yet from heaven.

How does this agree with the teachings of the earthly Jesus, who in preaching the Reign of God evidently saw it as something to be realized on earth, and taught His disciples to pray, "Your kingdom come, your will be done on earth as it is in heaven" (Mt 6:10), and promised to the hungry that they would have their fill (Lk 6:21)? It is clear that in Jesus' own teaching the Reign of God is to be a restoration of the earthly peace and justice which God had intended in the original creation and to which the Law and the prophets had continually recalled the people. And yet it is also a mysteriously transcendent life expressible only in parabolic symbols, an everlasting life with and in God.[117]

From these biblical data, it follows that Christian ethics has as its first principle a goal which comprises both an earthly and a heavenly order, since this goal is Jesus Christ Himself. To be a successful Christian is to be not only a disciple of Christ but to be incorporated in Him, a member of His community which lives by His life, and united to the Father through His union with the Father in the Spirit. The Reign of God is this union of all human beings in a single community of *agape* made possible by the vivifying Spirit of Christ. Since Christ is the New Adam who is both earthly and heavenly, because the "man from heaven" is the same who ascends to heaven (Eph 4:10), so the Reign of God is both earthly and heavenly, God's will "done on earth as it is in heaven" (Mt 6:10), and a Christian ethics includes the natural order and its natural law, but transcends them by a heavenly law of the Spirit. Christian ethics is a following of Christ, an *imitatio Christi*, which is necessarily the taking up of the Cross "daily" (Lk 9:23), but at the same time it is a life in the Risen Lord, a deification by his Holy Spirit of which it is possible for Paul to say, "All of us, gazing on the Lord's glory with unveiled faces, are being transformed from glory to glory into his very image by the Lord who is the Spirit'' (2 Cor 3:18).

What use can such an ethics make of the philosophical methodologies which we discussed earlier? Certainly it cannot neglect those deontological problems that concern the authority of God and obedience to His will as it is expressed in the Law, that is in the Old Testament Law as correctly interpreted and completed by Jesus (Mt 5:17-20), as well as it is expressed in the natural law of creation (Mt 19:4-9), and in legitimate human authority (Rom 13:1). Because of our creaturely finitude and also our sinfulness, we needed God's instruction to prepare us for the coming of Jesus, and although in him we now have the final Word of God, yet we still require instruction in his teaching through the ministries of church and state, whose authority he has confirmed. Jesus did not come to destroy the order of creation nor the social order, but to redeem and transform them into the Reign of God. He himself was obedient to this earthly order insofar as it was not distorted by sin, and he was obedient to the mysterious providence of his Father even when, as in the agony in the Garden (Mk 14:32-42), this will of God seemed beyond human endurance. The Christian, therefore, must also obey the moral law in all its forms, even when he or she cannot understand the wisdom on which it must rest in order to be law at all. This attitude of obedience, however, does not apply to those human "laws" which have the appearance and human sanctions of law but which in fact are a violation of the law of God (Acts 5:29), since obedience to God demands that we resist such "laws."

The permanent value of this deontological understanding of morality, however, is only preliminary, not ultimate. The Divine Father whom Jesus has revealed to us in his own Sonship (Jn 14:9) is not a despot who issues laws to demonstrate his power, but a wise and loving Father, whose will is not for himself but always for our happiness which is a share in His own beatitude. His will, therefore, is only the manifestation of his loving wisdom by which he guides our uncertain steps toward the true goal of our lives which is this share in his beatitude.

However, since we could not be open to this beatitude unless God had created us intelligent and free, his guidance of us is not only by the unconscious teleology of natural physical law, nor as for animals by the conscious but deterministic teleology of the laws of instinct, but by the law of the conscious, free teleology of reason, the natural moral law proper to us as intelligent creatures who can understand our own nature in both its individual and social needs, and by the interior guidance of the Spirit given us in Christ, by which we hear the Father's call to a union with himself in Christ, transcending even our created natures. God's pedagogy of the human race is to lead us in our childhood by law, but in our maturity by the inner guidance of the Spirit enlightening our own reasonable judgements of conscience which are a share in His Divine wisdom (Gal 3:24-25). Hence, Christian ethics must ultimately rest, not on a deontological, but on a teleological basis.

Such a Christian teleological ethics, however, cannot be a mere utilitarianism or consequentialism or pragmatism, because it is one of the most striking features of Jesus' moral teaching as seen in the Sermon on the Mount (especially Mt 6:1-18) that morality consists not merely in the external effects of an act, but first of all in the interior motive, in the intrinsic character of the free act. The good or harm done by an act is first of all its effect on the person making the free choice (which can be called a "consequence" only improperly). We become what we choose to do, so that the harm of lying is first of all that it makes the speaker a liar, whether his lie deceives anyone or not. "What comes out of the mouth originates in the mind . . It is things like these that make a man impure" (Mt 15:18-19).

Thus Jesus confirmed the central message of the prophets in their interpretation of the Law, "Obedience is better than sacrifice" (I Sm 15:22). This teaching is not deontologism, because the point of Jesus and the prophets is not that an act is good or bad because authority has so decreed, but teleologistic because its teaching implies that a human act is good or bad because it is measured by the right intention of the end. It is ultimately for the same reason that Proportionalism is not viable as a Christian ethic because it rejects the possibility that an act is *intrinsically* wrong by reason of its object, apart from the concrete circumstances in which this object is chosen (e.g. that a lie or the direct killing of the innocent are always wrong). Since morality begins with the intention of the object, if that object is of itself contradictory to the nature of the agent it cannot be made consistent with the moral good of the agent by the external circumstances in which it is performed.

Moreover, the ethics of Jesus cannot be adequately expressed in an ethics like that of Max Scheler or of Martin Heidegger, because such idealistic systems are not rooted in the concrete history of Jesus' life and the experience of his followers. Nor can Jesus' ethics be reconciled with cultural relativism, because Jesus came as a prophet to redeem all existing human cultures and all man-made ideologies from their sinful distortions of God's original order of creation. He wished to transcend the divisions of human culture and to unite the whole human family so that they might walk together toward one goal.

It is not sufficient, however, to assert against relativism the existence of transcultural needs. I have now to ask: What are these basic human needs which constitute the natural law or order of creation which I have claimed must be respected by any Christian ethics, since Jesus came first of all to redeem that creation from its fallen and frustrated condition? The goal of human life which is the first principle of any teleological ethics must include the satisfaction in an integral manner of these basic needs as the primary condition of a good human life and indeed of a Christian life, since "grace perfects nature." If Jesus is the existential embodiment of his own ethics and therefore its goal and first principle, these basic needs and their right fullfilment must be manifest in Him.

St. Thomas Aquinas reduced these basic needs to four: life, reproduction, truth, and society.[118] The scientific picture given in Chapter 2 of what it is to be human as interpreted in Chapter 7 and 8 provides us with six basic needs which amount to those of Aquinas by dividing his "life" into the need for food and security, and by adding the need for "creativity":

1) **Food:** Human beings are part of the animal world and therefore need appropriate nourishment, water and air.

2) **Security:** They also need bodily security from injury by natural forces, animals or other humans, especially so as to permit time for rest and sleep when they are defenseless. They also need physical freedom for movement and exercise.

3) **Sex:** Finally, in their animality humans need sexual reproduction by which the continuity of the species and of the evolutionary process is maintained. Moreover, in the specifically human form that reproduction takes, they need the stable relationship between man and woman, parents and children necessary to support this continuity of the species biologically and culturally.

4) **Information:** As thinking animals they need not only the opportunities to acquire sense knowledge but also the opportunity to learn intellectually for themselves and through communication by language with others.

5) **Society:** As humans they need society, not only to obtain the foregoing needs but also because human friendship and sharing of common goods is itself a need.

6) **Creativity:** Freedom from the pressures of the foregoing needs and other external social pressures is necessary for humans to be creative in the arts and science by which human culture advances: and especially is this necessary for human life to be open to transcend the human, to seek the Ultimate Totality.

396

Obviously an ethics requires a ranking of these needs, since sometimes in life some values must be given preference to others while attempting to integrate them all harmoniously without neglecting any. It is also clear that in assigning such priorities, two different kinds of priorities must be considered. On the one hand, the first items in this list are more necessary for human *survival*. We must have food, security, exercise, reproduction if we are to live at all. On the other hand, the latter items on the list have greater value for human self-realization. We do not live to eat, but eat to live, and human living consists in love and friendship, the search for truth, creativity and transcendence. Finally, we cannot simply sacrifice either half of the list to the other, because we cannot communicate socially, think, or create, except in and through our bodily existence and survival.

We have now to ask if Christian ethics is really concerned for these very mundane necessities of human life. Is it not rather a spiritual ethic in which these "necessities" become unimportant? Did not Jesus say, more spiritually than pragmatically, "O weak in faith! Stop worrying, then, over questions like, 'What are we to eat, or what are we to drink, or what are we to wear?'"? (Matt 6:31).

Notes

1. The *Humanist Manifesto II* (drafted by Paul Kurtz and Edwin H. Wilson), *The Humanist*, Nov. 1973: 28-36 affirms "a set of common principles that can serve as a basis for united action — positive principles relevant to the present human condition." The third of these principles is "that moral values derive their source from human experience. Ethics is *autonomous* and *situational*, needing no theological or ideological sanction. Ethics stems from human need and interest. To deny this distorts the whole basis of life. Human life has meaning because we create and develop our futures". It is not clear whether this is an acceptance of ethical relativism or not. On the one hand these principles are proposed as "common" and on the other they are purely human "creations". What prevents the adoption of many contradictory ethical standards which will set the human community at odds with itself?

2. Karl Popper, *The Poverty of Historicism* (New York, Basic Books, 1957). More commonly the term is used to apply to the philosophical position that history is the supreme form of knowledge to which all other forms of knowledge are reducible. A notable example is R. G. Collingwood, *The Idea of History* (Oxford, Clarendon Press, 1962).

3. The *Humanist Manifesto II* says "Traditional religions are surely not the only obstacles to human progress. Other ideologies also impede human advance. Some forms of political doctrine, for instance, function religiously, reflecting the worst features of orthodoxy and authoritarianism, especially when they sacrifice individuals on the altar of Utopian promises. Purely economic and political viewpoints, whether capitalist or communist, often function as religious and ideological dogma. Although humans undoubtedly need economic and political goals, they also need creative values by which to live." It is not clear just what "creative values" means.

4. See Chapter 3, pp. 85-88.

5. See Edward Schillebeeckx, O.P., *Revelation and Theology* (New York: Sheed and Ward., 1968), 2, 140-145, and J. M. Robinson, "Revelation as Word and as History", in J. M. Robinson and John B. Cobb Jr. ed., *Theology as History*, (New York: Harper and Row, 1967), pp. 1-100 which reviews the whole problem in Protestant theology, and Wolfhart Pannenberg, "The Revelation of God in Jesus of Nazareth", *ibid.*, pp. 1-1-134 and response to discussion pp. 221-276. However Paul Ricoeur, "Toward a Hermeneutic of the Idea of Revelation", *Harvard Theological Review* 70 (Jan.-Sept., 1977): 1-37 shows that history is only one mode of revelation.

6. See Claude Tresmontant, *A Study of Hebrew Thought* (New York: Desclée, 1960), pp. 17-38. On *Qoheleth's* reversion to cyclical time which some scholars believe indicates pagan influences see R. B. Y. Scott, *The Way of Wisdom in the Old Testament*, (New York: Macmillan, 1971), pp. 176-181.

7. See A. Luneau, *L'histoire du salut chez les pères de l'Église* (Paris: Editions du Cerf, 1964) for the relation of Augustine's thought on history to that of the other Church Fathers. On his own thought on history see J. N. Figgis, *The Political Aspects of St. Augustine's City of God* (London: Longmans, 1921), Chapter III; J. O'Meara, *Charter of Christendom: The Significance of the City of God* (New York: Macmillan, 1961) and G. L. Keyes, *Christian Faith and the Interpretation of History* (Lincoln, Nebraska: University of Nebraska Press, 1966).

8. See Bernard McGinn, *Visions of the End: Apocalyptic Traditions in the Middle Ages* (New York: Columbia University Press, 1979).

9. *Summa Theologiae* I, q. 1, a. 2 ad 2. Yet Aquinas frequently raises questions about why certain events of salvation occurred when they did, e.g., I-II, q. 98, a. 6; II-II, q. 174, a. 6; III, q. 1, a 5-6; III, q. 43, a. 3 etc.

10. *Physics* IV, c. 12, 220b 5-14; cf. *Meteorologica* I, c. 14, 351a 19 sq.

11. Aristotle, De Caelo, 1, 270b 1-25. For discussion see my *Aristotle's Sluggish Earth* (River Forest, IL: Albertus Magnus Lyceum, 1958) (*New Scholasticism*, 32, 1958): pp. 10-31, 202-234.

12. *Ibid.*, p. 39, note 29; cf. Aquinas' commentary on the *De Caelo* I, lect. 7, no. 6 and *Summa* I, q. 32, I ad 2 where he says, "Just as in astronomy it is hypothesized that there are excentrics and epicycles because, granted such a hypothesis, the observed phenomena of the celestial motions can be saved, nevertheless this explanation is not sufficiently demonstrated, because the phenomena might also be saved by some other hypothesis."

13. See Chapter 5, pp. 160-162 and, 6 pp. 208 with references. On this period of historical writing see Harry Elmer Barnes, *A History of Historical Writing*, 2nd rev. edition (New York: Dover, 1963), pp. 120-147. On the limitations of this polemical historiography see Cyriac K. Pullapilly, *Caesar Baronius: Counter Reformation Historian* (Notre Dame, Ind.: University of Notre Dame Press, 1975), pp. 144-178.

14. For an overview of the philosophy of history see John Edward Sullivan, O.P., *Prophets of the West* (New York: Holt, Rinehart and Winston, 1970). In what follows I have found the following works especially helpful: Paul Tillich, *The Interpretation of History* (New York: Charles Scribner's Sons, 1936); Karl Jaspers, *The Origin and Goal of History* (New Haven: Yale University Press, 1953); Reinhold Niebuhr, *The Self and the Dramas of History*, (New York: Charles Scribner's Sons, 1955); Jacques Maritain, *On the Philosophy of History* (New York: Charles Scribner's Sons, 1957); Karl R. Popper, *The Poverty of Historicism* (New York: Basic Books, 1957); Emil Fackenheim, *Metaphysics and History*, Aquinas Lecture, 1961, (Milwaukee: Marquette University Press, 1961); Grace E. Cairns, *Philosophies of History: Meeting of East and West in Cycle-Pattern Theories of History* (New York: Philsophical Library, 1962); R. G. Collingwood, *The Idea of History* (Oxford: Clarendon Press, 1962); Karl Löwith, *Nature, History and Existentialism* (Evanston: Northwestern University Press, 1966); Henri Marrou, *The Meaning of History* (Baltimore: Helicon, 1966); Eric Voeglin, *Order and History* 4 vols. (Baton Rouge, Louisiana State University Press, 1974); George G. Iggers, *New Directions in European Historiography* (Middletown, Conn: Wesleyan University Press, 1975); Isaiah Berlin, *Vico and Herder* (New York: Viking Press,

1976); Edgar V. McKnight, *Meaning in Texts: The Historical Shaping of a Narrative Hermeneutics* (Philadelphia: Fortress, 1978). Especially helpful is Pitrim A. Sorokin, *Social Philosophies in an Age of Crisis* (London, Adam and Charles Black, 1952) in which he reviews the attempts of Nikolai Danilevsky, Oswald Spengler, Arnold J. Toynbee, Walter Schubart, Nikolai Berdyaev, F. S. C. Northrop, Alfred L. Kroeber, and Albert Schweitzer to find patterns in history. See also Sorokin's, *Social and Cultural Dynamics*, 1962, 4 vols., (New York-Cincinnati: American Book Co., 1937-41) vol. 2, Chapter 10 "Problem of Linear, Cyclical, and Mixed Conceptions", pp. 351-384 and his "Reply to My Critics" in Philip J. Allen, *Pitrim Sorokin in Review*, (Durham, N.C.: Duke University Press, 1963), pp. 371-496.

15. For example the work of Jaroslav Pelikan, *Historical Theology* (New York: Corpus Books, 1971) and of Yves Congar, O.P., *After Nine Hundred Years* (New York: Fordham University Press, 1959).

16. As an exellent example of honest dealing with a scandal in the history of the Church see Edward H. Flannery, *The Anguish of the Jews: Twenty-three Centuries of Anti-Semitism* (New York: Macmillan, 1965).

17. See Jeffrey Burton Russell, *The Devil: Perceptions of Evil from Antiquity to Primitive Christianity* (Ithaca, N.Y., Cornell University Press, 1977) and *Satan: The Early Christian Tradition, (ibid.,* 1981). In the latter work Burton asks, "What, according to the historical theology of concepts is one to believe regarding the Devil? We should be willing to face the problem of evil squarely without trying to dodge it intellectually. We should be open to the possibility of the existence of an evil spirit or spirits beyond humankind. The metaphysical assumptions of our present age may lead many to prefer to interpret the diabolical in terms of depth psychology, arguing that the demonic exists within the human mind, or perhaps collectively among human minds. But on no account is one entitled to dismiss the idea of the Devil as irrelevant." p. 226. I will discuss this topic further in Chapter 11.

18. *Vatican II: The Conciliar and Post Conciliar Documents.* ed. by Austin Flannery, O.P., Vatican Collection, vol. I (Northport, N.Y.: Costello, 1975); *The Church in the Modern World (Gaudium et Spes)* p. 903-1014, Chapter 111 "Man's Activity in the Universe", nn. 33-39, pp. 933-938.

19. Chapter 7, pp. 264 f.

20. "Everything in the world, indeed, the world itself, is contingent; it might have been other than it is. Thus the clarity, distinctness, and logical adequacy of thought does not determine existence; logical necessities are merely metaphysical possibilities. Another principle, not merely logical, is needed to explain why this world is as it is, and is not another of the infinitely many possible worlds. This new principle for the determination of this particular world is the principle of the best possible (or the optimum). It applies to God's creation of this world and it applied to every thoughtful choice of the best possible decision out of the many possible decisions in a particular situation." p. 170. The empirical content of knowledge has two sources — sense perception for the external world, and internal perception, or reflection for man's knowledge of his own mental processes. But the external perception is not intuitive; it is part symbolic and limited by the *materia prima* of sense, while internal perception is not thus limited and gives reality immediately. Therefore man's knowledge of the external world is phenomenal, science deals not with reality but with 'well-found' appearances. Man's knowledge of himself, on the other hand, is the basis of metaphysics, for reality consists of many series of appetitive perceptive events such as he finds himself to be." p. 170; Leroy E. Loemker, *Struggle for Synthesis: The Seventeenth Century Background of Leibnitz's Synthesis of Order and Freedom* (Cambridge, Mass: Harvard University Press, 1972). See *Candide* in *Voltaire: Romans et Contes,* Bibliotheque de la Pléiade (Bourques: Gallimard, 1954).

21. Raymond E. Brown, *The Critical Meaning of the Bible*, (New York Ramsey, N.J.: Paulist Press, 1981).

22. See Paul Ricoeur's article cited in note 5 above.

23. See James A. Sanders, *Torah and Canon*, Philadelphia, Fortress, 1972 and Bevard S. Childs, *Biblical Theology in Crisis* (Philadelphia, Westminster, 1970); *The Book of Exodus*, same publisher, pp. 42-46 and *Introduction to the Old Testament as Scripture* (Philadelphia: Fortress, 1979, pp.71-83).

24. "The multiplicity of senses [in Scripture] does not result in equivocation, since . . . these senses are not multiplied because one word signifies many different things; but because *the things signified by the words can themselves be signs of other things*. And thus no confusion results in Sacred Scripture, since all the senses are founded on one sense, namely the literal sense; from which alone, not from what is said allegorically, can [theological] arguments be drawn". *Summa Theologiae* I, a. 1, a. 10, ad 2. Thus for Aquinas the persons and events of Biblical history are used by God to communicate meaning.

25. See Henri de Lubac, *Histoire et Esprit* (Paris: Aubier, 1950) and M. Van Esbroeck, *Herméneutique, structuralisme et exégèse* (Paris, Desclée, 1968), 101-199 on the medieval notion of the four senses of Scripture — one literal and three spiritual.

26. See Benedict Otzen, Hans Gottlieb and Knud Jeppsen, *Myths in the Old Testament* (London, SCM Press, 1980), especially Otzen pp. 58-61; J. W. Rogerson, *Myth in Old Testament Interpretation* (Berlin: Walter de Gruyter, 1974), pp. 174-189; and Avery Dulles, *Myth, Biblical Revelation and Christ*, (Washington, D.C.: Corpus Papers, 1969.)

27. See H. Christ, article "Cross, Finding of the Holy", *New Catholic Encyclopedia*, vol. 4, pp. 479-482.

28. For references see Chapter 7, note 3.

29. Jacques Monod, *Chance and Necessity: An Essay on the Natural Philosophy of Modern Biology* (New York: Knopf, 1971), pp. 169-180.

30. The nature-culture controversy was renewed by Edward O. Wilson in his *Sociobiology: The New Synthesis* (Cambridge, Mass.: Harvard University Press, 1975) and *On Human Nature* (same press), 1978 and vigorously attacked by many such as M. Sahlins, *The Use and Abuse of Biology* (Ann Arbor: University of Michigan Press, 1980). See also Richard Dawkins, *The Selfish Gene* (New York: Oxford University Press, 1976); Arthur L. Caplan, ed., *The Sociobiology Debate* (New York, Harper, Colphon Books, 1978); S. Gregory, A. Silvers and D. Sutch eds., *Sociobiology and Human Nature* (San Francisco; Jossey-Boss, 1978); Alexander Rosenberg, *Sociobiology and the Preemption of Social Science* (Baltimore: John Hopkins University Press, 1980). Recently Wilson has reformulated his theory as "gene-culture coevolution" in *Genes, Mind and Culture: The Coevolutionary Process* (Cambridge, Mass: Harvard University Press, 1981) and with Charles J. Lumsden, *Prometheus Fire: Reflections on the Origin of Mind*, same publishers, 1983. See Theodosius Dobzhansky, *The Biology of Ultimate Concern*, (New York: American Library, 1967) and Robert A. Wallace, *The Genesis Factor* (New York: William Morrow and Co. in association with Publisher's Inc., 1979) for examples of ethical applications. Marvin Harris, *Cultural Materialism: The Struggle for a Science of Culture* (New York: Random House, 1979) sums the controversy up as follows:

"The disagreement between sociobiologists and cultural materialists on the issue of human nature is a matter of the contraction versus the expansion of the postulated substance of human nature. Cultural materialists pursue a strategy that seeks to reduce the list of hypothetical drives, instincts, and genetically determined response alternatives to the smallest possible number of items compatible with the construction of an effective corpus of sociocultural theory. Sociobiologists, on the other hand, show far less restraint and actively seek to expand the list of genetically determined traits whenever a plausible opportunity to do so presents itself. From the cultural materialist perspective, the proliferation of hypothetical genes for human behavioral specialties is empirically as well as strategically unsound, as I shall show in the next section". p. 128.

32. See quote Chapter VIII, note 46. Aristotle uses the term *historia* chiefly as meaning a roughly organized collection of facts, as in his *History of Animals,* a usage carried over in our term "natural history". He does, however, show a strong historical sense in his use of history in the *Politics,* and the *Athenian Constitution,* the only survivors of his many politico-historical researches.

33. See Reijer Hooykaas, *The Principle of Uniformity in Geology, Biology, and Theology: Natural Law and Divine Miracle* (Leiden: Brill, 1963).

34. See the collection of texts with an introduction by H. P. Rickman, Wilhelm Dilthey, *Pattern and Meaning in History* (New York: Harper Torchbooks, 1961).

35. R. G. Collingwood, note 14 above.

36. See note 2 above.

37. See Hegel's, *The Philosophy of History* (translated by J. Sibree) (New York: Dover, 1956) and *Phenomenology of Mind* (translated by J. B. Baillie) (New York: Harper and Row, 1967) and the discussion in Sullivan, *Prophets of the West*, note 14, above, pp. 38-47. Also J. N. Findlay, *Hegel: a Re-Examination* (London: Allen and Unwin, 1958); James Collins, *God in Modern Philosophy* (Chicago: Henry Regnery, 1959), pp. 201-237; Karl Löwith, *From Hegel to Nietzsche* (New York: Holt, Rinehart and Winston, 1964); Walter Kaufmann, *Hegel: A Re-Interpretation* (Garden City, N.Y.: Doubleday Anchor Books, 1966) (a strongly Humanist view); Emil L. Fackenheim, *The Religious Dimension in Hegel's Thought* (Boston: Beacon Press, 1967); André Leonard, *Le foi chez Hegel* (Paris, Desclée, 1970).

38. David Hume, *Dialogues Concerning Natural Religion*, ed. by N. K. Smith (Indianapolis: Bobbs-Merrill, 1964).

39. See Richard E. Palmer, *Hermeneutics: Interpretation Theory in Schleiermacher, Dilthey, Heidegger and Gadamer*, (Evanston, IL: Northwestern University Press, 1969), pp. 118-121.

40. See Thomas C. Anderson, *The Foundation and Structure of Sartrean Ethics* (Lawrence, Kansas: The Regents Press of Kansas, 1979) who shows (pp. 57-60) that even in Sartre's ethics there is teleology and the concept of certain universal human needs which must be satisfied. John Passmore in his *The Perfectibility of Man* (New York: Scribners', 1970) p. 140 shows that Pico della Mirandola (d. 1494) in his famous *Oration on Human Dignity* anticipated Sartre's idea that man has no determinate nature before he makes a choice, and Carolus Bovillus (d. 1553) had put forward the same idea (p. 104). Originality in philosophy is hard to come by!

41. Peter Gay, *The Enlightenment: An Interpretation* (New York: Alfred A. Knopf, 1967), vol. 1, The Rise of Modern Paganism, pp. 31-211, Book One "The Appeal to Antiquity".

42. See Henri I. Marrou, *The Meaning of History* (Baltimore: Helicon, 1966), pp. 155-176 on the necessity in writing history for universal concepts, including the notion of "human nature".

43. For Luther on natural law see Paul Althaus, *The Ethics of Martin Luther* (Philadelphia: Fortress Press, 1972), pp. 25-34. For Calvin see Arthur C. Cochrane, "Natural Law in Calvin" in Elwyn A. Smith, ed., *Church-State Relations in Ecumenical Perspective* (Pittsburgh: Duquesne, 1966), pp. 176-217 and David Little, "Calvin and the Prospects for a Christian Theory of Natural Law" in Gene Outka and Paul Ramsey, ed. *Norm and Context in Christian Ethics* (New York: Scribner's, 1968), pp. 175-197. On the Biblical Foundations see Walther Eichrodt, *Theology of the Old Testament* (Westminster: Philadelphia, 1967), pp. 231-531, and C. H. Dodd, "Natural Law in the New Testament", in his *New Nestament Studies* (Manchester: Manchester University Press, 1967), pp. 129-142.

44. See Garry Wills, *Inventing America: Jefferson's Declaration of Independence* (Garden City, N.Y.: Doubleday, 1981), which argues that the philosophical basis of the notion of "self-evident" human rights was supplied by the Scottish school of "common-sense".

45. See Jacques Maritain, *The Rights of Man and Natural Law* (New York: Scribner's, 1943). On recent papal teaching on natural rights see Joseph B. Gremillion, *The Gospel of Peace and Justice* (Maryknoll, N.Y.: Orbis Books, 1976) pp. 10-13 and the encyclical *Pacem in Terris* of John XXIII, *ibid.* pp. 201-242.

46. The following philosophical, historical and theological works on natural law have been helpful: Mortimer J. Adler, *The Time of Our Lives: The Ethics of Common Sense* (New York: Holt, Rinehart and Winston, 1970), especially Chapter 17, "Presuppositions about Human Nature", pp. 185 ff.; Ross A. Armstrong, *Primary and Secondary Precepts in Thomistic Natural Law Teaching* (The Hague: Martinus Nijhoff, 1966); Oscar J. Brown,

Natural Rectitude and Divine Law in Aquinas (Leiden, Brill, 1981); Scott Buchanan, "Natural Law and Teleology" in John Cogley ed., *Natural Law and Modern Society* (Cleveland: Meridian Books for the Republic, 1962); pp. 82-153; M. B. Crow, *The Changing Profile of the Natural Law* (The Hague, Martinus Nijhoff, 1977); Illtud Evans ed., *Light on the Natural Law* (New York: Helicon, 1965); Josef Fuchs, *Natural Law: A Theological Investigation* (New York: Sheed and Ward, 1963); James M. Gustafson, "What is Normatively Human", *American Ecclesiastical Review* 165 (Nov., 1971): 192-207 (shows 5 different ways of arriving at answer); Stanley S. Harakas, "Eastern Orthodox Perspectives on Natural Law", *Proceeding of Annual Meeting of the American Society of Christian Ethics,* 1977, pp. 41-56; William A. Luijpen, *Phenomenology of Natural Law,* (Pittsburgh: Duquesne University Press, 1967); Jacques Maritain, *The Rights of Man and the Natural Law* (New York: Scribner's, 1943); D. Mongillo, "L'elemento primario della legge naturale in s. Tommaso" in L. Rossi ed., *La Legge Naturale* (Bologna, 1970), pp. 219-244; Johannes Messner, *Social Ethics: Natural Law in the Western World* (St. Louis: B. Herder, 1965); Jeremiah Newman, *Conscience versus Law: Reflections on the Evolution of Natural Law* (Chicago: Franciscan Herald Press, 1971); Yves Simon, *The Tradition of Natural Law* (New York: Fordham Press, 1965). The question of the origin of the term "natural law" raised by Helmut Koester, *"Nomos Physeos:* The Concept of Natural Law in Greek Thought" in Jacob Neusner, ed., *Religions in Antiquity: Essays in Memory of E. R. Goodenough* (Leiden, E. J. Brill, 1968), pp. 521-541 who denied it was Stoic and argued it originated with Philo and had Judaic roots: seems to have been settled by Richard A. Horsely, "Law of Nature in Philo and Cicero", *Harvard Theological Review* 71 (Jan.-Apr., 1978): 35-59 who shows that Philo and Cicero probably got it from Antiochus of Ascalon who got it from Zeno the founder of Stoicism who probably derived the notion if not the term from Aristotle, "The Stoics themselves, however, were responsible for the identification of the law according to nature with the universal reason and the constitution of the cosmos" (p. 39). The notion that God is the Legislator (found in Cicero and Philo) is, however, not typically Stoic but Platonic, and was probably revived by Antiochus.

47. See Barth's *The Epistle to the Romans,* (1928) translated by Edwyn C. Hoskyns, (Oxford: Oxford University Press), 1968, pp. 45-54. Barth admits that Plato and other pagans knew there was a divine, invisible reality, but reason without faith can only lead to erroneous and idolatrous notions of God. What reason tells us is that God is unknowable. On what this means for ethics see Robert E. Willis, *The Ethics of Karl Barth* (Leiden: Brill, 1971).

48. See James M. Gustafson, *Protestant and Roman Catholic Ethics: Prospects for Rapprochement* (Chicago: University of Chicago Press, 1978), pp. 80-94 and Ian T. Ramsey, "Towards a Rehabilitation of Natural Law" in his *Christian Ethics and Contemporary Philosophy* (New York: Macmillan, 1966), pp. 382-396. Also David Little, note 43 above.

49. *The Law of Christ* (Westminster, Md.: Newman, 1960) 2 vols. and *Free and Faithful in Christ,* 3 vols. (New York: Seabury, 1978).

50. Max Scheler, *Formalism in Ethics and Non-formal Ethics of Value* (Evanston: Northwestern University Press, 1973); cf. Alfons Deeken, *Process and Permanence in Ethics,* (New York: Paulist Press, 1974).

51. Many of the most important articles of this school of thought are collected in Richard A. McCormick, S.J., and Paul Ramsey, eds., *Doing Evil to Achieve Good* (Chicago: Loyola Press, 1978) and Charles E. Curran and McCormick, *Readings in Moral Theology,* No. 1: *Moral Norms and Catholic Tradition* (New York: Paulist Press, 1979). The two basic articles are Peter Knauer, S.J., "The Hermeneutic Function of the Principle of Double Effect", *Natural Law Forum,* 12 (1967): 132-162 and Josef Fuchs, S.J., "The Absoluteness of Moral Terms", *Gregorianum,* 52 (1971): 415-458. More extensive bibliography can be found in Servais Pinckaers, O.P., "La question des actes intrinsequement mauvais et l' 'proportionalisme,' " *Revue Thomiste* 82 (April-June, 1982): 181-212, and in Germain Grisez, *The Way of the Lord Jesus* (Chicago: Franciscan Herald Press), vol. 1, "Christian Moral Principles", Chapter 6: "Critique of the Proportionalist Method of Moral Judgement," pp. 141-172. Richard A. McCormick, S.J., clearly expounded the system in *Ambiguity in Moral Choice* (Milwaukee: Marquette

University Press, 1973) and has continued to answer his critics in the section "Moral Notes" of *Theological Studies*. These articles are published in *Notes on Moral Theology: 1965-1980* (Lanham, MD: University Press of American, 1981) and furnish what is probably the most extensive bibliography on the controversy. A very helpful analysis of the development of McCormick's own position is Lisa Sowle Cahill, "Teleology, Utilitarianism, and Christian Ethics", *Theological Studies* 42 (1981): 601-29 and "Contemporary Challenges to Exceptionless Moral Norms" in *Moral Theology Today: Certitudes and Doubts* (St. Louis: Pope John Center, 1984) pp. 121 ff.

52. Servais Pinckaers, O.P., "La question des actes intrinsequément mauvais" (note 51 above) argues that the "new moral theology" (proportionalism) is intelligible only as a reaction against the type of Post-Tridentine moral theology which was grounded in voluntarism and legalism and that it fails to make a radical break with this voluntaristic tradition. Richard McCormick has complained that Pinckaers fails to understand the issue which proportionalists are addressing, but himself fails to engage Pinckaer's critique. See "Moral Notes", *Theological Studies* 44 (March, 1983): 71-122, pp. 78-80.

53. On various meanings of the term "Law" in the Bible see Klaus Berger, article "Law in Karl Rahner ed. *Encyclopedia of Theology,* (New York: Seabury, 1975), pp. 822-830; and Gunnar Ostborn, *Tora in the Old Testament: A Semantic Study* (Lund, Hakan Ohlssons, 1945).

54. See E. Vernon Arnold, *Roman Stoicism* (1911): New York: Humanities Press, 1956, pp. 273-300 and Geneviève Rodis-Lewis, *La Morale Stoicienne* (Paris: Presses Universitaires de France, 1970), pp. 26-30 (especially on the concept of *homologia* or harmony between natural and moral law). Some recent authors are shocked by the way in which Aquinas makes use of the Stoic lawyer Ulpian's definition of the precepts of the natural law as "those things which nature teaches all living things" (*Summa Theologiae* I-II q. 94, a. 2 c.), but it should be noted that Aquinas quotes this legal saying only to affirm that the natural law is based on natural inclinations of human nature, *some* of which (reproduction, care of offspring etc.) are generic, while others are specific to human nature. Of course this implies, in Aquinas' philosophy, that generic traits are modified by the specific ones e.g., human sexuality differs from that of the brutes because it is the sexuality of a free and intelligent person.

55. On the history of this question of intrinsically evil acts see John F. Dedek, "Moral Absolutes in the Predecessors of St. Thomas", *Theological Studies* 38 (1977): 654-680 and "Intrinsically Evil Acts: An Historical Study of the Mind of St. Thomas", *Thomist* 43 (1979): 385-413. Dedek tries to show that the scholastics in their efforts to explain how God could have dispensed the patriarchs and others in Old Testament times to perform acts that the scholastics themselves consider to be intrinsically wrong, were in fact admitting the proportionalist thesis that the material or ontic character of an act *never* determines its morality absolutely. This argument suffers from two weaknesses: (1) He admits that the scholastics had no doubt that some acts are always wrong. Their embarassed attempts to deal with certain Biblical incidents, lacking the modern historical understanding of the Scriptures, should not be used to weaken this certitude. Rather we should admit that these *ad hoc* solutions were faulty; (2) Aquinas certainly, and the other scholastics probably, never maintained that the material or ontic character of the act as such determines the morality of an action but the *proximate intention* which renders that object formally a human, moral act. What they were insisting on was that there are some acts which even by their material nature cannot be proximately intended by a human agent without violating the nature of that agent, e.g., suicide. It is true that the scholastics, prior to the rise of voluntarism in the fourteenth century, also thought that *some* kinds of such intrinsically wrong actions would not be intrinsically wrong if commanded by God because God has rights over life and death which He can delegate to the human agent in a particular case; but as far as I know (and as far as goes the evidence cited by Dedek) they never thought that God could dispense the human agent to perform *any* kind of act, e.g., sodomy. Thus it would be fair to say that for the scholastics there were two types of intrinsically wrong acts: (1) those that were always wrong (2) those that were always wrong when undertaken on human initiative without the delegated authority of God.

56. *Summa Theologiae*, I-II, qq. 18-20.

57. On the ambiguities about the classification of moral theories see Richard A. McCormick, *Notes on Moral Theology* (see note 51 above), pp. 649-652; Charles E. Curran "Utilitarianism and Contemporary Moral Theology: Situating the Debates", *Louvain Studies* 6 (1977): 239-255, and *Transition and Tradition in Moral Theology* (Notre Dame: Ind.: University of Notre Dame, 1979), pp. 4-5. Also see Benedict Ashley, O.P., and Kevin O'Rourke, O.P., *Health Care Ethics*, 2nd ed. (St. Louis, Catholic Health Association, 1982), pp. 148-175 for an attempt to clarify this classification.

58. It should be noted that Aquinas admits the validity of *positive* laws enacted by legitimate human lawmakers, but he insists that in order to have moral force these must be "derived" from the natural law by way of determining it to particular circumstances; they cannot contradict natural law or have force independent of it. *Summa Theologiae* I-II, q. 95, a. 2.

59. See Guiseppe Angelini, and Ambrogio Valsecchi, *Disegno storico della teologia morale* (Bologna, EDB, 1972); pp. 106-132; Louis Vereecke, "Préface à l'historie de la théologie morale moderne", *Studia Moralia* I, Academia Alfonsiana Institutum Theologiae Moralis (Rome: Ancora, 1963), pp. 87-120; Stanley Hauerwas "Toward an Ethics of Character," *Theological Studies,* 33: (Dec., 1972): 698-715, and *Vision and Virtue* (Notre Dame, Ind.: Fides, 1974); *Character and the Christian Life* (San Antonio: Trinity University Press, 1975); and Alasdair MacIntyre, *After Virtue* (Notre Dame, Ind.: University of Notre Dame Press, 1982).

60. See Charles E. Curran and Richard McCormick, *The Distinctiveness of Christian Ethics, Readings in Moral Theology, No. 2* (Ramsey, N.J.: Paulist, 1980), for a collection of essays on this topic. Yves Congar, O.P., in "Reflexion et propos sur l'originalité d'une éthique chrétienne" *Studia Moralia 15* (Rome: Academia Alphonsiana, 1977), in answer to J. Fuchs, "Existe-t-il un morale chrétienne", *Recherches et Synthèses,* Morale 9 (Gembloux, 1972) says that while Fuchs distinguishes "categorial" and "transcendental" aspects of morality and sees in the last a special Christian "intentionality", Congar holds that we must insist not only on a difference in intentionality but also in ontology. Although the Creator and the Redeemer are one and the same and therefore the content of Christian ethics is natural, yet the Christian *exists* in a new way. Christian ethics or the "ethics of a Christian" is not merely moral, but theological and spiritual, a new life which is a Gift that cannot be reduced to a Law.

61. See *Summa Theologiae* I, q. 1, a. 8, ad. 2; q. 2, a. 2 ad 1; q. 62, a. 5 c; I-II, q. 99, a. 2 ad 1; III q. 71, a. 1 ad 1; etc. On the history of the axiom, see Bernhard Stoeckle, *Gratia supponit naturam* (Rome: Herder, 1962).

62. Chapter 7, pp. 276-280.

63. *Summa Theologiae* I-II, q. 99, a. 2 ad 1; q. 100, a. 1.

64. On Scheler's personal background see Deeken, *Process and Permanence* (note 50 above), and John N. Oesterreicher, *Walls are Crumbling* (New York: Devain-Adair, 1952), pp. 135-198.

65. "The Absoluteness of Moral Terms", note 51 above. For a critique of Fuch's argument see Benedict M. Ashley, O.P. and Kevin D. O'Rourke, O.P., *Health Care Ethics,* note 57 above, pp. 160-171 and the discussions by Grisez and Pinckaers in note 51 above.

66. The most extensive refutation of Proportionalism is to be found in Germain Grisez, *The Way of the Lord Jesus* (note 51 above) 1: 141-172. See also his "Toward a Consistent Natural-Law Ethics of Killing", *American Journal of Jurisprudence* 15 (1970): 65-96. Other critiques are Gustav Ermeke, "Da Problem der Universalität oder Allgemeingültikeit sittlicher Normen inner-weltlicher Lebengestaltung", *Münchener Theologische Zeitschrift* 24 (1973): 1-24; John R. Connery, S.J., "Morality of Consequences: A Critical Appraisal", *Theological Studies* 34 (1973): 396-414 and "Catholic Ethics: Has the Norm for Rule-Making Changed?", *ibid.* 42 (June, 1981): 232-250 and "The Basis for Certain Key Exceptionless Moral Norms in Contemporary Catholic Thought", *Moral Theology Today* (note 51 above) pp. 182 ff.; William E. May, *Becoming Human* (Dayton: Pflaum, 1975) Chapter 4 and *Human Existence, Medicine and Ethics* (Chicago: Franciscan Herald Press, 1977; "Ethics and Human Identity:

The Challenge of the New Biology", *Horizons,* (Spring, 1976): 17-37; Paul Quay, S.J., "Morality by Calculation of Values", *Theology Digest* 23 (1975): 347-364; Servais Pinckaers, O.P., "La question acts intrinséquement mauvais" (note 51 above); and Benedict Ashley and Kevin O'Rourke, O.P., *Health Care Ethics)* (note 51 above) pp. 148-175. Protestant critiques are Paul Ramsey, "The Case of the Curious Exception" in G. H. Outka and P. Ramsey, *Norm and Context in Christian Ethics* (New York: Scribner's, 1968, pp. 67-138 and Frederick S. Carney, "On McCormick and Teleological Morality", *Journal of Religious Ethics* 6 (1978): 81-107. An overview of Protestant views on this topic is supplied by Gene Outka, *Moral Theology Today:* (note 51 above) "The Protestant Tradition and Exceptionless Moral Norms" pp. 136 ff.

67. Louis Janssens, "Ontic Evil and Moral Evil", *Louvain Studies,* 4 (1972): 115-156. This terminology was an effort to improve on the traditional distinction between the moral object considered *materially* (physically) and *formally* as moral, by using the terms *ontic* and *ontological* derived from Heidegger. But the latter terminology has idealistic connotations because in this usage "ontological" implies a relation to the subject as *source* of meaning, while in the classic usage the subject is perfected by the moral object which is thus the source of meaning for the subject. Others use the term "pre-moral" for "ontic" but this misses the mutual causality of the material and formal aspects of the moral object.

68. "When is human action, or when is man in his action (morally) good? Must not the answer be: When he *intends and effects* a human good (value) — in the premoral sense, for example, life, health, joy, culture, etc. (for only this is recta ratio); but not when he has *in view and effects* a human *non-good,* an evil (non-value) — in the premoral sense, for example, death, wounding, wrong, etc., What if he intends and effects good, but this necessarily involves effecting evil also? We answer: If the realization of the evil through the intended realization of good is justified as proportionally related cause, then in this case only good was intended." (Fuchs, "The Absoluteness of Moral Terms", note 51 above). Fuchs refers to Peter Knauer (same note). Note that "life, health" etc., are human *goods* only in relation to the whole human person as by nature ordained to the ultimate goal of the person and hence as good (value) are not *pre*-moral, but related to morality as matter to form. It is only by distinguishing in an odd way the human value of the object from its morality that Fuchs' system works.

69. Joseph Fletcher, *Situation Ethics* (Philadelphia: Westminster Press, 1966); R. L. Cunningham ed., *Situationism and the New Morality* (New York: Appleton-Century, 1950) (see especially Fletcher's debate with Henry McCabe, O.P., pp. 35-87). See also the refutation of Fletcher by Paul Ramsey, *Deeds and Rules in Christian Ethics* (New York: Charles Scribner's Sons, 1967), p. 145 ff., and Harvey Cox, ed., *The Situation Ethics Debate,* (Philadelphia: Westminster Press, 1968).

70. See note 51 above.

71. See Ashley-O'Rourke, *Health Care Ethics* (note 57 above), pp. 157-175.

72. See note 50 above.

73. On the antecedents of Fuch's thought see Timothy O'Connell, *Changing Catholic Moral Theology* (diss. Fordham University) (Ann Arbor, Mich: University Microfilms, 1974), O'Connell stresses the influence of Scheler; that of Heidegger is seen in Fuch's stress on human historicity. It would seem that the thought of Karl Rahner has also played an important part in confirming Fuchs' revision of Thomistic ethics; as I have been informed by Fr. Terence P. Brinkman, a former student of Fuchs. Rahner did not write extensively on moral theology but see D. J. Dorr, "Karl Rahner's Formal Existential Ethics," *Irish Theological Quarterly,* 36 (1969), 211-229.

74. See Bernd Magnus, *Heidegger's Metahistory of Philosophy Amor Fati, Being and Truth* (The Hague, Martinus Nijhoff, 1970).

75. On the concept of "biologism" or "physicalism" see Charles E. Curran, *Transition and Tradition in Moral Theology* (note 57 above), pp. 3-43 and *Moral Theology: A Continuing Journey* (Notre Dame, Ind.: Notre Dame University Press, 1982), p. 144, and David F. Kelly, *The Emergence of Roman Catholic Medical Ethics in North America* (New York and Toronto: Edwin Mellen Press), pp. 421-429.

76. Aquinas' discussion of sexual sins, *Summa Theologiae* I-II, q. 154, is based throughout not on the sexual organs as such, but on the misuse of human sexuality as a human function having a God-given purpose and a mode of use intrinsically related to this purpose, but he understood the human body as designed by God to serve these needs of the human person; see I, q. 91, a. 3.

77. On the general problem see David Bidney, "Cultural Relativism", a part of the article "Culture" in the *Encyclopedia of Social Science,* vol. 3, pp. 542-547 and *Theoretical Anthropology* (New York: Columbia University Press, 1953); Clyde Kluckhon, "Ethical Relativity", *Journal of Philosophy* 52 (1955): 663-677; Helmut Schoeck and James W. Wiggins, eds., *Relativism and the Study of Man* (Princeton, N.J.: Van Nostrand, 1961), especially the article of Leo Strauss, "Relativism", pp. 135-157; Filmer S. C. Northrop and Helen H. Livingston eds., *Cross-cultural Understanding: Epistemology in Anthropology* (New York: Harper 1964); Richard Brandt, "Ethical Relativism" in Paul Edwards ed., *Encyclopedia of Philosophy* (New York: Macmillan and Free Press, N.Y., 1967) 3: 75-78; John Ladd, ed., *Ethical Relativism* (Belmont, Calif.: Wadsworth Publishing Co., 1973); and David Little and Sumner B. Twiss, *Comparative Religious Ethics: A New Method* (San Francisco: Harper and Row, 1978).

78. In addition to the work of Little and Twiss (see note 77) see Richard B. Brandt, *Hopi Ethics* (Chicago: University of Chicago Press, 1954), and *Ethical Theory* (Englewood Cliffs, N.J.: Prentice-Hall, 1959); John Ladd, *The Structure of a Moral Code: A Philosophical Analysis of Ethical Discourse Applied to the Ethics of the Navaho Indians* (Cambridge, Mass: Harvard University Press, 1957), and George Peter Murdock, *Culture and Society, Twenty-Four Essays,* "Cultural Relativity", pp. 144-51. Such detailed studies show that the main difficulty in establishing cross-cultural moral norms lies not so much in cultural variations as in obtaining reliable data and in the variety of opinions about morality *within* a culture, as well as the gap between professed norms and practice. See Derek Freeman, *Margaret Mead and Samoa: The Making and Unmaking of an Anthropological Myth* (Cambridge, Mass: Harvard University Press, 1983). In a review of this book by George E. Marcus, the reviewer notes that "Since Boas, systematic cultural and biological analysis of human behavior have been seen as theoretically complementary but practically separate research programs" ("One Man's Mead," *New York Times Book Review,* March 27, 1983, p. 3). In other words the problem is usually systematically avoided!

79. I found the following especially useful: Ronald de Vaux, *Le Genèse,* Jerusalem Bible (Paris, Ed. du Cerf, 1951); Dietrich Bonhoeffer, *Creation and Fall* (New York: Macmillan, 1959); Umberto Cassuto, *A Commentary on the Book of Genesis,* 2 vols. (Jerusalem Hebrew University Magnes Press, 1959); Gerhard Van Rad, *Genesis,* (Philadelphia, Westminister, 1961); Michael Schmaus and Alois Grillmeier, *Handbuch der Dogmengeschichte* (Freiburg: Herder, 1963), Bd. II; Walter Eichrodt, *Theology of the Old Testament,* (Philadelphia: Westminister, 1967); E. A. Speiser, *Genesis,* Anchor Bible (Garden City, N.J.: Doubleday, 1964); Jean Daniélou, *In the Beginning* (Baltimore: Helicon, 1965); Claude Tresmontant, *Christian Metaphysics* (New York: Sheed and Ward, 1965), pp. 45 ff.; Th. C. Vriezen, *An Outline of Old Testament Theology* (Oxford: Blackwell, 1970); Leo Scheffczyk, *Creation and Providence* (New York: Herder and Herder, 1970); Carroll Stuhlmueller, *Creative Redemption in Deutero-Isaiah* (Rome: Biblical Institute Press, 1970); Edmund Leach, *Genesis as Myth* (London: Cape, 1971), pp. 7-24; Claus Westerman, *Beginning and End in the Bible,* (Philadelphia: Fortress Facet Books, 1972); *Genesis I-II,* Biblische Kommentar Altes Test. (Neuricher-Vluyn: Neukirchener Verlag, 1974); *Creation* (Philadelphia: Fortress, 1974); G. J. Botterweck and Helmer Ringgren, *Theological Dictionary of the Old Testament,* (Grand Rapids, Michigan: Eerdmans, 1974), vol. 1, pp. 75-87, 88-98, 222-235, 267-284; 345-347; J. F. A. Sawyer, "The Meaning of 'In the Image of God' in Genesis I-II", *Journal of Theological Studies,* n. s. 25 (1974): 418-426; L. M. Pasyinya, "Le cadre littéraire de Genèse 1," *Biblica* 57 (1976): 224-41; Bruce Vawter, *On Genesis: A New Reading* (Garden City, N.Y.: Doubleday, 1977); Phyllis Trible, *God and the Rhetoric of Sexuality,* Philadelphia;: Fortress, 1978); Jean Richard, "Une nouvel essai théologique sur la création", *Laval Theologie Philosophique,* 38 (Feb., 1982), 77-80; Walter Brueggemann, *Genesis* (Atlanta: John Knox, 1982).

Helpful works dealing with Christian anthropology are: Jean Mouroux, *The Meaning of Man* (New York: Sheed and Ward, 1948), especially pp. 41-114 dealing with the nobility, misery and redemption of the body; E. C. Rust, *Nature and Man in Biblical Thought* (London: Lutterworth Press, 1953); Claude Tresmontant, *A Study of Hebrew Thought* (New York: Desclée, 1960) esp. 83-114; P. S. S. Le Troquer, *What is Man?* vol. 21 of *Twentieth Century Encyclopedia of Catholicism*, ed. by H. Daniel-Rops (New York: Hawthorne, 1961); Simon Doniger, ed. *The Nature of Man in Theological and Psychological Perspective* (New York: Harpers, 1962); C. J. de Vogel, "The Concept of Personality in Greek and Christian Thought" in John K. Ryan, ed. *Studies in Philosophy and the History of Philosophy* (Washington, D.C.: Catholic University of America, 1963), pp. 20-69; John A. T. Robinson, *The Body: A Study in Pauline Theology* (London: SCM Press, 1963); Werner G. Kümmel, *Man in the New Testament*, rev. ed., (London: Epworth Press, 1963); Israel I. Etros, *Ancient Jewish Philosophy* (Detroit: Wayne State University Press, 1964) Karl Rahner, *Hominisation: The Evolutionary Origin of Man as a Theological Problem* (New York: Herder and Herder, 1965); Joseph Endres, *Man as the Ontological Mean* (New York: Desclée, 1965); Leo Scheffczyk, *Man's Search for Himself: Modern and Biblical Images* (New York: Sheed and Ward, 1966); Josef Scharbert, *Fleisch, Geist und Seele in Pentateuch* (Stuttgart: Verlag Katholisches Bibel Werk, 1967); Hans Urs von Balthasar, *A Theological Anthropology* (New York: Sheed and Ward, 1967); Albert Gelin, *The Concept of Man in the Bible* (London: Geoffrey Chapman, 1968); Wolfhart Pannenberg, *What is Man?* (Philadelphia: Fortress Press, 1970); Arthur A. Vogel, *Body Theology: God's Presence in Man's World* (New York: Harper and Row, 1973); Hans Walter Wolff, *Anthropology of the Old Testament* (Philadelphia: Fortress Press, 1974); John F. O'Grady, *Christian Anthropology* (New York: Paulist Press, 1976); Robert H. Gundry, *Soma in Biblical Anthropology* (Cambridge: Cambridge University Press, 1976); James M. Childs, *Christian Anthropology and Ethics* (Philadelphia: Fortress Press, 1978).

80. See Carl G. Jung, *Symbols of Transformation*, vol. 5 (1956); *Two Essays on Analytical Psychology* , vol. 7 (1953); *Archetypes and the Collective Unconscious*, vol. 9 (1959), *Psychology and Alchemy*, vol. 16 (1953) in *Collected Works* (New York: Pantheon) with C. Kerényi, *Essays on a Science of Mythology*, vol. 9, part I, *ibid.*, and C. Kerényi, *Prometheus: Archetypal Images of Human Existence* (London: 1963); Jung et al., *Man and his Symbols* (New York: Dell, 1968); Eric Neumann, *The Great Mother: An Analysis of Archetypes* (New York: Pantheon, 1955). For critiques of Jung see Victor White, *God and the Unconscious* (London: Harvill Press, 1952) and Antonio Moreno, *Jung, Gods and Modern Man* (Notre Dame, Ind.: University of Notre Dame Press, 1970). For an extensive example of a Jungian reading of a text see Robert Donington, *Wagner's 'Ring' and its Symbols* (London: Faber and Faber, 1963). The use of Jung's comparative method of interpreting myths need not imply an unqualified acceptance of his psychological or psychotherapeutic system.

81. *De Genesi ad Litteram* Libri XII, PL 34, Book I, col. 245-262. *The Literal Meaning of Genesis*, Trans. by J. H. Taylor, S.J., *Ancient Christian Writers*, n. 41, 2 vols. (New York: Newman, 1982). See especially I, c. 18-21, vol. I, pp. 41-44.

82. See Andrew C. J. Alexander, "Human Origins and Genetics", *Clergy Review* 49 (1964), 344-353, and R. J. Pendergast, "Terrestrial and Cosmic Polygenism", *Downside Review*, 82 (1964): 189-198.

83. See Roger Shattuck, *The Forbidden Experiment: The Story of the Wild Boy of Aveyron* (New York: Farrar Straus Giroux, 1980).

84. See Karl Rahner, "Theological Reflections on Monogenism", *Theological Investigations* (London: Darton, Longman and Todd), 1, 229-296.

85. See M. Oldfield Howey, *The Encircled Serpent: A Study of Serpent Symbolism in All Countries and Ages* (Philadelphia: D. McKay, 1927); *Collected Works of Carl G. Jung*, vol. 20, General Index, (Princeton, N.J.: Princeton University Press, 1979), references under "Animals: Serpent", pp. 69-72; Weston La Barre, *They Shall Take Up Serpents*, (Minneapolis: University of Minnesota Press, 1962), pp. 53-112; and Charles P. Mountford, "The Rainbow Serpent Myths of Australia" in Ira R. Buchler and Kenneth Maddock eds., *The Rainbow Serpent* (The Hague: Mouton, 1978), pp. 23-98. Also Robert Graves and Raphael Patai *Hebrew Myths* (Garden City, N.Y.: Doubleday, 1964), "Primeval Monsters" pp. 47-53.

86. "That the song of Solomon was later universally allegorized by both church and synagogue as the love between God and Judaism or between God and the Church is as much a witness to Jewish and Christian freedom to ascribe sensuality to God as to confusions about sex in the acceptably pious Christian or Jew." p. 113, Marvin Pope's introduction to his commentary *Song of Songs*, Anchor Bible, (Garden City, N.Y.: Doubleday, 1977), pp. 93-232 with much data on various interpretations. Modern scholarship generally rejects the allegorical interpretation because (a) nothing explicit in the work indicates an allegorical interpretation; (b) its presence in the canon is sufficiently explained as a celebration of God's gift of faithful love and an exhortation to fidelity. Nevertheless, in my opinion the question remains whether once admitted to the canon this book needs to be understood in relation to the common prophetic use of the marriage metaphor to describe the Covenant. Recently, R. J. Tournay, O.P., *Quand Dieu parle aux hommes le langage de l'amour*, Cahiers de la Révue Biblique 21 (Paris: Gabalda, 1982) has argued that the *Song* refers *explicitly* to Israel's longing for a Messianic King.

87. It must be noted that *ruah* in the Old Testament is sometimes taken as masculine, and that in Hebrew the feminine gender is often used *impersonally*. Cf. Paul van Imschoot, *Theology of the Old Testament* (New York: Desclée, 1965), pp. 172-187. Also the elaborate study of Daniel Lys, *Rûach: Le Souffle dans l'Ancien Testament* (Paris: Presses Universitaire de France, 1962).

88. See Raphael Patai, *The Hebrew Goddess* (New York KTAV Publishing House, 1967) on the concepts of the Biblical *Shekina* or glorious presence of God, which is conceived as the bride of God (pp. 137-156) and the Kabbalistic figure of "Matronit" or Maternal Presence (pp. 157-206).

89. David Gonzalo Maeso "Punctualizaciones sobre Gn 2, 20-24; formacion de la primera mujer y concept del matrimonio" pp. 235-244 in *La Etica Biblica*, Semana Biblica Espanola, Madrid 22-26, Sept., 1969 (Consejo Superior de Investigaciones Cientificas, Madrid, 1971), who argues (1) the "great mystery" of Ephesians, 5:32, refers to the mystery of sexuality and this is symbolized by the sleep into which Adam falls; (2) we should translate "and the two will both cooperate in forming a new being" i.e., the child, as the finality of marriage.

90. See Chapter 8, pp. 315.

91. Carl G. Jung, "The Relations Between the Ego and the Unconscious" in *Collected Works* (New York: Pantheon), vol. 7 (1966), pp. 188-211.

92. On the "evil yetzer" see W. D. Davies, *Paul and Rabbinic Judaism*, 4th ed. (Philadelphia: Fortress Press, 1980), pp. 20-27.

93. On the various interpretations of the phrase "knowing good and evil" see Westerman, *Genesis*, note 79 above, pp. 330-333.

94. St. Thomas Aquinas, *Summa Theologiae*, I-II, q. 76.

95. See article "Original Sin" by Karl Rahner, *Encyclopedia of Theology*, ed. by Karl Rahner (New York: Seabury, 1975), pp. 1148-1155, and the following: N. P. Williams, *The Ideas of the Fall and of Original Sin* (London: Longmans, 1927); Julius Gross, *Geschichte des Erbsündendogmas* (Munich: E. Reinhardt, 1960-1971); A. M. Dubarle, O.P., *The Biblical Doctrine of Original Sin* (New York: Herder and Herder, 1964); A. Gelin and A. Descamps, *Sin in the Bible* (New York: Desclée, 1965); Piet Schoonenberg, S.J., *Man and Sin: A Theological View* (Notre Dame: Notre Dame University Press, 1965); Peter De Rosa, *Christ and Original Sin* (Milwaukee: Bruce, 1967); Paul Ricoeur, *The Symbolism of Evil* (New York: Harper and Row, 1967); Herbut Haag, *Is Original Sin in Scripture?* (New York: Sheed and Ward, 1969); John Bowker, *Problems of Suffering in the Religions of the World* (Cambridge: Cambridge University Press, 1970); Gerrit Berkouwer, *Sin* (Grand Rapids, Michigan: Eeerdman, 1971); Edward Yarnold, *The Theology of Original Sin* (Notre Dame: Fides, 1971); Maurizio Flick, *Il peccato originale* (Brescia: Queriniano, 1972); Henri Rondet, *Original Sin: the patristic and theological background* (Staten Island, N.Y.: Alba House, 1972); P. Grélot, *Péché original et rédemption a partir de l'epître aux romains* (Paris: Desclée, 1973); and Eugene Maly, *Sin: Biblical Perspectives* (Pflaum-Standard: Dayton, Ohio, 1973); George Vandervelde, *Original Sin: Two Major Trends in Contemporary Roman Catholic Reinterpretation* (Washington, D.C.: University Press of America, 1981). Vandervelde points out

that besides the explanation of original sin in terms of our historical "situatedness" (Rahner and Karl-Heinz Weger) a second more radical trend is to explain it in terms of *actual* sin, either in its *universality* (A. Vanneste) or its *depth* (Urs Baumann). My explanation in the text combines both tendencies.

96. See Jean Daniélou, *Holy Pagans of the Old Testament,* (Baltimore: Helicon, 1956).

97. See Davies, *Paul and Rabbinic Judaism* (note 92 above), pp. 111-146.

98. For a review of the question see Gustaf Aulén, *Jesus in Contemporary Historical Research* (Philadelphia; Fortress Press, 1976). On the notion of the *imitation of Christ* see Andreas Alpe, "De imitatione Christi in Novo Testamento", *Verbum Domini* 22 (1942): 57-63; W. F. Lofthouse, "Imitatio Christi", *Expository Times* 65 (1953-54): 338-342; E. J. Tinsley, *The Imitation of God in Christ,* (London: SCM Press, 1960); Joseph Fuchs, "The Law of Christ" in Enda McDonagh, Ed., *Moral Theology Renewed* (Dublin: Gill and Son, 1964), pp. 75-84; Louis Gillon, *Christ and Moral Theology,* (Staten Island, N.Y.: Alba House, 1967); James Gustafson, *Christ and the Moral Life* (New York: Harper and Row, 1968), pp. 150-187; Rudolf Schnackenburg, *Christian Existence in the New Testament,* (Notre Dame, Ind.: University of Notre Dame Press, 1968), 1, 99-128; E. C. Gardner, "A Critique of Christocentric Models of Ethical Analysis", *Religion in Life,* 39 (1970), 205-220; John Passmore, *The Perfectibility of Man* (New York: Scribner's, 1970) (attack on Christian notion of perfectibility); N. H. G. Robinson, *The Groundwork of Christian Ethics* (Grand Rapids, Mich: Eeerdmans, 1971), pp. 100-120; Charles-André Barnard, *Vie moral et croissance dans le Christ* (Rome: Gregorian University Press, 1973); David L. McKenna, *The Jesus Model* (Waco, Texas: Word Books, 1977). For a psychological study of the question of the "optimal personality" see Richard W. Coan, *Hero, Artist, Sage or Saint?* (New York, Columbia University Press, 1977), pp. 284-305.

99. Edward Schillebeeckx, *Jesus: An Experiment in Christology,* (New York: Seabury, 1979); and *Christ: The Experience of Jesus as Lord* (New York: Crossroads, 1981). See also Oscar Cullmann, *The Christology of the New Testament,* (Philadelphia, Westminster, 1959); A. J. Patfoort, *L'unité d'être dans le Christ d'après S. Thomas* (Paris: Desclée, 1964); Christian Duquoc, *Christologie: Essai dogmatique* (Paris: Editions du Cerf, 1968), 2 vols.; Piet Schoonenberg, *The Christ,* (New York: Herder and Herder, 1971); Karl Rahner and Wilhelm Thüsing, *Christologie: Systematisch und exegetisch,* (Freiburg: Herder, 1972); Jürgen Moltmann, *The Crucified God,* (New York: Harper and Row, 1974); Louis Bouyer, *The Eternal Son,* (Huntington, Ind.: Our Sunday Visitor Press, 1975); Walter Kasper, *Jesus the Christ* (New York: Paulist, 1976); Hans Küng, *On Being a Christian* (Garden City N.Y., Doubleday, 1976); Millar Burrows, *Jesus in the First Three Gospels* (Nashville, Abingdon, 1977); Jean Galot, *Who Is Christ* (Chicago: Franciscan Herald Press, 1981); Jon Sobrino, *Christology at the Crossroads* (Maryknoll, N.Y.: Orbis Books, 1978); Wolfhart Pannenberg, *Jesus: God and Man,* 2nd ed., (Philadelphia, Westminister Pess, 1981).

100. For the meagre but important extra-biblical information on Jesus see Guiseppe Riccioti, *The Life of Christ* (Milwaukee, Bruce, 1954), p. 78-92; and Walter Kasper, *Jesus the Christ* (note 99 above), pp. 65-71.

101. For Gautama Siddhartha see Edward J. Thomas, *The Life of the Buddha as Legend and History* (Barnes and Noble, 1952) and Richard Drummond, *Gautama the Buddha* (Grand Rapids: Eerdmans, 1974). For Muhummad see W. Montgomery Watt, *Muhammad at Medina,* 1956; *Muhammed at Mecca,* 1968; and *Muhammed Prophet and Statesman,* 1964 (all Oxford: Clarendon Press). See also Herlee G. Creel, *Confucius and the Chinese Way* (New York: Harper, 1960); Corlis S. Braden, *Jesus Compared: A Study of Jesus and other Great Founders of Religion* (Englewood Cliffs, N.J., Prentice-Hall, 1957).

102. For example, how different are the interpretations of Luther given by Hartmann Grisar, S.J., *Martin Luther, His Life and Works* (St. Louis: Herder Book Co., 1930) and Roland H. Bainton, *Here I Stand* (New York: Abingdon-Cokebury, 1950); and how novel the sympathetic biography of *Voltaire* by a Roman Catholic Alfred Noyes (New York, Sheed and Ward, 1936) or the interpretation by T. H. L. Parker of *John Calvin:* (Philadelphia: Westminister, 1975) as a representative of "Catholic Christianity".

103. On this point Catholic Christologists of very different points of view seem to agree; cf. Schillebeeckx (p. 71-76); Galot (pp. 20-30); Kasper (pp. 17-25); works cited in note 98 above.

104. On the *internum Spiritus Sancti testimonium* for the Lutheran tradition see Robert Preus, *The Inspiration of Scripture: A Study of the Theology of the Seventeenth Century Lutheran Dogmaticians* (Edinburgh: Oliver and Boyd, 1957). For Calvinism see *John Calvin, Institutes of the Christian Religion*, ed. by John T. McNeill, trans. by F. L. Battles, Library of Christian Classics, vol. 20-21 (Philadelphia, Westminister Press, 1960), vol 1, III, c. 7, pp. 78-81 and William Niesel, *The Theology of Calvin* (Philadelphia, Westminister, 1956, pp. 30-39.

105. Vatican I, Sess. III *Constitutio de fide catholica* cap. 3 (Denziger-Schönmester 3013) teaches that the Church is a moral miracle accessible to all as a certain and sufficient grounds of the rational credibility of the Gospel. Avery Dulles, *A History of Apologetics*, Theological Resources (New York: Corpus Books, 1971), pp. 244-246 points out that this line of argument is little stressed in current apologetics and that Vatican II makes little of it (p. 218). It seems to me, however, that the description of the Church in *Lumen Gentium* Chapter 1 holds up the Church as a sign to the nations. As a convert I myself testify that for me the moral miracle of the Church was my way to the Faith.

106. See Locke, *The Reasonableness of Christianity, Works* 10th ed. (London, 1801), vol. 7 and the essay of Richard Aschcroft, "Faith and Knowledge in Locke's Philosophy" in John W. Yolton ed., *John Locke: Problems and Perspectives* (Cambridge, 1969), pp. 195-223. Ashcroft emphasizes the importance of faith in Locke's system, but admits that the content of this faith is minimizing. For Jefferson see his *The Life and Morals of Jesus of Nazareth Extracted Textually from the Gospels*, ed. by Henry W. Foot (Boston, Beacon Press, 1951).

107. For an extreme view that Jesus provided no ethical teaching, except the love commandment, that remains relevant today see Jack T. Sanders, *Ethics in the New Testament* (Philadelphia, Fortress, 1975); cf. also R. Hiers, *Jesus and Ethics: Four Interpretations* (Philadelphia: Westminister Press, 1968).

108. See Reginald Fuller, "The Double Commandment of Love: A Test Case for the Criteria of Authenticity", in his *Essays on the Love Commandment* (Philadelphia: Fortress Press, 1975), pp. 49-56; Victor P. Furnish, *The Love Commandment in the New Testament* (Nashville: Abingdon, 1972); James M. Gustafson, *"Place of Scriptures in Christian Ethics: A Methodological Study"* pp. 121-146 and "The Relation of the Gospels to Moral Life" pp. 147-160 in his *Theology and Christian Ethics* (Philadelphia: United Church Press, 1974); Richard Hiers, *Jesus and Ethics* (note 107 above); John Knox, *The Ethic of Jesus in the Teaching of the Church* (New York: Abingdon Press, 1961); T. W. Manson, *Ethics and the Gospel* (New York: Scribner's, 1960); L. H. Marshall, *The Challenge of New Testament Ethics* (London, Macmillan, 1966), esp. Chapter VI, "The Validity of the Ethics of Jesus", pp. 171-215; John P. Meier, *Law and History in Matthew's Gospel* (Rome: Biblical Institute Press, 1976); John Piper, *'Love Your Enemies': Jesus' love command in the synoptic gospels and in the early Christian paranaesis* (Cambridge: Cambridge University Press, 1979); Rudolf Schnackenburg, *The Moral Teaching of the New Testament* (New York: Herder and Herder, 1965), pp. 15-167, and *Christian Existence in the New Testament* (Notre Dame, Ind.: University of Notre Dame Press, 1968); Ceslaus Spicq, *Agape in the New Testament* (St. Louis: B. Herder Book Co, 1963-1966), 3 vols., and *Théologie morale du Nouveau Testament* (Paris: Lecoffre, 1965), 2 vols., vol. 1, Appendix 1, "Vie Morale: Christ et Charité", pp. 381-393; John Yoder, *The Politics of Jesus* (Grand Rapids, Mich.: Eerdmans, 1972).

109. See George T. Montague, S.M., *The Holy Spirit: Growth of a Biblical Tradition* (New York: Paulist, 1978), especially pp. 349-362, and Raymond E. Brown, *The Gospel According to St. John*, Anchor Bible (Garden City, N.Y.: Doubleday, 1970), vol. 2, pp. 727-738 and Appendix V, pp. 1135-1144.

110. See Gustafson, *Protestant and Roman Catholic Ethics,* note 48 above, pp. 30-59.

111. Thus Eric W. Gritsch and Robert W. Jenson, *Lutheranism: The Theological Movement and its Confessional Writings* (Philadelphia, Fortress Press, 1976), sum up the position of the Augsburg Confession as Lutherans understand it: "(1) Good works must happen, not for merit before God, but for his praise. They are never a condition for justifiction; rather, they are the natural result of a grateful heart filled with the power of faith in the God who loves the ungodly.

410

(2) The new obedience is the call into a struggle between the "old" and the "new" Adam. Although sins are forgiven and sons of God are adopted, the conflict between good and evil has not yet come to an end. The Christian lives in an *interim* situation — between the Ascension and the Second Coming of Jesus. "Faith" — the relationship of absolute trust in what God did in Christ — determines what is "good" in this situation. Consequently, there are no absolute, eternal, or ethical norms by which the Christian is adjudged good or evil. There is only a faithful obedience to God through acts of love in the world.

(3) Lutheran "situation" — moral existence before God in anticipation of a new world — implies the use of reason when faith is active in love. Since Lutheran theology teaches the proper distinction between "law" (the work of God the Creator) and "gospel" (the work of God the Redeemer), there is an ethical dialectic between reason and faith; reason is able to create a limited "civil righteousness," as long as such righteousness is not confused with "spiritual righteousness" p. 140 f. "Thus the doctrine of justification by faith effects a particular *secularization* of morality" p. 147 in the sense that the motive for acting cannot be to gain a heavenly reward. "The radically proclaimed gospel frees me from moral egocentricity in that it does not merely tell me I *ought not* try to get anything out of my act for my neighbor, but that I *cannot* get anything out of it, that I will not in fact be rewarded at all." p. 146.

As a Catholic I would raise the following questions: (1) Granted that the Bible teaches the justification cannot be merited; why is the grace of God thought to be so weak that it cannot make the acts of the justified truly meritorious? (2) Why does "faithful obedience" exclude "ethical norms" which guide us in fulfilling God's will? (3) What is the biblical basis for saying that acts of love to be genuine must be *indifferent* to the advantage of the agent? When we act out of love don't we act for the *common* good of the giver and the receiver? To me it is a great paradox of Luther's thought that his conception of *sola gratia* seems to lead to the conclusion that although God's grace is able to get us into heaven, it is powerless to heal and elevate human nature until after death. If Jesus as man was able to live by the Spirit in this earthly life in a way that is worthy of God, why cannot the same Spirit which He has earned for us and which incorporates us in Him in faith, hope, and love also transform us? Granted that this transformation will not be completed until death, surely it *begins* from the moment of justification.

112. See note 69 above.

113. How far this is from St. Paul's intention is apparent from W. D. Davies; *Paul and Rabbinic Judaism*, note 92 above, pp. 111-146.

114. For a vigorous attack on the notion of the "canon within the canon' from a conservtive Lutheran point of view see Gerhard Maier, *The End of the Historical-Critical Method* (St. Louis: Concordia Publishing House, 1977) pp. 26-49, in which he reviews the book *Das Neue Testament als Kanon* (Goettingen, 1970) which collects 15 essays of different authors on the subject.

115. For a variety of interpretations besides W. D. Davies, note 92 above, see C. H. Dodd, *Gospel and Law: The Relation of Faith and Ethics in Early Christianity* (New York: Columbia University Press, 1951); Joseph A. Fitzmyer "Pauline Theology" in R. E. Brown, J. A. Fitzmyer and Roland E. Murphy, *The Jerome Biblical Commentary* (Englewood Cliffs, N.J., 1968), esp. 157-166; Victor Paul Furnish, *Theology and Ethics in Paul* (Nashville: Abingdon Press, 1968); William M. Longsworth, "Ethics in Paul", *The Annual of the Society for Christian Ethics*, 1981, pp. 29-56 (with bibliography); S. Lyonnet, "St. Paul: Liberty and Law" in C. L. Sam, ed. *Readings in Biblical Morality* (Englewood Cliffs, N.J.: Prentice Hall, 1967); J. M. Myers, *Grace and Torah* (Philadelphia, Fortress Press, 1975); Rudolf Schnackenburg, *The Moral Teaching of the New Testament* (See note 108 above), pp. 261-306; M. F. Wiles, *The Divine Apostle: The Interpretation of St. Paul's Epistles in the Early Church* (Cambridge: Cambridge University Press, 1967).

116. See W. D. Davies, *Paul and Rabbinic Judaism* (note 92 above), pp. 136-146.

117. On what Jesus meant by the "Reign of God" see Rudolf Schnackenburg, *God's Rule and Kingdom*, (New York: Herder and Herder, 1962), pp. 77-113.

118. *Summa Theologiae* I-II, q. 94, a. 2. See also Germain Grisez and Russell Shaw, *Beyond the New Morality*, 2nd ed. (Notre Dame: University of Notre Dame Press, 1980), Chapter 7, pp. 69-79 who refer to Peter A. Bertocci and Richard M. Millard, *Personality and the Good: Psychological and Ethical Perspectives* (New York: David McKay, 1963), pp. 157-172; A. H. Maslow, *Motivation and Personality* (New York: Harper and Row, 1954), pp. 80-106 and William K. Frankena, *Ethics*, 2nd ed. (Englewood Cliffs, N.J.: Prentice-Hall, 1973), p. 88, authors of very dissimilar points of view — for very similar lists.

PART IV
A PROCESS THEOLOGY
OF THE BODY

Chapter 10

Ethics of Co-Creative Stewardship

I: Ecology

A Christian ethical reflection on basic human needs must begin by looking to the Bible for the inspired light it casts on the human condition. We are immediately confronted by the difficulty that modern exegetical scholarship often seems to leave us with a vast amount of information and speculation about the various religious traditions, literary materials, and compositional stages represented in the canonical texts rather than a unified theological message. The best attempts at "biblical theologies" do not so much present us with a synthesis as with an account of a development of religious themes and the *diverse* theologies dealing with these themes to be found in the Bible, with the emphasis on "diverse."[1]

Nevertheless, this difficulty should not be exaggerated. Once we have accepted the historicity of human existence, it is no shock to discover that God has revealed himself to us in a thoroughly historical manner. The message of his Word has been conveyed to us in the Bible, The Book, not through a single unified sermon, let alone a theological treatise, but through a whole library of works of very different types, written by dif-

415

ferent authors for different purposes and at different times, many of them consisting not simply of the work of one author but of the combination and repeated revision of the work of many authors. Yet it remains the task of theology to listen to this chorus of voices in dialogue and even in debate, and to attempt to hear the Word of God in its absolute unity speaking through this polyphony. Consequently, it behooves theologians, especially myself, to remember humbly that our neat syntheses will always be inadequate to the richness of this inspired dialogue.

Fortunately, in the matter of the anthropology to which biblical ethics must be referred we have already found out in the last chapter that the *Book of Genesis,* especially in its opening chapters, provides us with the fundamental text. The final editor of this book supplies us with an account skillfully woven together out of several traditions of the creation of humanity and of the causes of our present human condition, as a necessary introduction to the Torah.[2] Since the Torah forms the fundamental statement of biblical ethics on which the Prophets and the Writings, drawn from other traditions and reflecting other ethical points of view, comment or to which they form the canonical complement; and since Jesus himself as the eschatological prophet gives the Torah his final interpretation, which St. Paul and the other New Testament writers simply aim to expound and apply, these first narratives of *Genesis* are the classical foundation of any theological anthropology and ethics. In the last chapter I tried to show that the story of Adam and Eve presents us with a profound anthropology which is consistent both with what we now know to have been the evolutionary history of our species, and with the ethical doctrine of a universal moral law rooted in human nature as that of a bodily being capable of spiritual acts of intelligence and free decision. In this chapter I will first re-read that story with a view to discovering what light it casts on human responsibility for our natural environment and for the building of human society.[3]

Genesis 1-3 tells us that God created an orderly universe of great variety. In order to crown this material universe (the question of a purely spiritual creation is passed over, except for the mystery of the speaking serpent), God created man and woman in his own image. In Chapter 4, I have already related the development of patristic interpretation of this fundamental biblical theme.[4] In brief it can be said we are images of God because God by his wisdom made and rules the world, and by the intelligence he has given us we are called to cooperate in this government of the world as God's ministers or stewards. This is why the Yahwist in the second chapter of *Genesis* pictures God placing the first human not in the desert wastes but in a garden oasis, "to till and to keep it" and why he has God bring the animals before Adam to be named. Just as God created the world by his word (Priestly account, Chapter 1), so Adam by naming the animals shows that he too understands their natures and therefore can control them. Although Adam was made out of the dust of the earth as were the animals, he has received a share of God's own wisdom by the inbreathing of God's spirit. As Adam comes to self-consciousness and perceives his difference from the other creatures, God gives to him Eve as a companion and promises them a family in the blessing "multiply and fill the earth and take possession of it" (1:28)[5], thus manifesting the essentially social nature of humanity. Moreover, God provides in the garden the "tree of life" (2:9), so this human family can escape the mortality to which their origin from the earth makes them liable. Nothing is forbidden to them, except the fruit of the "tree of the knowledge of good and evil", that is, the denial of their own creaturehood.[6]

Eve chooses to deny her humanity by listening to the "wisdom" of the earth (the serpent) rather than the true wisdom of God. Acting independently of her partner Adam instead of in consultation with him, she seeks to become completely autonomous as a goddess (2:4-5) apart from Adam and from God, seeking the immediate gratification of her senses and her intellectual curiosity (2:6) without concern for the true goal of human living, friendship with other human beings and with God. Then, no doubt frightened (the text leave us to guess this) by a sudden sense of her own isolation, she persuades Adam (who, no doubt also from fear of losing her more than losing God), allows himself to be persuaded to join her in her claim to divinity.

The sin of Adam and Eve was not that they wanted to be divine, since they were created in God's image, but in seeking "to know good and evil," that is, to determine for themselves apart from the wise order of God's creation what is for good and for evil. This claim amounts to a rejection of their own humanity which is a participation in God's divinity, but only a sharing, not an independent possession. In this they fooled themselves, for in fact their choice of autonomy was an enslavement to the "wisdom of the earth," that is, to what St. Paul calls, "the law of the members" that is "at war with the law of my mind" (Romans 7:23). In denying their humanity they entrapped themselves in a self-destructive contradiction, enmity to each other and to themselves, bodily and spiritual death.

This "wisdom of earth" which is in fact foolishness, irrationality, disorder, chaos, "the waste and void" of the abyss (1:2), the absurdity, darkness or Non-Being apart from God's wise act of creation bringing Being, light, meaning, order, plenitude, life into existence, is symbolized by the serpent; because (as we saw in the last chapter), in Mesopotamian myth and throughout the Scriptures the original chaos is portrayed as a dark ocean which is also a great dragon.⁷ Adam and Eve, by taking this path of folly, of a false freedom that is really enslavement to the blind war of created forces which, once the order that harmonized them is destroyed, attack eath other, begin to feel the inevitable consequences of their folly. First, they feel the war within themselves, the loss of personal integrity (the intrinsic evil of their act) manifest in their sense of shame for their naked bodies (3:7), their guilt before God their Maker (3:10), and their dishonest accusations and excuses (12-13). Second, they begin to feel the loss of their control over nature manifest in God's declaration of the accursed condition into which they have fallen.⁸ The woman finds herself burdened by childbirth which subjects her to slavish dependence on the man (3:16), and the man finds himself burdened with endless toil, no longer in a fruitful garden but in an infertile desert, to support himself and his family. And at the end of it all — death (3:3), instead of immortal life in the Kingdom of God.

In the subsequent chapters of *Genesis* the Yahwist shows how in His mercy God held back the full consequences of human sin. The human race did not perish because God gave Eve children and men learned to make a living, domesticating animals and cultivating the fields, inventing tools of work and instruments of music; yet at the same time learning to shed the blood of animals for food and then of men for envy's sake. They engaged in war, intercourse with demons,⁹ and indulgence of their brutish passions, until God in disgust released the chaos of the great flood to cleanse the earth, while saving the remnant of Noah's family and giving them the rainbow promise that the race would never utterly perish. After the flood God tolerated on earth the present ambiguous "human condition" in which we find ourselves. The race spread over the whole earth, built empires, and magnificent works of art and engineering (the tower of Babel), but these noble achievements were also the cause of the division of the human family into warring nations no longer able to understand each other's language. In the midst of this renewal of chaos, God revealed his mercy once again to Abraham. God chose a special people and began to prepare them as his witnesses to bring true wisdom (Torah) back into the world of folly, thus restoring the original Covenant between Himself and humanity.

This wonderful biblical theology is not an interpretation of mere myths. The Biblical writers are, in fact, concerned to "demythologize" the traditional narratives which they retell, leaving only a few traces of their mythological origins.¹⁰ Rather it is a theological interpretation of human *history* in which the common experiences of humankind are expressed in typical narratives, but these experiences are historically real — all too real. The essential message is that in creating us in His image,

418

God has made us free and intelligent, sharers in the governance and development of His creation; but this intelligent power over creation is not autonomous, it is a *stewardship*.

This important term "stewardship" is taken from Jesus' Parable of the Unjust Steward (*oikonomos*) (Luke 16:1-8), but is perhaps better illustrated from the Parable of the Talents (Matthew 25:14-30) which Matthew significantly places at the very end of Jesus' preaching, just before the account of the Last Judgement. This latter parable tells us that God has given us his gifts not merely that we may preserve them, but that we may invest them and return a profit to the owner. Thus God is not envious of human creativity (as were the gods of Greek myth in their wrath at Prometheus for giving fire to man). The Biblical God requires us to use his gifts creatively. The writers of *Genesis* show us that sin is the foolishness of seeking to be wise apart from God's wisdom, trying to take power into our own hands, a power that proves destructive. Jesus added to this teaching his own lesson, that neglect to use the power given us by God in accordance with the wisdom He has given us is also sin.

Christian ethics, therefore, begins with an affirmation of human responsibility to use God's gifts, our own talents and the resources of the earth, with creative freedom, with wisdom that chooses the right means to the right goal, not with the foolishness of the Prodigal Son who wasted his heritage in riotous living without thought of the consequences to himself or others (Luke 15:11-32). The responsibility is stewardship. It implies a respect for the sacredness of the order of creation, of our own bodies, and of the earth which was created to be an Eden.

This biblical theology raises three questions about stewardship? (1) Can we sacrifice the environment to achieve human goals? (2) Can we sacrifice the physical integrity of the human person to achieve higher spiritual values? (3) Can we sacrifice individual persons to the goals of the community?

As to the first of these questions I have shown in Chapter 8 what is also strongly affirmed by the *Genesis* account of creation, and what is agreed to by philosophers as different as Aquinas[11] and Kant: human persons, because they are *persons* endowed with intelligence and freedom, transcend the sub-human order. As Kant says, "human persons must be treated as ends, not means," that is, they are the subjects of moral rights.[12] On the contrary, some recent writers have attempted to extend the concept of moral rights to animals,[13] as religions which teach the transmigration of souls have done since ancient times. Some authors also have accused Christianity of being inferior in this respect to Buddhism, and even of being the source of our modern ruthless technological exploitation of nature because of the words of *Genesis,* "Be fruitful and multiply; fill the earth and *subdue* it" (1:28).[14]

A better translation of the verse of *Genesis* is "fill the earth and *possess* it"; there is no implication of brutal mastery or exploitation.[15] In fact *Genesis* pictures humans living in Eden at peace with the animals, both feeding only on plants (1:29). The "dominion" of man over animals (1:28) is that of guidance, like a shepherd and his sheep, of St. Francis and the birds, or like that portrayed by Isaiah as the condition in the Reign of God when "the wolf shall be the guest of the lamb . . . with a little child to guide them (11:6-9; cf. 65:25). It is only after the Flood when God has resigned himself to tolerating human sinfulness that He permits humans to eat flesh meat, while still forbidding them to eat blood (9:4), lest they lose respect for the sacredness of life. Throughout the Torah there are numerous provisions concerning the care of animals and of the land (e.g. "You must not muzzle an ox when it is treading out the corn", (Deut. 25:4), and the provision that domestic animals are to share in the sabbath rest and the land lie fallow in the sabbatical year, (Exodus 23:10-13). The only biblical teaching that seems contrary to this general carefulness for the resources of nature is the command of the Torah to offer bloody sacrifices of animals to God (not to mention the killing of human beings in the Holy War). We must remember, however, first, that these sacrifices were offered in acknowledgment of sin and as thanks-givings and first-fruits in recognition that these creatures are gifts of a good God to us to be respected, not exploited; and second, that the Jews were always somewhat troubled in conscience by these religious customs which *Genesis* pictures as arising only after the Fall, which the prophets tended to down-play in relation to "spiritual sacrifices", and which in the New Testament are abolished.[16]

The traditional position of the Eastern and Western Church seems to be in accordance with such biblical teaching and also to be fully ecological. We may use sub-human creation for necessary human pur-poses as gifts of God to us, but always with care and reverence for the Giver. Even in the treatment of animals, because they share sentient life with us, we should be sparing in the use of flesh meat and kind to animals, because brutality to them inclines us to be brutal to human beings. Never-theless, we ought not to prefer their welfare to that of human beings, and we ought not to fall into superstitious worship of animals as have many peoples, an indication that they have lost sight of human dignity (Romans 1:22-23).

Thus the modern ecological movement, in spite of its occasional ex-aggerations, is in profound harmony with Christian ethics.[17] In the Sermon on the Mount (Matt. 6:26-30), Jesus praises the beauty of the lilies of the field and the care-free life of the birds of the air and says that God's providence extends to them too, even if we are of more value; thereby implying that under that Providence we are to live in harmony with nature, not to exploit it, just as the Old Testament had taught.

The second question: Can we sacrifice the physical integrity of the human person to achieve higher spiritual values? has been made more urgent (as I showed in Chapter 1) by the fact that medical technology has introduced new procedures by which the radical remaking of the human body becomes possible. To date the standard teaching of Catholic moralists has been that it is not justifiable to suppress or destroy any of the basic functions of the human body, except to save the life of the whole body (the "principle of totality and integrity").[18] Of course this principle does not forbid reasonable cosmetic surgery nor the removal of parts of the body which are non-functional or whose function is not essential to normal human living (e.g. anticipatory appendectomy). Nor does it forbid the charitable donation of redundant organs such as one kidney or a regenerable part (skin grafts, blood transfusions), provided the risks to the donor are not disproportionate. Nor does it apply to the human corpse (even if vascular circulation is artificially sustained), although the corpse deserves respectful disposal.

Questions about the application of this principle arise because recently some ethicists have proposed to extend this principle of totality to justify the suppression even of normal and essential functions when these functions, although they are not life threatening in themselves, may occasion some physical or even merely psychological or sociological damage to the person. Thus they would justify a woman having herself sterilized in order to avoid a pregnancy when such pregnancy might result in risks to her physical or psychological health or even her economic security, although her reproductive system was normal and pregnancy could be prevented by abstention from intercourse. I will discuss the morality of contraceptive sexual activity later in this chapter, but for the present the issue is whether an extension of the principle of totality and integrity to cover cases of this type is legitimate.[19]

Those who defend this extension argue that the traditional view of the principle of totality neglected the good of the whole person and considered only the value of biological, bodily life and organic integrity. If it is permissable to sacrifice an organ to save the life of the body, why is it not permissable to do so to save psychological health? And why is it not even more justifiable to sacrifice bodily integrity to save psychological health? For example, why is it not permissable for a woman to undergo sterilization to avoid a pregnancy that might risk a nervous break-down when abstinence might risk a marital break-down? The case is indeed difficult, but it cannot be solved by equivocating on the term "totality." In the case where a part is sacrificed to save the life of the whole person, the "totality" in question is the person as *living,* and the part sacrificed is not *essential* to this totality, else its sacrifice would kill. But in the case presented, a part *essential* to the "totality" (to the person not merely as living but as functioning integrally with *all* his or her basic functions) is sacrificed not for the sake of perserving the integrity of this whole (which in fact is lost by this sacrifice of an essential part), but for the sake of another part deemed of higher value. Thus in the first case the "totality" in question is the person as living, in the second case it is the person as not merely living but functioning integrally. Again, in the first case a non-essential part is sacrificed for the sake of the whole; in the second an essential part (essential that is to integrity) is sacrificed not for the sake of the whole but for the sake of another part, with injury to the whole.

Therefore, if we are to follow the axiom that "art perfects nature", we must take the principle of totality and integrity in the traditional sense which requires us to fulfill a double responsibility: (1) to preserve the life of human persons, even when sometimes this can be done only at the sacrifice of bodily integrity; (2) to preserve the human person in all its essential functions, even when sometimes this can be done only at considerable risk and suffering. The first of these obligations has priority over the second, but the second is also inviolable.[20]

The current concern over ecology, which it would seem would give strong support to this principle of totality and integrity, nevertheless has led many to support contraception, sterilization, and even abortion on the ground that the "population explosion" is the greatest threat to our human environment, greater even than nuclear war.

No one can reasonably deny that the regulation of population growth is a serious social problem, especially in poor countries, and that responsible parenthood is a moral duty. But it is by no means certain that this problem can be solved by attempts to persuade the poor to practice contraception. Governments in underdeveloped countries have found it necessary to reinforce such persuasion by coercive methods and have rendered their policies effective only by encouraging abortion on a huge scale. The basic condition for effective population control which respects human freedom and the right to life of all human beings even the unborn is a vigorous effort to raise the standard of living, so that the poor are willing and psychologically able to plan their lives, including the planning of the number of their children by whatever method they prefer.[21] Given this condition, the issue then becomes a free moral decision as to the methods to be used (excluding abortion). Thus the advocates of natural methods as morally and medically superior to contraception, and sufficiently effective, have a right to make their case without being accused of being enemies of the people. Certainly this advocacy of natural family planning is consistent with the respect for the human body and for the total human person understood non-dualistically. [22]

The third question, as to whether we can sacrifice the good of the individual to that of the community, arises because some have wished to extend the principle of totality from the parts of the human person to the parts of society. I will postpone the consideration of this problem to the last section of this chapter.

II: Food

1. The Stewardship of Food and Drink

After thus establishing the fundamental principle of the stewardship of the resources by which basic human needs can be satisfied, I turn to the most basic need of all, the need for food and drink, which can serve as the paradigm of our other human needs. Commonly ethics has neglected this first of all ethical problems, the problem of eating and food-sharing. Yet, as depth psychology has shown, the most basic experience we have of the world is hunger, and the most basic human relation is to the mother who nursed us. What we eat and how we help others to eat is not a mere question of technology nor of expediency, but strictly a moral question, the most fundamental of all.

Eating and drinking are the primary *pleasant* experiences, and hunger and thirst the most recurrent pains, even for the healthy animal. Although these pleasures and pains are directed teleologically to the preservation of the individual, yet human eating as human also has a social character. Sociobiology sees in the food-sharing behavior of primates the foundation for the social character of humanity. [24] This food-sharing among humans also takes on aspects of recreation, celebration and communication, so that wisdom (*sapientia* is from the verb *sapere* to taste and we are *homo sapiens*) is thought of as a kind of "food for the mind."

In *Genesis* the first sin is the eating of the forbidden fruit of the tree of the knowledge of good and evil, which some believe is "carnal knowledge" [25] and which is probably more generally the choice to determine for oneself what is to count for good and bad, moral and immoral. The first murder takes place as a result of God's acceptance of Abel's sacrifice rather than Cain's (4:1-16). In other words, God eats with Abel (although, paradoxically it is flesh meat, until then not used by humans), and refuses to eat with Cain (who offers wheat) because Cain's motives are not right. [26] During the Exodus, the people are commanded to keep the Passover meal as a perpetual memorial of their liberation from Egypt

(Exodus 12:15-28), and God feeds them with a heavenly food and drink to prepare them for their entrance into the land of milk and honey (16:1-17:7). In the light of this tradition, Jesus instituted the Eucharistic meal in bread and wine as the fundamental act of Christian worship replacing the bloody sacrifices, an act of communal identification, of historical remembrance, and of prophecy of the future messianic banquet in the Reign of God. In that beautiful parable which gives us what is perhaps Jesus' most perfect picture of his Father, he pictures the father of the two sons who provides daily food for his children, who rejoices with the banquet of the fatted calf over the return of the prodigal, and wishes his other son to join in without resentment (Luke 15:11-32). In other parables Jesus speaks of Himself as the bridegroom of the wedding feast (Mark 2:18-21: Matt. 25:1-13).

It would seem to follow that greediness (gluttony) and drunkness are the primordial sins because they form the pattern of *addiction,* that enslavement to immediate gratification (Paul speaks of those "whose God is their belly", Phil. 3:19), which cuts off human beings from any truly human life, as we see so painfully in the alcoholic and drug-addict.[27] Even sexual indulgence takes on this aspect of addiction by which sexuality is reduced to a kind of greediness, a regression to the undisciplined demand of the infant for immediate satisfaction. In the parable, the prodigal who has wasted his inheritance in the pursuit of pleasure is reduced to keeping the swine and to eating what even they will not touch. Again the Gerasene demons of madness and death plead to be allowed to enter into the swine who rush to their destruction (Mark 5:1-20), and Jesus tells His preachers, "Do not cast your pearls before swine" (6:6) to indicate the blindness to truth and all spiritual values that arises from these vices.

Connected with this ethics of food is the prominent Old Testament theme of ritual uncleaness, which sometimes seem to be confused with sinfulness, no doubt because of the close psychological association (made famous by Freud) between a child's toilet training and its moral development. It was necessary for Jesus to distinguish sharply between sin and dirt, "Do you not see that nothing that enters a man from outside can make him impure? It does not penetrate his being, but enters his stomach only and passes into the latrine" (Mark 7:18). Nevertheless, the anthropologist Mary Douglas in commenting on the great concern of the Torah for ritual purity shows that such purity has a profound social function, since it serves as the basic paradigm and symbol of *order.*[28] The unclean is the chaotic, the disordered, that which does not fit into the categories of sane human existence, and is therefore the first notion of moral evil, although not identical with it. The dietary and other so-called taboos of the Old Testament, according to Douglas, were a way of instructing Israel in a theology, a world-view and value system, which although at first sight may seem arbitrary, when rightly understood proves to be good political theology.

In freeing us of these ritual requirements Jesus and St. Paul did not intend to leave us in a state of anomie (lawlessness), but meant to free the true religion of Israel from its particularism so that it might be the universal religion in which "all foods are clean" (Mark 7:19; Acts 10), precisely because the world-view and value system of Christianity is no longer veiled in symbols, but stands forth clearly in Jesus himself.

Of course there can also be sin not only in over-indulgence, but in the *neglect* of proper nutrition and care of the body, and in the injustice that deprives others of their necessary food and drink. Recently, there has been a remarkable development of medical ethics or bioethics concerned with the problems of human responsibility for the proper care of the health of our bodies as individuals and as a society. Too often this takes the form of debates about somewhat rare problems that arise from modern medical technology, such as heart transplants. An ethics of health-care should not begin with the problems of cure but of the proper care for health and the development of the body. Physical education ought not to be directed toward winning games and breaking records, but to the development of a truly useful and delightful body. In the Gospels, Jesus is pictured first of all as a healer, but the theological implication of this is that the root of human sickness is sin, not in the sense that God sends sickness directly as a punishment, nor that sin rather than germs directly produces pathology, but in the sense that it is our human failure to use natural resources and human creativity to produce a healthy environment, a healthy social order, and a healthy life style which has left us liable to physical and psychological disorders. The healing and feeding miracles of Jesus and the saints are not a substitute for our development of medical technology, but a rebuke to our failure to promote that development fully.[29]

Moralists (leaving aside some Manichaean ascetics and some extreme hedonists) have generally agreed that in all these questions the basic principle is that put forward by the Greeks of *moderation* (*sophrosyne, temperantia,* sobriety), and this moderation is inculcated both in the wisdom literature of the Old Testament (e.g. Sirach 31:13, 17) and in the moral instructions in New Testament epistles (e.g. 1 Thess. 4:4-8; Titus 2:1-10). Christian asceticism has emphasized fasting as a means to free us from addiction and has made this a feature of the Christian liturgy in preparation for great feasts. Yet it has also rejected many Manichaean ideas such as the notion that wine is the work of the devil. Some of the saints practiced an extreme asceticism. Yet by hard experience, many of

425

them also came to see that some of this was a mistake against which they warned their followers, as being easily infected with sadomasochism and reflecting not only a denial of the goodness of creation and the dignity of the human body, but also a hatred of others turned in upon the self.[30]

So *temperantia* or moderation is an indispensable foundation of Christian life and, as St. Thomas Aquinas showed, adds a special beauty to morality.[31] John Cassian tells us that the Desert Fathers considered the fight against gluttony as the first great step in Christian spiritual development,[32] no doubt because gluttony is the archetype of all enslavement to the body, in which the body in its animal character comes to be master of human freedom instead of being the beloved servant of the person's intelligence and free will. In a Christian ethics, sensual pleasure is a true good when it is the proper overflow and celebration of truly human activities, but it becomes evil when it becomes the goal rather than fruit of the attainment of some more permanent goal.

In the Sermon on the Mount, Jesus counsels us first to seek the Reign of God, without undue concern for food, drink, or clothing, because God will provide for these needs by his ordinary providence (Matt 6:25-34). This implies, of course, that we respect that providence by the reasonable use of the natural resources with which He has supplied us.

2. The Christian Ideal of Poverty

In Christian ethics and the parallel ethics of Buddhism there is marked emphasis on the ideal of poverty, that is, on freedom of spirit from great concern for material possessions, as urged in the Sermon on the Mount. While ancient Judaism did not greatly stress personal asceticism, it did develop a theology of the poor (*anawim*) which persisted in the rabbinical tradition.[33] A similar admiration for the simple life is found in Islamic tradition.[34] Central to Jesus' own manner of life and teaching, this ideal of the simple life carried over into the tradition of monasticism, in the mendicant movements of the Middle Ages, among the Calvinists (especially the Puritans) and certain sects of the Radical Reformation, and was even borrowed by Marxism. Among Humanists it has had its advocates such as Rousseau and Thoreau, but is generally countered by the theme of the progress of technology and the abundant life.

This ideal of poverty has two distinct aspects which are often confused. First, poverty is seen as an identification with the members of society who are the "have-nots." Second, it is seen as a means of detachment from material possessions so as to free the Christian to pursue spiritual values. The former of these motives is clear both in the Old Testament and in the teachings of Jesus, but the latter is by no means absent

426

either, as the passage from the Sermon on the Mount referred to above demonstrates. The influence of the Greek ideal of the philosopher who, like Socrates, was so absorbed in the pursuit of wisdom that he neglected all material concerns, caused this second notion of poverty to overshadow the first. In St. Francis of Assisi, however, the greatest medieval advocate of poverty, both ideals are united in the notion that detachment from material goods makes possible that charity which enables us to identify with the lowliest leper.[35] It must be admitted, however, that the subsequent history of the Franciscan mystique of poverty shows that if it is romanticized it can become an end in itself.[36]

A truly Christian attitude toward material possessions would insist first of all that these possessions are not evil but good gifts of God to us. Furthermore, by reason of our God-given creativity we ought to perfect these gifts through technology and the fine arts so there might be enough to supply the needs of all. "The man who has been stealing must steal no longer; rather, let him work with his hands at honest labor so that he will have something to share with those in need." (Eph. 4:28). The monks of the Middle Ages showed us that a life of poverty can also be one of invention and of artistry. As Zen monasticism has also taught us, a simple life-style need not be either ugly or drab, since the respect for simple things and the natural qualities of materials can lead to beauty of the highest order.

Therefore, "poverty" in the sense of the deprivation of material necessities is not a good but an evil, since it frustrates God's design to supply us with our needs.[37] In a world, however, which has been devastated by human greed and neglect, the majority of humankind live in this evil poverty, so that the Christian, following the example of Jesus himself, has a duty to remember Lazarus at the rich man's gate (Luke 16:19-31). Often the words of Jesus to Judas, "The poor you always have with you" (John 12:8), are quoted out of context and as if they were a prophecy or as if Jesus was saying, "There will always be poor people, so why try to overcome poverty?" In fact Judas, who was a thief, was wholly insincere in rebuking the woman for anointing the feet of Jesus with precious perfume; nevertheless, he was right in saying the poor should be cared for. Jesus did not deny that, but rebuked him for not seeing that the woman *was* caring for one of the poor, namely Jesus who was about to die. The Christian must always see Jesus in the poor (Matt. 25:31-46), and give them the same reverence and loving care he would give the Lord, which means that if Christians have riches and power they must share these with the powerless in such a way that the powerless are helped to help themselves and to achieve control over their own lives.

427

This identification with the poor does not mean that the more fortunate ought to forgo the opportunities which their resources make possible for them to achieve that level of knowledge and culture which will enable them to be of real service to society, since without this knowledge and creativity the society will never be productive enough to supply the needs of all. Such a life of an "aristocracy", however, must not be one of hedonism or idleness, but of intellectual and social work, and ought to be marked by a great moderation, so as not to set up a false goal for the poor of achieving the same luxury. Moreover, such social leaders must be ready to share their power with the poor as they are freed and prepared for the same participation in social decisions.

For everyone, whether rich or poor in origin, the need for an asceticism of detachment from material possessions in favor of a concern for other persons and a devotion to God is an essential mark of the Christian life.[38] It is false to say that the poor should be "content with their lot" if this means that they should be discouraged from seeking social justice and those things needed for a thoroughly human life, but it is true enough if it means that there are human advantages in living in a simple way that should make us never envy the rich whose spiritual danger is great (Matt. 19:16-30). The rich can escape this danger only if they take their responsibility of stewardship seriously. They can pass through the eye of the needle only if they use their wealth to overcome social injustice to the poor.

Jesus announced the coming of the Reign of God by saying,

Blest are you poor; the reign of God is yours.
Blest are you who hunger; you shall be filled.
Blest are you who are weeping; you shall laugh.
(Luke 6:20,21)

In the form in which Jesus spoke these words they were not directly given the moral sense which they have in Matthew's version,[39] but were an announcement that God intended to wipe out poverty, hunger, and death and here and now was beginning to accomplish what He had promised. This accomplishment is the work assigned to Christians, begun by Jesus in His miracles of multiplying the loaves and raising the dead, but for us to finish not only by miracles but by human creativity. In this we must learn from the Humanists who accuse us of waiting for miracles to do what we already have the intelligence to do if we would try.

III: Security

1. Stewardship and Aggression

I began this ethical discussion by considering our need for food because this is the most fundamental biological need of a positive kind. The need for security is negative, defensive. We cannot survive without protection from the weather, from animals, and, alas, from other human beings. Together with this need for security can be classed our need for the most elementary kind of freedom, namely, the physical freedom for movement and exercise. We cannot live well in a prison of restrained movement, although the ascetical tradition of the cloister and hermitage shows that the voluntary restriction of movement can have value in freeing the spirit of the restlessness of the body, if the contemplative is prepared to withstand the depression and *acedia* (boredom) this can also produce.[40]

Humanism generally sets a very high value on human "freedom", meaning by this, first of all, freedom from external coercion, as well as from such threats as capital punishment as a means of law enforcement. Nevertheless, it is obvious that although modern society has succeeded in reducing some forms of violence, others have actually increased in the form of total global war, genocide, terrorism, crime and use of torture. We must ask, therefore, what is the ethics of defending oneself and one's community against violence, and whether violence can be used for this defense?

Some sociobiologists believe that aggressiveness is genetically programmed in the human race, especially in males.[41] Because proto-man was a relatively weak and defenseless animal, males who had a high degree of aggressivity survived better than the less aggressive to pass on their genes. Can it be that modern men engage in apparently irrational violence because they are really psychologically obsolete, adapted to fight off enemies that no longer exist?

Freud also taught, partly as a result of observations during World War I, that we are primordially motivated not only by the Pleasure Principle, but also by a Death Wish or tendency to regress to the peace of the womb. Later, however, Freudian theory came to regard this second drive not as a Death Wish but as Aggression, the drive to destroy whatever frustrates the Libido or Pleasure Principle, a drive that can express itself in violence against others, or be turned back on oneself in depression or suicide.[42] Historically Aristotle and the scholastics had already identified these two drives, the first of which they called "the concupiscible appetite" and the second "the irascible appetite." In fact the distinction

429

was recognized in still earlier times by the myth coupling Venus and Mars, the goddess of love and the god of war.[43] Depth psychologists today generally agree that the source of neuroses is not only the frustration of the libido but also the lack of harmony between libido and aggression. The aggressive drive can become libidinized, so that people get pleasure from sadistic and masochistic practices, and in our male-dominated society with its *macho* ideal male sexual potency is often identified with violence and sadism.[44]

To conceive this innate aggressive or irascible drive, however, as merely destructive is itself a destructive error. Our enemy is not order but rather disorder. Consequently, aggression need not be aimed at destroying order, but its real teleology is aimed at overcoming existing disorder so as to create order. In mythical terms it ought to be the war of the Olympian gods against the Titans of chaos and the outcome should be the creation of a just and beautiful cosmos. The creative desire to tame an unfriendly environment, to manifest human power in great works of engineering and architecture, to create great musical compositions and works of art, to explore the world and space, to bring truth to the ignorant, even to discipline and remold oneself, all are expressions of our aggressive drives. While aggression as such is not creativity, since creativity has its source in intelligence and will, yet it is this bodily appetite for struggle and victory which lends power to our creative projects.

2. The Christian Ideal of Martyrdom and Non-Violence

Aggression under the control of reason can be used not merely in offense but in defense, not only of ourselves but others, and it can even take non-violent form in "passive resistance" which can demand enormous courage and endurance. Thus the martyr needs the courage given by his or her aggressive emotions to match the victorious patience of Jesus on the Cross. In the *Book of Revelation* the Risen Christ is at once "the Lamb who was slain" and the conquering "Lion of Judah" (5:1-14). Thus it is a mistake to think that in our modern culture we no longer need a high degree of aggressiveness just because we are no longer beset by wild beasts or savage tribes. Although we often express our aggressive drives in war and crime, or in the mock-war of spectator sports and TV violence, we can also use these drives in more truly human ways in the struggle to create order and beauty, to solve difficult problems, and to endure with courage the sufferings of modern life which may be no less than those which primitive men and women had to face.

The two great religious leaders who most opposed violence were Gautama the Buddha and Jesus the Christ. The Buddha opposed violence

430

because it is the enemy of that tranquillity of the mind through which we must pass to Nirvana, and thus entangles us ever deeper in that world of illusion which we falsely come to believe has the reality of an enemy, when in fact our only enemy is the illusion of the reality of our own self.[45] A somewhat similar view is to be found in the writings of the Stoics. Jesus, however, taught non-violence by teaching us to love our enemies (who are real enough), since only by love and not by violence can free human beings be transformed into our friends. He himself endured the violence of the Cross, but by his steadfast refusal to seek any revenge, or to withdraw from the service of his enemies, he converted many of them to friends and will convert many more.[46]

At first sight this attitude of Jesus to non-violence seems contradictory to the Old Testament notion of God as Yahweh Sabaoth, the Lord of Hosts, the warrior God who commands his people to undertake holy wars of destruction against his and their enemies (Deut. 20:1-12).[47] A more careful and historical reading of the Scriptures shows that the warlike God must be understood as the way God *appeared* to a primitive people who could not yet conceive of a God of power, or of loyalty to such a God, except in terms of total war against their enemies and against compromise with false religion. In *Genesis* we are shown that in the beginning God never created a world of war, but one of peace, and that bloodshed is man's invention, not God's. Through the course of the Old Testament as Israel is educated by God through its sufferings, its exile in Babylon and its persecution by the Greeks in the time of the Maccabees, it becomes clearer and clearer that the truest picture of God is not to be found in the warrior fighting to defend his country but in the martyr who is God's truest warrior.[48] Thus in the endurance of his passion and his prayer for those who killed him, Jesus is the final outcome of the highest Jewish morality.

Jesus lived at a time when his people were oppressed by a harsh, alien and idolatrous government, yet Jesus included among his followers both men who had been instruments of its oppression like Matthew-Levi, the tax collector, and men like Simon the Zealot (and perhaps Judas Iscariot) who had been "freedom fighters" (terrorists) against that government. Some have tried to make out of Jesus himself a zealot, but his repudiation of force as a means to promote his teaching and even to defend himself seems historically certain.[49] At the same time, neither he nor John the Baptist demanded pacifism as a condition of repentance, since they accepted soldiers among their followers; and it is clear both from the teachings of St. Paul (Romans 13) and St. Peter (1 Peter 2:13-17, assuming that this epistle, even if not authentic, represents Petrine tradition), that they did not interpret Jesus as denying the right and duty of governments to use police force to maintain order and by implication to use military force in defense of the nation.[50]

In the course of history the Church came not only to support defensive wars, but offensive ones whose purpose was to advance Christian civilization or to vindicate the supposedly rightful claims of kings, and thus to canonize warrior saints, to approve the establishment of military religious orders, to preach crusades, to wink at forced conversions, to approve the persecution and expulsion of Jews, and finally to institute the Inquisition which used torture in its proceedings and handed over the recalcitrant to the state on which it urged the use of capital punishment to maintain national religious unity. Yet in all these cases the Church justified its decisions only on the basis of the duty of public officials to maintain justice by the least amount of force necessary, and only when force might establish a more stable peace. That this principle was often applied dishonestly, fanatically, or foolishly cannot be denied, but it is also undeniable that serious Christians strove to be true to the principles of the theory of just war in an attempt to oppose unjust war.[51] The notion that war and the use of force are justified simply by *raisons d'état,* that "men were made to fight," or that the enemy can be treated as "non-persons", views that have been advocated by some Humanists and Marxists in our time, were never accepted in Christian ethics.[52]

Jacques Maritain has very well shown that the Christian attitude to the use of force is that it can sometimes be obligatory in order to defend the common good and human rights, but that even in such cases it brings with it evil side-effects which will still remain to be overcome after the unjust aggression has been successfully repelled.[53] Consequently, when we rely chiefly on force to achieve public order and justice, the evil effects tend so to accumulate that soon the remedy becomes as fatal as the disease it was intended to cure. Thus most wars and revolutions have ended by doing more harm to both sides than the good they accomplished. On the other hand, if we rely chiefly on non-violent and constructive means, doing good to our enemies where they have done us harm, these efforts also tend to accelerate, since good actions are fruitful of opportunities for still more good actions. The conclusion is that force should be used only when absolutely necessary to prevent an immediate injustice and then with the greatest moderation, while our chief efforts should go into constructive, not military actions. Hence it was that Jesus wished his Church to use peaceful means only, while not denying to the state the right and the duty to use the force necessary to stop unjust aggression.

Both pacifists and militarists will say that this complex position is unrealistic. Pacifists will point out that the use of force for good purposes

432

and with moderation always seems to end in total war. Militarists will point out that the Christian stand in the face of unjust powers is always read by them as weakness and therefore as an invitation to war. But genuine realism accepts the complexity of the human condition. War usually does more harm than good, and yet it is sometimes imperative to use force to protect the helpless.

The fundamental reason that Jesus' ethics of non-violence, with the qualifications just mentioned, is sound is that the human beings are free and yet are bodies. Because we are free we cannot be forced to submit in our hearts to external coercion, and will only abide our time until we can revolt against what we regard as tyranny. Yet because we are bodily beings we can be oppressed and killed. Consequently, we have an obligation to stop unjust aggression against the innocent insofar as this is possible without committing acts that are intrinsically wrong, but we must not fall into the illusion that the mere physical defeat of aggression can overcome enmity or bring true peace. True peace can only be the fruit of forgiveness and active love, as Jesus taught and proved by his own life.

We must, therefore, conclude that Christian ethics deals with the basic biological needs for food and security by a recognition that these needs generate inalienable rights of human persons, since we cannot continue bodily existence without food and security. The norms which it sets up for the protection of these rights are chiefly (1) moderation in the use of material possessions; (2) reliance primarily on spiritual rather than coercive means of obtaining security. As forms of ecological stewardship of the world, efforts by Christians to meet these needs for all human beings must be done creatively, but with respect for the natural order provided by God. We must cultivate our garden, in the sense not of Voltaire in *Candide*,[54] but of the author of *Genesis*.

IV: Sexuality And Sociality

1. Evolutionary and Other Meanings of Sexuality

In the foregoing I have dealt with the basic human needs that correspond to the biological requirements of human persons as bodily organisms for nourishment and security, except for the fact that individual humans die. Food-sharing and mutual defense are activities not only of individuals but are communal, yet they are directed to preserving the lives of individuals as such, while reproduction does not save individuals but the species. The need for the race to survive thus has an even deeper communal meaning than food-sharing and mutual protection. Hence it is sexual relations, not only of man to woman, but of child to parents, that found human community in which alone the other biological needs can be adequately met.

In our current culture the reproductive aspect of sexuality tends to be regarded as almost accidental, even out-moded by contraception and by prospects for test-tube reproduction. Sexuality is valued primarily as a way to satisfy sensual desires and to achieve intimacy liberated from the burden of child-bearing and child care. In Catholic thought this shift has been the source of great anguish and widespread revolt against traditional ethical norms and the efforts of ecclesiastical authority to maintain them.[55]

Yet if we grant that we are primary natural units, body-persons for whom biological needs condition all our spiritual needs, and that therefore these biological needs are not sub-human but authentically human, we cannot ignore the obvious fact that we share with all living things the need to reproduce, and that it is out of this need that human sexuality originates. We know today that the biological function of the sexual differentiation of any species is not just reproduction, since that could also be accomplished by asexual modes of procreation, but the recombination of genetic material in order (1) to achieve a rejuvenation of genetic transmission by a genetic balancing; (2) and perhaps more importantly to produce new genetic combinations from which natural selection can be made so as to promote further adaptation to environmental changes and thus to advance the evolutionary process.[56]

From this need to recombine genetic materials has arisen a kind of division of labor for which the sexes have been differentiated and adapted: the female providing the ovum which must contain nutritional material for the first stages of embryological development and which is therefore larger and less mobile, and the male providing the sperm which is small, mobile and able to seek out and penetrate the ovum. The ovum and sperm are equal in their genetic contribution to the new primary unit, except that the male contributes a Y chromosome in about half the cases of fertilization thus producing males approximately equal to what otherwise would all be females. In mammals this division of labor serves the function of permitting the female to gestate the egg for a longer period within her own body and then to feed it from her breasts for an even longer period so that embryological and infant development can be lengthened and thus a more complex organism, especially as regards the brain, can be built. This long period of gestation reaches it climax (allowing for differences in body size) in the human species which thus has the relatively longest period for the formation of the brain.[57]

Sexual differentiation also leads to a varying degree of dimorphism between the sexes adapting them to the reproductive and educative roles just described. In the case of the human species this dimorphism is moderate. The male is on the average physically larger and stronger (but perhaps less resistant to disease and shorter-lived) than the female, has more markedly differentiated cerebral hemispheres, resulting in some moderate differences in psychological aptitudes, and seems to have stronger aggressive drives. Nevertheless, males and females seem to be equal in fundamental intelligence, which makes functional sense in that both have to share substantially in the care of the children and in the transmission to the child of the cultural inheritance. Females, who as mothers must understand the needs of small children before they can speak, and who then play a large part in teaching them to speak, seem more sensitive than males to non-verbal clues in personal relations but also more facile in speech.[58]

Human sexuality is remarkable in that although the fertility of the female is cyclical, her readiness for intercourse is not, so that there is no restriction of intercourse to a season of the year. As a result, monogamous and relatively permanent bonding of man and wife is a basic feature of sexual behavior in all known human societies, and other varieties of sexual activity appear as variations on, not substitutes for this basic pattern.[59] This is biologically functional in providing for the prolonged period of child care necessary for the slowly developing human young. Nevertheless, what is unique to human parenting is not the mother-child relation which is more prolonged for humans but otherwise much the same as in other mammals, but the father-child relation which hardly exists among animals, yet is of immense importance for the human species.[60]

It is true that in the animal kingdom there are isolated examples where the male takes on something of the female's nurturing role (especially among birds), and it is true that among the primates there is often a kind of patriarchy of an old male over a harem of females, infants, and young males. These examples, however, are only dim anticipations of human fatherhood. The human paternal relationship is far more differentiated and long lasting, and is characterized by the father's responsibility to provide food and protect his wife and her children and to play an important role in the education of the child, particularly in its transition from childhood to maturity, and also by the father's normally intense psychological identification with his children. It is possible here, as in all human matters, to cite wide cultural variations in the parental roles, yet the fundamental pattern everywhere shows through the variations, and it seems that where these relations are weak the society is considerably weakened in its capacity to survive and develop, since these more extreme variations are found chiefly in small marginal societies.[61]

Thus it is hardly an accident, biologically speaking, that male "dominance" has been a common feature of all known societies, although again there are marked variations as to its mode and degree. Nor is this generalization invalidated by such special social arrangements as matrilineal inheritance of name or property in some societies, or by the rare occurence of polyandry.[62] The basis of this universal feature of human society seems to be that child-bearing and infant care render women more dependent on their males who are physically stronger and more mobile. The counterforce to this female dependence, however, is the fatherhood bonding of the male to his children, resulting in the institution of marriage which under various forms exists in all societies, and which ties the male psychologically to responsibilities for his children of which he would otherwise be physically free.

Polygyny is a wide-spread variation on monogamy in the human species and reflects a kind of compromise (resembling the arrangement among some primates) by which the male is bonded to a group of females and their offspring, but while this may be highly functional from the viewpoint of numerical reproduction, it seems obviously less functional than monogamy from the viewpoint of child-care and male participation in the education of the offspring.[63]

In spite of the rather clear biological origin of sexual differentiation in the human species, today many theories are put forward as to the truly *human* meaning of sexuality. The reason for these speculations is the apparent tension between the equality of man and woman as regards personhood and the apparent subordination of woman to man as regards reproduction. Can woman ever really hope to be treated fully as a person so long as she is disadvantaged by child bearing and rearing? Moreover, it seems to some that what is specifically human about sex is not the relation of parents to children who are not yet developed personalities, but rather the relation of the sexual partners. The feminist movement which often urges these considerations has been a characteristic feature of the romantic wing of Humanism from its beginning, partly, it would seem, becauses it symbolizes as no other reform could, the conviction that "biology is *not* destiny," since we can re-make ouselves culturally.[64]

Sigmund Freud put forward what can be called the *economic theory* of sexuality, according to which sex is the expenditure of a single fund of pyschic energy, the Libido or Pleasure Principle. Only in orgasm is there a total discharge of this energy. All other human activities, therefore, require to be energized by a diversion of libido through some kind of *sublimation*. This theory implies that individuals exist only to reproduce and raise their offspring to sexual maturity and that their other activities are mere means or by-products of this endless cycle. Freud, however, also

436

came to believe, as mentioned earlier in this chapter, that the death wish or aggression cannot be reduced to the libido and many Freudians today have gone still further in what they call "ego psychology" to admit still other sources of energy in the psyche.[65]

Carl G. Jung developed what can be called the *Androgyny Theory* of sexuality, according to which the psyche in all human persons has two aspects: the *animus* and the *anima*, the former of which is unconscious in the woman and the latter unconscious in the man. Consequently, no one is complete without union with a person of the opposite sex through whom the hidden half of the personality is awakened so as to prepare for the total individuation of the self which thus becomes consciously androgynous. Although this view is highly illuminating, as we saw in Chapter 9 in explaining the symbolism of *Genesis* 2-3, it is rejected by some feminists because it seems to assign to the woman the less rational and presumably more inferior half of human personality.[66]

A third theory, very popular today, is that sex exists not for reproduction (as Freud, and even Jung presupposed in their theories) but to satisfy a fundamental need of every human person to complete itself by *intimacy* or union with another person in a total sharing of self. Such totality is possible for body-persons only if it is expressed bodily through sexual intercourse as the most intimate possible union. For many who hold this theory, not only is the reproductive function of sex secondary, but it is even an obstacle since children obviously get in the way of intimacy.[67] The great weakness of this view is that it does not explain why human intimacy must take this rather peculiar form. Why should human beings have just two kinds of bodies, and why does their love have to be expressed in such an unhygienic and clumsy way? Obviously this theory makes sense only on the assumption that evolution was itself an essentially meaningless process which has left us with odd bodies which we have to make do for human purposes for which they are not really very suited, since they must be rendered infertile in order to be used for sex conveniently. Its strength as a theory, however, is that it does stress that although sex has a biological origin, in human beings it takes on new psychological aspects not evidently present in subhuman animals.

A fourth theory might be called *metaphysical*, since it sees the origin of sex not in reproduction but in an essential bipolarity of Being itself. God Himself, therefore, is bipolar and androgynous and this bipolarity is mirrored in creation by the differentiation of the sexes.[68] I will consider this notion in the next chapter, but will here only raise the question whether it is not more plausible that bipolar concepts of God (which are certainly found in many religions) are anthropomorphic rather than that sexuality, which is so obviously related to the mortality of individuals should be characteristic of the immortal God who can only be named the I Am.

A fifth theory is the *inspiration* theory according to which the function of sexuality is not physical reproduction, but the stimulation of human creativity to cultural advancement. This theory is as old as Plato, was wonderfully developed in the idea of medieval chivalry, and revived in romantic Humanism. Undoubtedly the experience of romantic love does awaken the person to aspects of his or her personality previously hidden, including intuitive and creative gifts. Don Browning in his work *Generative Man*, based on the psychology of Erik Erikson,[69] has shown how great is our need to transcend death by leaving children behind us, and that in many persons even this is not enough. They must also leave behind them novels, poems, paintings, scientific discoveries, or empires. Yet it seems reductive and narrowly Freudian to say that the wish to be creative is aroused only by sexual love. Creativity must be motivated, and this motivation can only come from love, but since human persons are intelligent and free they can love for other than sexual reasons.

An adequate theory of the meaning of human sexuality must integrate the true insights found in all these theories. The difference of the sexes is a division of labor which has arisen in earth history as the most effective way to guarantee the continuation of life, in a way open to the natural selection which promotes adaptation to the changing environment. In the human species this sexual differentiation is modified in view of the longer period of gestation and infancy required by the complexity of the human brain, and requires monogamous pair-bonding and the development of the role of the father. At the same time, because both male and female are equally intelligent free persons, it is necessary to develop cultural forms of marriage, varied from time to time and place to place, which can provide for the full development of both the female and the male primarily as persons and only secondarily as reproducing persons.

This development of human personality in the reproductive relationship means that the bonding and dedication to the child require a deep and intimate love of a specific, sexual kind in which the two personalities complement each other. In particular the human male needs to take permanent responsibilities in this relationship which do not arise from his simple role as impregnator, but in his continuing role as protector, food-gather, and educator of the children together with the mother. Since, however, human personality and not merely the continuation of the species, is the goal of reproduction, neither the man nor the woman can be entirely immersed in this sexual relation or in child-care. Family love may be the inspiration of wider activities, but these wider activities are

438

rooted not just in human sexuality but in human personality. Hence culture must not define either the woman or the man simply in terms of their family roles, but also in terms of their social roles.

2. The Future of Sex

The tension between the development of the individual personality, especially of the female, and the sexual role, has give rise in our time to the question as to whether we ought to use human intelligence to emancipate both men and women from this tension. In considering this question we should note that the protest against "Victorianism" has led some to treat sex as if it were a wholly positive aspect of human life, when in fact it is obviously highly problematic and ambiguous. If it were not for the fact that human beings are doomed to death, reproduction and consequently sexual differentiation would never have come into existence. No wonder that so much romantic literature and music climaxes in a *Liebestod.*[70] Certainly the first great epic of the west, the *Iliad* and the tragic dramas dwelling on its themes portray love as a kind of madness which destroys even the gods. The modern drama and novel, as well as rock and country music, tell us no other tale. The question therefore arises whether we would not be better off if we could eliminate human sexuality altogether, as we are eliminating its inconvenient fertility, and find other ways to fulfill even its hedonistic and romantic functions.

The result of such musings has been to lead some people such as Wilhelm Reich and Herbert Marcuse to call for a return to what Freud called "the polymorphous perversity of the child", that is, to release sexual satisfaction from any specific function and simply let it become the way, having a thousand expressions, of enjoying sensual release.[71] All existing civilization is thus regarded as oppressive and requiring a revolution. Along with such pleas has come a widespread denial (made official by the American Psychiatric Association)[72] that heterosexuality is any more "normal" than homosexuality. Some even assert that sexual intercourse is preferable to masturbation. The very notion of what is sexually "natural" seems to many meaningless. What is sexually good thus ceases to be a moral question and becomes purely a technical problem guided by personal preferences, provided that the partners give free consent and no serious physical injury results.[73]

This "liberal" attitude has become ethically plausible largely through the widespread conviction that just as our aggressive impulses are supposed to be atavistic, so our capacity to reproduce which was once necessary when infant mortality was very high, has become a liability in modern society whose very existence is threatened by the population explosion. It is urgent, it is said, to use human intelligence to separate reproduction from the necessary gratification of sexual appetites. What is less frankly dealt with is the fact that this logic should lead us to ask further whether the attachment of sensual release to sexual orgasm should itself be overcome. In fact, some today are finding drugs more pleasurable than sex. Does it not seem reasonable to suppose that in the future, by genetic engineering, we might produce unisex humans without sexual organs at all (thus eliminating all the many disorders of the reproductive system), but who would be able to achieve something better than orgasm with any other human being or by masturbation simply by pressing a button located in the middle of the forehead? How much that would simplify human life![74]

Such suggestions, really no stranger than proposing to have human beings conceived in a test-tube or frozen alive for future use, falter before the mystery of what it is to be human. Do we know enough about ourselves to be able to say whether such an elimination of sexuality would also destroy our humanity? What we do know is that human intelligence requires the body and bodily experiences. Our whole development as intelligent and free creatures is bound up with our experience of the world through our bodies and the actual human body is profoundly sexual. Freud seems to have been not far wrong in teaching that our whole system of feelings and of imagery, and even our lively curiosity about the world arise from the baby's first longing for its mother's breast, then for the mother's and its own body, and then for all the other relations of its body to other bodies.

If we were to eliminate or profoundly modify this development of human knowledge through sexual (in the broad sense) experiences, what would be the consequences for our capacity to know and to create? We cannot be justified in any profound change in human sexuality until we understand whether the consequences of this would be pervasively destructive of human personality. This is also why we cannot unthinkingly alter the basic parent-child relation by the ready use of artificial insemination and test-tube reproduction as easy answers to problems of infertility. Undoubtedly there is room for scientific study of sexuality and technological advancement of sexual health and functioning, but what are the limits set by stewardship?

3. A Creative Ethics of Sexuality

A Christian ethics of sexuality must be based on Jesus' own example, but Jesus was a celibate! So was his most prominent apostle Paul, and in the most explicit treatment of the sexual question in his *First Epistle* to the Corinthians St. Paul advises to Christians the free choice of the celibate life, the one he has himself chosen, as preferable to marriage because all Christians, married or unmarried, must keep in mind that

> . . .the time is short. From now on those with wives should
> live as though they had none; those who weep should live
> as though they were not weeping, and those who rejoice
> as though they were not rejoicing; buyers should conduct
> themselves as though they owned nothing, and those who
> make use of the world as though they were not using it, for
> the world as we know it is passing away. (7:29-31)

Consequently, married men or women are going to suffer from un-necessary trials (7:28) because their attention will be divided between the concern for their partners and "the pursuit of holiness in body and spirit" (7:34), while celibates will be "busy in the Lord's affairs" (7:32) and those alone. Paul here seems to be making another application of the Lord's words in the Sermon on the Mount when He said, "Seek first the kingdom, the heavenly Father's way of holiness, and all these things will be given you besides" (Mt 6:33). But we have also the words of Jesus to his disciples who complained of the strictness of his teaching with regard to divorce,[75]

> His disciples said to him, "If that is the case between man
> and wife, it is better not to marry." He said, "Not everyone
> can accept this teaching, only those to whom it is given to
> do so. Some men are incapable of sexual activity from birth,
> some have been deliberately made so; and some there are
> who have freely renounced sex for the sake of God's reign.
> Let him accept this teaching who can." (Mt 19:10-12)

And also,[76]

> "Everyone who has given up home, brothers or sisters,
> father or mother, wife or children or property for my sake
> will receive many times as much and life everlasting." (Mt
> 19:29).

And Jesus' statement to the Sadducees who did not believe in the resurrection,

> "When people rise from the dead, they neither marry nor
> are given in marriage, but live like angels in heaven."
> (Mk 12:25)

In the early Church and later in a more institutionalized way in both the Catholic and the Orthodox churches these teachings were taken as solid grounds for the encouragement of the celibate life as a practical expression of the reality of the coming Reign of God and a witness to married Christians that this reality must also be remembered in their own lives, where earthly cares might make it easy to forget. The churches of the Reformation have not denied that this is New Testament teaching, but have objected to its institutionalization under vows (which seem to them presumptuous) and have tended to the notion that after apostolic times this gift of God, like other special gifts, is rarely given. Today in the face of the psychological emphasis on the need for sexual expression for mental health, this New Testament teaching is also often explained away as reflecting a particular historical situation, namely the expectation of the imminent second coming of Christ which was never fulfilled, and such teaching, therefore, is irrelevant for us. It is impossible, however, to remove from the Gospel the essential teaching that the Reign of God is already present in the world, manifesting itself by an abundance of spiritual gifts so that every Christian in his or her individual life must live in constant expectation of its fulfillment as Jesus taught in the Parable of the Ten Virgins (Mt 25:1-13).[77]

How can this teaching of Jesus be reconciled with what has already been said about sex as one of the basic needs that define human nature? Is not celibacy a contradiction of this need and therefore *intrinsically* immoral? The Old Testament Law required all males to marry and beget children to strengthen the Chosen People. Women who were sterile were regarded as accursed. Yet by the time of the Qumran community some Jews had committed themselves to a celibate life.[78] The reason probably was that they were preparing themselves for the eschatological Holy War which they believed would usher in the Messianic Age, and they knew that the Law commanded warriors in time of battle to abstain from sex in order to remain in a state of ritual purity and to devote their full energies to battle. This was perhaps also reinforced by the institution in Israel of the nazarites who were ascetics also devoted to the Lord's work. It is not said in the Scriptures that the nazarites were celibate, but their abstinence from wine and from cutting their hair or shaving their beards was intended to express their withdrawal from ordinary affairs for the sake of their sacred mission, as in the case of John the Baptist[79] who was probably the earliest model for such Christian practices.

Every human being must eat in order to live. Sexual activity is absolutely necessary for the community to survive, but not for the individual, for whom its value is only relative. Consequently, it is not intrinsically wrong for an individual to forego sexual activity for the sake

442

of some greater value and for the good of the community, provided that in doing so he or she does not do injury to personal integrity. This proviso means that celibacy cannot justify (1) physical mutilation intended to desex the body; (2) psychological suppression of the person's masculinity or femininity, the attractions, feelings, imagery which form an integral part of human personality. The celibate remains a sexual person[80] because, as we have seen, the development of both the affective and cognitive life of the person has its libidinal sources which cannot be denied without psychological harm, and which require to be developed for full human maturity. Thus it is possible, although certainly not easy, for a person to achieve full psychological maturity and balance in a celibate life. In fact, people are not really ready for marriage until they have achieved a reasonable degree of maturity through a chaste single life.

Christians, therefore, are urged to consider dedicated celibacy as a possible life choice, and for those who are ready and willing for it, a preferable life choice because it will free them for a fuller dedication to the Holy War, not a war of violence and bloodshed, but the war against the violence and injustice which hold back the Reign of God. They are called not to be the only warriors in this spiritual battle, all Christians are called to that (Eph 6:10-17), but the front-line troops who give leadership and courage to the others.

What then of marriage? Some theologians today think that to stress the New Testament commendation of celibacy as preferable to marriage is to make married Christians second-class citizens, to treat marriage and active sexuality as unclean.[81] This is to misunderstand the very notion of the Reign of God so central to Jesus' teachings. The Reign of God is both an earthly and a heavenly, a temporal and an eternal reality. It is the love of God transforming this earthly life ("Thy kingdom come, Thy will be done on earth as it is in heaven"), and opening it up to everlasting life in the Trinity. Consequently, the Christian life and the Christian community in which it must be lived always have two inseparable and complementary aspects, forming a single age of past, present, and future time. On the one hand, the Gospel must sanctify earthly life, heal its sins, and make it fruitful in human culture. On the other hand, the Gospel must always point forward to the transformation of earthly life in the eternal Kingdom.

Consequently, Jesus and Paul, although celibate in order to serve their people and die for them, approved faithful sexual love and Paul severely reproved those forbidding marriage (1 Cor. 7:1-7; 1 Tm 4:3). Thus in Pauline teaching marriage is presented as a symbol or sacrament of Christ's love for his Church (Eph 5:25-33).[82] Marriage is the redemption of human sexuality from its fallen condition and its restoration to be the holy gift God intended it to be. Married love teaches the partners, their children, and the whole community something of Christ's love for his people, a lesson which lays the foundation in the life of every child to understand what true love can and should be. As such it is a sacred ministry in the Church which to despise would be sacrilege. Nevertheless, it is not the ultimate in love. It is still bound up with the world that is passing away, and it must eventually be transformed into a higher love which is eternal, the love which Jesus as celibate showed us, and which dedicated celibates choose as their portion even in this life.[83]

It should be added that what is in question is not only the religiously dedicated celibacy of vowed life. The commendation of celibacy also opens up to everyone the freedom to choose not to marry for reasonable motives, since the Christian community does not depend for its existence exclusively on physical fertility but also on conversion. Thus the commendation of the chaste single life is one aspect of the Christian contribution to the emancipation of women, since it means that the dignity of a woman as a person in no way depends on her being a mother or a religious.[84]

What, then, of those who freely choose to be sexually active? It is commonly said that traditional Christian ethics held that the only legitimate use of sex was for procreation, all other positive values being entirely secondary. It is true that this ethics has always insisted on two points: (1) sexual activity may not be deliberately separated from its natural teleological relation to procreation; (2) sexual activity may not be deliberately separated from the expression of committed love. The first separation violates the biological nature of sexuality, and the second its specifically human (*and* biological) nature. For Christian ethics all human acts must be motivated by the love of God and neighbor. This love is not a means to an end, it is the very end of human life. Consequently sexual relations which are not motivated by love and that specific kind of love which is in keeping with the nature of the two persons are an abuse.

Loving marrried couples do not have intercourse primarily to have children, but to express their love for each other, a love that seeks perfect union; but such love is not fully expressed in intercourse but in an intercourse which reaches its natural completion in creativity, in the begetting and rearing of children. There can be legitimate reasons why a couple in particular circumstances of health and of resources to care for the children ought to forego this completion of their love-making provided they do not so alter that love-making itself that it no longer has its natural, fully human character. That is why they may confine intercourse to the woman's known sterile periods, or may have intercourse even when they know that one of the partners is sterile.[85]

But why is it not ethical to separate love from procreation by contraception or in homosexual unions? Or procreation from intercourse as in artificial insemination or *in virto* fertilization? Or pleasure from love as in casual relations between consenting adults? Or pleasure from intercourse, as in solitary masturbation, mutual masturbation, sodomy and other perverse practices? Humanists today generally see no moral problem in any kind of sexual activity that is voluntary, and disapprove only of rape, of the molestation of the young who cannot give truly free consent, or of perversions that involve physical injury.[86] For the Christian, however, the teleology of sexuality for the human being is not itself a matter of free choice because sexuality is one of the defining characteristics of what it is to be human. Hence it is a goal set up for us to achieve in order to become fully ourselves. We are free in the way we achieve that goal, but it is the goal which determines what means to the goal are appropriate. To suppose that the goal is pleasure apart from committed love is to misunderstand what sex is all about and can in the end only lead to a distortion and frustration of authentic sexual fulfillment.

The masturbator, the frequenter of prostitutes, the promiscuous person obtain a momentary pleasure but do not realize themselves authentically as human sexual beings. Nor does the adulterer, the Don Juan, the homosexual lover, because their loves lack that permanent commitment to family life in which real sexual love grows to maturity. As for those who practice contraception or who seek to satisfy their longing for children by artificial means unrelated to their intercourse, they separate their love from its fruitfulness or bring into the world children whose right to be the fruit of the parental love-act has been denied them.[87] All these practices are an evasion of the true meaning of sexual love. For the Christian mindful of Jesus' own teaching, the only truly human sexual activity is that in a permanent marriage between man and woman, open to the transmission of life to the next generation and based on genuine self-giving love.

What then of the marriages in which such love does not exist? Christian love is something more than passion or even mutual attraction. When such attraction or passion do not exist a couple can still genuinely continue to love each other by seeking the good of the other, and no one in marriage has the right to demand more than that commitment of good will and fidelity.

445

This Christian ethic of sex which forbids sexual activity to all except permanently married people seems to Humanists utterly unrealistic and cruel.[88] Are we to condemn to loneliness and frustration young people who are not yet married or all those old or unattractive people who cannot find a partner willing to marry, or all homosexuals who cannot responsibily enter into marriage? Are we to condemn all those married people who find their situation intolerable and seek outside marriage for love, or wish to free themselves by divorce to find a better marriage for themselves or their children? Are we to condemn all those persons who can find satisfaction in their sexual lives only through pornography or other so-called perversions? What is moral about depriving people of harmless happiness in life or making them feel guilty or unworthy because they seek such happiness?

Such questions are based on a fundamental misunderstanding of what the Gospel teaches about sin.[89] Jesus taught that no one should judge others but only himself (Mt 7:1-6). He showed himself deeply compassionate to those others regarded as sinners, especially sexual sinners, and most particularly to women whom society judged with great severity (Jn 4:4-42; 8:1-11; Mk 14:3-9). At the same time He forbade divorce (Mk 10:1-12) and even a mental consent to extra-marital sex (Mt 5:27-32). How could one who so loved human beings just as they are in all their human longings, lay down so heavy a standard for them, especially since he condemned the Pharisees because "They bind up heavy loads, hard to carry, to lay on other men's shoulders, while they themselves will not lift a finger to budge them" (Mt 23:4)?

Jesus did not contradict himself in thus declaring the will of God in its purity while at the same time denouncing legalism, self-righteousness, and hypocrisy. He wanted to help us understand the real effects and consequences of our acts, so that we will be awakened from our delusions about where our real happiness is to be found. The misuse of our gifts, including the gift of sexuality, far from bringing us happiness, can in the long run only lead to our own personal disintegration and to great harm to others. For Jesus not to have told us the truth about ourselves would have showed a lack of compassion on his part. Moreover, Jesus does not warn us without at the same time giving us God's forgiveness for the past and strength for the future to take the difficult but sure road toward real happiness, not only by his interior gift of the Spirit but also by providing us the human ministry of the Christian community.

What really makes the difference between the Christian understanding of sexuality and the Humanist's is that the effects and consequences of extra-marital sexual behavior are differently estimated. The Christian sees the masturbator as acquiring a habit, difficult to break, which leads him or her to seek immediate sexual gratification rather than use of the gift of sexuality to express love and self-giving to another. The frequenter of prostitutes, the philanderer, the extra-marital or pre-marital cohabitor is exploiting another human being and making out of himself or herself a person who is not only self-indulgent but a deceiver. The adulterer is one who betrays the love and confidence of his wife or her husband or destroys another's marriage. The homosexual and those who seek perverse pleasures are handicapped persons who do not face honestly their own psychological abnormality and seek its cure or make the best of it without exploiting others or degrading their own dignity as persons.[90]

All these persons are hurting themselves and others in a way that defiles the deepest springs of human love and honest relations. Of course they are often driven to such behavior by psychological compulsions which may not be within their control. Even when they act freely, their freedom may be limited by their ignorance or by terrible emotional conflicts. They may often be more the victims than the offenders. That is why none of them are to be judged morally by us; but all need help for themselves and for the defense of those they injure even when they do not intend it. Still worse is the fact that these distorted relations are so often passed on to the next generation which is thus left still weaker in its efforts to be truly human in sexual life.

One of the great evils that flows from such abuses of sex is the oppression of women.[91] Today some feminists seem to believe that the emancipation of women is to be achieved by encouraging women to be as sexually "liberated" as are men in our society. Some advocate lesbianism, extra-marital affairs, and the use of contraception and abortion as means for women to be as free as men have been in most societies to seek uninhibited sexual satisfaction. They argue that men have insisted on virginity and chastity in women, and spread myths about the nature of female sexual response or even mutilated female children in order to secure their possession over the women they wish to exploit. They commonly accuse the Christian Church of aiding and abetting this exploitation.[92]

In a Christian ethics, I believe, we ought also to denounce the great evil of the oppression of women in most cultures, including most that have labeled themselves "Christian," but a Christian diagnosis and cure for this sinful situation is different than that of most Humanists. I have shown that the Bible recognizes that the reduction of most women to dependence on men because of the male's greater physical strength and the females burden of pregnancy and child care is a result of sin, just as is the crushing toil to make a living for most men (Gn 3:16-19). God intended in the beginning that woman and man live in complementary equality, having somewhat different roles in the family, but bound together in mutual dignity, respect, and love, because they are equally persons made in God's own image by their intelligence and freedom and their commission to share God's creative dominion of the world. The fact that male dominance, the enslavement of women, polygyny and irresponsible reproduction are historically so common in the human family is sinful and this sin must be overcome by Christian love and justice.

That men and women differ and have different gifts to use in the family and in the community is not itself an evil, but it is a grave injustice that women are not permitted to use their gifts fully lest this threaten the mastery which males have assumed. The source of masculine mastery is not superior intelligence but physical violence and sexual exploitation. This oppression continues because human cultures and their social structures favor violence and sexual exploitation as high values, and they disparage feminine values.

If this is the correct diagnosis, what is the remedy? Jesus taught, as we have seen, that the remedy for tyranny in this world is for Christians to imitate Him who said "The Son of Man has not come to be served but to serve — to give his life in ransom for many" (Mk 10:45, cf. Phil. 2:6-8). Jesus is male and this duty of imitation falls first on men. They must give up their claim to dominate, and they must accept that strict rule of chastity which Jesus stressed by his own celibacy. They must love their wives as their own bodies, their own selves, and must serve them and sacrifice themselves for them, as Christ did for the Church (Eph 5:25-33). Women must reciprocate in this mutual submission or self-giving, but the real burden of Christ's teaching falls on the husbands, as is shown by the fact that in forbidding divorce, his primary purpose was to forbid husbands to divorce their wives (Mk 10:2-12).

A fair reading of the history of the Christian Church will show that its constant effort has been to reform men, because until men change women will be oppressed. Of course these efforts have been far from successful, and sexism has continued in society, and from society has influenced many aspects of Church life. Yet a comparison of the results of Christian influence on the status of women and that of the other traditional religions shows the record of Christianity is *relatively* good. As for Humanism and Marxism, they have indeed promoted certain aspects of the emancipation of women more effectively than the traditional religions, but only at the cost of undermining the family, on whose health respect for human dignity ultimately rests.

Too often the churches have been confused into resisting changes that to be truly consistent with their own principles they should have favored, because these changes had been linked with others they could not consistently support. Thus many Christians have opposed the Equal Rights Amendment (obviously in itself demanded by simple justice) because so many of its advocates linked it with an approval of abortion and lesbianism.

A Christian ethics of sexuality is *creative* because it seeks to use human intelligence and freedom to enable human beings to use their sexuality well, rather than to submit to the widespread abuses built into our sinful world and encouraged by it. By a deeper exploration of the full meaning of sexuality in human life and its relation to the celibacy that witnesses to a greater and everlasting life, it seeks to make genuine sexual love more attractive than its distortions and perversions.

There is much in the culture which has developed under Christian inspiration which does glorify a true notion of sexual love, but it must be confessed that even the greatest Christian poet, Dante, although he managed to show in his *Inferno* the tragedy of perverted love, and in the figure of Beatrice the way that romantic love can lead to transcendent love, yet he does not give us a picture of married love in its own earthly beauty. It remains for Christian poets, novelists, artists, and filmmakers to show the world a truer vision of human love.[93]

Creativity also must be shown in finding ways to heal human sexuality at its roots through family therapy and a better emotional education of the child. Christian families, caught in the struggle between Christian ideals and the counter pressures of the world are often psychologically unhealthy, lacking genuine joy, filled with false guilt and distorted religiosity that makes God seem a tyrant rather than a loving Father, loving as a mother. To remedy this situation, we need to reform the social order so that it can give support to families rather than corrupt them.

Above all a Christian ethic will become creative when it leads couples to explore their love in a spirit of faith and prayer so as to discover in each other the authentic persons whom God created and by grace is redeeming. They must discover God in each other, and to do this they need the help of celibates who have used their freedom from earthly cares to discover God for themselves and thus to be able to encourage others less free to find Him in the very midst of their own pressured lives.

V: Information, Society, Creativity

1. Communication

The three basic needs so far discussed are common to all human beings and also to animals, although in human beings they take unique

449

form. The three remaining needs: information, society, creativity, listed at the end of Chapter 8 are specifically human, even if they are dimly anticipated in the animal world. They are explicit manifestations of human intelligence and freedom, and are so intimately linked that I will discuss them together.

Among animals, the only common social unit is that of mother and offspring and this chiefly among mammals. While insects sometimes exhibit remarkable social structures, these are of so different a type from what we recognize as a human society that the resemblance must be considered only analogical.[94] Satirists often compare totalitarian human societies to an ant-hill or a bee-hive in which the individual is totally subordinated to the survival of the species, but in human societies such ruthless subordination is clearly pathological and seldom lasts for long. Other animal societies are loosely bound together: schools of fish, flocks of birds, packs of wolves, colonies of prairie dogs. Even among the primates there seem to be only loose bands made up of the kinds of patriarchal "families" already described. Of course among primitive humans social organization is also rather weak, but wherever an economic surplus makes it possible, groups become highly organized even among pre-literates.[95]

As already discussed,[96] there are anticipations of language and of invention and the transmission of culture among animals and especially among our primate relatives, but these never rise to the level of what is so obvious a feature of human social life: the dependence of society on many kinds of invention and especially the inventions of modes of communication, which are not natural, but cultural. Animals of course need information, as they live by sensation, but since this information never rises to the level of self-consciousness,[97] it never becomes *truth,* which requires the judgment that I not only know but that I know reflectively that I know. It is only by this capacity for reflective truth that human beings have ethical problems, because they recognize goals as such and choose appropriate means to attain them. Moreover, in all our human activities, we ultimately act in order to take satisfaction in the results, and this satisfaction or enjoyment is an act of self-consciousness, of contemplation. We humans have as our most ultimate need, the need for truth, to know that all is well with us not merely in fantasy but in reality.

The truth which gives us ultimate satisfaction is not just any truth. Much of what we know is merely utilitarian (phone numbers for example), and of the things we enjoy knowing, many are not indispensable. The kind of knowledge which we absolutely need as valuable in its own right is our knowledge of those primary units which are ends in themselves, and these are *persons*[98]. To know ourselves and other persons is to know what is highest in creation and beyond it, for God is personal. Whatever else we know in science or history has value for us ultimately because it helps us to know ourselves and each other better. Persons, however ordinary and dull they may seem, are inexhaustible mysteries, so that we never need fear that we will be bored with others provided we can really get behind the masks that hide us from each other. We come to know ourselves only in knowing others in whom we see hidden aspects of our own nature manifested.

According to *Genesis* Adam needed Eve, someone human like himself, not only to know her sexually (although through this he discovered a hidden part of his own nature as feminine), but to discover himself as human. Nor would it have sufficed for him to know himself in Eve only. He and she needed children in whom they could recover (to take the story literally) what they had missed by not having a childhood, just as we need to recover the childhood we had but have forgotten. In fact we need the whole strange variety of the human family, of every race and every individual character to understand something of what is germinally contained in our own humanity.

Thus we humans are social animals, not in the way other animals have something of sociality, but because we have the need to know ourselves as the highest beings in our visible universe and as the only analogue by which we can make guesses about superhuman persons who may be citizens of our universe, or who may have created it.[99] Our need for society, therefore, is not confined to our needs to gather and share food, to band together to ward off enemies, or for reproduction. Above all, we need each other to explore the universe and by our mutual thought to come to understand it, and through it to know each other. In us, as Hegel said, the universe becomes conscious.[100]

It is odd that some scientist consider it arrogant of us humans to consider ourselves the most important items in the visible universe, more important than insects that may survive us, or quasars that shine so brilliantly. But we are important because we can be scientists and study insects and quasars, and insects and quasars cannot study us. Of course one might say, "What is so important about knowing the truth?" Aristotle said that people who ask why we admire beauty in the human body are blind.[101] Those who do not understand why it is so great a thing to be a truth-knowing animal, if they are scientists, are denying their own vocation. When we find other scientists in distant planets and if, as is usual in science fiction, they prove to know more than we, then it will be time for our scientists to be modest.

To know each other, however, we have to communciate, and that means we need to invent language. In fact it is generally conceded today (too facilely), that human thought itself is simply language. At least it seems certain that a child like Helen Keller was not able to develop mentally very far until she learned sign language. Current philosophy, whether of the British-American analytical school or the continental phenomenological school, has largely become a study of language and of hermeneutics. While this betrays an unfortunate tendency to idealism, it does make very clear that if it is to develop, human thought must not only reflect on itself but on the language in which it expresses itself and communicates itself.

What is communicated, however, is not only truth but attitudes. All human language has a feeling aspect most evident in rhetoric and poetry, but never absent even from scientific works which aim at perfect objectivity, not even in a mathematical treatise where "elegancy" of proof still reflects human taste. Human truth is always truth *for us*. To admit this is not to deny that truth is a correspondence of the mind with objective reality, but to assert that for us to pursue a truth we must first perceive and appreciate it as a value in our lives whether for its practical uses or its way of opening us up to the world and our own depths.

These attitudes which are communicated in language all rest on *love*. Because we love ourselves and others, we value what is good for ourselves and for them, and we fear and hate what is harmful. In the ethics of Jesus, the Great Commandment is the love of persons (Mark 12:28-34). To love someone with the God-like love of *agape* is not just to desire them for ourselves but to seek their good for themselves. Yet that good for human beings is ultimately their peace and union with each other and the common enjoyment of the truth about each other. It is communion and celebration — mutual praise and joy. As the *Gospel according to St. John* never tires of saying, the love of which Jesus speaks is always a love directed toward the light, toward truth. It is victory over darkness, over all that separates us and obscures our true selves from each other. The Reign of God which Jesus preaches is nothing less than this union of all created persons centering in God, knowing each other mutually by the Light of God who is Christ, and loving each other mutually by the Love of God who is the Spirit.

But how is this Reign of God to be realized in our human bodily condition in history? The Spirit is bringing this Reign to birth through every historical process that draws human beings together in brother-and-sisterhood. Therefore, this Reign is found first of all in families bound together in sexual and parental fidelity. Secondly, it is found in tribes, nations, states, empires insofar as these groupings are the valid expression of our social nature. Thirdly, it is to be found especially in the great world religions insofar as these religions tend to transcend national and cultural limitations and to unite the human family in mutual respect,

452

a tendency very clear in Buddhism, Christianity, and Islam, and in a secular way in Humanism and Marxism. Ultimately the Spirit is drawing the whole human race into a single community, especially through the modern technology of communication.

This work of the Spirit, however, is required for human beings to overcome their own natural limitations, especially as these have been aggravated by sin. Even apart from sin, as bodily beings we are separated from each other by space and time. We can communicate only through language of which the medium, however tenuous, is still material signs or symbols which are always inadequate to our thoughts, and how many and diverse are our human languages! Hence the process of coming to understand each other is laborious and dialectical. We can achieve it only by repeated corrections of messages half understood or misunderstood. Each generation, moreover, has the hard task of transmitting the cultural heritage to the next generation. It must overcome the inevitable resistance of the immature to the painful restriction of freedom required to go to school to its elders. This obedience is progressively more difficult as a child grows in his or her own understanding and freedom.

2. Government

The Old Testament contains not a little reflection on the problems of building a human community and organizing its government.[102] It presents us with a picture of patriarchal society in which government was not clearly distinguished from family (Genesis). Next we hear of the miseries of an oppressed people, a racial minority living under alien despotism (Exodus). Then we read of Moses' problems in organizing his people and the solution suggested to him by his father-in-law Jethro (Exodus 18:25-26):

> He picked out able men from all Israel and put them in
> charge of the people as officers over groups of thousands,
> of hundreds, of fifties, and of tens. They rendered decisions
> for the people in all ordinary cases. The more difficult cases
> they referred to Moses, but all the lesser cases they settled
> themselves.

Once the people had entered their land under the military dictatorship of Joshua, made necessary by Moses' death, they fell into a kind of anarchy in which they tended to lose their identity and take up the ways of their pagan neighbors until they again and again were in danger of losing their identity altogether, but:

> Whenever the Lord raised judges for them, he would be with the judge and save them from the power of their enemies as long as the judge lived. It was thus the Lord took pity on this distressful cries of affliction under their oppressors. But when the judge died, they would relapse and do worse than their fathers. (Judges 2:18)

> In those days there was no king in Israel; everyone did what he thought best. (Judges 21:25).

The last of these judges was the priest Samuel, under whose leadership the people began once more to accept God as their real ruler, but soon the people began to demand a king such as other nations had, because they saw monarchy as a way to national power and glory. Samuel warned them (I Sam. 8:1-22) that this would mean that the king could oppress them with taxation and enslavement. "When that day comes, you will cry out on account of the king you have chosen for yourselves, but on that day God will not answer you." (v.18). The people insisted, however, and Samuel anointed the warrior king Saul, whom a second source of more pro-royalist sentiments in *I Samuel* (9:1-10:16:14) depicts as guided by the Holy Spirit as long as he remained true to God. But Saul soon fell prey to evil spirits and was replaced by David. *II Samuel* and the *Books of Kings* paint for us with wonderful vividness the history of the Davidic dynasty: the unification of Israel by David in spite of the rebellion of his own son Absalom, the glory of Solomon's kingdom and the building of the First Temple, its decline through the wise king's luxurious folly and, under his son, its breakup because of unjust taxation into the two kingdoms of Judah and Israel. Then we are shown the history of these two kingdoms that never could get together again; their good kings and their bad; their attempts to reform in accordance with the Mosaic Law (which was codified as a constitutional limit on kingly power, but which the kings failed to obey in spite of the warnings of the prophets); and finally the destruction of Israel by Assyria, and the exile of the leaders of Judah by Babylonia.

After the Persian Cyrus permitted the return of the Jews to their land, the hopes of restoring the Davidic kingship faded, but under the governor Nehemiah and the priest Ezra (Chronicles, Ezra-Nehemiah) the nation was restored and the second Temple built. The country remained small and poor, but now it rigidly maintained its racial identity and came to see in the exact observance of the Law, interpreted more and more in a legalistic and ritualistic spirit, its one source of unity. This particularism, however, made it possible for Judaism to survive the occupation of the country by Alexander and the Seleucids and their attempts to Hellenize the Jews, and finally inspired the Maccabean Revolt led by a priestly family which achieved temporary independence for the nation. In the time of Jesus the country had fallen under Roman occupation with the high priests more or less collaborating with the oppressors. While the Zealots sought their overthrow, the Pharisees and the still more sectarian Essenes opposed both government and priests, and called for a reliance on the observance of the Law for the survival of the nation, rather than on the splendor of the cult of the Third Temple or on political rebellion.

Thus the teaching of Jesus on the Reign of God had a rich background of political experience and concern for social justice and peace, and should always be read in that context. In the two thousand years since his preaching, the Catholic Church has also had a long political history of efforts to survive and to influence the political system of Europe in particular. In the recent social encyclicals of the popes from Leo XIII on, an effort has been made to formulate this experience in the light of biblical teaching, as well as the sound results of political theory from Plato and Aristotle to the present.[103] Very briefly, certain principles of social justice emerge which I would formulate as follows:

First, the dignity of a human person as a primary natural unit endowed with intelligence and freedom is such that a community of persons exists for its members and not the members for the community. Hence the very significant biblical analogy of a Christian community to an organism or body (I Corrinthians 12), must not be understood univocally as it has been by totalitarian political theorists, of whom Plato unfortunately is the prime example.[104] This is one of the reasons it is so important to insist, as I did in Chapter 9, on the essential difference between a primary unit in which the parts exist only in the whole, and a system of such units in which the whole is unified only by secondary (although nonetheless real) relations. At the same time this primacy of persons should not be understood in an individualistic sense. The community exists for the good of persons but *the highest personal goods are common goods*.[105] Every person has two kinds of personal needs: (1) private needs for *material* things whose use demands that they be divided and distributed and hence used in a private way, e.g. my clothing, the food I consume; (2) common needs (which are just as personal, indeed *more* personal) for what are chiefly *spiritual* realities, which can only be achieved and fully enjoyed in common, e.g. truth, love, beauty, worship.

Second, a good human community should tend to *equality* in the distribution of common goods, not in the sense of arithmetic equality, but in the sense of distributive justice by which each member has his or her needs met to his or her capacity.[106] Not all have the same needs or the same capacities but all deserve their full share.

Third, the realization of these first two principles which Marx correctly stated (following Acts 2:42-47) "from each according to his ability, to each according to his need,[107] demands a division of labor or function because the members of a community are unequal in various ways in their talents and the development of these talents, so they contribute in a variety of ways to the common effort and the creation of the common good. As St. Paul says, (I Cor. 12:19), "If all the members were alike, where would the body be? There are, indeed, different members, but one body." This is the primary point of the body analogy, namely, the functional heterogeneity necessary for any whole made up of limited parts to have these parts differentiated so that the whole can engage in complex yet unified actions. Note, however, that this division is functional, not final. The goal of this complex common action is that all members should share in the results with distributive equality, i.e. to the fullest of their capacities. A society in which the functional division of labor becomes a rigid caste system, in which the function becomes the measure of the sharing of the common good, is an oppressive society.

Fourth, in order that this complex whole should be able to engage in common actions, it is necessary that there be a unifying authority. This authority can have two forms: (1) *substitutional,* as when a less competent person obeys one more competent in order that he may share the advantage of that competence (e.g. a patient follows the advice of a doctor, or a child obeys a parent); (2) *essential,* as when a group of equals obeys the common decision of the group not because this is necessarily more competent, but because the group cannot act together unless all agree to one course of action.[108] The second of these kinds of authority is necessary in every group of persons, no matter how mature and equal, because in matters of practical decision there often is no one way of acting which can be proved to be best by rational discussion. No matter how long a group debates about such problems, and no matter how good the will of all, consensus may not be reached. Therefore, lest the group be unable to act at all, there must at least be agreement as to how the deadlock is to be broken, whether by lot, by vote, by leaving it to one person, or to a committee to make the decision.

Fifth, in the human family, nature has to a degree determined a constitution by which decisions can be made apart from the personal qualities of the members. The man, by reason of his greater physical strength and relative independence of movement (not because of superior competence of judgment) is naturally able to determine what is to be done when he and his wife cannot agree (essential authority). The children by reason of their immaturity must obey the parents (substitutional authority).[109] In units larger than the family, however, there is no obvious natural order

456

of essential authority, and the substitutional authority is only with regard to particular functions, not to the action of the whole community. Consequently, there is no *one best form of government,* best, that is, in the abstract,[110] but only for particular historic communities. Each form of constitution has advantages and disadvantages, and all can become tyrannical if they fail to observe these other principles.

Sixth, an important part of the common good of persons without which their dignity as intelligent and free beings is not properly realized, is that they should participate in decisions as well as obey the decisions necessary for common action. Without such participation they do not make their full contribution to the common good, each from their own experience and talents; nor do they fully share in the creativity which is so important a part of the common spiritual good. Yet probably the most difficult of all social and political tasks is the achievement of this full participation. Historically in most societies the members tend to be more concerned with the problems of their family or their smaller group and willing to let others take much of the responsibility for community decisions. This results in the usurpation of power by a few who enslave the others and use government for private purposes, not the common good. Even when rulers seek the common good unselfishly, they may be tempted to replace essential authority by a patriarchal substitutional authority that reinforces the immaturity and irresponsibility of the citizens. If every citizen participated in community decisions according to his or her talents, tyranny and oppression would be impossible.

Seventh, in order to promote this participation in common decision it is necessary that the political order should conform to the principles of *subsidiarity* and *functionalism,* i.e. decisions should be made by those most concerned and most aware of the needs to be met. Thus the process of decision-making should be kept as close to the local community as possible. Only those decisions which directly affect the wider community should be made at higher levels of organization, although these higher levels must assist the lower to fulfill their proper functions, coordinating them and supplying for what they fail to do for their members. These interventions of higher authority, however, should not become ordinary lest subsidiarity collapse into patriarchal centralization, and should always be aimed at enabling the local communities to meet their responsibilities on their own. Subsidiarity applies to the retention of decision-making power at the local level (federalism). Functionalism applies to the retention of such power within functional groups such as professions, labor unions, industries, etc., rather than surrendering it to an omnicompetent state.[111]

These seven principles of social organization ought to be applied to particular organizations, to national states, and ultimately to the world community according to particular circumstances in a creative manner. Because we are body-persons who exist in the limitations of time and space, the achievement of this intricate system of communication of life demands the highest energies of intelligence and will to breach what appear to be insurmountable barriers.

VI: Ethics And History

In Chapter 9 and this chapter I have tried to show how a radical process philosophy taking full account of the findings of modern science manifest that we humans are primary units specifically characterized by intelligence and creative freedom and that consequently we are social and historical beings. A historically developing unit of this kind can be understood only in terms of his or her own individual biography in the social context of human history and the total context of cosmic history. This historicity of human bodily existence, however, by no means makes it impossible for us to work out an ethics of moral norms valid for the whole human community across space and time.

This universal morality, far from negating our human capacity to create new cultures and to advance technological control of nature, is what makes it possible for us to form such cultures in ways that will meet basic human needs. Hence, to answer the question raised in the first chapter of this book: "Can we create ourselves?", it must be affirmed that we can truly create ourselves if we understand this "creation" to be neither the destruction of nature nor the forgetting of history, but a perfecting in new ways of what has already come to be by constructive natural processes and a purification and incorporation of historical cultural achievements into our present and future selves.

But is such an ethics credible against the background of perhaps three million years of human history? That history shows us a very long time in which human beings lived in small scattered groups, leaving no memory but a few pitiful bones and instruments of bone and stone — but then we begin to get some idea of human personality through the cave paintings of 30,000 years ago. Then came the much more rapid but still very slow growth of neolithic village culture. There followed the great archaic, literate civilizations of Mesopotamia, Egypt, India, and China with their somewhat later and imperfectly literate parallels in Central America and Peru. Around 600 B.C., the great world religions arose and consolidated in the empires of China, India, Persia, Greece and Rome, each with its complex history, including the conversion of the Roman Empire into the

Constantinian Christian Empire of east and west, the rise of Islam as the rival of Christianity, the decline of Buddhism in India and the revival of new forms of Hinduism while Buddhism spread into east Asia and Japan arose as a major culture. In Europe, the discovery and colonization of the New World in the fifteenth century and the rise of modern science and technology in the seventeenth induced great cultural changes, as a result of which Christendom divided and yielded much of its cultural and political power to the new secularist religions of Humanism in the eighteenth and Marxism in the nineteenth centuries. [112]

An historian of markedly Humanist views, William H. McNeill, in his remarkable one volume world history *The Rise of the West* (1963) sees our present historical situation as one in which for centuries men have striven for greater power, economic, technological and military, stimulated by the rivalry of nations and cultures, but who now face the prospect that the last great rivalry between democracy (Humanism) and communism will end either in the destruction of humanity or in a world government. [113] Such a global sovereignty would almost certainly be controlled by an immense bureaucracy of experts remote from the lives of the billions of private citizens, yet its power might be checked by the countervailing power of modern communications and the pitiless gaze of the mass media. It would probably achieve a much greater stability than has resulted from the rivalry of nations, but it would also probably become much more conservative about allowing this stability to be disturbed by rapid technological changes. Hence, McNeill speculates, unless there is a nuclear Armageddon, human creative energies may shift from science and technology to the fine arts, to a more contemplative attitude toward life, much as Heidegger also prophesied.

Such a humanist interpretation of history still sees, as did the Enlightenment, the long line of history as progress toward scientific understanding and control away from ignorance, superstition, and powerlessness, but it is no longer confident of the inevitability of progress, nor assured that this always marks an increase in real human freedom. Since it is convinced that the great religions belong to the past, it compensates its disillusionment with the promises of science, by the substitution, typical of the romantic wing of Humanism, of esthetic contemplation for transcendence through divine revelation. A Marxist interpretation of the same facts would not differ greatly I believe, from McNeill's, except in the conviction that this world government will come about only through the universal communist revolution, since as long as capitalism remains, its internal contradictions will prevent social stability.

459

Can Christians interpret these same historical facts in any plausible way? Only if, as I have already argued in Chapter 9, we drop the rigid application of the "principle of uniformity" which Humanists use to exclude the possibility of unique historical events, can such a Christian interpretation receive a hearing. If this principle is applied *broadly*, however, the way becomes open. Just as Humanists today no longer accept a naive view of historical progress, so a viable Christian interpretation of history must abandon the triumphalism found in Bossuet's famous *Universal History,* in which the Christian Church is pictured as the only light in a world of error, first conquering its persecutors, and then under Constantine establishing a Christian Empire which has steadily advanced in its conquest of the world in spite of the opposition of heretics, schismatics, Islam and modern unbelievers. In fact, Bossuet's interpretation reflects his own heretical Gallicanism and differs markedly from Augustine's far more biblical understanding in *The City of God,* in which the ambiguity of history is explained in terms of a struggle between true and false love whose visible manifestations can be very deceptive.[114]

Following the lead of Vatican II, I would suggest something like the following.[115] A Christian looking at today's world sees the dominant rivals, Humanism and Marxism, for all their admitted accomplishments in scientific and technological progress and their efforts to liberate humankind from hereditary power in favor of popular government, as unable to meet the deepest human needs, which cannot be satisfied with material products to be consumed, nor with a scientific understanding that refuses to face ultimate questions, nor with a merely esthetic contemplation. In this respect modernity must progress by recovering for our time the transcendent wisdom of the world religions which cannot be regarded as obsolete, although these religions need to be freed by critical science and history from some of the burdens they still carry from their past. Christianity has no fear about facing this *aggiornamento,* whatever other religions may think about themselves, because the rise of modern science, as Alfred North Whitehead pointed out,[116] had its roots in Christian theology: in the conviction that faith must be reasonable, that God has made an orderly universe, that the human mind was made by God capable of objective truth, and finally that Greek philosophy which first established the goals and method of modern science is providentially in harmony with biblical revelation.

The Church has also accepted the historico-critical method in the study of the Bible, so that the evolutionary origin of the human species and the very long prehistoric age of humanity presents to her no insurmountable difficulty for faith. Once the modern species of humanity, gifted with some degree of self-conscience intelligence and freedom arose on

460

earth, they may have lived for two or three million years without written records like many pre-literate people today. For such people, the limits of human memory and tradition make history very narrow and the future little different than the present.[117] The accumulation of culture is very slow, and human consciousness is essentially confined within the present.

Consequently, for such generations it makes little difference to the meaning of life whether humans have been on earth a thousand or a million years; the drama of their life must be played out within their own horizon. Within this horizon, the revelation of God and the moral law is available to guide their living and give it essential meaning. No doubt too, God grants gifts of prophecy to the wise men of the tribes to recall them from their sins and hold up a promise of hope in which St. Paul and Christian theology see an implicit yet saving faith in the Christ to come.[118] Those who accepted this dim yet divine light through all those long ages (and today in the backwaters of the world) are like the biblical patriarchs. They are those of whom Paul says.

> When Gentiles who do not have the law keep it by instinct, these men although without the law serve as a law for themselves. They show the demands of the law are written in their hearts. Their conscience bears witness together with that law, and their thoughts will accuse or defend them on the day when, in accordance with the gospel I preach, God will pass judgement on the secrets of men through Christ Jesus (Rom 2:14-16).

Yet Christian theologians, unlike the romantic Humanist Rousseau, have never confused this prehistoric phase of human existence with Paradise or conceived it as an idyllic Theocritean pastorale.[119] Rather, *Genesis* 1-10 pictures this time between the Fall and the Building of the Tower of Babel as a time when the good Abels fell victim to the evil Cains, when human reversion to bestiality brought the race repeatedly to the brink of destruction, and when humanity in its technological poverty was helpless against the forces of nature such as the Flood. Through anthropology we have come to admire the true humanity and many noble traits of preliterate men and women and their "religion of the earth", but we also know that they are morally good and bad as we are, and that their feeble technologies leave them with a hard struggle for survival, a struggle which has in fact wiped out whole tribes and races, and left the survivers without the leisure needed for any great intellectual advancement.

The time came for human beings, led by God-given culture heroes, to break out of this long twilight. A concurrence of environmental and human factors which we do not fully understand and which the Christian can only see as God's providence, resulted in the rise of neolithic villages and then, after thousands of years, of the great civilizations of Mesopotamia, Egypt, India, China and the New World. The authors of *Genesis* could not untangle all the phases of this development which even now archaeology is only beginning to reveal to us,[120] and they knew nothing of India or China or the New World, but from what they saw in near-by Mesopotamia and Egypt, they knew the chief results of this advance. In the accounts of the later life of Noah, of the Tower of Babel, of the dispersal of the nations, and the narratives of the Patriarchs before Moses, they saw the broad picture of what had happened in the Near East, a progress to which India, China, and the New World simply provide parallels.

The biblical reflection on the rise of human civilization emphasizes its ethical ambiguity. On the one hand it is seen as negative, since it is viewed from the Jewish tradition of a small pastoral people living on the periphery of the great cities of Egypt and Babylonia, defensive of their own nomad identity. From this viewpoint civilization is a work of human pride and folly, rebelling against God by its huge architectural marvels, yet ending in the breakdown of human unity in the multiplicity of languages and cultures at war with one another, and overcoming this disunity only by the tyranny of empire and the rise of kings who made themselves gods and who fostered a priesthood with its lifeless idols.[121]

On the other hand, *Genesis* does not forget that we were created in God's image, that is, as co-creative god-like beings by our intelligence and freedom, who ought to use that intelligence to regain the dominion over the world lost through sin. For this reason in the following books of the Torah and in the *Books of Samuel* and of *Kings* we find that God meets this new manifestation of human pride not by destroying the race, nor by driving humanity back into primitivism, but by selecting a small and humble people to be his own, liberating them from the oppression of Egypt, and establishing them as his kingdom. Somewhat reluctantly he permits them their own human king, and warns them that their mission is not to rival the empires, but to witness to the empires that he alone is king. When, as he warned His people, their human kings also become tyrants and divide the people, he promises them through His prophets that someday the Christ will come,[122] the anointed king in whom his true reign will be established.

This prophetic criticism of the ancient civilizations which we read in the Bible is paralleled in world history by the rise of the great world religions, which in each of the major centers of civilization proclaimed the need of a profound moral and spiritual reform of the polytheistic

462

world-views which had arisen in these centers to justify the kingly and priestly tyrannies which the advance from neolithic to urban living had everywhere produced.[123] Zoroaster and Gautama the Buddha in Iran and India, Confucius in China, Socrates in Greece — these teachers presented themselves not as innovators but as restorers of the deepest spiritual traditions of their respective peoples. Not only do the ethical teachings of these great sages converge, but their world-views also all tend to transcend polytheism as an anthropomorphic projection of human needs and to reject magic as an acceptable means to control natural forces so as to meet these needs. Thus the Jewish prophets were not unique in their denunciations of polytheism and superstition. Even Humanists may admit only one deistic Supreme Being, and Marxists, for all their atheism, consider matter to be a single Supreme Principle which evolves dialectically through its own inner dynamism.

What seems specific to the Jewish prophets is their conviction that this Supreme Principle is the Creator in the strict sense of a personal being who creates the universe out of nothing by a *free* act of intelligent will. The Egyptian pharaoh Ikhnahton rejected polytheism but his one God was only the Sun who produced the universe out of pre-existent matter. Zoroaster was able to explain evil in the world only by admitting the existence of a Principle of Evil independent of the Good God. The great Eastern religions and the Greek philosophers seem never to have freed their concept of the One from the determinism of producing the universe by some kind of necessary emanation.[124]

The God of Abraham, Isaac, and Jacob, the Father of Jesus, however, creates the world *freely*. It wholly depends on him for every aspect of its being and he in no way depends on the world, yet he is present to it not merely by his power but by a personal presence of knowledge, love, and providence. Yet in thus trying to specify the unique role of Israel in world history, my intention is not to prove the inferiority of other religious traditions, but rather to leave open the possibility that as we of the Judaic tradition come to know these other religions better, we will discover that indeed their Supreme Reality is the God of Abraham and Moses.

Jesus as the last and greatest of the Jewish prophets and, as Christians believe, Emmanuel, God-with-us (Isaiah 7:14), revealed this Creator and Redeemer God at a moment when the great ancient world religions had come to their full maturity as typified in the Roman Empire of Caesar Augustus, to which the Han Empire in China and its adoption of Confucianism form a remarkable parallel.[125] The Christian Church claimed for Jesus' person and teaching a kind of universal significance which was predicted in Judaism for the Messianic king, but the Christians did not conceive this as an empire to replace existing governments. They accepted the view that during the time of the preaching of the Gospel the Church would exist alongside such empires, and its members would be loyal citizens while at the same time bearing witness even at the cost of their own lives to God as Jesus had done.[126] Confronted by the variety of cultures and by non-Christian pagan religions, the early Church Fathers, while rejecting the false gods of paganism, also attempted as had St. Paul, to show that the true, but unknown God was hidden even in their idolatries. Some of the Fathers were convinced that non-Christians could achieve the salvation that comes only through Christ as long as they did not sin against the light available to them, because it was the Logos who had become flesh in Jesus and this Logos is the "light that enlightens everyone who comes into this world" (John 1:9) from the first creation of humanity. Thus the revelation of God in Christ was not seen as exclusive of other religions, but as the glorious climax of God's revelation of himself in all times and places.

The history of the Christian Church has received many interpretations. In the Orthodox East the tendency has been to picture it as a triumphal expansion in the face of persecution by the pagan Roman Empire, followed by the conversion of that Empire which came to be the defender of the Church against outside foes, chiefly Islam. Thereafter, the Second Rome, now Byzantium, maintained orthodoxy through the first seven Ecumenical Councils of a united Church, after which the West broke away from loyalty to the Empire and the Church fell under the tyrannical domination of the popes. It was these same Roman popes who by the crusades prepared the destruction of Byzantium by Islam. Since that time the Orthodox Church has again lived in persecution, although some identified the Russian Empire with the Third Rome.[127]

The Reformed Churches, on the other hand, have generally seen the union of Empire and Church under Constantine as the beginning of a profound corruption of the Church, lasting until the Reformation to the point that the papacy appeared to some to be the Anti-Christ. The preservation of the faith during these long ages was attributed to the transmission of the Bible from which the Reformers were able to restore a true understanding of the Gospel. Subsequent history has been seen as a continuation of the preaching of the Reformed Religion throughout the world. The continued division of the churches and the course of secular history are seen simply as continuing illustrations of the sinfulness of humanity which constantly seeks to save itself through science and technology rather than through faith in God.[128]

Although Roman Catholic understanding of history has tended to accept the triumphalism of the Eastern Constantinian Establishment, it has

464

regarded the Eastern Church as schismatic both in its rejection of the legitimacy of the Holy Roman Empire and of the centralized papal power. This development of papal power has been justified as necessary in view of the inability of the Eastern emperors to defend the West against barbarian incursions, and then as a counter-balance to the overweening domination of Western emperors and the kings of France, Spain, and England. The Inquisition and the Crusades have been also justified as legitimate uses of secular power to defend Christendom against internal subversion and external attack by Islam.

The Protestant Reformation and the Enlightenment have been seen as successive stages in the modern revival of infidelity, the rejection of the authority of God in favor of subjective individualism and secularization reaching its culmination in the atheism of Communism. These losses of the Church have been viewed as offset by the strengthening of papal authority and of world-wide missionary extension of the Church.[129]

Vatican II, however, has led to a notable development in the Catholic Church's understanding of her own history, a development prepared by a deeper historical analysis in the light of the theological notion of "the development of doctrine" of which Cardinal Newman was so able an exponent.[130] Crucial to this new interpretation is our understanding of the Constantinian establishment. Was this the triumph of the Cross and the establishment of Christendom (the Orthodox view) or was it the corruption of the Gospel (the Protestant view)? The alternative to these interpretations which both now seem hard to justify theologically or historically, is to see in this Constantinian establishment a necessary but restricting phase of the Church's development. The Christian Church must work in the world to transform human society and to speak the Gospel in the forms of each culture and each age. Therefore, the efforts of the Church to translate the Gospel into the thought forms of the Roman Empire and to develop institutional forms that would be able to express Christian ethics in a concrete, historical way was not a corruption of the Gospel but its implementation.

At the same time, it must be admitted that in this effort to embody the Gospel in the Byzantine Empire and then in the western Carolingian Empire and in the feudal institutions of the Middle Ages, the full scope of the Gospel was constricted and narrowed. While this Christendom brought forth rich cultural and spiritual fruits that manifested the hidden potentialities of the Gospel, yet it also became identified with the time-bound and all too human European nations. Hilaire Belloc wrote a book called *Europe is the Faith*,[131] by which he meant both that Catholicism had created European culture and that it was historically wedded to it.

Certainly, very many features of the European class structure, its courtly rituals, its Roman law and Latin language, its use of thought control and force to maintain social identity (as reflected in the Inquisition) and to extend its empire (as in the Crusades) and its nationalism (as in the divisions of the Church) were assimilated by the Church in ways that often either limited or adulterated her own proper spirit. As a result such disasters as the schisms in the Church and the rise of Islam, of Humanism and Marxism in regions where for a time her influence was predominant, or the relative failure of her missionary efforts to become indigenized as in India, China, and Japan, or where they succeeded but without deep transformation as in Latin America, have again and again made Church history seem less a triumph than a long agony, a humiliating way of the Cross.

Yet the history of the Church is also a history of resurrection, of repeated reforms and renewals. In our own time, just when it seemed that Humanism and Marxism had made the wonderful advances of modern science and technology their own and exposed the obsolescence of all transcendent religion, the shortcomings of these two secular religions have themselves been exposed. At the same time the ecumenical movement has given us promise of a reunion of the Christian Church, and Vatican II has proposed a new self-understanding for the Church which may enable her to break free from her Europeanism and to become for the first time in her history culturally Catholic, able to bring the Gospel to all peoples in their own terms. The very notion of her mission is changing. On the one hand it no longer means an imperialistic subjection of the world to "Christian" political powers, nor does it mean the suppression of the other great religious traditions as devil worship, but an attitude of dialogue in the common search for God to which each tradition brings its truth. The Christian conviction is that in this convergence of truth the Christ will be be found at the center. At the same time, Christians know that they must continue to walk the way of the Cross. Already after Vatican II, new problems have arisen within the Church itself from the stress of change, and new enmity has appeared. In a special way this new relevance of the Church also means, as we see in liberation theology, a new engagement in the political struggles of the time, and hence of dangers perhaps even greater than those faced in the time of Constantine or Charlemagne or Charles V, or of the French Revolution.

In light of such a Christian interpretation of history, which of course is limited by our present horizons, the notion of a Christian ethic which transcends particular cultures by a co-creative stewardship of an earth whose dominion we share with God becomes again credible. It is precisely by the study of history that we can untangle what is perennially valid

466

in the ethical teaching of Jesus from its one-sided or even distorted exemplifications in various times and places. This teaching, moreover, maintains its own integrity since, as I have tried to show, its basic principles as exemplified in Jesus himself remain effective as answers to the deepest needs of human persons. The development since the seventeenth century of the scientific world-picture has deepened our understanding of these needs, and it is the task of theology to see that Jesus' answer to these needs is also more profoundly understood.

How theology is to accomplish this task is not easy to say. In the remaining chapters of this book I will try to suggest how the radical process interpretation of science which I have proposed in previous chapters, with its anthropology of the human person as essentially bodily and historical, can cast light on some of the basic problems of theology, especially human origins and human destiny. It is only in the perspective of a theological interpretation of what we know scientifically of this origin and destiny that a Christian ethics is possible today.

Notes

1. On recent debates as to whether there is one or many biblical theologies of the Old Testament see Gerard Hasel, *Old Testament Theology: Basic Issues in the Current Debate* (Grand Rapids, Mich.: Eerdmans, 1972). For a review of current types of Biblical theology see Wilfrid J. Harrington, O.P., *The Path of Biblical Theology* (Dublin: Gill and Macmillan, 1973).

2. On the unity of the Old Testament canon see Henri Clavier, *Les variétés de la pensée biblique et le problème de son unité* (Leiden: Brill, 1976), pp. 318-323. Also Ronald E. Clements, *Old Testament Theology: A Fresh Approach* (Atlanta: John Knox, 1978).

3. See Chapter 9, note 79 for exegetical works used in the following.

4. Chapter 4, pp. 113-114; 119-121.

5. Gerhard von Rad, *Genesis: A Commentary* (Philadelphia: Westminster Press, 1961), considers that the terms for human domination over the subhuman creation "are remarkably strong" implying "tread", "trample (as a wine press)" and "stamp". This has been taken to mean that Adam is to crush lesser creatures under his foot, but the image of stepping onto territory also symbolizes "taking possession" without any implication of destruction. This is more in keeping with the picture given in Chapter 2 of man as the guardian and cultivator of the Garden. See also Walter Brueggemann, *Genesis* (Atlanta: John Knox, 1982) p. 32 f.

6. According to Claus Westermann, *Genesis I-II: A Commentary,* trans. by John J. Sullivan, S.J., (Minneapolis: Augsburg, 1983), opinion on the meaning of "knowledge of good and evil" is chiefly divided into those who think it means the sexual mystery of mating and birthing (a divine, creative power) e.g., H. Gressmann, H. Schmidt, H. Gunkel, I. Engnell, J. Coppens, J. A. Soggin, J. McKenzie, and those who think it is a way of saying "*all* knowledge" e.g., Wellhausen, W. G. Lambert, Von Rad, etc. Westermann prefers the second opinion and says that the notion is a functional one applying to the human race rather than individuals, and indicates a desire for an unlimited knowledge that would give humanity autonomy apart from God.

I follow Westermann's view, but add that it does not contradict the first view, since for the ancients the sexual mystery was (as Freud also held) the paradigm of all other mysteries. More serious is the difficulty raised by Gnosticism (which seems to have in this regard somewhat influenced its arch-foe Irenaeus, see Chapter 4, pp. 113 above) and which has been adopted by the Jungians, that if Eve had not sinned, humanity would never have risen above an infantile "innocence", so that Eve's act was really a courageous risk necessary for maturity. In my interpretation, this maturity is symbolized by the tree of life which would have given true and god-like wisdom, free of all illusion. The tree of the knowledge of good and evil, on the other hand, gives a knowledge mixed with death and illusion. The contrast is between a true understanding and a sophisticated cynicism based on a "disillusionment" with life which is really the result of living with one's dead illusions, but not being really freed from them.

7. Chapter 9, note 85.

8. On the meaning of these "curses" see von Rad, *Genesis*, (note 5 above), pp. 91-94. They are etiological explanations of the natural miseries of human life to which humankind is liable unless they are prevented by God or by the full exercise of the human intelligence with which God has endowed us. When we fail to use God's gifts rightly, we fall victim to these miseries without any special act of punishment by God.

9. The belief in giants was fostered in the Near East by the remains of megalithic cultures and of the ruins of the great builders of Egypt and Mesopotamia. According to Umberto Cassuto, *A Commentary on the Book of Genesis,* 2 vols. (Jerusalem: Hebrew University Magnes Press, 1959) the narrative of *Genesis* 6:1-4 has as its purpose not to confirm the pagan legends that the giants were the offspring of demons and "the daughters of men", but to refute the notion that they could have been demi-gods, a notion offensive to biblical monotheism. The function of this narrative, therefore, is to attribute the corruption of humanity (symbolized by the shortening of the life-span) at least in part to devil-worship and the practice of divination and magic, not to perpetuate a mythical world-view.

10. "It is amazing to see how sharply little Israel demarcated herself from an apparently over-powering environment of cosmological and theogonic myths. Here [in Genesis 1] the subject is not a primeval mystery of procreation from which the divinity arose, nor of a 'creative' struggle of mythically personified powers from which the cosmos arose, but rather the One is neither warrior nor procreator, who alone is worthy of the predicate, Creator." von Rad, *Genesis* (note 5 above) p. 47.

11. "The common good is the end of single persons living in community, as the good of the whole is the end of each of its parts; but the good of each single person is not the end of any other person." *Summa Theologiae* II-II, q. 58, a. 9 ad 3. Cf. *Summa Contra Gentes* III, c. 25 where Aquinas argues that the common good of all created persons is to know and love God, and therefore no created person has another created person as its end.

12. "The ground of this principle is: *rational nature exists as an end in itself. . . .* The practical imperative thus should be as follows: *So act as to treat humanity, whether in your own person or in that of any other, always at the same time as an end and not as a means."* Immanuel Kant, *Groundwork of the Metaphysics of Morals,* trans. by T. K. Abbot, *Kant's Critique of Practical Reason and Other Works on the Theory of Ethics,* 6th ed. (London, 1909), p. 47. Although Aquinas and Kant agree on this principle, it means something different to each. For Aquinas all creatures are ordered to God as their final end and common good, but non-rational creatures attain this end by their service of rational creatures, i.e., as means. Rational individuals are subordinated not only to God as their ultimate good, but also to the created community of which they are a part. This subordination, however, is not that of a means to an end, but of members who share in a common good, which internally is the human life of the community and externally God Himself. Aquinas rests this view on a metaphysical analysis of the order of being, but this analysis in turn is grounded in an empirical description of the human world. Kant in idealistic fashion postulates that the rational nature exists as an end in itself by beginning with the *subject.* The notion of ends and means makes sense only if the thinking subject is an end for itself, and for this understanding to be rational it must be universalized, i.e., every subject must be considered equally an end in itself. But the existence of the noumenal self is for Kant a practical postulate, and not an objectively knowable fact. It is this Kantian way of founding ethics that has proved so vulnerable to moral relativism, although it was precisely his intention to defeat that relativism.

13. See Peter Singer, *Animal Liberation: A New Ethics for our Treatment of Animals* (New York: Avon, 1977).

14. Lynn White "The Historical Roots of Our Ecological Crisis", *Science* 155 (1967), 1203-7, provoked a considerable controversy by blaming the technological exploitation of natural resources on the influence of *Genesis* 1:28. This provoked a number of answers, e.g., Karlfried Froehlich, "The Ecology of Creation", *Theology Today*, 27 (1970): 263-276; Eric G. Freudenstein, "Ecology and the Jewish Tradition", *Judaism*, 19 (1970): 406-414; Gabriel Fackre, *Ecology Crisis: God's Creation and Man's Pollution* (St. Louis: Concordia Publishing House, 1971); Richard E. Sherrell, ed., *Ecology: Crisis and New Vision* (a symposium) (Richmond, Virginia: John Knox Press, 1971); Hans Jonas, "Technology and Responsibility" in James Robinson, ed. *Religion and the Humanizing of Man* (New York: Council on the Study of Religion, 1972), pp. 1-19; Cyril C. Richardson, "A Christian Approach to Ecology", *Religion in Life*, 41 (1972): 462-479; Lionel Basney, "Ecology and the Scriptural Concept of the Master", *Christian Scholars Review*, 3 (1973): 49-50; and George S. Hendry, *Theology of Nature* (Philadelphia: Westminister Press, 1980).

15. See note 5 above.

16. See Robert J. Daly, *Christian Sacrifice: The Judaeo-Christian Background Before Origen* (Washington, D.C.: Catholic University of America Press, 1978) pp. 157-171. Also Isadore Twersky, *Introduction to the Code of Maimonides* (New Haven: Yale University Press, 1980) pp. 390 f. and 418-424. Jacob Milgrom, *Studies in Cultic Theology and Terminology* (Leiden: Brill, 1983), especially "The Biblical Diet Laws as an Ethical System" pp. 104-118, claims that the whole tendency of the Priestly Tradition (which he argues is very old) is to inculcate reverence for life. Far from promoting blood-shed, the Torah strives to restrict it severely, permitting the taking even of animal life only under narrowly regulated conditions.

17. See David F. K. Standl-Rast, O.S.B., "What Can Theology Contribute to an Ecological Solution," in Richard E. Sherrell, ed., *Ecology* (note 14 above), pp. 125 ff., on the importance of the monastic tradition in East and West for a theology of ecology. Also cf. other references in note 14.

18. Benedict M. Ashley and Kevin D. O'Rourke, *Health Care Ethics: A Theological Analysis*, 2nd rev. ed., (St. Louis: Catholic Health Association, 1982), 37-45: 195-198.

19. *Ibid*, pp. 37-45.

20. See Chapter 9, pp. 364-365.

21. Charles E. Curran, "Natural Law and Contemporary Moral Theology" in *Contraception: Authority and Dissent* (New York: Herder and Herder, 1969), pp. 151-175, edited by him.

22. See Ashley and O'Rourke (note 18 above), pp. 272-276.

23. See Alfred L. Baldwin, *Theories of Child Development* (New York: John Wiley and Sons, 1968), pp. 305-390; E. James Anthony and Therese Benedek eds., *Parenthood: Its Psychology and Psychopathology* (Boston: Little Brown, 1970), especially Benedek's "Motherhood and Nursing" and "Fatherhood and Providing" pp. 153-183; and Myra Windmiller, Nadien Lambat and Elliot Turiel, *Moral Development and Socialization* (Boston: Allyn and Bacon, 1980), especially Windmiller's introduction and Terrence N. Tice, "A Psychological Perspective", in which he discusses three approaches: (1) Structural-Developmental (Piaget, Kohlberg); (2) Social Learning (Watson); (3) Psychoanalytic (Freud) to the study of child development.

24. See Edward O. Wilson, *Sociobiology: The New Synthesis* (Boston: Harvard University Press, 1975) who says that food-sharing is rare among the primates but a very important form of human behavior, pp. 206-208; 542; 551-553.

25. See note 6 above.

26. The text does not say why God accepted Abel rather than Cain's sacrifice. Von Rad *Genesis*, (note 5 above, p. 101) says it was because Abel's sacrifice was a blood sacrifice, but why God prefers this is left to His inscrutable will. Bruggemann (*Genesis*, note 5 above, pp. 55-64), attributes it also to God's free election and considers that Cain's sin is consequent and not antecedent to God's choice, and consists in his refusal to submit to God's inscrutable will.

469

27. See Adrian Van Kaam, *The Addictive Personality* (Chicago: Franciscan Herald Press, Synthesis Series, 1966). On the role of sensuous pleasure in Christian life see Cornelius Williams, "The Hedonism of Aquinas", *Thomist*, 38 (1974): 257-290 and John G. Milhaven, "Christian Evaluations of Sexual Pleasure", *The American Society of Christian Ethics: Selected Papers*, 1976, pp. 63-74.

28. Mary Douglas, *Purity and Danger: An analysis of concepts of pollution and taboo* (London: Routledge and Kegan Paul, 1966), pp. 41-57 and *passim*. See Milgrom, *Studies in Cultic Theology* (note 16 above), for a different approach to the question.

29. Some will say that in Jesus' time medicine and agriculture could not have reached the stage of progress that would have made the works he performed miraculously technologically possible. I am arguing that the slowness of human progress was not due to the lack of human intelligence or to some inevitable law of the rate of progress, but to the waste of these gifts in indulgence, warfare, and sloth through countless centuries of human existence, as the first chapters of *Genesis* seem to teach.

30. Karl Menninger, *Man Against Himself* (New York: Harcourt Brace, 1938) put up a strong psychoanalytic case for asceticism as a destructive form of self-hatred, a kind of prolonged suicide. Undoubtedly there is a pathological element in exaggerated asceticism, but the classic spiritual writers insisted on moderation and understood asceticism not as a war against the authentic self but against the false self which the psychoanalysts also seek to destroy. After all, psychoanalytic "working through" involves a quite rigorous mental and emotional asceticism.

"Intemperance is the vice most deserving of reproach for two reasons: first, because it is the most contrary to human excellence since it concerns pleasures we share with the brute animals . . . Second, because it is most contrary to the splendor or beauty of human nature, in that the light of reason from which comes the whole splendor and beauty of virtue is least apparent in intemperate acts.", *Summa Theologiae* II-II, q. 143, a. 4 c. See the dissertation of Cajetan Chereso, O.P., *The Virtue of Honor and Beauty* (River Forest, IL: Aquinas Library, 1960).

32. *The Institutes*, Book V, "On the Spirit of Gluttony" in Phillip Schaff and Henry Wace, eds. *Nicene and Post-Nicene Fathers*, 2nd Series, vol. 11, pp. 233-248; PL 49, cols. 201-266; CSEL 17: 78-113.

33. Albert Gelin, *The Poor of Yahweh* (Collegeville, Minn: Liturgical Press, 1964).

34. See W. Montgomery Watt, *The Formative Period of Islamic Thought* (Edinburgh: University of Edinburgh Press, 1973) pp. 9-37 on the Kharijites, a sect of Islamic fundamentalists who saw their ideal in the nomadic life of Muhammad and his first companions. The Sufi movement in Islam also proposed an ascetic way of life much like that of Christian monks. See Annemarie Schimmel, *Mystical Dimensions of Islam* (Chapel Hill, N.C.: University of North Carolina Press, 1975), pp. 35-38 and 98-186.

35. On the meaning of poverty in the Franciscan tradition see Duane V. Lapsanski, *Evanglical Perfection: An Historical Examination of the Concept in Early Franciscan Sources* (St. Bonaventure, N.Y.: Franciscan Institute, St. Bonaventure University, 1977).

36. On the controversies over poverty among the Franciscans see John R. H. Moorman, *A History of the Franciscan Order* (Oxford, Oxford University Press, 1968).

37. See the carefully nuanced answer of Aquinas to the question, "Whether to have common property diminishes the perfection of the religious state?" (*Summa Theologiae*, II-II, q. 188, a. 7.). He concludes that "with respect to poverty, any religious order will be more perfect to the degree that it practices the kind of poverty appropriate to its end" as a religious order. Thus poverty is a moral good only as it frees the person for a more human and Christian life.

38. See H. Wennick, *The Bible on Asceticism* (De Pere, Wisconsin: St. Norbert's Abbey Press, 1966), and Johannes B. Metz, *Poverty of Spirit* (Glen Rock, N.J.: Newman Press, 1968), and the article of H. Chadwick, "Enkrateia" in *Reallexicon für Antike und Christentum*, vol. 5, col. 347.

39. See Jacques Dupont, *Les Béatitudes*, rev. ed. 3 vols, (Bruges: Abbaye de Saint André, 1958-1973), 1: 343-345 and 379-381.

40. Cassian, *Institutes* (note 32 above) Books IX-X, pp. 264-274, PL 49, cols. 351-398; CSEL pp. 166-193.

41. See Konrad Z. Lorentz, *On Aggression* (New York, Harcourt Brace, 1966); Robert Ardrey, *The Territorial Imperative* (New York: Athenaeum, 1966) and refutation by Ashley Montagu, *The Nature of Human Aggression* (New York: Oxford University, 1976). See also Vernon H. Mark and Frank R. Ervin, *Violence and the Brain*, (New York: Harper and Row, 1970).

42. See Karl Menninger, *Love Against Hate* (New York: Harcourt Brace, 1942), and Erich Fromm, *The Anatomy of Human Destructiveness* (Greenwich, Conn.: Fawcett Publications, 1975), for discussion of the Freudian two-drive theory.

43. See Kenneth M. Clark, *The Nude* (New York: Random House-Pantheon, 1956), pp. 71 ff.

44. See Hendrik Ruitenbeck, *The Male Myth* (New York, Dell, 1967), and Joseph H. Pleck and Jack Sawyer, eds. *Men and Masculinity* (Englewood Cliffs, N.J.: Prentice-Hall, 1974), for analyses of the "macho image."

45. See H. Saddhatissa, *Buddhist Ethics: Essence of Buddhism*, (New York: George Braziller, 1970). The first of the Five Precepts forbids killing insects and all higher forms of life, pp. 87-93.

46. See Paul Hanley Furfey, *The Respectable Murders*, (New York: Herder and Herder, 1965), especially "Bombing of Non-Combatants", pp. 69-85; Harvey Seifert, *Conquest by Suffering* (Philadelphia: Westminster Press, 1965); Raymond Regamey, O.P., *Non-Violence and the Christian Conscience*, (New York: Herder and Herder, 1966); Thomas Merton, ed., *Gandhi on Non-Violence* (New York: New Directions, 1965), and *Faith and Violence* (Notre Dame, Ind.: University of Notre Dame Press, 1968); James W. Douglass, *The Non-Violent Cross* (New York: Macmillan, 1968); Joseph Comblin, *Théologie de la révolution* (Paris, Editions Universitaires, 1970); Helder Camera, *Revolution through Peace* (New York: Harper and Row, 1971); Jacques Ellul, *Violence: Reflections from a Christian Perspective* (New York: Seabury Press, 1971); James F. Childress, "Non-violent Resistance: A Bibliographical Essay", *Journal of Religion*, 52 (1972): 376-396; Jean-Michel Hornus, *It is Not Lawful for Me to Fight: Early Christian Attitudes Toward War, Violence, and the State*, trans. from the French by Alan Kreider (Scottsdale, Penn.-Kitchner, Ont. Herald, 1980 (1960).

47. See Evaristo Villar, "El Dios de la Biblia: Es violento?", *Biblia y Fe* 7 (Jan-Mar, 1981): 19-32, who shows that while Yahweh was originally conceived as a war-god, from the time of the exile the doctrine developed that the reign of the Messiah was to be a reign of peace. Moreover as Jacob Milgrom (note 16 above) has shown, the Priestly Tradition always emphasized God's opposition to bloodshed. On the concept of holy war see Ronald de Vaux, O.P., *Ancient Israel* (New York: McGraw-Hill, 1961) pp. 258-270. It pertains especially to the Deuteronomic tradition whose rhetoric requires to be interpreted in the same manner as Jesus' own words, "If your hand offends you, cut it off!" (Mt 5:30), as a way of saying hyperbolically that the temptation to compromise with the sin of idolatry must be avoided at all costs. It should be noted that the commands to carry on holy wars were probably written long after they had any practical implementation and, therefore, at the worst only as a *post factum* justification of the deeds of the ancestors in the remote past.

48. On the term *miles Christi* (soldier of Christ) for the martyrs see H. A. M. Hoppenbrouwers, *Récherches sur la terminologie du martyr de Tertullien à Lactance* (Nijmegen: Dekker and van de Vegt, 1961) pp. 144-151, and Adolf Harnack, *Militia Christi* (Tübingen: J. C. B. Mohr, 1905) pp. 4-46. In the Eastern Church many soldier saints are honored, some as martyrs.

49. S. G. F. Brandon, *Jesus and the Zealots* (Manchester: Manchester University Press, 1967), started the controversy over Jesus' political aims. Important replies to his claim that Jesus was a Zealot have been supplied by Oscar Cullmann, *Jesus and the Revolutionaries* (New York, Harper and Row, 1970), and by Martin Hengel, *Was Jesus a Revolutionist?* (Philadelphia: Fortress Press, 1971) with a useful bibliography pp. 28-41.

50. See Hornus, *It is Not Lawful* (note 46 above), for the controversies over these texts.

51. See F. H. Russell, *The Just War in the Middle Ages*, (Cambridge-New York: Cambridge University Press, 1975); Paul Ramsey, *The Just War: Force and Political Responsibility* (New York: Scribner's, 1968); J. T. Johnson, *Just War Tradition and the Restraint of War: A Moral and Historical Inquiry*, (Princeton, N.J.: Princeton University Press, 1981); William O'Brien, *The Conduct of Just and Limited War* (New York: Praeger, 1981).

52. See R. H. Bainton, *Christian Attitudes toward War and Peace* (New York: Abingdon Press, 1960) who, however, too closely identifies Catholic teaching with the Just War Theory to the exclusion of pacificism, on which see J. Bryan Hehir, "The Just-War Ethics and Catholic Theology: Dynamics of Change and Continuity" in Thomas A. Shannon ed., *War or Peace: Search for New Answers*, (Maryknoll: Orbis Books, 1982), pp. 15-39.

53. "On the Purification of Means" in *Freedom and the Modern World* (New York: Scribner's, 1936) pp. 139-192.

54. See Chapter 1, note 33.

55. The most authoritative statement on the Catholic teaching on sexuality is Vatican II "The Church and the Modern World" (*Gaudium et Spes*) nn. 47-52, as further expounded by Paul VI's encyclical *Humanae Vitae*, the "Declaration on Certain Questions Concerning Sexual Ethics" by the Congregation for the Doctrine of the Faith" of Jan. 22, 1976 (see discussion by J. McManus et al. in *Clergy Review* 61 (1976): 231-237; "The Message to Christian Families" of the 1980 Synod of Bishops, and John Paul II's encyclical, "The Role of the Christian Family in the Modern World" (*Familiaris Consortio*), Nov. 22, 1981. The points on which there is considerable dissent can be found in Anthony Kosnik, et. al, *Human Sexuality: New Directions in American Catholic Thought*, A Study Commissioned by the Catholic Theological Society of American (Garden City, New York: Doubleday, 1979 first published 1977). For some of the current theological issues see the symposium edited by John Y. Fenton, *Theology and the Body* (Philadelphia: Westminster Press, 1975).

56. William Etkin, "Reproductive Behaviors" in the book edited by him, *Social Behavior and Organization Among Vertebrates* (Chicago: University of Chicago Press, 1964), pp. 75-76 points out that sexual reproduction has two advantages: (1) gene diffusion; (2) accumulation of a pool of recessives ready to be adapted to a changed environment; and the second of these is probably the more important. See also George G. Williams, *Sex and Evolution* (Princeton, N.J.: Princeton University Press, 1974) and Edward O. Wilson, *Sociobiology* (note 24 above), pp. 315 f.

57. James A. Monteleone, M.D., "The Physiological Aspects of Sex," in *Human Sexuality and Personhood* (St. Louis: Pope John Center, 1981), pp. 71-85.

58. Jo Durden-Smith, "Male and Female — Why?", Quest/80 (Oct. 1980): 15-19, 93-98, on the research of Jerre Levy of the University of Chicago, who is reported as saying "It makes perfect sense to me to find in women this constellation of abilities, what I've called female intuition. For 99 percent of human history, we've lived as hunter-gatherers. The men have been hunters, loners, requiring pronounced visual skills and goal-direction. The women have lived together in groups, with children and the old. So it seems to me evolutionarily adaptive that women should have acquired different abilities — social, acculturative, nurturant ones that men, by and large, don't have. Second, this implies a sexual stamping, a genetic one. And I think that it's now becoming increasingly plain that the sexual stamping I'm talking about does indeed take place in the brain and does indeed start in the fetus. It is reinforced and magnified by our cultural institutions. But it is genetically based. It is part of our biological inheritance. And it is mediated by hormones." p. 94. See also Sid J. Segalowitz, *Two Sides of the Brain* (Englewood, N.J.: Prentice-Hall, 1982), pp. 158-66. For the more general question of sexual differences see Judith M. Barwick, *Psychology of Women: A Study of Bio-Cultural Conflicts* (New York: Harper and Row, 1971); B. F. Miller, E. G. Rosenburg and B. L. Stackowski, *Masculinity and Feminity* (Boston: Houghton-Mifflin, 1971) and Eleanor E. Maccoby and Carol N. Jacklin, *The Psychology of Sex Differences* (Stanford: Stanford University, 1975). What seems fairly well established is that males are physically larger and muscularly stronger (although perhaps less long-lived), tend to be more aggressive and physically active, and are more efficient at tasks requiring visual-spatial ability. Women tend

to be more nurturant, sensitive to personal non-verbal cues, more verbally expressive, and are perhaps longer lived. Sociobiologists generally interpret these differences as evolutionary adaptations to the reproductive needs of the species.

59. See George Peter Murdock, "The Universality of the Nuclear Family" in Norman W. Bell and Ezra F. Vogel, eds., *A Modern Introduction to the Family*, rev. ed., (Free Press, N.Y., 1968), pp. 37-44 and the attempts at refutation or qualification by M. E. Spiro, E. K. Gough, and Marion J. Levy, Jr., which follow in the same work. After reviewing this controversy Morris Zelditch, Jr., in Harold T. Christensen ed., *Handbook of Marriage and Family* (Rand McNally, Chicago, 1964), "Cross Cultural Analysis of Family Structures", pp. 462-500 sums up, "What is beyond dispute is that the nuclear family in many societies is so subordinated to the extended family that it is for many purposes not the paramount acting unit, and for some purposes not even relevant at all. But it has yet to be demonstrated that, except for the Nayar, the kibbutz, and perhaps the Minangkabau, there is not at least *one* purpose, aside from sex, with respect to which in most societies the nuclear family is at least *sometimes* the relevant acting unit." (pp. 478 f.). Perhaps more to the point is the conclusion of Clifford Kilpatrick, *The Family: Its Process and Institution*, (New York: Ronald Press, N.Y., 1963), p. 47: "The familial group known as the biological or nuclear family exists and is recognized in practically all the cultures of the world . . . Occasionally only a blurred glimpse of the nuclear family is obtained, imbedded as it may be in the polygamous family structure or perhaps in a kinship or household structure; *yet the vast majority of all the children in the world grow up experiencing distinctively intimate contact with their biological parents*" (italics mine).

60. "In contrast to the situation among birds, the contribution of the male mammal tends to be limited to that of insemination, and his social role is usually much reduced. His cooperation is generally not essential for the care of the young subsequent to fertilization. He commonly does not associate with the female and her young in any permanent manner. In some mammals, such as the cat and bear families generally, he is excluded by the female from contact with the young, being treated as their most dangerous enemy. We shall see that many of the behavioral characteristics of the mammalian male are related to the basic biological position of dispensability after insemination. Of course there is much variability among mammals, but it is only exceptionally that the male finds an important niche in the social relations of parent and child among mammals. From the point of view of natural selection, mammalian males may be characterized as expendable. This characteristic helps to account for the kinds of specialization in behavior and structures that we find in many mammals." William Etkin, *Social Behavior* (note 56 above), pp. 89 f. Thus it is all the more surprising that in the mammalian human fatherhood appears, since it does not exist even among primates; cf. Jane Beckman Lancaster, "Sex Roles in Primitive Societies" in Michael S. Teitelbaum, *Sex Differences*, (Garden City, N.Y.: Doubleday Anchor Books, 1976) p. 49.

Margaret Mead in Chapter 9, "Human Fatherhood as a Social Invention" in *Male and Female* (New York: William Morrow and Co., 1949), does not deny its biological basis or universality but says "When we survey all known human societies, we find everywhere some form of the family, some set of permanent arrangements by which males assist females in caring for children while they are young. The distinctively human aspect of the enterprise lies not in the protection the male affords the females and the young — this we share with primates. Nor does it lie in the lordly possessiveness of the male over females for whose favours he contends with other males — this too we share with primates. Its distinctiveness lies instead in the nurturing behavior of the male, who among human beings everywhere helps provide food for women and children — Analogies from the world of birds and fishes are far from man. Among our structurally closest analogues — the primates — the male does not feed the female. He may fight to protect her or to possess her, but he does not nurture her", pp. 188 f. She concludes, "Male sexuality seems originally focussed to no goal except immediate discharge; it is society that provides the male with a desire for children, for patterned interpersonal relationships that order, control, and elaborate his original impulses" (p. 229). I might add that fatherhood and marriage, if we can call them inventions, were probably invented by women! See also Peter J. Wilson, *Man the Promising Primate*, (New Haven: Yale University Press, 1981), who argues that human intelligence links the "primary bond" (mother and child) to the "pair bond" (male and female) and thus forms the family unit, and then by abstraction extends this notion of relationship to form other societies (kinship and then the state).

61. Anthropologists today generally admit the universality of "male dominance" in all known human societies. According to George Peter Murdock, a specialist in cross-cultural studies, *Culture and Society: Twenty-Four Essays*, (Pittsburgh: University of Pittsburgh Press, 1965), Essay 10, "Male and Female" pp. 199-231, the prevailing theory is that of Divale and Harris according to which male dominance is due to the superior ability of the male in physical combat. Otherwise mothers could have gained dominance by killing off baby boys, but they do not do so because this would lead to the destruction of the tribe in war. Peter Farb, *Humankind*, Boston: Houghton Mifflin, 1978 in a recent discussion of the problem (pp. 199-231) adopts this theory as the best. It is not contradictory to that of others who relate male dominance to the necessary dependence of the child on its mother, since it is this problem of pregnancy and child-care which chiefly renders the woman unable to engage in warfare. However, Karen Sachs, *Sisters and Wives*, (Westport, Conn. Greenwood Press, 1974), Chapter 2, "The Case Against Universal Subordination" pp. 65-95, argues that male anthropologists arbitrarily vary their criteria of "dominance" so as to deny the validity of every example of female dominance which feminists discover in the data. The chief argument of feminists, however, is that presented by Ann Oakley, *Sex, Gender and Society*, (New York: Harper, 1972), i.e., although male dominance has existed in all known societies, it need not exist in the future. Cannot women come to equal men in modern warfare where physical strength is no longer so important? Or better, can we do away with war? But I ask will women still not be restricted by child-care? Will contraception solve this problem? Or will women still *want* to be mothers and to care for their own children? My position is that the solution must lie not in trying to overcome male dominance as such, but to overcome the oppressive character of that dominance as Jesus did in the concept of authority as service. See the review of recent data by Mary Hotvedt in *Sex and Gender, A Theological and Scientific Inquiry*, ed. by Mark F. Schwartz, Albert S. Moraczewski, and James A. Monteleone (St. Louis: Pope John Center, 1983), pp. 144-177.

62. See the massive researches of Prince Peter of Greece and Denmark, *A Study of Polyandry* (The Hague: Mouton, 1963), who shows that it is rare, mainly found in Southern Asia (and for some reason which he does not attempt to explain, where Buddhism has been influential), and that it is usually due to exceptional economic conditions and is quickly abandoned when the marginal peoples who practice it are assimilated by more successful cultures. On the significance of matrilineality see Alice Schlegel, *Male Dominance and Female Autonomy: Domestic Authority in Matrilineal Societies* (no place given; HRAF Press, 1972). She shows its connection with economic forms, admits that it does not displace male dominance, but argues that it does increase the autonomy of women in some cultures (although the minority) in a remarkable degree.

63. Polygyny is the most widespread variation from monogamy, and according to Peter Farb, *Humankind* (note 61 above), is found in 75% of some 565 societies studied. Yet it is only a variation because so few marriages in these societies are polygynous. For example, in Islamic societies where four wives are permitted by law, probably less than 1% of the men have more than one wife (p. 409). Thus it is essentially a luxury for a few and a mark of their prestige, as it probably was for the Biblical patriarchs and the Jewish kings.

64. Christian theologians are attempting to assimilate the themes of the feminist movement as in three notable recent works by women; Phyllis Trible, *God and the Rhetoric of Sexuality* (Philadelphia: Fortress Press, 1978); Rosemary Radford Ruether, *Sexism and God Talk: Toward a Feminist Theology* (Boston: Beacon Press, 1983); Elisabeth Schüssler Fiorenza, *In Memory of Her: A Feminist Theological Reconstruction of Christian Origins* (New York: Crossroad, 1983). Much more radical is Mary Daly, *Beyond God the Father* (Boston: Beacon, 1973) and *Gyn-Ecology: The Metaethics of Radical Feminism* (Boston: Beacon, 1979). Some see radical feminism as already on the wane, Susan Bolotin, "Voices from the Post-Feminist

474

Generation" *New York Times Magazine,* Oct. 17, 1983, p. 28 ff., but unquestionably the issues which it has raised cannot be forgotten by theology.

65. See Daniel Yankelovich and William Barrett, *Ego and Instinct: The Psychoanalytic View of Human Nature Revised* (New York: Random House, 1970).

66. For an excellent defense of Jung from a feminist writer see Ann Belford Ulanov, *The Feminine in Jungian Psychology and Christian Theology* (Evanston: Northwestern University Press, 1971), especially pp. 139-167. The author rejects both the Freudian view that "anatomy is destiny" and the Neo-Freudian view (Horney and others) that sexual differences are purely cultural, in favor of a Jungian *symbolic view* that the *polarity* of the sexual archetypes is innate and transcultural, but that the expressions of these archetypes is culturally modified and often results in the *polarization* of men vs. women, i.e. in one-dimensional stereotypes which prevent both men and women from achieving full personal integration in an androgynous manner. In my opinion Ulanov's book is one of the best balanced presentations of the topic.

67. Andrew M. Greeley, *Sexual Intimacy* (Chicago: Thomas More Press, 1923), Chapter 7, pp. 129-148 argues strongly that children can contribute to true intimacy in marriage, against what he as a sociologist recognizes as a common attitude today.

68. Passing over the biological origins of masculine-feminine polarity, some philosophers and theologians have tried to find a basis for it in the very nature of God. Thus the Russian Orthodox theologians Vladimir Soloviev (d. 1900) and Sergei Bulgakov (d. 1944) developed a Sophiology in which the feminine Wisdom of God is the central feature. Cf. George A. Maloney, S.J., *A History of Orthodox Theology since 1453* (Belmont, Mass.: Nordland, 1976), pp. 61-65, and Paul Evdokimov, *La femme et le salut du monde* (Tournal-Paris: Casterman, 1958) especially p. 64-66; Helmut Theilicke, *The Ethics of Sex* (New York: Harper and Row, 1964) pp. 3-100, argues that the meaning of *Genesis* 1-28 is that since male *and* female are created in God's image, both are necessary to show us what God is like; and from a Jungian viewpoint, Ann Belford Ulanov, *The Feminine* (note 66 above) p. 293, takes the same line. It seems to me that we should not attribute sexual polarity to God, but rather should maintain that human nature, which is our best *analogy* to God, is only partially manifested in man and woman taken separately, because of the biological requirements of reproduction. In this sense the Greek Fathers were right in saying that the image of God in humanity is androgynous.

69. See Don S. Browning, *Generative Man: Psychoanalytic Perspectives* (Philadelphia: Westminster, 1973).

70. See Mario Praz, *The Romantic Agony,* 2nd ed., (New York: Oxford University Press, 1951) and (with caution) for the older background Denis De Rougemont, *Love in the Western World,* (New York: Pantheon, 1956).

71. Herbert Marcuse in *Eros and Civilization,* (Boston: Beacon, 1966), looks forward to a society in which non-repressive sublimation will become possible which can occur only if there is a certain "regression" to the polymorphous sensuality of the child. Gad Horowitz, *Repression* (Toronto: University of Toronto Press, 1971), reviews the theories of Freud, Reich and Marcuse and tries to defend Freud on the basis of a distinction between "basic repression" which is necessary for the achievement of adult genitality and "surplus repression" which is the unnecessary result of social oppression. See also Norman O. Brown, *Love's Body* (New York: Random House, 1966).

72. The history of the process by which this step was taken is detailed in Ronald Bayer, *Homosexuality and American Psychiatry: The Politics of Diagnosis,* (New York, Basic Books, 1981). See also Enrique T. Rueda, *The Homosexual Network* (Old Greenwich: Conn.: Devair Adair, 1982). On the present state of the question see D. J. West, *Homosexuality Re-Examined* (Minneapolis: University of Minnesota Press, 1977). The most extensive statistical study is Alan P. Bell and Martin S. Weinberg, *Homosexualities: A Study of Diversity among Men and Women* (New York: Simon and Schuster, 1978) and by the same authors with Sue K. Hammersmith, *Sexual Preference* (Bloomington, Indiana: Indiana University Press, 1981). The arguments used by some Catholics in defending homosexual behavior as ethically acceptable can be found in John J. McNeill, *The Church and the Homosexual*

(Kansas City, Kansas: Sheed, Andrews and McMeel, 1976). A sympathetic study of cases is Richard J. Woods, *Another Kind of Love* (New York: Image Books, 1978). A recent review of the theological problems is Bruce A. Williams, "Homosexuality and Christianity: A Review Discussion", *Thomist* 46 (Oct., 1982): 609-625. My own analysis of the data will be found in my essay "A Theological Overview on Recent Research on Sex and Gender, "in *Sex and Gender* (St. Louis: Pope John Center, 1983), pp. 1-47. This volume contains reviews of the most up-to-date data and theories on sexual identity and orientation by some of the leading researchers in the field (note 61 above).

73. Robert T. Francoeur, *Eve's New Rib: Twenty Faces of Sex, Marriage and Family* (Harcourt Brace Jovanovich, N.Y., 1972). In the "Epilogue" pp. 221 ff., he lists (1) Traditional monogamy, (2) Flexible monogamy, (3) Serial monogamy; (4) Trial marriages; (5) Polygamy for senior citizens; (6) Reversal of husband-wife roles in family, permanently or in rotation; (7) Unisex marriages; (8) Single parents; (9) Group marriages; (10) Retirement parenthood; (11) Temporary contractual marriages; (12) Informal marriages; (13) Limitation of marriages with children to professionally trained childraisers; (14) Third parents who work with others' children; (15) Legalized polygamy, polandry, bigamy, or triangular marriages for those who prefer them; (16) Multilateral marriages of bisexuals; (17) Celibacy, for religious or social dedication; (18) Non-sexual uni-sex partnerships ("odd couples"); (19) Celibate or celebrational marriage; (20) Other possibilities, as sexual "options", all of which he seems to regard as morally acceptable.

74. Current writers often leave the impression that Christianity has an exceptionally negative view of the dangers of sex. However, J. Duncan M. Derrett, *Jesus' Audience* (London: Darton, Longman and Todd, 1973), writes of the attitudes of the world into which Jesus was born as follows: "Sexual pleasure is regarded as inherently bad . . . the ancient civilizations (*except* the Western) concluded that because the sexual drive makes men hanker after other men's wives, and intrigue to get them, and since those women fell an easy prey to a determined admirer, the drive must be the work of Satan. The 'evil inclination' and Satan were virtually synonymous. Etiquette and society itself were largely aligned to the avoidance of sexual stimulation (to keep Satan's scope down), and a heavy taboo on all sexual activity, saving that directed to procreation, gave a particular character to the culture. It distinguished it profoundly and conclusively from the Greek. The Greeks must have thought the Jews collectively neurotic. Jesus' comment on Peeping Tom (Mt 5:29) would strike a Greek as crazy" (p. 132). No doubt the contrast between the Jews and the Greeks as regards sexual taboos was very great, but Derrett fails to note that the Greeks too had many anxieties about sexual behavior, as is obvious from the Greek myths, epics and tragedies. See the well known book of E. R. Dodds, *The Greeks and the Irrational* (Berkeley: California Press, 1951). Freud was closer to the truth in *Civilization and its Discontents* (Standard Edition: London, Hogarth Press, 1930, vol. 21, pp. 57-146), when he claimed that the advance of culture is connected with a sublimation (and repression) of the sexual instinct. The hope of Reich and Marcuse for a non-repressive sublimation seem Utopian as far as the historical evidence goes.

75. On the Scriptural view of sexuality the work of Derrick S. Bailey, *Sexual Relations in Christian Thought* (New York: Harper and Row, 1959), has been very influential, but is an example of special pleading. More useful discussions will be found in P. Grelot, "Le couple humain selon la Scripture," *La Vie Spirituelle,* Supplement 57 (1961): 135-198; J.J. Von Allmen, *Pauline Teaching on Marriage* (New York: Morehouse Barlow Co., 1963); W. J. Harrington, *The Promise to Love: A Scriptural View of Marriage* (New York: Alba House, 1968). On the divorce texts see Jacques Dupont, *Mariage et divorce dans l'Évangile Matthieu 19:3-12 et paralleles* (Bruges: Abbaye de Saint André, 1959), and Quentin Quesnell, "Made themselves Eunuch for the Kingdom of Heaven", *Catholic Biblical Quarterly* 30 (1968): 335-338. On the theology of marriage see Edward Schillebeeckx, *Marriage: Human Reality and Saving Mystery* (New York: Sheed and Ward, 1965); F. Böckle, ed., *The Future of Marriage as an Institution,* Concilium (New York: Herder and Herder, 1970); Mary R. and Robert E. Joyce, *New Dynamics in Sexual Love* (Collegeville: Minn.: St. John's University Press, 1970); Rosemary Haughton, *The Theology of Marriage* (Notre Dame, Ind.: Fides, 1971);

G. De Broglie, "La conception Thomist de deux finalités du marriage", *Doctor Communis* 30 (1974): 3-42. On the Jewish view see Robert Gordis, *Love and Sex: A Modern Jewish Perspective* (New York: Farrar Straus Giroux, 1978), who writes, "The Jewish attitude toward sex takes, as its point of departure, the fundamental principle that marriage and marriage alone is the proper framework for sexual experience" (p. 98), and strongly criticizes the Christian advocacy of celibacy (pp. 41-58). A recent study by a Catholic, Corrado Marucci, *Parole di Gesù sul divorzio* (Brescia: Morcelliana, 1982), again defends the opinion that the "Matthean exception" may after all have been a pastoral application to permit divorce and re-marriage, but solely in the case of adultery, as Paul permitted it in the case where the pagan partner refused to live with the Christian spouse.

76. Mark 10:29, as against Matthew and Luke (18:29), omits "wife" and D. E. Nineham, *The Gospel of St. Mark* (New York: Seabury, 1968), p. 276, comments "The absence of 'wife' from the list may well be significant. The early Christians may have regarded the marriage tie as so close (cf. above v. 9), that nothing short of martyrdom could justify the permanent breaking of it (cf. I Cor. 7:12 ff)". But why not rather suppose that "wife" is original and reflects the influence of Q, still earlier than Mark? While Mark's omission may reflect a concern on his part not to *seem* to contradict the Lord's words forbidding divorce, Matthew and Luke evidently did not see separation for the sake of the apostolate as forbidden by them.

77. See Rudolf Schnackenburg, *God's Rule and Kingdom* (New York: Herder and Herder, 1963), pp. 77-103, on the eschatological aspect of Jesus' teaching.

78. On Qumran celibacy see Geza Vermes, *Jesus the Jew,* (London: Collins, 1973), pp. 99-102, and *The Dead Sea Scrolls: Qumran in Perspective* (Cleveland: Collins-World, 1978), pp. 96-98, and 181 f., and the bibliography on p. 193. Vermez makes the important point that this celibacy was probably not so much the result of the desire to maintain cultic purity, but rather because of the rabbinic notion that Moses and the prophets abstained from sex permanently after their encounter with God, and that the Qumran people shared in this prophetic charism. The rabbis, on the other hand, opposed celibacy because their status as teachers of the people depended on the notion that prophecy had ceased after the return from Exile.

79. On the Nazarites see de Vaux, *Ancient Israel* (see note 47, above) pp. 465-467.

80. See Donald Goergen, O.P., *The Sexual Celibate* (New York: Seabury, 1975).

81. This attitude is naively expressed in William E. Phipps, *Was Jesus Married?* (New York: Harper and Row, 1970).

82. On the teaching of *Ephesians 5:32* that marriage is a "great mystery," see Heinrich Schlier, *Der Brief an die Epheser* (Düsseldorf: Patmos Verlag, 1963), pp. 252-280, and Markus Barth (from a Protestant point of view), *Ephesians,* Anchor Bible (Garden City, N.Y.: Doubleday, 1974), vol. 1, pp. 738-753.

83. For an effort to show that Jesus was not celibate see William E. Phipps, *Was Jesus Married?* (note 81 above). Phipps' case is chiefly based on the fact that rabbis were required to marry. Phipps' argument has had little success, see for example the review of J. Peter Bercovitz in *Perspective: A Journal of the Pittsburgh Theological Seminary* 12 (Fall, 1971): 271 f. On the biblical and patristic basis see Eugene H. Maly, "Celibacy" in *The Bible Today* 34 (Feb., 1968): 2392-2400; Donald W. Trautman, *The Eunuch Logion of Matthew 19, 12* (Rome: Catholic Book Agency, 1966); David E. Aune, *The Cultic Setting of Realized Eschatology in Early Christianity* (Leiden, E. J. Brill, 1972), especially pp. 215-219 on the asceticism in the Syrian Church; T.H.C. Von Ejik, "Marriage and Virginity, Death and Immortality" in J. Fontaine and C. Kannengiesser eds., *Epektasis: Mélanges Patristique offerts au Cardinal Jean Daniélou* (Paris: Beauchesne, 1972), pp. 209-235.

On celibacy as a requirement for ordained ministers, the standard history is Roger Gryson, *Les origines du célibat ecclésiastiques du premier au 5e siècle* (Gembloux: Paris, 1970). See also Jean Paul Audet, *Structures of Christian Priesthood* (New York: Macmillan, 1968) and Bernard Verkamp, "Cultic Purity and the Law of Celibacy," *Review for Religious,* 30 (1971), 199-217 The common account now is that in the early Church often even the bishops were married, but that gradually as a result of concerns for cultic purity, based on archaic notions that sex is unclean and evil, celibacy was enforced on bishops in the East and in the Middle Ages on all the clergy in the West, partly out of concern for the protection of ecclesiastical properties. Recently C. Cochini, *Origines apostolique du célibat sacerdotal* (Paris: Lethielleux, 1980), has given a very different interpretation to the known facts. According to Cochini, the example of Jesus and Paul established a powerful precedent which was reinforced by the Old Testament laws requiring the priests to abstain from marital relations at times of their liturgical service, not because sex was evil (since that is not the Old Testament attitude), but because the priest is an intercessor with God who must therefore transcend the concerns of mortal life and enter the divine sphere of eternal life. Since the priest of the Church so frequently celebrated the Eucharist this required permanent abstinence. Although it was common to elect married men of tried virtue (I Tim. 3:2; Titus I:6), there is evidence in the Scriptures (Matt 19:28), in the Fathers, and in Church legislation at least from the Council of Nicaea, that these elected bishops and priests were required to become celibate (separating from their wives in the East, living with them as sisters in the West). This tradition in some places and times fell into desuetude, but in the West was finally enforced by the reforming Popes of the Middle Ages. In the East in the seventh century it was legally allowed to lapse for priests, but never for bishops.

84. See Herbert Richardson, "The Symbol of Virginity" in Donald R. Cutler, *The Religious Situation: 1969* (Boston: Beacon Press, 1969), pp. 775-811. Richardson argues that in Hebrew and other ancient cultures, in order to overcome mother-dominance, men confined women and sex to privacy so as to leave a public space which was exclusively male. Friendship was homosexual. With the rise of the world religions the concern to emphasise the transcendent and eternal led to a negative attitude toward sex. Woman alone was a sexual being and was respected only if virgin. Friendship now could be heterosexual. Finally in modern culture, sexuality becomes a means of communication and friendship. Virginity is replaced by a period of pre-marital sexuality with a gradual sexual initiation so that sex can have a more pluralistic, differentiated combination of values.

85. See John T. Noonan Jr., *Contraception: A History of Its Treatment by the Catholic Theologians and Canonists* (Cambridge, Mass: Harvard University Press, 1965), on the controversy as to whether sterile persons were permitted to have intercourse (pp. 248-249, 290-292, 497-498). The Augustinian school tended to take the negative position, but Aquinas (without fully explaining his position) took the affirmative. It is consistent with his moral system to maintain that the marital act is justified simply as an expression of love, without the intention of procreation, provided that the nature of the act itself is not intentionally altered so as to destroy the ordination to procreation inherent in the nature of the sexual act. This is because for Aquinas, the physical character of the act, while it is not of itself formally determinative of the moral character of the act, enters into that determination as a material cause. The Augustinian school tended to attribute the morality of acts entirely to the spiritual *intention*. Curiously, this dualistic notion of morality has been revived by those theologians today who defend contraception by the proportionalist methodology.

86. See note 73 above.

87. See Benedict M. Ashley, "A Child's Rights to His Own Parents: A Look at Two Value Systems", *Hospital Progress* 61 (August), 47-49.

88. The notorious studies of A. C. Kinsey et al, *Sexual Behavior in the Human Male* (Philadelphia: Saunders, 1948), and *Sexual Behavior in the Human Female* (Philadelphia: Saunders, 1953), were methodologically faulty, but subsequent efforts by sexologists to determine empirically the actual range of sexual behavior have not greatly altered the picture which, after all, was not very surprising to theologians from their own pastoral experience. If, however, we are to draw from this data the conclusion that traditional Christian moral standards with regard to sex are unrealistic, must we not logically conclude the same as to those standards which apply to other moral matters? Would not empirical studies of behavior with regard to any of the Ten Commandments reveal a similar "morality gap"? Yet many who propose a relaxation of the Christian sexual code to make it more realistic, are also advocates of a very strict code of social justice and non-violence, and do not regard these demands as invalid because "unrealistic".

The historical and empirical study of actual moral behavior in relation to religious beliefs is still in the pioneering stage as in such works as Donald S. Marshall and Robert C. Suggs, *Human Sexual Behavior: Variations in an Ethnographic Spectrum* (New York: Basic Books, 1971); Geoffrey Parrinder, *Sex in the World's Religions,* (New York: Oxford University Press, 1980); and Vern L. Bullough and James Brundage eds., *Sexual Practices and the Medieval Church* (Buffalo, N.Y.: Prometheus Books, 1982), and is still often deeply biased by poorly analyzed pre-understandings; cf. Michel Foucault, *The History of Sexuality,* vol. 1, Introduction (New York: Pantheon, 1978). Moral Theology is in great need of such studies conducted with a more sophisticated methodology.

89. On the New Testament teaching on the nature of sin, see Ceslaus Spicq, O.P., *Théologie Morale du Nouveau Testament* (Paris: J. Gabalda, 1965), tom. 1, pp. 17-68 and 175-199.

90. It is curious that at a time when a great educational effort is being made to teach the public to respect the dignity and rights of the handicapped and to help the handicapped themselves to face their limitations realistically and without self pity, that the Gay and Lesbian Rights Movement systematically promotes a rhetoric of *denial* that homosexuality is a handicap and rejects that term. Instead the Movement attempts to sell the public on the idea that homosexuality is analogous to racial differences, a normal variation of sexuality. Yet to take the simplest consideration, is it not a handicap to be unable to achieve sexual fulfillment in a marriage based on a love that perpetuates itself in children? That this is a handicap is testified by the demands of some homosexuals to be allowed to adopt children.

91. Four main approaches to women's liberation are apparent in the vast, growing literature on the subject: (1) The Marxist approach which sees the oppression of women simply as one facet of the whole problem of capitalist oppression; (2) The liberal approach which sees this evil as a remnant of out-dated cultures to be overcome as racial discrimination is being overcome by appropriate legislation and education; (3) Radical religious views which blame it on the patriarchalism of the major religions and seek to remedy it by a return to a religion of "the Mother Goddess"; (4) Revisionist religious views which attribute it to a patriarchal distortion of the major religions and seek to remedy it by a purification of these religious traditions. There are of course writers who deny that such oppression exists, minimize it, or simply accept it as inevitable. My views are those of the revisionists, but I do not agree with the diagnoses or prescriptions put forward by some of these feminist writers.

92. See Mary Daly, *The Church and the Second Sex* (New York: Harper and Row, 1968), and *Beyond God the Father* (Boston: Beacon Press, 1974).

93. Authentic art must be rooted in the experience of the artist (at least his or her *imaginative* experience). The fact that artists, like all of us, are sinners, does not necessarily deny to them an experience of the beauty of chastity; but the fact that in modern society they often live in a social milieu which is ignorant of chastity and bigotedly blind to its beauty, does make it difficult for them to reflect in their art in any authentic and convincing way the nobility of chaste men and women. Yet the greatest artists of the past (notably Shakespeare), conscious of their own weakness and of the corruption of society, were able to do so.

94. Edward O. Wilson, *Sociobiology* (note 24 above), pp. 379-402, discusses what he calls the "four pinnacles of social evolution", i.e., the colonial invertebrates, the social insects, the social vertebrates, and humans, and shows their similarities but *radical* differences.

95. This dependence of social organization on its economic base is, of course, a fundamental principle of Marxism and is well argued for from an anthropological point of view by Marvin Harris, *Cultural Materialism: The Struggle for a Science of Culture* (New York: Random House, 1979).

96. See Chapter 2, p. 23 and Chapter 8, pp. 311; 325-328. Also John Deely, *Introducing Semiotics* (Bloomington: Indiana University Press, 1982), pp. 107-203 with bibliographical references.

97. See Chapter 8, pp. 327-328.

98. See references in notes 11 and 12 above.

99. I will say more about the problem of the knowledge of the Transcendent by analogy in Chapter 12.

100. See Chapter 6, note 53 on the various interpretations of what Hegel meant by this.

101. "When some one inquired why we spend much time with the beautiful, "That, 'he [Aristotle] said,' is a blind man's question". *Life of Aristotle* in Diogenes Laertius, *Lives of Eminent Philosophers* V, 20, Loeb Classics vol. 40-41, (Cambridge, Mass.: Harvard University, 1942), vol. 1, p. 463.

102. See Ronald de Vaux, *Ancient Israel* (note 47 above), pp. 91-163.

103. See Ernst Troeltsch, *The Social Teaching of the Christian Churches,* 2 vols. (New York: Harper, 1960); Jean-Yves Calvez, S.J., and Jacques Perrin, S.J., *The Church and Social Justice: The Social Teachings of the Popes from Leo XIII to Pius XI* (1878-1958) (Chicago: H. Regnery, 1961); Joseph Gremillion, *The Gospel of Peace and Justice: Catholic Social Teaching since Pope John* (Maryknoll, N.Y.: Orbis Books, 1976) and Roger Charles, S.J., with Droston Maclaren, O.P., *The Social Teaching of Vatican II* (Oxford: Plater Publications, 1982).

104. Karl R. Popper, *The Open Society and Its Enemies,* 5th ed., revised (Princeton, N.J.: Princeton University Press, 1966) attacked Plato as the source of totalitarianism. For replies and criticism see Ronald B. Levinson, *In Defense of Plato* (Cambridge, Mass.: Harvard University Press, 1953); John D. Wild, *Plato and His Modern Enemies* (Chicago: University of Chicago Press, 1953), and Renford Bambrough, ed. *Plato, Popper and Politics* (Cambridge: Helfer; New York, Barnes and Noble, 1967). While Popper is unfair to Plato, Aristotle long ago pointed out that Plato, by a too univocal understanding of the comparison of the unity of the state to the unity of an organism, opens the way to a totalitarian political philosophy; *Politics,* II, 2, 1261 b 10.

105. In the 1940's there was an instructive controversy on this problem among the Thomists. See Jacques Maritain, *The Person and the Common Good* (New York: Scribner's, 1947) who attempted to refute totalitarianism by means of a distinction between the "individual" (who is subordinated to the common good) and the "person" (who transcends the common good of human society). This was severely criticized by Charles De Koninck, *De la primauté du bien commun contre les personalistes* (Laval, Quebec: University of Laval, 1943) as an inadequate and unThomistic answer to totalitarianism. Fr. I. Th. Eschmann, O.P., attempted to answer De Koninck, "In Defense of Jacques Maritain," *Modern Schoolman,* 22 (May, 1945): 183-208, only to be rebutted by De Koninck, "In Defense of St. Thomas," *Laval Theologique et Philosophique,* 1 (2, 1945): 1-103 in an article which contains a useful collection of Thomistic texts. I am of the opinion that De Koninck was correct both as to St. Thomas' thought and in his solution of the problem.

106. Aristotle, *Nicomachean Ethics* V, 3, 1131aa 10 sq.

107. "In the higher phase of communist society, after tyrannical subordination of individuals according to the distribution of labor, and thereby also the distinction between manual and intellectual work, have disappeared, after labor has become not merely a means to live but is in itself the first necessity of living, after the powers of cooperative wealth are gushing more freely, together with the all-round development of the individual, then and only then can the narrow bourgeois horizon of rights be left behind and society will inscribe on its banners 'From each according to ability, to each according to his need.'" Karl Marx, *Critique of the Gotha Program* (New York: International Publishers, 1933), p. 13. See also Robert B. Fulton, *Original Marxism's Estranged Offspring: A Study of Points of Contrast and Conflict between Original Marxism and Christianity* (Boston: Christopher Publishing House, 1960), p. 81-85.

108. Yves Simon, *A General Theory of Authority* (Notre Dame, Ind.: University of Notre Dame, 1962).

109. "It is clear that in all known human societies, gender provides the basis for a fundamental division in social function. In no society to date do men take primary responsibility for the care of young children, nor do women take a principle role in organizing and implementing activities of offense or defense, although the degree to which the opposite sex is involved in these activities varies greatly from society to society. Even where the

division of labor by sex is minimal or nonexistent . . . this division occurs. Division of function, however, does not necessarily lead to stratification; rather, it can lead to balanced complementarity. Sexual stratification, then, is not panhuman, but rather poses a problem that must be explained for each society in terms of the forces to which it is responsive, and crossculturally in terms of variables that exist across societies. It is an enormously complex problem and a challenging one." Alice Schlegel, "Toward a Theory of Sexual Stratification" in the work edited by her, *Sexual Stratification: A Cross Cultural View* (New York: Columbia University Press, 1977), pp. 1-37. For Schlegel, the root of the tendency to male domination is found in the child-mother relation.

110. See Sir Ernest Baker, *The Political Thought of Plato and Aristotle* (New York: Russell and Russell, 1939), pp. 335-336 and 350-356.

111. On the Principle of Subsidiarity and Functionalism see Benedict M. Ashley, and Kevin D. O'Rourke, *Health Care Ethics* (note 17 above), pp. 121-127, pp. 196-197.

112. Arnold Toynbee, *A Study of History*, D. C. Somervell's authorized abridgement (London and New York: Oxford Press, 1946).

113. *The Rise of the West* (Chicago: University of Chicago Press, 1963), provides a unified view of human history, chiefly from the viewpoint of economics and technology.

114. See John O'Meara, *Charter of Christendom: The Significance of the City of God* (New York: Macmillan, 1961), and P. Barry, "Bossuet's Discourse on Universal History" in Peter Guilday, *The Catholic Philosophies of History* (New York: Kenedy, 1936), pp. 149-186, and Paul Hazard, *The European Mind, 1680-1715, (Cleveland: World Publishing, 1963) (reprint), pp. 198-216.*

115. *The Church in the Modern World (Gaudium et Spes)*, Dec. 7, 1965, Austin Flannery, O.P., ed., *Vatican II: The Conciliar and Post-Conciliar Documents* (Northport, N.Y.: Costello, 1975), pp. 903-1001.

116. "My explanation [of the rise of modern science] is that faith in the possibility of science, generated antecedently to the development of modern scientific theory, is an unconscious derivation from medieval theology." Alfred North Whitehead, *Science and the Modern World* (1925) (New York: Macmillan), 1962, p. 19.

117. On the value and limits of oral tradition among preliterate people see Jan Vansina, *Oral Tradition: A Study in Historical Methodology* (Chicago: Aldine Press, 1965).

118. See Robert R. Wilson, *Prophecy and Society in Ancient Israel* (Philadelphia: Fortress, 1980) on the various forms of prophecy in the Ancient Near East, pp. 21-252, and R. C. Zaehner, *The Comparison of Religions* (Boston: Beacon, 1958), Chapter 4, "Prophets Outside Israel", pp. 134-164.

119. See the foreword by Harry Levin to his edition of Irving Babbitt, *Rousseau and Romanticism* (Boston: Houghton Mifflin, 1919; Austin, Texas: University of Texas, 1977), and Rolf Tobiassen, *Nature et nature humaine dans l'Émile de Jean Jacques Rousseau* (Oslo: Presses Universitaires; Paris: Boyveau et Chevillet, 1961).

120. Ronald De Vaux, O.P., *Ancient Israel* (note 47 above).

121. Sigmund Freud, *Civilization and its Discontents*, newly translated and edited by James Strachey (New York: W. W. Norton, 1962).

122. S. Mowinckel, *He that Cometh* (New York: Abingdon, 1954); G. A. Riggan, *Messianic Theology and Christian Faith,* (Philadelphia: Westminster, 1967); Joseph Coppens, S.J.,, *Le Messianisme Royal* (Paris: Cerf, 1969); *Le Royauté, le règne, le Royaume de Dieu* (Leuven, Peeters and University Press, 1979) and *Le Fils de l'homme néotestamentaire,* same publisher, 1981; Joachim Becker, *Messianic Expectation in the Old Testament* (Philadelphia: Fortress, 1977) (expectation only late); Henri Cazelles, *Le Messie de la Bible* (Paris: Desclee, 1978).

123. The notion of the "axial period" in which the world religions developed was proposed by Karl Jaspers in his *The Origin and the Goal of History* (New Haven: Yale University Press, 1953).

124. "In the West we hold that man is a creature and his right attitude to God is one of creatureliness, and creatureliness expresses itself in worship and sacrifice. Neither Vedanta nor Buddhism accepts this. For the one, man is potentially the Absolute; for the other, it

is simply a question of putting an end to phenomenal existence in order to enter into a bliss which knows neither time or space". R. C. Zaehner, *The Comparison of Religons* (note 116 above) p. 103 f. See also Zaehner's, *Concordant Discord* (Oxford: Oxford University Press, 1970), Chapter 6, "The Birth of God", pp. 104-127. Although in Hinduism, Ramanuja opposed Shankara's non-dualism yet even for Vedanta the human self was part of God's "body", see Stuart C. Hackett, *Oriental Philosophy* (Madison: University of Wisconsin Press, 1979), p. 164. On Neo-Platonism see A. H. Armstrong, "Plotinus" in *The Cambridge History of Later Greek and Early Medieval Philosophy,* edited by him (Cambridge: Cambridge University Press) pp. 195-271, especially pp. 239-241.

125. See William H. McNeill, *The Rise of the West* (note 113 above), pp. 316-360.

126. Romans 13: 1-7 and I Peter 2:13-17.

127. See George A. Maloney, S.J., *A History of Orthodox Theology* (note 68 above), p. 23, 57 and *passim.*

128. See James Westfall Thompson, *A History of Historical Writing* (New York: Macmillan, 1942), 1, 473-646 and 2, 3-57.

129. See Josep Glozik, "The Springtime of The Missions in the Early Modern Period"; in Hubert Jedin and John Dolan, *History of the Church,* vol. 5 *Reformation and Counter Reformation* (New York: Seabury-Crossroads, 1980), Part Two, Section 4, pp. 575-614.

130. John Henry Newman, *An Essay on the Development of Christian Doctrine,* edited with an introduction by J. N. Cameron, (Baltimore, MD.: Penguin Books, 1974), and Jan H. Walgrave, *Unfolding Revelation: The Nature of Doctrinal Development,* (Philadelphia: Westminster, 1972), pp. 293-313.

131. The famous last line of Hilaire Belloc's *Europe and the Faith* (New York: Paulist Press, 1920), "The Faith is Europe. And Europe is the Faith" (p. 261), is misleading. Belloc's argument throughout his book is directed at proving that European civilization would never have been possible without the Christian Church. He does *not* argue that the Faith depends on European civilization, but only that Divine Providence prepared and used Greco-Roman culture to make it easier for the Gospel to gain a foothold.

Chapter 11

God's Fullness In Bodily Form

(Col. 2:9)

I: The Son of God

1. Christology From Below, From Above, From Without, From Within

An ethics of the sort sketched in the last chapter, rooted in our bodily existence, in our basic needs — both the needs for food, security, and sex that we have in common with the animals, and our needs for truth, society, and creativity which are specific to us as humans yet which depend on our animality — if it is a mere model created by human fancy, rather than something experienced in historical reality, still remains only an abstract ideal. For the Christian, this ethics has been historically realized in Jesus of Nazareth and hence it is only ultimately understandable in its fullness in the light of the great question, "But who do you say I am?" (Mt 16:15), i.e. the theological problem of Christology.

Recently there has been much discussion about whether Christology should be constructed "from above" or "from below." Should we begin with the human Jesus of Nazareth and then through what we know of his humanity seek to ask what it means to affirm his divinity, as Christians do when they confess that "Jesus is true man and true God;" or should we begin with the divine Christ, son of God whom we worship and then ask what it means to say that God became man, or was incarnate? In medieval scholastic theology it was usual to follow the latter method as Aquinas did when he began the Third Part of the *Summa Theologiae* with the question, "Was it fitting that God should become incarnate?" Today, however, when theologians are especially concerned to begin with the biblical data, it seems to many better to begin "from below" with the fact that the Gospels clearly present Jesus as a human being who suffered, died, and rose in his human body, and then to inquire as to how the early Church came to reverence this man as Lord and Savior, the true Son of God.

An outstanding example of a Christology from below is that of Piet Schoonenberg, S.J., who proposes to develop the so-called Chalcedonian formula of One Divine Person in two natures, one divine and one human (which seems to Schoonenberg to suffer from the grave difficulty that it makes Jesus less than a fully human person), by beginning with the evident fact that Jesus was a human person and then seeking some new way to express the way in which he manifests the Divinity.[1] Similarly Hans Küng has tried to give an "up-to-date positive interpretation" of the formula by first affirming unconditionally that Jesus was a man, and then explaining how he was "truly God" as follows:[2]

> The whole point of what happened in and with Jesus of Nazareth depends on the fact that, for believers, *God himself* as man's friend was present, speaking, acting, definitively revealing himself *in Jesus*, who came among men as God's advocate and deputy, representative and delegate, and who, as the Crucified raised to life, was confirmed by God. All statements about divine sonship, pre-existence, creation-mediatorship, and incarnation — often clothed in the mythological or semimythological forms of the time — are meant in the last resort to do no more and no less than substantiate the *uniqueness, underivability* and *unsurpassability* of the *call, offer, and claim* made known in and with Jesus, ultimately not of human but of divine origin and therefore absolutely reliable, requiring man's unconditional involvement.

That either of these theological efforts are really successful as new formulations of an ancient faith is not apparent. Edward Schillebeeckx has argued that the effort to construct a Christology strictly "from below" is not possible in view of the fact that in the New Testament, which is the source of the data for any such construction, both types of Christology were present from the beginning.[3]

In Chapter 9, in keeping with my own methodology of beginning always from our sense experience of the world, I presented Jesus as it were "from below" as a member of our human species, sharing fully our bodily existence, and subject to the laws of nature, physical and moral; but I assumed, without trying to analyze it, his claim to be in some special sense Son of God. How then am I to approach the question of his transcendent aspect? To speak of this as an approach "from above" is certainly not unbiblical. St Paul in speaking of Christ as the New Adam says,

> Scripture has it that Adam, the first man, became a living
> soul; the last Adam has become a life-giving spirit . . . the
> first man was of earth formed from the dust, the second is
> from heaven. (1 Cor 15:45,47).

Nevertheless, the metaphor of "below vs. above" or "lower vs. higher" easily lends itself to that Platonic dualism which we are struggling to escape. The Platonic tendency to apply this hierarchical model to every problem probably influenced Chalcedonian Christology and earlier may even have influenced this passage in Paul. It certainly is felt in Hellenistic-Jewish apocalyptic thought and later in Gnosticism.

But Paul in psychological contexts also uses another metaphor of "outside vs. inside": "We do not lose heart, because our inner man is renewed each day even though our outer man is being destroyed at the same time" (2 Cor 4:16); "My inner self agrees with the law of God, but I see in my body's members another law at war with the law of my mind" (Rom 7:22); "For who among men knows the spirit of a man except the spirit of that man which is within him?" (1 Cor 2:11). This metaphor can serve, it seems to me, as the point of departure for a Christology which is less Platonic if we keep in mind that the without and the within of a thing are not two parts but the very same thing seen from the surface and seen from the depth. Nor is this way of talking foreign to Jesus himself, whose emphasis in his ethical teaching was always on the interiority of true morality: "Do you not see that nothing that enters a man from outside can make him impure? . . . What emerges from within a man, that and nothing else makes him impure," (Mk 7:18,20; and see especially the Sermon on the Mount, Mt 5-7).

Of course we must not make the mistake to which we are very liable today, of confusing these interior depths with the "depths" of psychology. As I showed in Chapter 8, the unconscious psyche is really at the surface of psychic life, while the real depths of the human personality are to be found in the will and intelligence, and especially in that intuitive and free center which is the source of human creativity and transcendence, the *apex mentis* or *scintilla animae* of the mystics.[4] We must also keep before us that this "inner man" is not, in Platonic fashion, the real self inhabiting the human body. Rather it is the human being viewed from its center looking out to the periphery. The body is indeed the periphery or surface of the human being but it is a surface by which the interior communicates with the rest of reality, receiving information from that outer world and expressing itself to that world. It is, so to speak, the phenomenon of the inner noumenon, which reveals rather than conceals (although sometimes *per accidens* it does conceal). In every primary unit, as we have seen, there are inner and outer aspects; but in the human unit, because the interior is spiritual, transcending the physical, it spiritualizes the outer, bodily humanity, so that free human actions of the body, our words, our gestures, our deeds, our embraces are spiritual actions. No wonder that Jesus transformed others both bodily and morally by the touch of his hand (Mk 1:40-45; 5:25-34).

A Christology in these terms of the inner and outer man will search the Scriptures for those words and deeds of Jesus which manifest his own subjective experience and his personal self-understanding, both in his earthly and his risen life. Yet it must take fully into account that the Ressurrection was first of all an earthly event and that Jesus' continued presence in his Church in word, sacrament, and life, is not merely spiritual but a prolongation of his bodily existence and human experience. Some will object that if it is so difficult to arrive at the historical truth about the bare events of Jesus' pre-Resurrection life, it is mere fantasy to speak of his interior life. On the contrary, it can be replied that the New Testament writers were in some respects more concerned with the mind and heart of the Lord than with the details of his biography, consequently they disappoint the historian who wants to know when and where, but they are generous in communicating to us what Jesus thought most important to share with his disciples. Of course the inner life with which the evangelists were concerned was not Jesus' "psychology" in the modern sense of which the Gospels say only a little, but in the sense that St. Paul can say boldly,

> The spiritual man, on the other hand, can appraise everything, though he himself can be appraised by no one. For, "Who has known the mind of the Lord so as to instruct him?" But we have the mind (*nous*) of Christ. (1 Cor 2:16)

The "mind of Christ" is, of course, set forth most explicitly in the *Gospel according to St. John* where the author expands traditional

486

narratives and sayings into long poetic discourses which cannot be taken as literal reports of Jesus' own sermons but which nevertheless are not just "putting words into the mouth" of Jesus, but are an effort to convey the Johannine community's profound understanding of its Lord.[5] The author of the *Gospel according to St. Matthew* also composes discourses from traditional material such as the Sermon on the Mount, and we may even wonder whether the beautiful parables found only in Luke's Gospel such as that of the Good Samaritan (10:25-37), the Lost Sheep (15:1-7), the Prodigal Son (15:11-32), or the Rich Man and the Beggar (16:19-31) represent such efforts on his part. Mark alone avoids such expansion because he wishes to highlight the obtuseness of Jesus' hearers, yet in this indirect way he conveys very powerfully his sense of Jesus' identity through Jesus' own rejection of the many false interpretations others sought to impose on Him. As for St. Paul, he was convinced that he had "the mind of Christ" through his own profound spiritual experiences (1 Cor 2:16).

Thus a Christology in terms of the outer and inner man will put a special stress on what the New Testament has to tell us about the spiritual interpretation which the early community put upon their memories of the Lord as he had lived among them and shared his inner life with them. Of course there is a risk that in taking this approach we may end as did the Gnostics (possibly under the influence of the *Gospel according to St. John*),[6] when they claimed an *esoteric* (inner) Gospel in contrast to the exoteric (outer, public) Gospel of the Church. For this very reason, it is essential that the inner and the outer Jesus be not dualistically divided, but that whatever is attributed to "the mind of Christ" be rooted in historical events.

Edward Schillebeeckx in his *Jesus: An Experiment in Christology*, whatever reservations one might have about particular arguments in the work, has given us a "sketch" of a Christology which is thoroughly historical, both in its study of the biblical sources and its understanding of the person of Jesus through his humanity, and in how that humanity is revelatory of his divinity in relation to the Father and the Spirit.[7] It is essentially a Christology from without to within (although the author does not stress that point), and it strives to keep together the Jesus of history and the Christ of faith by recognizing that the historical Jesus cannot be isolated from the Christian community which he created and with which he shared his life. Schillebeeckx centers this Christology in the "Abba experience" of the earthly Jesus, that is, his self-consciousness. This consciousness was not a static, but a dynamic and ever deepening awareness of his intimate relation to God as Father, a relationship which was more than that of creature to creator, since it rested on the Father's own giving of Himself in the Spirit to his Son in a unique way. It was this relation experienced by Jesus that was the ground of his total self-giving culminating in the Cross, as servant and savior to all humanity, so that all the actions of his earthly life are revelatory, not only of his sonship, but also of the Trinitarian nature of God. Yet this experience is to be conceived in incarnational, bodily, and hence historical terms, since, although it is original and unique to Jesus, through him and only through him it becomes accessible to other human, bodily, historical beings as adopted sons and daughters of the Father in the Spirit.

Where does this experience come from? Every experience of a human person, granted all its originality, stands at the same time in a tradition of social experience, and is never a simple drawing upon some interior plenitude without any mediatory factors. Jesus' human self-consciousness, like that of every human being, was a consciousness in and of the concrete world of living encounter in which he was set — this, for Jesus, was the Jewish practice of piety, bred and fed by the synagogue, in a family where it was a father's duty to initiate the boys into God's revelation, the Law. Experience of the creator-God, the Lord of history, was partly nurtured in Jesus by the living tradition in which he stood; the living hand of God was apprehended in nature and the world of men. This creaturely consciousness, the living centre of which is the fact of God's lordship, is already noticeable in Jesus' basic message of God's rule. In contrast, however, to John the Baptist and the traditional prophetic legacy, he proclaims not God's eschatological judgement (although he does not suppress that), but God's approaching definitive salvation for man, with a determination that did not falter in face of death. Either this man lived in an illusion — as some are able to say, because after his death history went on its customary way — or we put our trust in him, partly on the strength of his career as a whole and the manner of his dying, a trust which is only possible in the form of an affirmation of God, namely, that God vindicates him. This latter, specifically Christian solution implies in the final instance, the affirmation: the salvation that is coming is this man Jesus himself, the crucified-and-risen One. There is really no middle way: Jesus' method of approaching salvation is either an illusion, or 'it is of God' (Acts 5:35-39); it is true, that is, a reality to be found nowhere else than in the risen Jesus himself.[8]

488

I leave to the reader this painstaking effort of Schillebeeckx and other recent Christologists and turn my attention to one aspect of what the New Testament tells us about the inner experience of Jesus which it seems to me such studies tend to neglect, because the present critical historical methods of exegesis (already discussed in Chapters 9 and 10)[9] find this material more an obstacle than a help in the quest for the historical Jesus, namely the accounts of *visions*.[10] Already in the authentic epistles of St. Paul we have references to his own visions (Gal 1:11-24; 2 Cor 12:1-6), and to the prevalence of prophetic experiences in the Corinthian community, phenomena he seeks to control but not suppress. Since the Fourth Gospel tells us that Jesus promised the abiding presence of the Holy Spirit of prophecy in the Church (Jn 16:13), and *Acts* narrates many such events (e.g. 7:55; 9:10-16; 10:1-6, etc.), it cannot be doubted that visions were an important feature of the life of the early Church. The *Book of Revelation* provides us with a canonical example of such a vision and the *Shepherd of Hermas* an early non-canonical one.[11] Such visions are inner, spiritual events, yet both the canonical and extra-canonical literature represent them as usually manifested in some way to the senses, often to more than one witness, and always as pertaining to the historical and not to the merely mythical order. Even if sometimes they can reasonably be interpreted as imaginative rather than sensible events, their imaginative rather than purely intellectual character humanizes them, and gives them an irreducible relation to bodily reality.

Even before Pentecost we read of the apparitions of the Risen Lord and of the angels at the tomb, and how, at the crucifixion, "Many bodies of saints who had fallen asleep were raised. After Jesus' resurrection they came forth from their tombs and entered the holy city and appeared to many" (Mt: 27:53). It is not suprising, therefore, that Jesus also is reported to have experienced visions, some of which were shared with others, and that his birth was announced in visions which presumedly were related to him as he grew up.

Current exegesis tends to deal with these accounts by reducing them to *theologumena*, defined by Schillebeeckx as follows.[12]

> In broad terms a "theologumenon" means an interpretation having (no more than) a theological value. But this unfamilar word is used only when it is meant to imply that a theological interpretation (a) is to be distinguished from a commonly recognized interpretation, normative for faith, and (b) is also distinguishable from a historically verifiable affirmation.

Schillebeeckx gives as an example of a theologumenon Matthew and Luke's claim that Jesus was born at Bethlehem. St. Paul, however, declares in *Romans* 1:1-4 that he, Paul, was

> . . . called to be an apostle and set apart to proclaim the gospel of God which he promised long ago through his prophets, as the holy Scriptures record — the gospel concerning his Son, who was descended from David according to the flesh but was made Son of God in power according to the spirit of holiness by his resurrection from the dead: Jesus Christ our Lord.

This comes about as close to a creed "normative for faith" as anything in the New Testament. Of course Jesus could be of legal Davidic descent even if He was not born in Bethlehem, but if the genealogies given by Matthew and Luke are to be dismissed as apologetic constructions[13] and their location of Jesus' birth in Bethlehem (as a result of the census requirements that his legal father, Joseph, register in the city of David his ancestor) explained as a theologumenon, what historical context can be found in the Bible for Paul's assertion? If Paul believed that Jesus was descended from David, his belief must have rested on some credible tradition, and what reason is there to think this tradition was any other than the one to which Matthew and Luke's narratives bear witness? Can we be satisfied to leave a matter of faith (or at least proximate to faith) suspended in a historical void? A hermeneutical method that arrives at such results is itself subject to a "hermeneutic of suspicion."

I recall to the reader the point I have already argued in Chapter 9.[14] For the systematic theologian, exegesis does not rest simply on the results of critical historical exegesis (although it should not neglect these results), but on the correctly interpreted assertions of the canonical text in its final form, assertions whose truth is guaranteed, not by historical evidence, but by the inspiration and consequent inerrancy of the text. But (and this unfortunately is sometimes overlooked in current discussions), these guaranteed assertions are sometimes *historical* assertions, and are therefore normative for faith not merely as abstract theological truths but as historical truths. Thus, exegesis has to determine not only whether or not Jesus' Davidic descent is historically verifiable by critical methods (perhaps it is not), but also whether the inspired author intended to assert that it was an historical fact normative for faith, or whether it was for him only incidental (as may be the case for Jesus' birth in Bethlehem, but can hardly be so for his Davidic descent). Too often exegetes today, having undermined the possibility of an historical verification of a biblical statement that appears to convey a historical truth, reduce it to a theologumenon without first deciding whether, even if it cannot be historically verified by the critical historical method, it can be verified as historically true because it is guaranteed by biblical inspiration. Of course this last point is a delicate one since it requires the exegete to distinguish between what the inspired text asserts as a matter of faith or contextual to it and what can be understood as merely incidental to the text's religious message or its literary dramatization. Nevertheless, the fact that an assertion is not historically verified by the critical historical method does not automatically permit the exegete to explain it as merely incidental or as literary dramatization.

Thus in discussing the visions related in the life of Jesus I do not intend to reject the results of current critical historical exegesis, nor to deny that the Catholic doctrine of biblical inspiration and inerrancy is perfectly compatible with a hermeneutic that recognizes in Scripture every honest mode of human communication from the most directly literal to the most indirect and imaginative.[15] Indeed I want to emphasize the value of symbolic interpretation in arriving at the inspired sense of Scripture. But I want to give special attention to the visions narrated in the synoptic gospels prior to the Resurrection (which will be discussed in the next chapter), because it seems to me that these are biblical ways of expressing interior experiences which could be communicated to us in no other way. At the same time I want to avoid the mistake of jumping rashly from the conclusion that these narratives express interior experiences to the very different conclusion that they are theologumena or that they are not rooted in historical fact or, finally, that they are always to be understood as literary constructs of the author intended to express in other terms some truth of faith already known independently apart from the events narrated.

Even from the view of critical historical exegesis it seems strange to me that exegetes are often so quick to interpret accounts of visionary experiences in the New Testament as literary constructs when we have so many parallels which can be historically verified in the lives of later religious communities.

We have, for example, extensive historical documentation on the life of St. Francis and his companions and of St. Catherine of Siena and her circle.[16] This documentation has been subject to something of the same kind of critical historical criticism as has the New Testament with the result that we can observe many of the same literary processes at work. Without a doubt these *legenda* have been shaped in part by the tendency to exaggerate and elaborate and dramatize the historic facts and to bring them in line with conventional literary models, just as critics find for the New Testament narratives. Yet when all these processes are allowed for, it remains historically certain that many of the visions recounted are not the invention of the writers but are reports of experiences which actually occurred to one or many individuals. We may wish to consider them purely subjective experiences of Francis or Catherine, but it is an objective historic fact that Francis and Catherine related these experiences to others and that sometimes others claimed to have shared in these experiences, so that they are not merely literary constructs of the *legenda* writer.

Why then should historical exegetes be so quick to interpret such a narrative, for example, as that of the Transfiguration not as an experience of Jesus which he related to his followers and in which perhaps some of them reported that they had shared, but as a dramatization of the early Church's faith in the Resurrection, retrojected into the life of Jesus by a literary device?[17] Is this perhaps a reluctance on the part of exegetes to attribute to Jesus, whom they revere, the characteristics of mystics of whom they are suspicious?

Visions are sometimes purely interior experiences, but they are not merely spiritual.[18] Rather they relate inner life to outer life, the spirit to the body, since they have to be communicated in symbols derived from bodily experience. Moreover, they are often experienced as actual, physical events, as miracles which can be publicly verified at least by the privileged witnesses. Thus all the miracle narratives of the New Testament also have a relevatory aspect, since Jesus did not perform miracles simply to astonish the crowds, but as revelatory signs. Since these miracles, however, were centered on the Reign of God rather than on Jesus' own identity, with which I am here concerned, I will pass over them and focus on those narratives of visions which seem to mark important steps in the development of his own self-understanding.

The chief of these visions are (1) those of Jesus' baptism and temptation in the desert (Mk 1:9-13; Mt 3:13-4:11; Lk 3:21-4:13); (2) the transfiguration (Mk 9:2-10; Mt 17:1-8; Lk 9:28-36); (3) the vision of the fall of Satan on the return of the seventy-two disciples (Lk 10:17-20); (4) the eschatological discourse (Mk 13:1-37; Mt 24:1-35; Lk 21:5-36); (5) the Last Supper (Mk 14:12-31; Lk 22:7-38); (6) the agony in Gethsemane (Mk 14:32-42; Mt 26:36-46; Lk 22:39-46). To these we must preface the visions related in the infancy narratives of Matthew 1-2 and Luke 1-2 which, of course are not visions experienced by Jesus himself (except perhaps Luke 2:41-52), but which are directed to establishing his identity.

2. The Annunciation of the Virginal Conception

Although Matthew is unequivocal about the virginal conception of Jesus by Mary through the power of the Holy Spirit, his purpose does not seem to be so much to present this as a revelatory sign as to provide an apologetic answer to the objection that the belief of Christians in this miracle, an accepted tradition in the circles for whom he wrote his gospel (evidently current or his argument would have lacked force), was contradictory to the claim that Jesus was a descendant of David and therefore a possible Messiah, a claim that was of supreme importance for his Jewish Christian readers, and which was attested already by St. Paul (Rom 1:3-4, cf. II Tim 2:8). Current scholarship no longer attributes the origin of this tradition to the prophecy of Isaiah 7:14 which Palestinian Jews probably did not understand as referring to such a conception, but believes that Matthew added this text to his source to further his apologetic and other theological purposes.[19]

Luke, independently of Matthew, takes up this same tradition, but no longer merely to counter the Christological embarassment it caused for Matthew. Instead for Luke this miracle is rich in positive Christological meaning. According to Fitzmyer,[20] Luke's Christological use of this tradition is primarily intended to assert that Jesus is truly Son of God by the power of the Holy Spirit, but secondarily it also has a Mariological significance, namely to picture Mary, the Mother of Jesus, as the model of Christian discipleship. Thus, if we leave aside this secondary theme, the question becomes "How does Mary's vision reveal the true identity of her son?"

Today many Christians find the notion that Jesus was miraculously conceived without a human father not only meaningless but contradictory to the first principle of any "Christology from below", namely, that Jesus was truly and fully human. Moreover, they see it as implying that sex is somehow evil or at least degrading, or they regard it as anti-feminist in its exaltation of virginity for women. Undoubtedly the patristic and medieval Platonic Christian Theology *did* exploit some of these dualistic themes in the interpretation of this vision, but the modern interpretations are even more anachronistic in attributing to Luke notions about sexuality foreign to the biblical mentality. Other sceptics point out that the virginal conception of Jesus is no adequate proof that he was the Son of God in the Chalcedonian sense, and that metaphysically speaking, it would seem that the Second Person of the Trinity could have become incarnate through a human father. But this objection also is raised from a point of view foreign to St. Luke.[21]

As shown in Chapter 10, for the ancient Jews as for most ancient peoples, sexual power was considered as the most fundamental kind of political power because on it depended the survival of the nation. To be sterile was to be accursed of God. Yet at the same time it was acknowledged that this power of sexuality was a gift of God which could not have its effect without His active intervention. In *Genesis* 4:1 we read of the very first human birth, "The man knew Eve his wife, and she conceived and bore Cain, saying, 'I have given birth to a man-child *with the help of the Lord.'*" Thus the origin of every human being, like that of the first man and woman, requires the creative action of God, as well as the sexual activity of the parents. Consequently in the case of the great leaders and prophets called by God for the salvation of His people, the Old Testament often dramatizes this intervention by relating that their mothers were sterile and (in the case of Abraham, the father was also impotent, Genesis 17:17). Only the direct intervention of God removes this incapacity (Isaac, Gn: 18:9-15; Jacob, Gn. 25:21; Samson, Judges 13:1-7; Samuel, 1 Sm 1:1-28 and in the New Testament, John the Baptist, Lk 1: 5-25).[22]

Therefore, Luke uses the tradition of Jesus' virginal conception not to prove that Jesus is the Son of God, since it is Jesus' resurrection alone which establishes this definitively, but to show that Jesus is called to a saving mission which *absolutely* (and not merely as a matter of degree) surpasses that of all the previous leaders and prophets of Israel. God has intervened in the call of all these his former spokesmen, but in Jesus his intervention is not merely a restoration of Israel's failing powers, but a new beginning. As St. Paul had taught that Jesus is the New Adam (1 Cor 15:21-49, Rom 5:15-17), so Luke remembers that Adam was made directly by God out of the dust of the earth by the power of the creative Spirit, and portrays Jesus as directly created by God out of the flesh of Mary, herself a type of Israel. In the words of St. Paul, "God sent forth his Son born of a woman, born under the law" (Gal 4:4), i.e. out of the flesh of the People of the Covenant. In this way Jesus was truly an Israelite, and yet the Israelites could not pride themselves on this claim as if it gave them an exclusive right to salvation. Luke has John the Baptist say, "Do not begin by saying to yourselves, 'Abraham is our father.' I tell you, God can raise up children to Abraham from these stones" (3:8). Thus for Luke, who is especially concerned in his Gospel to show the continuity of salvation history in the Old and New Testament and in the Age of the Church, and yet also to show the universal salvific significance of the Christ event, the virginal conception shows the continuing creative action of the Holy Spirit preparing the way for Jesus, culminating in his coming, and flowing on in the adoption of Christians from all nations to share in his Sonship.

The virginal conception for Luke's theology, therefore, is a revelatory sign whose significance is that the Sonship of Jesus in relation to God is of a different and higher order than any of the ways in which in the Old Testament men were called "sons of God", because he has been called by God from the beginning of his existence for an ultimate mission, the *effective* proclamation of the good news of salvation open to all. For this mission, Jesus has been conceived by no human power but by the Spirit alone and, as he announces in his first sermon in his home town of Nazareth, "The spirit of the Lord is upon me; therefore has he anointed me. He has sent me to bring glad tidings to the poor . . . to announce a year of favor from the Lord" (4:18; Is 61:1-2). Dependency on a human father for his life (whether this would have been metaphysically possible or not is no issue for Luke), would have negated or at least obscured the revelatory sign given by God to help us understand the deeper significance of these events if, like Mary, we treasure "all these things and reflect on them" in our hearts (2:19). It is noteworthy that this annunciation vision is Trinitarian in the sense that the message comes from God as from a father concerning one who is to be called Son by the power of the Holy Spirit. Many commentators see in the fact that the spirit is to "overshadow" Mary a reference to Genesis 1:1 where the creative "Spirit of God hovers over the waters" like a bird hatching its eggs.[23] The presence of the angel Gabriel (Dn 9:21), announcing the coming of the Reign of God) as well as of the angel hosts at the birth of Jesus (2:13), in counterpoise to Luke's account of the temptation of the adult Jesus by Satan (4:1-13), indicate the cosmic significance of this event which is to restore the whole of creation under God's reign.

The antecedent events (in Luke's narrative annunciation and birth of John the Baptist) and the subsequent ones (visitation of Mary to Elizabeth, the births of John and Jesus, the apparition of the angels to the shepherds, the presentation of Jesus in the temple, the prophecies of Simon and Anna, and the finding of the boy Jesus in the temple) — all these events involve visions, except the last. This last event, because of Jesus' reply to the question of his puzzled parents, "Why did you search for me? Did you not know I had to be in my Father's house" (2:49, the first words of Jesus in Luke's Gospel), also implies on Jesus' part already at the age of twelve a knowledge of his identity as Son of God in a unique sense which has previously been made known to the reader through these visions.[24]

Does Luke imply that Jesus knew this by some vision of his own, or that he had learned it from Mary? Against the latter is the fact that Luke adds "They [Mary and Joseph] did not grasp what he said to them." (2:50). It would seem, therefore, that for Luke, Jesus by reason of his Sonship possesses a hidden knowledge, yet this does not prevent Luke from picturing Jesus as "listening" and "asking questions" of his elders, and Luke adds "Jesus, for his part, progressed steadily in wisdom and age and grace before God and men" (2:52), while Mary observes and remembers all these events and trys to understand their full meaning (2:52). It seems clear, therefore, that Luke portrays Jesus as knowing his own identity

495

as Son of God from his early years, while to others, especially his mother Mary, this was also made known by revelatory signs, but signs whose full significance was still obscure to them, just as in Luke's Resurrection accounts (chapter 24), he represents the apostles as only gradually coming to an understanding of the meaning of the great event which they had witnessed but did not comprehend. Nevertheless, without the revelatory sign of the virginal conception we, as readers of Luke' account, could not understand the meaning of the boy Jesus' cryptic statement. Without Mary's witness to the miraculous character of her motherhood, the whole of Luke's account would seem to fail of its manifest intent to assert the ultimacy of Jesus' mission.

3. Baptism, Temptation, Transfiguration, Sacrifice

Mark, who omits any account of Jesus' birth or early years, portrays him as aware of his divine Sonship at the moment of his baptism in a vision of the opening of the heavens, the descent of the Holy Spirit as a dove, and the voice of God the Father saying, "You are my beloved Son. On you my favor rests" (Mk 1:11), and the other synoptics give much the same account (Jn 1:32-34 says that the Baptist shared this vision).

The most enigmatic feature of this vision is the appearance of "the Spirit desending on him like a dove" (Mk 1:10). Two explanations of this have been given: (1) the dove is "the spirit of God hovering above the waters" of *Genesis* 1:2; (2) the dove is the People of God (Ps 68:14) as the greatest creation of the Spirit.[25] Both senses are appropriate, since baptism is a "new creation" (2 Cor 5:17; Gal 6:15), and in Christ "everything in heaven and on earth was created . . . all were created through him and for him . . . he who is the beginning, the first-born of the dead" (Col 1:15-18); and on the other hand the People of God, the Church, is also born of the Spirit in baptism and incorporated in the Risen Christ. Hence in this vision Jesus becomes aware in some new way of his Sonship to God and his mission to preach the reign of God, and the renewal of God's people in the Spirit which illuminates him and empowers him for his mission. He can announce the coming of the Reign of God here and now and not, as did former prophets, merely in the remote future or even, as John the Baptist announced it, as imminent; because that Reign begins in himself, in his immediate relation of Sonship to God. As a boy he had recognized the temple as his rightful home, but now he begins to see that he himself is that temple, the dwelling place of God's Spirit on earth (Jn 2:21).

According to Mark, immediately after the baptism "the Spirit sent him out toward the desert. He stayed in the wasteland forty days, put to test there by Satan. He was with the wild beasts, and angels waited on him" (1:12-17). Matthew and Luke fill out this account by relating what this threefold temptation was. According to Luke, it was first a temptation for Jesus to satisfy his hunger by a miraculous conversion of the stones into bread, second to obtain earthly empire in return for worshipping Satan as a god, and third to call on the miraculous aid of angels by leaping from the pinnacle of the temple (no doubt in the sight of the worshippers gathered there). Matthew, whose order may be the more original, reverses the second and third miracle, but Luke wants to end with a return to the temple theme. These temptations all involve a distortion of what Jesus' Sonship and his mission really mean, and seem to show that just as his mother Mary had to meditate on the events she witnessed to understand their full meaning, so Jesus himself retired into solitude to reflect on the full implications of his baptismal vision, to clarify its meaning by rejecting the false interpretations offered him by the Father of Lies. As indicated in the last chapter these temptations relate especially to Jesus' human condition: the need for bread, the search for power, the ostentatious display of power reminding us of Satan's temptation of Eve, "you shall be as gods."

Thus prepared with a clear understanding of his mission, Jesus began to preach the coming of the Reign of God and to confirm its presence in his person by his miracles, including the casting out of demons. Soon, however, opposition against him mounted, taking the form of attributing his miracles to Beelzebul rather than to the Spirit of God (Mk 3:20-30). At this time, when the threat of death and the destruction of his mission began to loom, and in connection with his thrice repeated prediction to his disciples of his coming passion, Jesus along with his three most intimate disciples, Peter, James and John, experienced the vision of his transfiguration with the apparition of Moses (representative of the Law) and Elijah (representative of the prophets). Since Moses and Elijah were believed by the Jews, on the basis of certain interpretations of the Old Testament narratives, not to have come to an ordinary end but to have been assumed (Dt 34:5-6 and 2 Kgs 2:11), their presence with the glorified Jesus is a prediction of the Resurrection, but it also connects the Resurrection with the baptismal vision, because "A cloud came, overshadowing them, and out of the cloud a voice: 'This is my son, my beloved. Listen to him.' (Mk 9:7-8). Again we have the voice of the Father acknowledging the Son through the "overshadowing cloud", i.e. the Holy Spirit which "overshadowed" (same word) Mary in the annunciation. What is unique to this vision is the glorification of Jesus' body, especially of his face (Matthew and Luke, but not Mark), reminding us of the glory on Moses' face when he returned from the theophany of Sinai (Ex 34:30) where he too had been overshadowed by the cloud.

The sending forth of the seventy-two disciples (as distinguished from the Twelve) and their return is unique to Luke. When these seventy-two rejoice in their power over the demons, Jesus says, "I watched Satan fall from the sky like lightning. See what I have done; I have given you power to tread on snakes and scorpions and all the forces of the enemy, and nothing shall ever injure you. Nevertheless, do not rejoice so much in the fact that the devils are subject to you, as that your names are inscribed in heaven "(Lk 10:18-20). Here Jesus clearly speaks of a vision, and Luke immediately adds to this the famous "Johannine" passage which is probably from Q, but which Matthew places after Jesus' complaint about the lack of faith he has found in Chorazin and Bethsaida (Mt 11:25-30).[26] In Luke's version (10:21-22) it reads:

> At that moment Jesus rejoiced in the Holy Spirit and said, 'I offer you praise, O Father, Lord of heaven and earth, because what you have hidden from the learned and the clever you have revealed to the merest children. Yes, Father, you have graciously willed it so. Everything has been given over to me by my Father. No one knows the son except the Father and no one knows the Father except the Son — and anyone to whom the son wishes to reveal him.

Since Luke connects this saying so closely with the vision of Satan's fall by the theme of "rejoicing" it is possible that this vision was also in Q, although omitted by Matthew. In any case for Luke the vision of the fall of Satan is related to Jesus' "rejoicing" in the Holy Spirit as He joyfully praises the Father because of his own Sonship (again the Trinitarian motive as in the annunciation, baptism, and transfiguration). Here Jesus expresses the "Abba" experience in its fullest form in a way similar to the great discourses attributed to him in the Johannine Gospel. He has received "everything" from the Father, i.e. the dominion promised falsely to Jesus by Satan in return for worshipping him. It has truly been given Jesus by the Father, in return for his humility, his identification with the "little ones", because "No one knows the Son except the Father and no one knows the Father except the Son — and anyone to whom the Son wishes to reveal him." These words not only imply that the Father has revealed himself to Jesus so that he can reveal the Father to others, but also and more wonderfully that "no one knows the son except the Father", i.e. that the identity of Jesus is a mystery beyond all human comprehension, which can be known only by humble faith, by the "little ones." To use the technical language of theology, it is a mystery *essentially supernatural.*[27] The cause for joy is not the power to do miracles, but that the humble have been admitted to the secret of the saving plans of God, while the powers of evil have been cast down by these humble, powerless ones.

The eschatological discourse in the form which it takes in Luke (21:5-27), where a clear distinction is made between Jesus' description of the last judgment and his prediction of the more immediate destruction of the temple as a type of that last judgment, can also be called a "vision" because it is a vivid portrayal of events which Jesus could only have known through the prophetic Spirit. It is noteworthy also that according to Mark (13:3), Jesus revealed this vision to the same three of the Twelve (with the addition of Andrew) who had shared the transfiguration vision. The feature of this discourse which is important for our purpose is Jesus' reference to the Son of Man prophecy of Daniel 7:13-14 (Lk 21:27-28,36). To this can be added Luke's "little apocalypse" (17:20-37) which may be from Q and in which the Son of Man is also the one whose coming will bring history to its close. There are, of course, many problems and much controversy over these eschatological discourses and the authenticity of Jesus' use of the Son of Man title as applying to himself.[28] There can be no doubt, however, that Mark (14:61-62) represents Jesus at his trial before the Sanhedrin answering the question of the high priest, "Are you the Messiah, the Son of the Blessed One?" with the reply, "I am; and you will see the Son of Man seated at the right hand of the Power and coming with the clouds of heaven," and this identification by Jesus of himself with the Messiah, Son of God, and the Danielic Son of Man provides the most obvious and consistent explanation of the various synoptic uses of these terms because the Christian community so identified them.

Thus the synoptics represent Jesus at the end of his earthly ministry as convinced that he was Son of God in a unique sense. Moreover, this entails that he, along with the Twelve ("In my kingdom you will eat and drink at my table, and you will sit on thrones judging the twelve tribes of Israel" (Lk 22:30; Mt 19:28), will sit at the side of the Father and judge the whole world. Yet his way to this triumph will be one of suffering, as indicated by the lowly title Son of Man, which in Hebrew simply means a member of the human race.

The incidents at the Last Supper and the institution of the Eucharist are related by the synoptics with many variations, but they certainly contain a visionary element.[39] Jesus, with seeming clairvoyance, tells the apostles where to find a place for the passover (Mk 14:12-16), prophesies "I will drink no more of the fruit of the vine, until that day when I shall drink it new in the kingdom of God" (14:25), detects the plot of Judas, and predicts his desertion by the apostles and his denial by Peter. In the Fourth Gospel (13-17), we have the great discourse of Jesus in which he reveals his most intimate sense of his Sonship and promises the coming of the Holy Spirit. In the following scene of the agony in Gethsemane (Mk 14:32-42), Jesus withdraws with Peter, James, and John (the witnesses of the transfiguration), and prays "Abba, Father, you have the power to do all things. Take this cup away from me. But let it be as you would have it, not as I"; and to this account, common to all the synoptics, Luke adds, "An angel then appeared to him from heaven to strengthen him" and "his sweat became like drops of blood" (22:42-44). The imagery of the cup and bloody sweat seems to connect this incident with the symbolism of the eucharist related shortly before. Thus, even in this dark hour, Jesus is conscious of his Sonship and understands that the offering of himself body and soul as victim is the unavoidable way to the triumph of the Kingdom. The testing by Satan and the consolation of angels in his trial experienced at the beginning of his ministry is repeated at the end.

4. The Historical Authenticity of these Experiences

Are these synoptic narratives of the stages by which Jesus' Abba experience deepened rooted in historical fact or they are theologumena?

In approaching this problem, we must first note that only in one of these passages does there seem to be any explicit acknowledgement that Jesus is not fully aware of all details of his destiny. At the close of the eschatological discourse, Mark (13:32; Mt 24:36, but omitted by Luke) reports Jesus as saying, "as to the exact day or hour, no one knows it, neither the angels in heaven, nor even the Son, but only the Father." Does this contradict the Q statement (Lk 10:21-22; Mt 11:25-27), that "no one knows the Father but the Son"? It seems to me that the traditional interpretation of "nor even the Son", which avoids any such contradiction, is correct. Jesus is not declaring his ignorance, but only that to answer this question does not fall under the scope of his mission from the Father. The text limits his mission rather than his actual knowledge concerning which it says nothing. It does not even deny that Jesus had the beatific vision, since the angels enjoy that already, yet it is not their mission either to reveal "the day or the hour".

The evangelists are not especially interested in Jesus' psychological development, just as they are little concerned with exact chronology. Nevertheless, taken together, these vision texts are related to the main events of Jesus' ministry and seem to show a certain unfolding and deepening of his self-understanding as that ministry progressed, especially in the sense of its inevitable tragic ending. This development, however, is always an unfolding of the same identity, of Jesus' constant experience of his intimate relation to God as a unique Son to the Father in which also is implied a mission empowered by the Spirit. If we are to accept the infancy narrative of Luke as having a historical basis, we must also conclude that Jesus gave evidence of this awareness as a teen-ager and that its ultimate origin in his consciousness was fostered in him by the way he experienced the relation between his mother Mary and his legal father Joseph.

For Luke the virginal conception of Jesus is of deep theological significance since it provides a clue to the better understanding of all the visionary episodes that follow and for which it provides the basic pattern. From a psychological point of view (which is hardly that of Luke), this makes sense. The lives of many Christian saints and what we know of the religious psychology of children is clear evidence that some children have remarkable experiences of their relation to God and dreams about their call to His service.[30] It is also well known that children wonder about their real parentage and their true identity. Finally, it is clear that parents, and particularly mothers, often share their pre-natal hopes and dreams for their children with those children after their births, and that such impressions may lay the groundwork of any child's personal image. "Jesus' human self-consciousness, like that of every human being," as Schillebeeckx says in a passage already quoted, "was a consciousness in and of the concrete world of living encounter in which he was set — this, for Jesus, was the Jewish practice of piety, bred and fed by the synagogue, in a family where it was a father's duty to initiate the boys into God's revelation, the Law," but even in patriarchal Jewish families a mother might be a determining factor in her son's life, like the mother of James and John who pleaded with Jesus that her sons might be first in his kingdom (Mt 20:20-25). After all, it is often the mother's attitude to the patriarchal father that determines the relation of her sons to that father and his influence on them.

My point is not to attempt a psychohistory (dubious even when our information is extensive) of Jesus, concerning whose inner lifer we have only the limited data just surveyed, but to make clear that he must have had such a psychohistory and that it is important for our understanding of *his* self-understanding whether what Luke tells us is historical. Unfortunately, two very distinguished Catholic exegetes, Raymond E. Brown and Joseph A. Fitzmyer have recently concluded after careful study that it probably is not. Fitzmyer summarizes the situation as follows:[31]

> Whereas in the Pauline and probably pre-Pauline use of the formulation, the resurrection of Jesus was the moment when the title Son of God became attached to him, Luke pushes the christological affirmation back to the conception of

Jesus. What is involved here is the growing understanding of the early church about the identity of Jesus. Though first such titles as Son of God were attached to him primarily as of the resurrection (besides Rom 1:41; see Acts 13:33), the time came when early Christians began to realize that he had to have been such even earlier in his career, even though it had not been recognized. It is not so much that the "christological moment" . . . was pushed back as that there was a growth of awareness as time passed among early Christians that what Jesus was recognized to be after the resurrection he must have been still earlier. Luke, in affirming that Jesus was Son of God, not only at his conception, but through his conception, is representative of early Christians among whom such awareness was achieved. Still later, in the Johannine community, the awareness will grow into the idea of incarnation — a notion foreign to Luke (as to Matthew). This, then is the primary import of this passage, its christological affirmation: the announcement to Mary identifies Jesus to the reader of the Lucan Gospel as the Davidic Messiah and the Son of God.

Brown writes:[32]

In my book on the virginal conception, written before this commentary, I came to the conclusion that the *scientifically controllable* evidence [note: I mean the type of evidence constituted by tradition from identifiable witnesses of the events involved, when that tradition is traceably preserved and not in conflict with other traditions] leaves the question of the historicity of the virginal conception unresolved. The resurvey of the evidence necessitated by the commentary leaves me more convinced of that.

Both commentators agree that Matthew and Luke are independent witnesses of infancy traditions older than their gospels, which they date around 80 A.D. The two infancy narratives are very different and can be harmonized only very artificially, but they do agree on many points (Fitzmyer, p. 307, lists twelve) including the virginal conception. What was the origin of this tradition? Laurentin and others[33] have argued that it must go back to family traditions, but Brown is unconvinced. Our earliest witnesses to the facts of Jesus' public ministry, Paul, Mark and even John and the preaching in *Acts*, are silent on Jesus' infancy. The synoptics seem to show (Mt 13:54-58; Lk 4:31-32,36-37; Mk 3:21-22,31-35; 6:1-6, cf. also Jn 7:5), that the people among whom Jesus was reared had no knowledge of his extraordinary birth. Since Joseph was dead, only Mary could have provided this information, but the passages cited by Laurentin (Lk 1:66; 2:19; 51) about Mary pondering over these events are borrowed from Gn 37:11 and Dn 4:28 (LXX) where they only signify an effort to interpret dreams. In view of the inconsistency between Matthew and Luke's narratives, and the way in which both accounts seem constructed from Old Testament materials, it seems these accounts should be regarded not as history but as symbolic narratives whose purpose is theological, namely to communicate in a different way the truth of Jesus' divine sonship, his passion and resurrection, truths which in no way depend on the historicity of the virginal conception.

Neither Fitzmyer or Brown, of course, rejects the creedal statement "conceived by the Holy Spirit, born of the Virgin Mary", but Brown raises the question as to whether the essential meaning of this affirmation includes a physically miraculous birth.[34] Schillebeeckx is even more blunt: "the earlier life of Jesus, prior to what we can recapture, historically, of his ministry, is almost completely unknown to us. His birth in Bethlehem is a Jewish *theologumenon*, that is an interpretative vision and not a historical assertion".[35] These authors, however, do not mean thereby to question the doctrine of biblical inspiration since, as Brown correctly says, this doctrine admits of "non-historical theological dramatizations".[36] Consequently, Fitzmyer concludes his discussion by saying,[37]

> . . .this commentary is concerned with Stage III of the gospel tradition; there is, as far as I am concerned, no real proof for or against the fact of the virginal birth in Stage I. Christian belief in it is governed by factors other than what one can ascertain by careful exegesis.

But what might these factors be? It is well-known that since Vatican II, theologians are not inclined to appeal to extra-biblical tradition,[38] so it is difficult to see how if "careful exegesis" cannot find some basis for the virginal conception in the Scriptures, it can still be asserted as an article of faith. Certainly Brown and Fitzmyer are to be praised for their objectivity in following their method to its conclusons, even in the face of this theological difficulty, but as far as I can see they have not directly faced that difficulty and have simply passed it on to the systematicians.

To deal with this difficulty, I would first point out that Brown does not doubt that Luke himself believed this tradition, but Brown finds no problem in assuming that Luke may have been mistaken in this belief.[39] Of course Luke may have been mistaken about the historicity of the virginal conception, if in narrating it he did not intend to assert that its historicity was an element in the Good News and therefore to be believed by divine faith, as, for example, he clearly intends the historicity of the crucifixion to be. But if Luke intends to relate the virginal conception as an actual event through which God made his purposes known in a special way, then it cannot be reduced to a "non-historical dramatization", but the truth of Luke's admitted belief will be guaranteed by biblical inerrancy, whether we can establish by historical methods how Luke verified that fact or not. The real question, therefore, that a "careful exegesis" must ask (and I do not find that these authors very directly ask it), is whether Luke asserts the virginal conception as an historical fact to be believed by Christian faith. That he *does* seems to me to be evident from the fact that (as I have attempted to show in the foregoing) it is for him a *revelatory sign* which establishes something about Jesus' ultimate mission which could otherwise not be known or at least known so fully. It cannot be, therefore, merely a "non-historical dramatization" of some theological truth known in some other way (e.g. as a deduction from the truth of the resurrection), as these exegetes suggest.

How did Luke, guided by inspiration, certify to himself that the tradition about the virginal conception was reliable and therefore dare to use it in his gospel as a truth to be believed (while not hesitating to dramatize it with that literary freedom that is compatible with inspiration)? One can concede to these scholars that we can only speculate about the sources of this tradition, since by the very nature of the case our information about Jesus' early life is not as "scientifically controllable" as that about his public ministry and death. For apologetic purposes, arguments for the historicity of the virginal conception can never be very strong. Nevertheless, must we concede that the arguments against its historicity are as serious as these exegetes believe? "The real difficulty about a preserved family (Marian) tradition is the failure of that memory to have had any effect before its appearance in two Gospels in the last third of the first century," says Brown.[40] Yet if we admit Fitzmyer's point in the previous quotation that "there was a growth of awareness as time passed among early Christians that what Jesus was recognized to be after the resurrection he must have been still earlier," it is not hard to understand that the earliest witnesses avoided including in their writings elements of the tradition about which there was still little *public* understanding.

Paul's epistles give us almost no information about Jesus' life except his death and resurrection, not even his baptism by John, which is certainly a historical fact, both because that was not the purpose of the epistles and because Paul's own theology focuses simply on that great fact of the Paschal Mystery which seems to have been the content of his own conversion vision. As for Mark, exegetes today generally admit that

504

his brief Gospel is highly selective and that his selection of material is based on his theological purpose, a purpose which accounts for the fact that both the beginning and end of his Gospel are so abrupt. One of his theological concerns, according to some, seems to be to counter the dynastic pretensions of Jesus' relatives in the Jerusalem church.[41] If in fact this was the case, it is not surprising that he passes over family traditions. As for the argument that the Gospels show that the relatives and neighbors of Jesus did not believe in him until after the resurrection, the same is true to a degree even of the apostles who witnessed his miracles! We can be sure that if there were family traditions about Jesus' birth and early years, they would be taken seriously only *after* those who had heard them had become Christians following Pentecost, in which case they probably were mixed with much that was rumor and popular exaggeration, as such stories about a famous man's early years always are. Because of this fact, the early kerygma may well have avoided such topics as more embarassing then convincing. Was it not hard enough to convince people of the Resurrection, without trying to convince them that Mary was a virgin mother!

After the Christian community was well established the interest in such traditions increased, as can be seen by the apocryphal infancy narratives;[42] but the teachers of the Church also had a problem of sifting the wheat from the tares, and of course this could not be done by a modern critical historical method, any more than was possible in the formation of the canon of Scripture.[43] Luke's credit as an historian is not very high today among critical exegetes; yet as Fitzmyer himself has well shown,[44] he was not void of either historical or theological acumen. The Church in such matters, however, and Luke is very much a churchman, uses as its chief criterion (again witness the history of the canon), one which is excluded by the historical method as many modern exegetes practice it, namely the *analogia fidei,* i.e. the inner consistency of salvation history. Are not the probabilities in favor of the speculation that Luke verified the essential elements which he took from the traditions then current in the Church by accepting only what was so congruent with the informed Christian faith as he knew it that he was quite confident it could be taught as something to be believed? Did not the Church canonize his Gospel as a result of the same kind of process, and is it not the proper, final step of "careful exegesis" to determine just what the canonical text asserts to be historically true in the sense of a revelatory event that "really happened"?

In my opinion the most likely explanation of this tradition about the virginal conception of Jesus for which modern exegesis seems to provide no better solution, is that it originated with Jesus himself who learned it from his mother in an intimate sharing of religious experience which they enjoyed together from his childhood, as Augustine after his conversion came to share with Monica in an adult way. We have either to suppose that Jesus came to the Abba experience only after a conversion in youth or manhood, or we must suppose, as Karl Rahner has well argued,[45] that this experience was intrinsic to Jesus' self-consciousness and developed as he grew up just as does the self-consciousness of any other human being. As anyone's childhood consciousness of self-identity is *relational* — ''I am my mother's and my father's child'', so for Jesus this included a gradually deepening awareness of being Son of God and Son of Mary in the Spirit of faith, which is the reality expressed by the doctrine of the virginal conception, not simply as a theological proposition, but as an experienced, concrete, historical event.

It is true that the Gospels never record an occasion on which Jesus speaks about this to his disciples. Yet, as I have argued above, it is probable that he shared a number of his inner experiences with his more intimate followers, as the mystics have done with their confidants throughout the centuries. Some of these are recorded in the Gospels as visions, no doubt with a degree of literary stylization and dramatization. As we have seen, these visions center on the Abba experience, and in the light of such memories the Christian community or certain circles of that community (most probably the Jerusalem church) came to the conviction of the truth of rumors about Jesus' miraculous origin, rumors which gave rise to attacks which Matthew in his gospel is concerned to answer, but which Luke is more concerned to use theologically, only because guided by the inspiration of the Holy Spirit, he recognizes in them gospel truth congruent with better known elements of tradition. Yet Luke does not merely repeat those elements, but clarifies them by additional, authentic information.

II: The Body of Christ

1. True Man

The inner life of Jesus was not only expressed through his body but was rooted in his bodily life. If one accepts an Aristotelian epistemology, then Jesus' psychological development in which he "progressed steadily in wisdom and age and grace before God and men" (Lk 2:52) must be conceived in terms of his bodily interaction with the world about him. The medievals thought that since Christ is not only the New Adam but also the source and cause of the human elevation to eternal life, he was even in this life both *viator* and *comprehensor,* i.e. like us he was journeying to his final union with his Father, but that somehow he already possessed this union. Consequently, they distinguished three kinds of

506

knowledge possessed by him: (1) the immediate face-to-face vision of his Father, such as the Christian hopes to attain in heaven; (2) ordinary human knowledge acquired through experience and reflection such as we have; (3) an "infused" knowledge intermediate between these like that possessed by the prophets. Aquinas insists on Jesus' experiential knowledge, in addition to the other two kinds, but argues that Jesus was always a teacher rather than a pupil, i.e. he was a creative genius.[46]

Current theology, of course, affirms Jesus' experiential knowledge and seems to admit that he also had prophetic or infused knowledge, that he was inspired by the Holy Spirit, but has generally abandoned the notion that he had the beatific vision in this life, because it seems contrary to the gradual development of his own self-identity and knowledge of his mission, and his frank admission of the limitation of his knowledge. I have already pointed out that in fact there is actually little biblical evidence for any limitations on his knowledge,[47] but I will return to this point later. How are we to conceive the relation between his infused or prophetic knowledge and his experiential knowledge? In the Old Testament, prophetic knowledge seems ordinarily depicted as an enlightenment which comes to the prophet unexpectedly and only at times.[48] What we know of the great mystics, on the other hand, seems to show that there is a distinction between such occasional illuminations and a profounder union of the intellect and will that occurs in the highest mystical state when the mystic is habitually under the guidance of the Holy Spirit through the Spirit's permanent gifts.[49] Certainly, we ought to suppose that Jesus, in whom the prophecy of Isaiah (11:1-2) about the anointing of Emmanuel by the Spirit applied even before the Resurrection, was endowed from his baptism with the Spirit's gifts, since these gifts were given for his ministry. If Luke's comparison between Jesus and John the Baptist, whom Luke pictures as filled with the Holy Spirit in his mother's womb, is to guide us, *a fortiori* we should suppose that Jesus was so illumined from birth.

Yet in accordance with my Aristotelian epistemology I would prefer not to conceive this illumination in terms of Platonic innate ideas, but rather as an elevation or deepening of Jesus' ordinary, experiential knowledge. There is no need to think of the prophets as persons who were inspired only by hearing voices or seeing apparitions. The voices and visions described in Scripture (and many of those described in the lives of the mystics) may often be dramatizations of intuitions, creative insights which cannot be expressed except by such dramatization. In Chapter 8, I showed that creative intuition works by the capacity of the human intelligence, precisely as intelligence and not only as discursive reason, to attain to the essential in experiential data and to unify this essential information analogically. A Shakespeare or an Einstein uses the same experiential data which is available to all of us, but such geniuses see clearly the significance of that data which most of us miss. The prophet similarly has a heightened intuitive awareness of the meaning of the events of ordinary life, an awareness which is "inspired" in the sense that he rightly attributes this awareness to a gift of God which surpasses his own talents. Thus prophetic inspiration is *essentially* not the infusion of information in the prophet's mind but a strengthening of his intuitive capacities (the "light" of his mind) so that he perceives in his experience what otherwise would have escaped him. The Old Testament prophets, therefore, do not exhibit a knowledge of historic facts not known to their contemporaries but rather a deeper understanding of the providential significance of these facts.

Jesus' "infused knowledge", therefore, ought to be understood not so much in terms of his possession of all the information in an encylopedia, but rather as a profound penetration of his ordinary human experience, possible for him because of his intimate relation to God as His Father. A scientific genius who has hit upon a great idea, a clue to the laws of nature, begins to see that this clue will unravel countless mysteries of nature. Similarly Jesus, because "No one knows the Father except the Son — and — and anyone to whom the Son wishes to reveal him" (Lk 10:22), had the clue to the ultimate mysteries of nature and history in his profound understanding of who God really is.

As we have seen, Jesus' knowledge of God as his Father was with him from childhood, from the breast of his mother, just as every child develops in self-consciousness through its relation to its parents. If Luke is right, it was Mary's own unique experience of motherhood that provided the special milieu for Jesus' unique Abba experience, yet this uniqueness was not something that separated Jesus from the ordinary course of human experience, but rather rooted him more profoundly in it. It opened up for him the inner secret of human history which is within each of us, but hidden from us.

Jesus' experience was that of a male child in the town of Nazareth at the time of the Herods and the first Caesars of the occupying Roman Empire. The more we know about that period the more we realize that although Nazareth was an insignificant village, it was in a Galilee which was a point of intersection of many cultures, and we need not suppose that either Jesus or his family were ignorant of the great world. He probably knew three languages, perhaps four.[50] But his daily life for his first thirty years was that of a carpenter's family and he himself worked with his hands at the bench.[51] His education was that of a pious Jew, i.e. he could read and write but his learning was strictly in Torah, the Scriptures, in which Joseph, and no doubt some local teacher, must have instructed him. He must have been acquainted with both the religious and political controversies of the day, including the criticism by the Pharisees of the Sadducean party of the official priesthood, the opposition of the Zealots to the Roman oppressor, and the existence of the Essene and other sectarian movements which attacked both the religious and political leaders of the country and raised apocalyptic expectations. The one exceptional fact about Jesus himself which seems certain, was that unlike other young Jews, he did not marry.[52] There seems no likely explanation of this except that as in the case of some Essenes and Paul of Tarsus, this was connected with deep eschatological religious expectations. Thus Jesus' experience was in some respect limited, yet it provided for him the major ingredients for reflection on the human condition.

Jesus' celibacy is a special problem for us today. D.H. Lawrence questioned that a man who had not had the courage to enter into sexual intimacy could ever really understand what it is to be human.[53] Moreover, feminists today ask whether it is possible that a male, a member of the oppressive sex, could ever be a liberator of women from their age-old oppression. For Christian feminists it is a real problem that salvation comes not through a Daughter of the Mother-God but through the Son of the Father.[54] Is it not the final sexist outrage that God became *man*! Yet it is generally admitted by feminists that of all the great religious leaders, Jesus showed himself the least sexist in his attitudes toward women.[55]

It is certainly inadequate to say, as is commonly said, that God became present in the world through a male simply because in a patriarchal society no one would have listened to a woman. In my opinion the answer is rather to be found in the fact that the incarnation was a *kenosis,* an emptying. The pre-pauline hymn sung in the earliest Christian communities reads:

> Though he was in the form of God, he did not deem equality with God something to be grasped at. Rather, he emptied himself and took the form of a slave, being born in the likeness of men. He was known to be of human estate, and it was thus that he humbled himself, obediently accepting even death, death on a cross! Because of this, God highly exalted him and bestowed on him the name above every other name (Phil 2:6-9).

Thus the early Christian communities understood the incarnation not as exaltation but as debasement. Jesus was the Suffering Servant who had said to his apostles,

509

> "You know how among the Gentiles those who seem to
> exercise authority lord it over them; their great ones make
> their importance felt. It cannot be like that with you.
> Anyone among you who aspires to greatness must serve the
> rest; whosoever wants to rank first among you must serve
> the needs of all. The Son of Man has not come to be served
> but to serve — to give his life in ransom for the many" (Mk
> 10:42-45).

If the Savior had been a woman, this debasement would have lacked all
significance precisely because in patriarchal society, as feminists point
out, all women are already slaves. The Gospel message could have been
proclaimed only by a male who freely surrendered all that is characteristic
of male oppression of which the principal traits, as I showed in Chapter
10, are violence and sexual exploitation. Jesus, by renouncing the revolutionary violence of the Zealots and the sexual power associated with it
(witness the harems of David and Solomon!), and making himself con-
temptible in the sight of other males as a *pacificist celibate*, struck at the
very roots of sexism, the primordial form of all oppression as we see from
Genesis 4:19-24. Thus Jesus was not Daughter but Son. Later I will discuss
why biblically God is called Father rather than Mother. It is only when
we think of Jesus as a figure of power rather than of servanthood that
his maleness seems an injustice to women.

Because Jesus was a Jew and his experience was Jewish, his whole
teaching is profoundly in continuity with the Old Testament. The distortion by some theologians of St. Paul's teachings about the Christian Church
as the "true Israel" has led them to picture Jesus as abolishing Judaism
to replace it with a new religion. This Neo-Marcionism reflects once more
that Platonic dualism which deformed Jewish Christianity to produce
Gnosticism. In Chapter 10, I attempted to show that Jesus' own ethics
is a completion of the ethics of the Old Testament and his wisdom the
flowering of Jewish wisdom. To the degree that historic Judaism has
remained true to its own deepest sources and developed them in the course
of its long history, Judaism and genuine Christianity have not grown apart
but tend to converge, because they have evolved from the same fundamental principles.[56] Modern Judaism at its most authentic is not a dead
religion but pulses with the same life that leads Christianity to constant
repentance of its own infidelities to God and renewal of the faith of
Abraham, Isaac, and Jacob. It is only a perversion of Christianity which
has encouraged persecution of the Jews, and a distortion of Judaism that
has resulted in the notion that Judaism has more in common with
Humanism than with Christianity.

510

We cannot properly understand the humanity of Jesus unless we see it as Jewish humanity. Yet for Jesus and for Paul this rootedness in their own people and its irrevocable covenant with God was the source of that universalism which is so evident in the great prophets.[57] The Jews were chosen to be the servants of Yahweh to lead all people to worship the one true God. The great prophets sometimes spoke of the gathering of the nations in the Jerusalem temple as if this universalism was one of an empire of the Jews over the defeated nations, but they also sometimes transcended this conception and envisaged it not as the reign of one nation but of God equally over all, and it was this higher understanding which flowered in Jesus.

We can say, therefore, that God in assuming human nature in his Son Jesus, also *assumed all humanity,* female as well as male, in and through him (Eph 2:11-22). Jesus, as a bodily being, is of one flesh and blood with all humanity, and as its teacher, its embodied Wisdom, he has drawn from all of human experience throughout history. The history of the Jews as Chosen People is unique, and yet in its uniqueness it is also the paradigm of all nations, as it remains today in its suffering, its endurance, and its never failing creativity. This is the reason that for the Christian, the claim of Humanism to be the religion of humanity seems like an effort to usurp the claim of Judaism and of the Christianity which grew out of Judaism, yet remains profoundly united to it, to be that religion of humanity. And the same goes for Marxism, in many ways so Jewish in its eschatology, as Ernst Bloch has shown.[58]

2. The Christian Church as the Body of Christ

If the body of Christ is all humanity, how can St. Paul say that the Christian community, in his day a tiny Jewish sect opening itself up to Gentile proselytes, is the "body of Christ"?

St. Paul speaks of the human body in many ways, depending on the context and his audience,[59] but fundamentally for him as for the Bible generally, the term refers to the substantial reality of the human person. Moreover, again in biblical categories, individual human persons are always members of a living community, a *corporate personality* from which they derive their life, but to which they also contribute so that its very existence depends on their existence.

> God has so constructed the body as to give greater honor to the lowly members, that there may be no dissension in the body, but that all the members may be concerned for one another. If one member suffers, all the members suffer with it; if one member is honored, all the members share its joy. You, then are the body of Christ. Every one of you is a member of it (I Cor. 12:24-27).

511

Jesus, in obedience to the Father, has suffered death through his body, but through the power of the Spirit of the Father, has been raised again to bodily existence. In thus being raised, he has also raised with him all those human beings doomed to death by their sins, yet now incorporated in Jesus by faith, a faith signified in baptism which is an enactment of Jesus' death and resurrection. Paul understands his own life as a life *in Christo.* "I have been crucified with Christ, and the life I live now is not my own; Christ is living in me. I still live my human life, but it is a life of faith in the Son of God, who loved me and gave himself for me" (Gal 2:19-20). In Paul's conversion vision he heard a voice saying," 'Saul, Saul, why do you persecute me?' 'Who are you sir?' he asked. The voice answered, 'I am Jesus, the one you are persecuting.'" (Acts 9:5), thus identifying the community Paul was attacking with the risen Jesus. And Paul can say, "Even now I find my joy in the suffering I endure for you. In my own flesh I fill up what is lacking in the sufferings of Christ, for the sake of his body, the church." (Col. 1:24)

No wonder then that Paul in his moral instructions as regards food and sex can say.

> Do you not see that your bodies are members of Christ?
> Would you have me take Christ's members and make them
> the members of a prostitute? God forbid! . . . You must know
> that your body is a temple of the Holy Spirit, who is within
> — the Spirit you have received from God. You are not your
> own. You have been purchased, and at a price. So glorify
> God in your body (I Cor 6:15-20, 19-20).

Finally, he sees the eucharistic meal, which is the chief act of worship of the Christian community, as an incorporation into Christ's body,

> Is not the cup of blessing we bless a sharing in the blood
> of Christ? And is not the bread we break a sharing in the
> body of Christ? Because the loaf of bread is one, we, many
> though we are, are one body, for we all partake of the one
> loaf (I Cor 10:16-17).

Paul understands this sacrament so physically that he attributes sickness in the community to an unworthy eating and drinking of the Eucharist (11:30). Thus for Paul Christian life is an incorporation in Christ by which not merely individuals, but the whole community become the very body of Christ living by his life, suffering with him, rejoicing with him, and worshiping the Father in him in the power of his Holy Spirit. Hence sin is an offense against this community and against the Christ whose body it is. Evidently Paul likes this kind of language precisely because it is so physical, ontological, real, as contrasted to any merely mental or moral terminology which would imply that the relation of the Christian to Christ and the community is simply a matter of ideas and attitudes.[60]

What is the relation between this Pauline conception of the Church as the Body of Christ, i.e. the visible presence of Jesus in the world following his ascension, and the Johannine and Lukan notion of the Risen Christ's continued presence with his people through the Holy Spirit? Certainly Paul's language is more incarnational, but he too sees the Holy Spirit as giving life to this body and distributing His gifts to the various members (I Cor 12:12-31). A more puzzling question is what is the relation between all these conceptions of the Christian community and Jesus' central theme of the Reign of God? Every one knows the famous saying of Loisy, "Jesus announced the coming of the Kingdom of God and what came was the Church." The truth in this saying is that although the Kingdom did come in Jesus himself, it was resisted by the world and will continue to be resisted until its final triumph. In the meantime, Jesus continues to invite men to enter before it is too late into the Father's kingdom, in and through the witness of those who have accepted him, that is, through his Church. This Church shares his servanthood in order that some day it may triumph with him, yet it does not now triumph, but serves and suffers. The worst of its sufferings are its own internal divisions, its corruptions, its lukewarmness, all the evils which were already evident in the seven churches of Asia to which the *Book of Revelation* was sent in the name of John the Evangelist. Although it is the Body of Christ and the sign of his kingdom, unlike him its members are still sinners.

At the same time, this presence of Jesus in his visible and suffering Church would not be real or effective unless the Church of the New Covenant was to remain essentially faithful to its mission of preaching the gospel, through the power of Christ's Holy Spirit and throughout the whole course of its history. Under the Old Covenant a Remnant remained faithful, but the New Covenant is far superior to the Old, not because Christians are better than the patriarchs but because Christ is present to his Church in the fullness of the Spirit. The notion of some Protestants, that in the Middle Ages the Church was corrupted to its core and was restored by the Reformers to faith by means of the Bible, can hardly recommend itself today to serious theologians. What would guarantee the Bible to us as canonical, if the Church had for a time lost its faith?

Recently, Hans Küng has argued that the Church is "indefectible in faith" but not "infallible."[61] To me it seems that this distinction is pertinent only if we succumb to Küng's philosophical difficulties about admitting that any proposition in human language (even the inspired words of Scripture) could be infallibly true, even by the *potentia absoluta* of God! If indeed the Church is unfailing in its faith, then as a community it must somehow know with the certitude of faith what that faith teaches, and that is "infallibility" as Vatican I and II understood the term. The Eastern Church also agrees to the infallibility of the Church, although it has its own theology of how this infallibility is manifested. What we can concede to Küng is that this certitude of Christian faith is by no means of Cartesian clarity, but that it is a dark, sometimes almost desperate, clinging to the Person of Christ crucified and risen; yet it is a faith that knows in whom it believes — what Jesus is and what he taught and commands that the Church teach. It can be as positive as Paul was when he said, "I assure you, brothers, the gospel I proclaimed to you is no mere human invention. I did not receive it from any man, nor was I schooled in it. It came by revelation from Jesus Christ" (Gal 1:11-12).

But granted that Jesus by the power of His Spirit gives life to the Church and perserves her radically faithful to himself until her work is done and He returns, granted that the Body of Christ makes Jesus visible in our earthly, bodily world of human experience and history, need that Body be an *institution*? Contemporary people are profoundly disillusioned with institutions, which seem always to become tyrannical, reactionary, bureaucratic barriers to life and progress. The Christian Church, marginal as it is to modern societies dominated by Humanism and Marxism, to many seems to be among the most reactionary of institutions.

I think we must recognize that in our present epoch of pluralism, where the older religions have all been pushed into marginal positions and where Humanism and Marxism seem to be making good their claim that religions of transcendence are obsolete and incompatible with modern science and technology, these older religions manage to survive only by one of two strategies: (1) withdrawal into closed and static institutions; or (2) a kind of "decomposition" or deinstitutionalization and privatization. Before Vatican II, the Catholic Church relied on the former strategy. With Vatican II she is experimenting with the latter strategy in which organized lines of command are loosened and the symbolic marks of identity are toned down. The result is a considerable loss of clear visibility as an institution, but in compensation, a greater *presence* in the world.

Nevertheless, even if we accept, as I believe we must in the main, this second strategy, we must be realistic about its grave risks and not be led into the foolish error of thinking we can dispense wholly with ecclesiastical structures. St. Paul wrote two letters to a Corinthian Church which was in the process of making this mistake. Amusingly enough, some recent writers have used that church of enthusiasts as the norm of authentic church life, and have seen the growing institutionalism evident in so much of the New Testament as "early Catholicism", the beginning of corruption. [62]

514

Did Jesus himself *institute* an organized church? Any valid answer to that question must preserve the essential tension implied in the very notion of "thy kingdom come . . . on earth as it is in heaven," in the Cross of Christ, in the ascension of the Risen Christ to the Father to send the Holy Spirit, in the "filling up of what is lacking in the sufferings of Christ, and in the delay of the Parousia. Jesus announced the actual beginning of a Reign of God which is eternal and which, therefore, transcends earthly conditions. The Christian community, however, lives in time, in earthly conditions in a world still only beginning to be redeemed, where the social conditions of life in full keeping with the standards of the Kingdom are lacking. Consequently, Jesus himself advises his followers to be "as clever as snakes and as innocent as doves" (Matt 10:16). St. Paul says to the Corinthians,

> I tell you, brothers, the time is short. From now on those with wives should live as though they have none; those who weep should live as though they were not weeping, and those who rejoice as though they were not rejoicing; buyers should conduct themselves as though they owned nothing, and those who make use of the world as though they were not using it, for the world as we know it is passing away. I should like you to be free of all worries (I Cor 7:29-31).

In these remarks Paul was echoing the thought of Jesus in the Sermon on the Mount which proposes a morality of detachment from earthly concerns, of poverty, openess, and non-resistance, which to many has seemed totally impractical. The most obvious example of this tension is that Jesus and Paul are celibates and Paul warmly encourages celibacy in the Corinthian church, yet neither forbids marriage, although Jesus declares that "When people rise from the dead, they neither marry nor are given in marriage, but live like angels in heaven" (Mk 12:25). In the fulfilled Kingdom all will be celibate, but in the time of the Church, marriage remains a necessary and holy institution.

It is an evasion of this intrinsic tension, however, to conceive of the Church in its pilgrim condition as a "compromise with the world," an adulteration, or corruption. Such ecclesiologies, including, it seems to me, Luther's Two Kingdom Doctrine[63] are other reflections of the Platonic dualism that has so affected Christian theology. The Church is conceived (as some Gnostics conceived it),[64] as an *ideal* reality in the heavens, and the historic Church as a faint and distorted reflection of the reality. This hyper-spiritual and docetic conception cannot do justice to St. Paul's thoroughly realistic and incarnational theme of the Body of

515

Christ. This living and visible body is the real Church, but in the process of growth and maturation. As the Pauline *Ephesians* 4:14-16 says,[65]

> Let us, then, be children no longer, tossed here and there, carried about by every wind of doctrine that originates in human trickery and skill in proposing error. Rather, let us profess the truth in love and grow to the full maturity of Christ the head. Through him the whole body grows, and with the proper functioning of the members joined firmly together by each supporting ligament, builds itself up in love.

This developmental ecclesiology, of course, by no means excludes that the Body does suffer from serious sickness and not only matures, but often requires healing.

If we accept such a realistic view of the Church, then just as Christian marriage remains subject to the natural law, physical and moral, although it is transformed into a sacrament of eschatological significance; so also the Christian community as a whole remains subject to those laws and to the principles of social and political organization, although it cannot be reduced to a merely political institution. As anarchism is a dubious political theory, so is any conception of the Christian community which reduces it to a "movement" or to a society operating by pure "consensus." Before we ask whether Jesus founded an institutional church, let us recall some of the basic ethical principles discussed in Chapter 10 that govern any community made up of human persons,[66] and then consider how Jesus embodied these in his own community, and gave them a new meaning in view of the special life and mission of that community.

The first of these basic ethical principles was that the community exists for its members, yet their highest good is not private and individualistic but consists in spiritual values which are enhanced by being shared in common. Jesus' concern for the most lowly and neglected members of the community, even its outcasts, is so evident as to require no argument. It is the very theme of the beatitudes that the Reign of God is open to all, especially to the poor who are its favored citizens, "the little ones."[67] But this principle is also embodied in his unique example and teaching that all are to be servants one to another, as he showed them by washing the apostles feet at the Last Supper, and by the teaching of Mark 10:42-45 already quoted above. "The Son of Man has not come to be served but to serve — to give his life in ransom for many," is the Christian form of this fundamental principle.

The second of these principles was that this service of persons must be realized by an ever growing equality in order to achieve the distributive justice of "to each according to their need."[68] This is also embodied in Jesus' beatitudes in which it is promised that "Blest are you poor; the reign of God is yours. Blest are you who hunger; you shall be filled" (Lk 6:21). What is most remarkable, however, is that for Jesus even those who are unworthy or less worthy are to receive forgiveness and to be included in the distribution of benefits. This is the meaning, for example, of the Parable of the Prodigal and the Dutiful Sons (15:11-32), and the Parable of the Laborers in the Vineyard (Matt 20:1-16), of whom those who work only an hour get the same pay as those who work all day. In these parables Jesus recognizes the apparent injustice of this generosity but points out that in the Kingdom everyone gets *more* than he or she deserves.

The third principle is that to achieve this abundance for distribution there must be a division of labor within the community. It is here that St. Paul, writing to the Corinthians whose enthusiasm about their new found freedom had led them to a kind of anarchy, insisted on the analogy of the Body of Christ in which, Paul says (I Cor 12:14-18):

> The body is not one member, it is many. If the foot should
> say, "Because I am not a hand I do not belong to the body,"
> would it then no longer belong to the body? . . . As it is,
> God has set each member of the body in the place he wanted
> it to be. If all the members were alike, where would the body
> be? There are, indeed, many different members but one
> body.

It is essential that this analogy used by Paul be understood within the context of the previous two principles, or the result will be a univocation which would support a totalitarian notion of society in which the individual exists only for the society as a part of the body exists for and can be sacrificed for the organism.[69] That Paul is no totalitarian is evident from the conclusion he draws from this analogy (12:24-26):

> "God has so constructed the body as to give greater honor
> to the lowly members, that there may be no dissension in
> the body, but that all the members may be concerned for
> one another. If one member suffers, all the members suffer
> with it; if one member is honored, all the members share
> its joy.

The fourth principle is that unified action by any human society demands authority and obedience at least in their essential function of arriving at common agreement on action, and usually also of substitutional authority by which less competent members profit from the guidance of more competent, at least with respect to particular decisions.[70] That Jesus exercised such authority, however gently, is perfectly evident from the Gospels, and that the apostles were no less insistent upon their own authority is plain from the example of St. Paul. Jesus defended the authority even of his opponents, "The scribes and Pharisees have succeeded Moses as teachers; therefore, do everything and observe everything they tell you. But do not follow their example" (Matt 23:1-3, because they themselves were not obedient to the Law in their own personal conduct). Yet both Jesus and Paul made clear enough that this obedience is owed only when authority is used for the common good within its proper limits, and therefore refused to obey commands that transgressed a higher law. In Acts 5:30, St. Peter is portrayed as saying, when faced with the Sanhedrin's demand that he cease to preach, "We ought to obey God rather than men."

The fifth principle is that the form of government of a society is open to human creativity, since biology has not provided any one order of social relations as it has done, to a degree, for family relations. As for these family relations, it is clear enough in St. Paul's teaching that the man is the head of the family as Christ is head of the Church (11:3, obviously a controversial point today, which I have already discussed in Chapter 10).[71] Paul's teaching implies no inferiority in women since he recognizes that as baptized they are equal in dignity to their husbands, but it does imply a functional division of roles which can be solved in different ways in different cultures as long as the fundamental relations set by the nature of sexuality itself are not contravened.

As for the question of the best form of government for the larger society, it is never raised in the New Testament, but is implied by this very silence, since the early Church was indifferent to the form of governmental authority as such (Rom 13:1-7; I Pt 2:13-17), on the grounds that God gave authority to existing governments for the public good, "Let everyone obey the authorities that are over him, for there is no authority except from God, and all authority that exists is established by God . . . Rulers cause no fear when a man does what is right, but only when his conduct is evil" (Rom 13:1,3). It is clear from this last sentence that what Paul has in mind is obedience to just laws, not to ones that command wrongdoing. This principle, therefore, ought not to be understood in the extreme dualistic sense sometimes given Luther's Two Kingdom Doctrine according to which authority is to be passively obeyed unless it commands what is against the spiritual freedom of the Church.[72]

The sixth principle requires that the members of a society participate in the decisions that affect the common good and themselves. It would be disingenuous to claim that the Gospels show us a Jesus who acted "collegially" with his apostles. Nor is there any indication that St. Paul governed the churches he founded according to some kind of "democratic process." The prophetic authority of both related them to their followers by substitutional authority as father to children, leader to followers, teacher to disciples. Moreover, it would be anachronistic to read back into these communities the concern for democracy typical of modern systems of government based on universal literacy and rapid communications, not to mention the psychological strategies of modern administrative and communication theory. Yet in the Gospel there is the deepest root of the principle of democratic participation, a root often ignored by modern theories, namely the inviolability of the individual conscience. Jesus' Sermon on the Mount, as we saw in Chapter 10, focuses on what is especially characteristic of Christian ethics, namely, the *primacy of the interior motive,* "No man can serve two masters" (Matt 6:24). This insistence on the purity of motives means that ultimately the individual stands accountable for his or her actions before God, not before any human tribunal.

Humanism borrowed this Christian principle of the inviolability of the individual conscience and built on this principle its own basic dogma of the autonomy of the individual, a dogma so brilliantly defended by Immanuel Kant;[73] and concluded from it that it is degrading to the human person to obey anyone, not even God, except oneself. This exaggerated individualism leads to grave antinomies as in Rousseau's theory of the general will, which has so often been the excuse for majoritarian tyranny or even dictatorship. It forces democratic theorists to claim that "the people" all think alike in order to justify the necessity of public action or to hypostasize the law so as to avoid admitting that obedience not only to laws but to persons is ethically necessary. For the Christian, however, it is no degradation to be a servant of another for the sake of the common good. Nor is the inviolability of conscience contradictory to obedience to God and to authorities established by God. Paul Tillich has developed this theme well by showing that the polarization of autonomy and heteronomy is solved by *theonomy.*[74]

Once it is seen that it is because of the proper autonomy of the individual that participation by all in matters that affect the common good is an essential element of the common good, since without it the members of the community never attain maturity of conscience and responsibility, then it becomes obvious that a Christian community built on obedience must also understand obedience not as a blind and mechanical subjection to authority, but as a free and intelligent cooperation with authority. It means also that authority must be so exercised as to make possible growth in participation in decision by every adult member of the community. This true authority *is* evident in the Gospels by the way in which Jesus drew the Twelve more and more into his confidence inspite of their slowness to understand, and gave them a share in his authority. In Luke's Gospel, Jesus not only chooses the Twelve and commissions them, but he then (Chapter 10) chooses the Seventy-Two who are similarly commissioned. It seems obvious that this Seventy-Two (or as many MSS read "Seventy") is an allusion to the action of Moses (Ex 18:13-27; 24:1) in dividing his authority not only among seventy minor judges but among "officers over groups of thousands, of hundreds, of fifties, and of tens."[75] Jesus promises the Twelve that in the Reign of God they will sit on thrones beside him judging the nations (Matt 19:28), and St. Paul applies this to the action of the whole community in governing its affairs (I Cor 6:1-3):

> Do you not know that the believers will judge the world? If the judgment of the world is to be yours, are you to be thought unworthy of judging in minor matters? Do you not know that we are to judge angels? Surely, then, we are up to deciding everyday affairs.

The seventh and last principle is that of subsidiarity and functionalism which is simply a corollary of the previous principle, a way of insuring maximum participation while maintaining unity of action.

3. Ministry

It appears, therefore, that these principles of organizing any community are recognized by Jesus and by Paul, but did they lead to some definite constitution for the Church? Or, since the fifth principle states that the form of government of larger social groups is in no way determined by the nature of the human person but is open to human creativity, did Jesus leave this entirely up to the Church to vary in different times and places? The Roman Catholic Church seems to have taken the position that Jesus himself (whether immediately or mediately through his Spirit) instituted the three-fold hierarchy of bishops, presbyters, and deacons with the bishop who is successor of St. Peter as head of the college of bishops.[76] In spite of reservations about the exact meaning of the primacy of the successor of St. Peter, the Eastern Church seems to agree. Recently, however, serious questions have been raised even by Catholic scholars about this ecclesiology. If there is no one best form of secular government, but that is best which human creativity devises for a certain place and time, why should Jesus have fixed on a universal and enduring church an unchangeable form of government which was bound to be less suitable for the changing times and expanding global community? Moreover,

520

modern biblical scholarship seems to find no clear evidence in the New Testament for the notion that Jesus established a priesthood, let alone a definite form of Church government. Rather, it appears that the early Church experimented with many forms of government and that the idea of a continuation of the cultic priesthood of the Old Testament was quite foreign to its life.[77]

Certainly we must recognize that Jesus did not write a constitution for his church the way that St. Benedict wrote a rule for his monasteries. Jesus' whole ethical teaching, as we have seen, accepted the existing structures of Judaism and even those of the Roman government. His purpose was not to abolish these structures, but to infuse them with a new and transforming life. There is no evidence that Jesus himself intended to abolish the Jewish priesthood, although by his prediction of the destruction of the temple, he clearly anticipated that the rejection of his preaching by the leading priests would necessarily bring about a new religious situation. He left to his closest followers, the Twelve, under the guidance of the Holy Spirit he promised to them, to meet this new situation in a creative way. Yet, it is also undeniable that he not only instructed them carefully in the principles of the carrying on of his ministry (Matt 10:1-42), but he also gave them a model in his own conduct as to what the fundamental principles of that ministry must be.

Comparative study of religion makes evident that for any religion which claims a transcendent source, it is essential that the basic symbols of religion should be *given* from above, not created by the community whose life and unity depend on these symbols. Humanism and Marxism, which have attempted to be man-made religions with symbols that are self-created, not given, have discovered how difficult it is to touch and to move their devotees at the deepest level of human personality and to achieve a unity of action. That is why art and propaganda take on such central importance in these secular religions as they strive to create their own "myths." As I indicated in Chapter 9, they seem to have succeeded in some measure: the Humanists by adopting the myth of inevitable evolutionary progress, and the Marxists by the eschatological myth of the final revolution.

Today, however, they are faced with the problem (I do not predict that they will fail to solve it more or less satisfactorily) of growing scepticism about these myths. Christianity, too, suffers in our age from the problem of revising its myths to make them credible to our age without emptying them of their original meaning. The Roman Catholic Church at Vatican II took the perilous step of revising its liturgy and restricting its sacred language. To some Catholics this appeared a fatal step, but the Council was convinced (I believe rightly), that it could make the Church more Catholic, not so much by abandoning its sacred language, but by going back to the roots of its traditional symbolism in the Scriptures. The language of the Church is not Latin, nor even Greek or Hebrew,[78] but the symbolism of the Bible which is at once "mythical" and historical; that is, it is rooted in the archetypal symbols of common human experience derived first of all from the human body and familial relations, and in the paradigmatic story of a people, a story that reaches its climax in the life and person of the Church's Founder.

The essential difference between good secular and good ecclesiastical government is not with respect to the principles I have already discussed, but consists in the fact that a religious community is primarily symbolic and sacramental, while a secular community is primarily functional and only secondarily symbolic. Certainly symbolism plays an important role in secular government. We have only to consider the symbolic importance of the British monarchy or even of the United States Presidency to see that. Yet it is generally recognized that the desacralization of secular government, i.e. the lessening of its symbolic character in order to increase its functionality, has been a positive cultural achievement for which Humanism and Marxism can rightly take much of the credit.

Vatican II acknowledged that this "separation of church and state,"; when it does not merely mean the substitution of Humanism and Marxism for Christianity as the established religion, is not incompatible with Christianity but actually favorable to it, because it frees the Church for its own proper mission.[79] Only secular government has the vocation from God and the means at its disposal to bring about social justice, peace and the order of the common good; and the Church cannot of itself undertake or achieve that task. It is the Church's mission to speak for Christ in announcing the coming of the Reign of God, which Reign demands of earthly government that it fulfill its responsibilities of justice, but which also reminds us that all earthly power is passing away to make room for the eternal reign of Christ, the only true king.[80] Thus the Church is above all a *sign*, a sacrament that points beyond this life while making real the presence of the Risen Christ within this life.

The early Church, as early as in the *Epistle to the Hebrews,* came to see in Jesus the summation of all the different kinds of leadership it had experienced in the past, although it seems he took to himself only the title Son of Man and Servant, speaking not of his own authority but of the Reign of God his Father.[81] Jesus was Messiah, shepherd and king of the flock of Israel, but he was also a teacher, a prophet, Wisdom itself, and he was priest, a priest greater than Aaron, although he was not a member of the Jewish priesthood.

Every high priest is taken from among men and made their representative before God, to offer gifts and sacrifices for sins . . . One does not take this honor on his own initiative, but only when called by God as Aaron was. Even Christ did not glorify himself with the office of high priest; he received it from the One who said to him, "You are my son; today I have begotten you"; just as he says in another place, "You are a priest forever, according to the order of Melchizedek." In the days when he was in the flesh, he offered prayers and supplications with loud cries and tears to God, who was able to save him from death, and he was heard because of his reverence [humble submission]. Son though he was, he learned obedience from what he suffered; and when perfected, he became the source of eternal salvation for all who obey him, designated by God as high priest according to the order of Melchizedek (Heb 5:1, 5-10).

This capital text shows clearly that Jesus' priesthood, although it summed up and confirmed whatever was contained in the Aaronic priesthood, was far superior to it. As Yves Congar has shown,[82] the old Jewish cultic priesthood, like all the priesthoods of other religions was an offering of God's subhuman gifts in acknowledgement of humanity's dependence on the Creator and humanity's sinfulness before Him. Even human sacrifice was the offering by one human of another sinful human victim as a substitute. But Jesus' priesthood is the offering of himself in his perfect (and therefore sinless) love to the Father in intercession for all sinful humanity, even the enemies who killed him.

Too often the meaning of this sacrifice has been distorted by a one-sided emphasis on the fact that the ancient sacrifices were immolations, destructions of a victim, and that they were conceived as expiations, i.e. as payment of a debt to an offended sovereign. Such language is actually used in the Bible and Jesus himself uses it in some of the parables, but in a notably *ironic* manner.[83] His own conception of God as Father, for example in the Parable of the Prodigal and Righteous Sons, makes clear that God is not concerned about collecting debts owed to him. He is always ready to forgive, and certainly could never have taken pleasure in seeing his own Son suffer vicariously in order to satisfy God's hunger for vengeance. The essence of sacrifice is not the destruction of a victim, but the offering of a gift symbolic of the gift of self, which is the only thing a God of Love could ever desire from us. It is only a human projection to picture God as demanding payments and revenge. God the Father did not will that his Son should die, but he did permit sinners to kill His Son. Jesus acceded to this decision in order that he might give proof *to us* (not to his Father who needed no such proof), that his and the Father's love for us, even as his rebellious enemies, is without limit.

The supreme act of Jesus' priesthood was this offering of himself on the Cross, summing up all the other acts of love in his whole life, but at the Last Supper he also provided the sacrament of the Eucharist in order that his community throughout all ages might *remember* this sacrifice, not merely in the sense of recalling an event of the past, but in the sense of participating in that permanent reality which perdures in the heart of the Risen Lord, his everlasting self-giving to the Father in the Spirit. The Christian community as the Body of Christ co-offers itself to the Father with, in, and through its Head and is offered by him with himself in one single act of sacrificial love, a love ready to endure death itself to prove its fidelity, since nothing is as strong as death except love (Sg 8:6). This act of love is salvation, because "The man who continues in the light is the one who loves his brother; there is nothing in him to cause a fall" (1 Jn 2:10), and nothing "will be able to separate us from the love of God that comes to us in Christ Jesus our Lord" (Rom 8:39).

Through baptism the Christian community is reborn by faith in this love of God, which it continually renews in the most perfect possible act of worship that is the eucharistic sacrifice of Christ the only priest. The Christian community, therefore, in offering itself in Christ in its daily life as well as in its sacramental worship, its sacrifice of praise and thanksgiving, is a priestly people. "You are 'a chosen race, a royal priesthood, a holy nation, a people he claims for his own to proclaim the glorious works' of the One who called you from darkness into his marvelous light'" (1 Peter 2:9; Ex 19:6). This priesthood is "royal" and also prophetic since it "proclaims the glorious works" of God. Catholics are grateful to the Reformation for re-emphasizing this universal priesthood of the faithful which certainly is part of Catholic tradition, but which has often been obscured by the pride of the clergy. It is important, however, to note that it is a *corporate* priesthood. It is as a community, the Body of Christ, that the laity shares in his priesthood, not as isolated individuals. Furthermore, this corporate priesthood is in behalf of the world, that is, the Christian community exists to serve not itself but the total human community, to witness to it the message of the Gospel, (its prophetic function) to provide it an example of love and care (its kingly or shepherding function, which is not that of domination but of service), and to intercede for it in prayer (its priestly function).[84]

Thus the corporate priesthood of all Christians is especially directed *outward* to the world. The laity, that is the whole Christian people taken

524

together apart from intra-church distinctions, lives in the world, takes part in its daily life, and has a joint ministry to it.[85] Each member of the church shares in this ministry according to their particular gifts and life situations. In a church where every member was fully matured and active as a Christian so that all lived in the harmony of the Spirit, only a minimum of internal structure would seem necessary to the Church's performance of its ministry, and in the earliest hours after Pentecost no doubt this was the situation. But the Christian community is a fully human community which strives to be "catholic" in the way that Jesus was, i.e. it is concerned especially to include the poor and neglected, the outcasts of other human communities. It can never be elitist or sectarian, because the Spirit of Christ compels it to give the world an example of what the Reign of God must be, a net that takes in every kind of fish. "The reign of God is also like a dragnet thrown into the lake, which collected all sorts of things" (Mt 13:47). This means that it is faced with the basic problems of organization which I have outlined: it must provide for the guidance of its members toward Christian maturity in the faith and it must hold this vast variety together into a single living cooperating body.

Jesus himself faced this problem and chose twelve among his followers for special attention and education, while he preached to the multitude in parables. Even among the twelve he drew three, Peter, James, and John, into special intimacy, and, as all four Gospels attest, as his death approached he appointed Peter as their chief.[86] This could only mean that he willed the twelve to share in his prophetic, shepherdly, and priestly ministry in a special way different from the way in which the whole community shared in it. What is that difference? As we have seen, every community must have a division of labor, and this includes relationships of authority if it is to act as an organic community. Hence the leadership in the Christian community, like that of Jesus himself, is directed not only outward but inward, to the building up of the community. Only if the leaders build up the Body of Christ can that Church be strong enough to carry on *its* priestly ministry to the world.

Perhaps nothing is so confused at the present moment in the minds of Christians as to just what is the role of the clergy. Is the task of the clergy to help Christians save their souls, or is it to carry on a campaign of conversion of new members, or is it to transform the social order? Certainly Jesus came to transform the whole world, but he chose to do this mediately through the community which he formed. On the one hand he preached to all, but on the other (and one gains the impression from the gospels that he shifted his emphasis as he met with increasing opposition), he gave much time and energy to forming the Twelve through whom he hoped his work could continue and spread. We must remember that Jesus spoke of the limits of his own ministry, "My mission is only to the lost sheep of the house of Israel" (Mt 15:24), yet he hoped, as did all the prophets, that through Israel the whole world might come to know God.

Thus it seems that we should say that the Church as a *whole* is outwardly directed toward the entire world, yet to be able to carry on this mission, it requires an internal division of labor by which some have as their primary concern the building up of the Church, and these are its ministerial priests or clergy. Their ministry or service, however, does not stop short with simply strengthening the Church interiorily, but must also be one of inspiration and guidance of the whole people in its outward ministry. Yet the clergy can never substitute for this ministry of the whole People. When the clergy tries to do what the laity has been called to do, rather than to prepare, inspire, and guide them in doing it, the clergy becomes the Church and the people are reduced to mere clients of the clergy.

This division of labor, by which some members of the People of God are given responsibility to build the community and enable the various charisms of all to be exercised harmoniously, requires both substitutional and essential authority. Therefore, these ministers must be designated by those in the community already invested with authority derived from Christ to whom alone this authority properly belongs. The notion that the Church can exist effectively under the leadership of self-chosen charismatics is absurd, as the experience of the Jews under the judges makes clear to anyone who reads the Bible. Samuel, with many warnings about the abuses that might follow, solved that problem not by looking for some one on whom the Spirit of the Lord had fallen, but by a ceremonial or sacramental anointing of Saul and then of David, by which this empowerment by the Spirit could be made evident to all the people. The Gospel according to St. John portrays the Risen Lord as breathing the Spirit upon the Twelve (20:21-23), and in *Acts* we read of various acts of designation by the laying on of hands, whose specific purpose may not be altogether clear, but which evidently are intended for this same general purpose of establishing publicly recognized ministries. This is what the Christian churches understand as *ordination* of a person to carry on the functions of the Twelve.

In the churches of the Reformation, the tendency has been to see this ordination primarily as an authorization to preach the Gosepl, although some special role in the administration of the sacraments is usually also included. This reflects the Reformation stress on the conversion of the individual to Christian faith and its fear of emphasizing the further action of grace in santification in forming a community of love. Paradoxically, this tends to surrender the mission of the church to the clergy, who then alone are actively engaged in the work of converting others through preaching. In order to counteract this tendency, it then becomes necessary to urge the laity themselves to become preachers or at least bible distributors. On the other hand, in Catholic churches the tendency of the clergy has often been "to do it all", reducing the laity to purely passive roles.

What then is the specific role for which the priest is ordained? It seems to me that it is a mistake to answer this question primarily in terms of *function*, because of the essentially sacramental character of the Church. The priest is first of all a *sign* of the presence of Christ in the Church as its Head, or to use another metaphor used by Jesus himself, its Bridegroom (Mt 25:1-13; Eph 5:22-33). Such a sign is necessary to the life of the community to indicate the source of its unity and its continuity. As Jesus was the sign around which the Church originally gathered, so the priest stands *in persona Christi*. For this reason, the primary personal quality demanded of a priest by his office is not any special talent or competence, but *fidelity* to Christ and to the community for which he has been ordained. He is wedded to that community, as Christ to the Church, and willing to give up his life rather than to leave his post as committed to the Church. Every member of the Church, of course, is committed by baptism to Christ and therefore to his or her fellow members of the Church, but lay Christians do not assume responsibility for the building up of the community so much as for witnessing Christ to the world. It is the priest, who as a special realization of his baptismal commitment, assumes for Christ a commitment to the preservation of the Church and its up-building.[87] Like the captain of a ship on whom the responsibility for the ship devolves even if others desert, so the priest is committed never to desert the Church but to marshal its remaining forces for survival.

Because the priest makes this commitment, he becomes a sign of the Church's unity and this is indicated by the traditional theological thesis that ordination, like baptism, is unrepeatable because it confers a "character" which is permanent.[88] A priest designated merely to carry out a function would cease to be a priest when he ceased to perform that function; but because ordination relates not merely to function but to the very existence of the priest as person, as one who has surrendered himself to stand for the unity and permanence of the community, it transforms that person ontologically, not by raising him *above* the baptized in another realm of existence, but by locating him *within* the community of the baptized in a special relation to the other members.

In my opinion, this is the reason why in the Eastern Church it was decided to ordain to the episcopacy only men who if married, lived apart from their wives or who had dedicated themselves as celibates and that in the West, this discipline was extended to all priests. This discipline, which has always been difficult to enforce, which the churches of the Reformation abandoned, and which today is again under wide attack even in the Catholic Church, is said by some to have originated in notions, derived from the Old Testament, about cultic purity. While this reason undoubtedly contributed to the development of the discipline of celibacy for the clergy, it was itself only a reflection of the more fundamental conviction that Jesus was the model of priests, and that what St. Paul recommended to Christians as preferable should be observed at least by those who symbolized the Bridegroom of the Church, dedicated wholly to her.[89]

Schillebeeckx[90] has recently argued that since all Christians have a right to the Eucharist, and since the exclusion of married men from the priesthood has resulted in too few priests to meet this need, therefore the Church should make celibacy optional. The assumption underlying this argument obviously is that if married men were permitted to be ordained, it would greatly increase the number willing to commit themselves to priesthood. The Christian people, however, have a right to the Eucharist only if they themselves generate from their own ranks men willing to make this commitment. After all, it is not bishops who produce priests but Christian families who inspire their sons with the desire to make this commitment. Now it may be true that sometimes married men, after providing for their own families, through their experience as fathers come to the desire to commit what remains of their lives to minister to the community. It was from such men that the early Church drew its priests (1 Tm 3). It is also possible that it may become necessary for the church to develop a two-tiered system of celibate priests who function somewhat as St. Paul did in the founding of communities and their supervision, while permitting the ordination of married men as less well-trained and part time priests in local churches. Yet what is most needed in the Church is not numbers of priests but committed priests, since commitment (and not sacramental functions merely) is what defines priesthood. Celibacy for priests (as distinct from that for religious), is so effective a sign of their total commitment to the Church and of its eschatological witness in the world that it remains for the Christian Church the ideal provided by Jesus himself, and it seems to me very probable that to obscure this ideal by making celibacy merely optional, far from solving our acute problem about priestly vocations, would worsen it.

The priesthood fails to attract young men today not because it requires total dedication, signified by celibacy, but because it seems to lack that dedication and urgency which would make it stand out in a secularized world. The abandonment of required celibacy would only make priesthood seem still more functional, just another profession among others and an ill paid one at that. As for the comparison of the Catholic clergy with the lower married clergy of the Eastern churches and the Protestant clergy, without in any way wishing to impugn their merits, I believe that this comparison makes it clear that celibacy has been a powerful factor in maintaining a *level* of discipline and channeled energy, as well as a practical faith in the transcendent realities of Christian faith in the Catholic clergy, which demonstrates its practical value.

Because the priest is first of all committed to stand as a sign in *persona Christi*, he has certain functions which he may or may not be called on to perform. These are the same functions which the laity shares with Christ: those of the king, prophet, and priest; but the priest's task is to *facilitate* the community's performance of these functions. Thus the function of prophecy is not limited to priests. Every member of the church takes part in some measure and in some mode in proclaiming the Gospel to the world, but the priest of the community has the ultimate responsibility to see that this is done, and rightly done. Even if he never himself preaches or is even incompetent to preach, as priest he must see that the faithful are instructed by others in the Gospel, and that his people bring this Gospel to the world. In the same way the kingly role of governing the Church and of extending this care for the needs of human beings into the world by charity and social action can be shared in many ways by the lay members of a local church, but it falls to the priest to call the faithful to these tasks, to coordinate them, and to make sure that the community remains one in charity. This he does, not primarily because of his talents as an organizer, but because of his constant concern for the survival and growth of that community and the expansion of its mission.

In medieval theology the nature of priesthood was deduced from the fact that it is the priest who alone can "confect the Eucharist and offer it in sacrifice." Undoubtedly the priest's role in the Eucharist is his supreme activity, but of course in this offering he does not act alone. The Eucharist (even when the priest celebrates alone) is always the offering of the whole people of God. But because the priest is the *sign* of the whole community, it must be he who leads that community in its offering. In fact, in so doing, he is an essential element of the Eucharistic sign, since it is he who acts *in persona Christi*, representing visibly the action of the Lord at the Last Supper. Of course the Risen Lord is present in the Eucharist substantially and personally in the sign of the consecrated bread and wine, but this sign cannot be significant, and therefore, cannot be effective of the real presence of Christ except through the consecratory thanksgiving spoken by the committed head of the community, i.e. by the priest. In thus acting *in persona Christi*, the priest is not arrogating to himself of role of power but of servanthood, identifying himself with the Crucified as the victim who lays down his life for others (Jn 15:13).

Because the priest's primary role is sacramental, the fact that historically the Church has ordained only males is not to be attributed to patriarchal sexism, but is rooted, as all symbolism must be, in the human body. Today it is sometimes objected that if the sex of the priest is sacramentally significant, then so should be his race, age, and other physical characteristics, since Jesus was a Jew, thirty years old, etc. But these differences are relatively superficial, since they do not determine the basic human relations on which archetypal symbolism is based, while sex does, since it is the relations within the family that determine our whole archetypal symbol system whose meaning is common to all humanity.

Far from conforming to the sin of sexism, this traditional restriction of the priesthood to males, when priesthood is rightly understood, is a powerful remedy for sexism. I have shown earlier in this chapter that the incarnation of God in a man rather than a woman can best be explained as a rectification of the primal injustice of the oppression of women by men, because in this incarnation God takes the role of servant, not of master, thus overthrowing the whole system of values based on physical force, and by coming into the world through a virginal conception, he overthrows male lust and sexual exploitation of the woman, setting her free to be wholly herself, fully a person in her own right. The Church calls only men to enact this servant role on behalf of the People of God who, as I will show more fully in a later section, are sacramentally represented by a woman, the Virgin Mary, and by all Christian women. Christ was thus Bridegroom of the Church in that *complementary equality* (or equalizing complementarity) which is God's original intention for the relation of man and woman[91] and this bridal relation requires symbolically that the priest acting *in persona Christi* at the eucharistic wedding banquet, as indicated in the miracle of Cana (Jn 2:1-12), should be male.

Such arguments at the present moment fall on deaf ears as far as most Christian feminists are concerned.[92] They see them as rationalizations of male domination of the Church. No doubt they are right in saying that men have used the priesthood to obtain that kind of power which leads to sexual and other forms of oppression. Is it not inevitable that human sinfulness will manifest itself again and again in the distortion and misuse of the very means established by Christ to remedy sinfulness? Has not the preaching of the Gospel been used so as to convert Christ's message of liberation into one of imperialism?

530

The mistake of this type of feminism, it seems to me, is that it seeks to judge the life of the Church primarily in terms of earthly government. It is very true that the Church must give an example of social justice and observe the principles of organization which I have already described, but these principles must be applied in view of the special end of the Church which, we have seen, is primarily sacramental and not, as in the case of secular government, primarily the establishment of a just social order. The Church is not inconsistent because she makes use of a symbolism in her life which only designates men for certain offices, while at the same time she works in the world to urge civil government to open all of its offices to both sexes indifferently, because the meaning of "office" in the two cases is quite different. No one considers the theater discriminatory because certain roles in acting, ballet, or the opera are given to men and others to women, because it is recognized that here the sex of the person, their physique and voice and manner, are highly relevant to the esthetic effect. Feminists are making a "category mistake" when they apply the concept of "equal rights" in the same way to a sacramental ministry where symbolism is of the essence and to civil functions where competence is the only essential factor.

Nor does the argument hold that St. Paul says "There does not exist among you Jew or Greek, slave or freeman, male or female. All are one in Christ Jesus," (Gal 3:28-29), because here Paul is speaking specifically of baptism, the sacrament by which all become members of the Body of Christ, not of that division of offices within the Body of Christ to which ordination pertains. Similarly, it is not valid to argue that since women are created in God's image equally with men and are also "other Christs " by reason of their baptism, that they are equally qualified for the symbolic role of priest, since when it is said that the priest acts *in persona Christi*, what is refereed to is not the fact that Christ is the image of God, but that he is the Crucified, the Servant, and Victim which requires that he put aside that power and glory which is sinfully claimed by men in their role of oppressors. Women, as the wrongfully oppressed, cannot symbolically fulfill that kenotic role as effectively as men.

I do not intend this argument as a demonstration based on revelation that the Church lacks the power validly to ordain women to priesthood. The Church has plenary power to adapt the sacraments pastorally to changing needs as long as their substantial meaning is not changed.[93] She has used this power in the past, for example, to give up for a time the administration of baptism by immersion, and to restrict the chalice to priests. Since Vatican II we recognize that these modifications, made for pastoral reasons that seemed valid at the time, probably did more harm than good because they impoverished the symbolism of the sacraments,

531

but they were undeniably a legitimate exercise of the Church's pastoral authority and did not invalidate the sacraments. In the same way, it is conceivable that the Church in our times may be forced by feminist demands and the shortage of priest to ordain women validly, but it would be an impoverishment of the universal symbolism of the sacraments made in the interests of our own distorted culture and, I believe would have to be corrected later in order to maintain the true meaning of the Christian priesthood in its fullness.

What I have said about priesthood helps to explain also the teaching of St. Paul in 1 Corinthians (11:1-16 and 14:33-36) as well as in the Pauline instructions on the family (Eph 5:1-6-9; Col 3:18-21: 1 Tm 2:8-15), concerning the role of women which have given so much scandal to Christians working to overcome the sin of sexism. In Chapter 10, I attempted to show that these texts must be understood as expressing, sometimes with a midrashic type of argument with which we are not very comfortable, the fundamental doctrine of the *complementary equality* of the sexes, not the inferiority of woman. In the context of ecclesiology, however, they have an additional significance.

St. Paul wished to dampen the eschatological extremism which threatened the early church. While strongly recommending celibacy, because "the time is short" (1 Cor 7:29), he also wanted to maintain the institution of the family because it is necessary to the continued existence of the Christian community during that time of expectation whose duration he never claimed to know (1 Thes 5:1-5), and of whose imminence he came to feel less and less sure (Phil 1:20-26). Consequently we find in Ephesians 5: 22-33, that Paul or his followers had begun to apply the symbolism of Christ as Bridegroom of the Church to the family so as to see in the husband a sacramental sign of the presence of Christ in the home, and to draw from it the conclusion that the husband should not exercise his headship of the family in a dominative way, but in service and care, sacrificing himself for his wife and children as Christ did for the Church. The priesthood of the father of the family[94] was in keeping with what was noblest in the old patriarchy, in opposition to what was sinful and tyrannical in it.

If the principle that the Christian family is the fundamental cell of the Church was to be maintained, then it was essential that in the Christian assembly it should be the husband who spoke for the family. Hence it was that Paul forbade women to speak in Church, not because they did not have the gift of prophecy, which they sometimes had, but that they might not publicly undercut the authority of the family in the Church embodied in the natural representative of the family, the father. Obviously in other times and cultures such a pastoral rule might be unnecessary and inappropriate. It does not in itself stand as the reason for not ordaining women, although it has often been used for this purpose;[90] yet back of it is the same fundamental symbolism which does provide the permanent reason that only men are called to priesthood.

So far I have spoken of "priesthood" in the Church. What was the origin of the three-fold hierarchy? The New Testament furnishes us only scanty information about the situation in the first century, and to attempt to make out of this evidence a case either for a single system of government throughout the Church or for the "many forms of Church government" of which some recent exegetes speak as confidently as did the older ones about a single system,[95] seems to me historically insecure. We can, however, be certain of a few things. First, Jesus himself chose the Twelve with St. Peter as their chief, gave them a special preparation distinct from that of the other disciples, and shared his mission with them.[96] Second, after the resurrection and ascension, the Twelve (reduced to Eleven by the desertion of Judas), did not hesitate to chose others to assist in their work, including the acceptance of Paul who claimed an authority equal to theirs.[97] Third, the Church of Jerusalem was presided over by James, and then by Simon; while in the churches founded by Paul he continued to act with full authority under the title of an "apostle" or one sent by an existing local church to found other churches.[94] Fourth, by the beginning of the second century, many of these same churches had resident heads who were called bishops (*episkopoi*) and this soon became universal until the Reformation, and even continued in some of the reformed churches.[98]

These resident bishops were for the local church the kind of priest I have described, i.e. the man committed to preserve the unity of the Church as a sacramental sign. They were recognized as such by the Roman government, which commonly chose them to be martyred in order to destroy the Church. To say that this priesthood was instituted by Jesus himself is to claim, with solid reasons, that the churches guided by the Holy Spirit recognized in these leaders the presence of the same Jesus who had gathered the Twelve about him and sent them forth.[99]

Traditionally the text of Acts 6:1-7 was supposed to describe the origin of the diaconate in the Church. Exegetes today reject this identification,[100] but the operative words seem to be the statement which Luke attributes to the Twelve, "It is not right for us to neglect the word of God in order to wait on tables" i.e. to take care of the distribution of food to the poor. These words express the principle which the Church early acted upon, namely, that the bishops needed assistance in those corporal works of mercy which Jesus himself had performed miraculously, but which the bishops found interfered with their own performance of their primary task of the spiritual works of mercy. Why then should they not have used their authority to choose such assistants, and sometimes also to share with them even some of the spiritual functions of preaching and baptizing?[101] What is notable, however, is that the deacon did not symbolize the unity of the Church, but its diversifed ministry. It was not so clear, therefore, that this task could not be performed by women, and there were women deaconessess in some places.[102]

The example of Jesus and the Twelve, however, gave rise to another fundamental feature of church order, the presbyterium or council of elders. Already there was both Jewish and Greek precedence for such a council.[103] If Jesus himself had always acted with a band of men whom he treated as brothers and equals while clearly holding authority as their teacher, and if he had provided for the continuation of this brotherhood by the appointment of Peter as their head, did this not provide the Church with a clear principle that the priest who symbolized the unity of the community and was totally committed to its life should strengthen himself with a council of men who were not simply assistants as were the deacons but who shared this responsibility with him? This council, therefore, was also made up of priests equal as priests to their head and able to function in his place when necessary and yet subordinate to him.

Before Vatican II there was controversy as to whether episcopacy was a distinct order from that of priests, but the Council made clear that it is only the bishop who has the plenitude of priesthood.[104] The logic of this is evident if we define priesthood not simply in terms of the function of presiding at the eucharist, but in terms of the relation to the church as sign of unity, commitment, and authority, of which presidency at the eucharist is a symbolic summation but not the complete implementation. In the beginning it seems the presbyters seldom so presided, but they came to do so as the local church was subdivided into parishes.

Thus the threefold hierarchy traditional in the Church is an expression of the model given by Jesus himself, and it is difficult to imagine how this model could be better expressed. What it leaves open is a still greater diversification of ministries. St. Paul (1 Cor 12:28) says, "God has set up in the church first apostles, second prophets, third teachers, then miracle workers, healers, assistants, administrators, and those who speak in tongues." Since he was himself certainly an apostle, and also a prophet and teacher, it seems obvious that these different gifts are not exclusive of each other. The important point is the primacy of the apostle, since Paul as an apostle asserts his authority and then proceeds to regulate the exercise of the other diversified gifts. These gifts are functional, and of course the work of the apostle is also a function, but what gives primacy to the apostle as Paul saw his own role was his ultimate responsibility for his churches.

534

> You must endure a little of my folly. Put up with me, I beg
> you! I am jealous of you with the jealousy of God himself,
> since I have given you in marriage to one husband, pre-
> senting you as a chaste virgin to Christ. My fear is that, just
> as the serpent seduced Eve by his cunning, your thoughts
> may be corrupted and you may fall away from your sincere
> and complete devotion to Christ (1 Cor 11:1-3).

Paul's imagery here is significant. He does not dare to speak of himself
as the bridegroom of the Church but he clearly implies that he stands,
as it were, in the place of God who was spouse of the People and of Christ,
the bridegroom of the Church, and he thinks of the Church as the archetypal woman, Eve, Bride of the New Adam, Bride of the Covenant.
Granted the presence of the apostle-bishop committed to the unity of
the Church as head, the diversity of gifts among the people is not divisive
but enriching, as Paul shows in 1 Corinthians 12.

St. Thomas Aquinas, contrary to some later theologians, taught that
the diaconate can be subdivided by the Church into many "minor
orders."[105] Thus there seems room in the essential structure of the Church
to institute many forms of ministry open to women, although ordination to these is unnecessary, except when for particular reasons there
needs to be a public recognition of their function within the Church. For
the action of the People of God in its ministry to the world, no special
commission is required beyond baptism and confirmation, except when
it is essential that these ministries explicitly express the unity of the
Church, e.g. when a theologian is called to teach in the name of the
Church, rather than as a private scholar.

III. The Mother of God

1. The Mother of Jesus

Joseph Fitzmyer, S.J., to whose commentary on Luke I have previously
referred, shows that in Luke's Gospel Mary is representative of the model
disciple, the one who truly believes.[106] It is generally said that the special
veneration given to her in both East and West, which the Reformation
tended to regard as at best an exaggeration and at worst idolatry, was first
given official status by the Council of Ephesus in 431 which canonized
her title of *Theotokos*, "Mother of God", not primarily to honor her but
to refute the heresy of Nestorius who had distinguished the human person of Jesus from the divine person of Christ. Of course well before that
in the so-called Apostle's and Nicene Creeds, she had been honored as

the Virgin Mother of Jesus. [107] The Middle Ages developed an ever more elaborate theology of her virtues and "privileges", and in the Roman Catholic Church this culminated in the dogmatic definitions, invested with the full weight of infallibility, of the belief that she had been exempt from original sin from the first moment of her personal existence in preparation to be the Mother of God (1854), and that when her earthly life was completed she was immediately assumed body and soul to be present with her Risen Son. (1950)[108]

Vatican II, without going so far as a new dogmatic definition, declared Mary to be the "Mother of the Church." The Orthodox object to such declarations by the Bishop of Rome apart from an ecumentical council in which their bishops participate, although their own traditions to the truth of these beliefs was part of the grounds for their definition by Rome. For Protestants, however, and for some Catholic theologians, it is a source of amazement that Catholics can accept as truths of revealed faith an account of Mary of Nazareth which seems so far beyond what the Scriptures tell us, and even further beyond anything in Scripture that can be historically verified. It seems to them like the acceptance of myths as dogmas.

In my opinion, the Catholic understanding of the role of Mary in the plan of salvation is, as it were, a summary of the theology of the body and its historic development in the Church shows how the guidance of the Holy Spirit has overcome the dualistic influences of Platonism on Christian thought. To understand in what sense this Mariology is based on the Scriptures, however, requires a special hermeneutic, which I have already employed to a certain extent in discussing the visions narrated in the New Testament as providing us with a Christology from within, but which requires a few more words of justification which will recall and develop certain points already made in Chapter 9.

Any transcendent religion, such as that proclaimed in the Bible must of necessity be symbolic, because the realities which it reveals surpass the human mode of knowing which is always derived from *bodily* experience. The symbols used are taken from this bodily experience, and the archetypical ones are derived from the human body itself and the familial relations which arise from the sexuality of the human body-person, because these are primordial in our experience, and transcultural and transhistorical in their universality. Symbols are always ambiguous and polysemic so that the meaning of any given symbol must be contextually determined by its relations to other symbols within a symbol system.[105] It is the task of theology to clarify this ambiguity by attempting as well as it can (recognizing all the while that the symbol is inexhaustible), to reduce the symbol to explicit *analogical* terms. For example, the biblical

536

expression "Lamb of God" is interpreted as "sacrificial victim", but here too the notion of "sacrifice" is used only analogically.

The term "symbol", however, should not be understood in the idealistic sense in which Paul Tillich and other modernist theologians use it, to imply that the symbol receives its meaning from the one who interprets the symbol rather than from the one who uses the symbol to reveal. In Tillich's theology, God does not really reveal Himself to us by biblical symbols, but rather, we are stimulated by such symbols to project on the unknowable One our notion of God which serves as a heuristic, unifying symbol for our own life experiences in a Kantian manner, just as for Bultmann the historical reality of Jesus is unimportant as long as He serves as a paradigm for our own existential decisions. But, we read in the Bible that God really does speak to us, using, however, the only language which we can understand, i.e. in symbols which require analogical interpretation.[109]

In this latter sense, it is acceptable to speak of the Bible as a system of religious "myths", i.e. of narratives which require a symbolic interpretation, but not in the sense that a "myth" is contrary to "history." Because God is Lord of history, He can communicate with us through historical events which are at the same time symbols, e.g. the Exodus was an historical event with a symbolic, religious significance. From this coincidence of history and myth, a strange consequence follows which is scandalous to the modern historian, namely, that in the case of historical events whose essential revelatory meaning is known, it becomes possible to determine the historicity of details by their consistency with this essential meaning, just as from the general tenor of a text it becomes possible to restore its lacunae. Consequently the writers of the New Testament, once they had determined that Jesus was the fulfillment of the Old Testament prophecies, shaped their narratives to bring out this correspondence in details with a confidence that the modern historian who views history as in itself quite meaningless would regard as pure fiction. I do not mean by this that the New Testament writers simply invented these details but that when they discovered them in floating traditions they selected those that seemed appropriate with the confidence that they must be the true traditions because they "made sense."

This same process continues in the Church as a community of tradition and of common faith. Meditating on the data of tradition in light of the clue to its meaning given by those articles of faith already explicit in the official preaching of the Church, the community gradually draws out of the symbols of this tradition meanings which to the historian appear new, but to the believer are seen as already germinally contained in these symbols and guaranteed by the *analogia fidei*, the mutual interrelation of all the elements of the Faith. Thus once it was seen that Mary of Nazareth was chosen by God as the virgin mother of the Son of God (and this is clearly asserted by Luke and Matthew in the canonical Gospels), then for believers it gradually became evident that a woman chosen by Divine Providence for this role in cosmic history must have been prepared by God in such a way as to be entirely fitted for her role.

Of course historians ordinarily would sin against the very nature of history which, as we saw in Chapter 9, is essentially *contingent*, full of incongruities, if they attempted to deduce a historic fact from some notion of what *should* have happened, what would be fitting or appropriate. Such deductions presuppose metaphysical necessity, or at least the necessity of determined natural law, while the contingency, or pure facticity of history lacks such necessity. While it may appear fitting, and even statistically probable that a great and good man had a great and good mother, we know very well that in a given case his mother may have been a bad one indeed. Nevertheless, in the case of the great, revelatory events of history, the Christian knows that a higher necessity is operative, namely God's infallible will to fulfill His promises and to communicate His message unambiguously to humanity. Consequently, it is possible in such cases to be sure that what *should* have happened *did* in fact happen.

Thus, as John Henry Newman showed so well,[110] the Catholic understanding of biblical revelation permits a "development of doctrine" under the guidance of the same Holy Spirit who inspired the writing of the Bible in the first place, not by new revelations, but by the unfolding of the meaning contained in the biblical symbols. Some of the efforts to explain this development have erred by supposing that what was in question was an abstract logical deduction of a new article of faith from antecedent articles, or from one article of faith and another having metaphysical certitude, or that it was an explication of what was already implicitly contained in some abstract concept.[111] Such theories are too rationalistic and fail to take into account the symbolic (rather than the univocally conceptual) language of revelation. The explication of the implicit meanings in a symbol takes place not by logical deduction but by the exercise of the poetic imagination, aided by empathetic affectivity, resulting in more penetrating intutions which discover new analogical relations.

The development in the Church of Mariology is a prime example of this kind of intuitive, symbolic reflection, and it first of all took place not in the study of theologians but in the devotional life of the people and among the poets, musicians and artists who invented the liturgy and constructed the churches, as well as among the contemplatives. It is the fruit of meditation, prayer and human anguish and need. This process is already at work in the New Testament in the infancy narratives of Matthew and

Luke and in Marian symbolism of the Johannine writings;[112] and the apocryphal infancy narratives of the second century show that it was more widespread than its comparatively small space in the New Testament might indicate. As this develpment continued, it enriched itself from contact with the myths of Hellenistic culture and in the Middle Ages with those of Celtic and Germanic barbarism. The Reformers were shocked to realize how much of this popular Marian devotion could not be explicitly documented in the Bible, and this scandal increased in the seventeenth century, as theology more and more fell under the spell of Cartesian rationalism. Today, fortunately, as a result of the romantic reaction to rationalism, we are less allergic to mythical thinking. Nevertheless, academic theological elitism still finds itself very uncomfortable with popular religion, which it strives to "demythologize."

The Magisterium has always viewed these Mariological developments with caution and only as theologians have sifted through these popular devotions and the critical historians and exegetes eliminated the fantastic, has it been willing to give them liturgical, let alone dogmatic approval. It took a pope of the temperament of Pius IX to dare to move toward a dogmatic definition of the Immaculate Conception, over which the Councils of the Church had long hesitated, or an event like World War II with its Holocaust to move Pius XII to declare the dogma of the Assumption. In both cases, what was in question was not the refutation of a heresy (as in most dogmatic definitions), but the displaying of an ancient symbol in a new light to speak to the hearts of the people in a language more penetrating than theological pronouncements. In both cases, what was symbolized in Mary was the Church. Pius IX saw in the lovely Immaculata the truth that the Church was infallible in its faith in the face of modern unbelief which came to a climax in his time, and Pius XII hoped that the radiant Assumpta would be a rainbow of hope to an age of death and despair. Vatican II, by declaring Mary the Mater Ecclesiae summed up its own work in trying to give the Church a new self-understanding.[113]

I have quoted Fitzmyer as saying that Mary is presented by Luke as the model disciple of Christ who says, "I am the servant of the Lord. Let it be done to me as you say" (1.38). This is echoed again in Jesus' saying, "My mother and my brothers are those who hear the word of God and act upon it" (Luke 8:21) and his answer to the woman in the crowd who cried out "Blest is the womb that bore you and the breasts that nursed you!" — "Rather . . . blest are they who hear the word of God and keep it" (Luke 11:28). Thus Mary's role in salvation history if not simply a biological but a human one. It is through her faith and obedience that the Son of God has become one of us from her body. I have argued earlier in this chapter that Luke also wishes to show that this faith of Mary is not to be understood merely as her own act as an individual, but that she speaks for the whole of faithful Israel from Abraham through all the centuries of its suffering and trial. She is a corporate personality in whom the patriarchs before the Covenant and all the people of the Covenant say to God at the critical moment, "I am the servant of the Lord." More than that, Luke, who stresses that the Savior has come not just for the Jews but for all humanity, traces the genealogy of Jesus all the way back to Adam, whom he calls "son of God" (3:38), so that in Mary the whole of the human race which has not rejected the light given to them speaks with one voice.

If all of humanity and all the Chosen People speak in Mary in faith, then this faith is not merely hers as an individual, but the faith of her people and of all graced humankind. It is the Spirit of God who speaks in her, and what she says is not merely her word but the prophetic word. In speaking it by the Spirit, she conceives the Word of God in her soul and in her body in a new act of creation. In her mere human lowliness she can say

> "My being (*psyche*, i.e. the human body-person
> as living with human life)
> proclaims the greatness of the Lord,
> My spirit (*pneuma*, i.e. the faith that fills her)
> finds joy in God my savior,
> For he has looked upon his servant in her lowliness;
> all ages to come shall call me blessed.
> God who is mighty has done great things for me,
> holy is is his name (1:46-49).

As lowly, nothing before God, she is the original chaos or nothingness out of which God created the primordial sinless world, culminating in the New Adam. This whole divine work of the new creation which culminates in the New Adam, who is Son of God in even a greater sense that the first Adam was, has taken not merely the symbolic seven days of creation (which today we understand as the millions of years of evolution), but the whole long history of the human race and the tragic history of the Chosen People to complete. Mary is the new garden of Eden which God has prepared as the dwelling place of the New Adam. As such God intended her to be the Chosen People if they had been faithful. All that we read in the prophets about the future glory of Jerusalem ready for the coming of the Messianic King could not have failed, because God cannot fail. He flowered in Mary and will triumph definitively at the end of history.

Therefore, in spite of the many infidelities of which the Jews through their prophets accuse themselves throughout the long course of the Old

Testament, nevertheless the Spirit of God was infallibly accomplishing its work in the Remnant. In Mary, this work had to come to its perfection so that she could speak the word of faith with *perfect* love, or rather that the Spirit could speak through her without any hindrance. She is thus the masterpiece of God's creative love prepared for God's own entrance into His world. As she is the original Paradise, so also she is Mount Sinai, she is the Tabernacle of the desert, the Ark of the Covenant, the first, second, and third Temple, the city of Jerusalem, all those symbols of the Old Testament which express the presence of God in the midst of his people. And all that is said in the Law about the sanctification of this resting place of God had to be and was fulfilled.

To deny this fulfillment in Mary would be to deny the whole trajectory of the Old Testament which shows us the Spirit of God moving through history, irresistibly preparing a place for the Presence. Because the Church cannot deny that this work of God in preparing a place for His Son (not merely a physical place, but the human heart to receive Him in perfect faith) was surely accomplished and will be perfected, it has to confess that Mary as a person was never touched by original sin, since if there had been in her any trace of sin, even in its effects, she could not have spoken the perfect word of faith, she would have (in the terms of the Old Testament) been ritually impure.

To many this doctrine of the immaculate conception of Mary seems to remove her from ordinary humanity, but this is precisely to misunderstand what the doctrine means. All humanity descended from Adam receives its existence in the state of original sin in the sense I have earlier explained,[114] but it has not remained in that state. From the beginning the Spirit of God has been converting those who have not rejected His light back to God in faith, but their conversion is imperfect and they continue to struggle with sin, as the Bible shows in its narrative of human failures. Yet the Bible also shows that this work of conversion by the Spirit is progressive, an education and disciplining of the human race, a gradual restoration which could have been completed in the Jewish nation if as a whole it had received Jesus as its Lord in his first coming, and will nevertheless be completed in the Reign of God at his return. But it is already *proleptically* (in anticipation) completed in his first coming in Bethlehem when the Reign of God had its epiphany. It was in a lesser but real sense also completed in the formation of the Covenant with Israel, and in each son of the Covenant in circumcision. More truly, it is completed in each Christian at baptism when original sin in its essence is taken away, although not in all its effects.

Historically it was only in the person of Mary that this preparatory work of grace was complete, which is to say, that she was spared original sin in its essence and in its effects. What is done for every Christian in baptism was done for her in absolute plenitude, not after birth but in her very coming into existence in the womb. In teaching the virginal conception of Jesus, Luke argues (symbolically not dialectically), that if in the Old Testament the Spirit gave Israel its saviors by a miracle that removed the barrenness of their mothers, so the birth of the ultimate Savior must have been by the still greater miracle of a virginal conception. By the same symbolic argument it follows that if God prepared the prophets from the womb to speak the word of God, He must have prepared the prophetess who was to speak that Word in its actual presence from the very moment of her personal existence. Thus Mary Immaculate is not an isolated event, but the Daughter of all humanity in its conversion from sin through the power of the Spirit.

But does not this make Mary somehow a *duplication* of Jesus? Of course she is created by grace in his image as every Christian is reborn in Christ; but her special role is to be the model disciple, the perfect Christian in whom that image of Christ has its perfect reproduction, firstly because she is the culmination of the whole Old Testament which had to be ready to receive Emmanuel, the Temple of his presence, and secondly because she is the symbol of the Church, the Reign of God, the New Jerusalem where he will dwell forever. She is the past and the future of which Jesus is the present.

We must ask ourselves however, how Mary in her perfection of faith is related to Jesus, who is the object of that faith? This is the same as to ask how Israel was related to God in the Covenant and how the New Israel is related to Christ. *Revelation 21:1-22-5* sums up all the symbols of the Old Testament by showing us the heavenly Church as the New Jerusalem descending on earth as the New Creation, the New Garden of Eden, the New Jerusalem and calling this Church "The Bride of the Lamb." *Ephesians 5:22-33* uses the same imagery of the Church as the Bride of Christ and adds to it the metaphor of Christ as Head and the Church as Body. These texts obviously reflect the meditations of the early Church on the symbols of the Old Testament reinterpreted in the light of symbols used by Jesus in his parables. Basic to them all is the Genesis symbolism of the Garden of Eden and Adam and Eve, which I discussed at length in Chapter 10. Although nowhere does the Bible call Mary the New Eve, the unfolding of this symbolism is found in the Fathers of the Church as early as St. Justin Martyr and St. Irenaeus in the second century.[115]

Once again I hear the objections of Christian feminists. Is it not sexist thus to identify Mary the Woman with the nothing out of which the world was created, the garden prepared for the Man Adam, the Eve who fell into sin, the Jewish people who proved unfaithful, the Temple which is only a building with Jesus who is a person, etc.? Is not this symbolism one grand put-down of womankind in relation to mankind, which is always given the superior aspect of the symbolic duality? To this objection I grant that when this symbolism is misinterpreted by sexists it does become a put-down, but that is true of every symbolism, which by its analogical character can always be reversed to mean the opposite of what was intended. The only way to prevent this from happening is by a correct reading of the symbol.

The hermeneutic clue to the right reading of this symbolism which renders it liberating and not oppressive is to be found, I believe, in the explanation already given of why God become incarnate as a male and not as a female, namely, to show men that true manhood is not to be achieved by violence and rape but by a self-giving love that accepts the role of servant in relation to the woman. Significantly, St. Matthew in his infancy narrative gives us the portrait, in few but sufficient words, of Joseph, the man called by God to live in virgin marriage with Mary, protecting her and her son in faith.[116] It has taken even longer in the Church for the meaning of this symbolism of St. Joseph to be understood than that of Mary; but it is no accident that in our times this devotion has been given more official recognition.[117] No doubt it was from St. Joseph that Jesus learned to be a true man who, as many feminists admit, was the only great man in history who was not a sexist. On the other hand, the meaning of Mary's virginity in her motherhood is not that she was sexless, but that she achieved her full realization as a person *independent* of any man. As we have just seen, that full actualization as a person did not consist merely in being a biological mother, but arriving at perfect faith, i.e. at the heights of mystical contemplation.

In the contemporary discussion of the ordination of women, it is assumed by many that in the Christian Church the highest status is that of the priest, or the bishop, or the Pope. Therefore, it is argued, the Church puts down women because it excludes them from the highest place. The trouble with this argument is that the assumption is false. In authentic Christian understanding the highest place in the Church does not belong to the clergy as such, but to *contemplatives*. The clergy is ordained to ministry which is part of the *active* life of the Church, but Christianity is in agreement with the Greek view of both Plato and Aristotle that the contemplative life is superior to the active life. The writings of many of the Church Fathers is filled with laments that they were compelled to leave the monastery to become bishops out of a sense of duty to the Church. It is Luke again who gives us the story of Martha and Mary (10:38-42), to make the point that Mary who listens to the word of God has chosen the better part, rather than Martha who performs a ministry.[118]

As woman of faith who "treasured all these things and reflected on them in her heart" (Luke 2:20), who is a prophetess (1:46-55) filled with the Spirit, Mary is also the embodiment of the Wisdom figure of the Old Testament (Proverbs 8, etc.). This Wisdom is higher than reason, it is the creative intelligence,[119] the loving wisdom which is described as feminine in order to contrast it with the lower discursive and manipulative wisdom which seeks power and technical control, and is therefore masculine. It is the wisdom which is the gift of the Holy Spirit, leading to mystical contemplation and which is found in such women as St. Catherine of Siena and St. Teresa of Avila, whom Paul VI declared Doctors of the Church. Mary has that higher contemplative wisdom which is superior to the wisdom of theologians. In the contemplative life, not only are women equal to men, but as many spiritual writers have pointed out,[120] they have a natural aptitude for contemplation commonly lacking to men. Plato represents Socrates as having to go to the woman Diotima to learn the highest secrets of wisdom,[121] and in all the great male theologians of the Church we find a great devotion to Mary. In the lives of not a few, we find that it has been the influence of a woman which has led them on from their theological rationalism to enter into the mystical way.

If we return to the notion of the complementary equality of the sexes developed in Chapter 10, we recall that Adam and Eve were created to complete each other so that both could equally attain the perfection of human personhood; the man learning through the woman to discover his own femininity and the woman through man her own masculinity. Adam was head of this family, Eve its body. In any theology colored by Platonism (and in his discussion of woman even St. Thomas Aquinas, the least Platonic of theologians, shows traces of such influence),[122] to say that femininity is related to masculinity as form to matter, as body to head, is obviously a subordination of inferior to superior, and it is this understanding that has given these biblical expressions a sexist interpretation. If, however, we understand the relation of matter to form and of body to soul in the manner developed in Chapters 7 and 8, a different interpretation emerges.

In primary units existing in process, the relation of matter and form is one of mutuality, in which each is superior and inferior to the other in different respects. St. Paul, thinking biblically, tried to express this point when he wrote to the Corinthians (I Cor. 11:7b-13):

> A man . . . is the image of God and the reflection of his glory. Woman, in turn, is the reflection of man's glory. Man was not made from woman but woman from man. Neither was man created for woman but woman for man. . . . Yet, in

the Lord, woman is not independent of man nor man independent of woman. In the same way that woman was made from man, so man is born of woman; and all is from God.

Thus Paul sees in the *Genesis* narrative that man is an image of God and woman an image of man since she was made out of his side to be his partner. In all these respects woman seems inferior to man. Yet, on the other hand, men are inferior to women in that they are born from their mothers by the power of God (Eve said, "I have given birth to a man-child with the help of the Lord" (Gen. 4:1). Hence both men and women are dependent on each other and on God. The superiority of the man is not with regard to personhood but with regard to the masculine role in the family and the superiority of the woman is not with regard to personhood but with regard to the feminine role in the family. In matters that do not pertain to the family they are equal.[123]

Yet, although in the family each sex has superiority with regard to its own role, does it not remain that the man's role itself is superior to the woman's role? Is not that expressed by saying that he is the head and she the body? Not at all, if we keep in mind the profound mutuality implied in the matter-form analogy. Form is superior in that it gives actuality to the matter while the matter is a passive principle; but the matter is superior to the form in that it pre-exists the form and perdures even after the form is corrupted. Thus although the form gives essence and also actual existence to the matter, it can do so only because the matter pre-exists every new form it receives. Moreover, while the form and matter mutually limit each other, the matter always remains as an *infinite* capacity for new forms, it is always open to the Divine Creativity, while form, as it specifies, also limits and renders finite.

I must emphasize that in all this discussion we are speaking of symbols, of metaphors, of the right use of analogy, and analogies must not be reduced to univocities. In every analogy, the difference is greater than the similarity. Consequently, it is *more* true that women are the *equals* of men as persons, than that they are their inferiors as matter to form, head to body; or conversely it is more true that men are equal to women as persons, than that they are inferior to them because form can only exist in matter or the head can only live and know in dependence on the body. In the individual person, man or woman, these sexual relations are secondary, although substantial and significant, because they indicate analogously the division of labor which naturally exists in the family, and out of which we derive the categories and symbols by which all our human thinking is shaped.

Thus it becomes clear that the presence of God in the human world could not have been merely through the man Jesus, but required also a woman, Mary. The image of God would not have been fully manifested to us only in Jesus as a man, but requires a manifestation in a woman. Why then did not God become incarnate in a human pair, an Adam and Eve? Perhaps that would have been metaphysically possible, but it is not so difficult to understand why in fact this was not what God chose to do in order to manifest the Divine Presence in the world. That Presence is in time but it is for eternity. It points beyond this world of birth and death, and therefore beyond the sexual relationship to a spiritual relationship. Consequently, the manifestation of God in one human being pointed more clearly to that eternal unity of all persons in the unity of God. I have indicated already what seems to me the reason that a male humanity was chosen rather than a female one, not to exalt masculinity but to humble it, since the Incarnation is not for God's sake but for the sake of the service of humanity in its collectivity, which symbolically is the Woman, "the mother of all the living", the New Eve (Gn 3:20).

By the *spiritual* marriage between the New Adam and the New Eve, a union which in its virginity and its identification of the bridegroom with the son of the bride completely transcends the realm of sexuality and of physical birth and death, the Christian Community or Church is raised to complementary equality with Emmanuel, God Present. In the *Song of Songs* this mutuality of true love is shown when the Lover begs, "Open to me, my sister, my beloved, my dove, my perfect one" (5:2), and so Jesus and Mary are as brother and sister, as Adam and Eve were equally created in God's image (Gn 1:27). It is true that the bridegroom is God and the bride is only human, but the bridegroom owes his humanity to the bride, and the bride owes her deification through grace to the bridegroom. In perfect love there are only equals in mutual self-giving. Thus the union between Christ and His Church, made up of all the faithful, is a union in one flesh and one Spirit.

Therefore, these symbols, when rightly interpreted and deeply felt in their true meaning, far from promoting sexism, can only teach the love of God for his people and their love for him, and therefore for each other; a love which liberates each person to be fully her or himself, because it is not merely an erotic love, seeking one's own satisfaction, but an agapetic or generous love seeking the happiness of others in service to them.

2. True God

In the forgoing attempt at a Christology from within-and-without, that is, in terms of the relation of the self-awareness of Jesus to his bodily existence, while assuming without discussing the Christian dogma that in being true man Jesus is also truly the Son of God in some absolute sense that makes him Emmanuel, God present in the world, I have tried to show how his Presence was manifested in his human experience and the experience of the early Church which saw, and heard, and touched him. I have not yet dealt with the question of how this real human being could be simply and without qualification God Himself, yet not God the Father or the Holy Spirit, since the Abba experience clearly implies that he is truly son and yet not the Father. Moreover, in the visions we have studied, the Spirit is also experienced by Jesus as dwelling in himself and empowering him with a unique plenitude, yet as a person distinct from himself and the Father. This mystery must be explored more fully in the last chapter of this book in the light of the Resurrection, Ascension, and Pentecost, where it was most fully revealed.

At this point of the discussion I will only note three sources of confusion in Christology which I hope the data of this chapter have shown to be misconceptions. First, many people read the traditional formula, "Jesus was a Divine Person in two natures, one divine and one human" to imply that because Jesus was a Divine Person he therefore lacked something of human personhood. Leaving aside the various metaphysical theories as to exactly how the *hypostatic union* can best be expressed, it is evident that any such formula must be understood in a sense consistent with the Scriptural teaching that Jesus was like us in everything but sin (Heb 4:15). Therefore, whatever is *positive* in the human person, including what is *positive* about personhood itself must be attributed to Jesus. What is often not appreciated is that in the language of Greek philosophy, "person" contains not only the positive connotations of primary existence (the primary natural unit of Chapters 7 and 8), and of intelligence and freedom, but in the case of created persons (and particularly of body-persons), it also implies something *negative*, namely a limitation or closure of being by which one person is enclosed from communication with another, a barrier which can be overcome only by an effort of reaching out and establishing contact through language or physical interaction. It is this negative limitation or closure separating the human person from direct communication even with God (although God is so much closer to the depths of the person than one created person is to another), which is lacking in the human reality of Jesus in relation to his divine personhood, and only this. Obviously this lack of limitation to his reality does not in any way make him less a human person, but it does make it possible for him to be more than a human person. Hence in being a Divine Person he is as much *human* person as any of us, but He is infinitely more a *person* than any created person can be.

Second, great difficulty is felt in the notion that if Jesus was both God and man there would have been in him two centers of self-consciousness, which seems contrary to the notion that a person is by definition a single such center. Such a difficulty seems based on a Cartesian conception of consciousness and fails to take into account the obvious fact of experience that in a single person there can be several levels of awareness. All of us experience simultaneously in ourselves both a level of sense awareness and of intellectual awareness, and today it is generally conceded that we have a subconscious or peripheral awareness. I have argued in Chapter 8 that within intelligence there is a distinction between our super-consciousness or creative intelligence and our rational ego and its verbalized consciousness. When we speak of "self-consciousness" we ordinarily refer to the unification at the level of the ego of these kinds of awareness; but our deeper self-identity is at the level of the super-consciousness.

In the writings of mystics we find that they are very intensely aware of these levels of consciousness, and can speak of the experience of deep and peaceful union with God in the midst of great sensory suffering and even of intense anguish of the practical ego. Certainly Jesus must have experienced these interior depths of his human soul. If it is possible for a human person to have a unified self-consciousness which includes several levels of awareness, what is the difficulty in supposing that as Divine Person Jesus also experienced within his divine self-awareness a human self-awareness at a lower level?

The third difficulty is how to understand the way in which God is related to *this* particular individual humanity in a way different from His relation to all humanity. If God is immanent in every human being, how was He present in Jesus Christ in an absolute sense? In Indian Vedanta, the goal of mysticism is to discover one's authentic self (*atman*) and then to realize that this authentic self is identical with the universal Self.

In the next chapter I will attempt a reply to this difficulty, and I will stress the importance for the Christian world-view of God's free *creation* of the world *ex nihilo* and the infinite difference between God and creatures that this implies. Consequently, the union between a human being and God must be a union of *grace* by which God bridges this infinite abyss to unite the creature to its Creator. Since this wonderful union is accomplished by a *free* act of God, it is not strange that He has chosen freely to accomplish it by the mediation of incarnation in a single member of the human race, Jesus of Nazareth, and through him as head of a new spiritual race to adopt all humanity. This God had done by creating this humanity of Jesus as his own by which he is *personally* present in our world as one of us. Anything less than this *personal* presence would fall short of his promises and his generosity. This is why such a Christology as that of Hans Küng, quoted at the beginning of this chapter, according to which Jesus is said to have been "God's advocate and deputy, representative and delegate", even if he was commissioned to present God's "call, offer, and claim"; in its "Uniqueness, underivability and unsurpassability", falls so infinitely short of saying who Jesus really is.

Notes

1. Piet Schoonenberg, S.J., *The Christ: A Study of the God-Man Relationship in the Whole of Creation and in Jesus Christ* (New York: Herder and Herder, 1971). See Steven Pujdak, S.C.J., "Schoonenberg's Christology in Context", *Louvain Studies* 6 (1976-77); 338-353 for a defense of Schoonenberg. For other recent Christologies see Chapter 9, note 99.

2. Hans Küng, *On Being a Christian* (Garden City, N.Y.: Doubleday, 1976).

3. Edward Schillebeeckx, *Jesus: An Experiment in Christology*, (New York, Seabury, 1979); *Christ: The Experience of Jesus as Lord* (New York: Crossroads, 1981).

4. See Chapter 8, p. 314.

5. Raymond E. Brown, S.S., *The Gospel According to St. John*, The Anchor Bible (Garden City, N.Y.: Doubleday, 1966), Introduction, pp. xxviii-xl, discusses Bultmann's theory of a Revelatory Discourse Source and rejects it.

6. See Brown, *Ibid.* pp. lii-lv.

7. Note 3 above.

8. *Ibid.* p. 658 f.

9. See Chapter 9, pp. 352-355 and Chapter 10, pp. 418 f. and Raymond E. Brown, *The Critical Meaning of the Bible,* (New York: Paulist, 1981).

10. Thus Rudolf Bultmann, *The History of the Synoptic Tradition* (New York: Harper and Row, 1963), says of the account of the baptism that "Admittedly it would not do to psychologize it and talk of a 'call' story and reckon its content as a calling vision" (p. 247) because it says nothing of Jesus' inner experience, commission or answer to a call. "The legend tells of Jesus' consecration as messiah, and so it is basically not biographical, but a faith legend." p. 248. The Temptation is a "scribal haggada" (p. 253-259), the Transfiguration is a retrojection of a Resurrection account (pp. 259-261), and the Agony in Gethesemane a faith legend (pp. 267-268). These judgements are made on form-critical grounds, but it is questionable whether the literary models which the evangelists use for their narratives are a sufficient criterion on which to judge the nature of the materials cast in these forms.

11. See E. Earle Ellis "Prophecy in the New Testament Church and Today" in J. Panagopoulos, *Prophetic Vocation in the New Testament and Today* (Leiden: Brill, 1977), pp. 46-57. J. Reiling, "Prophecy, Spirit and Church" in the same volume pp. 58-76. Also Robert Joly in his edition of Hermas, *Le Pasteur,* Sources Chrétienne no 53, 2nd ed., (Paris: Cerf, 1968), pp. 46-54, and J. Reiling, *Hermas and Christian Prophecy* (Leiden: E. J. Brill, 1973), pp. 5-20. While the students of *The Pastor* generally consider that it is not the record of actual visions, but that it uses the vision form as a literary device as Dante was to do later, the recorded visions of the saints make it clear that "imaginary" visions and literary imitations are not easy to distinguish by internal criteria.

12. *Op. cit.* p. 752.

13. See Raymond E. Brown, *The Birth of the Messiah* (Garden City: N.Y.: Doubleday, 1977), pp. 57-95, on a comparison of the Matthean and Lukan genealogies. This work of Brown, along with his *The Virginal Conception and the Bodily Resurrection of Jesus* (New York: Paulist, 1973) and the report *Mary in the New Testament* (New York: Paulist, 1978) prepared by him, Joseph A. Fitzmyer, S.J., and two Lutherans Karl P. Donfried and John Reumann constitute one of the most thorough recent investigations of the Scriptural foundations of the creedal doctrine of the Virgin Birth.

14. See Chapter 9, pp. 352-355.

15. See Pierre Grelot, *The Bible Word of God* (New York: Desclée, 1968), pp. 82-99 and 112-138 on the literary forms in Scripture.

16. For St. Francis see Marion H. Habig ed., *St. Francis of Assisi: Writings and Early Biographies (Omnibus of Sources)* (Chicago: Franciscan Herald Press, 1972); Omer Engelbert, *Saint Francis of Assisi*, 2nd English ed., revised and augumented by Ignatius Brady and Raphael Brown (Chicago: Franciscan Herald Press, 1976) and John R. H. Moorman, *The Sources for the Life of St. Francis of Assisi* (Manchester: Manchester University Press, 1940), republished Farnborough: Gregg Press Limited, 1966). For St. Catherine see Raymund of Capua, *The Life of Catherine of Siena*, translated, introduced and annotated by Conleth Kearns, O.P., (Wilmington, Delaware: Michael Glazier, 1980), especially pp. lx-lix on the critical controversies with regard to this life and the bibliography pp. lxxv-lxxxix.

17. See Bultmann, *History of the Synoptic Tradition* (note 10 above) pp. 250-261. Vincent Taylor, *The Gospel according to St. Mark* (London: Macmillan, 1952), pp. 386-387 lists numerous authors for and against this theory. More recently Hugh Andersen, *The Gospel of Mark* (London: Oliphants, 1976), p. 22 f concludes that we simply don't know what lies behind this account.

18. See Karl Rahner, S.J., *Visions and Prophecies* (New York: Herder and Herder, 1963) for the theology and psychology of visionary revelations. See Juan Arintero, *The Mystical Evolution in the Development and Vitality of the Church* (St. Louis, B. Herder Book Co., 1951), 2, pp. 304-33, for discusion based on the spiritual writings of Christian mystics.

19. See Brown, *The Birth of the Messiah*, note 13 above, p. 143-153 on Isaiah 7:14. A helpful recent article on the whole prophecy is Joseph Jensen, O.S.B., "The Age of Immanuel", *Catholic Biblical Quarterly*, 41 (1979): 220-239.

20. Joseph A. Fitzmyer, S.J., *The Gospel According to Luke I-IX*, Anchor Bible (Garden City, N.Y.: Doubleday, 1981), pp. 334-355, see the up to date bibliography pp. 352-355. Some works that have been especially helpful are: José M. Boyer, S.J., "Cómo conciben los Santos Padres el misterio de la divina maternidad: La Virginidad, clave de la maternidad divina." *Estudios Marianos* 8 (1949): 185-256; René Laurentin, *Court Traité sur la Vierge Marie*, 5th ed., (Paris: Lethielleux, 1967); M. J. Nicolas, O.P., *Theotokos: Le Mystère de Marie*, (Paris: Desclee, 1965); F. M. Braun, O.P., *Mother of God's People* (Staten Island: Alba House, 1967); Jean Galot, "La conception virginale du Christ", *Gregorianum* 49 (1968) 637-666, and "La filiation divine du Christ", *Ibid.*, 58 (1977) 239-274; Anton Vögtle, "Offene Fragen zur lukanischen Geburts-Kindheitsgeschichte" in his *Das Evangelium und die Evangelien* (Düseldorf; Patmos Verlag, 1971), pp. 43-56; Pierre Grelot, "La naissance d'Isaac et celle de Jésus", *Nouvelle Revue Théologique'*, May, 1972: 462-487 and June, 1972: 561-585; *Jesus et sa Mère d'après les récits Lucaniens de l'enfance et d'après Saint Jean* (Paris: Gabalda, 1973), Robert L. Humenay, "The Place of Mary in Luke: A Look at Modern Biblical Criticism", *American Ecclesiastical Review*, 168 (May, 1974): 291-303; Antonio Ammasari, "La Famiglia del Messia", *Bibbia et Oriente* 19 (1977): 195 ff.; John A. Saliba, S.J., "Virgin Birth and Anthropology," *Theological Studies*, 36 (1975): 428-454; Antonio Salas A.S.A., *La Infancia de Jesus (Mt 1-2) Historia o Telologia?* (Madrid: Ed. "Biblia y Fe", 1976); James T. O'Connor, "Mary, Mother of God and Contemporary Challenges", *Marian Studies* 29 (1978): 26-43; Giuseppe Danieli, "Maria e i fratelli di Gesú nel Vangelo del Marco", *Marianum*, 40 (1978): 91-109; I. De la Potterie, S.J., "La Mere et la conception virginale du Fils du Dieu", *ibid.* pp. 41-90, and Lucien Legrand, *L'Annonce à Marie: Une apocalypse aux origines de l'Évangile* (Paris: Cerf, 1981), René Laurentin, *Les Évangiles de l'Enfance du Christ,* 2nd ed. (Paris: Desclée, 1982) extensive bibliography pp. 549-528.

21. Current scepticism about the Virginal Conception is epitomized by Hans Küng, *On Being a Christian,* (Garden City, N.Y.: Doubleday, 1976), who, after asserting that this doctrine "does not belong to the center of the Gospel'', p. 456 (which can be conceded, in my opinion, only with qualification, considering its place in all the ancient creeds), goes on to express what he considers "acceptable" for faith today as follows: "Although the virgin birth cannot be understood as a historical-biological event it can be regarded as a meaningful *symbol* at least for that time. It would symbolize the fact that, with Jesus who closes and surpasses the Old Covenant, God has made a truly new beginning; that the origin and meaning of his person and fate are ultimately to be understood, not from the course of the world's history, but from God's action in him" . . . "This new beginning can be proclaimed also *today* without the aid of a legend of a virgin birth, which more than ever is liable to be misunderstood in modern times. No one can be obliged to believe in the biological fact of a virginal conception or birth. Christian faith is related — even without a virgin birth — to the crucified and still living Jesus manifest in his unmistakability and individuality." p. 456. Küng is confident that the meaning of Jesus is "unmistakeable" even after we have eliminated from the Gospel narrative all the elements which the "modern mind" finds it difficult to accept; but after all the Scriptures are *God's* way of accrediting His message to all ages, including our own. Will our apologetic strategies be successful if we substitute our own arguments for those God Himself has supplied? Surely the failure of liberal Protestantism to win over the "modern mind" ought to have taught us a lesson.

22. Brown, *Birth of the Messiah* (note 13 above) p. 155-159 with Table VIII, p. 156 showing literary form of Biblical annunciations of birth. See René Laurentin, *Structure et theologie de Luc I-II,* (Paris: J. Gabalda, 1957). Also M. Allard, "L'Annonce a Marie et les annonces de naissance miraculeuses de l'Ancien Testament," *Nouvelle Revue Théologique* 78 (1956): 730-733; and J. P. Audet, "L'Annonce à Marie", *Revue Biblique* 63 (1956): 346-374).

23. *Ibid.*, p. 290 f., and 327 f., and Fitzmyer, *op. cit.* p. 351.

24. *Ibid.*, p. 437 and 443 f., and René Laurentin, *Jésus au temple: Mystère de Pâques et foi de Marie en Luc. 2: 48-50* (Paris: Gabalda, 1966), pp. 38-72.

25. See Vincent Taylor, *Gospel According to St. Mark,* (note 17 above), p. 160 gives the various opinions about the meaning of the Dove and prefers an allusion to Genesis 1, 2. A. Feuillet, "Le Symbolisme de la Colombe dans le récits évangeliques du baptême", *Recherches de Science Religieuse* 46 (1958: 524-544), argues that the principal meaning is that the People of God of whom Jesus is the Head, from whom the whole life of the Spirit derives, is the Dove. Fitzmyer, *Luke I-IX,* p. 483 f., relying on L. E. Keck, "The Spirit and the Dove", *New Testament Studies* 17 (1970-71): 41-67 concludes that the meaning of the symbol is uncertain.

26. See L. Howard Marshall, *The Gospel of Luke,* New International Greek Text Commentary (Grand Rapids, Eerdmans, 1978), pp. 432 and 439, and M. E. Boismard, O.P., in P. Benoit, O.P., and Boismard, *Synopse de Quatres Evangiles* (Paris: Cerf, 1972), tom. 2, Note 110-111, pp. 169-170. "Q" is the hypothetical source from which in the Two Document Hypothesis *Matthew* and *Luke* derived the material common to them but which they did not find in Mark. If this Two Document Hypothesis is valid, Q is older even than Mark. However, recently not a few scholars reject this theory.

27. Thus miracles of healing, for example, are only supernatural in their *cause* (God) and their *mode* (instantaneous) but not in their *essence,* since what they produce is a natural state of health, while the miracle of conversion to the Faith is essentially supernatural since its effect (Christian faith) transcends all created powers and modes of existence. See Reginald Garrigou-Lagrange, O.P., *De Revelatione* 5th ed. (Rome: F. Ferrari, 1950), pp. 185-197.

28. For an up to date bibliography on this very controversial question see Maurice Casey (a pupil of C. K. Barrett), *Son of Man* (London: SPCK, 1979), pp. 241-259. Casey himself holds that Jesus sometimes used this idiomatic expression as a way of applying general statements of a more or less proverbial character to Himself. When this explanation does not fit a Gospel use, Casey holds that the passage is inauthentic and is the result of a custom which arose in the Church of using the expression as a Christological title.

29. All four evangelists picture Jesus as acting during these crucial hours as a *prophet* possessed of superhuman awareness of His situation, and particularly of the future. Paul in I Cor. 11:23-26 does not portray Jesus as prophesying, but says "Every time, when you eat this bread and drink this cup, you proclaim the death of the Lord *until he comes*." "Until he comes" seems to allude to the words of Jesus transmitted by all three sympotics, "I will never again drink of the fruit of the vine until the day when I drink it new in the reign of God" (Mark 14:25, Matthew adds "drink it with you" 26-29 and Luke gives it for both the bread and the cup). In John, Jesus promises "Within a short time you will lose sight of me, but soon after you will see me again" (16-16), and adds the promise of the coming of the Paraclete.

30. For a thoroughly studied example see René Laurentin ed., *Lourdes: Dossier des documents authentique* (Paris: Lethellieux 1957-1960), 6 vols., and *Lourdes: Histoire authentique*, same publishers, 1962-1964, 6 vols; and his biography *Bernadette of Lourdes* (Minneapolis: Winston Press, 1979). Is it not significant that Laurentin who has so carefully studied the life of Bernadette approaches the Infancy Narratives with a greater readiness to accept their historically than many exegetes?

31. Fitzmyer, Luke (note 20 above) p. 340.

32. Brown, *Birth of the Messiah* (note 13 above), p. 527.

33. René Laurentin, *Structure* (note 22 above). Manuel Miguens, O.F.M., *The Virgin Birth: an Evaluation of the Scriptural Evidence*, (Westminster, MD., 1975), and John McHugh, *The Mother of Jesus in the New Testament* (Garden City, N.Y.: Doubleday, 1975) (see bibliography pp. 475-498 and list of opponents and defenders of the historicity of the virginal conception pp. 456-458), and Laurentin, *Enfance du Christ,* (note 20 above) accept the notion of a family tradition and are critical of Brown. McHugh, "Exegesis and Dogma: a Review of Two Marian Studies: with a reply by Raymond E. Brown", *Ampleforth Review* 1 (1980): 43-60 raised the question how the Church came to know the truth of the Virginal Conception if, as Brown holds, it cannot be established as an historic fact from Scripture. In his reply Brown says, "I believe that Mary remained a virgin because the Roman Catholic church teaches this doctrine with an authority that is usually classified as infallible. Yet from a historical-critical reading of the New Testament I find no decisive evidence one way or the other about this issue in New Testament times." "Rather, theologians reflected on the role of Mary in God's plan of salvation, and the church through the guidance of the Holy Spirit gradually accepted" these Marian doctrines (p. 57). But the problem pointed out by McHugh remains: How did the church come to be certain of a historic fact by "theological reflection" without a historic tradition as its basis? And if this historic tradition is not to be found in the Scriptures, must we resort to extra-biblical tradition only?

I appreciate Brown and Fitzmyer's exegetical honesty, but are they equally hard-headed in facing the theological objections to their exclusive dependence on the historical-critical method? Laurentin, "Exègeses réductrices des Évangiles de l'enfance", *Marianum* 41 (1979): 76-100, severely criticizes Brown for his exclusive dependence on one methodological approach which Laurentin regards as somewhat out-dated. Brown, in his reply to McHugh, on the other hand, points out that McHugh tends to rely on French exegetes against whom Brown quotes German exegetes. I tend to agree with the critics who believe that Brown has too much confidence in his method, but in the text I have attempted to show that even if Brown is correct in his analysis (which I do not think is the case) still the doctrine of the historicity of the Virginal Conception has a basis in the Scriptures from which the Church arrived at it by a theological reflection (which seems to be what Brown intends, although I do not find that he is clear on the matter). Other reviews of Brown's book are R. H. Fuller *Catholic Biblical Quarterly* 40 (1978): 116-120 and M. Bourke p. 120-124; Jacques Dupont, *Commentaria Liturgica* 58 (1977), p. 545 f.; L. H. Marshall, *Evangelical Quarterly* 51 (1979): 105-110; Gerald Collins, S.J., *New Testament Abstracts* 22 (1978): 85 f.; X. Léon-Dufour, *Recherches de Science Religieuse* 66 (1978): 128-131; *Revue des Sciences Philosophiques et Théologiques* 62 (1978): 99-105; cf. 60 (1976): 311-315; Francis L. Maloney, *Salesianum* 40 (1978): 684-686; H. Wansbrough, *Tablet* 232 (1978): 304; Neil J. McEleney, *Theological Studies* 39 (1978): 771 f.; Manuel Miguens, O.F.M., *Communio* 7 (1980): 24-54; Jerome Quinn, *Biblical*

Theology Bulletin 10 (1980): 134-136; Leopold Sabourin, S.J., *Religious Studies Bulletin* 10 (July, 1980): 134-136; James T. O'Connor, *Marian Studies* 29 (1978): 24-43.

34. "I do not imply that Matthew and Luke presented the virginal conception only as a symbol and were indifferent to what really took place. I think that both of them regarded the virginal conception as historical, but the modern intensity about historicity was not their's. For them, the primary importance of the virginal conception was theological, and more specifically christological." p. 517, *Birth of The Messiah* (note 13 above) ". . . I avoid the term 'virgin birth' because of its creedal implications. The ancient creeds tend to speak of Jesus' being 'born of the Virgin Mary,' and while the authors of those creeds certainly thought that Jesus had been virginally conceived, it is not clear that what they were proposing as a matter of Christian faith was the biological manner of Jesus' conception . . . 'Born of the Virgin Mary' echoes the Pauline formula 'born of a woman born under the Law' (Gal 4:4) and is designed to underline the historicity of *Jesus* — a much wider question than the historicity of his conception without a human father." *Ibid*, p. 518.

35. Schillebeeckx, *Jesus* (note 3 above) p. 115.

36. See Brown's discussion in *Birth of the Messiah* (note 13 above), pp. 525-531.

37. Fitzmyer, *Luke* (note 20 above), p. 342.

38. Since Vatican II and the *Dogmatic Constitution on Divine Revelation (Dei Verbum)*, Catholic theologians generally emphasize the unity of Scripture and Tradition and avoid interpreting the Council of Trent's defense of "Scripture and Tradition" as witnessess of revelation against the Reformation formula of "sola scriptura", as if Trent defined that there are matters of faith which are not *somehow* contained in Scripture but only in Tradition. It is necessary to insist, however, that Vatican II did not attempt to settle this question. To show that some truth is not certainly contained in Scripture is not equivalent to proving that it is not a matter of Catholic faith; while to prove that it *is* contained in Tradition is equivalent to proving that it *is* a matter of Catholic faith. The classical example is that the canon of Scripture is certainly contained in Tradition and is a matter of faith, but it is hard to show that it is in any way contained in the Scriptures themselves.

39. "I do not imply that Matthew and Luke presented the virginal conception only as a symbol and were indifferent to what really took place. I think that *both of them regarded the virginal conception as historical*, but the modern intensity about historicity was not theirs." Brown, *op. cit.* p. 517 (my italics). The question, therefore, is: "Did Matthew and Luke *assert* the virginal conception to be historical?" Brown is not explicit on this point. If they did intend this assertion, then theologically speaking, the doctrine of inspiration and inerrancy of Scripture certifies the historicity of the virginal conception, even if we cannot in a "scientifically controllable" way discover the sources of their information. Their inspired witness is certain in a way that historical reconstruction of traditions never can be. In my opinion the theological weakness of Brown's discussion is a failure to address this point of exegesis explicitly.

40. *Ibid.*, p. 526.

41. S. G. F. Brandon, *Jesus and the Zealots* (Manchester: Manchester University Press), 1967, pp. 274-279. Hans von Campenhausen, "The Authority of Jesus' Relatives in the Early Church," in von Campenhausen and Henry Chadwick, *Jerusalem and Rome* (Philadelphia: Fortress Facet Books, 1966), pp. 1-20, undertakes to refute the notion that Jesus' relatives constituted a kind of "Christian caliphate", but points out that this idea, orginated by Harnack, has been accepted by many scholars such as Johannes Weiss, Rudolf Knopf, B. H. Streeter, and more recently H. J. Schoeps.

42. The *New Testament Apocrypha* are available in the work of E. Hennecke and W. Scheemelcher, 2 vols. (Philadelphia: Westminister, 1963-1965). The most important infancy Gospels are the *Protoevangelium of James* and the *Infancy Gospel of Thomas*, vol. 1, 363-417. The former is usually dated to the middle and the latter to the end of the second century,

43. See article (with bibliography) "Canonicity" by James C. Turro and Raymond E. Brown, in Raymond E. Brown, S.S., Joseph A. Fitzmyer, S.J., and Roland E. Murphy, O. Carm., *The Jerome Biblical Commentary* (Englewood-Cliffs, N.J.: Prentice-Hall, 1968), pp. 515-534.

44. Fitzmyer, *Luke* (note 20 above), pp. 14-17.

45. Karl Rahner, S.J., "Dogmatic Reflections on the Knowledge and Self-Consiousness of Christ" (which contains useful bibliography) in *Theological Investigations* (Baltimore: Helicon Press, 1966), 5: 193-218, and the article "Jesus Christ" in the *Encyclopedia of Theology* edited by him (New York: Seabury, 1975), pp. 751-772.

46. Aquinas, *Summa Theologiae* III, q. 12, a. 3.

47. The Biblical difficulties with respect to the existence of the beatific vision and in-fused knowledge in Jesus are collected in William B. Most, *The Consciousness of Christ* (Front Royal, Virginia: Christendom Publications, 1980). Most defends the classical scholastic teaching on the subject and proposes solutions of these difficulties.

48. On the psychological range of prophecy as a psychological phenomenon see David L. Petersen, *The Roles of Israel's Prophets*, Journal for the Study of the Old Testament, Supplemental Series, no. 17, (Sheffield: University of Sheffield, 1981), pp. 20 ff. Petersen concludes, however, that ecstatic trance was not a main feature of Israelite prophecy (p. 30).

49. See Arintero, *The Mystical Evolution* (note 18 above), pp. 170-251 for descriptions of this state taken from the writings of Christian mystics.

50. That Greek was common knowledge in Galilee in Jesus' time is shown by Martin Hegel, *Judaism and Hellenism* (London: SCM Press), 1, pp. 586 f. and Séan Freyne, *Galilee from Alexander the Great to Hadrian* (Wilmington, Delaware and Notre Dame, Ind: Michael Glazier and University of Notre Dame, 1980), pp. 139-145. That Jesus could speak Hebrew as well as Aramaic is shown by Jean Carmignac, *Recherches sur le "Notre Pere"* (Paris: Letouzey and Ané, 1969), pp. 3-35. That Jesus may have known some Latin appears also from the fact that the title of the Cross put up for the public was in Hebrew or Aramaic, Greek, and Latin (Luke 23:38 and John 19:20).

51. Brown, *Birth of the Messiah* (note 13 above), pp. 537-541. The term translated "carpenter" has the wide-sense of "builder".

52. William E. Phipps, *Was Jesus Married?*, (New York: Harper and Row, 1970) argues that Jesus was married because rabbis are required to marry; but Qumran is proof that in Jesus' time celibacy for religious reasons was possible. Moreover, there was a Rabbinic belief that Moses and the other prophets were celibate, from the time of their encounter with Yahweh. For a refutation of Phipps by a Protestant theologian see George Wesley Buchanan, "Jesus and Other Monks of New Testament Times", *Religion in Life* 48 (1979): 136-142.

53. See George A. Panichas, *Adventures in Consciousness: The Meaning of D. H. Lawrence's Religious Quest* (The Hague: Mouton, 1964).

54. Marilyn Chapin Massey and James A. Massey, *Feminists on Christianity* show how feminists such as Letty M. Russell first tried to argue that it is Jesus' personhood that is of religious significance and his maleness a mere historical accident, but that the more radical, such as Mary Daly, *Beyond God the Father* (Boston: Beacon, 1973), have come to believe that His maleness cannot be ignored but inevitably gives religious support to sexism, so that feminists must abandon Christianity and seek the religion of the "Goddess" or some new religion whose symbols are androgynous.

55. "In all four Gospels, Jesus is never reported as asking or speaking to women in a derogatory fashion. He always treated them as equals, individuals, persons. He testified in his practice and doctrines that he saw woman, as created by God, equal to man, endorsing the view of *Genesis* 1:27: 'So God created man in his own image, in the image of God he created him; male and female he created them.' " Alicia Craig Faxon, *Women and Jesus*, Philadelphia: Pilgrim Press, 1973, p. 11.

56. See J. M. Oesterreicher ed., *Brothers in Hope* (vol. 5 of *The Bridge: Judaeo-Christian Studies)* (New York: Herder and Herder, 1970). Several of the theologians collected in Alan T. Davies, ed., *Anti-Semitism and the Foundations of Christianity* (New York: Paulist, 1979), discuss the vocation of Israel in its own right and the convergence of Christianity and Judaism as each follows its own call.

57. On the problem of reconciling the fact that Jesus confined his preaching to his own people with his universalism see Joachim Jeremias, *Jesus' Promise to the Nations* (Studies in Biblical Theology No. 24) (London: SCM Press and Naperville, IL: Allenson, 1958), and T. W. Manson, *Only to the House of Israel: Jesus and Non-Jews* (Philadelphia: Fortress, 1964).

554

58. Ernst Bloch, *Des Prinzip Hoffnung,* 2nd ed. (Frankfurt: Suhrkamp, 1959), 2 vols., and *Atheism in Christianity: the Religion of the Exodus and the Kingdom* (New York: Herder and Herder, 1972).

59. Robert Jewett, *Paul's Anthropological Terms: A Study of their Use in Conflict Situation* (Leiden: Brill, 1971). See also John A. Robinson, *The Body: A Study in Pauline Theology* (Chicago: Regnery, 1932), and Robert H. Gundry, *Sōma in Biblical Theology with Emphasis on Pauline Anthropology* (New York: Cambridge University Press, 1970).

60. See Jerome Murphy-O'Connor, *Becoming Human Together,* (Wilmington, Del.: Michael Glazier, 1977), who shows how the Reformation reading of Paul in terms of the justification of the individual fails to perceive that Paul's whole understanding of the Gospel is *corporate.*

61. Hans Küng, *Infallible?* (Garden City, N.Y.: Doubleday, 1971); *Fehlbar? Eine Bilanz* (Einsiedeln: Benziger Verlag, 1973); and his introduction to August Bernhard Hassler, *How the Pope Became Infallible* (Garden City, N.Y.: Doubleday, 1981). On the controversy see Karl Rahner, S.J., ed., *Zum Problem Unfehlbarkeit: Antworten auf die Anfrage von Hans Küng* (Freiburg-Basel-Wien; Herder, 1971); John J. Kirvan, ed., *The Infallibility Debate* (New York: Paulist, 1971); *The Küng Dialogue: a documentation on the efforts of the Congregation for the Doctrine for the Faith and of the conference of German Bishops to achieve an appropriate clarification of the controversial views of Dr. Hans Küng (Tübingen)* (Washington, D.C.: United States Catholic Conference, 1980); Leonard Swidler, ed., *Küng in Conflict* (Garden City, N.Y.: Doubleday, 1981).

62. Thus Hans Küng, *The Church* (New York: Sheed and Ward, 1967). Küng on p. 19 states the principle that the ecclesiologist must consider and balance *all* the New Testament evidence, the Church of Corinth becomes for him the paradigm of the original Christian communities. For reactions to Küng's treatise see Y. Congar's review of the book in *Revue des Sciences Philosophiques et Théologique,* 53, (1969): 693-706 with Küng's reply, pp. 175-221. Also H. Haring and J. Nolte ed., *Diskussion um Hans Küng Die Kirche* (Freiburg-Basel-Wien, 1971). Contrast with Küng, Louis Bouyer, *L'Église de Dieu: Corps du Christ et Temple de l'Esprit* (Parish: Cerf, 1970).

63. See Paul Althaus, *The Ethics of Martin Luther* (Philadelphia, Fortress, 1972), pp. 43-82.

64. See Pheme Perkins, *The Early Church and the Crisis of Gnosticism,* (Theological Inquiries) (New York, Paulist, 1980), pp. 191-204, and Elaine Pagels, *The Gnostic Gospels* (New York: Random House, 1979), Chapter 5, "Whose Church is the 'True Church'?", pp. 102-118.

65. On the controversy as to the authenticity of *Ephesians,* see Markus Barth (who defends it) in his commentary on the work in the Anchor Bible (Garden City, N.Y.: Doubleday, 1974), 1, pp. 36-50. Also Ceslaus Spicq, O.P., article "Pastorales (Épîtres) Authenticité" in *Dictionnaire de la Bible* ed., by H. Cazelles and A. Feuillet (Paris: Letouzey et Ané), Supplement, col. 50-73, and Stanislas de Lestapis, S.J., *L'Énigme des Pastorales de Saint Paul* (Paris: J. Gabalda, 1976). A vicious circle arises from the fact that the chief argument for dating the Pastorals after the death of St. Paul is the fact that they seem to show a more structured and conservative church, but their supposedly late date is also used as a proof that the hierarchical structuring of the church is a late development. John A. T. Robinson, *Redating the New Testament* (Philadelphia: Westminister, 1976), pp. 67-85, points out that there is no really solid reason for doubting that the development of a highly structured church took place very rapidly, even within the lifetime of St. Paul.

66. See Chapter 10 pp. 455-458.

67. See the magisterial work of Jacques Dupont, *Les Béatitudes,* rev. ed., 3 vols. (Bruges: Abbaye de Saint André), 1958-1973.

68. See Chapter 10, pp. 456.

69. For Paul the unity of the Christian community is not merely "spiritual" in the sense of non-material; it extends to the whole persons, including the bodies, of its members. Rather, it is spiritual in the sense that it raises this community to a new mode of existence and this mode of existence is one of true *freedom.* Such freedom is free of all totalitarian deper-

sonalization and coercion, but is communitarian in that out of love each member seeks to serve rather than to dominate the others. That this does not contradict the principle of authority and leadership may seem a mystery, but it is exemplified in Paul's own firm but loving leadership of the churches in his care.

70. See Chapter 10 pp. 456.

71. See Chapter 10 pp. 435-439.

72. See note 63 above.

73. See Brendan E. A. Liddell, *Kant on the Foundations of Morality* (Bloomington, Ind.: Indiana University Press, 1970), pp. 180-234.

74. See Paul Tillich, *Systematic Theology* (Chicago: University of Chicago Press), 1951-63, vol. 1 pp. 83-86; 147-149 and vol. 3, pp. 249-265; 274 f.

75. Some manuscripts say 72. The Rabbis thought that the Gentile nations of Genesis 10 were 70 or 72 in number. Also the elders appointed by Moses (Nm 11:16) were seventy in number, or with Eldad and Medad (Nm 11:26), seventy-two. Luke, who alone of the evangelists mentions these seventy or seventy-two disciples probably intended to compare Jesus to Moses, who found it necessary to share his authority with other subordinate authorities. See G. B. Caird, *The Gospel of St. Luke*, Pelican Gospel Commentaries (New York: Seabury, 1963), p. 144.

76. Vatican II, *Dogmatic Constitution on the Church* (Lumen Gentium) Nov. 21, 1964. Chapter 3, "The Church is Hierarchical", nn. 18-29.

77. The common view among exegetes today seems to be that Jesus commissioned the Twelve but that the traditional church structures are developments subsequent to the Ascension which can be attributed to Jesus only by the mediation of His Holy Spirit. See Raymond E. Brown, *The Critical Meaning of the Bible* (note 9 above), pp. 124-146, who concludes, "This survey shows that the manner and exercise of supervision varied greatly in the different places and different periods within the first century or NT era. Only at the end of the century and under various pressures was a more uniform structure of church office developing . . . Many of us see the work of the Holy Spirit in this whole process, but even those who do must recognize that the author of *I Clement* is giving overly simplified history when he states (I Clem. 42) that the apostles (seemingly the Twelve), who came from Christ, appointed their first converts to be bishops and deacons in the local Churches" p. 146. Brown seems confident that his laborious and debatable reconstruction of history from admittedly sparse, disparate and ambiguous data is more credible than the plain statement of Clement who lived in the New Testament period.

78. The value of a "sacred language" for liturgy should not be overlooked. According to Carmignac, (see note 48 above), Jesus himself accepted the Jewish use of Hebrew rather than the vernacular Aramaic in composing the Lord's prayer. Nevertheless, the decision of Vatican II in permitting the vernacular is in line with the practice of the early Church, which adopted the languages of the nations rather than imposing Hebrew.

79. Vatican II, *Declaration on Religious Liberty (Dignitatis humanae)*, Dec. 7, 1965.

80. Pope John Paul II in *Redemptor Hominis,* March 4, 1979, stressed the share of the laity in the three-fold office of Christ, including His Kingship, which is a kingship of service and of self-mastery. The public media have continually misrepresented his thought in charging that the Pope condemns *political* activity by the Catholic clergy and religious and that he is inconsistent in engaging in political questions himself. The Pope insists over and over again that it is the duty of all Christians to concern themselves with matters of peace, justice, and human rights; but he also insists that they should do so in a manner proper to the vocation of each. It is proper to the laity to do this by engaging in all the roles that belong to citizenship, but to the clergy and religious in a manner consistent with their total dedication to the specific mission of the Church, which requires that they not become identified with any political party or ideology or the exercise of public power.

81. See Schillebeeckx, *Jesus* (note 3 above), pp. 475-499, for the problem of the New Testament community in formulating Jesus' identity and authority. He concludes that at least there was agreement that He was the *eschatological prophet* promised in the Torah. Of course

556

every true prophet had divine *authority*, i.e., the right to proclaim, "Thus says the Lord," and for the eschatological prophet this meant the *ultimate* revelation of God, the definitive establishment of God's Reign.

82. Some of the most profound work on the theology of priesthood has been done by Yves Congar, O.P., whose views have undergone a considerable evolution. This is well traced in the dissertation of Richard J. Beauchesne, *Laity and Ministry in Yves Congar, O.P., Evolution, Evaluation, and Ecumenical Perspectives*, Boston University, 1974 (Xerox University Microfilms, Ann Arbor, Mich.), with a prefatory letter by Congar highly commending the analysis. Congar has gradually come to a strong emphasis on the universal priesthood understood as the spiritual offering of the entire life of the Christian community *within* which the ministerial priesthood is called to serve by fostering and coordinating the universal priesthood through its preaching, shepherding, and leading in liturgical worship. Beauchesne believes he finds a convergence of the views of Congar with those of Küng, but it seems to me their two theologies are based on different principles. See also Albert Vanhoye, S.J., *Prêtres anciens, prêtre nouveau selon le N. Testament* (Paris: Seuil, 1980) and Jean Galot, S.J., *Theology of the Priesthood* (San Francisco: Ignatius, 1984).

83. Henri Cormier, *The Humor of Jesus* (Staten Island, N.Y.: Alba House, 1971), in a popular rather than a scholarly way shows how much in Jesus' teachings as actually set down in the Gospels (Cormier does not enter into critical problems) is permeated with various types of humor. Since the Old Testament and Rabbinic literature are also permeated with touches of humor, usually of a didactic and ironic character, it would be strange if Jesus had not made use of this important method of teaching.

84. The explicit teaching on the participation by all the baptized in the threefold office of Christ (prophet, king, and priest) was one of the chief features of Vatican II. See *Dogmatic Constitution on the Church (Lumen Gentium)* of Nov. 21, 1964, Chapter 2, 10-13, and *Decree on the Apostolate of Lay People (Apostolicam Actuositatem)*, Nov. 18, 1965, "In the Church there is a diversity of ministry, but unity of mission. To the apostles and their successors Christ has entrusted the office of teaching, sanctifying, and governing in his name and by his power. But the laity are made to share in the priestly, prophetic, and kingly office of Christ; they have, therefore in the Church and in the world, their assignment in the mission of the whole People of God" (no. 2), Austin Flannery, O.P., ed., *Vatican Council II: The Conciliar and Post Conciliar Documents* (Northport, N.Y.: Costello Publishing Co., 1975), p. 768. John Paul II in *Redemptor Hominis* (see note 79 above) especially stresses this teaching of the Council.

The historical problems of (1) how the structure of church offices developed from the Twelve to the three-fold hierarchy in the Eastern and Western Churches which the Council of Trent declared to be instituted by Christ; and (2) how this three-fold office was "sacerdotalized" in the manner of the Old Testament priesthood, when the term "priest" is never used of the officers, but only of Christ; and the theological problem of the relation between the "priesthood of the faithful" and the "ministerial priesthood", which Vatican II reaffirmed but declared to be one of kind and not merely of degree, has stimulated a vast literature. I have found the following especially helpful: Jean Colson, *Les Functions Ecclésiastique au deux premieres siècles* (Paris: Desclée de Brouwer, 1956); *L'Episcopat Catholique* (Paris: Cerf, 1963), and *Ministre de Jésus Christ ou Le Sacerdoce de l'Évangile* (Paris: Beauchesne, 1966); *Prêtres et peuple sacerdotal* (Paris: Beauchesne, 1969); Pierre Grelot, *Le ministère de la nouvelle alliance* (Paris: Cerf, 1967) (note specially pp. 114-121 on the sacramental "character"): A Descamps, ed., *Le prêtre: Foi et Contestation* (Responses Chretiènnes) (Paris: Duculot, 1969); James A. Mohler, S.J., *The Origin and Evolution of the Priesthood* (Staten Island, N.J.: Alba House, 1970); Raymond E. Brown, S.S., *Priest and Bishop* (Paramus, N.Y.: Paulist, 1970); Herbert Vorgrimmler, ed., *Der Priesterliche Dienst* (Quaestiones Disputatae) (Herder: Freiberg-Basel-Wien, 1970-1973), 6 vols.: André Lemaire, *Les Ministères au origines de l'Église: Naissance de la triple hiérarchie: évêques, presbyteres, diacres* (Paris: Cerf, 1971) (his conclusons are summarized in a shorter work *Ministry in the Church* (London: SPCK, 1977, pp. 1-80); *Les Ministères* is one of the most useful works on the subject because it analyzes each of the New Testament texts relating to ministry (the conclusions drawn are more debatable); Hans Küng, *Why Priests* (Garden City, N.Y.: Doubleday, 1972), with review by Raymond E. Brown, "The Changing Face of the Priesthood," *America,* May 20, 1973, pp. 531 ff.; Jean Delorme, ed., *Le Ministère et les ministères selon le Nouveau Testament* (Parole de Dieu) (Ed., du Seuil: Paris, 1973) (the exegetical section pp. 16-280 is especially good); David Power, O.M.I., *The Christian Priest: Elder and Prophet* (London: Sheed and Ward, 1973) (see pp. 89-129 on the sacramental character); Leopold Sabourin, S.J., *Priesthood: A Comparative Study* (helps to define the unique meaning of Christian priesthood) (Leiden: Brill, 1973); Aurelio Fernandez, *Sacerdocio Común y Sacerdocio Ministerial: Un Problema Teologica,* (Bruges: Ed. Aldeco, 1979) (note bibliography pp. 242-261); Charles R. Meyer, *Man of God* (Garden City, N.Y.: Doubleday, 1974); Robert Zolitsch, *Amt und Function des Priesters* (Freiberg-Basel-Wien; Herder, 1974); André Feuillet, *The Priesthood of Christ and His Ministries* (Garden City, N.Y.: Doubleday, 1975); Donald W. Wuerl, *The Catholic Priesthood Today* (Chicago: Franciscan Herald Press, 1976) (discusses the Synod of 1971 on the ministerial priesthood); Manuel Miguens, *Church Ministries in New Testament Times* (Westminister, MD: Christian Classics Press, 1976); Alexandre Faivre, *Naissance d'une hiérarchie; Les premiers étapes du cursus clérical,* (Paris: Beauchesne: 1977) (post New Testament developments); Albert Vanhoye, S.J., *Prêtres anciens, prêtre nouveau* (note 82 above) and Jean Galot, *Theology of Priesthood* (note 82 above). Two important works by Protestant authors are, E. Schweitzer, *Church Order in the New Testament* (Naperville, IL: Allenson, 1961) and Hans von Campenhausen, *Kirchliches Amt und Geistliche Vollmacht* (Tübingen, Mohr, 1953).

Most of these works are well acquainted with the available data, but it is obvious that the quite different interpretations to which they come on many points reflect different attitudes. Those concerned to make ecumenical relations "easier", those who dislike centralized authority in the Church, those who desire to stress the continuity of Catholic church order to sustain Trent, those who favor or oppose the ordination of women all find support for their views in the interpretation of what must be admitted to be rather meager data. The fixed point in these historical reconstructions remains the explicit testimony of Ignatius of Antioch to the existence of the three-fold hierarchy in a considerable area at the beginning of the second century (c. 110). Consequently attempts have been made recently to disprove the authenticity of the Ignatian letters by R. Weijenborg, *Les lettres d'Ignace d'Antioche* (Leiden: Brill, 1969); Robert Joly, *Le dossier d'Ignace d'Antioche* (Brussels: Ed. de Université Libre, 1979); J. Ruis Camps, "Las Cartas autenticas de Ignacio, el obispo de Sirio", Revista Catalana de Teologia, 2 (1977), 31-149, but these attempts (which do not agree among themselves) do not seem to have succeeded; cf. Charles Kannengiesser, "L'affaire de Ignace d'Antioche", *Recherches de Science Religieuse* 67 (1979): 599-623 and W. Schoedel, "Are the Letters of Ignatius of Antioch Authentic?" *Religious Studies Review,* 1980: 196-201.

85. The *Dogmatic Constitution on the Church,* mentioned in the last note, declares of the laity ("all the faithful except those in Holy Orders and those who belong to a religious state approved by the Church"), "Their secular character is proper and peculiar to the laity. Although those in Holy Orders may sometimes be engaged in secular activities, or even practice a secular profession, yet by reason of their particular vocation, they are principally and expressly ordained to the sacred ministry. At the same time, religious give outstanding and striking testimony that the world cannot be transfigured and offered to God without the spirit of the beatitudes. But by reason of their special vocation, it belongs to the laity to seek the kingdom of God by engaging in temporal affairs and directing them according to God's will. They live in the world, that is, they are engaged in each and every work and business of the earth and in ordinary circumstances of social and family life which, as it were, constitute their very existence. There they are called by God that, being led by the spirit of the Gospel, they may contribute to the sanctification of the world, as from within like leaven, by fulfilling their own particular duties. (n. 31) (Flannery, note 84 above).

86. See *Peter in the New Testament:* A Collaborative Assessment by Protestant and Roman Catholic Scholars, ed., by R. E. Brown, K. P. Donfried, and J. Reumann (Minneapolis: Augsburg Publishing House, 1973), pp. 158-168, with a bibliography pp. 169-178, for an effort by Lutherans and Catholics to agree on the role of St. Peter as head of the Twelve.

87. The scholastic theology of priesthood defined the priest as one who can offer the Eucharistic Sacrifice and this definition is still evident in Vatican II in the *Decree on the Ministry and Life of Priests (Presbyteriorum ordinis)* Dec. 7, 1965, but it creates difficulties when it is observed that in the *Dogmatic Constitution on the Church,* contrary to this scholastic theology, the episcopacy is declared to possess "the fullness of the sacrament of Orders" (n. 21), although they have no more power to offer the Eucharistic Sacrifice than does a presbyter. It is evident, therefore that the offering of the Eucharist, whether by a bishop or priest, expresses something in which the bishop and priest share unequally, namely, the *responsibility* as shepherd to a local church. This responsibility the bishop has absolutely and in its fullness, the priest relatively and as a co-worker with the bishop. What is to be said then of the ordination of bishops who do not in fact have the responsibility of a local church; or the very large number of priests (especially in religious orders) who do not have parochial office? Undoubtedly this can be an abuse. It seems justified, however, that bishops and priests should be freed of responsibility to a local church in order to serve the needs of some larger segment of the church or the church as a whole in some properly priestly function, provided that his *principal* commitment remain the service of the church and not some secular work. Thus it is primary commitment to the mission of the Church, which receives its supreme expression in the leadership of the Eucharistic Sacrifice with and on behalf of the Christian community, which is sealed by Holy Orders and which constitutes the ministry of priests as specifically distinct from that of the priesthood of the laity which it promotes. The building up of the Church is *not* the principal responsibility of the laity (although they must share in it), but rather principally the laity share in the priesthood of Christ and the Church by fulfilling the responsibilities they have in the world's work and thereby sanctifying that world.

88. Although the scholastic theology of the sacramental "character" impressed by ordination may require rethinking, there is no doubt that Vatican II continued to teach that ordination is not merely the assignment to an office or function, but an *ontological* transformation of the person by which he is constituted permanently in a new mode of existence. "Because it is joined with the episcopal order the office of priests shares in the authority by which Christ himself builds up and sanctifies and rules his Body. Hence the priesthood of priests, while presupposing the sacraments of initiation, is nevertheless conferred by its own particular sacrament. Through that sacrament, priests by the anointing of the Holy Spirit are signed with a special character and so are configured to Christ the priest in such a way that they are able to act in the person of Christ the head," *Decree on the Life and Ministry of Priests,* no. 2. (Flannery, note 87 above). The document insists that this does not separate the priests from the rest of the baptized, but unites them to the laity as the shepherds to the flock.

89. For references on the origin of priestly celibacy, see Chapter 10, note 83.

90. Edward Schillebeeckx and Johannes B. Metz, *The Right of the Community to a Priest* (Concilium no. 133) (New York: Seabury, 1980), especially Schillebeeckx, "The Christian Community and its Office-Bearers", pp. 95-134 and Schillebeeckx, *Ministry: Leadership in the Community of Jesus Christ* (New York: Crossroad, 1981) and the critique by Pierre Grelot, *Eglise et ministères* (Paris: Cerf, 1983) It can be granted, certainly, that the Christian community has a need and therefore a "right" to participate regularly in the Eucharist and other sacraments, to the ministry of the Word, and to a pastor; but it also has the responsibility to call forth from its own members those truly qualified to be priests. To argue as Schillebeeckx does that the requirements of sex, of celibacy, and of education which the Church has come from long experience to regard as necessary qualifications of priesthood are the chief obstacles to the community providing priests for itself, overlooks the possibility that the real obstacle are not these qualifications but a failure on the part of the *laity* to accept their responsibilities to the community. Is it not modern individualism and hedonism, resulting from the influence

of Humanism, which have so eroded the laity's sense of the need for truly qualified priests that is the basic cause of the shortage of priestly vocations which normally are the fruit of Christian marriages? If this is the case, then the relaxation of the qualifications for priesthood will only acerbate this spiritual illness. If we really believe that in the future the laity must participate much more actively in the work of the Church (and to this I heartily agree), then they must also come to see *their* responsibility (not the bishops) to provide candidates for the priesthood who are totally dedicated in a way that will make a spirit of humble service and sacrificial celibacy possible. Perhaps nothing will awaken them to this sense of responsibility so effectively as experiencing the hunger for the Sacraments which the growing shortage of priests and resistance of the hierarchy to a lowering of the qualifications for priesthood is beginning to produce.

91. See Chapter 9 pp. 379-381.

92. The bibliography on this controversial topic is vast and growing. See John H. Morgan and Teri Wall, *The Ordination of Women: A Comprehensive Bibliography (1960-1976)* and Bernhard Asen, "Women and the ministerial priesthood; an annotated bibliography", *Theology Digest* 29 (1981): 329-342. Much of the research favorable to women's ordination was done by Else Kahler, *Die Frauen in den Paulischen Briefen*, (Zürich: Gotthelf Verlag, 1960), and Ida Raming, *The Exclusion of Women from the Priesthood: Divine Law or Sex Discrimination* (1973) (Metuchen, N.J.: Sacrecrow Press, 1976), To which should be added Roger Gryson, *Le ministères des femmes dans l'Église ancienne* (Gembloux: Duclot, 1972); *The Ministry of Women in the Early Church* (Collegeville, Minn,: Liturgical Press, 1976). The chief points of history that remain unresolved are (1) In the time of St. Paul did women exercise priestly ministries which were then suppressed? (2) Were the "deaconesses" of the early Church *ordained* in the strict sense? (3) Did the exclusion of women from ordination arise out of theological or merely cultural reasons? Eric L. Mascall in his short essay *Women Priests* (London: Church Literature Association. 1972), made the case against; while Haye van der Meer, *Women Priests in the Catholic Church? A Theological-historical Investigation* (Philadelphia: Temple University Press, 1974), made the best case for it. Most of the other literature seems to depend largely on these authors. From a Protesant point of view one of the best pro works is Paul K. Jewett, *The Ordination of Women* (Grand Rapids: Eerdmans, 1980) (see appraisal of official Roman Catholic position, pp. 85-97). On Oct. 15, 1976, the Congregation for the Doctrine of the Faith issued a declaration *On the Question of the Admission of Women to the Ministerial Priesthood* with a *Commentary* (Washington, D.C.: United States Catholic Conference, 1977) in the negative as regards ordination to priesthood or episcopacy and passing over the question of the diaconate. This somewhat peremptory official stand on a question which had not been under public discussion for long was probably motivated in part by ecumenical considerations — the strong opposition of the Orthodox churches and the hope that the Anglican Church might not commit itself to the ordination of women. It was received, however, with indignation by some theologians, especially in the United States; cf. Catholic Theological Society of America, *Research Report: Women in Church and Society* edited by Sara Butler (Mahwah, N.J.: Darlington Seminary, 1978); Carroll Stuhlmueller, C.P., ed., *Women and the Priesthood: Future Directions, A Call to Dialogue from the Faculty of the Catholic Theological Union at Chicago* (Collegeville, Minn.: Liturgical Press, 1978); Leonard and Arlene Swidler, *Women Priests: A Catholic Commentary on the Vatican Declaration* (Ramsey, N.J.: Paulist, 1977), and by others more favorably, cf. *The Order of Priesthood: Nine Commentaries on the Vatican Degree Inter Insigniores* (Huntington, Ind.: Our Sunday Visitor Press, 1978). Other statements which I have found helpful are World Council of Churches, Department of Faith and Order, and Department on Cooperation of Men and Women in the Church, Family and Society, *Concerning Ordination of Women,* 1964 (See the statements of the Orthodox Nicolas Chitescu and George Khodre, pp. 57-64); National Conference of Catholic Bishops, Committee on Pastoral Research and Practice, *Theological Reflections on the Ordination of Women* (Washington, D.C.: 1972); Arlene Swidler, *Woman in a Man's Church From Role to Person* (New York: Paulist, 1972); Emily C. Hewitt and Suzanne R. Hiatt, *Women Priests: Yes or No* (New York: Seabury, 1973); George W. Rutler, *Priest and Priestess* (Ambler, PA: Trinity

Press, 1973); George H. Tavard, *Woman in Christian Tradition* (Notre Dame, Ind.: Notre Dame University Press, 1973); pp. 211-229. Robert J. Heyer, *Women and Orders* (Paramus, N.J.: Paulist, 1974); Michael P. Hamilton and Nancy S. Montgomery, ed., *The Ordination of Women: Pro and Con* (New York: Morehouse-Barlow, 1975) (see specially the articles by Leonard Swidler, "Roma Locuta, Causa Finita?" pp. 3-18; Elisabeth Schüssler Fiorenza "The Twelve" (argues that the Twelve are not the origin of the Church's hierarchy) p. 114-122; the Declaration and its Commentary are printed, pp. 37-49; 319-337 and also the report of the Pontifical Biblical Commission to the Congregation for the Doctrine of the Faith which report concluded that the New Testament does not settle the question; Bernard Lambert, O.P., *Can the Catholic Church Admit Women to Priestly Ordination* (Boston: Daughters of St. Paul, 1976) (in my opinion this inconspicuous essay is one of best defenses of the Declaration); David M. Maloney, *The Church Cannot Ordain Women to the Priesthood* (Chicago: Franciscan Herald Press, 1978); *An Expanded Vision, The Detroit Ordination Conference,* proceedings ed., by Anne Marie Gardiner, S.S.N.D. (Paramus, N.J.: Paulist, 1976); two studies by Fran Ferder, *Called to Break Bread* and by Maureen Fiedler, *Are Catholics Ready?* (Mt. Ranier, MD: Quixote Center, 1975), on 100 women who "feel called to priesthood in the Catholic Church" and the views of the laity; Peter Moore ed., *Man, Woman and Priesthood* (London, SPCK, 1978); Elizabeth M. Terlaw, *Women and Ministry in the New Testament* (Ramsey, N.J.: Paulist, 1980); Michael McFarlene Marrett, *The Lambeth Conference and Women Priests* (Smithtown, N.Y.: Exposition Press, 1981); and Elizabeth Schüssler Fiorenza, *In Memory of Her: A Feminist Theological Reconstruction of Christian Origins* (New York: Crossroads, 1983). On this last work see the review by Phyllis Trible, *New York Times Book Review,* May 1, 1983, p. 28-29.

93. See M. Amaldoss, S.J., *Do Sacraments Change? Variable and Invariable Elements in Sacramental Rites* (Bangalore: Theological Publishers in India, 1979), with an important bibliography pp. 127-51. The author holds that in favor of cultural adaptation in the missions the Church can, for example, replace the use of bread and wine in the Eucharist with the form of food and drink common to a given culture.

94. In the Old Testament the patriarchs are the priests of their family. Thus Abraham, Isaac, and Jacob offer sacrifice for their wives, children, and servants; as does the pagan Job. This primordial priesthood of the father is part of the order of creation and is not negated by the later rise of a priesthood that serves the larger community. This is implied I believe in *Ephesians* (5:22-33) when the husband is compared to Christ "who gave himself up" for his bride the Church, i.e., the father of the family *sacrifices* himself for his family, not only in the sense of suffering for them, but in the more inclusive sense of offering his whole life, both its joys and his sorrows, for them.

95. See Brown, *The Critical Meaning of the Bible* (note 9 above), pp. 124-146.

96. See *Peter in the New Testament* (note 86 above), for discussion.

97. It is clear from *Galatians* and the two epistles *To the Corinthians* that for Paul it is the Gospel of Christ which has absolute authority, but he also claims that as an apostle he is bound to obey the Gospel and to demand obedience to it from the community, not to merely represent the opinions or wishes of the community. While he seeks to follow Christ in acting as a servant of the community, this service demands that he call them to responsibility to the Gospel, and that if necessary that he excommunicate those who refuse obedience (5:1-13) both for the good of the community and in hope of the repentance of the offender. Such claims to spiritual authority seem to some today "undemocratic" and "oppressive". Others attribute them to St. Paul's "paternalism" or "patriarchalism", and believe it unworthy of adult Christians to be treated this way, as if they were children. An important study by Pedro Gutierrez, *La Paternité Spirituelle selon Saint Paul* (Étude Biblique) Paris: J. Gabalda, 1968 shows that St. Paul *did* think of himself as a spiritual father, participating in the Divine Fatherhood which was central to Jesus' own experience of God as *Abba*. But the New Testament sense of paternity is not that of a tyrant God but is precisely that God is *not* a tyrant but a loving, wise, and patient Father, who far from being "paternalistic" is concerned above all that His children grow to spiritual maturity and share in His own dignity and authority. Such a father is a servant of his children in the truest sense, that he seeks their good, not his own power.

561

98. See Walter Schmithals, *The Office of Apostle in the Early Church*, Nashville-York: Abingdon, 1969, and Lemaire, *Les Ministères* (note 84 above), pp. 179-184. An episcopal form of church order remains not only in all the ancient eastern Churches separated from Rome, but also in churches of the Reformation, notably in some Lutheran churches (as those of Sweden), in the Church of England, the worldwide churches of the Anglican Communion, and some of the offshoot Methodist churches. The churches having a presbyterian form of polity, nevertheless, find it necessary in practice to select some form of presidency from the presbyterium. Finally, the congregational churches have pastors who fulfill all the functions of a bishop in a local church. Thus the problem of an ecumenical church order is not so much the question of the episcopate as one of a unified order that transcends the local and national churches. Would this constitute a problem if the principles of collegiality and subsidiarity recognized by Vatican II were thoroughly implemented so as to foster rather than extinguish full participation in church life at the grass-roots?

99. Lemaire, *Les Ministères* (note 84 above) concludes that the first clear evidence of resident bishops is with Ignatius of Antioch at the beginning of the second century but it was general by the second half of that century (p. 199). It is difficult, however, to deny that Clement was a resident bishop at Rome c. 96, or James "the brother of the Lord" at Jerusalem. That local churches were able to survive without at least a "chairperson" at the head seems to me a myth of modern scholarship.

100. Current exegesis generally rejects the traditional view that Acts 6:1-7 recounts the institution of the diaconate (see Lemaire, *Les Ministères,* note 84 above, pp. 49-58) on the grounds that the Seven are not called deacons in the text, that Stephen and Phillip preach and evangelize, etc., and prefers to see in the Seven a group of presbyter-bishops appointed to administer the community of Greek-speaking Christian Jews independent of the Twelve who administer the Aramaic-speaking Christian Jews of Jerusalem. As such their appointment constitutes an important step in the transmission of authority from the Twelve to local communities (Brown, *The Critical Meaning:* note 9 above, pp. 124-146). In my opinion this explanation remains rather speculative in view of the fact that the text seems to give as the reason for the institution of the Seven not pastoral care in general but the specific function of "the daily distribution of food" and "waiting on tables." Nor is this explicit reason given by the text negated by the further accounts of Stephen and Phillip, two members of the Seven, preaching and evangelizing, since no doubt at that period of church history preaching and evangelizing, although the primary task of the Twelve and other "apostles," was not restricted to them but could be engaged in by any Christian. Perhaps Jean Colson is right in his conclusion to his study *La Fonction Diaconale aux origines de l'Église* (Paris: Desclée de Brower, 1960): "If then the Seven whose institution is reported in Acts 6 were installed by the apostles in a 'presbyteral' function, it is doubtless because in their group of Hellenists the 'episcopopresbyteral' function very quickly and very early differentiated into the 'sacerdotal' function on the one hand and the 'levitical' or 'diaconal' function on the other. Thus if the Seven were perhaps not the first deacons, they were probably at least at the origin of the diaconate in the Church." p. 46.

101. Lemaire, *Les Ministères* (note 84 above), pp. 49-58. On the role of the deacon see Norbert Brockman, *Ordained to Service: A Theology of Permanent Diaconate* (Hicksville, N.Y.: Exposition Press, 1976), and James Monroe, *The Diaconate: A Full and Equal Order* (New York: Seabury, 1981).

102. See Roger Gryson, *The Ministry of Women in the Early Church* (Collegeville, Minn.: Liturgical Press, 1976) To argue, as some do, from the scattered New Testament references to women as "apostles," "coworkers" with the apostles, or "ministers" to the conclusion that they exercised the functions we now consider proper to bishops or priests; or to argue from the liturgical rites associated with the deaconesses that they were "ordained" in the sense of having the functions now proper to deacons (when the evidence shows their functions were much more restricted), seems to me anachronistic. In the early Church terminology was fluid and the notion of "ordination" and its liturgical expression not theologically defined. Consequently this type of evidence remains ambiguous. The question stands: Were women officially deputed in a sacramental manner to perform the functions which the Church now regards as defining the episcopacy, presbyterate, and diaconate? The Declaration of 1976 took the negative position as regards the first two functions, and did not treat of the third.

103. Lemaire, *Les Ministères* (note 84 above), pp. 17-30.

104. Vatican II settled the controverted theological question as to whether the episcopate is a distinct order from priesthood (cf., *Dogmatic Constitution on the Church (Lumen Gentium)* n. 21, conveying the plenitude of the priestly office.

105. St. Thomas Aquinas, *Summa Theologiae* III Supplement, q. 37, a. 2-3. In post-Tridentine theology two factors (1) historical information that the minor orders had not always existed in the church; (2) an overly strict interpretation of the teaching of Trent that the sacraments are instituted by Christ, led theologians to deny that the minor orders are sacramental. St. Thomas had a more flexible concept of the power of the Church over the sacraments, although more strict than that of current theology.

106. Joseph Fitzmyer, *Luke* (see note 20 above), pp. 237, pp. 340-341.

107. See Hilda C. Graef, *Mary: A History of Doctrine and Devotion* (New York: Sheed and Ward, 1964), 2 vols, vol. pp. 101-160.

108. On the "Marian definitions" see Thomas A. O'Meara, O.P., *Mary in Protestant and Catholic Theology* (New York: Sheed and Ward, 1960), pp. 41-108 and 259-300, and Gustave Thils, *L'Infailliblité Pontificale* (Recherches et Syntheses) (Gembloux: Duculot, 1968); Narcisus García Garcés, "De pietate populari relate ad mariologiam," pp. 219-256, and Adolfus Kolping, De phaenomina interventionis multiplicis magisterii recentionis in re mariologica" p. 331-352 in *De Mariologia et Oecumenismo,* Rome: Pontificia Academia Mariana Internationalis, 1962.

109. See Pierre Grelot, *The Bible Word of God* (New York: Desclée, 1968), pp. 86-88; 209-213 on Biblical symbolism.

110. *An Essay on the Development of Christian Doctrine,* ed., with introduction by J. N. Cameron (Baltimore, Md.: Penguin Books, 1974).

111. Jan H. Walgrave, *Unfolding Revelation* (Philadelphia: Westminster, 1972), pp. 293-313.

112. If we accept the usual chronology as used by Brown and Fitzmyer then it is clear that between the silence of Paul, Q and Mark, and the infancy narrative of Matthew and that of Luke, there must have been a remarkable enlargement and deepening of the Testament community's understanding of the role of Mary in salvation history; to which must be added the symbolic use made of her figure in *John* and *Revelation* as the type of the New Testament Church itself. If we are sceptical of this chronology, it is still evident that in the New Testament Church there were different theological attitudes toward her and that a "high Mariology" was current in some circles before the end of the New Testament age. Thus the later development of Mariology which appears so unbiblical to many Protestants was a part of the historical theological trajectory already traced out in the Bible.

113. See the "Promulgation of the Dogmatic Constitution on the Church (Lumen Gentium)" by Paul VI, *Acta Apostolicae Sedis,* 56: 1007-1018.

114. See Chapter 9, p. 385.

115. See Lino Cignelli, O.F.M., *Maria Nuova Eva nell Patristica Graeca,* (Assisi, Studio Teologico "Porziuncola", 1966).

116. Recent theologians tend to regard Josephology as a pious aberration resting on almost no theological data. See, however, Henri Rondet, S.J., *St. Joseph,* (New York: Kenedy, 1956), and Boniface Llamera, O.P., *St. Joseph* (St. Louis, B. Herder Book Co., 1961) for treatises on the subject by excellent theological scholars.

117. See the article by F. L. Filas, S.J., "Joseph, St., Devotion to", *New Catholic Encyclopedia,* vol. 7, pp. 1108-1113. The decree of John XXIII inserting a commemoration of St. Joseph into the Eucharistic Prayer is found in *Acta Apostolicae Sedis,* 54: 873.

118. "The story is not meant to exalt the contemplative life above the life of action, but to indicate the proper way to serve Jesus; one serves him by listening to his word rather than by providing excessively for his needs", L.H. Marshall, *The Gospel of Luke,* New International Greek Text Commentary: Grand Rapids, Mich. 1978. But is not "serving Jesus by listening to Him" a very good definition of the contemplative life? It has been also pointed out that what Luke had in mind in relating this incident was to make clear that while in Rabbinic Judaism the study of Torah was a male privilege, for Christianity the study of Torah

or Wisdom as embodied in Jesus himself was open to women equally with men.

119. On the many faceted Biblical concept of wisdom see Roland E. Murphy, *Seven Books of Wisdom,* Milwaukee: Bruce, 1960, pp. 143-156.

120. See Juan Arintero, O.P., *The Mystical Evolution in the Development and Vitality of the Church* (St. Louis: B. Herder Book Co. 1951), vol. 1, 296-303. Arintero's use of the old-fashioned term "weaker sex" is only to high-light his conviction that in the highest activities of life, those of contemplation, women are the "stronger sex," e.g., the Virgin Mary and Mary Magdalen.

121. *Symposium,* 201.

122. It is easy to gather a considerable dossier of "sexist" statements from the works of St. Thomas Aquinas, but they must all be read in the context of his unequivocal teaching that men and women are *essentially* equal and alike created in the Divine image. Most of them occur in his exposition of Aristotle or in quoting Aristotle as an authority and reflect Aristotle's biological theories, rather than any theological misogynism. It is as anachronistic to cite them in the present feminist controversy, as to cite his defense of "delayed hominization" in the abortion controversy, and yet the latter is a frequent tactic of those who favor "free choice." On the historical context see E. Power "The Position of Women" in G. G. Crump and E. F. Jacob, *The Legacy of the Middle Ages* (Oxford: Oxford University Press, 1962).

123. See Stephen B. Clark, *Man and Woman in Christ,* (Ann Arbor, Mich., 1980) pp. 619-666 for an argument why, at at least in modern cultures, the problems of family leadership, of church leadership, and of leadership in secular occupational and political society should not be treated as identical, but require different kinds of answers.

Chapter 12

The Spiritual Body

(I Cor. 15:44)

I: Glorious Wounds

1. The Dead Body of Christ

In the last chapter we considered Jesus' developing understanding of himself as Son of God as this is expressed in the chief vision narratives in the synoptic gospels. Intrinsic to this development was his call to proclaim the Reign of God and his ever growing awareness that obedience to this call would inevitably lead him to the Cross.

The infancy narratives of both Matthew and Luke picture Jesus as born and reared under the shadow of death. Matthew tells how the child, before he was two, narrowly escaped the massacre of the infants of the region of Bethlehem by the paranoic tyrant Herod, as well as the perils of his parents' flight into Egypt. Luke says nothing of all this, but relates how Mary, when she presented her first-born in the Temple soon after his birth, was greeted by the prophet Simeon who told her,

> This child is destined to be the downfall and the rise of many
> in Israel, a sign that will be opposed — and you yourself shall

be pierced with a sword — so that the thoughts of many hearts may be laid bare (2:34-35)

Furthermore, Luke (2:41-52) recounts how at the age of twelve the boy was lost for three days and frantically sought by his anxious parents.

Current exegesis sees in these stories legendary constructions whose theological purpose is to retroject the passion of Jesus to his earliest years.[1] The details of Matthew's narrative certainly show the processes that produce legends at work, yet if we make allowance for this, there is nothing incredible about the Herod we know from Josephus being frightened enough by astrological predictions to order a small atrocity, nor in Jesus' parents taking refuge over the border to escape. The real difficulty is why Luke ignores this event, as Matthew ignores the events which Luke relates. Unless, however we attribute their narratives to pure fiction on their part, we must suppose that both Matthew and Luke selected from among the stories about Jesus' infancy known to them those which suited their particular theological purposes. It is not even improbable that Luke knew the traditions used by Matthew, but ignored them because they did not fit in well with his very careful scheme of contrasting the births of John the Baptist and of Jesus, yet took care to include in his own story no less than 12 different facts in Matthew's account[2] and to provide parallels to other points of particular theological significance. Thus in place of Matthew's foreshadowing of Jesus' passion by the massacre and flight, Luke's story of Simeon's prophecy and of the three day loss of the child perform the same theological function.[3]

Whatever one concludes about the historicity of these narratives, it is evident that the two evangelists wished to make clear that the whole life of Jesus was under the shadow of the Cross. To reduce this merely to a literary retrojection is certainly a mistake. We know enough of the political and religious situation of Palestine at the time of Jesus' birth to know that he must have grown up in an atmosphere of the terrible foreboding which we find in the extracanonical apocalyptic literature. It was a time of oppression and nascent revolution, of terrorism and atrocity, and of feverish anticipations of the Day of the Lord. The traditions of the deaths of the prophets and the martyrdom of the Maccabees made it clear that anyone assuming the role of a prophet or witnessing to a doctrine different than the official line of the Jewish or Roman authorities was in constant peril, and that the punishment might be atrocious.

If we suppose, as I argued in the last chapter, that the virginal conception was an historic fact, and that it shows us that Mary was a person in whose religious conscience was summed up all the faith of Israel —

566

and therefore both its righteous anger and its fears, as Luke depicts them in the *Magnificat* — we ought also to suppose that Jesus was raised in an atmosphere which was from his infancy one of high religious tension. The house in Nazareth was no doubt one of domestic peace, but its windows were not closed to "wars and rumors of wars." How could Jesus have preached the coming Reign of God as he did unless he had first spent the years of his youth filled with an ever deepening concern over the agony of the world around him? Palestine was then as now a "trouble spot" of the world in which no one could escape seeing the miseries of the human condition, and Jesus must have viewed this human condition through the eyes of his fervently religious parents.

Moreover, as his resolution grew in him to engage himself more directly in this situation and to play an active part in the world conflict, his widowed mother must have known it, and grown increasingly anxious about what the outcome for her son would be; yet in her own zeal she must have refused to try to hold him back. All this is guesswork, yet it is in accord with what the four evangelists tell us, and any other supposition would seem to make the psychology of Jesus inexplicable. If we take seriously an epistemology which says that human knowledge arises from bodily experience rather than from an inner life dualistically isolated from the body, we must understand Jesus in terms of his historical context and his family origins.

All four evangelists tell us that the work and the fate of John the Baptist played a critical role in Jesus' own mission. The Baptist had preached a message far less disturbing than that of Jesus, yet he was beheaded for it. According to the Fourth Gospel, when Jesus came to John to be baptized and thus to take part in the general act of conversion and renewal to which the Baptist called the whole people, John cried out "Look! There is the Lamb of God who takes away the sin of the world!"[4] thus predicting from the very beginning of Jesus' ministry that Jesus would end as a sacrificial victim. Mark, after relating the Baptist's execution by Herod (6:14-29), tells us that Jesus, much to the shock and denial of the Twelve, predicted his coming death and resurrection three times (8:31-33; 9:30-32; 10:32-34). The Fourth Gospel portrays Jesus as predicting his own crucifixion as early as his dialogue with Nicodemus (3:13-21) even before the imprisonment of the Baptist (3: 22-30) and as alluding often to it in his later discourses. Thus there is no reason to doubt that throughout his ministry Jesus was constantly aware of his probable martyrdom.

As the fatal time approached, in spite of his favorable hearing by the crowds of common folk, Jesus' anguish over his failure to awaken the faith of the leaders of Israel grew deeper. In striking parables (of the vineyard tenants, Mt 21:33-46; of the wedding banquet, Mt 22:1-14) he predicted the fatal outcome, lamenting,

> "O Jerusalem, Jerusalem, murderess of prophets and stoner
> of those who were sent to you! How often have I yearned
> to gather your children as a mother bird gathers her young
> under her wings, but you refused me. Recall the saying, 'You
> will find your temple deserted.' I tell you, you will not see
> me from this time on until you declare, 'Blessed is he who
> comes in the name of the Lord!'' (Mt 23:37-39)

And he saw not only the disaster of the fall of Jerusalem but in a vast
apocalyptic vision (Mk 13:1-37) the death of the nation, the death of the
world.

There is a notorious difficulty about whether the Last Supper was
legally the Passover meal, since according to the Fourth Gospel the date
for Passover, at least as it was reckoned in Jerusalem, was one day later.[5]
Nevertheless, the synoptics are probably correct in considering it to have
been Jesus' intention to celebrate that festal meal with the Twelve and
in so doing to reveal to them his own understanding of what he was convinced was about to happen to him — betrayal by one of the Twelve to
his enemies and his torture and death at their hands. He said to the
Twelve, "I solemnly assure you, I will never again drink of the fruit of
the vine until the day when I drink it new in the reign of God" (Mk 14:25),
words which they may have understood as a vow, but came later to
understand as a prophecy.[6] Consequently, he wished to give them a way
of commemorating the time of his presence with them after the separation he saw was coming, and

> Then, taking bread and giving thanks, he broke it and gave
> it to them saying: "This is my body to be given for you.
> Do this as a remembrance of me." He did the same with
> the cup after eating, saying as he did so: "This cup is the
> new covenant in my blood, which will be shed for you."
> (Lk 22:19-20)

These words, which in their essential features are historically well
attested not only by the synoptics but by St. Paul (1 Cor 11:23-26),[7]
show us the final phase of Jesus' developing self-understanding. No better commentary on this could be offered than that of the profound
theologian who wrote the *Epistle to the Hebrews:*

> On coming into this world, Jesus said, "Sacrifice and
> offering you did not desire, but a body you have prepared
> for me; Holocausts and sin offerings you took no delight
> in. Then I said, 'As is written of me in the book, I have come
> to do your will, O God.'''. First he says, "Sacrifices and

offerings, holocausts and sin offerings, you neither desired nor delighted in" (These are offered according to the prescriptions of the law.) Then he says, "I have come to do your will." In other words he takes away the first covenant to establish the second. By this "will," we have been sanctified through the offering of the body of Jesus Christ once for all (10:4-9; the quotations are from Psalm 40 and 1 Sam 15:22).

Jesus in this moment clearly understood himself as a priest, not in the merely ritual sense, for the Passover was a domestic act, but in the fundamental sense of one who has been called by God to offer sacrifice because he is totally committed to the people for whom he is called to make this sacrifice. Moreover, this sacrifice, unlike those of the Old Law, is not merely symbolic of an interior sacrifice, but is that interior sacrifice expressed in the offering of his own body.

As mentioned in the last chapter, the essence of sacrifice is not the destruction of a victim, nor is it primarily an expiation of guilt by the acceptance of punishment, but it is simply an offering of a gift in acknowledgement to God that He alone is the source of all gifts.[8] Moreover, the offering to God of external gifts can only be symbolic of self-offering to God in love, since all that God really wishes of us is ourselves. His greatest gift to us is the gift of our very being and above all our free will, our love, and only this total return of ourselves to Him can be a proper response to the creative love by which He has brought us into being and guides us to union with Him. God has never sought vengeance for human sin. All biblical expressions that seem to imply this must be interpreted in light of the supreme principle that God is Love. What He has sought is our love, which completely restores all damage ever done by sin and even renders it a *felix culpa*.

2. Why Did He Die?

Why then did Jesus die? Too often, especially in late medieval times and the subsequent theology of the Reformation, the death of Jesus has been understood as if God were a wrathful despot who, when he was insulted by His rebellious subjects, delighted in taking bloody vengeance on them. Therefore, He had to have his feelings soothed by beholding his innocent Son tortured and killed in their place.[9] No wonder Humanists rebelled against such a blood-thirsty Deity! This type of theology arose by an exaggerated stress on certain biblical metaphors used originally in view of a primitive culture in which blood feuds were still a part of life, but which by the time of Jesus had already been repudiated by the prophets and by the Pharisaic rabbis. That such metaphors appealed so strongly to late medieval culture probably reflects the rise of absolute monarchies in that troubled period, and constituted a reversion to an obsolete conception of God.

W. D. Davies has shown[10] that the essential idea of the Old Testament sacrifices was the acknowledgement that God is the Giver of Life. The proper role of the priest was not to kill the victim, but to *offer* it. What did this offering signify? It was the return to God of a gift taken from the gifts which he had given (a "first-fruit"), as an act of thanksgiving. An exchange of gifts between human beings is always a sign of an exchange of persons. It signifies a bond or *covenant* between them. By such a covenant God binds himself to care for his people, and the people bind themselves to obey him with an obedience which is itself an act of love because its purpose is not to enhance the well-being of God, which is impossible, but to cooperate in God's care of his world rather than to attempt to frustrate that care. Thus sacrifice does not benefit God, but it does benefit those who offer themselves to him, because it draws them into a closer relation with him and makes them partners in his creative work for the world.

This was the meaning of all the sacrifices of the Old Law whose purpose was to deepen the covenant of the Chosen People with God, and thus to further God's plan for the restoration of the whole world to its original splendor at creation. Why then was the typical Old Testament sacrifice bloody? Why did it require the death of the victim? The answer seems to be that it was not death as such that was important in the ritual sacrifices (cereals and incense were also offered) but the *consumption* of the victim. A relation of love between persons requires a mutual exchange of personal life, a communication of selves. Such a communication, as I have shown in Chapter 10, is really possible only at the spiritual level, but our bodies participate in it sacramentally. Since, however, at the physical level there cannot be a real sharing of a common good, it must be symbolized by the giver *losing* what he gives through its consumption by the one to whom it is given. The archetype of such an act is the *sharing of food*, as shown in Chapter 10. We exchange food and I eat what you give me and you eat what I give you. Thus we "eat each other," so that I am immolated to you and you to me.

Even sexual intercourse, which is the second archetypal symbol of mutual relationship, is a kind of mutual eating (the "two become one flesh"), so that the basic social unit of the family is founded on the sacraments of *torus et mensa*, "bed and board." Thus the first meaning of the sacrificial immolation is that God and His people sit down together at a common meal, exchange and consume the food, and thus "consume each other", i.e. are united with each other. So evidently Jesus conceived the Last Supper, as the messianic banquet which was also a wedding banquet in which he was the bridegroom and Israel was his bride. When in the *Gospel of John* we read that the Beloved Disciple reclined on the bosom of Jesus we should not be scandalized that this is symbolic of the Church as virgin bride, for at that moment the Twelve were the Church.

Sacrifice in this first sense is a gift, an offering of self in exchange for God's prior offering of Himself to us, and the immolation involved is essentially not destructive but unitive, although as a physical sign it involved a destruction. Such a sacrifice, which is fundamental, is a confirmation of the Covenant and it is also a "sacrifice of praise" (Heb 13:15), a *berakah*, or *eucharist,* i.e. a *thanksgiving*, since thanksgiving is simply the acknowledgement of a gift and praise is the acknowlegement of the giver. It is also a "blessing", because to bless someone means both to give a gift (as God's blessings to us), or to acknowledge a gift (as we "bless God" for His gifts).

We, however, are not worthy to sit down with God at His table, to offer ourselves to Him. The prophets say that Israel was an unfaithful wife. Jesus gives the rule, "If you bring your gift to the altar and there recall that your brother has anything against you, leave your gift at the altar, go first to be reconciled with your brother, and then come to offer your gift" (Mt 5:23-24). Before the one who has broken the Covenant can sit down to eat with God, he must first repair the damage he has done, not to God who cannot be injured, but to God's creation which God loves, and especially to other children of God. The prophets constantly declaim against those who suppose they can offer a sacrifice acceptable to God without conversion and reparation.

W. D. Davies shows that to avoid this error, the rabbis taught that the sacrifices do not obtain forgiveness for *deliberate* breaches of the Commandments, but only for offenses committed in ignorance or at least from mere negligence.[11] For deliberate sins (the only kind that Christians recognize as capable of being *mortal* sins that would destroy the union of the believer to God), the only remedy is repentance ratified by the sacrifices of the Day of Atonement. Under the influence of the prophetic teaching that "God desires mercy, not sacrifice" (Hos 6:6), and that "Obedience is better than sacrifices" (1 Sam 15:22), by Jesus' time the Pharisees, alienated by the Sadducean priests, no longer greatly emphasized the sacrificial system. Thus according to Davies, St. Paul's own rabbinical background shows through in his greater emphasis on the more fundamental covenantal aspects of Jesus' self-offering than on the cultic metaphors found chiefly in *Hebrews.*[12]

Yet undoubtedly the notion of "sacrifice for sin" is prominent both in the Old and New Testaments, and the immolation involved is a *destruction*, but what is to be destroyed is not the sinner but his sin. The New Testament brings out forcefully the teaching of Ezechiel 33:11, "As I live, says the Lord God, I swear I take no pleasure in the death of the wicked man, but rather in the wicked man's conversion that he may live" (Lk 15:1-32). Conversion, therefore, not sacrifice, is the way to the forgiveness of sins; but conversion is a kind of sacrificial immolation in which the sinner gives up his sinful acts and ill-gotten goods, and attempts to repair the damage he has done. The resolution to do this is symbolically enacted by the immolation of a victim who represents the offerer's sinful or false self. The sinner must "die to sin" so as to "rise to life" (Rom 6:11). What God wills, therefore, is not death but life. Thus in the Old Testament, the scapegoat laden with the sins of the people who was sent into the desert to die (Lv 16:20-28) was a symbol of the people's resolve to destroy their own sinful ways. The holocausts or total sacrifices (Lv 4:1-6:10) symbolized the fact that the sinner was not ready for communion with God in the thanksgiving and peace offerings until his sin was totally destroyed. Thus the holocaust was the sinner's acknowledgement that while God had a claim on him, he by reason of his sin had no claim on God. Finally, the significance of the offering of blood, which was the supreme act of the priest, was that for the Jews, blood was the principle of life,

> Since the life of a living body is in its blood, I have made you put it on the altar, so that atonement may thereby be made for your own lives, because it is the blood, as the seat of life, that makes atonement. That is why I have told the Israelites: No one among you, not even a resident alien, may partake of blood (Lv 17:11-12).

Thus the blood was the equivalent to the "breath of life" (Gn 2:7), or the soul or spirit by which God as spirit was present in the human being, perserving him or her in existence. By offering to God the blood of an animal, the priest symbolically offered to God the very heart, life, mind, and spirit of the People, their innermost being, in the hope that God would give that life back to the People rather than allow them to fall into death as the result of their own foolish violation of the Covenant. In order to teach this reverence for life as the supreme gift of God, the People were solemnly forbidden to drink blood or to eat meat with the blood in it. This symbolism therefore, does not mean that God delights in suffering and death, but rather that He is the giver and restorer of life. Yet there is a sense in which God is an avenger; what he avenges is the violence by which one human being takes the life of another, of his brother whose "blood cries from the ground" (Gn 4:10) as Abel's did.

In view of this understanding of sacrifice, first of all as a thanksgiving offering confirming the covenant, and second as symbolizing the conversion from sin, the action of Jesus at the Last Supper becomes evident. The Passover Supper was a joyful celebration of God's deliverance of his People from slavery, constituting them a new community wedded to him by the Covenant. Jesus celebrated it (perhaps even out of season) to reassure the Twelve that whatever befell him, and even if they betrayed him, yet the promises of God would be fulfilled and the Reign of God would certainly come. At the same time, Jesus wished to prepare them to understand the meaning of his death in the plan of God. It was not that the Father wanted his own son to die, but that his Father had called him to witness to the truth about God, the truth of God's love for his people, even if this witnessing led to his own son's death. In fact, God willed to permit that death in order that it might be manifest to all that Jesus was the Suffering Servant foretold by the prophets, one whose word could be trusted because he was a martyr acting only for the sake of truth. Since the authorities had accused Jesus' of being an agent of the devil, a blasphemer and deceiver, no other proof than his righteous death could ever have made absolutely clear that he was a true witness to the Father.

The Jews had already developed the view that a good man can merit for others, just as a bad man can implicate others in his sin. Such a notion was congenial in a culture where the notion of *corporate personality*, of which I have already spoken, was accepted. This notion is difficult for us, because we live in a culture which is radically individualistic. If everyone is morally responsible only for his own actions, how can one person be responsible for the sins of another, or one atone for those of another? The rabbis had already perceived the ethical danger in this notion.[13] Had not Ezekiel (chapter 18) already made clear that everyone is responsible before God only for his or her individual deeds? Yet it is also clear that as human beings, we form a community in which the actions of one necessarily affect the actions of another. I have already shown in Chapter 9 and 10 that the doctrine of original sin does not mean that the individual is guilty before God because of the sins of his or her ancestors, but that individuals are nevertheless victims of the sins of their ancestors, suffering some of the inevitable consequences of those destructive acts, but personally sharing in their moral guilt only when they freely conform to the same pattern. In the same way we benefit from the good actions of others, even when we have not deserved it, but we can take no moral credit for this unless we ourselves freely choose to follow their good example.

3. Merit

How then is it possible for one person to merit for another or to expiate the sins of another? Since all conversion from sin and all doing of good works that repair the damage of sin and create the Reign of God can only come from the grace of God, from His gift of the life-giving Spirit, the fundamental way that one person helps another to be rid of sin and to rise to a new life is by intercessory prayer. In the Old Testament Moses is pictured as the great intercessor for his sinful people (Dt 9:7-29), and *Hebrews* says,

> In the days when he [Jesus] was in the flesh, he offered prayers and supplications with loud cries and tears to God, who was able to save him from death, and he was heard because of his reverence. Son though he was, he learned obedience from what he suffered; and when perfected, he became the source of eternal salvation for all who obey him, designated by God as high priest according to the order of Melchizedek (5:7-10).

and then goes on to show that Jesus as Risen Lord continues this intercession forever. Intercessory prayer, however, is effective primarily because of the relation of love between God and the one who prays, according to the rule "the fervent petition of a holy man is powerful indeed" (Jas 5:16). Consequently, the prayer of Jesus, "perfected" through "obedience", is universally effective because he is the innocent Son of God perfectly united to his Father. Thus what obtains the salvation of the world is not Jesus' sufferings and bloody death as such, but the obedience and love which were perfected and manifested through his bodily sufferings and death. This efficacy is not merely one of example, but of real empowerment, because it obtains the coming of the Holy Spirit to convert sinners and to unite them in Christ as his Body actually at work in the world.

Thus the sacrificial act of love by which Christ offers himself to the Father in intercession for us completes the sacrifices of the Old Law and makes them no longer necessary, since what they signified symbolically, it actually achieves, namely the perfect revelation of God's love to the world manifested in the love of his Son for the Father and for the world, even for his and the Father's enemies. Yet there is needed in the Church which Jesus founded some way to supply for its members what the temple cult had supplied for the Jews, namely a way to express sacramentally both the renewal of the Covenant and conversion from sin in an act of thanksgiving and reconciliation. This act can not be a new act as if something can be added to the total offering of Christ himself on the Cross, but it can be a *anamnesis* or "remembering" of that same act, by which it becomes actual in the lives of all those who freely participate in that one, sufficient offering of Christ.

For the theologians of the Reformation it seemed that the Catholic understanding of the Mass or Eucharist was a blasphemous denial of the once-for-all sufficiency of Jesus' sacrifice so clearly taught in the Scriptures (Heb 9:23-28). They were misled both by the decadent theology and the distorted piety and liturgical practice of the late middle ages. The Eucharist is a remembering of an event which need never and can never be repeated, but which is not something merely of the past. It commemorates a permanent change in the relation of the world to God brought about by Christ's death and resurrection; a change, however, which must be effectively actualized in individuals by their conversion to Christ and sanctification in him.

All the Christian churches recognize that the salvation which Jesus brought by his suffering and death on the Cross is made efficacious in the individual only by faith; but most of them, even those of anabaptist persuasion,[14] believe that it can also be efficacious in children who have not yet reached the age when they can personally believe, if they are somehow included in the Church by the faith and intercession of their parents or other believers, whether this takes the form of infant baptism or not. The Catholic Church at Vatican II made perfectly clear that this possibility of salvation extends also to those who are not visible members of the Church, provided that they freely accept the grace of Christ coming to them by the mysterious vivification of his Holy Spirit which is at work in the world in other religions (even in Humanism and Marxism) which do not recognize who Jesus is.[15] The reason for this will become more evident later on, but certainly Jesus' intercession was for all of humankind, and must be somehow efficacious for all who do not knowingly and willingly turn away from the Father who has offered Himself to us through His Son.

What is this salvation which comes through Jesus' sacrifice?[16] Clearly it cannot be something which affects us merely passively, since its primary effect is reconcilation with the Father and the beginning of a new life with him as he originally intended for us in creation. Such a reconcilation and new life must be a free and personal activity on our part, although not merely from our own powers. The saying of St. Paul, "I have been crucified with Christ, and the life I live now is not my own; Christ is living in me" (Gal 2:19-20) is not to be understood as if St. Paul had become a passive automaton moved by an external power, but in the sense that in him the "old man" (Rom 6:6), or sinful self, had been destroyed in the sacrificial death of Christ by repentance, so that the new man, his true self as God created him to be in nature and in grace, has begun to live in union with Christ and by his Spirit. Thus Paul adds, "I still live my human life, but it is a life of faith in the Son of God, who loved me and gave himself for me" (Gal 2:20).

Salvation is a new life in accordance with the Reign of God, and as such is an activity in the world which is a share in Jesus' own activity, a life of intercessory prayer for others and of efforts to bring justice and peace on earth and the union of all human beings in one community, looking forward to the eternal life of that community in Christ. Such a life, however, as St. Paul learned, also implies that we must share in Christ's sufferings (Phil 1:18-26). As Jesus was himself perfected through suffering, learning an ever deeper and wider love, so we must learn through suffering for the sake of the Gospel. The Father permitted his Son to suffer only that Jesus might in this way manifest the Father's perfect love of the world, and so God would not permit us to suffer even the smallest trouble if it were not for the same reason.

For the Christian, therefore, all the evils of life can become, when lived as Jesus' lived them in obedience and love, not obstacles to life, but opportunities for growth in a deeper life of faith and love. Through them as love grows, it purifies us of the distortions which have resulted from sin, both personal and social. Today it has become difficult for us to understand the Christian attitude toward suffering, because it seems to us that suffering is destructive. Yet we are quick to realize that persons undergoing psychoanalysis cannot arrive at personal wholeness and true freedom except by a terribly painful process of coming face to face with their neuroses. Such a grappling with the unconscious is only part of a much deeper and more comprehensive process of growth, in which those who would be whole and free must encounter the truth about themselves and the world of injustice in which we live.

For a free person to change is not itself painful, but to be changed passively by another tends to be painful because it goes against our essential freedom. Yet even self-change becomes painful when we meet in ourselves resistance to change because of the remaining effects of sin within us. Among these effects are prejudices, habits, neuroses, diseases rooted in our body which by its materiality is subject to determinisms which cannot be simply altered by a change in our will or our reason. That is why, in all religions, it is recognized that for spiritual growth a certain *asceticism* is necessary by which our spiritual control over our bodies, imagination, and feelings is enhanced. We have only to think of the severe discipline of meditation practiced in Eastern religions to recognize this necessity. The Jews and early Christians practiced fasting, sexual abstinence, vigils for prayer. The contemplatives of the desert

576

engaged in astonishing "mortifications", and in the middle ages flagellation was adopted by the western monks.[17]

Such asceticism, with its bodily sufferings, is unquestionably a necessary part of the Christian life and is still a duty of every Christian,[18] since without it it is hardly possible that there can be consistent and sustained spiritual growth; and because it is the *easiest* form of the suffering which is an essential part of the imitation of Christ who said, "If anyone wishes to come after me, he must deny his very self, take up his cross and follow in my steps. Whoever would preserve his life will lose it, but whoever loses his life for my sake and the gospel's will preserve it" (Mk 8:34-35). It is easiest because being active, it is not contrary to our freedom. It requires, however, to be regulated lest in the effort to discipline the body, the spirit become more disorderly. The great spiritual guides have always warned ascetics: (1) self-mortification can itself become a preversion either because it is eroticized as in sado-masochism, or worse because it becomes a kind of suicide, an aggression against the self; or still worse it becomes a source of pride and self-righteousness; or finally, it can so injure physical and mental health as to leave the person with no energy for any good activity; (2) consequently, it must be moderated as a means to an end and adjusted to the condition and advancement of the individual.

Nevertheless, because it is the easiest form of purification, every Christian must undertake it in order to be strong enough to bear what is far harder, namely the discipline which comes not from one's own choice and activity, but from without in the form of persecution, opposition, frustration in one's attempts to do good, or from the sickness of the body and from psychological illness as it arises from the body, or from unconscious determinisms. Because these must be endured passively and against our freedom, they are suffering in the strict sense, and are very difficult to accept. Nevertheless, only by our accepting what we cannot change out of the love of God, who only permits this that we might grow closer to Him, does this passive suffering become a source of purification and growth.

Finally, there is the form of suffering of which the mystical writers speak as "the dark night of the spirit", which is also passive, but which arises from the very presence of God in his human temple. Only the Spirit of God is able to enter into the very depths of the human spirit and to purify and vivify it at its very source.[19] This suffering, which the mystics tell us is the most terrible of all (the men of the Old Testament believed that to see God means to die), is a suffering of love because it arises from the intense desire to be united with Christ and the agony felt because of the remaining darkness of sin (the sin of the world, if not their own), which still holds them back.

Even after such purgation, suffering still remains. St. Paul writes the Philippians (1:24-26):

> For, to me, "life" means Christ; hence dying is so much gain. If, on the other hand, I am to go on living in the flesh, that means productive toil for me — and I do not know which to prefer. I am strongly attracted by both: I long to be freed from this life and to be with Christ, for that is the far better thing; yet it is more urgent that I remain alive for your sakes.

Some have seen in the first part of this statement the influence of the Platonic idea that the soul is imprisoned in the alien body. However this may be, Paul's thought is not really Platonic, since to be with Christ for Paul is to be incorporated in the *body* of the risen Christ. "Living in the flesh" here means being still involved in a sinful world and burdened with all the anxieties of loving and serving this world — Paul's "Leaving other sufferings unmentioned, there is that daily tension pressing on me, my anxiety for all the churches" (2 Cor 11:28) — and it is this suffering which brings the Christian to ultimate perfection, to the closest likeness to Jesus who suffered not for his own sins but from the terrible anguish of seeing his brothers and sisters destroying themselves and each other.

Thus we may believe that Jesus himself arrived at his own fullness of love for the Father and for us only on the Cross itself when he was able to say, "Father forgive them; they do not know what they are doing" (Lk 23:34), and to recite Psalm 22, "My God, my God, why have you foresaken me," which expresses the ultimate temptation to doubt the love of the Father, but which concludes

> To him my soul shall live,
>   my descendants shall serve him.
> Let the coming generation be told of the Lord.
>   that they may proclaim to a people yet to be born
>   the justice he has shown. (30c-32)

or as Luke has it, "Father into your hands I commend my spirit" (Ps 31:6), and John "Now it is finished" (19:30), words of ultimate trust.

All four evangelists take pains to assure us that although the Twelve deserted Jesus in this fatal hour, his mother and some of the women did not (Mk 15:40-41; Mt 27:55-56; Lk 23:55-56; Jn 19:25-27). The significance of this seems to be to show, as Luke emphasizes, the fulfillment of God's covenant in the Remnant that awaited the coming of the Messiah, i.e., that there has always been a community of faith. In the Fourth Gospel, Jesus on the Cross entrusts his mother, symbol of the church, to the "beloved disciple", symbolic of the ministers of the Church,[20] a Church which shares the passion of her Lord. It is also this Fourth Gospel which relates the piercing of the side of the dead Christ from which flowed blood and water (18-31-37), signifying, it would seem, the sacraments of baptism and the eucharist.[21] The Fathers saw in this event the birth of the Church as the New Eve from the side of the New Adam, sleeping in death; and it is from this also that the devotion to the Sacred Heart of Jesus arose.[22]

As the wounded body of Jesus is wrapped in linen and laid in the rock-hewn tomb (recapitulating Luke's narrative of how the new born child was swaddled and laid in a manger in a cave[23]) his bodily existence seems finished and we stand before the mystery of death which seems the last word about what it means to be, as we are, mortal bodies.

II: The Immortal Body

1. Life After Death?

Do we in any way survive death? In the time of Jesus the only school of thought which altogether rejected an after-life was the Epicureans. They hoped thereby to free humanity from its fear of death, which they believed was chiefly a fear of punishment after death fostered by priests for their own profit. Some exegetes see a trace of Epicureanism even in the biblical book of *Qoheleth*.[24] While the Epicureans were a relatively small school, the much more influential Stoics also came close to denying survival since they taught that the soul is only a spark of the divine, impersonal energy which fills the world and will be reabsorbed at death into that cosmic fund of energy, or destroyed in the cyclically recurrent world conflagration.[25] The chief proponents of the immortality of the soul were the Pythagoreans who taught reincarnation, and Plato and his school, who accepting the Pythagorean doctrines, defended them by well worked-out arguments to prove that since the soul is immaterial it is eternal, pre-existing the body, descending at intervals into the body, and then returning to its heavenly home, but never ceasing to exist. Aristotle argued for a much closer relation of soul and body, but agreed that at least the intellectual element in the human being cannot be destroyed by any material process. Just how this intellectual element can remain as individuated or operate without the senses from which it draws its data, Aristotle never seems to have been able to answer. No doubt one of his difficulties was that he, like Plato, believed the world to be eternal. Consequently, if every individual soul survives death there would now be in existence an *infinite* number of souls. Later the Averroistic Aristotelians were to interpret Aristotle as teaching that there is only one world intelligence which is immortal and which is the sole agent of intellectual knowledge in human beings, but that there is no individual human soul that survives.[26]

Biblical thought, although its later authors were somewhat influenced by these Greek speculations, went back to the much more primitive beliefs which underlay all the Near Eastern and Mediterranean cultures, including those of the Greeks as is evident in Homer. The belief in some kind of survival, for which we have evidence clear back to Paleolithic times,[27] seems to have been universal. Its fundamental form was that the human individual survives death (at least for an indefinite time) as a *ghost*, a kind of shadow of the bodily self, which dwells in the tomb or the underworld (the Hebraic *Sheol*), and still has need of the clothing, servants, domestic animals and even food and drink which it needed in this life. Often this life was pictured as a meaningless existence in the dark. The living feared that the ghosts might return, or that they would suffer still more if their funeral rites were neglected or they were deprived of offerings of food and drink. In Sumer, however, and especially in Egypt, royalty believed they could continue to live a life of luxury and enjoyment by means of richly furnished tombs, magical rites, and (in the case of Egypt) mummification of the body.

The notion that this underworld might be a place of judgment and of punishment and reward also was elaborated and became, as the Epicureans complained, an obsessive preoccupation of the living. Consequently, the place reserved for the righteous was more and more pictured as a place of happiness which might still be in the underworld (as the Greek Elysian Fields), or in order to distinguish it clearly from the place of punishment might be located in the heavens. The blessed dead, therefore, were venerated as immortal gods (although even the immortality of the gods was often considered to be merely relative); sometimes as companions of the gods of the underworld such as the Egyptian Osiris or the Greek Pluto or Hades, and sometimes, especially in the case of royalty or heroes, as companions of the heavenly or Olympian gods.

The biblical writers generally shared belief in survival of the individual as a ghost in Sheol.[28] Mitchell Dahood has pointed out evidence in the Psalms and other biblical books that although Sheol was commonly pictured as a place of darkness where positive human activities, especially the worship of God, was impossible ("It is not the dead who praise the Lord, nor those who go down into silence", (Ps 115:17); nevertheless, by others it was pictured more positively as it had been in the older Canaanite literature.[29] By the time of the Pharisees, the rabbis taught that at death there is a judgment and the shades of the unrighteous go to a place of punishment in Sheol called Gehenna, and the just to a place of happiness called Paradise, like the garden of Eden. It is evidently to this that Jesus himself refers on the Cross, when according to Luke's account he said to the good thief, "I assure you: this day you will be with me in paradise."

(23-43).[30] It is not strange then that a late Old Testament work, *Wisdom*, written in a Hellenistic atmosphere, seems to accept the Platonic notion of the immortality of the soul (which of course is not quite the same as that of a ghost) without question (3:1-9).[31]

Nevertheless, the Platonic dualism of perishable body and immortal soul could not be successfully fused with Jewish monotheism. Scripture teaches that the human person is made in God's *image*, but also teaches that it is *made*, created out of nothing in its entirety. It would seem, then, that what has been made out of nothing can also return to nothing.[32] Moreover, since the body has been created by God and vivified by the infusion of the "breath of life" (Gn 2:7), the body is also good and inseparable from the human person. The solution to this puzzle seems to have arisen not from abstract speculation but from the bitter experience of the Jewish people.

It is generally agreed that it was the Maccabean period which witnessed a crucial development of Jewish thought about the next life as a result of the martyrdom of many Jews faithful to the Law of God in the face of Hellenistic persecution.[33] Surely a just God would restore to full bodily life those who had yielded their bodies to torture and death for His Name's sake! This is most nobly expressed by the mother (symbolic no doubt of the whole of faithful Israel and foreshadowing in Christian symbolism Mary at the Cross), who said to her youngest of seven sons faced with the choice of apostasy or martyrdom:

> "I do not know how you came into existence in my womb; it was not I who gave you the breath of life, nor was it I who set in order the elements of which each of you is composed. Therefore, since it is the Creator of the universe who shapes each man's beginning, as he brings about the origin of everything, he, in his mercy, will give you back both breath and life, because you now disregard yourselves for the sake of the law" (2 Mc 8:22-23).

This conviction that God would *resurrect* the faithful (note that nothing yet is said of the unrighteous) was adopted by the Pharisees, by the Essenes, and clearly affirmed by Jesus himself, although the Sadducean priesthood retained their conservative scepticism (Mt 22:23). All three synoptics report Jesus' argument for this doctrine:

> "As to the raising of the dead, have you not read in the book of Moses, in the passage about the burning bush, how God told him, 'I am the God of Abraham, the God of Isaac, the God of Jacob'? He is the God of the living, not of the dead" (Mk 12:26-27).

It is notable that Jesus was here hard pressed to find in the Torah the grounds for his position (since only the Torah had indisputable canonical authority at that time), but did find it by a profound intepretation. God's covenant with the patriarchs, antedating even that with Moses and Israel, establishes a relationship between God and his faithful which God cannot permit death to destroy. Since for the Bible the anthropology of *Genesis* is normative (as in the words of the Maccabean mother), this covenant can be fulfilled only if God gives back to the faithful dead their lives as body-persons, not merely as souls.

Does this mean that the Pharisaic and Christian doctrine of the resurrection of the body *excludes* the older Jewish notion of ghostly survival or the more developed Platonic doctrine of the immortality of the soul? Oscar Cullmann has convinced many theologians that it does.[34] He understands St. Paul at one time (1 and 2 Thes) to have expected the Lord to return in the lifetime of the first generation of Christians and later, when this expectation was disappointed, to have come to believe that the dead would "sleep" until the final judgment when all would be resurrected. H. Lassiat[35] has argued that the early Church Fathers, notably St. Irenaeus, in their catechesis based on apostolic tradition, taught that while all souls, both those of the just and the unjust, are *immortal* in the sense that they survive death, they are not *incorruptible*, hence the souls of the unjust in hell are eventually destroyed by their sins while those of the unjust are perserved in perpetual life by the Holy Spirit. According to Lassiat, the notion of the natural immortality of the soul entered Christian catechesis through the influence of Platonism in the second century apologist Athenagoras and became influential chiefly through Origen in the third century, in whose thought it was connected with the soul's pre-existence to the body and transmigration from one body to another. Lassiat urges theologians to abandon the notion of the soul's natural immortality as alien to the Bible and to apostolic tradition.

I have already shown in Chapter 3 and 4 how Platonism came to dominate Christian theology in both East and West until the time of St. Thomas Aquinas in the thirteenth century, so that the doctrine of the natural immortality of the soul (but not its pre-existence or transmigration) generally prevailed. With the adoption by Aquinas of the Aristotelian conception of the soul as the form of the body, the question arose whether this doctrine could still be sustained. Yet theologians continued to accept it as a matter of faith, while raising doubts about whether the classical Platonic philosophical proofs were valid. Both the Averroist Aristotelians (who tended to the notion of a single intelligence for all human beings) and the Nominalists denied the certainty of these arguments.

Aquinas' own position was nuanced and not without its difficulties. He rejected the Platonic arguments for the immortality of the soul, but attempted to assimilate the elements of truth contained in them as they were formulated by St. Augustine.[36] Aquinas accepted and further developed the Aristotelian argument that the human intelligence cannot be an act of a material organ. Furthermore, he established against certain "Augustinian" opponents that this immortal intelligence is a faculty of the unitary human soul, and not a soul distinct from the form of the body. Against the Averroists he argued that each human person has an individual soul and is not merely a participant in some universal intelligence. Finally, he concluded that this individual human soul cannot be destroyed by the material processes which destroy the human body, and hence it must survive the body, yet in this state of separation from the body it cannot be said to be a human person but only a "partial human substance."[37]

Such a "separated soul," obviously cannot function in a completely human way, since for any Aristotelian, human intellection is essentially dependent on the bodily senses (how Aquinas attempted to solve this difficulty I will discuss later). Hence Aquinas concluded that the separated soul retains a *transcendental* relation (i.e. a relation inherent in its very nature rather than added to it) to the body which it once possessed. Hence although death is "natural" to the human person in the sense that the matter of the body is liable to destruction and the soul cannot prevent this, yet resurrection is also "natural" to it in the sense that the teleology of the soul cannot be achieved without the body.[38]

The great commentators of the Thomistic school strongly defend this position, except for one of the most eminent of all, Cardinal Cajetan, who was led, probably by Averroistic influences, to doubt the conclusiveness of these arguments.[39] The influence of Averroism in the Renaissance and especially the denial by Pomponazzi that the immortality of the soul could be philosophically established, led the Fifth Lateran Council in 1513 to declare *de fide* that the human soul is the form of the body, individuated in each person, and that it is immortal. While it is commonly said by theologians today that this definition leaves open the question whether the soul is immortal by nature or by grace,[40] it is difficult for me to see how this can be reconciled with the clear *intent* of the decree which states that the error which it seeks to remedy includes the opinion of "some philosophers who teach that at least according to philosophy" the soul is mortal and one for all humans. Since, however, the church does not determine philosophical questions except as these are intimately related to revealed truth, it must be asked why the church should be concerned to defend the "natural" immortality of the human soul, when (as H. Lassiat argues),[41] the Bible and the early Church were concerned chiefly with that immortality which is a participation in the life of the Risen Christ.

In answering this question two preliminary points should be made. First, if the immortality of the human soul can be established by human reason, this is of great apologetic value since it confronts unbelievers with a great natural mystery: the paradox of a human destiny which transcends this life and human control; and thus inclines them to be open to hear the Word of God which alone can provide a solution to this inescapable mystery. Second, if we follow the lead of Aquinas, what philosophy can demonstrate is not the immortal life with God of which the Scriptures speak, but only the fact that the material processes which destroy the body cannot destroy the soul. What the life of the soul after death can be remains shrouded in mystery, and the Platonic claim to say what the ultimate destiny of the soul can be is rejected as sheer guess work. Thus Cullman's view, that the philosophical doctrine of the immortality of the soul detracts from the Christian doctrine of our absolute dependence on the grace of God for eternal life in the true sense of that term, applies to the Platonic doctrine rather than to that of Aquinas.

Granted these limitations on the significance of any philosophical demonstration of human immortality, the connexion of this doctrine with faith is to be founded, as the later Greek Fathers tell us,[42] on the strictly theological truth that since we are created in God's image with intelligence and free will, we are *open* to receive the gift of participation in God's life if God freely wills to bestow it on us, i.e. we are not only bodily beings like the animals but we are spiritual beings, persons. Yet if we are truly spiritual beings even by nature, how can we be entirely subject to that kind of death which is produced by bodily processes? Even before the efforts of Greek philosophy to provide philosophical proofs of the immortality of the human soul, the conviction that the intelligence and freedom of the human person must survive death was already common and is present in the biblical notion of Sheol.

Since, however, this restricted notion of the natural immorality of the soul leaves entirely mysterious what kind of life without the body this might be, it is understandable that many theories, more imaginative than philosophical, were purposed, and that most of these fell short of a sufficiently radical notion of spirituality, i.e. non-materiality. Thus, they pictured the separated soul as a ghost or shadow and its life either as one of minimal life in the dark, or on the other hand, as a projection of earthly pleasures. The great contribution of Plato to the history of human thought, carried over in more cautious form by Aristotle, was to attempt a truly philosophical understanding of non-material spiritual existence, freed of merely imaginative elements.

584

Thus, in contradistinction to the views of H. Lassiat already quoted, I have taken the position in Chapter 4, that the transition from Jewish-Christian theology (still evident in St. Irenaeus) to Platonic Christian Theology was not on the whole a loss, but a great gain for the right understanding of the Gospel, although it undoubtedly did result in certain incidental losses with which this book has been very much concerned. Although Plato's purified understanding of spiritual life led to a certain neglect of the Biblical understanding of the value of matter and the body, it also led in Christian theology to a greater understanding of the reality of the spiritual order, without which, in the long run, the right understanding of the human person as a whole, and therefore of the body itself, is not possible. The early Fathers such as Justin and Irenaeus struggled with this problem. Origen went too far to the Platonic side by adopting transmigration and pre-existence (which after all are remaining Pythagorean mythical elements in Plato, rather than the essence of his doctrine). With the Cappadocians in the East and Tertullian and Augustine in the West, a more satisfactory view begins to emerge which the Church came to accept as reconciling *all* the elements of biblical teaching in a consistent whole. To attempt, as Lassiat does, to return to the theories of the early Fathers as if these were "apostolic tradition", is to violate the principle of legitimate homogeneous doctrinal development.

2. The Body Transformed

The Christian view of life after death, therefore, is based on two fundamental biblical teachings. The first is that we are created in God's image and, therefore, that somehow we are not meant in God's creative intention for death, which is the result of sin and not of God's will for us. The second is that at the end of history, all those who have been incorporated in Christ, the New Adam, will live forever with him in our total personalities as body-persons, because he already has attained this immortal life in his humanity.

In trying to understand this view, therefore, we must work backward from the final consummation of all things to our present condition and the problem of individual salvation. Clearly the Bible focuses on the final coming of the Reign of God, not on a detailed explanation of the here and now, which has remained the task of theology.

In Chapter 2 we saw that as far as science is concerned, our universe is probably doomed to entropic death. Long before this cosmic death our earth will become uninhabitable. Humanity will survive only if we develop a technology that can take us elsewhere in remote space, a prospect which, although not impossible, according to present knowledge is not very probable. If we escape to other planets, we may discover that on them exist other humanoids, intelligent persons like ourselves. Yet although this idea is very appealing because it delays somewhat our having to face the ultimate questions, it rests on no secure foundations. Thus, some scientists argue wistfully that if life evolved on our earth, the uniformity of natural processes would seem to make it probable that it has also evolved on many other planets throughout our vast universe. Other scientists, however, point out that since evolution is a historical process which can go in many directions, it is just as probable, or even more so, that in the whole universe, only here on our own earth has intelligent life evolved.[43] Certainly our recent explorations of the solar system seems to make certain that we are alone in the only region of space which is ever likely to be accessible to us except by "remote control." Thus science leaves room for many possible scenarios for the end of human history, but it at least makes it very probable that the end is inevitable.

Today some Christian theologians see nothing more definite in the eschatology of the Bible than an indefinite hope for some indefinite future for human life, without specifying whether this includes life after death, or whether this future is anything more than an indefinite extension of our present history with its ups and downs. For them the "salvation" promised by the Gospel can only mean for "modern man" some kind of hope to live his or her individual life on earth in a more fully satisfying way. They seem to think that the whole Bible has to be interpreted in the light of the saying of Qoheleth (9:7):

> Go, eat your bread with joy and drink your wine with a merry heart, because it is now that God favors your works. At all times let your garments be white and spare not the perfume for your head. Enjoy life with the wife whom you love, all the days of the fleeting life that is granted you under the sun. This is your lot in life, for the toil of your labors under the sun. Anything you can turn your hand to, do with what power you have, for there will be no work, nor reason, nor knowledge, nor wisdom in the nether world [Sheol] where you are going.

Those indeed are inspired words, as well as the bitter wisdom of long human experience, but they are not the Good News.

Fundamentalist theologians, on the other hand, find in the Scriptures a very detailed scenario of an ever worsening earthly situation, a final, apocalyptic battle between good and evil, and then the return of Jesus in glory on the clouds to destroy his enemies and welcome his faithful into "the city four square" of gold and pearls.

In Chapter 9, in discussing the Christian interpretation of history, I stated my opinion that in fact the Scriptures provide us not with a single eschatological scenario, but with a spectrum of possibilities ranging between two extremes: (1) the world may get worse and worse so that at last the Risen Lord will return to judge its utter wickedness and to rescue a tiny Remnant of the faithful and take them with him into eternal glory; (2) the whole of humanity in a global community of brother and sisterhood may come to accept the Gospel and to live it by transforming our earth into a "millenial" manifestation of the Reign of God, so as to be ready to welcome the returning Lord when he comes to complete this triumph of his Gospel by elevating the earthly kingdom to the realm of eternal, risen life, as the Bridegroom coming to meet his Bride already adorned for their nuptials. Or there may take place any one of an indefinite number of intermediate scenarios. The one which will be actually realized in history depends of course on the predestining will of God, but it is God's will that *it also depends on us*, on our freely exercised stewardship.

God has truly called us to cooperate freely in the full establishment of his Reign, and he prefers that we all share in building that kingdom on earth. Yet if we freely reject his invitation, we will not share in that reign. If only a small Remnant remains faithful then the end will be an apocalypse. Who can say that this apocalypse is not even now at hand in the form of universal nuclear destruction? If it is delayed, it is only because of the prayers and work of God's servants, many of whom may be doing his will in other religions, even Humanism and Marxism, by following such light as has come to them in the world darkness. Yet in view of all the possibilities of the Gospel which have not yet been realized in history, may it not be that God will remain patient still for many centuries?

To my understanding, such an eschatology which takes seriously the many Biblical warnings to be awake and expectant of the end, yet which also takes seriously God's generosity in calling us to follow him freely, takes more adequately into account the various themes of biblical revelation than do fundamentalist or existentialist heremeneutics. An application of the hermeneutic which I advocate to the individual leads again to a spectrum of Christian biographies that range from the life of the Blessed Virgin Mary, who has already joined her Son in resurrected glory, to last hour converts like the good thief on his cross. In between are included every possible narrative of faltering saints and repenting sinners, which in fact we see in the history of the church in the past and the experience of the church in the present, and which great Christian writers like Dante and Chaucer have depicted so variously and vividly.

Great theological problems about these scenarios arise, however, and must be explored: (1) Can the process of spiritual perfection and growth, which seems so incomplete in the lives of most Christians at death, be completed after death? (2) What is the condition of those who have completed the Christian journey at death, between death and resurrection? (3) What is the fate of children who die before they are capable of a free choice of God? (4) Are those who die as enemies of God capable of repentance after death, and if not, what is their destiny?

It cannot be said that the Scriptures provide us with any explicit solutions to these questions, except perhaps to the last, but even for it the Scriptural data seems subject to wide differences of interpretation. No doubt the vagueness of biblical revelation itself says something, namely, that we ought not to be distracted by such questions from the fundamental task of living for God's Reign here and now in view of the certain hope of resurrected and eternal life with Christ, of which the Bible does speak with definitive clarity. Our freedom makes it possible to reject partially or totally the offer of God's grace, but that grace also makes it possible for us in our freedom to follow God in total fidelity. Whatever our spiritual biography, if we are to triumph with Christ, we must suffer with him through the trials of life and come to glory marked by his wounds.

Current exegesis, however, has gone very far to reduce the bodily resurrection of Jesus himself to a purely "spiritual" event.[44] This reduction is possible because the earliest New Testament documents, namely St. Paul's epistles, give no details of how the resurrection took place, although Paul plainly says,

> If Christ was not raised, your faith is worthless. You are still in your sins, and those who have fallen asleep in Christ are the deadest of the dead. If our hopes in Christ are limited to this life only, we are the most pitiable of men (1 Cor 15:18-19).

Yet he does not seem to distinguish the reality of the Risen Christ from what he beheld in his vision on the road to Damascus, a vision, which as he relates it in Galatians 1:1-20 (Paul's only account, except such allusions as 1 Cor 9:1 and 15:8; compare Luke's three versions in Acts 9:1-19; 22:1-6; 26:9-18)) might have been a purely interior experience. Moreover, in the chief text in which Paul attempts to answer the question, "How are the dead to be raised up? What kind of body will they have?", he answers by saying that it will be a "spiritual body" *(pneumatikon)* (I Cor. 15:44).

When we turn to the Gospels for some real understanding of the meaning of Jesus' resurrection as an objective event publicly witnessed

588

we find that *Mark*, generally considered to be the earliest Gospel,[45] although it speaks of the empty tomb and promises that Jesus will appear to his disciples in Galilee, does not, in its original ending, given any account of this appearance. Although in *Matthew* the story is filled out with more details (some of them, such as the earthquake, seemingly legendary), it is only in *Luke-Acts*, rather late documents of doubtful historical value, and in *John* (possibly influenced by *Luke* and highly dramatized), that we find a detailed account of the empty tomb, the appearances of the risen Christ in Jerusalem (for the Fourth Gospel also in Galilee), and his ascension into heaven after forty days. Between these gospel account there are numerous inconsistencies, while the silence of the earlier accounts about so many details found in the later ones is suspicious and evidently influenced by apologetic considerations.[46]

As a result of these undeniable problems, some Catholic exegetes wish to distinguish a *tomb tradition* (connected with later pilgrimages of the early Christians to the tomb site) and an *apparition tradition* (like that to Paul), and to see the latter as the original, the former as an apologetic dramatization.[47] Schillebeeckx, on the other hand, argues that even the apparition accounts are not certainly original, so that what actually happened is hidden from us, although we can be sure at the least that whatever happened it was sufficient to convince the Twelve and the other early Christians, on whose witness the credibility of the New Testament and the Christian faith rests, that Jesus in his complete humanity is still alive with God.[48] This conviction is then said to be essentially a faith-conviction in a transcendent event which cannot be called a part of human history as such. Similar positions are taken by many Protestant exegetes, although Wolfhart Pannenburg has undertaken to argue that Christian faith requires that the resurrection of Christ can be established by modern critical methods as an historical event.[49]

Pannenberg's effort is certainly of interest for theology in its apologetic function. It can be reasonably argued to non-believers that the faith of the Church in the historicity of the resurrection of Jesus cannot be adequately explained either as delusion or fraud. Nevertheless, it is clear from the New Testament itself that the Risen Christ was seen only by those who came to believe, and that what they experienced was something different from what they witnessed in the restoration to life of the daughter of Jairus or of Lazarus, since in several of the accounts they do not immediately recognize Jesus or he appears or disappears suddenly in a para-normal fashion. Consequently, it seems necessary to say that this event was intended by God as a direct communication to the Church, on whose testimony alone it must stand or fall as historically real.

In my opinion, the same difficulties about the critical historical method of biblical exegesis which I raised with regard to the current treatment of the infancy narratives can be raised about the treatment of the resurrection narratives. Indeed, exegetes following this method tend to class the infancy and resurrection narratives (both absent from Mark) together as the latest level of the gospel tradition and to explain both as largely theological constructs. Oddly, there seems an inconsistency in the way in which these exegetes first make extensive use of the supposition that the evangelists very freely selected from the traditions they knew in view of their theological purposes and presented them with wide literary freedom, and second, treat the texts as if omissions in earlier accounts are solid arguments *ex silentio* that the presence of material in later texts is not of early origin, or conclude from disagreements in details between different accounts (inspite of substantive agreements between them), that these diverse accounts cannot be historical.

If we grant that the New Testament writers are highly selective of the material known to them in view of their theological and literary purposes, why should we be surprised that St. Paul should not speak of the empty tomb in epistles which do not aim at narration? Why should we then pass over Mark's witness to the empty tomb and make much of his omission of any narration of the Jerusalem appearances, when it seems that this omission is required by his special theological orientation and the literary scheme he has chosen to convey it?[50] Finally, why should so much weight be given to differences between the accounts of Matthew, Luke, John, and the canonical conclusion of Mark, differences which, difficult as they may be to harmonize exactly, relate only to secondary matters and are no greater than we might expect in oral traditions modified by theological and literary bias, when all four accounts are in substantive agreement not only on the resurrection, but also on the empty tomb and the appearance of the Risen Christ to the faithful women and to the apostles both in Jerusalem and in Galilee (with the exception that Luke omits the Galilee appearances, an omission also easily explained by literary reasons)?

I do not claim to dispose so lightly of all the legitimate exegetical problems which have been raised about these resurrection texts but only to state my opinion that these difficulties are not such that they reduce these narratives to "non-historical dramatizations" of minimum use in developing a theology of resurrection. Rather, every detail remains theologically significant, provided that we do not read into these texts more than they really say. Certainly the results of modern exegesis do not give us solid warrant to doubt that the apostles saw, heard, and touched with their bodily senses the Risen Christ, who presented himself to them after his historic death a few days later in Jerusalem and Emmaus and then in Galilee, in an objective manner in the total reality of his humanity. The guarantee of the truth of this assertion is the witness of the inspired Scriptures as interpreted by the living faith of the Church, *independent* of, although not excluding, apologetic arguments based on an evaluation of the New Testament as a set of historical documents and of the attempts of critical historical exegesis to reconstruct the sources and compositional stages of these documents.

Certainly the most curious feature of the resurrection accounts is the failure of the witnesses, emphasized in them all in spite of the apologetic difficulties this raises, to recognize him immediately. Moreover, the accounts of Luke and John emphasize that he appears suddenly and disappears mysteriously, yet that he also ate and drank with the Twelve and Luke, that he "was not a ghost", i.e. that he was tangible, and that his body retained the wounds of the crucifixion. None of the accounts claim that his body was glorified as it had been in the transfiguration vision.

In the Eastern Church, the incident of the transfiguration, with its mysticism of light or "glory" plays a central role in spirituality. In the theology of St. Gregory Palamas (d.1359), the essence of God will forever remain hidden from created intellects, but He can be experienced even in this life and more perfectly in heaven through his "uncreated energies", which are distinct from his essence, yet communicate in his divinity as the rays of the sun communicate in its splendor. Christ's glory, hidden during his earthly life, was for a time revealed in his human body to the bodily eyes of Peter, James, and John at the transfiguration.[51] Obviously, this theology raises many theological questions, not only about how the notion of the "energies" can be reconciled with the divine Unity, but how such energies could be manifest to bodily eyes; but it has at least this great merit, that it assigns to the body of the Risen Christ an important function of mediating the Divine Glory to human eyes and to the bodies of the resurrected faithful an important function in the beatific vision. Western theology has tended to be a *theologiae crucis* (and Lutheranism has gloried in radicalizing this type of theology) rather than a *theologia gloriae*, but how can the Cross of Christ be separated from his risen glory?

The Risen Christ, however, did not appear in glory to the witnesses of his resurrection. The fact that he was at first not recognized, does not seem to mean that he looked different than formerly, but rather that the witnesses "could not believe their eyes." To Magdalen he seemed a gardner (Jn 20:15), to the disciples at Emmaus a traveler (Lk 24:16), but in neither case could his appearance have been extraordinary. This difficulty in recognition was on the part of the beholders, it would seem, rather than in Jesus' actual objective appearance. Today it is often said that the resurrection was not a "re-animation of a corpse", but these accounts seem to say that it was. Yet the sudden appearances and disappearances and finally the ascension as Luke portrays it (24:50-53; Acts 1:1-11), suggest that it was more than that.

Exegetes point out that it is only Luke who represents the ascension as occurring at the end of forty days while in the Fourth Gospel it appears to take place on the very day of the resurrection (Jn 20:17-23).[52] Since, however, in John 21 (according to most exegetes), we have an appendix relating the appearance of Jesus in Galilee some time after the day of the resurrection, it would seem that the earlier edition of this gospel telescoped the two events for literary rather than historical reasons. The tradition related by St. Paul (1 Cor 15:1-11) certainly seems to imply that there was a considerable time during which the Risen Lord appeared to various witnesses, so that while Luke's "forty days" may be symbolic, his account of the ascension as the final appearance to the Twelve at a considerable interval after the Passover is perfectly credible.

Nevertheless, the fact that in these resurrection appearances Jesus presented himself much as he had been in life, does not tell us what the risen body is to be in glory, anymore than it asserts that the transfiguration was a permanent transformation. In both cases we are dealing with a transitory phenomenon. Indeed in the transfiguration, we come closer to knowing what the glorified body will be eternally than we can from the resurrection appearances, yet even here we are only told that his face and *clothing* shone like the sun. How Jesus appeared to St. Paul on the road to Damascus we are not told in any of the four biblical accounts already referred to.

The only text of Scripture which really provides us with a theological basis for understanding the risen body is St. Paul's grappling with the problem in 1 Cor 15:35-58. Richard Kugelman, C.P., very precisely analyzes this text into the following points.[53] First, Paul lays down as a principle that God provides each kind of living thing, plants, fish, birds, men and even the heavenly bodies with the kind of body appropriate to it; nor does death prevent this, as is evident from the fact that to the (apparently) dead seed in the ground, God provides a new leafy and fruitful body. Next he applies this to the resurrection of the dead human body in four antitheses which illustrate the new qualities by which it is transformed: (1) "What is sown in the earth is subject to decay, what rises is incorruptible"; (2) What is sown is ignoble [unhonored], what rises is glorious"; (3) "Weakness is sown, strength rises up"; (4) "A natural body is put down, and a spiritual body comes up."

Saying nothing more, at least explicitly, about the second and third of these antithesis, Paul elaborates on the fourth and the first in that order. As to the fourth he says that the first man Adam, made from the earth, became "a living soul" ("psyche" in the sense of a body living with natural life like an animal lives); while the last Adam, the Risen Christ, comes not from earth but from heaven and has become "a life-giving spirit" in whose likeness Christians will be resurrected.

As to the first antithesis, Paul says that corruptible flesh and blood cannot inherit the kingdom of God. Consequently at the last judgement the present corruptible and mortal bodies, even of those who are still alive at that time, must become incorruptible and immortal. He then breaks out in a hymn of praise and thanksgiving for Christ's victory over death, and urges the Corinthians to persevere "in the work of the Lord" in order that they may share in this triumph.

Although St. Irenaeus seems to have distinguished "immortality" from "incorruptibility" and have taught that the souls of the unrighteous are eventually corrupted or annihilated by their punishments in hell,[54] it is doubtful that St. Paul intends two different qualities by these terms, since he merely seems to use one to clarify and give emphasis to the other. The term "glorious" here is contrasted to *atimia*, "without honor," hence does not indicate any intrinsic quality such as "shining" but simply means "honorable." As for "strong," the term merely negates the weaknesses of mortality and hence adds little to the term "immortal." Thus St. Paul's description of the resurrected human person reduces to saying that it is not subject to death (immortal) and that it is "spiritual." Evidently it is immortal because it is spiritual, so this word *spiritual* or *pneumatic* is our only clue.

As is well known, for St. Paul the term "spiritual" is not the opposite of "bodily" or "material," but is used much as Jesus uses the term "Reign of God", to indicate that order of things which is in accordance with the will of God in contrast to the disorder of things as alienated from God by sin.[55] This restoration of the created world to conformity with God's will is the work of the Spirit of Christ which incorporates all things into the "body" of Christ, i.e. the Reign of God in which Christ rules until the judgment is concluded:

> Just as in Adam all die, so in Christ all will come to life again, but each one in proper order: Christ the first fruits and then, at his coming, all those who belong to him. After that will come the end, when, after having destroyed every sovereignty, authority, and power, he will hand over the kingdom to God the Father. Christ must reign until God has put all enemies under his feet, and the last enemy is death . . . When, finally, all has been subjected to the Son, he will then subject himself to the One who made all things subject to him, so that God may be all in all (1 Cor 15:22-26,28).

Here "every sovereignty, authority, and power" indicates the rule of sin in the world, i.e. every power opposed to God's will.

Thus the first thing we can say about the resurrected body is that it is restored to the condition it had "in the beginning", that is, as God created it in the first man and woman (or men and women) before they sinned. Because of the modern evolutionary theory of human origins, we should add to this that what is in question is not precisely the historically original condition of the human body, but rather that body as it would have developed to its most perfect condition if sin had not intervened and modified this course of development. This does not mean simply the *natural* state of the human person as Reformed and Jansenistic theologians often said, ignoring the distinction between human nature as such and its super-natural endowment with the life of grace and the charisms appropriate to it. It is true that this distinction between nature and super-nature is not explicit in the Scriptures nor altogether clear in the writings of the Fathers, and this is what led the Reformers and the Jansenists (and some modern Catholic theologians)[56] to deplore it as a scholastic invention, but in my opinion this distinction was a great achievement of medieval theology which greatly assists us to do full justice to the rich but unsystematic data of Scripture.

The anthropology of *Genesis* not only describes humanity as created in the image and likeness of God in the sense that the human person is superior to the animals by reason of intelligence and free will (i.e. superior by *nature*), but it also pictures Adam and Eve as living in intimate friendship and direct communication with the Creator, living in a garden perfectly adjusted to their needs, and with the choice before them of the forbidden tree or, ultimately, the opportunity to eat of "the tree of life" i.e. to become immortal. These images symbolize in a most vivid way that humanity was created not only as a part of the order of nature but also in the *grace* of God, that is, as being prepared for a life surpassing anything that created nature can by right expect, but which can only be a free gift of God in no way necessitated by natural requirements.

Nature to be sure, is itself a gift, a grace, but it is a limited gift, although in its limitations still open to the further generosity of God. But "grace" in the strict theological sense of the term is not the gift of nature, but the far greater and unlimited, infinite gift of eternal life in intimate union with God. This is what the serpent in his evil wisdom understood when he tempted Eve — "you shall be as gods" — knowing that in fact God wished Eve and Adam to become "as gods" by his gift, but not by theft, because theft destroys the very nature of a gift and produces not a god but an idol.

594

Super-nature is not to be understood either as destructive of human nature, or as a mere addition to it, a second story added to the first. The difference between super-nature and nature is best understood *relationally*, not substantively, by analogy to an important distinction we make in ordinary inter-personal relations. When I meet a stranger and recognize her or him as human, there is established between us a relation of recognition and acquaintance which is certainly personal, but still very objective and remote. If later we become close friends, this original relationship is not destroyed, nor is the relation of friendship simply added on to it. Rather the first human relation of human person to human person has been vastly *deepened*. The original relation was *open* to this deepening and not in the merely negative sense of a "non-repugnance", but in the positive sense that it provided the grounds for a possible development. Yet it could not of itself *demand* that we become close friends, since the mutual gift of friendship must by its very essence be a free gift that cannot be demanded.

Analogously, before sin, human nature was established in a certain personal relation to God as intelligent and free creature to its Creator, and this relation established a positive openness to God's gifts exceeding the limits of nature. In fact *Genesis* indicates by the images just mentioned, that from the beginning God deepened this natural relation to one of intimate friendship, and as the Greek Fathers say, by *theosis* (divinization), or as the New Testament says "the gracious gift of life" (1 Pt 3:7) by which Christians are "sharers of the divine nature" (2 Pt 1:4). Thus in saying that the resurrected body is "spiritual", Paul is teaching that it is a body wholly fitting to a human person in whom this relationship of intimate friendship to God has been fully accomplished by his or her perfect incorporation into the body of Christ, as a full citizen of his Kingdom ready to be subjected "to the One who made all things subject" to Christ, "so that God may be all in all."

3. Guesswork

In Chapters 2, 7 and 8, we saw that the present human body is adapted to serve the human brain by supplying it with the data of thought and to execute its commands in changing the relations of the human person to the environment in time and space, in modifying that environment itself by various kinds of artistic and technological actions, and above all in serving as a medium of communication between persons so as to form the human community. We also saw that the brain itself does not perform intellective or free acts, but is the necessary consubstantial instrument of such acts, and participates in them more and more perfectly as the human spirit comes to achieve more and more perfect control over the body. Finally, we saw that the limitations of this body put limitations on the human mind, which, however, can be in large measure overcome by human technological creativity, provided that this technology is wisely used so as not to destroy the essential bodily structure and thus make creative thought and freedom impossible. It follows, therefore, that the "spiritualization" of the body signifies that these bodily limitations will be overcome in a way that technology can partially accomplish, but which will surpass the possibilities of human art as grace surpasses nature.

First of all, this means that the resurrected body will be freed from those defects contrary to the full development of human nature which have been the result of the sinful condition of the world (original sin), including death. I have already shown[57] that these effects of sin in the long history of human development have been very profound, so as we now experience it, the human body may be very far inferior to what it would have been through biological and cultural evolution if these had been under the control of human beings living by the new law of love and using their intelligence with wise stewardship. In principle, all the diseases and defects to which our bodies are liable, all the new lives lost by spontaneous or induced abortions, all those stunted by genetic defect or in childhood by a bad environment, all those injured and twisted in adult development by social injustices and vicious customs, could have been and even now can be remedied by the right use of human creativity.

Even death is natural to us only in the sense that our bodies are liable to suffer it from the natural and man-made forces in the environment, but it is unnatural in the sense that as intelligent beings, we have in principle the power to find ways to forestall these forces and to remedy their effects, so that the dream of earthly immortality is not beyond the power of human technology, at least as long as human beings can maintain their earthly existence in the cosmic current of entropy. Thus if human history extends a long way into the future, the science fiction dreams of a human race far more perfect in body, and therefore in mental capacities than at present, may be realized if we courageously undertake the task. In the most hopeful scenario, therefore, when Christ comes again he will find a humanity already "immortal" and physically spiritualized in the likeness of Christ by its own creative efforts through a wise stewardship of the nature given it by God under the guidance of the Holy Spirit everywhere at work in our world.

Second, we must ask, what then will remain for the transformation of the human body to adapt it not merely to immortal but to *eternal* life in the vision of God, since this transformation is a gift which can only be given us by God through Christ at his final coming into the world?

596

It seems clear that in eternal life, some bodily functions will cease to have any use. Jesus said, "When people rise from the dead, they neither marry nor are given in marriage, but live like angels in heaven" (Mk 12:25). Whatever we think about "angels" (a point to be discussed in the last chapter of this book), it is clear that Jesus means that the life of the resurrected will not require procreation to multiply and perserve the species, because the human community will be completed and deathless. It follows also that the human person will not require food, drink, air or any other biological in-put for its maintenance. What will remain therefore, can only be those functions of the body necessary for thought, freedom of communication and modification of the environment. Since most technological modifications of the environment are directed to supplying these biological needs, it would seem that the needs for any kind of action except for the sake of communication will also be eliminated. Thus the role of the resurrected body will be to serve as the instrument of human thought, love, and the communication of thought and love to other persons.

Does this mean that in eternal life the enjoyment of sex and of food and drink and of the other sensual and esthetic delights will be eliminated? Christians have scorned Muslims because the Quran seems to promise sensual delights in heaven, but is this contempt more Platonic than honest?[58] Although Jesus says we will be "as angels", he also did not hesitate to speak of the Messianic wedding banquet. Furthermore, in Chapter 10 I stressed the profound humanness of sexuality and of our need for food and drink as well as the artistic creativity of the human spirit. To eliminate all these seems to deprive our humanity of much of its richness.

This difficulty can perhaps be overcome if we consider the point made in Chapter 10, that what makes these biological activities specifically human is that they are also *symbols*. They form a language by which we human persons communicate and draw closer to each other in community. The family meal or food and drink shared by friends are a way of gift-giving that expresses friendship and mutuality. Sexual intercourse is the language of love which takes its ethical character, as Paul VI taught in *Humanae Vitae*, from its "meaning". Finally, artistic activity is a way of expressing and communicating thought and love through artifacts that are a medium between persons.

It is not strange then that the risen Jesus ate with his friends and permitted Thomas to touch his body to assure himself of the Lord's reality. To be sure he did not permit Mary Magdalen to touch his feet, "Do not cling to me, for I have not yet ascended to the Father" (Jn 20:17), but do not those very words gently suggest that in the future, when she herself has risen to immortal life, the Magdalen will embrace her Lord? As for artistic activity, Jesus' art was in the wonder of his miracles and the poetry of his parables, and the risen Christ repeats the miracle of the miraculous haul of fish (Jn 21:11) and develops the parable of the good shepherd (21:15-17).

We can suppose, therefore, that the sensuous joy of human banqueting, sexual play, and artistic expression remain in heaven as forms of communication freed of whatever binds them now to the limitations of bodily existence which Paul calls "corruption" (*phthora*, decay). Eating and drinking as we know them are delightful, but they involve all the "decay" (metabolism) of the digestive and eliminative processes as well as the danger of excessive weight-increase, surfeit, and lethargy. Sexual intercourse is a pretty sweaty and messy business, embarassingly linked to the organs of elimination and menstruation, and connected with all the problems of pregnancy, not to mention the limitations of male potency and the brevity of the whole affair. These limitations impose the exclusiveness of the sexual relation, confining it normally to heterosexual relations and to one partner. Finally, artistic activity in this life is, as every writer or artist knows, an exhausting and frustrating mode of expression in which the materials, whether they are marble or merely musical sounds or words, are exceedingly refractory to the form we seek to give them. In eternal life, therefore, such forms of communication must be freed of so much that burdens them now that it is difficult to imagine what they might be like.

Since, however, the communicative value of food and drink is not to be found in the consumption of material objects whose energy will not be needed by perfect and immortal bodies, let us imagine the banquet of heaven rather as a symposium in which our food is the delight in being together and exchanging the gifts of ourselves to each other, of which the food and drink at a banquet is only the symbol. The sensuous pleasure and inebriation will be felt in our bodies wholly enlivened by our spirits by each other's presence, even as now sometimes we are filled with delight just to see our friends, a delight that is not merely of the spirit but of the body enlivened by the spirit. "You make me feel alive again just to be with you."

Yet human intimacy is never satisfied merely with looks and words as the medium of presence. We want to touch, to embrace, to kiss, and to become one flesh. That desire for bodily union as the expression of the union of mind and heart is qualified by the physical structure and basic human relations founded on this structure, so that incestuous and homosexual relations are contrary to our natures,[59] and the exclusivity of monogamy is ethically necessary. In immortal life, however, it would seem that this desire to be physically as well as mentally united with *all*

598

other human beings would prevail. Obviously such intercourse would not take the form of the genital intercourse proper to this life and so conditioned by the biological ordination to procreation. Perhaps it is best imagined simply as an *embrace*, the union of two persons through the sense of touch, the most basic and existential of our senses,[60] by which we become immediately aware in our bodies of the intimate presence of another, but an embrace that involves the contact not merely of the arms, the lips, or the genitals, but of the other person to ourselves in our mutual *totality*, a total contact or compenetration impossible in this life, accompanied by the intense sense of physical well-being, of intense aliveness that is sometimes experienced in sexual orgasm, but also in certain drug-induced, or esthetic or natural "mystical" states — an ecstacy in which the bodily limits that seem to enclose the self are opened up so that one person can give him or herself totally in body and spirit to another, and receive the other in return, as in Eucharistic communion. Curiously, the Spanish mystic Maria of Agreda,[61] in her revelations, asserted that the resurrected bodies of the blessed are able to "compenetrate" i.e. to occupy each other and in doing so they share a common bliss.

This heavenly embrace, this true "kiss of peace", can be shared between any two of the dwellers in glory. Does this mean that the relations of this life will no longer exist? The answer to this depends on the more general question, to be discussed in a moment, of how our present life is continuous with the next. At this point it suffices to say that there seems no reason that within the community of the kingdom there cannot be a network of more particular relationships, just as there is now. Our parents, our friends, our partners in marriage will remain close to each of us in special ways. While there can no longer be differences in bodily age between those who have all arrived at their full development, there will still remain the complementarity of the sexes, not indeed as a need of man for woman and woman for man to be complete in their humanity, for each of the blessed will have achieved wholeness, but in the sense that each individual will retain his or her personal identity, of which sexuality is an essential feature.

Third, we are thus brought to the chief question, which is why the body remains necessary to the wholeness of the person. Granted the antidualistic, Aristotelian epistemology I have adopted throughout this work, the answer to this question must be found in the nature of human intellection and freedom, namely, the fact that it is the least possible of all created intelligences in that it can know and make free choices only in dependence on bodily sense organs and a bodily brain. Does this mean that the soul which survives the body would be "asleep"? Thomas Aquinas found a way to avoid such a conclusion without falling into Platonic dualism. Aquinas admitted that Plato was right in supposing that if the human self was the soul, then it ought to be "transparent" to itself, i.e. fully self-conscious by a direct, immediate intuitive vision of its own essence. In fact, however, we do not really experience any such intuition because the human self is not a soul inhabiting a body but a body whose form is the soul. The closest we can come in this life to any such intuition is in that natural mystical experience in which our normal *indirect* self-awareness is intensified by introverted meditation or some esthetic experience in which we have an existential awareness of our *otherness* from the material world of our natural experience.[62] Such an experience is negative, but it would seem that when the soul has been deserted by its body through death, its energy would be withdrawn within its purely spiritual activities and the result would be a direct, *positive* intuition of the very essence of the personal self. This intuitive self-consciousness would include the knowledge, on the one hand, that it has been deprived of the body, necessary for it to know the world outside itself, including the body essential to its own personal wholeness, and on the other, the existence of its Creator reflected in his image in its own limited being.

This theory of Aquinas (which he developed as his solution of the famous problem of the fate of unbaptized children, who Augustine had been compelled by the logic of his anti-Pelagian campaign to consign to hell) with its notion of a *limbo* (threshold) where the souls of these innocents might dwell in a state of natural happiness, has little theological support today.[63] It is generally supposed that even before baptism, the grace of Christ reaches the child through the mediation of the prayers of their parents and the Church.[64] Nevertheless, it seems to me that this theory is still significant in helping us understand the resurrection, because it helps us to understand exactly why the soul needs to get its body back. Evidently it is not in order to be awake and self-conscious, since Aquinas' analysis seems to establish that the soul after death has a self-awareness that far surpasses any self-knowledge that we can possess in this life.[65]

Aquinas also saw no great difficulty in supposing that the separated soul is able to know other separated souls and the angels in somewhat the same way that angels, as pure spirits, know one another by the simple will to communicate.[66] Further, the separated soul can know something of God through His image impressed in the very substance of the soul in its creation.[67] But it would seem that without the aid of the senses, the soul could not acquire new knowledge of the material world from which it was separated, nor even remember its experiences in that world. Death, as the Greeks had thought, would indeed be Lethe, the River of Forgetfulness. The separated soul would still possess the abstract knowledge it had gained in its former life, but no remembrance of the singular experiences from which it had drawn these abstractions. As for the souls of unbaptised children in Limbo, their minds would remain blank of all knowledge of the material world.

Aquinas did not see how this conclusion could be reconciled with the Scriptural exegesis of his day, which read certain of the parables of Jesus as describing the separated souls remorsefully remembering their past sins and suffering physical pain as well. Nor did it fit well with his theory of Limbo. He hypothesized therefore that just as the angels have knowledge of the material world through an illumination from God in the form of innate ideas, so the separated souls receive a similar illumination which is preternatural but not a revelation in the strict sense.[68] Yet Aquinas acknowledges that while this theory of illumination might provide for the souls in Limbo, it was unsatisfactory as an explanation of how souls might remember their earthly past. Such illumination must be received according to the condition of the receiver, and while in the case of the angels such knowledge is received innately and is therefore perfect, in the case of the human soul it is preternatural and therefore imperfect and can furnish only an abstract, confused and general knowledge of the material world which does not extend to knowing or remembering concrete experiences.[69]

Aquinas concluded, therefore, without detailed explanation, that it must be that the separated soul knows only those concrete singulars which it knew or to which it was connected by acts of the will in its previous earthly existence. It cannot learn anything new about the material world, or perform any new free acts. His great commentator, John of St. Thomas, explains this by saying[70] that the human intelligence, in knowing a universal also knows the singulars from which it has been abstracted, but only by a reflex reference to the sensible images of the singulars supplied by the organs of sense. In the separated soul this reflex reference is no longer possible; nevertheless the *determination* of the intellect to this reference remains. Aquinas posits that this determination must be sufficient to make it possible for the intelligence, even without the assistance of the senses, to recall the singulars which it had once known experientially.

How are we to evaluate this theory? First it must be remarked that modern scriptural exegesis would not think it necessary to conclude, as Aquinas did, that Jesus in the Parables of the Sheep and the Goats and of Lazarus and the Rich Man intended to instruct us on the condition of the soul before its reunion with the body. Nor as I have indicated, has the theory of the Limbo of unbaptized infants been defined by the Church. Consequently, it is possible that the separated souls exist in a state of forgetfulness of their past. But even if Aquinas is correct, and memory traces remain in the separated soul as aspects of the transcendental relation which

601

it certainly retains to the body which it will receive back in the resurrection, these traces must, without the body which actually endured these experiences and bore (as it were) their scars, be lacking in existential vividness. Human memory as we experience it in this life is not merely a contemplation of the perished past, but it is often the past *relived*, because that past remains existentially enfleshed in our bodies. Our past is built into us as body-persons, as in Rome one sees the remains of pagan temples incorporated in the fabric of Christian churches.

This need of the separated soul to recover its past in fullness, as well as to be opened up to new knowledge, to its eternal future, provides us with the principle by which we can consider what the resurrected body needs to be. What we need to recover from the earth is whatever minimal physical structure is necessary to be our "memory bank" which will make it possible for us to relive our past. In brief, *our resurrected body is our living past*. In view of the discussion in Chapter 9 of the essential *historicity* of human nature, it is clear that our personal identity, our existential personality develops throughout our earthly life. We *become* ourselves through a life of experiences and relationships, and if we were to lose these and retain only the abstract understanding we have acquired, we would have lost a great part of our real selves. It is here that the Christian view of life stands in marked contrast to the ahistorical outlook of neo-Platonism or the Vedanta and of Buddhism, in which the goal is to escape all that is individual and particular. When the Risen Lord appears to doubting Thomas he shows Thomas his *wounds*, the marks of his earthly life even in its bitterest hours. So it will be with us in regaining our bodies; we will regain the memory of our past. Moreover, all that has been irretrievably sunk in our unconscious will, no doubt, then become available once more to memory.

Is there not much, however, that we would like to forget? No, because the Christian knows that "God makes all things work together for the good of those who have been called according to his decree" (Rom 8:28). Much of our lives now seems to us to be aimless, accidental, absurd, disastrous. As we grow older sometimes we have a little understanding of events which formerly seemed to us without meaning, but the story still seems without a clear plot. To Jesus' human consciousness on the Cross his life must have appeared a foolish and pitiable dream, summed up in the words, "It is finished" (Jn 19:30), cynically understood.

Thus we need a body to remember our earthly life. While God may miraculously supply these memories without the body for a time, the perfection of the human person requires its body to do this naturally. We do not need a body for the vision of God since God can reveal Himself to us immediately. Nor will this fullness of the vision of God exclude all other knowledge, since the light of God cannot be destructive like the blazing light of the sun, but healing and strengthening. Will we not then still wish to understand our past lives and the whole of human history in the light of God's loving purpose in which at last every dark puzzle will receive a splendid answer?

It must be remembered that although all the blessed will see God face to face, no created finite intelligence can wholly *comprehend* God (it is in this sense that the Eastern Fathers and St. Gregory Palamas can be understood when they say that "the essence of God is forever inaccessible") since He is infinite Being.[71] Thus each of us will posses the inexhaustible plenitude of God to the degree that each has loved him. This love, however, has grown with the experience, sufferings, and personal efforts of each life, so that memory, dependent on individual bodies, will provide, as it were, the *horizon* of our vision of God. The child who dies in infancy will have a limited possession of God proportionate to its small life compared with the possession of God by the adult who has served God through a long or at least an intense life.[72]

Heaven, therefore, will not be, as some fear, a bore; because we will never cease to reflect over our own lives and the lives of others that form the whole of history which we will share one with another. Through this reflection on history we will plunge ever more deeply into the unlimited riches of God. Our human past will become luminous with an infinite future which is God himself, a future in which there is "process" forever, but a process which is all gain and never any loss, because it is the spiritual process of finite minds contemplating Infinity.[73]

To retain and use our memory, therefore, we need a genuine material body, but no longer one which has constantly to repair itself and furnish energy to our "data-storing-and-processing system", the brain. It does not seem that we would need any of our other organs except the brain, since we need no new information through sense organs from a universe whose history is complete, nor do we need the organs that keep the nervous system alive and in repair, nor do we need to move the body about in order to get from one place to another if we can communicate more perfectly without such transportation of a large mass of matter, nor, finally, do we need to transform a completed world.[74] Moreover, our brain might be considerably less bulky than it is at present, marvelously compact as it is, once it is freed of its dependency on our physiological system and is free to do nothing except perform its essential data-storaging-and-processing.[75] We have recently discovered how a huge computer system can be reduced to a tiny chip imprinted with an electric circuit.

Might we not imagine that our brain might be reconstructed out of some configuration of a field of radiant energy, a system of energy exchanges, of waves of light? Light is a *material* action in a field with space and time dimensions, and is thus truly the action of a body, but it is body in which the passivity and potentiality of matter has been actualized to its highest level.[76] It is no accident that the visions of mystics are described in terms of light, and that artists have conceived the images of saints as figures of pure light dancing to music in the ocean of the sky. Such a body, freed of all which in our present bodies is left over, as it were, from the long process of evolution — a scaffolding no longer necessary for the completed building — and refined to that spatio-temporal pattern of pure energy, is not unlike the dreams of some science-fiction writers. Thus in Arthur C. Clarke's famous *2001*, the super-human beings who left behind them the Monolith on the moon had evolved to the point that "their bodies were empty space".[77] We have seen, however, in Chapters 2 and 8 that "empty space" is really a field of energy and therefore, still a material being. If we must guess, then it would seem that this guess provides a reasonable idea of St. Paul's "spiritual body", yet it would be truly *my* body because it would retain for each one of us the whole of that individual's earthly experience in all its concreteness and particularity.

It can be objected, of course, that the Risen Christ did not seem to appear in this fashion, but rather exactly as he had been in earthly life. To which I would reply that he presented himself to his disciples exactly as they had known him because he retained in his spiritual body the remembrance of what he had been and suffered and could still communicate this to them under whatever aspect he wished. Again, some will object that such a body is not identical with the one in the tomb. Aquinas long ago replied to this difficulty by pointing out that since matter is pure potentiality it receives its identity from its form.[78] That is why the body I now possess is the same I had as a child, although all the matter which was then in it has long ago been replaced. It retains its identity because it has always been informed by the same soul. Consequently, when the soul takes to itself a body in the resurrection, it is of no consequence what matter it uses provided that it gives to it the same information that it gave to its body before death. Thus the risen body will be identical with that which died not by the identity of all the anatomical structures, but by the identity of the embodied memory of its whole biography.

How will such spiritualized body-persons communicate with one another? Aquinas supposed that angels as pure spirits could communicate by a simple act of the will by which one intelligence would become transparent to another.[79] It would seem, however, that these body-persons, no matter how "dematerialized" in the sense just explained, would still

604

have to communicate by a bodily language, but this might take the form of "thought waves" which would be some form of radiant energy, perhaps even that still undiscovered form (about which I remain sceptical), that some believe is evidenced in the para-normal events called "telepathy" or "clairvoyance."

4. The Intermediate State

When will the resurrection of Christians take place? Development of the elaborate doctrine of the Catholic Church on the topic (and to a somewhat lesser degree of the more conservative Eastern Churches), which at the Reformation was largely rejected in the name of strict adherence to explicit New Testament teaching, took place by a series of logical steps which, however, all have their germ in biblical texts.[80]

The point of departure was, of course, the resurrection of Jesus himself and the conviction expressed by St. Paul,

> Are you not aware that we who were baptized into Christ Jesus were baptized into his death? Through baptism into his death we were buried with him, so that, just as Christ was raised from the dead by the glory of the Father, we too might live a new life. If we have been united with him through likeness to his death, so shall we be through a like resurrection (1 Cor 6:3-5).

From this basic doctrine, some drew the conclusion that after baptism they were already living the resurrected life and would never die, but would still be alive at Jesus' second coming. This encouraged the expectations of an immediate parousia and led some to give up work, reject marriage, and even to indulge in licentious behavior. In the *Epistles to the Corinthians* Paul takes pains to correct these extravagant expectations, and to insist that since the exact time of the parousia has not been revealed, it is necessary for Christians to continue their earthly lives with sober responsibility and discipline.[81] Moreover, he assured the Thessalonians (1 Thes 4:13-18) that those who were still alive at the parousia would have no advantage over those who had died, because all alike would be judged and transformed by the resurrection conversion of the body to the spiritual state already described.

There was evidently also anxiety among these Christians about their relatives who had never been baptised, so that a custom arose (at least in the Corinthian church, 1 Cor 15:29), of proxy baptism for them (as practiced even today by the Mormons). Significantly, St. Paul does not repudiate this custom but seems to approve it when he uses it as a valid argument for the reality of the resurrection. That such concerns were felt throughout the Church is clear from the evidence of speculation about the fate of the righteous who had lived before the first coming of Christ, not only of the righteous Jews but of Adam and Eve, Abel, Enoch, Noah, and others, i.e. of the righteous dead of the whole human race. Thus in the First Epistle of Peter we read (3:18-21):

> [Christ] was put to death insofar as fleshly existence goes, but was given life in the realm of the spirit. It was in the spirit also that he went to preach to the spirits in prison. They had disobeyed as long ago as Noah's day, while God patiently waited until the ark was built. At that time, a few persons, eight in all, escaped in the ark through the water. You are now saved by a baptismal bath which corresponds to this exactly.

The meaning of this text is controversial.[82] Why would Christ have preached to disobedient souls who had already been condemned? The passage must somehow be connected with several clearer texts on which are based the article of the Apostle's Creed "he descended into hell", namely, Rom 10:6-7; Eph 4:8-10, Heb 13:20, etc. In these texts a contrast is drawn between Christ's descent into the grave (Sheol) and his ascent to heaven. As we have seen, Sheol was the place of the dead awaiting the resurrection. There were also speculations in the apocalyptic literature, according to which the souls of the dead were located not under the earth but in successive spheres of the heavens though which Christ passed as he ascended. Thus in Ephesians 4:9-10 we read the comment on a verse of the psalms,

> "He ascended" — what does this mean but that he had first descended into the lower regions of the earth? He who descended is the very one who ascended high above the heavens, that he might fill all men with his gifts.

Thus it was believed that Jesus in spirit descended after death into the region of the dead ("hell", or Greek *Hades,* in the creed simply means "the grave" or Sheol), or alternatively passed through the heaven of the spirits as he ascended after the resurrection, in order that he might preach the Gospel to those who had lived too soon to hear him so that they might repent and be saved.[83] Obviously the theological motive of these speculations was to explain how the universal salvific will of God ("For [God] wants all men to be saved and come to know the truth". 1 Tim 1:4) could be applied to those who had never heard the Gospel. This explanation had also to be reconciled, however, with the conviction that there is no more opportunity for repentance after death, a conviction that seems implied in Jesus' own parables of Lazarus and the rich man (Lk 16:19-20), and of the rich man who built a place to store his harvest (Lk 12:13-21). Thus it appeared that Jesus' visit to the dead was not to call the unrighteous dead to repentance but fully to enlighten the righteous dead and to release them from the prison of the grave.

There were other speculations about Enoch "who walked with God; and he was seen no more because God took him" (Gn 5:24); Moses, who died but was buried mysteriously so that "to this day no one knows the place of his burial" (Dt 33:6); and Elijah, who was taken up into heaven in a fiery chariot (2 Kgs 2:11).[84] Presumably these, and perhaps other holy ones, had been "assumed" into some secret place until the Day of the Lord. Such notions seem reflected in Matthew's report that at the crucifixion of Jesus "The earth quaked, boulders split, tombs opened. Many bodies of saints that had fallen asleep were raised. After Jesus' resurrection they came forth from their tombs and entered the holy city and appeared to many" (27:53).

What is notable in all these stories is that there is a wavering between the conception of the dead as still existing in the tomb with their bodily remains, or "assumed" bodily into some mysterious place of waiting, or as disembodied spirits dwelling in one of the lower spheres of the heavens; and that it is unclear exactly how salvation is to come to them. What is certain, however, is that this salvation must come through Christ and will lead to the believers sharing in his resurrection.

If Paul did not repudiate proxy baptism for the dead *a fortiori* it is unlikely that he repudiated praying for them, since we have clear evidence that such prayers were already part of Jewish practice and were soon accepted by the Christians, as the catacomb inscriptions plainly attest. In 2 Maccabees 12:38-46, it is related how Judas Maccabaeus was deeply troubled to discover that some of his soldiers who had died fighting for the Law of God were found to be wearing idolatrous amulets which they had taken as booty. Consequently

> he took up a collection among all his soldiers, amounting to two thousand silver drachmas, which he sent to Jerusalem to provide for an expiatory sacrifice. In doing this he acted in a very excellent and noble way, inasmuch as he had the resurrection of the dead in view; for if he were not expecting the fallen to rise again, it would have been useless and foolish to pray for them in death. But if he did this with a view to the splendid reward that awaits those who had gone to rest in godliness, it was a holy and pious thought. Thus he made atonement for the dead that they might be freed from this sin (43-46).

607

But what was the purpose of these prayers of Judas? Would these men who were martyrs for the Law have been condemned without the sacrifice? It is not clear. And what was the purpose of the Christians' prayers for the dead? It is unlikely in view of the character of the catacomb inscriptions that they were for the repentance of sinners after death. Most likely, the thought was that the time of awaiting the resurrection was long and the living wished somehow to console and strengthen them in their waiting.[85]

It was the development of the veneration of the martyrs, then of the Blessed Virgin and other saints that seems to have led to the conviction that at least some of the dead were not merely awaiting the resurrection, but had already attained the full glory of the Lord, leaving the question of their bodies unclear, since in fact their tombs and the relics of their bodies were highly honored.[86] Once this conviction was established, it brought into question the condition of the righteous dead who nevertheless were known to be less exemplary in their faith and lives than the martyrs and saints. Encouraged no doubt by the formalization of the practice of public penance and by the ideals of ascetic discipline fostered by the hermits of the desert, the view developed in both the Eastern and Western Church that for those dying in Christ, but who were yet far short of spiritual perfection, this time of waiting must be one of purification. The Eastern Church never developed a very concrete view of the process, but in the West it was elaborated into the notion of a Purgatory conceived as a place of purifying suffering and pictorialized as a place of fire not unlike the hell of the damned. Yet, as can easily be seen by contrasting the picture painted by Dante in his *Inferno* from that in his *Purgatorio,* the vast difference of meaning of the two states was deeply felt. While the *Inferno* is a place of darkness, stench, and cursing; the *Purgatorio* is a place of dawning light, of solemn music, of prayer, leading up to the Earthly Paradise.

In recent theology, there is an effort to rethink these various themes rooted in the New Testament and developed by Christian experience, but never entirely coherently.[87] First of all, it is fundamental to understand the conversion and baptism of the Christian as an incorporation into the Body of Christ by his life-giving Spirit. In a very real sense, eternal life for the Christian in his or her whole person, soul and body, begins at this moment and is fed by the Eucharist, which is a pledge of the resurrection. Yet, as St. Paul warned the Corinthians, there still remains a time of testing in the faith during which the Christian must "work for your salvation in fear and trembling" (Phil 2:12) in the Lord's service. Such a life lived in union with Christ leads progressively toward holiness, not as something self-possessed, but as an ever greater openness to his grace and culminating with sharing his own suffering and death, not merely symbolically as in baptism, but experientially.

Death for the Christian is not an interruption of life. The whole Christian life has been one of vigil, of expectation of the Reign of God and of labor that this Reign might come in its fullness at last. At death, some are ready to share with Christ the vision of his Father face-to-face in glory, and with him to make intercession to the Father for those who are still on their journey.

For some, however, this journey is not yet completed because they have in life resisted the purifying grace of God, running from the Cross which was given them to carry daily. Yet they have died in Christ, trusting in his mercy and repentant of mortal sin. Consequently, they must still undergo the purifying action of the Spirit which they too much avoided in this life. This purifying action is essentially the same as that experienced here, except that it must be endured passively, while in this life it was possible to accept it cooperatively and actively. Consequently, they are far wiser who seek Christ with all their strength in this life as St. Paul did, so as to have their necessary task completed at death. Yet by the principle of the "communion of saints" by which we are to "help carry one another's burdens" and in that way "fulfill the law of Christ " (Gal 6:2), the living can intercede for the dead and obtain for them a quicker purification, just as those already in glory intercede along with Christ for the living. Thus the prayers for the dead avail for those who died in Christ, yet have not attained to the fullness of his glory.

This process, however, need not be considered merely something that affects immortal souls, separate from the body. We can avoid some of the problems we have already noted about the separated soul if we suppose that resurrection itself is a process which begins at death and is completed in the parousia of the Lord when all the dead come finally to the resurrection and universal judgement. By the power of God, the soul after death begins to reconstitute its "spiritual' body by beginning to recover its own history, requiring as we have seen some sort of spatial-temporal structure so subtle that it may be present and undetectable even here on earth, or may be located elsewhere in our vast universe — the old guesses about whether it was in the earth or in the heavens are as good as any we can make. I am inclined to take seriously the popular belief of many primitive people that some of the dead linger about the places in which they lived and others move more freely.

For those who die still very attached to this life, the gradual recovery of their memory and their growing understanding of what their life has really meant must undoubtedly be the source of great suffering, like that we all suffer at times in this life as we confront ourselves, sometimes on the psychotherapist's couch, and which the saints endure more intensely in their spiritual dark nights, but which is probably most intense of all for these who have died, because of the growing light of the Spirit which illumines them. The prayers of the living greatly assist this healing process because of the love which empowers them.

As for those who have already achieved the full vision of God, we need not suppose that the resurrection process is always complete in them, since as in this life the soul follows the action of the body, so in the next life it is the soul that goes first, and we can suppose that for them too there is a kind of waiting for the complete reconstitution of their bodies. But in Jesus himself this process was complete in the ascension (I do not say at the resurrection since, as we have seen, there is something transitional in his resurrection appearances), and Christian meditation on the role of the Blessed Virgin came to the conclusion, formally defined by Pope Pius XII in 1950, that in Mary too this resurrection has already been completed.[88]

The doctrine of Mary's Assumption flows from the same understanding of her role that led to the conviction in the New Testament of the virginal conception of Jesus and of Mary's own freedom from original sin in preparation for that great act of faith. Because she is the New Eve and also the Church, the Bride of Christ as a corporate personality in whom the work of God has been summed up, she too must anticipate the resurrection. She is the Church as it will become at the end of history, just as Christ is its and her Head. The definition of the Assumption calls this a "singular privilege", but this does not, as far as I can see, necessarily exclude the possibility that others of the saints have also attained to the completion of this process of resurrection through her intercession with her Son.[89] Thus it would seem that we can conceive the Church as the Body of Christ in every stage of completion from baptism to glory in a continuous process of resurrection to new life.

5. The Self-Condemned

No Christian doctrine, it would seem, has been so much responsible for the development of Humanism as a counter-religion to Christianity as the doctrine of hell. Pierre Bayle, the Calvinist sceptic who furnished the Enlightenment with so many of its best anti-Christian arguments, is said to have implied that it was very difficult to distinguish the God of Calvin from the Devil[90] As unjust as this certainly is to Calvin, he applied to the themes of Augustinian theology a pitiless logic which exposed some of its most shocking features. Does it not seem to many that what Christians believe is that God set up a code of draconian laws and then created some poor human beings to break these laws so that he could then punish them eternally, while gloating over them as they suffered tortures far worse than any human sadist could inflict, while rejoicing in the flattery of other subservient human beings who slavishly kept the laws in order that they too might look down on their fellows suffering in hell with smug self-satisfaction like Roman decadents at the gladiatorial games?

I leave it to Milton "to justify the ways of God to men", and content myself with asking what the view of human salvation which I have been advocating implies about the destiny of those who knowingly and of their own free will reject the salvation offered to all by Jesus on the Cross. In Chapter 10, it was argued that human freedom is what makes it possible for human beings to love themselves and others, but it also makes it possible for them to love themselves with a false love (which is *objectively* self-hatred) by which they become closed to a true love of others. Such a false love is what is called "mortal sin" because it is a serious breach of the fundamental commandment to love God and neighbor as oneself and therefore makes impossible union with the God who is the source of all love, life, and truth. It is self-alienation in the profoundest sense of the word.[91]

Such a knowing and deliberate closure of the self to God, to others, and to one's own self-fullfilment is effected by any free decision to do serious harm to another human being or to oneself, either by commission or omission. Once this road of alienation is chosen there is no turning back unless God himself converts the human heart. Yet as long as the person lives, such conversion is always possible, and in a thousand ways God seeks out the sinner to turn him or her back on the true road, especially through all the ways that other human beings offer their love and help, knowingly or unknowingly, as agents of God's love. Because we do not know the heart of anyone, we can never judge that another person is "in mortal sin." Only they and God can know. Outward signs may suggest this alienation, but they can be deceiving, since the greatest "saint" might be a deceiver and the greatest sinner, even a Hitler, or those responsible for Jesus' death, might be simply crazy or acting in error. "You can know a tree by its fruit" (Mt 7:20) is a rule given us by Jesus, but he combined it with "Do not judge" (Mt 7:1).

Consequently the efforts of some of the Fathers of the Church to find in Scripture a warrant for saying what proportion of the human race will be saved (the resistance they met to the Gospel led them to conclude that the majority of mankind would be lost), or the efforts of others such as Origen and even St. Gregory of Nyssa to argue from the mercy of God that all or almost all will be saved, are in vain.[92] If my argument about

611

the spectrum of scenarios furnished by the Scriptures for the course of history has any merit, it is an open possibility that almost all or almost none will be saved. The reason for this is not that the mercy of God is limited, but that because of his mercy God has really made us free. Hence, the number of those who will be saved or damned is up to us, both personally and collectively, since we cannot only accept salvation for ourselves or reject it, but we can work effectively by prayer and service to help others achieve salvation or we can lead them into temptation.

To say that the number of those who will be saved is truly up to us, is not contradictory to the doctrine of predestination and election in that high form advocated by Thomas Aquinas,[93] although clearly it *is* contradictory to the form in which it was defended by Calvin and Jansenius. Aquinas teaches that in His eternity, God knows the outcome of the world drama, including who will by His grace accept God and neighbor, and who by their own refusal of grace will reject them. Nevertheless, we must not imagine this predestination as if our fate is already determined at this present point of time in the way the world is made or in our personal characters as they here and now exist, since God's knowledge does not depend on the present make up of the world or on the character of human agents. God does not know the future by *predicting* it from the present situation.

God's freely willed knowledge of the world is creative of the world, and his knowledge of the individual is creative of the individual, of the individual's freedom, and of the individual's free use of his or her freedom. God knows my free acts by enabling me to be free in each of those acts, and this freedom can permit me to sin freely in choosing to separate myself from his love, just as it can cause me to do good freely in choosing to cooperate with his love. God's predestination is his loving will to enable some persons freely to do good, and to *permit* others freely to do evil, but always under the condition that in his wisdom he can bring a greater good out of the evil for the sake of those who freely wish to do good, as when he permitted some to murder Jesus his Son in order that this would make clear to all who wished to understand how to love one's enemies. We do not know in fact that God has permitted any particular human being in history to persist in mortal sin to death, but *if* He has done so, it is in order to make way for some greater good for others. Nor is there any way that we can know, short of a special private revelation, whether in fact he has permitted some to go their own way into eternal death.[94] Of course many persons would like some guarantee that no matter what they do, God in his mercy will in the end convert them to himself. They often say that what really worries them is that they cannot bear the thought that anyone else is damned; but really, who can worry about the damnation of others unless they are also worried about themselves? Because we are free, we cannot ask for any such guarantee for ourselves or for anybody else. What I have a right to ask is that (1) God will always efficaciously help me, even until the last breath of my life, if I want and keep asking his help; (2) God will listen generously to my prayers for the salvation of others.

What are the consequences of dying in mortal sin, that is, dying still freely committed to alienation from my true self, my neighbors and my God? The great early theologian Origen proposed to adopt the ancient theory of reincarnation so that the soul might return again to this life in another body and have another chance, and so on indefinitely until at last, supposedly, all would repent and be saved.[95] Philosophically the notion of reincarnation is unacceptable in the non-dualistic anthropology I have been proposing, because in such a view of human existence the soul is individuated by the body, and the notion that one soul could occupy different bodies seems excluded.

Yet in view of ecumenical dialogue with the great religions of the East, I would suggest that their doctrine of reincarnation may be reinterpreted without losing its essential truth, and that so reinterpreted it is not inconsistent with Christian doctrine, and may even be an important ground of agreement. This essential truth seems to be the idea of *karma* or moral responsibility.[96] On the other hand, the notion of the passage of an identical personality from one body to another seems rather to be a *mythical* way of formulating the doctrine of *karma*. It can hardly pertain to its essence, since in these religions empirical personality is considered ultimately unreal.[97] Buddhism denies that there is a "self" which passes from one body to the other, and non-dualistic Vedanta maintains that the cycle of lives is in reality only illusory, since the true self is always identical with the Absolute. The essential purpose of this mythical formulation of the doctrine of *karma*, therefore, is to maintain that human salvation depends not only on the actions of individual lives but on the corporate *karma* of all humankind.[98]

In Christian doctrine each person lives but a single life, but he or she comes into life burdened by the effects of the sins of all who have lived before, yet also able to profit from all the hard-won wisdom of the past. Our salvation is the freeing of the self from error and its establishment in the full light of truth. Moreover, the doctrine of the communion of saints insures that whatever good we do in our lives will be of profit to all those who live in Christ, just as our sins will burden all who follow us. Thus the law of *karma* is fulfilled.

What then are we to say of the apparently reliable reports of those who believe they recall their former lives? May we not explain this as the emergence from the unconscious of the images of significant persons, derived through many channels of our cultural tradition, who have unconsciously affected us by shaping our lives as ideals with which we identify, and which in dreams or fantasy appear to be our true selves?[99]

The Church, although it has accepted the fact that we come to be in a world profoundly conditioned by the behavior of others in the past and that we influence the destiny of others in the future, has not accepted the myth that the soul passes from one body to another; instead, it teaches that through Christ an opening is given to each individual (with the exception of those who never achieve normal reason) to achieve the freedom to decide to live for the common good or to enclose his or her self in self-destructive selfishness. Through a non-historical reading of the Bible, many have come to think of God as "condemning" sinners and "punishing" them on the model of a human judge, who if he were compassionate, might refuse to condemn or to decree punishment. But it would be much truer to say that God never condemns or punishes anyone; they condemn and punish themselves.[100] We recall the words of Jesus to the adulteress, "Nor do I condemn you. You may go, but from now on, avoid this sin" (Jn 8:11). God's judgment on sin is the absolutely truthful revelation which takes place in the conscience of each of us, prevented by death from any longer hiding the truth from ourselves, of our honest responsibility for what we have done. In that clear light of God, God's compassion will be evidenced so that we will not blame ourselves for anything that was really not our fault. We will need no better advocate than God himself, who will make every excuse for us that is possible. We will see how much of the ill we did was *not* our fault, but the fault of our ancestors, of society, of our ignorance, our psychological compulsions, and our human weakness. But after God in his compassion has showed us how much in our conduct was not our responsibility, then each of us will be faced inescapably with our *real* sins, the harm to ourselves and others, the hatefulness in the face of God's love and others' love for us for which *I* and no one else in the universe was responsible. My own conscience will condemn me with absolute justice.

Nor does God have to add any punishments to our crimes other than the consequences which follow from them.[101] While we cannot exclude the possibility that God in his providence sometimes miraculously strikes down a sinner, or sends natural disasters to strike him as a warning to him and to others; it is improbable that this is the ordinary way in which sin is punished. Actions are evil because they injure the agent and others, and it is this harm which is the punishment that exactly fits the crime, "the eye for an eye, the tooth for a tooth" (Ex 21:24).

614

This is clear enough when we think of sins against oneself (if I overeat and drink it is I who suffer from overweight and cirrhosis of the liver), but what about the harm we do to others? This question is, of course, one of the great questions of the Old Testament: "Why do the wicked prosper" (cf. Psalms 37, 49, 73). It is here that the Indian notion of *karma* is relevant. The first effect of every sin against the neighbor is an injury to the moral integrity of the sinner. Plato rightly said that in a profound sense no one can injure us except ourselves.[102] The liar often does great harm to others. Sometimes he does no harm to others at all. But he always harms himself, because our actions form our character, the full development of ourselves as persons. In the case of mortal sin, this harm to the self is a fundamental re-orientation of life from the search for union with the human community and with God to living enclosed on oneself. Moreover, since we are by nature social and open to persons, to live enclosed on oneself is to be imprisoned within a *false* self, to become a living self-contradiction. It is in this state of self-contradiction, which if persisted in until death takes away our freedom to repent of our disastrous decision, that *damnation* consists.[103] God does not damn anyone, nor can the Church, but only the individual who damns him or herself.

Hell, therefore, is not a place of punishment into which an angry God hurls us to be tortured; but a state which the damned have chosen for themselves and in which they torture themselves and each other.[104] As the blessed are resurrected by the recovery of the memories of their lives, and in the light of God come to rejoice in all that has ever happened to them and to the world because it has ended by making them what they are, citizens of the Reign of God; so the condition of the damned must be understood as the free rejection of all this. The resurrection of the damned is the process of recovering the memories of their lives, and in the light of what God created them to be (i.e. the image of God in their own souls) perceiving what they made of themselves and of others in whose damnation they have shared. The "eternal fire" of hell (Mt 18:8-9) of which Jesus himself spoke must be understood as this terrible self-inflicted remorse. No doubt it also has a physical aspect, not in the sense that God tortures the damned, but in the sense that resurrected sinners feel their spiritual anguish throughout their whole body-persons, just as the blessed feel the delight of their spirit sensuously.[105]

Our understanding of hell, therefore, must be freed of every notion that God tortures his enemies or delights in their suffering. But some good people will still object that they could never be happy in heaven if they know that the damned are suffering, even if what they suffer they deserve; and that a good God himself must either make these people repent or at least annihilate them so their suffering will come to an end. I deeply respect such persons, and know that in trying to answer them, I may simply make their scandal worse. Nevertheless, we need to realize that we are not talking about sinners for whom we feel compassion because we know that they are really not so bad after all since there are many extenuating circumstances in their lives, but only about those who are *really* guilty of persisting in the will to do serious harm to themselves or others. Nor are we talking about "first time offenders", but about people who persist to the very end of their lives in evil, no doubt after God and many others have again and again forgiven them and tried to help them. Nor, finally, are we talking about persons who any longer have the possibility of repentance.

It is this last point which is perhaps the most puzzling. Why is it that after death the righteous cannot sin, nor the unrighteous repent? The answer given by Aquinas still seems the most reasonable.[106] He pointed out that a pure spiritual intelligence (as he conceived angels to be), would have such a clarity and completeness of knowledge that when they would make a free decision it would be irrevocable, because no new factors would ever appear, either from the side of their objective knowledge or their subjective attitudes toward it, which could make them change their minds. That is why in Scripture there is never any hint that the devils will repent.[107] It is precisely the fact that human beings have the lowest type of intelligence which requires the assistance of the body, that makes it possible and necessary for human beings in this life to grow in knowledge and to make repeated renewals of their free decisions, because from day to day, and even minute to minute, new aspects of a problem come to mind and our feelings about them vary with our moods and bodily states. This variability of the human bodily condition is what makes us liable to temptation to give up our good commitments, but also opens up for us the psychological possibility of repentance. At death, however, the separated soul is in the condition of the angels, committed forever to decisions already made.

Why, then, when we begin to get our bodies back do we not begin to be open once more to sin and repentance? To answer this, we must recall that when by the power of God the soul begins to recover its body as a spiritual body, that body follows and is adapted to the condition of the soul, just as in life the soul followed and was adapted to the condition of the body. Consequently, the reconstitution of the body of the self-condemned is in accordance with the commitment which they made to themselves and persisted in until death. The body of the evil ones has become an expression of their evil, just as the body of the blessed has become the expression of their goodness. If we reflect on this we will

616

begin to see, I believe, why it is not possible for the blessed to have "sympathy" or "compassion" with the damned. Human compassion or sympathy is to "feel with" another human being who is essentially like ourselves. We sympathize with them because we can "feel myself in his place." This is why we feel our sympathy cool when we begin to think that another is himself "heartless," "cold", "inhuman." Yet for anyone we know in this life, even a Hitler, we can always suppose that there is something human, vulnerable in that person, at least at moments. But the damned, if there are any such, will have become by their own choice simply and irredeemably evil.

Dante faints away with pity at the sight of the misery of Francesca da Rimini and Paolo because he feels in himself that he could have loved as they did, but this was his projection, not a true understanding of what a real sin of adultery is, namely the heartless rejection of the love of one's spouse for the sake of one's own false satisfaction.[108] Perhaps few "adulterers" really subjectively commit that sin. Who knows? But only such truly responsible sinners are in hell. Some have wondered at the understanding Jesus' must have had of hell when in the Parable of Lazarus and the Rich Man (Luke 16:27-28) he attributes to the Rich Man condemned to Hades an apparently charitable concern to save his own brothers from coming there. But perhaps what is implied is not that the Rich Man cares for his brothers but that he fears that his own remorse will be increased if his example brings them to his own fate.[109] What we can be sure of us is that no one in hell feels any real pity except for themselves, and that will make any pity for *them* empty.

Why then may it not be that God will annihilate the wicked, at least after a time of punishment commensurate with their sins? After all, the harm they did could only have been finite, why should their punishment be infinite? H. Lassiat argues[110] that for St. Irenaeus and most of the early Fathers, this was answered by saying that although the wicked were resurrected it was only for a temporary punishment proportionate to the sins of each, and then these sins caused their total annihilation. He supports this by showing that although there are a good many texts in Scripture which speak of the punishment of the damned as "eternal" (e.g. Mt 25:46), this term in Greek does not necessarily mean "everlasting" but only "terminal."[98] In my opinion, however, the development of the doctrine of everlasting punishment in both Eastern and Western churches, was the positive result of an ever clearer understanding (as we have seen under Platonic and Aristotelian influences) of the spirituality of the human person created in God's image and therefore of the "immortality" of the soul, i.e. its incorruptibility by any natural process, even by sinful acts which can disorder the soul, but not destroy it.[111]

Lassiat seeks to escape this conclusion by holding that the soul is not immortal by nature, but is sustained in being only by the constant concurrence of God's creative activity.[112] I grant, of course, that no contingent being can continue in existence without God's constant concurrence, but this means that the annihilation of a spiritual being requires on God's part a free act of annihilation. Lassiat again seeks to avoid this conclusion by saying that the damned person is the one who withdraws himself from God, not God from the damned; but this reply presents impossible metaphysical difficulties, since just because God's action is always primary, no action of a creature can bring itself into existence nor can it remove itself from existence. Thus, Lassiat's hypothesis would force us to see God as one who freely annihilates some of his creatures to save them from the consequences of their own free acts.

Perhaps the matter is somewhat cleared up when we realize that the damned do not *will* to be annihilated but continue to will to exist. Of course we imagine that if their remorse is so terrible, they will cry out to cease to exist as do some suicides. But the choice of suicide is possible only to those who are somehow confused in mind while the damned know themselves pitilessly. Their very torment consists in their state of self-contradiction by which they will themselves to exist, in fact to be gods, who exist necessarily; and yet they have freely willed their own self-frustration. It would be an injustice on God's part to take back from them the existence he has given them to do with what they will.

But does not the mystery remain how God and the blessed can be perfectly happy knowing that the damned continue to exist? Perhaps the answer here lies in the *relativity* of time. We have seen that in modern physics time is localized so that what is present for one locality can be past for another. Might we not say that for the blessed the damned are wholly in the *past*? Their story, horrible as it is, is over as far as the blessed is concerned. Moreover, the blessed can think of this past with joy, not because they enjoy revenge on their enemies or take delight in suffering, but because it is an evil from which they have escaped, yet which, having been endured and overcome, adds to their triumph. As for God, in his eternity the unending condition of hell is seen not as something which he has willed, but which he has permitted in permitting his enemies to live the life they have chosen. He cannot change this situation except by being false to the covenant made with even these creatures whom he knows have removed themselves as far from him as they can.[99b]

I have one final suggestion. Perhaps we can think of the perpetuity of hell as a state of *asymptotic* existence. In geometry an asymptote is a curve that constantly approaches a line, but which will never touch it, no matter how far the curve is extended. May not the everlasting existence of the damned be a kind of "fading away", just as the past recedes from us forever, a "fading away" of awareness, a kind of spiritual entropy? Perhaps this is what some of the Church Fathers intended by their hypothesis of God's merciful "mitigation" of the pains of hell.[113]

Where then will heaven be? And hell? Our present universe is doomed by natural law to entropic death in which matter and energy will become so diffused and of such low intensity that all communication and action will cease. But the heaven of the redeemed and resurrected persons will be one of intense intercommunication in which all life is shared. St. Paul says (Rom 8:18-23):

> I consider the sufferings of the present to be nothing compared with the glory to be revealed in us. Indeed, the whole created world eagerly awaits the revelation of the sons of God. Creation was made subject to futility, not of its own accord but by him [God or Adam? the exegetes differ] who once subjected it; yet not without hope, because the world itself will be freed from its slavery to corruption and share in the glorious freedom of the children of God. Yes, we know that all creation groans and is in agony even until now. Not only that, but we ourselves, although we have the Spirit as first fruits, groan inwardly while we await the redemption of our bodies. In hope we were saved.

He thus teaches that the earth, and indeed the whole material creation was cursed in Adam's sin (Gn 3:17-19), because it had been created to be a good environment for humanity and thus was frustrated in its original purpose when humankind turned away from God and began to exploit the world for evil ends. Yet Paul assures us that the cosmos will finally attain its God-given purpose when humanity turns back to God. May this not mean that the entropy and growing diffusion of the universe will be reversed, and that all its parts will again come into close communication with each other, so that our dark world with its scattered nebulae and wandering stars will become brilliant with the interplay of intense light radiating in every direction, centering on this earth where God became incarnate and where the People of God, now become the community of all humankind, will undergo their transformation to the state of spiritual bodies?

Yet we need not exclude the possibility that God has created other earths with intelligent, embodied beings. Will they then also come into communication with each other? Will they then receive the Good News from us? Or may the Gospel already have been made known to them, not by Jesus of Nazareth, but by their own prophets, or even by the Word who became man in Jesus Christ, but who has also become flesh in another nature than ours? May it be that some of these earths have had a different history than ours? That in some, sin has never entered, while others like our earth have fallen into their own tragic history and now await redemption? May God have created some intelligent embodied beings purely in a state of nature, while having elevated others, like ourselves, to the order of grace? In the infinite creativity of God, all such possibilities are open.

Yet whatever God had done in the rest of the universe, it still remains true that Jesus of Nazareth remains Lord of the whole bodily universe, because in that universe he is the personal presence of the Son of God, the creative Word of God than whom nothing can be greater. I myself do not find it incredible, although wonderful indeed, if it turns out that God has become incarnate *only* in our world, because as he chose to become human, the least of all intelligent natures, so too he may have chosen our earth as the least of all the worlds.

Notes

1. See Joseph A. Fitzmyer, *The Gospel According to Luke I-IX,* Anchor Bible (Garden City, N.Y.: Doubleday, 1981), pp. 418-433 for commentary and current bibliography.

2. Fitzmyer, p. 307, gives the most complete list of the details of the infancy narratives shared by Matthew and Luke and refers to the studies of J. Schmid, Léon Dufour, G. Schneider, and R. E. Brown. Fitzmyer remarks "It has, of course, been popular to harmonize these accounts: Luke 1, Matthew 1, Luke 2:1-38, a postulated return to Bethlehem, Matthew 2. But with what right, apart from pious speculation? The harmonization tends to obscure the individual thrusts of the two narratives, and does not summon up great credence for them." (p. 308). Certainly the different purposes and principles of selection from the tradition of the two evangelists should not be obscured. But what is wrong with "harmonization"? Ultimately, all historical interpretation involves a harmonization of the disparate data. A choice always has to be made between a type of interpretation which excludes all data difficult to harmonize and one which includes all data that *can* be harmonized. The decision must be grounded in our general trust in the sources. In the case of the Bible, Christian exegesis ought to rest on the doctrine of inspiration and hence to favor trust in the tradition, unless its purpose is *apologetic,* in which case it can reasonably assume a sceptical bias for the sake of argument. In the case of purely secular history, it seems to me obvious that historians make similar decisions and cannot avoid doing so.

3. For example, Brown points out the parallel between the shepherds in Luke and the magi in Matthew (*op.cit.* p. 411-412).

4. See Raymond E. Brown, *The Gospel According to John,* Anchor Bible (Garden City, N.Y.: Doubleday, 1966), 1, 58-63. The most important interpretations of the Lamb are as the apocalyptic lamb, as the suffering servant, and as the paschal lamb. Brown gives support to all three.

5. *Ibid.,* vol. 2, pp. 555-6. J. Jeremias, *The Eucharistic Words of Jesus,* rev. ed., (New York: Scribner, 1966), is the great defender of the view that the Last Supper was the Passover. P. Benoit, O.P., "The Holy Eucharist", *Scripture* 8 (1956: 97-108), shows that the Supper even in John's account resembles a Passover meal. Brown's conclusion is "We suggest then that, for unkown reasons, on Thursday evening, the 14th of Nisan by the official calendar, the day before Passover, Jesus ate with his disciples a meal that had Passover characteristics.

The Synoptics or their tradition, influenced by these Passover characteristics, too quickly made the assumption that the day was actually Passover; John, on the other hand, preserved the correct chronological information.'', p. 556.

6. Jeremias, *The Eucharistic Words* pp. 237-255.

7. *Ibid.,* p. 186-203.

8. Aquinas defines "sacrifice" as follows: "Something is properly called a sacrifice which is performed for the sake of that honor which is due to God alone in order to please him." (*Summa Theologiae* III, q. 48, a. 3 c.) Furthermore, he explains that "Through the sacrifices were represented the raising of the mind to God which the offering of a sacrifice stimulated. Now to the raising of the mind to God pertains that a man acknowledge that all that he possesses he has from God as from a first principle, and should order to God as his ultimate goal; and this is represented by offering and sacrifices . . . And among all the gifts which God has given to the human race, after it had fallen into sin, the greatest is that He gave His own son . . . and therefore the greatest sacrifice is that by which Christ offered himself to God." (*Ibid.* I-II, q. 102, a. 3 c). For the Biblical development of the concept see Robert J. Daly, *Christian Sacrifice: The Judaeo-Christian Background Before Origen* (Washington, D.C.: Catholic University of America Press, 1978).

9. This issue was effectively raised by Gustaf Aulén in his well known *Christus Victor* (New York: Macmillan, 1969). The result has been that many current Christologists simply pass over the problem of "satisfaction," but that this is too facile is shown by Romanus Cessario, O.P., *Christian Satisfaction in Aquinas: Towards a Personalist Understanding* (Washington, D.C., University Press of America, 1982).

10. "The principle laid down in Lev. 17, 11 was fundamental. 'For this life of the flesh is in the blood: and I have given it to you upon the altar to make an atonement for your souls; for it is the blood that maketh an atonement for the soul' . . . By the outpouring of blood life was released, and in offering this to God the worshipper believed that the estrangement between him and the Deity was annulled, or that the defilement which separated them was cleansed." W. D. Davies, *Paul and Rabbinic Judaism,* 4th ed. (Philadelphia: Fortress Press, 1980), p. 235.

11. *Ibid.* pp. 254-259.

12. *Ibid.* pp. 226-284.

13. *Ibid.* pp. 58-85.

14. George Hunston Williams, *The Radical Reformation* (Philadelphia: Westminister, 1962), pp. 134-136; 300-314.

15. See *Decree on Ecumenism (Unitatis Redintegratio),* the *Declaration on the Relationship of the Church to Non-Christian Religions (Nostra Aetate)* and the *Degree on Missionary Activity (Ad Gentes Divinitus.)* In the last it is said, "So, although in ways known to himself God can lead those who, through no fault of their own, are ignorant of the Gospel to that faith without which it is impossible to please Him (Heb. 11:6), the Church, nevertheless, still has the obligation and also the sacred right to evangelize." n. 7. *Vatican Council II: The Concilar and Post-Concilar Documents,* Austin Flannery, O.P., ed. (Northport, N.Y., Costello, 1975), p. 821.

16. On theories of atonement and redemption see Vincent Taylor, *The Atonement in New Testament Teaching,* London: Epworth, 3rd ed., 1958 (emphasizes that "reconciliation" is the basic concept); Gustaf Aulén, *Christus Victor: An Historical Study of the Three Main Types of the Idea of Atonement* (note 9 above); (Objective or Anselmian, Subjective or Abelardian, and Reconciliatry which Aulén attributes to Luther); Philippe de la Trinite, O.C.D., *What is Redemption?, Twentieth Century Encyclopedia of Catholicism* no. 25, (New York: Hawthorne, 1961); Jean Galot, S.J., *La Rédemption Mystère de l'Alliance* (Paris-Bruges: Desclée de Brouwer, 1964); John R. Sheets, S.J., *The Theology of the Atonement* (Readings) (Englewood Cliffs, N.J.: Prentice-Hall, 1967); A. Büchler, *Studies in Sin and Atonement in the Rabbinic Literature of the First Century* (New York: KTAV Publishing House, 1967); Sam K. Williams, *Jesus' Death as Saving Event: The Background and Origin of a Concept* (Harvard diss.) (Missoula, Montana, Scholars Press, 1975) (believes it was of Hellenistic origin); and Romanus Caesario, O.P., *Christian Satisifaction in Aquinas* (note 9 above).

17. See articles of J. De Guibert and M. Olpha-Gallard, *Dictionnaire de Spiritualite,* (Paris: Beauchesne, 1932 —) vol. 1: 936-960 and 964-977 and P. Bailly, vol. 5: 392-408.

18. See Chapter 10, pp. 425-426.

19. "Since we find that we are so seriously wounded and vitiated and since our evil is so extensive and deep-rooted that it penetrates the most intimate depths of our nature, it follows that all our mortifications and abnegations are not sufficient of themselves to 'purge out the old leaven . . . of malice and wickedness' and convert us into 'the unleavened bread of sincerity and truth' (I Cor. 5: 7 f). Nor can they cure our vices, root out our evil inclinations, put in order all those things that are disordered, and restore us to our primitive rectitude and purity. In addition to our practice of mortifications we must abandon ourselves to the divine action so that the fire of the Holy Spirit may purify and renew us to the extent that it is necessary for perfect union with God." Juan G. Arintero, O.P., *The Mystical Evolution* (St. Louis: B. Herde ·, 1951), 2, 63.

20. See Brown, *The Gospel According to John* (note 4 above), 2, 922-927, "By way of summary, then, we may say that the Johannine picture of Jesus' mother becoming the mother of the Beloved Disciple seems to evoke the OT themes of Lady Zion's giving birth to a new people in the messianic age, and of Eve and her offspring. The imagery flows over into the imagery of the Church who brings forth children modelled after Jesus, and the relationship of loving care that must bind the children to their mother. We do not wish to press the details of this symbolism or to pretend that it is without obscurity. But there are enough confirmations to give reasonable assurance that we are on the right track." p. 926.

21. *Ibid.* 2, 944-956. Brown, Barrett and Dodd all incline to the historicity of the incident. Brown thinks the primary significance which the Evangelist sees in this is *theological* i.e., the blood and water stand for the coming of the Spirit in Jesus' place. The sacramental significance, according to Brown, is secondary and only probable, although supported by Cullmann and Bultmann, and that it is more probable the reference is to Baptism.

22. *Ibid.,* 2, 949. Brown finds little support in the text for this symbolism but says that it is as old as the fourth century and received the approval of the Council of Vienne (1312). The patristic evidence is in E. C. Hoskyns, *The Fourth Gospel,* ed., F. N. Davey, 2nd ed. (London: Faber, 1947), pp. 534-35.

23. See J. Duncan M. Derrett, "The Manger: Ritual Law and Soteriology", in *Studies in the New Testament* (Leiden: Brill, 1977) 2, 48-53.

24. The influence of Epicureanism on *Qoheleth* was proposed by T. Tyler (1899) and is frequently repeated. It is rejected, however, by Robert Gordis, *Koheleth: The Man and His World* (New York: Schocken), 3rd ed., 1968, pp. 51-58; Koheleth doubts but does not *deny* immortality. He may be somewhat influenced by Hellenism but is strictly in line with Jewish tradition.

25. See Chapter 4, pp. 000. See also R. Noyes, Jr., "Seneca on death", *Journal of Religion and Health,* 12 (3) 1973, 223-240, for a recent sympathetic account of the Stoic attitude toward dying, and J. N. Sevenster, *Paul and Seneca* (Leiden: Brill, 1961), p. 218-240, for a comparison of Stoic and Christian attitudes to death.

26. See Chapter 5, pp. 160 and 170.

27. See E. O. James *Prehistoric Religion* (London: Thomas and Hudson, 1957), pp. 17-33 and Mircea Eliade, *A History of Religious Ideas* (Chicago: University of Chicago Press, 1978) 1, 3-28.

28. See Nicholas J. Tromp, *Primitive Conceptions of Death and the Nether World in the Old Testament* (Rome: Pontifical Biblical Institute, 1969).

29. Mitchell Dahood, S.J., aroused a good deal of controversy by his Anchor Bible commentary on the Psalms (Garden City, N.J.: Doubleday, 1966), xxxvi; 2, xxvi f., and 3, xli-lii by arguing that the Psalms early and late, reflect a belief in a blessed future life, already present in Caananite religious poetry. His thesis has not won general acceptance. Moses Greenberg in his article "Resurrection", *Encyclopedia Judaica,* Jerusalem: Macmillan, 1971, 14, 96-103 disagrees with Dahood on the ground that Dahood interprets phrases which refer to long life on earth and survival in one's progeny as meaning personal immortality (p. 98).

30. On this verse see G. B. Caird, *The Gospel of St. Luke,* Pelican Gospel Commentaries (New York: Seabury, 1963), p. 252 and Barnabas M. Ahern, C.P., *Proceedings of the Catholic Theological Society of America, 1961,* 3-22, especially 9 f.

31. See A. M. Dubarle, O.P., *Les Sages d'Israel* (Paris: Cerf, 1946), pp. 190-197, points out that the author of *Wisdom* never makes clear whether he has immortality of the soul or resurrection in mind. David Winston, *The Wisdom of Solomon,* Anchor Bible (Garden City, N.Y.: Doubleday, 1979), argues (p. 25 ff.), that the author is clearly influenced by Platonic ideas of the soul.

32. Aquinas argues, however, that as only God can create from nothing only God can annihilate but that to annihilate would be contrary to God's wisdom and goodness. "Since God is the creator of nature, he does not subtract from things that which is proper to their natures. Now since it has been shown that it is proper to intellectual natures to be perpetual, God does not subtract this from them. Hence intellectual substances are in every respect incorruptible." *Summa Contra Gentes* II, c. 55; cf. *Summa Theologiae* I, q. 9, a. 2.

33. According to Helmer Ringren, *The Faith of Qumran* (Philadelphia: Fortress, 1963), pp. 148-151 and 164-166 and Geza Vermes, *The Dead Sea Scrolls* (Philadelphia: Fortress, rev. ed., 1981), pp. 186-188 the Essenes certainly claimed some kind of immortality and probably believed in the resurrection but this latter is not clear in the documents we possess. On the Pharisees and Sadducees see Greenberg, "Resurrection" (note 29 above). On present Jewish attitudes see Jack Riemer, ed., *Jewish Reflections on Death* (Schocken Books, New York, 1975).

34. Oscar Cullmann, *"Immortality of the Soul or Resurrection of the Dead"* (London, Epworth: New York: Macmillan, 1958). This essay, originally published in 1955 has been extremely influential even in Catholic circles, but its argument assumes, as George W. E. Nickelsburg Jr. points out in *Resurrection, Immortality and Eternal Life in Intertestamental Judaism* (Harvard Theological Studies xxvi) (Cambridge, Mass.: Harvard University Press, 1972), pp. 177-180, a uniformity of opinion in Judaism on the after life which did not exist. Furthermore, in my own opinion, it suffers from too sharp a distinction between Jewish (biblical) and Greek conceptions, and from a failure to allow for a legitimate development of doctrine in the Church. See also Joseph Moingt, "Immortalite ́de l'âme et/ou re ́surrection" in *Lumie ́re et Vie* 21 (1972) 65-78; the whole number is devoted to this problem. Also Pierre Benoit, O.P., "Resurrection: At End of Time or Immediately after Death" in *Immortality and Resurrection,* ed., by him and Roland Murphy, vol. 60 (New York: Herder and Herder, 1970), pp. 113-114; Claude Tresmontant, *Le proble ̀me de l'âme* (Paris: Seuil, 1971). For a study on the terminological question see Daniel Lys, *Ne ̀phesh: histoire de l'âme dans la re ́ve ́lation d'Israel au sien des religions proche-orientales* (Paris: Presses Universitaires de France, 1959). In all this discussion the following points must be kept in mind: (1) Many different conceptions of after-life existed simultaneously in Near Eastern and Jewish culture without any systematic attempt to choose among them until Greek philosophy began to force such a choice; (2) Biblical thought insists that our life here and hereafter is *totally* dependent on God; (3) Jesus' resurrection settled the question for Christians as regards the fact of resurrection but the question of immortality of the soul (provided that this immortality still depends on God as does every created entity) remained vague in New Testament times. Cullman and others who follow his view seem to think that the doctrine of the natural immortality of the soul is contradictory to the creaturely dependence of the soul on God, but this is not necessarily the case. See also Josef Pieper, *Death and Immortality* (New York: Herder and Herder, 1969), and Pierre Grelot, *De la mort ̀a la vie e ́ternelle,* Lectio Divina 67 (Paris: Cerf, 1971).

35. H. Lassiat, *Jeunesse de l'e ́glise, la foi au 2ᵉ sie ̀cle,* 2 vols. (Paris: Mame, 1979) reviewed by M. Spanneut in *Me ́langes de Science Religieuse,* 37 (1980), 118-121. Refuted by Adelin Rousseau editor of Irenaeus in *Sources Chre ́tiennes,* "L'e ́ternite ́ des peines de l'enfer: et l'immortalite ́ naturelle de l'âme selon saint Irene ́e," *Nouvelle Revue The ́ologique,* 99 (Nov-Dec., 1977): 834-864 and answered by Lassiat "L'anthropologie d'Irene ́e",

ibid., 10 (May-June, 1978): 399-417. Rousseau's strongest point is *Contra haereses* V, 7, 11 where Irenaeus seems to say formally that the soul is naturally immortal (cf. also V, 4, 1 and V, 13, 3) but see III 20, 1, where Irenaeus distinguishes between immortal and incorruptible (II, 34, 2-4, III, 8, 3 and IV, 4, 1).

36. Eugène Portalié, S.J., *A Guide to the Thought of Saint Augustine* (London: Burns and Oates, 1960), pp. 145-151. The *De Immortalitate Animae* and the *De Quantitate Animae,* PL 32: 1021-1080 are translated by John J. Macmahon, S.J., in *Fathers of the Church,* ed. Ludwig Schopp (New York: Cima, 1947).

37. *Summa Theologiae* I. q. 89, a. 1-4; cf. also III *Sent,* dist. 31, q. 2, a. 4; IV, dist. 50 q. 1, a. 3-4; *Contra Gentes* II, 8, 81; *De Veritate,* q. 19, a. 1-2; *De Anima,* a. 15, 18, 20; *Quodlibet* III, q. 9, a.1.

38. "In the creation of human nature, God gave something to the human body beyond what was due to it from its natural principles, namely a certain incorruptibility through which it was conveniently adapted to its form, in order that, just as the life of the soul is perpetual, so the body through the soul was able to live perpetually. Even if it was not natural as to the active principle, yet it was in a way natural from the order to the end, that namely the matter might be proportioned to its natural form, which is the end in relation to the matter . . . Considering the creation of human nature, therefore, death is something which happens to man *per accidens* because of sin." *Summa Contra Gentes* IV, c. 81.

39. See Chapter 5, pp. 160 and 170.

40. Thus Jörg Splett in an article "Immortality" in Karl Rahner, ed., *Encyclopedia of Theology* (New York, Seabury, 1975), says, "Just as the description of the body-soul relationship by the magisterium in terms of hylomorphism do not erect Aristotelian philosophy into dogma, so too the official declarations on eschatology (D 530, DS 1000), and against Averroism (D 738, DS 1440), do not give any particular philosophical explanation of immortality any status as part of the faith. While rejecting false interpretations of the faith, they used a ready-made philosophical terminology as a handy way of allowing the faith to find self-expression" (p. 688).

41. See note 35 above.

42. See Chapter 4, pp. 000. and Jaroslav Pelikan, *The Shape of Death: Life, Death, Immortality in the Early Fathers* (New York: Abingdon, 1961).

43. See Chapter 7, pp. 256-258.

44. The following have been helpful: Karl Barth, *The Resurrection of the Dead,* New York: Arno Press, 1977 (1933); J. G. Davies, *He Ascended Into Heaven: A Study in the History of the Doctrine* (London: Lutterworth Press, 1958); Francis X. Durrwell, *The Resurrection* (New York: Sheed and Ward, 1960); David M. Stanley, S.J., *Christ's Resurrection in Pauline Soteriology* (Rome: Pontifical Biblical Institute, 1961); C. F. D. Moule, ed., *The Significance of the Message of the Resurrection for Faith in Jesus Christ* (Naperville, Allison, 1968 (essays by Willi Marxsen, Ulrich Wilckens, Gerhard Delling and Hans-Georg Geyer); Pierre Benoit, O.P., *The Passion and Resurrection of Jesus Christ* (New York: Herder and Herder, 1969); Edward L. Bode, *The First Easter Morning: The Gospel Accounts of the Women's Visit to the Tomb of Jesus* (Rome: Pontifical Biblical Institute, 1970); C. F. Evans, *Resurrection and the New Testament,* (Naperville, IL: Allenson, 1970); Willi Marxsen, *The Resurrection of Jesus of Nazareth* (Philadelphia: Fortress, 1970); Reginald H. Fuller, *The Formation of the Resurrection Narratives* (New York: Macmillan, 1971); Gerald O'Collins, S.J., *The Resurrection of Jesus Christ* (Valley Forge, Penn,, Judson Press, 1973); Béda Regaux, *Dieu l'a ressucité* (Paris: Duculot, 1973); Charles Kannengiesser, *Foi en la résurrection: Résurrection de la foi* (Paris: Beauchesne, 1974); Xavier Leon-Dufour, *Resurrection and the Message of Easter* (New York: Holt, Rinehart, Holt and Winston, 1974); Pierre Guibert, *Il ressucite le troisiéme jour* (Paris: Le Centurion, 1975); George Eldon Ladd, *I Believe in the Resurrection of Jesus* (Grand Rapids: Eerdmans, 1975) (see his "harmonization" of the accounts pp. 90-93); Leo Scheffyczk, *Auferstehung: Prinzip Christlichen Glauben* (Einsideln: Johannes Verlag, 1976); J. Cantinat, C. M., *Réflexions sur la Résurrection de Jésus d'après Saint Paul et Saint Luke* (Paris: Gabalda, 1978); Edward Schillebeeckx, *Jesus: an Experiment in*

624

Christology (New York: Seabury, 1979), pp. 399-438; James D. G. Dunn, *Christology in the Making* (Philadelphia: Westminster, 1980) (up-to-date bibliography pp. 354-403).

45. The priority of Mark on the basis of the Two-Documentary Hypothesis remains the common working hypothesis of exegetes. See a defense of it in modified form by Joseph A. Fitzmyer, S.J., "The Priority of Mark and the 'Q' Source in Luke," in his *To Advance the Gospel* (New York: Crossroads, 1981), pp. 3-40 and his recent discussion in his commentary on Luke (note 1 above) pp. 63-97. Nevertheless, it is being attacked from many sides as his bibliography pp. 97-106 indicates, notably (but not exclusively) by those who defend the "Griesbach hypothesis" that Mark is a conflation of Matthew and Luke; see William R. Farmer, *The Synoptic Problem* (Dillsboro, N.C.: Weston North Carolina Press, 1976); and "Modern Developments of Griesbach's Hypothesis". *New Testament Studies* 23 (1976-1977) 275-296). Hans Herbert Stoldt, *History and Criticism of the Marcan Hypothesis,* (Macon, Georgia: Mercer University Press, 1980); and Bernard Orchard, *Matthew, Luke and Mark,* 2nd ed., (Manchester, England: Koinoia, 1971). The significance of this controversy for theologians is that for the present at least it is not feasible to construct theology on any one system of the historical reconstruction of the Gospel tradition, even the more probable, but it is necessary to take into account alternative possibilities.

46. In the works referred to in note 44 above C. F. Evans (1970) and Willi Marxsen (1968 and 1970) are examples of exegetes who conclude that the accounts of the Resurrection cannot be harmonized without violence to the texts and that consequently it is impossible for us to know "what really happened". On the other hand Pierre Benoit (1968) and Béda Rigaux (1973) are examples of scholars who believe that with due allowance for the different points of view of the writers and the traditional literary devices which they use, such a harmonization is easy enough within the normal limits of historical reconstruction.

47. Some Catholic authors such as Xavier Léon-Dufour, (note 44 above) seem to accept the more extreme view that the "empty tomb" tradition is entirely secondary to the "apparition tradition" of which Paul's conversion experience becomes the model, and then have the problem of arguing that these "apparitions" were not merely *subjective*. Marxsen (*ibid.*), when confronted with St. Paul's assertion that "If Christ has not been raised, your faith is worthless. You are still in your sins" (I Cor. 15:17), by a torturous "hermeneutic" process comes to the conclusion that what is important is not whether Christ has been raised, but whether we have faith and hope in the future! What seems to underlie all this theological maneuvering is a philosophical idealism which is reluctant to admit that *physical* facts such as the empty tomb or the bodiliness of the Risen Christ as shown in his tangibility and his eating with his disciples can be in any way essential to the spiritual act of faith. The Gospel writers were not idealists.

48. Schillebeeckx, *Jesus* (note 44 above), pp. 71-76, is certainly right in holding that even if we admit, as we must that the "Jesus of history" cannot be opposed to the "Christ of faith" (because we have no way of knowing the Jesus of history except through the witness of those who knew and believed in Him); nevertheless, it is still important to try to arrive at as critical a picture of the Jesus of history as we can, both for apologetic reasons, and also as a means of purifying our faith from illusory elements. I would want to add, however, that theology is founded on the witness of the apostles and not on our attempts at historical reconstruction whose function, as Schillebeeckx is well aware, have only an apologetic or critical function, not the function of grounding theological assertions.

49. Wolfhart Pannenberg, *Jesus-God and Man* (Philadelphia, Westminister, 1968), pp. 88-105. See Ricardo Blasquez Perez, *La Resurreción en la Christologia de Wolfhart Pannenburg* (Victoria: Editorial ESET, 1976), and Gerald O'Collins, "Is the Resurrection an 'Historical Event'?", *Heythrop Journal* (1967): 381-387. O'Collins concludes that the Resurrection is a "real, bodily event" but not an "historical event." His argument is that the resurrected body exists in eternity outside space and time. As I will show in the next chapter, this can't be granted only in a qualified sense. O'Collins does not seem to consider the possibility that the risen Jesus even, if his proper state is outside history, might not be able to enter into history (i.e., into space and time) at will. This is what the Gospels seem to relate, i.e.,

the Risen Lord willed to sit down and eat with the Twelve at Emmaus, in the supper room, on the sea-shore. Why not?

50. On the different explanations of Mark's theological aim see Seán P. Kealy, C.S.Sp., *Mark's Gospel: A History of its Interpretation* (New York: Paulist, 1978).

51. See Chapter 4, p. 124.

52. Pierre Benoit, O.P., *Jesus and the Gospel* (New York: Herder and Herder, 1973), vol. 1, pp. 209-253, on "The Ascension" distinguishes between a *spiritual* ascension simultaneous with the Resurrection which consists in the reunion of Jesus with his heavenly Father, and its *manifestation* to the Apostles in some visible event which concluded the period of the various resurrection appearances to them. Luke's "forty days" is a biblical convention. Benoit, however, is not asserting that Luke's account is a mere construction.

53. In *The Jerome Commentary,* ed., by R. E. Brown, S. S., Joseph A. Fitzmyer, S.J., and R. E. Murphy, O. Carm. (Englewood Cliffs, N.J.: Prentice Hall, 1968), Vol. 2, No. 51-86, p. 274. See also William F. Orr and James A. Walther, *I Corinthians,* Anchor Bible, Garden City, N.Y.: Doubleday, 1976, pp. 316-354.

54. See note 35 above.

55. See Burger Albert Pearson, *The Pneumatikos-Psychikos in I Cor.* (Society of Biblical Literature Dissertation Series No. 12, 1973). Also Orr and Walther, *I Corinthians* (note 53 above), pp. 341-349.

56. Henri de Lubac, *Surnatural: Études historiques,* (Paris: Aubier, 1946), and *The Mystery of the Supernatural,* (New York: Herder and Herder, 1967). The difficulty with de Lubac's famous attack on the notion of "the state of pure nature" was that without this concept it seems impossible clearly to defend human nature in its own intrinsic wholeness. De Lubac is too much influenced by the Augustinian tradition with its fear that assigning a genuine autonomy to the human being will make humanity independent of God; cf. J. Nicholas, O.P., *Les profondeurs de la grace,* (Paris: Beauchesne, 1969), pp. 334-397.

57. See Chapter 9 pp. 385-86.

58. Some Islamic commentators of the *Quran* interpret the promise of sexual pleasure in heaven literally, while others regard it as symbolic. See Jane Idleman Smith and Yvonne Yazbeck Haddad, *The Islamic Understanding of Death and Resurrection.* (Albany: State University of New York Press, 1981), Appendix B, "The Special Case of Women and Children in the Afterlife," pp. 157-182. On the general attitude of Islam to sex see G. H. Bosquet, *L'Éthique sexuelle de l'Islam* (Islam d'hier et d'aujourd'hui, no. xiv) (Paris: G. P. Maisonneure et Larose, 1966), especially pp. 191-195.

59. See Chapter 10, p. 445.

60. *Summa Theologiae* I, q. 76, a. 5 c., argues that all the senses are founded on the sense of touch (cf. q. 91, a. 3, ad. 1m). I, q. 84 a. 7 and 8 shows that without reference to *actual* sense knowledge, it is not possible for the human intellect to make a perfect judgement. That is why in the dreaming state, when the mind is out of *contact* through sense with the really existing world and supplied only with images which are abstracted from concrete existence, that it thinks illogically and cannot distinguish between the real and the merely mental. No wonder, then, that to assure ourselves we are awake we *pinch* ourselves!

61. Mary of Jesus of Agreda, *The Mystical City of God,* translated by Fiscar Marison (George J. Blatter) (Chicago: The Theopolitan Press, 1902), vol. 3, *The Transfixion,* no 760, p. 732 f.

62. See the remarkable essay of Jacques Maritain, "The Natural Mystical Experience and the Void" in his *Redeeming the Time* (London: Geoffrey Bles: The Centenary Press, 1943), pp. 225-255.

63. On Limbo see Ch.-V.Heris, O.P., "Le salut des enfants morts sans baptême", *Maison Dieu* 10 (1947): 86-105 who discusses Cajetan's solution (which goes back at least to Jean Gerson) but prefers the view that they are saved through death as a "quasi-sacrament"; George J. Dyer, *The Denial of Limbo and the Jansenist Controversy* (Mundelein, IL.: St. Mary of the Lake Seminary, 1955); and *Limbo: An Unsettled Question* (New York: Sheed and Ward, 1964); Vincent Wilkin, S.J., *From Limbo to Heaven* (New York: Sheed and Ward, 1960); Bertrand Gaullier, *L'état des enfants morts sans baptême d'après Saint Thomas d'Aquin*

(Paris: Lethellieux, 1961) (defends existence of Limbo; note especially pp. 141-142 on Trent); and Dom Edmond Boissard, *Réflexions sur le sort des enfants morts sans baptême* (Paris, Ed., de la Source, 1974), who argues for the universal salvation of all unbaptized infants even if we cannot explain how grace is mediated to them.

64. The most recent official document, *Instruction of the S. Congregation for the Doctrine of the Faith on Infant Baptism,* Oct. 20, 1980, *The Pope Speaks* 26 (1981): 6-19, simply repeats the obligation to baptize infants and says "As for infants who have died without baptism, the Church can do nothing but commend them to the mercy of God", as in fact she does in the funeral rite designed for them (Rite of Funerals, 1969, n. 82). The fact that the Church thus officially prays for the unbaptized proves that if the view of theologians who hold that unbaptized children are saved by the prayers of their parents and the Church is not certain, it certainly is not heterodox but at least probable.

65. *Summa Theologiae* I, q. 89, a. 2 c. and ad 2; and *Quaestio Disputata De Anima* a. 17.

66. *Summa Theologiae* I, q. 89, a. 2.

67. *Ibid.*

68. *Ibid.* a. 4.

69. *Ibid.* a. 3 and 4.

70. *Cursus Philosophicus*, Ed., by P. B. Resier, O.S.B., (Turin: Marietti, 1937) III, Naturalis Phil. iv, q. x, a. iv-v, pp. 322-339.

71. See Chapter 4, and Vladimir Lossky, *The Vision of God* (London: Faith Press, 1963) for a detailed history of the notion of the inaccessibility of the Divine Essence according to the Greek Fathers. For this tradition God is known only in his "uncreated energies."

72. The fact that the unborn child has never had the opportunity to learn to love God through the experiences of life and therefore may have less capacity to love and know Him in the beatific vision adds to the horror with which Christians have always regarded the crime of direct abortion.

73. See George A. Maloney, S.J., *The Everlasting Now* (Notre Dame, Ind.: University of Notre Dame Press, 1979), pp. 132-140, develops the idea of an "evolving heaven." He goes too far, in my opinion, when he argues (pp. 109-131) that it may be that God's mercy will overcome the resistance to Him even of those in Hell, since this neglects the firmly established teaching of the Church that human destiny is finally determined in our present life.

74. Nevertheless, it does not seem that the eschatological state would exclude the artistic (as distinguished from the merely practical) exercise of human skills. Traditionally, the blessed dance and sing. Why may they not also engage in artistic play in the arrangement and rearrangement of the elements of the transfigured universe, but without the physical toil and frequent frustration which artists now experience?

75. I suggested in Chapter 8 that the human brain even now may well be at the limit of physical evolution in its compactness and complexity of intra-communication. Yet it is still very much involved in directing the physiology of the body, a task from which it will be relieved in the resurrected state where it will be needed for only two tasks (1) providing the intellect with sense data; (2) directing the body to carry out the commands of the will. Moreover, it appears that our brains in their present state include a good deal of redundant material to provide for new learning and for substitution in case of trauma. Perhaps, therefore, the resurrected brain will be much more efficient and "spiritualized" than our present brain, wonderful as it is.

76. On the so-called "light metaphysics" of Robert Grosseteste (d.1253) see Étienne Gilson, *History of Christian Philosophy in the Middle Ages,* (New York: Random House, 1955), pp. 262-263. Throughout the Middle Ages, the notion of light as the spiritualization of matter was very influential. Art historians often connect it with the Gothic style of architecture with its emphasis on great expanses of stained-glass.

77. Arthur C. Clarke, *2001: A Space Odyssey* (New York, New American Library, 1980).

78. *IV Sent.* dist. 44. q. 1, art. 1: *Summa Theologiae* III Supplement 79 (81) a. 1-3.

79. *Ibid.* I, pp. 106-107.

80. See E. J. Fortman, S.J., *Everlasting Life after Death*, (Staten Island, N.Y.: Alba House, 1976), pp. 95-218. For traditional Protestant views see G. C. Berkouwer, *The Return of Christ*, (Grand Rapids, Mich.: Eerdmans, 1972), pp. 32-64; and for those of an Adventist see Freeman Barton, *Heaven, Hell and Hades*, (Charolotte, N.C.: Advent Christian General Conference, 1981), pp. 30-80.

81. See W. D. Davies, *Paul and Rabbinic Judaism*, (Philadelphia: Fortress, 1980), pp. 111-146 and T. J. Deidun, *New Covenant Morality in Paul* (Rome: Biblical Institute Press, 1981). Deidun sums up, "Paul's rejection of the Mosaic Law is based not on the mere fact that it was *law*, but on the fact that it was *not* Christ", p. 258.

82. Exegetes still disagree on the meaning of 1 Peter 3: 18-22, but the trend seems to be to interpret this in terms of apocryphal literature and to see it as Jesus' *ascent* to the upper regions of the air inhabited by evil spirits to whom Jesus declares their defeat and condemnation, thus emphasizing Jesus' victory over the evil powers through baptism. Thus *with variations*, Ernest Best, *1 Peter* (New Century Bible, London: Aliphants, 1961), pp. 135-147; Bo Reike, *The Epistle of James, Peter and Jude*, Anchor Bible, (Garden City, N.Y.: Doubleday, 1964), pp. 106-115; Charles Perrot, "La descents aux enfers et la prédiction aux morts", in *Études sur la première lettre de Pierre*, Lection Divina no. 102, (Paris: Cerf, 1980), pp. 231-246; J. N. D. Kelly, *A Commentary* on the *Epistles of Peter and Jude* (London: Adam and Charles Black, 1969). However, Donald Senior, C.P., *1 and 2 Peter*, New Testament Message (Wilmington, Del.: Michael Glazier, 1980), pp. 67-73, believes that it indicates a second chance to repent for those who died in the flood.

83. See Wilhelm Mass, "He descended into Hell", *Theology Digest* 30 (1982): 43-48.

84. See Vincent Taylor, *The Gospel According to St. Mark*, (London: Macmillan, 1952), pp. 390, and Heinrich Baltenweiler, *Die Verklärung Jesu* (Zürich: Zwingli Verlag, 1959, pp. 69-82).

85. On prayers for the dead in the early Church see J. H. Wright, article "Dead, Prayers for", *New Catholic Encyclopedia*, vol. 4, pp. 671-673, and J. H. Crehan, S.J., article, "Dead, Prayers for", *Catholic Dictionary of Theology* (London: Nelson), 2, 156-160.

86. See Hippolyte Delehaye, S.J., *Les origines du cult des martyrs*, 2nd ed., (Brussels: Société des Bollandistes, 1933), W. H. C. Frend, *Martyrdom and Persecution in the Early Church* (New York: New York University Press, 1967), and Victor Saxer, *Morts, martyrs, reliques en Afrique chretiénne aux premiers siécles* (Paris: Beauchesne, 1980). An amusing study of the extremes to which this could go is Patrick J. Geary, *Furta Sacra: Thefts of Relics in the Central Middle Ages* (Princeton, N.J: Princeton University, 1978).

87. On purgatory see Fortman, note 71 above, pp. 125-142; Martin Jugie, *Purgatory and the Means to Avoid It*, Westminister, MD, 1950; *La purgatoire profond mystère* (Bibliothéque Ecclesia 40; a collection of essays by various authors, Paris: Arthème Fayard, 1957) and Elmar Klinger, article "Purgatory" in Karl Rahner, ed., *Encyclopedia of Theology*, (New York: Seabury, 1975), pp. 1317-1320.

88. Pius XII, Apostolic Constitution *Munificentissimus Deus*, Nov. 1, 1950, *Acta Apostolicae Sedis* 42, 1950: 753-771. See special number of *The Thomist* 14 (1951) for theological analyses.

89. "In teaching her doctrine about human destiny after death, the church excludes any explanation that would deprive the Assumption of the Virgin Mary of its unique meaning, namely the fact that the bodily glorification of the Virgin is an anticipation of the glorification that is the destiny of all the other elect," *Letter on Certain Questions Concerning Eschatology*, May 17, 1978, *Origins*, 9 (August 2, 1979): 131-133, p.133, n.6. The Assumption of Mary is "singular" and of a different order than that of others in at least that (a) it has a unique foundation, i.e., the divine motherhood; (b) it is the reward of her total freedom from sin which she alone shares with her Son; (c) its term is the highest share in the glory of her Son granted to any mere creature. But its singularity need not imply that she alone among the redeemed was resurrected before the eschaton.

628

90. Daniel Walker, *The Decline of Hell* (Chicago: University Press of Chicago, 1964), pp. 195-201 (note on page 197 refers to Bayle's *Oeuvres Diverses,* t.III, p. 807). Henry More held the same view, p. 147. For a fascinating discussion of how the Enlightenment changed the attitude to death, hell and heaven see John McManner, *Death and the Enlightenment* (New York: Oxford University Press, 1981).

91. *Summa Theologiae*, I-II, q. 72, a. 5; q. 87, a. 3-5 and q. 88, a. 1. See my essay, "The Development of Revealed Doctrine: Sin, Conversion, and the Following of Christ," in *Moral Theology Today: Certitudes and Doubts* (St. Louis, MO: Pope John Center, 1984), pp. 51-57, on the nature of mortal sin.

92. See Chapter 44, pp. 000, Hans Urs van Balthasar in his preface to *Origen*, Classics of Western Spirituality (New York: Paulist, 1979), p. xiv, sums up: "Though it must be admitted that his [Origen's] doctrine of apokatastasis (the salvation of all) was not securely anchored (though he had recourse to numerious scriptural texts), this in no way implies that it was incorrect: the Christian may hope for the salvation of all men, but may not forestall the judgement of God." Wilhelm Breuning, "Zur Lehre von der Apokatastasis", *Communio* (Internationale Katholische Zeitschrift) 10 (1981): 19-31 points out that Origen is *tentative* in his teaching of universal salvation and that what Constantinople II condemned was what it called Origenism not Origen's own teaching as such.

93. Aquinas, *Summa Theologiae* I, q. 22, a. 4 shows that God's providence does not impose necessity on either contingent events or free acts, but makes them respectively to be contingent or free, and in q. 23, a. 1, he shows that predestination is part of providence but does not remove free will (cf. a. 3 ad 3). This predestination follows from God's election and this from his *predelectio* or loving choice of those who are to be saved, not by their own merits, but by His grace (a. 4). Those who are not saved (if there are in fact any such, although certainly Aquinas supposes that there are), are not saved only because of their own free refusal of grace.

94. See M. and L. Becqué, *Life After Death* (Twentieth Century Encyclopedia of Catholicism) (New York: Hawthorne Books, 1960), pp. 175-181.

95. See *Origen on First Principles* (Paul Koetschau's text translated with introduction and notes by G. W. Butterworth) (Gloucester, Mass.: Peter Smith, 1973), Book II, Chapters VIII-IX, pp. 120-146.

96. "Karma: in Buddhism, Hinduism and Jainism, the law of the deed and the effect, according to which the life circumstances of an individual in any given phase of the cycle of births are an effect of his deeds and character in previous lives," p. 209, Stuart C. Hackett, *Oriental Philosophy*, (Madison, University of Wisconsin Press, 1979).

97. Charles A. Moore with Aldyth V. Morris, ed., *The Status of the Individual in East and West*, East-West Philosopher's Conference, 1964, Honolulu: University of Hawaii Press, 1968.

98. For an extensive discussion of transmigration of the soul as a possible alternative to the traditional Christian view see John H. Hick, *Death and Eternal Life*, New York: Harper and Row, 1976, a very interesting work which I did not come across until after completing this book. Hick's proposals, however, amount very much to Origen's *apocatastasis*, and require the abandonment of the teaching that death ends the possibility of conversion. For a view favorable to the notion, see Geddes McGregor, *Reincarnation as a Christian Hope* (Totowa, N.J.: Barnes and Noble, 1982).

99. According to Jung, "The collective unconscious is a common psyche of a super-personal kind whose contents are not acquired during the individual life time", p. 3, as distinguished from the Freudian unconscious which is relatively superficial and whose contents are the result of individual experience. Antonio Moreno, O.P., *Jung, Gods and Modern Man*, (Notre Dame, Ind.: University of Notre Dame Press, 1970), pp. 1-30.

100. Such texts as "Better for you to enter the kingdom of God with one eye than to be thrown with both eyes into Gehenna where 'the worm dies not and the fire is never extinguished.' " (Mark 9:48), or "Out of my sight, you condemned, into that everlasting fire prepared for the devil and his angels!" (Matthew 25:41), do not determine whether God *causes* the punishments or merely declares the inevitable consequences of the actions

of the wicked. Allowing for the rhetoric of these passages, they are easily susceptible of the latter interpretation, i.e., that the sinner condemns himself to suffer the effects on himself of his own actions — aliention from God and the torment of remorse which has not only spiritual but physical effects on the resurrected body.

101. See Fortman, *Everlasting Life* (note 80 above), pp. 81-110, for a discussion of the nature of eternal punishment.

102. *Gorgias* 469 C.

103. *Summa Theologiae* I, q. 64, a. 4 ad 2.

104. Hell as the torment which sinners administer to one another by mutual hatred and accusation is pictured by Jean Paul Sartre in his play *No Exit*. Might we not imagine it also as the continuation of the kind of world which unrepentant sinners have chosen to fashion for themselves, a world which might superficially appear a utopian paradise, but which interiorly would burn with the fire of unendurable absurdity and meaninglessness?

105. See Fortman, *Everlasting Life* (note 80 above), pp. 172-174.

106. Since according to *Summa Theologiae* I, q. 89, a. 4, the separated soul had only imperfect knowledge of singulars, it would seem that it lacks the capacity for practical deliberation which is required for free choice. As for the resurrected, since the resurrected body will be conformed to the state of the soul at death and it will no longer be subject to physiological variations, the resurrected person will resemble the angels who, because they lack bodily mutability and imperfection of knowledge are immutable in their "fundamental options' (I, q. 13, a. 6 ad 3; q. 64, a. 2). It is generally held by theologians that it is *de fide* that mortal sin or conversion are impossible after death, in conformity with the profession of faith of Michael Paleologus of the Council of Lyons II (1274) on the particular judgement (Denzinger-Schönmetzer, 856-858). Cf. Antonio Rudoni, *Escatologia* (Turin: Marietti, 1972), pp. 126-137.

107. According to Aquinas, *Summa Theologiae* I, q. 64, a. 2 c; II-II, q. 14, a. 3, 5m and III q. 52, a 6, 3m; q. 86 a. 1 c., the devils are confirmed in evil because of the perfection of the angelic intellect which makes every free choice absolutely decisive. In any case such texts as Matthew 26:41, 'Then he will say to those on his left: 'Out of my sight, condemned, into that everlasting fire prepared for the devil and his angels.' ", and the absence of any biblical reference to repentance by the devils seem conclusive, if we do not dismiss the New Testament understanding of the angels as mere myth or allegory.

108. *Inferno*, Canto V, 73 ff.

109. See Gerhard Kittel (G. W. Bromley, translator and editor) *Theological Dictionary of the New Testament*, (Grand Rapids, Mich: Eerdmans, 1964), articles of Joachim Jeremias, "Hades," vol. 1, pp. 146-149 and "Gehenna", vol. 1, pp. 657-658.

110. Lassiat, note 35 above.

111. If it be granted that the Bible never explicitly settles the question of whether the human soul is naturally immortal, what is the ground for the Church's favoring this doctrine? Is it merely the result of the influence of Platonic (and perhaps Aristotelian) philosophy on the Church? I believe its Biblical basis is the teaching that we are created in God's image, i.e., possessed by nature of intelligence and free will. Reflection on the meaning of this (a reflection no doubt assisted by Platonism), led the Church to an understanding of the distinction between the human spiritual order and the merely material order, which distinction necessarily implies that the human soul cannot be destroyed by the material processes of death, and thus survives them.

112. Lassiat (see note 33 above), admits that even if the soul is naturally immortal, it still requires the constant concursus of God as the source of all existence in order to perdure. Thus, for a Christian who believes in the utter contingency of all creation, to maintain the natural immortality of the soul is not to assert (as the Greeks did) that the soul is divine and possessed of a life independent of the Divine Life; but only to assert that God can and has created contingent creatures who are not liable to be destroyed by any natural force. Denial that God can so create can be sustained only if it can be proved that the notion of an everlasting creature is metaphysically contradictory.

630

113. See the article of A. Michel, "Mitigation des peines de la vie future", in *Dictionnaire de Théologie Catholique* (ed. A. Vacant, E. Mangenot and E. Amann) (Paris: Letouzey et Ané, 1929), t.10 (2) cols. 1997-2009. This idea may have Jewish origins in the notion of a Sabbath rest even for the damned.

Chapter 13

The Glorified Body

("The God of all grace, who called you into
everlasting glory in Christ" I Peter 5:10)

I: The God Revealed In Jesus

1. Father, Son, and Spirit

In Chapter 11 I discussed Jesus' own self-understanding in his "Abba experience" and in the last chapter the way in which this self-understanding was completed in his death and resurrection. My emphasis was on the problems connected with his bodily existence: the way in which his human body and earthly environment contributed to this understanding of himself as uniquely the Son of God and on how the death of his body and its reconstitution as (in St. Paul's term) a "spiritual body" can be interpreted in the light of his Sonship, a Sonship in which by grace all are called to share. The task of every theological investigation, however, is not complete until we ask, "How does God reveal himself in these created realities and events? What does our long meditation on the human body and its material environment tell us about God?" Some great theologians have claimed, of course, that theology is simply the study of

632

human salvation; but it is more than that, since human salvation is not just our escape from the miseries of our condition but is the *vision* of God, our entrance into the glory of his Reign. Theology is not just an anthropology, but a *praise* of God.

Who then is that God whom Jesus and his community the Church in their bodily, historical existence reveal? I say "and his community" because, as we have seen, it is in the community of faith that the significance of Jesus' life and teaching becomes fully manifested through a long process of development in historical experience which will not be completed until history reaches its goal in the Lord's second coming. In the face of modern agnosticism and atheism, of Humanism and Marxism, any effort to answer that question will have a hard time even getting a serious hearing. Hans Küng in his *Does God Exist?* has made a courageous effort to face all the objections raised by "the modern mind" from the time of Descartes and Paschal to Feuerbach, Marx, Nietzche, Heidegger and Wittgenstein and to provide the kind of answer he believes has a chance of at least making sense to our contemporaries.[1] His conclusion is that although the existence of the God whom Jesus called Father cannot be *proved* by rational argument, it can be shown to be "reasonable" to believe in this God's existence because the only alternative is to accept the nihilism or absurdism to which, as Nietzche showed, modern agnosticism had led. Thus Küng rejects the efforts of philosophers like Descartes to demonstrate God's existence, and accepts the road to that conclusion opened by Paschal, which demands that the believer take the risk of believing in something of which he cannot be theoretically certain but which our practical living in the hope that it all makes sense demands of reasonable persons of good will.[2]

Whatever one may think of Küng's non-demonstrative argument, Vatican II in confirming the solemn definition of Vatican I in its exposition of the text of St. Paul who said, "Since the creation of the world, invisible realities, God's eternal power and divinity, have become visible, recognized through the things he has made" (Rom 1:20), a text which summarizes one of the great themes of the whole Bible,[3] has assured us that God, prior to his revelation of himself in Jesus, has revealed himself even in creation, in a way that goes farther than Küng has dared to go. What holds Küng back, it seems to me, is that in order to meet the "modern mind" (read "the Humanist or Marxist mind" that dominates academia and the media), he feels compelled to leave unquestioned (or at least not radically questioned) the Cartesian point of departure which characterizes this "modern" thought.[4] "Modern" thought continues to move within the limits of a fundamental *idealism* with its dualistic anthropology, which I have tried to show in this book is inconsistent with the Christian world-view and in no way necessitated by the rightly interpreted results of modern science.

Here I only intend to sketch without detailed argument what seems to me a more adequate way of dealing with the "God problem", proceeding, however, not from an apologetic starting point as does Küng (although I do not deny the legitimacy of apologetics), but rather from what Jesus has revealed about God to the question whether what he has revealed can be reasonably believed and *in part* demonstrated from our historical and scientific experience of our visible, material world. To take this route is not to try to substitute reason for faith (since faith guides reason and transcends it), nor is it to deny that faith requires that we take the "risk" of believing what we do not and cannot see for ourselves. But this "risk" is not that God's revelation is insufficient to give certitude to open minds, a certitude of reason as regards the existence of the Creator and a still greater certitude of faith that this Creator has become personally visible to us in Jesus Christ. Of course these objectively certain truths are received by us with varying degrees of subjective certitude according to the depth of our own reasoning and the humility of our own faith.

As Küng's historical survey shows, modern atheism and agnosticism repeat the two basic difficulties about the existence of God which the medievals tried to answer,[5] namely: (1) science gives a sufficient explanation of human experience without the hypothesis of a God; (2) the existence of evil in the world refutes the existence of a good God; but have added a third and characteristically modern difficulty: (3) if God is transcendent to human experience, then whether he exists or not, there is no way to know him. This last objection also sometimes takes the form of denying that "God language" even has any meaning except the expression of feelings.[6]

Contemporary process theology has tried to answer these difficulties by positing a *finite* God who cannot prevent all evil in the world, yet can be known by us, and is necessary for a complete scientific explanation of the world because he is not absolutely transcendent to the world but is its immanent unifying principle (panentheism.)[7] The radical process philosophy which I proposed in Chapters 7 and 8, however, seeks to retain the notion of the absolute transcendence of an infinite Creator as entirely compatible with modern science and as the only one compatible with the Bible and with Jesus' own acceptance of the God of the Old Testament who creates the world from nothing. Consequently, I must deal with these difficulties in my own way.

I have already answered the first of them by showing that the data of modern science can be given different interpretations including a theistic one.[8] I have also tried to answer the second difficulty in Chapter 10 on history, by showing that the evil in the world can be explained: (a) as the responsibility not of God but of human persons who have misused their stewardship of God's creation; (b) as tolerated by God for a time in order to permit these persons to exercise the freedom of such stewardship, only with a view to God's ultimately bringing a greater good out of the evil. It is a paradox that Humanists and Marxists are so reluctant to accept this theistic type of explanation, when their own world-views force them also to place the whole responsibility for correcting the evils in the world on human shoulders.

The third difficulty, however, remains to be addressed. It has been given many philosophical forms, but the one which has probably been most widely influential on both Humanism and Marxism has been that of Feuerbach, (d.1872)[9] which is itself rooted in the critical philosophy of Immanuel Kant, rightly regarded as the most influential of modern philosophical systems. Kant had contended that because human reason can be validly applied only within the limits of human experience, there can be no theoretical knowledge of the existence or nature of God; but he had also tried to save God as a necessary *postulate* of his philosophical system because without such a postulate, the unity and consistency of a world-view which also included an ethics could not be maintained. Feuerbach, accepting the impossibility of a theoretical knowledge of God, went beyond Kant by explaining that the term "God", as it still functioned in the Kantian world-view, need not be considered to refer to any extra-mental reality but was simply a *projection* of the ideal human self. The importance of this proposal was that on the one hand it explained why historically most humans, including most philosophers, have believed there is a God, and on the other opened up the possibility that modern men "come of age" could learn to do without a belief in the reality of such an ideal.

I concede at once that our notion of God *must* be a projection of the ideal human self, since all our concepts must be derived from human bodily experience. The question, however, is whether this kind of projection is justified, i.e. whether it is ontologically, existentially grounded, as, for example, repeatedly in the history of science certain ideal models of physical entities have been demonstrated to be approximately correct descriptions of actually existing realities. Before I deal with that question, however, it is well to ask just what this projection of an ideal by Jesus himself in his Abba experience was really like, since, as I have already noted, the problem for many of our contemporaries takes the form of saying that "God-language is meaningless".[10] In other words, they cannot imagine what a God might be like. The anti-theistic attacks of Humanism and Marxism, often bitterly satiric and contemptuous, have painted a picture of the Christian God which is manifestly absurd or revolting, and this distorted image has been reinforced by the survival

635

of out-worn devotional symbols of popular religion which have become so repugnant to our contemporary sensibilities that the message they once conveyed is now misread. No wonder that the first step of Vatican II (which as yet has only partially succeeded, but of which we ought not despair), has been the effort to restore or create more adequate symbols suitable to our times.

Jesus himself seems to have accepted the Old Testament image of God in the form that it had taken in the Deuteronomic literature, the Prophets, and the Wisdom writings. This God is an unlimited Power which created the universe in its entirety and controls its activity and destiny absolutely through the whole of its history, and which is transcendent to time and space. Apart from this Power nothing can exist or act, yet It remains utterly independent of what It has created and controls. While in the Old Testament other images of God inconsistent with this radical monotheism can be found, as the Bible unfolds in a record of doctrinal development, these divergent images were more and more overcome and replaced by an uncompromising and consistent conception. Yet this God of that radical monotheism in which Jesus grew up and from which he never departed was also conceived as radically *personal*, not merely as a remote, Absolute or nameless Power, but as the Person confronting every created person in what Martin Buber has taught us to call "the I-Thou relation", speaking to such created persons in their own language and expecting them to reply to Him, even question or bargain with Him.

How difficult it is to maintain the tension in this Jewish idea of God is shown by eastern religions in which God is usually regarded as the Absolute which may, of course, manifest itself to human beings in the form of personal gods or avatars of a personal character, but only *provisionally*, because to those who would really know It, all such personal images must be left behind. Even in Islam every tendency to personalize Allah becomes suspect and heterodox.[11] As for the ancient nature religions, or the deism of the Enlightenment, or the so-called "civil religion" of modern urban culture, God fades away into a *deus otiosus*, a "Supreme Being" or even (as in the recent and enormously popular movie *Star Wars*), "the Force." On the other hand the human need to project a more humanly personal image of God has again and again in the history of religion ended in polytheism, or the multiplication of buddhasatvas or saints whose veneration pushed God into the background. The heart of Jewish prophetic religion, however, which is central to Jesus' experience and teaching, is the uncompromising maintenance of this tension between the absolute transcendence of The Power and his real, intrinsic Personhood, confronting the worshipper in a two-way communication, in dialogue.

The personality which Jesus attributes to this personal God is that manifested in the Abba experience. Jesus speaks to God as "Abba, Dearest Father" and in the Parable of the Prodigal Son (Lk 15:11-32), we find vividly portrayed how Jesus understood that word "Father."[12] The father in this parable freely gives to one son his share of the patrimony in order that this young man may achieve the independence of adulthood; and when that son wastes his patrimony to his own destruction, the father runs out to greet him with love, rejoicing without a word of reproach, receiving him back not as a servant but once more as his beloved son. To the older son such behavior seems inexplicable as (if we are honest) it probably seems to us. Is not this implausible notion of God a projection by Jesus of what he himself wished to be to his people, and in fact proved himself to be on the Cross when he prayed, "Father, forgive them; they do not know what they are doing" (Lk 23:34)? It is the highest realization of true humanity.

Yet there is something unique in Jesus' understanding of God as Father, namely, the *intimacy* of his own sonship. It is precisely this aspect of Jesus' teaching which Jews and Muslims have never been able to accept because it seems to contradict radical monotheism. According to Mark, at Jesus' trial before the high priest he was formally interrogated: "Are you the Messiah, the son of the Blessed One?" and Jesus answered, "I am; and you will see the Son of Man seated at the right hand of the Power and coming with the clouds of heaven" (14:61-62). Although this claim by Jesus to be the Son of God in an unqualified sense that made him in the eyes of the Sanhedrin guilty of a blasphemy worthy of death is most explicit in the Fourth Gospel, exegetes recognize that it is found in the synoptic tradition as well, and of course is a fundamental theme of the Pauline epistles.[13]

Somehow Jesus experienced himself as standing in an I-Thou relation to the Father from whom as Son he had received his entire being, yet as having received from the Father not merely some share in being but the *totality* of the Father's "glory." "Everything has been given over to me by my Father. No one knows the Son but the Father, and no one knows the Father but the Son — and anyone to whom the Son wishes to reveal him". (Mt 11:27, the Q document).[14] "Whoever has seen me has seen the Father. How can you say, "Show us the Father'? Do you not believe that I am in the Father and the Father is in me?" (Jn 14:10).

This is mysticism of a unique kind, since generally speaking, the mystics of all religions speak of their union with God either as "losing oneself in God," as in the non-dual forms of Vedanta, or, as in the dualist form of Vedanta or in bahkti devotional cults of a *complementary* union symbolized as that of man and woman.[15] Jesus, however, experienced this union as that of son to father, in which the son is distinguished from the father precisely because he has received the totality of the father's being. It might be objected that this is like the dual Vedanta of Rāmānuja, for which the human self unites with the absolute Self and yet remains in an I-Thou relation with it. However, Rāmānuja did not presuppose a radical monotheism as did Jesus, and hence for the Hindu, the human side of the dialogue was not a creation but an emanation from the Absolute to which it simply reunited although it is not reabsorbed.[16] Jesus' self-understanding, on the other hand, has to be read in the context of the doctrine of creation, in which the fact that the Son has taken to himself a created human nature manifests that even at the level of divinity, the Son is not the Father who has not been thus incarnated.[17]

Moreover, in Jesus' baptismal vision, the essential historicity of which I defended in Chapter 11,[18] and in Jesus' subsequent identification of himself as the New Temple (Jn 2:21), as well as in his promise to send the Spirit to his Church after his ascension to the Father (Acts 1:4-9), we find that in this experience of union with the Father, Jesus found the union between them to be a unity in the Holy Spirit. Jesus experienced the Spirit as a gift to him from the Father, and yet also as given him to transmit to his church. Therefore, the Spirit whom Jesus sends to be teacher in his place is, like Father and Son, a person.[19]

Thus the mystical Abba experience of Jesus is truly a revelation of the One God as a Trinity of Persons, who remain absolutely one because they are distinguished only by the total communication of one single life and being. The early Church struggled to find formulas that would do some justice to Jesus' own expression of His intimacy with the Father in the Spirit.[20] Recent theology has attempted to revise and improve on these admittedly inadequate formulae, but it appears to me that to date the results have not been very successful, not because improvement is not possible, but because the instruments commonly employed have been merely philological, or if philosophical, have been some anti-metaphysical system that reduces the biblical data to psychological categories.

What did Karl Barth, for example, gain in expounding the Trinitarian doctrine by dropping the word "person" because it was non-biblical, only to substitute for it the equally non-biblical (and Spinozistic) "mode of existence" which fails to do justice to the terms "Father" and "Son"?[21] Schoonenberg and others have emphasized that the Scriptures speak only of the "economical Trinity" i.e. the manifestation of the Trinity in the Incarnation and the sending of the Spirit to the Church, and wish to leave the "ontological" Trinity beyond theological discussion as so much metaphysical speculation. The view, however, that the Son exists eternally with the Father independent of his manifestation in the Incarnation is explicit in many texts of Scripture, some of which are pre-Pauline.[22]

Karl Rahner and Yves Congar[23] have taken the more meaningful view that since God has revealed himself to us through history, inevitably the "ontological Trinity", i.e. what God is in himself, in his inner life, has become known to us through the "economy", God's dealings with us. Indeed this "economy" would have failed of its very purpose if it had not made God known to us as he really is ontologically.[24] In the events of Jesus' life, his true identity in relation to his Father and the Spirit became evident to him in his human understanding, and he has invited us to share in this understanding. "No one knows the Father but the Son — and anyone to whom the Son wishes to reveal him" (Mt 11:27). Certainly God was known to the Jews by his hidden name of Yahweh, yet he can be known as Father in the profoundest, inner meaning of that name only by the Son in the Spirit, and only by the Son through the Spirit can the Father thus be known by us.

The Fathers of the Church found the Trinity reflected first of all in the human person as the *image* of God, and Augustine developed this theme by finding this Trinitarian image in the intercommunication of memory, intelligence, and freedom in the human soul.[25] It should not be denied, however, that this Trinitarian image is also reflected in the human body to the degree that the body visibly manifests the human psyche. The human body is first of all a natural unit whose relative permanence and historicity manifest our spirit as it is *memory*, a being-in-process which comes to be integrally in history, and this manifests the Father who is the source of all being and Lord of history. The human body is human because it is *intelligent*, and this intelligence is manifested in its articulated anatomy as an organism which supplies the intelligence with sense data and executes its commands, and this corresponds to the Son, the Word of God who took human form in the Incarnation, sharing our human way of knowing, our bodily experiences, and teaching us how to understand them. Finally, the human body is *free*, expressive of the human will, because it is not limited to the fixed patterns of animal behavior, but participates in creativity both in its sexuality and in its artistic and technological activities, and this corresponds to the dynamism of the Spirit.[26]

Thus Jesus, in his own constant remembrance of the Father, in his own knowledge of himself as the New Adam, crowning creation and revealing the Father to the world, and in his love in the Spirit by which he willed the world to be transformed into the Reign of the Father, manifests the Trinity to us. The Trinity is also manifested in history: the past (memory), the present (self-aware actuality), and the future (eschatology), not merely as it is considered as a linear succession of events, but as these dimensions of past, present, and future are immanent in every moment of history as each moment is a unique manifestation of God. This long history is recapitulated, as St. Irenaeus said,[27] in Jesus, as first Adam, as the New Adam, and as the Victor at the end of time — the Bridegroom of the New Jerusalem, the Lion-Lamb who alone can open the sealed book of history (Rev 5:1-14).

In all these ways Jesus made the ineffable mystery of the Trinity present, immanent in human history, but this does not mean that by so doing he obscured the divine transcendence. In the Old Testament no book speaks more profoundly of that transcendence than the *Book of Job* in which through suffering, Job learns that "I have dealt with great things that I do not understand, things too wonderful for me which I cannot know" (Jb 42:3). Human mystical experience necessarily is negative. It requires entrance into divine darkness, and this darkness is not absent from Jesus' life, even if we grant with the medievals that he always enjoyed the beatific vision throughout his life. On the Cross, Jesus drained the cup of death to its dregs, somehow experiencing abandonment by the Father more profoundly than any other creature, *"Eloi, eloi sabacthani"*. Nor did Jesus experience this "emptying" only in that last moment, since the synoptics make clear that He lived always before the Cross.[28]

Whatever we can say of God, in the attempt to express what he is, must be said in *analogical* terms derived from ordinary bodily experiences, and hence is necessarily projective.[29] Such analogical terms must (1) be based on a causal link between the realities we experience in this world and God as their ultimate and necessary cause; (2) imply a comparison based on the axiom that "the effect is (at least imperfectly) like its cause," since whatever positive is in the creature must be found in God in its plenitude; and (3) also imply a denial of the adequacy of any concept of God.

As we have just seen, the Bible, especially as it helps us enter into the Abba experience of Jesus, provides us with the fundamental concepts for describing God by this analogical method, and it is the task of the theologian to develop these analogies with philosophical finesse so as never to attribute to God the defects and limitations of creatures, and yet to draw out of what we know of creatures all the positive riches which they contain and to attribute all these to the Creator so that our idea of him may not be too jejune. The idea of God which Jesus gives us has all the ethical riches of Jesus' own holy humanity, which provides us with the best analogy by which we can come to know his divinity, and the Father from whom he receives that divinity in the Spirit.

The question then becomes, "Is this vivid and profound concept of God, built up by projecting all the wonder of creation and of history,

640

and especially the wonder of the man Jesus in his life of profound wisdom and sacrificial love proven on the Cross, — is this concept rooted in reality, or is it a figment of the idealizing human imagination?

2. Does the Christian God Exist?

Since Immanuel Kant attempted his "Copernican revolution" in philosophy by changing the definition of truth from "the conformity of our minds to reality" to "the conformity of reality to our minds", modern philosophy, no matter how critical it has become of Kant's own "critical philosophy," has accepted at face value Kant's denial that the analogical method of Christian theology provides us with any objectively valid knowledge of the unseen God, because according to Kant we have no warrant to extend concepts taken from our bodily experience to a supposed realm of being transcending bodily experience. Kant's reason for this denial was that he was convinced that the first condition of analogy, namely, that there is a causal link between the things of our experience and a supposed First Cause which transcends our experience, can never be established.[30]

It is taken for granted by many contemporary theologians that Kant's refutation of the medieval arguments for the existence of God was conclusive, or at least that it has historically become so widely accepted in modern culture that it is unthinkable to return to older views. Since the medieval arguments were only precise philosophical formulations of informal arguments found in the Bible and in patristic tradition, this capitulation to modern prejudices amounts to undercutting the Christian world-view at its roots. The valiant efforts of some Christian theologians to find some new way to justify the Christian conviction of the reality of God while supinely accepting Kant's position seem to me as futile as they are industrious. I have already pointed out that Vatican II reaffirmed the definitive teaching of Vatican I that:

> God, the beginning and end of all reality, can be known with certitude by the light of human reason from created things, for "since the creation of the world, invisible realities, God's eternal power and divinity, have become visible, recognized through the things he has made" (Rom 1:20). Nevertheless, it has pleased His wisdom and goodness to reveal Himself and the decrees of His eternal will to the human race by another and supernatural way, as the Apostle says, "God spoke in fragmentary and varied ways to our fathers through the prophets; in this, the final age, he has spoken to us through his Son." (Heb 1:1)

641

Historians of philosophy have shown that Kant knew the medieval proofs only indirectly through the Cartesian tradition, and many critics have exposed the serious flaws in his refutation of these proofs,[31] yet both Humanists and Marxists continue to refer to this refutation as if it were one of the most solid achievements of philosophy. It would seem that this is a good example of what Thomas Kuhn in his well-known theory of scientific revolutions[32] calls a "paradigm shift", namely the wide-acceptance of a new theory not because the old has been decisively disproved, nor even because the new is more probable than the old, but because the new theory seems more consistent with some broad shift in cultural orientation due to many and varied, but mostly non-scientific factors. As Humanism developed in the eighteenth century, Kant's views seemed very plausible at a time when Christian theism was yielding to deism. By eliminating any theoretical knowledge of God Kant undercut theism, but by also providing *practical* arguments for the postulation of God's existence, he was able to remain a deist, a position which he regarded as useful both for the defense of the Newtonian conception of a universe governed by invariable natural physical laws, and for the defense of the French Revolution and the rights of man based on a universally self-evident natural moral law.[33] Humanism, of course, was eventually to abandon both Newtonian physics and the natural moral law, but it has not abandoned its trust in Kant's explosion of the medieval proofs of God's existence.

If we are seriously to ask ourselves about whether the traditional proofs or their Kantian refutation has the better case, we cannot assume either the Christian or the Humanist world-view as given, nor prefer one to the other on the grounds either of "tradition" or "modernity", without simply begging the question. Yet neither can we seek some "neutral ground" which hermeneutics shows us it is impossible to find. What can be done is to compare the positions of Kant and Aquinas (to take the latter as a good example of the classical view), to see where they agree and where they disagree, and then to evaluate the disagreements in the light of our own experience. Of course the very notion of "experience" is not neutral, but each of us has access to realities which we cannot honestly deny.[34]

Kant and Aquinas *agree* that the *a priori* (from cause to effect) ontological argument (which Aquinas knew in its Anselmian form and Kant in its Cartesian and Leibnizian form) is invalid because it illicitly concludes from a *mental* concept of God (a Feuerbachian "projection" such as the one I developed in the first section of this chapter) to the *real* existence of God. They agree, therefore, that any proof to be valid must be *a posteriori* (from effect to cause). To eliminate this second possibility Kant

642

proceeds to show that human reason in its theoretical use cannot attain to any reality that cannot be experienced in our world of time and space.

He tries to do this by showing that when the human reason attempts to transcend time and space it falls into *antinomies* or contradictions which betray that it is really talking nonsense. The four antinomies Kant presents are: (1) "The world is finite" vs. "the world is infinite, in space and time;" (2) "A continuum is infinitely divisible" vs. "a continuum is composed of indivisible parts;" (3) "Human behavior is determined by natural law" vs. "human behavior is free;" (4) "The world is caused by a Necessary Existent" vs. "every existent is contingent." Medieval philosophy was well acquainted with these antinomies and proposed solutions to all of them with which Kant shows little acquaintance.[35]

It is the last of these antinomies which is directly relevant to Kant's refutation of the *a posteriori* arguments for God's existence. His disproof amounts to the claim that whether we attempt to prove the existence of God as efficient cause of the world (the cosmological argument) or its final cause (the teleological argument), we are surreptiously making use of the admittedly invalid *a priori* ontological argument, because we are still using the definition of God as the Necessary Existent which is just what has to be proved. In fact, Kant points out (and Aquinas also agrees[36]) we cannot define God.

Again Aquinas agrees with Kant that our *practical* reason requires the existence of God as the ultimate ground for that hope for final justice which is the condition of any rational ethics, and also that the idea of God as the final cause of the world is needed as a regulative idea to give transcendental unity to our world-view.[37] Finally, Aquinas and Kant agree that reason is dependent on sensible experience for valid *existential* knowledge.

The real point of difference between them is that for Kant, theoretical reason, although dependent on sense experience for the content of its thinking, is able by self-reflection to act as "pure reason" in establishing *a priori* the conditions of human knowledge. Thus for Kant our valid knowledge is reduced to constructing a world-picture out of the data of experience in accordance with these necessary conditions derived from the structure of the human mind. Hence such a universal principle as the "principle of causality" is not derived from the objects of our experience, but from the *a priori* conditions of thought.[38] Aquinas, however, does not posit the possibility of any such "pure reason." For him mundane human intelligence has no direct knowledge of itself, but only of Being-in-process (*ens mobile*). Hence the principle of causality is not derived from the structure of the mind but from an analysis of Being-in-process.

Aquinas' analysis leads to the conclusion that the Being-in-process experienced by us cannot exist without a First Cause which is Being-*not*-in-process. Such a Cause (as Kant also agrees) cannot be conceived by us by a positive concept, since it is other than anything we experience, but (and this possibility Kant ignores, no doubt because of his Cartesian background in which only "clear and distinct" concepts were considered as philosophically legitimate) we can conceive of such a First Cause by a positive-negative concept, i.e. by *analogy* to the causal agents known to us by experience. Such an analogy is no mere comparison of ideas, but is grounded in the knowledge that Being-in-process could not be without a cause other than itself.[39]

Thus the difference between these two positions is not a question of their logical consistency, nor of fact, but of *epistemological standards*. Our choice therefore, has to be between Aquinas' Aristotelian epistemology in which all knowledge is empirically derived, and Kant's epistemology in which a Platonic element (of course considerably modified), namely, the *a priori* categories derived from pure reason, still survives. Today many philosophers think they can retain Kant's critical standpoint while not succumbing to idealism if they reduce his categories from the status of universal structures of human reason to "hypotheses" by which we tentatively arrange the data of science, and which are only pragmatically justified as guides to further empirical research. But this modification, far from saving Kant's position, reduces it to the Humean scepticism which it was Kant's great aim to escape.[40]

If, on the other hand, we accept Aquinas' Aristotelian epistemology because it resists both the scepticism of Hume and the *a priori* idealism of Kant, and because it is more radically critical than either Kant or Descartes since (as I argued in Chapter 7[41]) it goes behind the *cogito* to Being-in-process, then we can certify that the classical *a posteriori* proofs of God's existence are valid. Of course they are not of the mathematical type which Descartes demanded (and which seem still to be demanded by writers like Küng when they reject any hope of "demonstrating" God's existence and settle for some sort of Paschalian wager). They are of the sort we have whenever we conclude that contingent facts must have an explanation and we try to form some notion of their cause by a closer observation and analysis of the effects.

Once we are convinced that our world-in-process is not self-explanatory but manifests the existence of a Necessary Existent which is not in process but which is the ultimate cause of our world-in-process and its processes, then we can with good reason "project" all our ideals of God derived from our experience as an attempt to describe, very imperfectly, what that God is like. Such projections will be valid to the degree

644

that by analogy they are purified of their negative aspects, while whatever is positive in them is preserved. The First Cause must be the ultimate source of whatever these projective concepts have of positive being, yet it cannot be the source of whatever is negative in them, because negativity is always some aspect of contingency, of the incompleteness of process.

Such efforts to "describe" God do not function in the *a posteriori* proofs of God's existence as "real definitions" on which the demonstrative force of the proofs turn (as Kant thought when he asserted that they implicitly presuppose the *a priori* ontological proof based on a real definition of God as the Necessary Existent) but only as "nominal definitions" by which we identify the First Cause whose existence is proved with what we popularly mean by the word "God." But once we have established that the First Cause does in fact exist, then such descriptions become "real", although they remain inadequate descriptions of the Reality they seek to portray.

If we now look back over the road we have been traveling, we see what the words of St. Paul, "since the creation of the world, invisible realities, God's eternal power and divinity, have become visible, recognized through the things he had made" (Rom 1:20) must mean for us today, enlightened as we have been by the remarkable advance in the scientific understanding of our world and ourselves outlined in Chapter 2. As this advance in scientific understanding has proceeded since the seventeenth century, its theological implications have become more and more obscured, not by the scientific discoveries themselves, but by the Humanist interpretation of these discoveries in terms of Kantian epistemology. Kant seemed to have closed the path by which we could learn more about God theoretically through His creation. The only way left open to Kant was ethical, the way of "values" separated from "facts", the way of subjectivity but not of objectivity. The romantic wing of Humanism eagerly pursued this way only to end in the cultural relativism discussed in Chapter 10, since values which are not grounded in human needs, in human nature as part of cosmic nature, are arbitrary.

Since, however, this Kantian obscurantism has exposed itself historically as groundless, the way again lies open for us to see in the scientific world-picture that Book of Nature, as the medievals called it, in which God reveals himself to us. Every one of the marvelous discoveries of science is the Creator speaking to us as creatures who can understand what he is saying. The Book of History, which also can be read by us with much sharper eyes than formerly, also reveals God, and in more human ways than the Book of Nature. At the center of this history stands Jesus, the New Adam, clue to the evolution of nature and to the course of history.

645

II: Persons Without Bodies

1. Do Angels Exist?

The early Church praised the Risen Lord in a hymn that St. Paul quotes:[43]

He is the image of the invisible God, the first-born of all creatures. In him everything in heaven and on earth was created, things visible and invisible, whether thrones or dominations, principalities or powers; all were created through him, and for him. He is before all else that is. In him everything continues in being. It is he who is head of the body, the church; he who is the beginning, the first-born of the dead, so that primacy may be his in everything. It pleased God to make absolute fullness reside in him and, by means of him, to reconcile everything in his person, both on earth and in the heavens, making peace through the blood of his cross (Colossians 1:15-20).

This text confronts us with a topic which has to do with our human bodily condition and its theological significance only by way of *contrast*, namely, "Are there invisible creatures? What are they?" Yet I cannot pass over this difficult topic, because it seems to me that the current theological embarassment with it is also a result of anthropological dualism. When theology forgets the human body and concentrates on human subjectivity, it also finds it can say nothing about the angels except that they are myths symbolizing the providence of God which no longer has meaning for our scientific age. This neglect, however, only results in the angels reappearing in secularized forms in science fiction and pseudo-scientific research on paranormal psychology.

The Bible speaks in many places about intelligent creatures who are superhuman, some good, some evil.[44] It must be admitted that what it says about them is fragmentary and often very puzzling. Nowhere do the Scriptures provide an angelology like the anthropology presented in *Genesis* 1-3. In the Pentateuch they appear as messengers of God who often seem to be apparitions of God himself. In the Prophets and Psalms they appear as forming the court of God, singing his praises and executing his commands. In the biblical writings of the Persian period they begin to take more definite shape probably under the influence of Zoroastrianism which had a rather elaborate angelology.)[45] In *Job* we find Satan as a dramatic figure, a kind of prosecuting attorney in God's court of judgement, and in *Tobias* the detailed story of Raphael as a guardian angel.

Finally, in *Daniel* and the non-canonical apocalyptic literature angels are everywhere the powers behind the scenes of history, working the rise and fall of empires.

In the New Testament it is plain that belief in angels was an important part of the world view of the contemporaries of Jesus, and of Jesus himself, as is evident both from the vision experiences I have discussed in Chapter 11 and in his miracles of the casting out of demons. Some theologians today, however, would dismiss Jesus' references to the angels on the ground that after all, he was a man of his own time and spoke in the terms of his culture. But the existence of angels was not a universal or unexamined belief in Jesus' time. As is well known, the Sadducean party among the Jews, which included the high-priest, positively repudiated both belief in the resurrection and in the existence of angels, probably on the ground that these doctrines were not clearly taught in the Pentateuch. The Pharisees, however, and also the Essenes, strongly defended both doctrines.[46] Jesus clearly sided with the Pharisees when he was questioned by the Sadducees about the resurrection and answered,

> You are badly misled, because you fail to understand the Scriptures or the power of God. When people rise from the dead, they neither marry nor are given in marriage but live like the angels in heaven (Mk 12:24-26).

Of course in this text, the principle question is about the resurrection, but it is so closely coupled by Jesus with the doctrine of angels that it certainly appears that he is rebuking the Sadducees for their blindness to the Scriptural teaching on a transcendent spiritual world under God peopled by angels and by the risen dead.

Why has this doubt about the existence of angels arisen among contemporary theologians? As early as *Humani Generis* in 1950, Pius XII protested against this denial, and Paul VI required that the famous *Dutch Catechism* be revised to give explicit support to this traditional doctrine, which it had managed very neatly to pass over.[47]

Several factors seemed to have contributed to these doubts: first, Teilhard de Chardin's laudable efforts to bring evolutionary theory into theology called attention to how difficult it is to fit angels into an evolutionary scheme. If the universe emerged from matter where did the angels come from?[48] Second, modern biblical scholarship made theologians keenly aware that sound hermeneutics requires at least some "demythologization" of biblical imagery, and the angels and devils seem likely candidates for such decoding. Third, apologetic concerns to make the Gospel more credible to our contemporaries seem to demand that we free ourselves from unnecessary baggage.[49] Even if angels are a matter of faith, they evidently are not very high in what Vatican II called "the hierarchy of truths." Finally, there is the ecumenical concern to pare away "creeping infallibility" by which theologians have too often labeled doctrines *de fide* which in fact cannot be proved to have been solemnly defined by an ecumenical council or a pope.

As to the last point, it has become common to argue that while the creeds of the church have frequently included some such formula as "I believe in one God, creator of all that is visible or invisible," such formulae only say that *if* there are invisible beings they also are God's creatures, but do not formally assert their actual existence. [50]

It is a poverty-stricken theology, however, which eliminates all doctrines not infallibly defined by ecumenical councils or popes. Before such definitions are possible, their substance must already have been included in the ordinary teaching of the Church, and it is the business of theology to attempt to discriminate and preserve these irreformable elements in the Church's ordinary teaching from the reformable elements which it also undoubtedly contains. In this case, the existence of angels is so deeply embedded in the Church's liturgy and in the writings of the Fathers and in many non-definitive but important documents of the councils and popes that *prima facie* a biblical exegesis which would treat angels as mere metaphors for God's providence or devils as metaphors for human sin, requires to be supported by very serious arguments, not merely by apologetic convenience.

I would offer three positive reasons for believing that the biblical references to angels and devils is to be understood as an inspired assertion that they exist as intelligent, non-human creatures. First, it is not accurate to say that the existence of angels and devils is merely peripheral to the message of the Gospel. While it is certainly true that their existence is not one of the *articles* of faith, i.e., the principle mysteries of the Gospel; nevertheless, they have an intimate connection with the central doctrine of the victory of Christ over sin.

Louis Bouyer has shown that the biblical understanding of history necessarily involved the existence of angels. [51] In the present age God does not rule the world directly but through the "powers and principalities", who are both good and evil, and even the good angels are limited in wisdom. Such powers rule the nations, and God directly rules only his own people, Israel. The nations often worship the angels as their gods. Only in the Age to Come will God establish his Kingdom over all nations through the hegemony of Israel. This Old Testament view is taken over in the New Testament in the belief that the Messianic Age has begun with Jesus, who is by his death victor over the "principalities and powers", but this victory will be finally realized only at the Last Judgment, when

Christ's reign over all nations will be perfectly established and he will turn over this universal reign to his Father. This view of the world in which the angels are dominant until Jesus subordinates them to his own power is quite prominent in the theology of St. Paul[52] (Gal 4:8-9; 14; 1 Cor 6:3; 11:10; Col 1:16: Col 2:10,15,18), although admittedly his main concern was to assert the primacy of Christ, as is evident also for the author of *Hebrews* 1:5-14.

Since the emphasis of the New Testament is on this victory of Christ over whatever powers there may be, it can be argued that the existence of angels as one of these powers is an entirely secondary question which can just as well be relegated to the theological scrap-heap. On the contrary, I would contend that to deny the existence of these super-human powers is to minimize the New Testament teaching on both the human need of salvation and the magnitude of Christ's victory. Jesus is represented by Mark (1:13) as aided by angels in his temptation and by Luke (22:43) in his agony in the garden, and by Matthew (26:53) as no longer asking for their aid in his final submission to arrest; yet Mark (8:38) also shows that Jesus expected to share his victory with the angels,

> If anyone in this faithless and corrupt age is ashamed of me
> and my doctrine, the Son of Man will be ashamed of him
> when he comes with the holy angels in his Father's glory.''

The author of *Ephesians* (6:10-12) explicitly teaches that Christian victory over sin is a victory not merely over human weakness but the angelic powers:

> Draw your strength from the Lord and his mighty power.
> Put on the armor of God so that you may be able to stand
> firm against the tactics of the devil. Our battle is not against
> human forces but against the principalities and powers, the
> rulers of this world of darkness, the evil spirits in regions
> above.

Thus to eliminate the angels is also to eliminate Jesus' and the early church's understanding of what the battle is all about. The Good News is a message not only of Christ's victory over human sin but also over cosmic evil of which human sin is only a part.

A second reason for not reducing the biblical references to angels to the status of pure myth can be drawn from the fact that the modern scientific world-picture, although it has profoundly modified the ancient cosmology with which the New Testament belief in angels was closely connected, has not (contrary to the all too facile assumptions of revisionist theologians) eliminated a cosmological argument for the existence of superhuman intelligences as necessary factors in a consistent modern cosmology.

According to Greek astronomy all the natural processes on earth are ultimately caused by the action of the heavenly bodies embedded in celestial spheres which turn perpetually.[53] Since the ancients were well aware that a perpetual motion machine is not possible, the problem became how to explain this unlimited source of energy which kept the spheres moving, perhaps forever. From this originated the famous proof of the existence of a Prime Mover which is not itself a physical entity but a spiritual Intelligence. But the ancients did not immediately suppose that this Prime Mover was the One God. Instead they assigned to each of the spheres a separate intelligence. The unity of the universe was then explained by saying that the motion of all the spheres was coordinated by the largest or outer sphere, the Primum Mobile, and it was moved by a Supreme Prime Mover, the One God who directed all the other intelligences so that they act in harmony. The Jewish and Christian (and later the Muslim) theologians, with some hesitation, identified these intelligences with the biblical angels.[54]

If we now look to modern cosmology, we see of course that this picture has been profoundly modified. We still attribute most of the processes on the earth to energy coming from the sun, and we attribute the existence of the sun to galactic processes and so on back to the original Big Bang. But most scientists today do not see the universe as a steady-state system which could go on perpetually as the ancients thought. The universe is running down, although in our region of space at least, evolution is going on as I showed in Chapter 7. The new problem that arises, as I argued there, is to discover what accounts for this counter-current to entropy which has led to the production of the most complex and unified kind of primary units which we know in the universe, namely human beings, including the scientists who have discovered these natural processes. I showed that cosmic evolution cannot be explained simply by natural laws, although of course they produce each of the processes involved, but must be explained *historically* as a very long sequence of events produced by unique concurrences of natural forces. This sequence could have turned aside at any point and is guided by no known natural law. Of course it is possible to say that the actual outcome, the emergence of intelligent life, is just an accident bound to happen sooner or later in infinite time. The universe actually revealed by science, however, does not exhibit processes that go on infinitely. We cannot avoid, therefore, asking for the explanation of what has in fact happened.

Hence today, in spite of the vast changes in our scientific views, we still face the same problem the ancients faced: the need to posit the existence of superhuman intelligences guiding natural events. Nor should we immediately conclude that only one superhuman intelligence, namely God, is involved, because the history of evolution does not *prima facie* exhibit perfect unity. The ancients concluded from the imperfect simplicity of the astronomic patterns, especially from the inclination of the ecliptic and the complexity of the motions of the planets ("the wanderers") that there was not just one Prime Mover, but also many subordinate and relatively independent although coordinated prime movers.[55] Similarly, we might well conclude from the fact that the pattern of evolution is not smooth but conflictual, even sometimes revolutionary or dialectical, that it should not be immediately attributed to One Intelligence but to many lesser intelligences, working sometimes at cross-purposes, which is just the biblical notion of the "principalities and powers" who are under God but who do not always agree with each other.

If this seems too speculative, let us note how uncomfortable to the modern mind it has seemed to suppose that we human beings are the only intelligences in the universe.[56] To understand this discomfort, which has resulted in the proliferation of science fiction fantasies about life in other worlds and in perfectly serious efforts of scientists to communicate with other humanoids, we should note that one of the modes of creative thinking that has paid off richly in science, although of course it always requires testing against the evidence, is *extrapolation* or pattern thinking. For example, Mendelejeff's periodic table was based on a symmetrical arrangement of known elements according to their properties, but it contained blanks.[55] Eventually it was possible to fill in these blanks by the discovery of new elements. Again, the table of possible kinds of crystalline structures was first worked out mathematically from known types and the blanks were eventually all filled in by new discoveries.[57] Our evolutionary view of the world presents us with a great variety of kinds of primary units from atoms to the most complex of living forms. We are always looking for "missing links" to complete this pattern. Whenever we find a new type of living thing we immediately suspect that we will soon discover that it has "radiated" in a number of genera and species adapted to the various possible environmental niches.

Therefore, when we discover that in our visible universe there is a type of organism, the human species, which introduces a wholly new principle of behavior, namely abstract, symbolically expressed, creative thought, we naturally conjecture that the very limited exemplification of this type of life found only in the single human species cannot be the only one. If we also accept that the world has been created by a God who is an infinite intelligence, we are even more struck by the immense gap that lies between these two extremes of mental power, the human and the divine. Undoubtedly this gap must still puzzle us today, just as it puzzled ancient people the world over, and even more so because we have greater awareness both of the wonderful scale of natural forms and of the vast differences between the human beings who have attained to scientific understanding and technological control of the world and the other animals.

2. Encountering the Angels

But if such superhuman beings exist, why do we not encounter them or they us? This is indeed a powerful argument against their existence if they are humanoids, i.e. embodied minds. As we have seen, even Plato and Aristotle associated the intelligences very closely with "astral bodies" as if they were the souls of the planets and stars. The Bible represents the angels anthropomorphically and the Church Fathers tended to think of them as having ethereal bodies. The medieval scholastics as late as St. Bonaventure hesitated to deny some type of corporality to them, less the distinction between God as pure Spirit and any of his creatures should be lost. It was St. Thomas Aquinas who decisively took the position that angels are pure spirits, intelligences so perfect, that unlike human beings, they have no need of bodies with which to think. [58] Aquinas no longer feared a pantheistic confusion between God and the angels, because for him that difference depended not on the distinction of form from matter, but on the famous "real distinction" between essence, which only is in potentiality to existence, and existence itself, which all creatures receive from God who alone is Existence itself identical with his own Essence. Angels are pure spirits, but nevertheless they are contingent, not necessary beings.

In order to understand the implications of the notion of a created intelligence so powerful as not to require a body, it is useful to describe the model of an angel which Aquinas developed. It is traditional to mock the scholastics as engaged in futile discussions of such questions as "How many angels can dance on the head of a pin?", but such mockers ought to read a little modern mathematics with its learned discussions of hyperspaces. [59]

According to Aquinas' model, an angel is a created primary unit having intelligence and free will independent of a body. [60] It knows by means of innate ideas in a purely intuitive, non-discursive manner. Angels differ from one another only in the power of their intelligence. The more powerful their minds, the fewer innate ideas they require and the more perfectly integrated is their knowledge. [61] Because they have no bodies, there cannot be many individuals in a species, but each individual angel is specifically different from every other. The highest angel has a vast comprehensive and detailed knowledge of the universe and of God through[62] one (or at least very few) innate ideas. Consequently, the condition of the highest angel's mind approaches the eternity of God's ever present Now.[63] The lowest angel, on the other hand, comes close to the human condition in that its thought must move from one of its many ideas to another in order to gain a comprehensive view of things, and thus it lives in a time which is spread out into past, present, and future like our own. Yet none of the angels is in our kind of time determined by the continuity of physical processes, since angelic time is determined by the succession of thoughts each of which is discrete (we might say "quantized").

All of the angels, with different degrees of clarity and precision depending on their specific rank in intelligence, know the whole material and spiritual universe in its fundamental order.[64] But since the universe is in historical process and this depends on a certain indeterminism in nature and on free will, the angels cannot know the exact course of future history by means of their innate ideas. For Aquinas, who accepted a steady-state universe, this was less significant than for us who believe in an evolutionary universe. We would have to say that the angels in this model might know the general direction God intended for evolution, but would have to await its actual determination in history to know the details. The Fathers of the Church, relying on such texts as "We have become a spectacle to the universe, to angels and men alike" (1 Cor 4:9), and "Into these matters angels long to search" (1 Peter 1:12), were convinced that even the good angels were astonished at the way of God in history as the events of the Bible unfolded, and the bad angels were baffled by it.[65]

The angels, according to Aquinas, communicate with each other not by any kind of external language, but simply by an act of the will which makes one transparent to another. Yet if they do not wish to communicate, no other angel, but only God, knows what they are thinking. Because in the power of their intelligences they form a perfect linear hierarchy, the lower angel has nothing to say to its superiors that they do not already know except its own free acts, the expression of its personal needs. Thus the lower only "pray" to the higher. An angel, on the other hand, can teach any angel lower than itself, not so much by giving it new information (although of course commands to act can be relayed downward), as to help it *synthesize* its ideas in the light of the superior's more unified knowledge.[66] All the angels pray to God who is infinitely superior to any of them, and they know of his existence and his nature from their study of his creation. It is only to the good angels that God by *grace* has revealed himself in the beatific vision as the Trinity. But by *nature* all the angels know through innate ideas, and therefore do not require a body to learn as we do. Nevertheless they do need the objective existence of the material universe as part of creation, since it is in the mirror of the total universe, material and spiritual, that they enjoy a natural knowledge of God. Moreover, since the greatest revelation of God has taken place not merely in creation but in the drama of human history culminating in the Incarnation, Cross, and Resurrection, the angels need our little world to understand the ways of God and thus to know God himself better. For the angels as for us the Cross is the supreme revelation, short of the face to face vision of God, of who God is: God as humble, servant love. [67]

The angels, however, are not only contemplative, but play a role as God's messengers and ministers in the carrying out of God's plan for the universe. [68] As messengers they communicate God's revelation down the hierarchy of spiritual beings and finally to human beings in the material world. As ministers they guide the course of nature and of history according to the plan of God. Just as our human spiritual intelligence has control over our material bodies, so angels can affect natural processes. Aquinas believed, as do modern physicists, that natural laws are statistical only. Consequently, the angels can influence the course of natural events without disturbing the laws of nature by an influence which is so subtle that it would ordinarily not be detectable. Because angels are spirits, they do not themselves occupy space, but because they are finite they cannot act everywhere in space at once. More powerful angels can control a large volume of space at one time, lesser angels smaller volumes. But by an act of the will an angel can transfer its "virtual" presence i.e. the focus of its efficient energy, from one region of space to another without any passage through the intervening space. Angels, therefore, do not dance on pins, but they can dispute for the control of a territory. [69]

Although the angels have power over the physical world and therefore over the human body and brain, they cannot directly affect either the human intelligence or free will, just as they cannot know what another angel is thinking or force its will. Only God can enter into the depths of created spirits, angelic or human. Because angels have free will, it is not strange that some have sinned. [70] Indeed, because of their wonderful intelligence and power of will, they were tempted even more than human beings to set themselves up as rival gods, preferring their own autonomy to citizenship in God's community of love with his creatures. A rebellious angel is one who rejects the plan of God and wishes to run the universe in its own way. Many of the fathers believed that the great test of the angels, before which many fell, was the shock of learning of God's plan to become a human being.[71] This incredible act of humility on the part of God the Son, the Greatest becoming the least out of love, was too much for Lucifer, to accept — and he led a rebellion against God the Father who in the Holy Spirit had commanded this act out of love for fallen humans. Certainly the action of the fallen angels has as its aim to win over the allegiance and worship of human beings to themselves as gods, as is shown in the temptation of Jesus himself.

Perhaps this Thomistic model suggests why the angels do not appear to communicate with us. Angelic communication would normally take the form simply of a subtle action on the human brain by which we are assisted in our own thinking to *synthesize* the information we have gained by our ordinary human learning processes in such a way as to arrive at more secure and penetrating insights. An evil angel might use such clarification of ideas as a means to tempt us, while the good angel would use it as a means to lead us to a fuller understanding of truth. Sometimes such insights might be so dramatic as to constitute actual "visions", or they could take the form of "apparitions' to our bodily senses. An angel would have the power actually to produce paranormal events, but not a true miracle which would require divine power. It must be assumed that God through the good angels so maintains the ordinary course of natural events that even paranormal events are greatly restricted.

The third line of argument for the existence of angels flows directly from this last consideration. The action of angels will be most evident, not in the disturbance of the natural order, but in the course of history which is determined not so much by natural law as by chance and human free will. The good effects of the angels should be seen in the fact that God's plan has been carried out, as is evident in salvation history, in spite of the obstacles set by the bad angels, and the existence of evil spirits will be evident principally in the way this plan has been frustrated. I would suggest that what is most diabolic is not open and militant atheism, but the efforts of the evil ones to gain control over whatever is good in the world, and to turn *good against good*. What is most clearly evidence of the action of good angels, on the other hand, will be found in whatever has been remarkably novel in human history in the promotion of human creativity and the good stewardship of God's gifts.[72]

Of course we must not attribute to angels what can be explained either by human creativity or by human selfishness and pride. But are there not events in human history which seem to transcend the intelligence, good or bad, of the human actors? How can we explain the Holocaust?[73] Or on the other hand, how can we explain the "Greek Miracle" and all that it has meant for human civilization?[74] The Holocaust was not the product merely of Adolf Hitler, who could never have come to power without the concurrence of many forces which were not under his control or any human control. Nor was the Greek Miracle the work of one or two geniuses or any planning committee. It was the flowering of many cultural forces that were somehow synthesized to form an ideal environment for creative thought. In Humanism, the Romantics attempted to explain such remarkable events by the *Zeitgeist*. Why not by angels?

Thus the belief in the existence of angels, so deeply embedded in the biblical world, still is a rational possibility in ours, as a way of understanding the evolutionary character of the world and the drama of human history. It does not exempt us from trying to investigate the natural and human causes of these phenomena, but it does leave open a wider view of creation. Modern man and woman are oppressed by the seeming *absurdity* of the world. May it not be that this absurdity is the result of trying to explain the world in merely human terms?

I would suggest that the doctrine of the angels illuminates several current theological dilemmas, of which I will mention three. One is the problem of *physical* evil in a world which *Genesis* tells us was created "very good". The Victorians were horrified at Darwinianism because it seemed to reveal a *cruel* universe, "red in tooth and claw," in which all that is good and beautiful about us has been the product of vicious warfare, "the survival of the fittest."[75] Modern evolutionary theory has softened this picture considerably, because we have found that symbiosis and cooperation between life forms also contributes importantly to evolution,[76] but it still remains true that living things prey on each other, and that the origin of new species often entails the extinction of the old. Is it possible that if angels guide evolution, the fallen angels are responsible for some of its tragedies?

In attempting some answer to this problem, it is essential first to recall that physical evil is radically different from moral evil. A universe which was monolithic, i.e. a single natural unit, would be a very simple and uninteresting universe; but in any universe made up of many and specifically different kinds of units, like the one of which we are actually some of the units, there is necessarily a degree of conflict, since each finite unit will follow its own teleology without coordination with others. To unify and coordinate such a universe requires governance from inside or outside, and if from inside then this intrinsic governance itself requires to be governed from outside by the Creator on which it depends for its very existence and operation.

If, therefore, we consider that the intrinsic governance of the universe is the responsibilty of created intelligences, the angels, then it follows that their governance, since it is that of powerful but limited intelligences, will be open to a certain degree of conflict; just as in human governance, since

656

no created intelligence human or angelic can be omniscient. The angels know from God what his plan for the universe is, but it seems likely that he leaves to them a considerable degree of freedom in how they execute that plan, just as he does to us. If this is so, then it is possible for angelic intelligences to disagree, and this is just what we find the author of the *Book of Daniel* (Chapter 10) believed, since he writes of the angels of the nations debating with each other.[77]

Apart from sin, however, the angels, inspite of the finitude of their intelligences, no doubt would have lived in harmony under the governance of God. It was the fall of some of the angels in their free choice to be gods in their own right that has disturbed the universe to its foundations. It would seem, therefore, that the more we come to understand through scientific investigation the course that the evolution of the universe has actually taken, we may see that it could have followed a very much smoother path toward the production of our earth suitable for human intelligence and of the evolution of humankind than in fact it has. Already we perceive many crises in that immense history where the outcome was perilous. Think only of the time when meteors and planetoids were raining down heavily on the earth and the moon, leaving the moon with the scars it still bears. How did the earth survive that crisis? Thus, it may well be that the history of evolution has reflected the civil war among the cosmic intelligences.

A second problem that could be illuminated by the doctrine of angels is the proper understanding of the history of the Christian Church.[78] This history has been marked by veritable "mysteries of inquity" such as the ravages of antisemitism in the name of Christ, the rise of gnosticism, the secularization of the Church by the Constantinian establishment, the long wrangles over Christology that eventually led to the division of the Eastern and Western Churches, the imperialistic violence of the Crusades, the Inquisition, the Great Western Schism, the Renaissance corruption of the papacy until it appeared as Anti-Christ, the divisions of the Reformation period, the employment of the Christian missions as an instrument of imperialism, the defection of priests and religious following Vatican II, etc. What characterizes these spiritual disasters within the Christian community is the fanatical war of the good against the good. How can it be that Christians have been martyred by Christian saints? That the missions to bring Good News of love to the Indians of the New World ended by murdering and enslaving them? The much neglected *Epistle of St. Jude* and *II Peter* show that in the early church, the distortion of the Gospel by antinomian Gnosticism (or proto-Gnosticism) led to a discussion of the role of the good angels in protecting the Church and of the evil ones in producing internal dissensions and moral perversion. The remedy for

657

such problems, however, is not inquisitions and witch-hunts,[79] which often turn out to be more demonic than the heresies and abuses they seek to correct with fanatical zeal, but an increase of disinterested charity, care for moral integrity, and hard-headed intellectual work to distinguish truth from illusion

> Beloved, grow strong in your holy faith through prayer in the Holy Spirit. Persevere in God's love, and welcome the mercy of our Lord Jesus Christ which leads to life eternal. Correct those who are confused; the others you must rescue, snatching them from the fire. Even with those you pity, be on your guard; abhor so much as their flesh-stained clothing. (Jude: 20-23).

The third current theological problem to which the doctrine of angels is relevent is ecumenism. How does the Bible understand the pagan (i.e. non-Jewish) religions?[80] It does not regard them simply as errors, but rather the worship of false gods *who actually exist.* In asserting that Yahweh is the One True God of the universe, biblical thought does not therefore deny reality to the gods of polytheism. Rather it regards them as the worship of idols which in themselves are powerless but which represent angels who are by no means powerless. The victory of Christ over the "powers and principalities" is a victory over these gods who are false gods, but gods nevertheless.

Does this mean that all religions except Judaism and Christianity are devil worship? Certainly the Church Fathers often speak as if this were the case,[81] but what they have in mind are the tyrants who claimed to be gods, the practitioners of magic and fraudulent prophecy, those who encourage moral perversity, hatred and dissension, all those evils which corrupt human society and which are sometimes fostered by religion (even by Judaism and Christianity as I have just said). Yet the Bible also recognizes that the good angels are at work in the world, and, as many recent authors have brought out, the Bible recognizes righteous men and women outside God's Chosen People. The *Book of Job* is the story of a man who pleased God but was not a Jew, and whom God allowed Satan to tempt.[82] In this book Satan is presented as a servant of God whose role is to test the authenticity of human religion. Consequently we can see in non-Christian religions the working of God through his ministers, some of them willing, perhaps others, like Balaam,[83] unwilling but compelled by truth. Behind these human ministers are good angels who illuminate the minds of religious leaders, so that in a Buddha, a Confucius, a Socrates, a

Mohammed, we see prophetic men guided by good angels to raise the level of human wisdom and morality among the peoples to whom they were sent. At the same time we see the work of evil angels in the way that such noble truth often became distorted and was used as a weapon against other religions. St. Paul says that even the Old Law was imperfect because although it was given by God it was given through the mediation of angelic messengers:

> It was given in view of transgressions and promulgated by angels, at the hand of a mediator . . . The law was our monitor until Christ came to bring about justification through faith. (Gal 3:19, 24)

Only in Christ is God's perfect law made known to us, because in him God is present in person. In ecumenical dialogue with other religions, therefore, we can (1) admit that God has spoken in them through his angels, as he has done for us also in the Old Testament; (2) admit that in all religions, including Christianity, the work of the devil is apparent whenever injustice, hatred or contempt for other human beings, or uncritical credulity is evident; (3) claim nothing for Christianity that would give it the right to dominate others, but only to communicate to others the loving and humble Christ whom we have come to know and want others to know.[84]

Undoubtedly in any such theological theory of the role of angels in the creation and redemption of the universe we can only reason analogically from the world in which we live to a larger realm of intelligences superior to our own and give them credit and debit for aspects of our visible world which seem to exceed explanation in terms of natural laws or human foresight. The dangers in such speculation of course lie in (1) the temptation to neglect scientific research to explain what science can explain; (2) fascination with an unseen world which could lead us to neglect what God has clearly revealed about himself to indulge our own fantasies. Nevertheless, there is also danger in misreading the biblical view of the creation in such a way as to dismiss as literary metaphor much of its teaching which gives us a deeper insight into the cosmic drama. My purpose in the foregoing discussion has been to widen our theological horizons, not by neglecting modern science, but by showing that it can be read so as to suggest the existence of a vaster universe than its own field of exploration.

659

III. The Eucharistic, Transformed World

1. Eucharistic Transformation

Before the Reformation the whole Christian Church, East and West, believed that the eucharistic consecration has an ontological effect by which the bread and wine, while remaining in every respect phenomenally unchanged, are noumenally changed into the living Body and Blood of the Risen Lord, so that he becomes personally present to his Church in time and space in each of its local assemblies throughout the world. While Luther among the Reformers continued to defend this traditional position, much of the Protestant world questioned it in various degrees. In the twentieth century Teilhard de Chardin carried this traditional view a step further by suggesting that this marvelous transformation of the material bread and wine is the beginning of the Reign of God by which the whole universe will be transformed into risen glory. He relied especially on the Romans 9: 19-23 in which St. Paul speaks of the way in which "all creation groans and is in agony even until now" as it "awaits the revelation of the sons of God" as we too "groan inwardly while we await the redemption of our bodies."[85]

Certainly St. Paul connects the reception of the Eucharist with the resurrection of our bodies. He says that it is "a proclamation of the death of the Lord until he comes" (1 Cor 11:26). He warns the Corinthians that "whoever eats the the bread or drinks the cup of the Lord unworthily sins against the body and blood of the Lord" (11:27) and even attributes the sickness and death of some to such unworthy reception. In subsequent chapters he insists that the Church which shares the Eucharistic meal "is the body of Christ" (12:27) vivified by the Holy Spirit, and that because this Christ has risen, the members of his Body will also be raised from the dead to share in the Reign of God (15:20-28). In the previous section I showed that for the New Testament, the Risen Christ is not merely the head of the church but of the whole universe, both the visible and the still vaster invisible spiritual realm, including those evil cosmic intelligences over whom he has achieved victory through the Cross.

> After that will come the end, when, after having destroyed every sovereignty, authority, and power, he will hand over the kingdom to God the Father. Christ must reign until God has put all enemies under his feet, and the last enemy to be destroyed is death . . . When finally, all has been subjected to the Son, he will then subject himself to the One who made all things subject to him, so that God may be all in all (1 Cor 15:24-26, 28).

Thus the Eucharistic worship of the community looks forward to that total transformation of the universe which is the Reign of God in its fullness. By incorporating the members of the Church ever more perfectly into the living Christ, it begins that incorporation of the whole universe into what Teilhard called the Cosmic Christ, that is, the universe as the temple in which dwells the entire assembly of persons, angels, and humans beings, who are headed by the Risen Jesus, Son of God, Son of Mary and united to him in the Holy Spirit in the everlasting Kingdom of the Father.[86]

The victory has been achieved by the Cross, that is by Jesus' act of sacrificial love, by which he permitted himself to suffer the hatred of sinful humankind in order to manifest the forgiving love of the Father for all his enemies, and thus win for them the gift of the live-giving Spirit. Consequently, the Eucharist is a remembrance of Jesus' sacrificial death, the shedding of his blood, the giving of his very life that the universe might rise from death to life once more.

How then are we to understand this eucharistic transformation? In recent theology there have been attempts to improve on the traditional theology of the Eucharist in two respects: (1) an effort to free it from what might seem out-moded philosophical explanations and terminology which have often proved an obstacle to ecumenism and to the modern mind; (2) an effort to explain it more personalistically, more as the presence of Jesus in personal relation to his people, rather than in terms of an objective or thing-like reality.[87] The very term "real presence" raised difficulties because the word "real" is derived from the Latin *res*, "thing," which seems to derogate from Jesus' personal presence.

This difficulty is increased by the Council of Trent's use of the term "transubstantiation", because in modern philosophy "substance" is also understood as an "object" standing in opposition to the subjective realm of human experience. Recently some Catholic theologians have thought that this dilemma might be overcome by replacing the term "transubstantiation" with "transignification", understood in the light of Heidegger's analysis according to which the ontological reality of entities is not to be found in their noumenal character but in the meaning ("signification") which they have in our human life.[88] According to this new theory, what had been bread and wine become ontologically the Body and Blood of Christ, when through the eucharistic prayer of the priest they come to signify the presence of Christ to the believing assembly.[89] While this solution has been generally judged as orthodox in intention because it attempts to go beyond a merely Zwinglian, symbolic theory of the Eucharist by affirming a real change in the bread and wine, it has seemed to many difficult to see how in fact this "ontological" change is anything more than symbolic, since on this point, at least, Heidegger's philosophy reflects its origins in German idealism.[90]

In my opinion, this controversy again confronts us with the persistent influence of Platonic dualism in its idealistic form, which dislikes any theological doctrine which is too "physical", i.e. which attributes a positive role to matter in its mind-independent objectivity. If, on the contrary, we insist on a realistic understanding of the Incarnation according to which it can honestly be said that "The Word was made flesh," then it is important not to "spiritualize" the Eucharist in the manner of idealism. We human beings are given eternal life by actually becoming the Body of Christ, i.e. by sharing his total existence, including his existence as part of our bodily world, and by rising to new life in him by a genuine resurrection of our body. I have tried to show in the last chapter that although the risen body is, according to St. Paul, "a spiritual body", it remains truly matter although transformed matter. Consequently, in the same way our understanding of the mystery of the Eucharist must somehow retain the genuine *materiality* of Christ's presence, otherwise its relation to the incarnation and resurrection will become obscured.

I do not intend to propose a new theory of the Eucharist, but only to examine the canonized term "transubstantiation" to see what it means in the context of a theology of the body which takes modern science fully into account. One difficulty that has been frequently raised against this term was that "transubstantiation" as explained by scholastic theology, meant the total conversion of the substance of the bread into the substance of Christ's body and of the substance of the wine into his blood; but how, it was asked, can this be reconciled with modern science which would deny that either bread or wine are true "chemical substances" since they are artificial mixtures of many substances? The issue, however, is not whether the bread (or the wine) is one substance or many, but whether one or many, it is converted into the one substance of the humanity of the Risen Christ as that is a substantial body of flesh and blood. Thus after the consecration of the bread or the wine the whole substantial resurrected humanity of Christ is present although effectively signified first as body and then as blood.

It might be asked in what way the risen body, since it is "spiritual" in the sense discussed in the last chapter, can have "blood"? We have seen that the essential meaning of the identity of the mortal and the risen body is that the risen body preserves in matter the memory-traces by which the risen person retains the self-awareness of his or her own past. Thus Jesus in his risen body must retain the memory of his earthly flesh and blood existence with all that is implied by blood — vitality, feelings, suffering.[91]

662

Certainly the eucharistic revisionists are correct in saying that we must not conceive the eucharistic presence in a merely objective manner. The real presence is first of all a *personal* presence, the Risen Christ coming to his people to give himself to them sacrificially as he gave himself to the Twelve at the Last Supper. Yet we must also remember that this self-giving in the case of a human body-person is the giving of something that is material. It is the giving of one's body. Sexual union is a very personal act, the mutual donation of persons, yet it means the giving over of oneself as first of all a *thing*, a body to another as that other's possession. It implies an element of passivity, surrender, vulnerability, defenselessness, precisely because the human lover is not just a spirit but a spiritual body. We have seen that in the gift giving of food-sharing, which is an even more primitive experience than sexual union, the food which is consumed symbolizes the very body of the host who feeds us in friendship as it were on his or her own body. To share a drink together is to arouse the primordial memory of being fed at one's mother's breast on the milk which is from her very substance. Thus in trying to understand how intimately personal Christ's presence is in the Eucharist, we ought not to think of this merely as the presence of someone facing us in dialogue, but we must retain the fact that he is present to offer himself as our food and drink.

Indeed, if in the older Eucharistic devotion there was a certain distortion of the full meaning of the Real Presence, it was not so much in its depersonalization as in its *visualization*. Medieval devotional practices which arose at a time when the reception of communion was infrequent, stressed gazing on the host at the elevation, or in the monstrance, or reposed in the tabernacle.[92] People became accustomed to kneeling before the tabernacle and imagining themselves kneeling before Jesus and looking into his face in a kind of silent dialogue. The element of *touch* by which bodily existence is primarily known hardly entered into such devotions, although it is the essential element of communion: "Take this, all of you, and eat it." Fortunately the Church has corrected this by again placing the *primary* emphasis not on the veneration of the reserved sacrament, but on holy communion.

If then we begin from the fact that in the Eucharist the Risen Lord is present to give himself to us as our food and drink, i.e. as our very life, beginning, as human life begins, from our material existence and ending in our spiritual self-consciousness, but never losing its existential reality rooted in our material bodilyness, we will begin to see why the Eucharistic conversion is a *substantial* conversion. Today the very notion of "substance", so attacked by Lock, Hume, and Kant, seems meaningless to many scientific-minded persons who would say "What is a human being but what a human being does?'' and "What is a chemical 'substance' but its properties which we can observe, measure, or alter?'' Even in an Aristotelian epistemology where all knowledge must be derived from sensation, it would seem that for us a reality is what we can sense — the phenomena, not some inaccessible noumenon. I would suggest, however, that modern science does show us in quite dramatic ways that the notion of a "substance," as distinguished from its directly observed properties or behavior, is the fundamental reality. In Chapter 7, I showed that the only interpretation of the data of modern science which is coherent with the self-consciousness of the scientific observer is one which ascribes primary reality to *primary natural units*, which alone exist as such, and which alone have properties and behaviors to be observed, through which observed properties and behaviors they are known by interaction with human observers who are themselves also primary natural units.

One of the dramatic ways in which this distinction between a "substance" or primary natural unit and its "accidents" or observable characteristics is made evident is in the case of any living organism. Genetics and embryology have shown us that a minute cell, the fertilized zygote, will develop into a mature adult organism totally different in size, appearance, behavior and in almost every observable characteristic we may name, and yet remain identically the same unit throughout this amazing biography. The term "substance" simply indicates that recognizable unity which makes an organism a living system-in-process. Other examples could be chosen from the way in which even an atom receives its identity, its properties, and its behavior from its nucleus, which is not directly observed but which governs the whole structure and "function" of the atom.

Thus the paradoxical words of Jesus when he handed his disciples the bread and the cup and said, "This is my body," "This is my blood" were obviously not true if what he intended was to make a statement about what they saw, touched, and tasted, for what they saw, touched, and tasted was only bread and wine. Since he was serious, it was obvious that what he said had to be taken in some non-obvious, prophetic, *parabolic* sense. The Twelve were certainly used to his speaking in this mysterious and parabolic manner. Undoubtedly these words and this action were symbolic, sacramental, a sign which required to be interepreted, to be understood "spiritually", not "carnally" (1 Cor. 3:1). It is not strange then that St. Augustine, with his neo-Platonic bent, and influenced by the Fourth Gospel with its stress on the need for a spiritual, faith-understanding of events, should have set in motion a theological tradition which at the time of the Reformation led to the purely "spiritual" or symbolic understanding

664

of the Eucharist by Zwingli and the Radical Reformers, which Luther never accepted but which penetrated deep into most of the churches of the Reformation.[93]

Yet biblical parables and symbols are seldom mere "symbols" in the weak sense of that term, but in the strong sense of an epiphany[94] of something greater and more powerful than the rather ordinary appearance at first suggests. In this case the solemnity of the occasion and the strangeness of the words demanded of the Twelve that they look deeper into the mystery. We have no way of knowing how much of it they then grasped, but we do at least know that by the time St. Paul related the story to the Corinthians, he spoke in the very realistic terms I have quoted earlier, warning them that to eat and drink this food could mean their bodily death, or their resurrected life![95] It is the Deuteronomic language of the Covenant of life and death: "This cup is the new covenant in my blood." As the Church continued to receive and meditate on this mystery, in spite of the temptation to "spiritualize" (in the improper sense of that term) this sacrament in accordance with Hellenistic ideas, as the great St. Augustine tended to do, the more it understood this mystery realistically.[96]

Jesus was saying to the Twelve and to all who would remember him, that his coming death would take its meaning from the fact that then and there and always he had come to give himself entirely in his whole being to them and to others as the manifestation of the Father's unfailing and forgiving love for his sons and daughters. The bread and wine to every appearance were only bread and wine; but in truth, the truth of life and death, the Twelve by this eating and drinking were entering into total union, body and soul in the Spirit, with Jesus in his total humanity and divine Sonship. The only way that those words could be unambiguously true (a truth, however, which could be grasped only by their faith in him and all that he said and did), was if that seeming gift of bread and wine was in fact the gift of himself in his total, existential and therefore bodily, and sacrificial reality, that is, *substantially*, in *primary* reality.

Thus in receiving the Eucharist, the continuous tradition of the Church tells us that we receive the Lord *substantially*, in his primary reality, which implies that the primary reality of bread and wine have been converted by the Lord's creative Word into himself, so that we too, by consuming them under these sacramental appearances, which remain unchanged in order to perform their symbolizing function, begin to be transformed in him, incorporated ever more profoundly into his body the Church.

But is it not nonsensical to speak of this bodily presence of the Risen Lord in our space and time when He is also present "in heaven", and when obviously he does not exist in any determinate spatial relation to the dimensions and positions of the bread and wine? To put it naively (and we need to keep hold of our naive sense of the reality of things bodily, their tangibility), how can an adult man be present in a small bit of bread?

Obviously we are talking here, just as when we talked about how the risen body could be truly a "spiritual body", about a state of affairs beyond our experience, to be understood only by analogy. Yet I trust that my discussion of a possible understanding of the nature of the risen body has made clear that the substantial reality of the human body is not to be identified with its various states, but with that "information" which was present in the genetic code at the first moment of our conception and in our human memory, by which the historical development of the unique identity of each person reaches it full actualization. It is that "information" which survives death in the spiritual part of man, but which again requires embodiment in the resurrection in some matter, although in a much more actualized matter than we possess in our present mortal state.

Hence it should be maintained that the *substance* or primary reality of the Risen Christ is not as such located in any time or space, but becomes so located by relating itself to particular times or spaces. Yet Luther's theory that Jesus is present in the Sacrament by reason of his divine omnipresence is insufficient, since for his humanity as such to be really present in a place it must have a special relation to that place.[97] Thus the Risen Christ entered into the room with the apostles by taking on such a special spatial relation. In the Eucharist, however, the localizing relation is established not by occupying a volume of space but by being signified as present by the continuing sign of the phenomenal bread and wine, a signification which they have acquired by the historic words of Christ (or some equivalent of them) repeated by the priest acting *in persona Christi*. Hence in the Eucharist the Risen Christ is bodily present, offering himself to his people sacrificially, yet his bodily presence is of a unique, sacramental kind — total presence through a bodily sign, a total presence which is itself noumenal, substantial, bodily, but not phenomenally so.

The fact that the Eucharistic Prayer of the Church effects this real, earthly presence of the Risen Christ in his sacrificed humanity by an ontological, substantial conversion, and thus also effects the real tranformation of sinful humanity into the Church, the Body of Christ, as the hidden mystery accessible only to faith, and of the transformation of the entire universe, leads us to consider more precisely what this transformed universe will be like.

2. The Trinity Centered Universe

In the text of St. Paul already quoted we read, "When finally, all has been subjected to the Son, he will then subject himself to the One who made all things subject to him, so that God may be all in all" (1 Cor 15:28). In some oriental religions the conception is that as the phenomenal world has emanated from the One, so at the end of every cosmic cycle it will be reabsorbed into the One. In the neo-Platonsim of Plotinus all things emanate from the One and return to it by contemplation in a timeless process.[98] Most Humanists would probably accept the inevitability of the ultimate entropic death of the universe, although some may accept a kind of cyclic eternal return. When St. Paul, however, says "that God may be all in all" he is not thinking either of absorption of the universe in God nor its annihilation.

Because God has *created* the world by a free act of love, he has created it to be itself, yet to share as fully as is possible for something necessarily finite in his infinite life. Unlike the universe of Plotinus, moreover, the Christian universe is essentially historical. It moves through time to a goal which transcends time, but, as I have shown in discussing the meaning of the resurrection, its history is perserved in eternity and constitutes, as it were the completed text by meditation on which the infinite depths of Eternity will be forever explored.

Eternity, the immortal Now is, of course, God Himself. Only in God is there a plenitude of Life which is gathered together in absolute simultaneity where there is no before or after, no spreading out of existence, no undershadow of potentiality or unrealized possibilities, because he is pure Act, pure Existence. Recently a number of theologians have proposed to substitute for this traditional notion a *di-polar* God who has an antecedent and a consequent nature, the former an unordered collection of eternal possibilities, and the latter the undying memories of all these possibilities as they have been realized in the endless history of the universe.[99] Such a God, it is argued, is not remote from the universe but immanent in it, forever actualizing himself by inducing the universe to actualize more of the possibilities which lie in his antecedent nature so that he can become conscious of them. But the Christian God is the Now which is the *future* of his creation, drawing it out of the darkness of nothing into the full light of his eternal presence. He is never remote from his creation, because in creating he communicates his very life to the creature, "Through him all things came into being and apart from him nothing came to be. Whatever came to be in him, found life" (Jn 1:3-4a).

Since the economic or historic Trinity is the manifestation in time of the ontological or inner Trinity, the New Testament interprets human history as a movement of creation toward the full revelation of the inner life of God. In the Old Testament God was experienced as Creator and Lord of History. Indeed, some exegetes would say that He was first known as Lord of History, the God of the Exodus who made himself known to the Jews as their Liberator, and from this encounter they came to see that he was also the Creator, the One and Only God.[100] As such he was understood as the absolutely First or Unbegotten, the Principle or Source of all that is, and as such having no name but Yahweh, a name so great it could not even be pronounced, and meaning simply the I Am (or according to others the One-Who-Causes-to-be or "I am what I am" or "I am the one who is truly God") who cannot be imagined or represented in any way. Yet He is ever present in every event of history, all-knowing and all-caring, and filled with anger at every injustice done to his people.[101]

The Old Testament is also concerned to understand how this hidden, yet ever present God is manifested in His world, and hence there arose the notion of Wisdom, not as a second God nor yet merely as another name for God himself, but as a personification of his presence in creation. In Chapter 11 I showed that Wisdom is given a feminine personification to indicate the covenant of love between God and his universe, and yet to indicate that the universe is not a mere emanation from God, but stands before him as a covenant partner. Because in fact the universe has become subject to injustice, to sin, an injustice which extends from the angels to man and has frustrated even the guiltless sub-human world, it needs to be redeemed, restored and brought to its originally intended fulfillment, which is called both in the Old and New Testament the Reign of God.

Jesus as a Jew worshipped this One God, Creator of the Universe and Lord of History, but he experienced with this God that unique Sonship which permitted him to call God "Abba", and to proclaim that the Reign of God was actually beginning in himself in whom the prophetic promises of God's coming personal presence in the world were fulfilled. Luke presents him as making this claim in the synagogue of Nazareth, soon after in his baptismal vision he had heard the voice of God saying "You are my beloved Son. On you my favor rests" (3:22), and seeing the Holy Spirit descend upon him as a dove, symbol of the Chosen People to whom he was being sent, and also soon after the temptation in the desert when he began directly to confront the evil cosmic powers who actually dominate the world and prevent the coming of God's Reign. After reading the scroll of the prophet Isaiah in which the coming of the one who by the power of God's spirit will "announce the year of favor from the Lord", when every injustice will be righted, Jesus says, "Today this Scripture passage is fullfilled in your hearing" (4:21). Such a claim astonished his hearers, and as he continued to make it quietly but firmly throughout his preaching, even in Jerusalem, it led finally to his condemnation for blasphemy and his abandoment to the Romans to be crucified. He made this claim not to aggrandize himself or even to call attention to his own identity, but precisely to assert that the promises of God were being fulfilled and the Reign of God beginning, not in the way so many had expected them, by the political liberation of the Jews and their rise to world empire under the Messiah as king, but by a spiritual transformation of the world initiated by the revelation of God in his Son and in his Spirit.

The reconciliation of this Good News with the radical monotheism of the Old Testament, which Jesus himself accepted, was of course utterly mysterious; and we can hardly be surprised that the scribes and Pharisees who were the accredited theologians of the time (Jesus himself said, "The scribes and the Pharisees have succeeded Moses as teachers; therefore, do everything and observe everything they tell you" Mt 23:2), were unable to accept it. Mark has Jesus say to them that somehow their theology is faulty since they cannot explain why in Psalm 110, David addresses the Messiah as "Lord". "If David himself addresses him as "Lord" in what sense can he be his son?" (Mk 12:37), implying that the Messiah is "Lord" in some greater sense than they have imagined. In the Fourth Gospel Jesus is presented as also arguing from Psalm 82 in which the judges of Israel are called "gods", that his own claim to be God's Son ought not be *prima facie* rejected as blasphemy without a more open-minded effort to understand a prophetic announcement, since he has proved by his miracles that he has a right to speak as a prophet in the name of God (Jn 10:31-39).[102]

The early Church itself had to continue to struggle with this paradox through the first five hundred years of her existence until the Trinitarian dogmas had been decisively formulated, and Unitarianism was again to appear with certain sects in the Radical Reformation and to be renewed with the deism of the Enlightenment. The original Unitarianism of Faustus Socinus (d. 1604) accepted the Reformation *sola Scriptura*, and based itself on the fact that the New Testament presentation of the Lordship of Jesus fails to reconcile this with monotheism in any systematic way.[103] The closest approach is in the Fourth Gospel, which is written in such lofty and shifting "mystical" language that the matter of fact mind feels compelled to regard it as so ambiguous as to require sober interpretation. Today, however, most Unitarians would no longer affirm the inspiration or inerrancy of the Bible, and would admit that the New Testament

669

exhibits numerous Christologies some of which are high and some low, and would prefer to demythologize the high Christologies as products of a mythologizing age.

Yet if we accept the Fourth Gospel, as did the early Church in canonizing it in its preaching and teaching, we find that without contradicting either the synoptics or St. Paul, it does provide the grounds for a reconciliation between the two apparently conflicting aspects of Jesus' own teaching: his acceptance of the One God of Israel and his claim to be the unique Son of that God through whom his kingdom begins on earth in the power of the Spirit.

The Fourth Gospel in its famous prologue (which may be from a source antedating the final edition of the Gospel[104]), shows that the Word of God has always been present with God, is in some sense identical with God ("the word was God", but without the article used with God in the ordinary Old Testament sense), yet distinct from God, since he is "with" God and creation and redemption take place "through" him. It is the Word which "became flesh" and whose "glory" the Church has witnessed, "the glory of an only Son coming from the father, filled with enduring love" (1:14). Later in this Gospel, Jesus, this Word made flesh, repeatedly uses of himself the divine name of the "I am" (8:24, 28; 8:58), and asserts that "The Father and I are one" (10:30), and that "the Father is in me and I in him" (10:48) and that "whoever has seen me has seen the Father" (14:9), although he also says "The Father is greater than I" (14:28) in explaining why he must return to the Father. Finally, Jesus promises that when he returns to the Father, at his intercession the Father will send the Holy Spirit to take his place as teacher, who will guide and empower the Church, and who evidently dwells with the Father as the Paraclete or Advocate of the Church (Jn 14:26), as Jesus is himself also its Advocate (14:12-13).[105]

When we turn back from the Fourth Gospel to St. Paul we find in the text from *1 Corinthians* with which this section began, the prophecy that at the end of history "all things "including" every sovereignty, authority, and power", (which for Paul includes all the angels) will be subject to the Risen Christ who will then "subject himself to the One who made all things subject to him, so that God may be all in all." In other words, Paul sees Jesus as Lord of the universe, yet subject to the Father who has "made all things subject to him [Jesus]". This of course raises the question by what right Jesus has become Ruler of the Universe as the vicar, as it were of God. The Pauline answer of course is that he has won the right to this by his obedience even to the Cross,[106] but this still does not say why the obedience of one man should merit the salvation of the world.

670

The great hymn of *Philippians* 2:6-11, quoted by Paul adds that "God highly exalted him and bestowed on him the name above every other name", because "Though he was in the form of God, he did not deem equality with God something to be grasped at" but "being born in the likeness of men" "he humbled himself, obediently accepting even death on a cross", i.e. it was because he was by right God's Son, yet humbled himself for the salvation of others, that he is now exalted to the Father's right hand. Finally, in a parallel hymn in *Colossians* 1:15-20, it is said of Jesus that "He is the image of the invisible God, the first-born of all creatures. In him everything in heaven and on earth was created, things visible and invisible, whether thrones or dominations, principalities or powers; all were created through him and in him".

Although some doubt that St. Paul wrote *Philippians* and especially *Colossians*, these epistles certainly represent developments in the Pauline tradition, and the two hymns in question are earlier than the epistles themselves.[107] Thus the Johannine community is not alone in its conclusion that the exaltation of Jesus as Lord implies that he is Son of God in his very person, and this is why his self-humiliation and sacrificial death has its universal significance. As for the Holy Spirit, we have already seen that Luke supports the Fourth Gospel in its teaching that, after his ascension to the Father, Jesus promised to send the Spirit to take his place, which can only imply that the spirit is equal to the Son and like him has dwelt in eternity with the Father.

Thus it is necessary to conclude that the New Testament writers, although they always begin from the "economy", speak of the Three Persons in terms that are incomprehensible unless the Son and the Spirit exist with God in eternity from before the creation of the world till after the end of history, and that God as Father has shared with them the totality of his power over the creation. In the functional mode of thinking characteristic of the Bible, according to which "as a thing acts, so it is", this can only mean that the Son and Spirit who share the power of God in its totality, although they are somehow *from* or *with* Him, are Divine without qualification. As John 1:1 says of the Word, it was "with" *the* God" (i.e. the one God), and it "was God" (without the article).[108]

Thus the One God in eternity is accompanied by his Son and his Spirit who share totally in his power, yet are somehow distinct from him. Why does this not contradict monotheism? The only answer that the New Testament gives (and it suffices for those who are willing to accept the word of Jesus on the basis of his life and his resurrection witnessed by the Church), is that Jesus experienced in his lifetime the reality of his distinction *from*, yet identity *with* the Father. He experienced that he was truly the Father's Son, yet by a relation which surpasses that of human sonship as the heavenly Father surpasses in fatherhood any human father. "Do not call anyone on earth your father. Only one is your father, the One in heaven" (Mt 23:9). Jesus' experience of this distinction and identity in earthly life was completed in his risen life. He knew and revealed to his apostles that the Father had given to him *all* the life the Father has, that eternal, uncreated life to which alone belongs the name "I am." "God highly exalted him and bestowed on him the name above every other name" (Phil 2:9).

Notice that the problem here is not really that of the "pre-existence" of the Son, as it is usually termed, but of the *co-eternity* of the Son with the Father, a notion difficult to express in biblical language which is so resolutely historical and narrative.[109] It is put plainly enough by the author of the *Epistle to the Hebrews*:

> In times past, God spoke in fragmentary and varied ways to our fathers through the prophets; in this, the final age, he has spoken to us through his Son, whom he has made heir of all things and through whom he first created the universe. This Son is the reflection of the Father's glory, the exact representation of the Father's being, and he sustains all things by his powerful word. When he had cleansed us from our sins, he took his seat at the right hand of the Majesty in heaven, as far superior to the angels as the name he has inherited is superior to theirs (1:1-4).

The common view is that the author of this work was not Paul, but that it certainly must be dated before A.D. 96 (since it is quoted in (*I Clement*), and possibly before A.D. 70, since he speaks as if the Temple is still standing without any reference to its destruction.[110] In this passage we see (1) the coming of the Son has been foretold for many ages: (2) his coming marks the "final age"; (3) this Son has received the whole universe as his inheritance, evidently by reason of his obedience; (4), but it is rightfully his because all things were created by the Father "through" him in the first place; (5) the Son is the perfect image of the Father; (6) and continues to be operative as the one who "sustains" the universe; (7) it is he who has cleansed the Church from sin in his earthly life; (8) and has now taken full possession of his rightful inheritance with the Father; (9) because his "name" (i.e what he really is) is superior to all created things. Thus the earthly, temporal life of the Son (the economy), is seen as a manifestation of his *eternal* relation to the Father with whom he totally shares the same life and power.

What is said of the Son must also be said of the Spirit, because both come from the Father, and the Spirit in the economy fills the same role as the Son in teaching and empowering the Church, but while the Son "became flesh", the Spirit abides in the Church as in his Temple, as its life principle and source of its holiness and its active gifts (1 Cor 6:19:12). Exegetes have often been troubled that St. Paul sometimes speaks as if the Son and the Spirit are the same ("The Lord is the Spirit" 2 Cor 3:17). Paul often, however, also uses triadic formulae (e.g. 2 Cor 1:21-22). Thus he certainly distinguishes the Holy Spirit from the Son, but does not hesitate to speak of both in the same terms, except that the Spirit is never identified with the earthly, suffering Jesus. Back of all this is the Old Testament notion of Wisdom as the epiphany of God in the world. For the New Testament, this supreme epiphany is in Jesus Christ and then in the Spirit whom he has sent to the Church, so that this Wisdom is now distinguished into the Son and the Spirit, to both of whom the attributes of the Old Testament Wisdom are attributed, although in different ways. There is then never any doubt that both the Son and the Spirit are related to the same God as the ultimate and sole Principle of all things; but finally there is the conclusion that God has shared his *total* life and power with the Son and the Spirit so that they exist with him in a way that transcends the before and after of time, in eternal life. The economy, therefore, becomes an epiphany of the Eternal and Uncreated into our temporal world for its redemption from sin and its return to what God had originally created it to be:

> Jesus, who was made for a little time lower than the angels,
> that through God's gracious will he might taste death for
> the sake of all men (Heb 2:9).

The Sonship of Jesus (and the equivalent role of the Spirit) can be reconciled with radical monotheism by recalling that since in the Old Testament the Wisdom of God is described as distinct from God yet as his epiphany in creation, therefore it should be possible also to affirm that his Wisdom is distinct from him in eternity, without derogating from his absolute primacy and unity, because the Son and Spirit have nothing they have not received from the Father, yet they have received from him the total plenitude of his being. That such a total communication by God of himself in eternity is possible is, of course, absolutely beyond human comprehension, but the Jewish faith in God has always acknowledged that he exceeds all human comprehension. We must simply adore him and cry out with the angels and all creation, "Holy, Holy, Holy" (Is 6:3).

This mystery beyond human understanding is however, a light by which we must finally understand our universe and ourselves. "In your [uncreated] light shall we see [created] light" (Ps 36:10) by which we see everything else. It's essential meaning is that God is a God of the *communication* of his existence in truth and love. His absolute transcendence and otherness must, therefore, never be understood as if it implied his isolation or aloofness from the universe, or his jealousy of his perogatives as the One. When the Old Testament portrays him as saying "I am a jealous God" (Dt 4:24) the meaning is that God is jealous of the love of his bride, his people, and cannot tolerate that she should love other false gods, but not that he is unwilling to share his life with all who will receive it. Just the contrary, it is because he wishes that all should come to share in his glory that he cannot endure that they should seek life where it cannot be found — in delusion. The cunning of the serpent in saying to Eve, "No, you shall not die; for God knows that if you eat [of the forbidden tree] you will be like God" (Gn 3:4), was precisely that God forbade us *that* tree of death in order that we might eat of the other tree of life and share in his (God's) immortal existence, divine wisdom, and unending love.[111]

If we now return to Jesus' Abba experience, and ask how it could be possible that "the Word became flesh" (Jn 1:14), we are faced with the difficulties raised by some contemporary theologians about the Chalcedonian formula, namely that "one and the same Son and only-begotten Word of God, Jesus Christ is "true God and true man", consubstantial with the Father in deity and consubstantial with us in humanity", and thus dual in "nature" but not in "person"[112]. Nevertheless, now we can perhaps see better why this formula, however it may have been influenced by the thought-forms of its own time, is completely consistent with the New Testament faith and by no means meaningless to our own age and its understanding in the light of modern science and psychology of what it is to be human.[113]

The Trinitarian understanding of God as a God of communication tells us that God is not only personal, but personal in such a way that even within his eternal life there is an eternal process of communication and love which constitutes a community of persons, each of whom possesses the totality of the divine life so that this life is not divided among them (then it would not be totally given), but remains identically one life in the Father as Principle, in the Son as God's Wisdom, and in the Spirit as God's Love. Since the Father is Person, so are Son and Spirit each a Person because the Father has given them all that He is and all that they are, and these Three Persons are one God because they are distinguished from each other only by *relations* of communication, a communication which does not separate but absolutely unites them. Finally, these "relations" are not something added to the Persons, but are their very Personhoods, since these are Persons characterized precisely by their being for *other*

674

persons: the Father being father because he is the Principle of Son and Spirit, the Son being son because he receives all that he is from the Father, and the Spirit being spirit because he too receives all that he is from the Father but in a way that unites the Father to the Son is mutual love so that in this order of communication he is the *Third* Person, not the Second.[114]

To speak of the Father, Son, and Spirit as "persons" is not literally biblical, but it preserves the fundamental biblical truth of Jesus' experience, since for him the Father was a father and he was a son. In modern language no term but "person" can preserve that sense of two existents who know and love one another in such a way that they confront each other face to face in order to be united with each other ever more perfectly through the communication of life, of understanding, and of love. The very term in which this word "person" originated, *prosopon* or "actor's mask", indicates the confrontation of two *faces*, and it indicates first the dilemma that the face of another "masks" who that actor really is; it is a surface hiding an abyss; but it also indicates that the actor speaks *through* the mask to express, reveal, and *communicate* thought and feeling, an inner life. All of us experience this dilemma of face to face confrontation with another person in which we feel our own personhood most keenly. His or her face is both a concealment and a revelation; it distinguishes us, yet it unites us. The Jews conceived of God in a most intensely personal way, yet his face remained hidden from them by its terrible, blazing light, to see which might mean death. In some sense Moses had "seen God face to face" on Sinai in thunder and storm (Ex 33:11), but this was imperfect, since God says to Moses "You may see my back, but my face is not to be seen" (23). Jesus, however, as the final prophet, experiences God in absolute intimacy, an intimacy which surpasses even that of Adam, since *Genesis* represents God as "speaking" to Adam but never describes Him. For Jesus, on the other hand, while the visions which we discussed in the last chapter also speak of the "voice" of the Father, they also describe the "heavens as opened" and the "Spirit descending and remaining" on the Beloved Son, thus proclaiming that no barrier remains between the Father and Jesus, although the mention of the "voice" perhaps suggests that this intimacy on earth would still reach its full manifestation only at the Ascension.

In discovering himself as unconditionally the Son of God, Jesus, therefore, discovered that as *this person*, he stood in relation to God as Son to whom all the Father's life and power had been communicated. Such a total communication, as we have noted the New Testament writers came to realize, could only be *eternal*, so that Jesus found himself as person the Eternal Son. Did this mean a disappearance of his own consciousness that he was human? By no means. I have stressed through this book that we do not discover ourselves by first looking at our subjectivity, but at our body. I am who I am first of all because I have a body that feels itself to be itself and to be over against, yet in contact with other bodies. It is only little by little that I come to see that my body is special in that it is not an unfeeling body like the earth and the rocks or the wind and the air, but a living, feeling body like the animals, yet different from the animals in that it is a speaking and therefore thinking body; and that finally, as Adam, I come to recognize other human beings as other than myself, yet able to communicate with me.

Jesus knew himself as *flesh* first of all, and then as human and then as *more* than human, yet always flesh and always human. In discovering himself in relation to God, he discovered as we all discover that we are not only human, we are *creatures*. Our being is a gift, not something we made, and yet it is a gift which we now possess as ours, yet only as God sustains us. Jesus found himself to be related to God in a human creaturely way, but also uniquely as the Son of that God, whose humanity had been given him that he might be the way to God for others, as Adam must have come to see that he had been given life to be the parent of all who were to come from him. I would think, with Karl Rahner,[115] that in thus discovering who he was, Jesus realized that he had somehow always known this in the depths of his being, as each of us, as we discover ourselves growing up, realize that since we first knew anything, we knew who we were without knowing it explicitly. Jesus had always known from the dawn of his memory that he was Son of God and yet had to learn the full meaning of this fact through human living and suffering. Even in dying on the Cross, there remained a darkness that was lifted only when he rose again on the third day and ascended to the Father whence he had descended, had "become flesh." "He who descended is the very one who ascended" (Eph 4:10).

The modern difficulty that the Chalcedonian formula cannot account for, the fact that "two natures" seems to imply "two centers" of consciousness, which the "one person" seems to contradict, disappears if we consider that in Jesus, as in us, there is one center of consciousness surrounded by several concentric circles, or levels of awareness.[116] At the center of my consciousness is a very dim and hidden awareness that "this is me" surrounded by several levels of subjective and objective awareness, terminating in the sense of my own body touching other bodies. From time to time, one or the other of these levels becomes activated as the focus of psychic energy and attention, but they never lose their roots in my ultimate sense that "this is me." In Jesus this "me" was precisely the sense that he was the son of the Father become "this flesh".

676

As he was the Second Person of the Trinity, in his divine knowledge of himself was included the eternal knowledge that he was willing at a certain time in history, in obedience to the Father's will, to become flesh, to become fully human in history. This divine consciousness was thus at the center of Jesus's human consciousness and included all the levels of his human consciousness, so that at the Last Supper it was God who said, "This is my body given for you", and who on the Cross said "I thirst", and "My God, why have you forsaken me", that is, "Father who are always with me in eternity, how can you leave me to die in time?" On the Cross it was really God who died for us in his humanity, while always living in his divinity. If, *per impossibile*, he had died also in his divinity (as the Death of God theologians claim), this death would not have been so completely sacrificial as it was, but would have been in part the inevitable capitulation to death that all of us must undergo. The marvel of the Cross is that it was truly the Divine Son, living eternally, who with absolute freedom willed to permit that revelatory death of his perfect human nature, which was just as truly his nature as was his divinity.

I believe that the ultimate stumbling block for many theologians today is that it is so hard for us to be convinced that the Son truly *became flesh*, not in a dualistic sense of putting on human nature as a garment, but in the fullest sense that the Divine Son made human nature truly and unqualifiedly *his* nature, his mind, his body as truly as his divinity is his, and that both must remain so. In the back of our minds sticks the notion that when we worship the Lord in prayer, especially in the Eucharist, that what we worship is the Divine Son merely hidden in the garment of human nature. Or for some theologians of unitarian tendencies, that what we really worship is God, announced to us by Jesus. But what we worship is the Divine Son made flesh. "In Christ the fullness of deity resides in bodily (*somatikōs*) form" (Col 2:9) In Jesus God is a body!

3. The Transformed World

Holding firmly in faith to that fact, that God the Son is God the man, and that in his risen *body* he has become the head of the whole material and spiritual creation, the doctrine of the redemption of the universe follows. This redemption is what Jesus in all his preaching called "the Reign of God" and it began with him through his Church empowered by the Holy Spirit on Pentecost Day. The great objection of the Jews against the identification of Jesus as the Messiah is that in the Messianic Age, justice and peace will reign on earth, yet while Jesus came and went, holocausts continue to occur. It seems to me that we must take this objection of our Jewish brethren with absolute seriousness.[117] Jesus is not the Messiah if his preaching two thousand years ago did not truly inaugurate justice and peace on earth, a justice and peace which must begin with our daily earthly and bodily realities.

Yet, as I showed in Chapter 10, Jesus was completely in the tradition of the prophets when, as in the Sermon on the Mount, he taught that the coming of this reign must be first from within, from a conversion of the heart. At the same time he pointed out that the leaders of the people should not lay heavy burdens of morality on the weak without making sincere efforts to help them carry these burdens. Thus the kingdom of justice and peace cannot come by an apocalyptic act by which evil is destroyed and righteousness created, but only by a human cooperative effort, empowered by the Spirit, of obedience to the Law of God in which the leaders of the people change social structures so as to make individual conversion easier through example and encouragement. Jesus himself gave the example of this by refusing to assume dominative powers, or resorting to violent revolution (with its establishment of a new tyranny), which so much appealed to his times as to ours. On the other hand, he did not preach a merely individual conversion, but the change of the whole social order, without which individuals find the life of virtue so exceedingly burdensome.

The first example and witness of this transformation of the earth is the Christian community, the Church. Jews, thinking of the many persecutions, deportations and ghetto imprisonments throughout history which the Church has seemed to support, find it a mockery to speak of the Christian Church as a sign of the coming of the Messianic age. Yet the Holocaust and the continuing anti-semitism of the Soviet Union are beginning to expose the fact that it has been Satan who was the inspirier of anti-semitism under the guise of a revenge for the death of Jesus, who sought no such revenge and forbade it to his followers. The Holocaust, far from being the work of the Church, was the work of those forces which rejected the non-violence of Christ in the name of human pride and violence, a pride and violence which had deluded certain church leaders (among them great saints) to think as Peter did in the Garden of Gethsemane that the Church can perform its mission with the use of the methods of empire, only to be rebuked by Jesus, "Put back your sword where it belongs. Those who use the sword are sooner or later destroyed by it" (Mt 26:52). It is well known that this refusal of force in its mission has always been a teaching of the Church, although the Church has not been consistently faithful to it in practice. The death of so many Christians along with the Jews in the Holocaust led in Vatican II to a firm rejection of the anti-semitic delusion so contradictory to the real nature of the Church.[118]

Does this mean, therefore, that Christians must withdraw from political and other earthly activities which necessarily involve the acquiring of power and its use to prevent evils and to implement ideals? The New Testament assumes that the Christian community will not withdraw from the existing social order, but will live in it in spite of the morally mixed character of its institutions. St. Paul's advice to the Christian slave-holder Philemon on sending back to him the run-away slave Onesimus (who, be it noted, went voluntarily), expressed the Christian spirit in its astonishing "impracticality":

> I had wanted to keep him with me, that he might serve in your place while I am in prison for the gospel; but I did not want to do anything without your consent, that kindness might not be forced on you but might be freely bestowed. Perhaps he was separated from you for a while for this reason: that you might possess him forever, no longer as a slave but as more than a slave, a beloved brother, especially dear to me; and how much more than a brother to you, since now you will know him as a man and in the Lord (Phlm 1-16).

This impracticality aims at the radical change of every social institution by the witness to and practice of a new conception of what it is to be human, of the dignity of the human person which mandates not only that all are brothers and sisters, but *slaves* to one another.

After two thousand years, this conception is still new. Slow as its reception has been in the Church and after that in the world, it nevertheless has had its profound effects, so that even the enemies of the Church find no better argument against her than her inconsistencies in implementing it fully, and find it necessary to build their own secular religions on some declaration of the rights of man, or to raise the slogan of revolution in the name of the abolition of every form of slavery.

As I have argued earlier, there is no guarantee in the Gospel that this slow progress will finally achieve a breakthrough and attain a universal kingdom of peace and justice on earth by human effort. Only Jews who have assimilated Enlightenment faith in inevitable progress interpret the coming of the Messianic Age in this sense. For Christians and Orthodox Jews alike, it remains a possible and mandatory human task, under the favor of God, to prepare the way for the Messiah by an earnest striving to transform the earth according to the Law of God, but both know that the coming of the Messiah can ultimately be the work of God alone, and that it will occur even in spite of human failure.

Does this mean, therefore, that Christians lack the powerful motivation to work for social justice that Humanists and Marxists have, for whom this is humanity's only hope and sole responsibility? This is the constant accusation against the Church, that it is "the opiate of the people", encouraging passivity by waiting for God to do it all. But what is the strength of the motivation provided by Humanism and Marxism, which can only promise the possibility of justice in the remote future after this generation and many others have been sacrificed to a distant dream? Such motivation has proved effective only when the people were whipped up to believe the revolution was immediately at hand, or had their hopes raised for a time by an influx of new technology which produced a higher standard of living, before the cost of these "advances" in violence, misery, and depletion of the environment were exposed. Christianity experienced such millenial expectations in its early years, but has proved that it can survive when such hopes are again and again destroyed by the stubborn resistance of the evil forces in the universe and the hardness of the human heart.

Nevertheless, we must keep before ourselves the real possibility offered by the Gospel in its power to transform the human heart that we can achieve universal peace and brotherhood, and by the wise stewardship of the earth and the creativity with which God has endowed us we can come closer and closer to the reign of God on earth. The possibilities of modern technology and of therapy for the ills of the human psyche which science can make available to humanity are not only immense, but exceed anything we now imagine. There is not one single human evil that is not in principle able to be overcome by the right use of human creativity, except the evil of the human will; and for that the Gospel provides the total cure, not merely by individual faith, but by the faith and love and prayer of the Christian community.

The greatest obstacle to the acceptance of the Gospel at first sight seems to be the existence of rival religions, including Humanism and Marxism, but since in these too the Holy Spirit is at work drawing all human beings toward the light, the Gospel also has the power to reconcile these in a community of love where all share without envy.[120]

Let us imagine, however, not the worst but the best scenario in which the Gospel is preached to the whole world, accepted and lived in a Eucharistic community where Jesus Christ is truly King over the whole human family, and there are no other "kings", but only people who serve one another. By the wise use of technology, earth has become Paradise once more, no longer as in *Genesis* a watered oasis in a desert, but a garden-earth. The human family has world government, but a government where

680

the principle of subsidarity is carried out, so that there are many cultures adapted to the different natural regions of the earth and perserving each one its long history, yet constantly sharing with each other their special gifts. The Church is distinguished from the State in the sense that it is primarily contemplative, directed to awaiting for the coming of the Lord in meditative anticipation, while the State is concerned with the active affairs of earth; yet both live by the Great Commandment of love and service. The population of the earth is controlled by the fact that many prefer the celibate state in order better to live contemplatively, and only those marry who have a special vocation for the raising of children.

Because the length of life has been greatly increased and old age freed of disease, many older people become contemplatives. The chief employment of human resources is no longer in the production of material necessities, but in scientific research for the sake of understanding the world and the human self more profoundly and thoroughly, and thus achieving a fuller and richer idea of the true humanity in Christ Jesus in whom all creation is summed up and through whom alone the Triune God can be better known. If death comes after a long life, it is a peaceful rest in anticipation of the resurrection. Because most people have been aided by the good society in which they live to have achieved perfection in this life, most will immediately pass to the vision of God and need only wait for the reconstitution of their bodies and the completion of the number of all those who are to see God for the consummation of their life's journey.

This earthly City is filled with great activity, because until now the human race had hardly begun to use its creative talents. [121] The wonders of the universe and the drama of history are being opened up in a way that makes previous ages seem dark and ignorant. The question of the existence of other humanoids elsewhere in the universe is being solved, and the world of the angels is becoming accessible to human communication because a perfect humanity has become open to their helpful influence. With this ever increasing sharing of cosmic truth, the human vision of God is no longer abstract but concretely rich, profound, the bright dawn of the eternal beatifying vision.

A remnant of evil remains (the Anti-Christ), as once there was only a remnant of the good. Its source, however, has been exposed in the ever increasing fury of the devils against the City of God they see being built on earth. Consequently, the security of this City as it awaits the coming of its Lord still requires the ceaseless vigil of the contemplatives, who through their prayers will be enlightened concerning the strategems of the evil angels and will warn their fellow citizens to be on guard.

In due season, the Lord will come to bring history to its joyful end, its Sabbath rest. All the dead from the beginning of human history will appear in their transformed bodies and form a single community in the praise of God. Those still living will be transformed with them. The earth on which the supreme historic event of God's appearance in the flesh first took place will not be destroyed, but it too will be transformed into a place of light in which the bodies of all human beings will be spiritualized as the highest form of matter, in patterns of light retaining the whole past history of the universe. In this mirror of the past, humankind, and with them the angels (who have been awaiting the final revelation of God's creative and redemptive love[122]) will at last share with Jesus the final illumination which came to him in his resurrection and ascension. In the mirror of the past the blessed will begin to see God, the eternal Future, shining forth in the Trinitarian community which will be revealed as the center of the entire material and spiritual universe.

Entering into this Trinitarian Community, created persons will find that there will unfold for them an unending abyss of new wonders. Yet the history of the world will also remain forever preserved in the risen bodies of human beings, and in the transformed universe through which they will know themselves in eternal dialogue with one another in the Spirit, together with the Son before the Father, who will be "all in all."

In this scenario, whose actualization depends on the openness of our free wills to God's grace, it can be truly said in answer to our original question, "Can we create ourselves?" that yes, we can create ourselves as faithful stewards of God's gifts, who by His grace have shared in the bringing of His universe to its surprising completion, like the novel and unpredictable resolution of a uniquely great symphony in which the most daring discords have found their perfect resolution.

Notes

1. Hans Küng, *Does God Exist? An Answer for Today* (Garden City, N.Y.: Doubleday, 1980). See also James Thrower, *A Short History of Atheism* (London: Pemberton Books, 1971); Hans Urs von Balthasar, *The God Question and Modern Man* (New York: Seabury, 1967), and Claude Tresmontant, *Les problèmes de l'atheïsme* (Paris, Seuil, 1972), and Patrick Masterson, *Atheism and Alienation: A Study of the Philosophical Sources of Contemporary Atheism* (Notre Dame, Ind.: University of Notre Dame Press, 1971). For the case for atheism see Anthony Flew, *God and Philosophy* (New York: Harcourt, Brace and World, 1966) (discussed by Masterson p. 118-123); Kai Nielsen, "In Defense of Atheism" in H. E. Kiefer and Milton K. Munitz, eds., *Perspectives in Education, Religion and the Arts* (Albany, N.Y.: State University of New York Press, 1970), pp. 127-156 and Ernst Bloch, *Atheism and Christianity* (New York: Herder and Herder, 1972).

2. "To sum up: What are we pleading for? We are pleading with Descartes and his followers decidedly for *critical rationality,* but with Pascal and his followers equally decidedly against an *ideological rationalism",* p. 124. So far so good, but when Küng finally comes to his own argument for the existence of God, he rejects any "logically conclusive proof" and appeals to a "free decision" based on a "fundamental trust in reality". This trust, he argues, is "rationally justified" by an "inward rationality, which can offer a fundamental certainty. In the accomplishment, by the 'practice' of boldly trusting in God's reality, despite all temptations to doubt, man experiences the reasonableness of his trust, based as it is on an ultimate identity, meaningfulness and value of reality, on its primal ground, primal meaning, primal value" (p. 574). This seems to me to be Pascal's wager in a more wordy form: *If* we freely choose to believe in God's existence, we reap pragmatic advantages greater than from choosing to live as atheists.

3. On Romans 1:20, see Karl Barth, *The Epistle to the Romans,* 6th ed. (Oxford: Oxford University Press, 1933), pp. 45-48, who argues that what the nonbelievers can know of God is His "invisibility"! For exegetes who (for the most part) admit some genuine knowability see John Murray, *Epistle to Romans,* New International Commentary, Grand Rapids, Mich: Eerdmans, 1959, pp. 20-42; William Barclay, *Letter to the Romans,* Philadelphia: Westminster, 2nd ed., 1957, pp. 19-20; Ernest Best, *The Letter of Paul to the Romans* (Cambridge: Cambridge University Press, 1967), pp. 20-21 (God reveals himself through the creation); D. M. Coffey, "Natural Knowledge of God: Reflections on Romans 1:18-20", *Journal of Theological Studies* 21 (1970): 674-91; Matthew Black, *Epistle to the Romans.* New Century Bible (London: Marshall, Morgan and Scott, 1973), pp. 50-51 (God reveals himself); Ernest Käsemann, *Commentary on Romans* (Grand Rapids: Mich.: Eerdmans, 1980), pp. 38-44; W. D. Davies, *Paul and Rabbinic Judaism* (Philadelphia: Fortress Press), 1980, pp. 27-29.

4. Küng, *Does God Exist?* (note 1 above), pp. 24-41, correctly criticizes Descartes (*ibid.* pp. 24-41) for his mathematicism and identification of truth with "clear and distinct concepts." He also point out that the *cogito ergo sum* can easily turn into the *cogito ergo non sum* of existential despair when the abstract clarity of thought is confronted by the dark obscurities of the concrete human condition (p. 39). But he fails to engage in that criticism of the *cogito* itself as the starting point of philosophy which Cornelio Fabro, *God in Exile: A Study of the Internal Dynamic of Modern Atheism from its Roots in the Cartesian Cogito to the Present Day* (Westminister, MD.: Newman Press, 1968), has shown to be essential in understanding present atheism. Küng, however, does lay the "blame" for Vatican I's declaration of papal infallibility at the door of the alliance of Neo-Thomism and Cartesianism. Can it be that Küng himself rejects the possibility of infallible propositional truth because he identifies propositional truth with conceptual clarity of the Cartesian type? For Aquinas, however, the proposition "God exists" is *certain,* but by no means *clear.*

5. St. Thomas Aquinas, *Summa Theologiae,* I, q. 2, a. 3, objections 1 and 2.

6. On the God-language problem for a review of opinions see Edward Cell, *Language, Existence and God* (Nashville: Abingdon Press, 1971); Malcolm L. Diamond and Thomas Litzenburg, Jr., *The Logic of God: Theology and Verification,* (Indianapolis: Boss-Merrill, 1975), and Stuart C. Brown, ed., *Reason and Religion* (Ithaca, N.Y.: Cornell University Press, 1977). For some positive approaches from different points of view see Ian T. Ramsey, *Religious Language,* (London: SCM Press, 1957) and *Christian Discourse* (London: Oxford, 1965), and criticism by Brian Hebblethwaite "The Philosophical Theology of I. T. Ramsey: Some Reflections," *Theology,* Dec., 1973, pp. 639-640; Frederick Ferré, *Language, Logic and God* (New York: Harper, 1969); Jean Ladrière, *Language and Belief* (Notre Dame, Ind: University of Notre Dame Press, 1973), especially the Conclusion, pp. 187-202, and Kenneth H. Klein, *Positivism and Christianity: A Study of Theism and Verifiability* (The Hague: Martinus Nijhoff, 1974).

7. See Charles Hartshorne, *The Divine Relativity* (New Haven, Yale University Press, 1964).

8. See Chapters 7 to 10.

9. Ludwig Feuerbach, *The Essence of Christianity* (1841) trans. by G. Eliot (New York: Harper, 1957).

10. See references in note 6 above.

11. "In spite of the 'Excellent Names,' there is a rooted, and honorable uneasiness for Islam about all such thoughts . . . those other involvements of the Divine with the human, namely the terms of invocation, are also outside his inaccessible character. They can still, however, be used. For they are religiously indispensable, despite their theological untenability . . ." The worshipper used the Names of God "without implying 'how'" and "without representational intentions," Kenneth Cragg, *The House of Islam*, 2nd ed., Encino and Belmont, (California: Dickenson, 1975), p. 9; also Cragg's *The Mind of the Qur'ān* (London: George Allen and Unwin, 1973), pp. 129-145; 163-182 and Helmut Gätje, *The Qur'an and Its Exegesis*, (Berkeley: University of California Press, 1976), pp. 146-163, on the vision of God.

12. See Chapter 11, p. 508.

13. See Reginald H. Fuller, *The Foundations of New Testament Christology* (New York: Charles Scribner's, 1965), pp. 114-115. Fuller concludes "For although there is no indubitably authentic logion in which Jesus calls himself the "Son," he certainly called God his Father in a unique sense.", p. 115.

14. See Richard A. Edwards, *A Theology of Q* (Philadelphia: Fortress, 1976), pp. 106-107.

15. See Troy Wilson Organ, *The Hindu Quest for the Perfection of Man* (Athens: Ohio University Press, 1970), Chapter ix, "The Way of Devotion", pp. 248-297 (cf. pp. 88-97).

16. On Rāmānuja's system John B. Carman, *The Theology of Rāmānuja* (New Haven: Yale University Press, 1974), and see Surendranath Dasgupta, *A History of Indian Philosophy* (Cambridge: Cambridge University Press), 3, 165-388 (contrasted to that of Sankara, 2, pp. 1-227). For Rāmānuja the Scriptures reveal that God creates for play and for the good of creatures (p. 296), but the created souls and world are God's *body* (pp. 156-158 and p. 297).

17. See Walter Kasper, *Jesus the Christ* (New York: Paulist, 1976), pp. 170-172, and Jean Galot, *Who is Christ?* (Chicago: Franciscan Herald Press, 1981), pp. 78-83.

18. See Chapter 11, pp. 487-489; 496-497.

19. See Gonzalo Haya-Prats, S.J., *L'Ésprit Force de l'Église* (Paris: Cerf, 1975, pp. 82-92), and George Montague, S.M., *The Holy Spirit: Growth of a Biblical Tradition* (New York, Paulist, 1976), pp. 353-362, on the personhood of the Holy Spirit in the New Testament.

20. On the early Christologies see Alois Grillmeier, *Christ in Christian Tradition* (London: Mowbrays, 1975), vol. 1; J. N. D. Kelly, *Early Christian Doctrines,* 5th ed., (San Francisco: Harper and Row, 1978); and Bernard J. Lonergan, S.J., *The Way to Nicaea* (London: Darton, Longman and Todd, 1976).

21. See *Church Dogmatics* (Naperville, IL: Allenson, 1969), 1:354-368, for Karl Barth's refusal to use the term "Three Persons", of the Trinity. See also Karl Rahner, S.J.,*The Trinity* (New York: Herder and Herder), pp. 103-114, who has similar difficulties. For an answer to these hesitations see William J. Hill, O.P.,*The Three-Personed God: The Trinity as a Mystery of Salvation,* (Washington, DC: Catholic University of America Press, 1982), pp. 111-148 and pp. 268-272. Note also Hill's defense of the personhood of the Holy Spirit, pp. 287-302. See also Galot, *Who is Christ* (note 17 above), pp. 279-285.

22. See below pp. 670-673 with references.

23. On the economic and ontological Trinity see Rahner, *The Trinity,* (note 21 above), pp. 21-23 and Hill, *Three-Personed God* (note 21 above), pp. 149-184; Also Yves Congar, O.P., *I Believe in the Holy Spirit,* 3 vols. (New York: Seabury; London: Geoffrey Chapman, 1983) 3:11-18, 34-44, and Joseph Bracken, S.J., *What Are They Saying About the Trinity?* (New York: Paulist, 1979), pp. 5-24.

24. Rahner, *The Trinity,* pp. 99-102.

25. On the Augustinian Trinity see John Edward Sullivan, O.P., *The Image of God: The Doctrine of St. Augustine and its Influence* (Dubuque: Iowa: Priory Press, 1963).

684

26. The metaphor of the Head is often used of the Father, the Face of the Son, and the Heart of the Spirit. It might be better said that the human body as a total vital organism reflects the Father; as engaged in cognitive activities, the Son; and as engaged in affective activities, the Spirit. Thus a dancer in balanced repose manifests the Father; in movements that indicate awareness and thought, the Son; in movements that indicate giving and receiving, the Spirit. The structure of the human body as it is adapted to these functions of vital repose, of alertness, and of action manifests these functions statically.

27. See Chapter 4, p. 113.

28. See for the threefold prediction of the coming Passion, Matthew 16:21-23; 17:22-23; 20:17-19 and parallels. But Matthew has also indicated this awareness as occuring earlier, as in 12:39. It is of course impossible to rely on the chronology of these events, but the intention of the Evangelist is clear, namely, to show that Jesus, from very early in His ministry, was aware of how it would end. This theme was beautifully developed by Louis Chardon, O.P. (d.1651) in *The Cross of Jesus,* 2 vols. (St. Louis: B. Herder, 1957).

29. On the Thomistic use of analogy see William J. Hill, O.P., *Knowing the Unknown God* (New York: Philosophical Library), 1971, p. 111-144. For an attack on this methodology see John S. Morreall, *Analogy and Talking About God* (Washington, D.C.: University Press of America, 1978). See also B. Mondin, article "Analogy" in *New Catholic Encyclopedia* 1:461-468. The Church Fathers at least *implicitly* follow this method in their frequent apologetic arguments for the knowability of God by human reason. For references to some of these arguments see M. Chossat, *Dictionnaire Théologique Catholique,* article "Dieu", 4(1): 879-881.

30. Germain Grisez, *Beyond the New Theism* (Notre Dame, Ind.: University of Notre Dame Press, 1975), pp. 94-180.

31. See Küng, *Does God Exist?* (note 1 above), pp. 536-551. Küng shows that modern science has not accepted Kant's view that "all basic categories of the world can be traced back to pure human subjectivity", but it has accepted that "there is no purely objective, but only a subjectively colored objective knowledge" (p. 543); but Küng also grants that "concerning knowledge of God, Kant is right in principle to appeal not to 'theoretical' but to *'practical' reason,* which is manifested in man's actions." (p. 544). Also Grisez, *Beyond the New Theism* (note 30 above), *loc. cit.*

32. See Thomas S. Kuhn, *The Structure of Scientific Revolutions* (Chicago: University of Chicago Press, 1970), with discussion by William A. Wallace, O.P., *Causality and Scientific Explanation,* 2 vol. (Ann Arbor; University of Michigan, 1974), 2: 238 ff.

33. On Kant's practical attitude toward religion see Chapter 3, pp. 70 f.

34. For a discussion of the many senses in which the word "experience" is used in current theology see Gerard O'Collins, S.J., *Fundamental Theology* (New York: Ramsey, Paulist Press, 1981), pp. 32-52.

35. "Kant had reflected long and deeply, ever turning his thoughts on this side and that, but in the philosophical field he had not read widely and not with the proper attention to the texts. He was not lacking in a general knowledge of the history of philosophy, both ancient and modern, and he understood also how to use this knowledge aptly" Friedrich Paulsen, *Immanuel Kant: His Life and Doctrine* (New York: Frederick Ungar, 1963) (an author very sympathetic to Kant). That the "history of philosophy" in Kant's day included little on medieval thought is clear from Guiseppe Micheli, *Kant, Storico dell Filosofia,* Padova: Antenore, 1980. For Kant's background see Lewis W. Beck, *Early German Philosophy* (Cambridge, Mass: Harvard University Press, 1969), pp. 426-456 and Ernst Cassirer, *Kant's Life and Thought* (New Haven: Yale University Press, 1981), pp. 12-77.

36. Aquinas: "This proposition 'God exists' in itself is self-evident, because the predicate is identical with the subject, for God is His own existence (as will be shown later); but since we do not know the definition (*quid est*) of God, this proposition is not self-evident to us", *Summa Theologiae* I, q. 2, a. 1 c; Kant: "People have at all times been talking of an *absolutely necessary* Being, but they have tried, not so much to understand whether and how a thing of that kind could even be conceived, as rather to prove its existence. No doubt a verbal definition of that concept is quite easy, if we say that it is something the non-existence

of which is impossible. This, however, does not make us much wiser with reference to the conditions that make it necessary to consider the non-existence of a thing as absolutely inconceivable", *Critique of Pure Reason* Chapter III, Section 4, F. Max Müller trans. (New York, Doubleday Anchor, 1966).

37. For Aquinas see *Summa Theologiae,* q. 44, a. 1 and a. 4, where he shows that God is the efficient and final cause of all creatures and their actions; Kant: "The supposition, therefore, which reason makes of a Supreme Being, as the highest cause, is relative only, devised for the sake of the systematical unity in the world of sense, and a mere Something in the idea, while we have no concept of what it may be by itself." *Critique of Pure Reason,* III, Appendix 2.

38. See Grisez, *Beyond the New Theism,* (note 30 above), pp. 152-160.

39. The criticisms of the classical theistic proofs are often based on the assumption that they are simply "arguments from analogy" in the sense of comparison (e.g., this seems the assumption of Humphrey Palmer in his book *Analogy: A Study of Qualification and Argument in Theology* (New York, St. Martin's Press, 1973). The point which is missed is that such arguments by "analogy of proportion" (as scholastics called them) are used in the theistic proofs only after the establishment of an "analogy of attribution" i.e., an *a posteriori causal* proof of the existence of the First Cause. Once we know the existence of the cause from the existence of its effect, we can then use the principle "nothing can be in the effect that is not equally or more perfectly in the cause" to argue by analogy of proportion to the *nature* (not the existence) of the First Cause.

40. "It is a curious fact that while almost no philosopher today regards the philosophy of Immanuel Kant as a defensible system, practically everyone assumes that proofs for the existence of God are definitely shown to be impossible in Kant's great work, the *Critique of Pure Reason,*" Grisez, *Beyond the New Theism* (note 30 above), p. 137.

41. See Chapter 7, pp. 273-274.

42. See Thomas C. O'Brien, *Metaphysics and the Existence of God,* (Washington, D.C.: Thomist Press, 1960) for an excellent analysis of the many misunderstandings of the Thomistic proofs.

43. See Eduard Lohse, *Colossians and Philemon,* New Century Bible (London: Aliphants, 1974), pp. 41-61, and Ralph P. Martin, *Colossians and Philemon,* Hermeneia (Philadelphia: Fortress, 1968), pp. 55-66.

44. A delightful and thought provoking book which will introduce the sceptical reader to the angels is that of my old teacher who first introduced me to this subject, Mortimer J. Adler, *The Angels and Us,* New York: Macmillan, 1982, written from a purely philosophical and speculative point of view (see the bibliography pp. 191-197). Rob Van der Hart, in a small book *The Theology of Angels and Devils,* Theology Today Series n. 36 (Notre Dame, Ind.: Fides Press), gives a good introduction to the biblical material. The classical theological treatment of Aquinas is in the *Summa Theologiae* I, qq. 50-64 and qq.106-114, along with *Summa Contra Gentiles,* II, cc. 46-55 and 91-101, and the opusculum *De Spiritualibus Creaturis* and the Disputed Question *De Substantiis Separatis.* The history of the doctrine of the Church on the subject can be found in Georges Tavard, in collaboration with André Caquot and Johann Michel, *Les Anges* (Histoire des Dogmes, tom. 2, fasc. 26) (Paris: Cerf, 1971). Helpful are the articles "Angels and Angelology" in the *Encyclopedia Judaica,* vol. 2, pp. 956-978; Jean Danielou, S.J.,*The Angels and Their Mission: According to the Fathers of the Church* (Westminster, MD: Newman, 1957); Michel L. Guérard des Lauriers, O.P., *La péché et la durée de l'ange* (Rome: Desclée, 1965); *Anges, démons et êtres intermédiaires* (3rd colloquium of the Alliance Mondiale des religions, 1968), ed., by Maryse Choisy and Bernard Grillot, (Paris: Labergerie, 1969); Ladislas Boros, *Angels and Men* (New York: Seabury, 1977), and Claus Westermann, *God's Angels Need No Wings* (Philadelphia: Fortress, 1979).

45. See S.G.F. Brandon, *Religion in Ancient History* (New York: Scribner's, 1969), Chapter 24, "Angels: The History of an Idea" pp. 354-371; and the article from the *Encyclopedia Judaica* referred to in the previous note.

46. See Chapter 12, pp. 580-582.

47. The Declaration of the Commission of Cardinals appointed to examine *A New Catechism* ("The Dutch Catechism") by Paul VI is printed in *Acta Apostolicae Sedis* 60 (1968): 658-91, and the revisions which implemented this report are published in Edouard Dhanis, S.J., and Jan Visser, C.SS.R., *The Supplement to A New Catechism* (London: Burns and Oates, 1969). The Declaration required that "The Catechism must teach that God created along with the visible world in which we live, a realm of pure spirits whom we call angels (see for instance Vatican I, Constitution *Dei Filius,* ch. 1; Vatican II, Constitution *Lumen Gentium* nos 49, 50). It must also be explained that the soul of each man, since it is spiritual (cf. Vatican II, Constitution *Gaudium et Spes,* no. 14), is created directly by God (see for instance the Encyclical *Humani Generis,* AAS 42 (1950) p. 575)." This text and revisions are found on pp. 3-5. The *General Catechetical Directory* issued by the S. Congregation for the Clergy, 11 April, 1971, in n. 51 includes the angels in its listing of the "most important elements" of catechetical instruction, and the S. Congregation for Divine Worship issued a study in French "Christian Faith and Demonology," June 26, 1975, which presents an extensive discussion on the Biblical and traditional basis for the liturgical recognition of angels and devils. These two latter documents can be found in *Vatican Council II: More Post Conciliar Documents,* Vatican Collection, vol. 2, ed., by Austin Flannery, O.P., (Collegeville, MN.: The Liturgical Press, 1982), pp. 529 ff. and pp. 456 ff., respectively.

48. Claude Cuénot, one of the best informed Teilhardians in *Teilhard de Chardin et la pensée Catholique* (Colloque de Venise sous les auspices de Pax Romana) (Paris: Seuil, 1965) edited by him, in one of the discussions answered someone who inquired how Teilhard fitted the angels as pure spirits into his system by saying "Il n'a intégré les anges, dans sa phénoménologie, ni dans son énergetique ni dans sa mystique" . . . "pour lui, l'Esprit est toujours lié à certaine matérialité," p. 101. Paul Chauchard, *Man and Cosmos: Scientific Phenomenology in Teilhard de Chardin,* (New York: Herder and Herder, 1965) in a section on *Christ, The Angels and Satan,"* pp. 125-128, seems to agree with this summation, but argues (without much explanation) that perhaps the angels could be considered as "co-evolving" with the material world.

49. In current exegetical discussion of the Resurrection (see Edward Schillebeeckx, *Jesus: An Experiment in Christology* (New York: Seabury, 1979), pp. 329-345) scholars generally seem to take it for granted that the angels at the tomb are mere literary devices to provide spokesmen for the evangelists' faith-proclamation. But is this really the evangelists' intention? Or are they reporting a detail of the tradition that goes back to the original witnesses, not only out of fidelity to that tradition, but also because they are concerned to show that for these witnesses the veil between the spiritual and human world was for the moment withdrawn to manifest the *cosmic* significance of the event?

50. See Paul M. Quay, S.J., "Angels and Demons: The Teaching of IV Lateran", *Theological Studies* 42 (1981): 2045, for a convincing argument that the existence of angels *is* a defined doctrine. Also the argument in "Christian Faith and Demonology" (note 47 above), argues to the same effect.

51. "Les Deux Économies du Gouvernement Divin: Satan et le Christ" in *Initiation Théologique* par un groupe de theologiens (Paris: Cerf, 1952), pp. 504-535. This rather obscure article of Bouyer is a very important study on the biblical and patristic conception of the historical role of the good and bad angels, a notion almost entirely neglected in modern soteriology.

52. Wesley Carr, *Angels and Principalities: The Background, Meaning and Development of the Pauline Phrase hai archai kai hai exousiai* (Cambridge: Cambridge University Press, 1981) has raised serious questions about how much of a role this notion of the domination of the world by evil angels plays in Paul's own soteriology, as against Cullman (see also the essays of Bouyer cited in note 51 above), but this does not deny its influence on the teaching of the New Testament as a whole, nor on that of the Church Fathers.

53. Aristotle, *Metaphysics* XII (Lambda), Chapter 8 argues that besides the first mover of the whole universe, there must be either 55 or 47 subordinate prime movers, each of which is associated (as a soul or as separate?) with one of the celestial spheres hypothesized by the Greek astronomers to account for the motions of the planets and the sun and moon. For the influence of this theory see the great work of Pierre Duhem, *Le Système du Monde,* 10 vols. (Paris: Hermann et Fils, 1913-1959), Chapter 4 and *passim.*

54. See Étienne Gilson, *History of Christian Philosophy in the Middle Ages* (New York: Random House, 1955), p. 215, on the theory of the Arabian Avicenna (Ibn Sina, d.1937), the Jew Moses Maimonides (d.1204), p. 230, and the Christian Pseudo-Grosseteste (13th century), p. 273.

56. Chapter 7, pp. 254 f.

57. See René Taton, ed., *A General History of the Sciences: Science in the Nineteenth Century* (London: Thomas and Hudson), pp. 284-286, on the periodic table. Another example is that of classification of the symmetries of crystals by René Haüy and others, see pp. 304-310. On the general method see Mary B. Hesse, *Models and Analogies in Science* (Notre Dame, Ind.: University of Notre Dame Press, 1966).

58. On earlier views of the angels see the article "Anges" by A. Vacant and others, in the *Dictionnaire Théologique Catholique*, 1: 1189-1272.

59. See Adler, *The Angels and Us* (note 44 above), pp. 114-118, on "Angelology and Mathematics."

60. *Summa Theologiae* I, q. 50.

61. *Ibid.* I, q. 102, a. 1.

62. *Ibid.* q. 55, a. 3.

63. *Ibid.* q. 10, a. 45.

64. *Ibid.* I, q. 57, a. 1 and 2.

65. *Ibid.* I, q. 57, a. 3 and 5.

66. *Ibid.* I, q. 106, a. 1.

67. *Ibid.* I, q. 58, a. 6 and a. 7.

68. *Ibid.*, I, q. 110.

69. *Ibid.*, I, q. 52.

70. *Ibid.*, I, 63.

71. "What, according to the historical theology of concepts, is one to believe regarding the Devil? We should be willing to face the problem of evil squarely without trying to dodge it intellectually. We should be open to the possibility of the existence of an evil spirit or spirits beyond humankind. The metaphysical assumptions of our present age may lead many to prefer to interpret the diabolical in terms of depth psychology, arguing that the demonic exists within the human mind, or perhaps collectively among minds. But on no account is one entitled to dismiss the idea of the Devil as irrelevant", Jeffrey Burton Russell, *Satan: The Early Christian Tradition* (Ithaca, N.Y.: Cornell University Press, 1981). See also Russell's earlier, *The Devil: Perceptions of Evil from Antiquity to Primitive Christianity*, same publisher, 1977 and *Satan*, a special number of *Études Carmélitaines* (Paris: Desclée de Brouwer, 1948); Trevor O. Ling, *The Significance of Satan: New Testament Demonology and Its Contemporary Relevance* (London: SPCK, 1961) and Rivkah Schärf, *Satan in the Old Testament* (Evanston, IL: Northwestern University Press, 1967).

72. See Jane Chance Nitzche, *The Genius Figure in Antiquity and the Middle Ages* (New York: Columbia, 1975) on the ancient conception of the "genius" or "guardian angel" who guides human activity. Interestingly enough, this was often symbolized by the *snake* and by the *cornucopia* (symbolizing gifts or talents), pp. 8-9. On Socrates' *daemon* and the Platonic theory of genius, see Ottomar Wichmann, *Platos Lehre von Instinct und Genie, Kantstudien* no. 40 (Berlin: Von Reuther and Reichard, 1917).

73. See John T. Pawlikowski, *The Challenge of the Holocaust for Christian Theology* (New York: Center for Studies on the Holocaust, Anti-Defamation League of B'nai Brith, 1978); Arthur Allen Cohen, *The Tremendum: A Theological Interpretation of the Holocaust* (New York: Crossroads, 1981), and Henry James Cargas, *When God and Man Failed: Non-Jewish Views of the Holocaust* (New York: Macmillan, 1981).

74. See Joseph Needham, *The Grand Titration* (London: Allen and Unwin, 1969), "Poverties and Triumphs of the Chinese Scientific Tradition", pp. 14-54 for a comparison of the Greek with the Chinese tradition of science which is the best parallel for the "Greek Miracle".

75. See William Irvine, *Apes, Angels and Victorians,* (New York: McGraw-Hill, 1955), and James R. Moore, *The Post-Darwinian Controversies 1870-1900,* (Cambridge: Cambridge University Press, 1979). Moore (pp. 348-350) shows that Huxley preferred the Calvinist pessimistic evaluation of human nature to that of Liberal Protestantism and that, indeed, his evolutionism was a kind of "secularized Calvinism."

76. See Chapter 7, pp. 256-257 and Chapter 10, pp. 429-430.

77. "The angels are ruled in their own actions according to divine decree. Now it sometimes happens that in different kingdoms or among different men contrary merits or demerits are found, so that one is subjected to another or made superior, and the angels are not able to know what the order of the Divine Wisdom as to this may be unless God reveals it, so that they must need to consult the Wisdom of God. Thus, therefore, insofar as they consult the Divine Will with regard to these merits which are contrary and conflicting with each other, the angels are said to "resist" each other (Daniel 10:13), not because they are of contrary wills, since they all agree in fulfilling God's decree, but because the matters concerning which they consult God are conflicting." *Summa Theologiae* I, q. 113, a. 8. Cf. also *Sent II,* d. 2, q. 1, a. 5 and *De Veritate* q. 8, a. 2 ad 5.

78. See Bouyer, *Les Deux Économies* (note 51 above).

79. See Charles E. Hoskin, *The Share of Thomas Aquinas in the Growth of the Witchcraft Delusion* (Philadelphia: University of Pennsylvania (Thesis), 1940); Julio Caro Boroja, *The World of the Witches* (Chicago: University of Chicago Press), pp. 242-257; Alan C. Kors and Edward Peters, eds., *Witchcraft in Europe 1100-1700: A Documentary History* (Philadelphia: University of Pennsylvania Press, 1972), Jeffrey Burton Russell, *Witchcraft in the Middle Ages* (Ithaca, N.Y.: Cornell University Press, 1972) (important annotated bibliography pp. 345-377). Gregory Zilboorg, *The Medieval Man and the Witch during the Renaissance* (New York: Cooper Square Publishers, 1969) gives a psychoanalytic explanation. Russell (pp. 265-287) gives the broadest and most balanced explanation comparing the witch craze to the anti-Communist and other conspiracy theories of our own time and showing that all these movements reflect a high level of anxiety in a society beset by disturbing social changes.

80. See Jean Daniélou, S.J., *Holy Pagans of the Old Testament,* (Baltimore: Helicon, 1957). On the pagan view of Christianity in antiquity see Johannes Geffchen, *The Last Days of Greco-Roman Paganism* (Amsterdam: North Holland Pubs., 1978).

81. See, for example, St. Augustine, *The City of God,* Bk. VIII, c. 14-24, *The Fathers of the Church:* (New York: The Fathers of the Church, Inc., 1950), pp. 45-70 on pagan idolatry as devil worship.

82. Daniélou, *Holy Pagans* (note 80 above).

83. *Numbers* 22-24.

84. See Yves M.J. Congar, *Dialogue Between Christians,* (Westminster, MD: Newman, 1960), pp. 53-156.

85. See Robert L. Faricy, S.J., *Teilhard de Chardin's Theology of the Christian in the World* (New York: Sheed and Ward, 1967), pp. 129-138 and Piet Smulders, S.J., *The Design of Teilhard de Chardin* (Westminster, MD: Newman, 1967), pp. 235-256.

86. See Christopher F. Mooney, S.J., *Teilhard de Chardin and the Mystery of Christ* (New York: Harper and Row, 1966) and the very careful, critical study of Richard W. Kropf, *Teilhard, Scripture and Revelation: A Study of Teilhard de Chardin's Reinterpretation of Pauline Themes* (Rutherford: Farleigh Dickinson University Press, 1980).

87. See Colman E. O'Neill, O.P., *New Approaches to the Eucharist,* (Staten Island, N.Y.: Alba House, 1967).

88. Heidegger chiefly developed the ideas which seem to have influenced this discussion in *What is a Thing?*, translated by W. D. Barton and Vera Deutsch (Chicago: Regnery-Gateway, 1968) and *The Origin of the Work of Art,* translated by David F. Krell, ed., *Martin Heidegger: Basic Writings* (New York: Harper and Row, 1977). These works are critically discussed by Thomas Langan, "The Problem of the Thing", in John Salliss, ed., *Heidegger and the Path of Thinking* (Pittsburgh: Duquesne University, 1970), pp. 105-115.

89. See Edward Schillebeeckx, O.P., "Transubstantiation, Transfiguration, Transignification" in R. Keven Seasoltz, O.S.B., *Living Bread, Saving Cup* (Collegeville, Minn.: Liturgical Press), pp. 175-189. Schillebeeckx, p. 180, attributes the initiation of this discussion to J. Möller and to the influence of phenomenology. See also Joseph M. Powers, S.J., *Eucharistic Theology* (New York: Herder and Herder, 1967), pp. 111-179.

90. "Car si tout le *Seiendes* est recupéré dans le *Sein,* il n'y a plus aucun motif de repousser les thèses de l'idéalisme absolu. Seule la résistance radicale, opposée prétendûment par l'existence brut à projection de l'intelligible, éloigne Heidegger de l'idéalisme absolu.", Alphonse de Waelhens, *La philosophie de Martin Heidegger* (Louvain: Nauwaelaerts, 7th ed., 1971).

91. Aquinas, *Summa Theologiae* III, q. 54, a. 4 (quoting St. Augustine and the Ven. Bede), suggests five reasons that Christ retained the wounds of his passion in his risen body: (1) as signs of his victory; (2) to confirm the faith of his disciples in the reality of the resurrection; (3) to show them to the Father in intercession on our behalf; (4) to assure us of his continued mercy toward us; (5) to show to the condemned at the last judgment as proof of the justice of their sentence. He also quotes St. Augustine (*The City of God,* 22, chapter 19) as of the opinion that the saints will also retain the wounds of their martyrdom "not as deformities but as signs of dignity" and the "beauty of virtue."

92. Hubert Jedin ed., *Handbook of Christian History* (New York: Herder and Herder, 1970), 4:570-574 and Nathan Mitchell, O.S.B., *Cult and Controversy: The Worship of the Eucharist Outside Mass* (New York: Pueblo Publishers, 1974), pp. 129-200 and 367-423.

93. See George Hunston Williams, *The Radical Reformation,* (Philadelphia: Westminster, 1962), Chapter 3, "The Eucharistic Controversy Divides the Reformation 1523-1526", pp. 85-117.

94. See Mircea Eliade, *The Sacred and the Profane,* (New York: Harcourt Brace, 1968), pp. 116-125, on the notion of "epiphany."

95. "Every time, then, you eat this bread and drink this cup, you proclaim the death of the Lord until he comes! This means that whoever eats the bread or drinks the cup of the Lord unworthily, sins against the body and blood of the Lord. A man should examine himself first; only then should he eat of the bread and drink of the cup. He who eats and drinks without recognizing the body, eats and drinks a judgment on himself. *That is why many among you are sick and infirm, and why so many are dying.* If we were to examine ourselves, we would not be falling under judgment in this way; but since it is the Lord who judges us, he chastens us to keep us from being condemned with the rest of the world" (I Cor. 11:27-32).

96. See Eugène Portalié, S.J., *A Guide to the Thought of St. Augustine* (Chicago: Regnery, 1960), pp. 247-260, on Augustine's eucharistic theology.

97. See Werner Elert, *The Structure of Lutheranism,* (St. Louis: Concordia Publishing House, 1974), pp. 313-321.

98. See R. T. Wallis, *Neo-Platonism* (New York: Scribner's, 1972), pp. 61-72, on the concept of *emanation* in Plotinus.

99. Some important contributions to this discussion are Leonard J. Eslick, "God in the Metaphysics of Whitehead" in Ralph M. McInerny, ed., *New Themes in Christian Philosophy* (Notre Dame: University of Notre Dame Press, 1968); Eric Mascall, *The Openness of Being: Natural Theology Today* (Philadelphia: Westminster, 1972), Chapter 10, "God and Time"; Burton Z. Cooper, *The Idea of God: A Whiteheadean Critique of St. Thomas Aquinas' Concept of God,* (The Hague: Martinus Nijhoff, 1974); Schubert M. Ogden, "The Temporality of God" in his *The Reality of God and Other Essays* (New York: Harper and Row), pp. 144-163; William J. Hill, O.P., "Does the World Make a Difference to God" *Thomist* 38 (1974): 146-64; W. Norris Clarke, S.J., "A New Look at the Immutability of God" in R.J. Roth, S.J., ed., *God Knowable and Unknowable* (New York: Fordham University Press, 1973), pp. 43-72.

100. See Carroll Stuhlmueller, C.P., *Creative Redemption in Deutero-Isaiah* (Rome: Biblical Institute Press, 1970).

101. "Anger and similar affects are attributed to God by the similarity of their effects; for as it is a proper effect of an angry man to punish, so God's punishments are metaphorically called his anger." *Summa Theologiae* I, q.3, a. 2 ad 2. In q. 21, a. 1 ad 3, Aquinas shows that God acts justly in punishing and rewarding both to manifest to us that he is wise and good and also out of respect to creatures themselves in rendering to them "what is due to each according to the measure of its nature and condition."

102. "Therefore, although the Johannine description and acceptance of the divinity of Jesus has ontological implications (as Nicaea recognized in confessing that Jesus Christ, the Son of God, is himself true God), in itself this description remains primarily functional and not too far removed from the Pauline formulation that "God was in Christ reconciling the world to Himself" (II Cor. v, 19)," p. 408, Raymond E. Brown, S.S., *The Gospel According to John,* Anchor Bible (Garden City, New York: Doubleday, 1966), 1:400-412. Brown notes, however, that the ontological implication is made explicit in the prologue hymn of the Gospel.

103. On Socinus see Earl Morse Wilbur, *A History of Unitarianism* (Cambridge, Mass: Harvard University Press, 1945-52) 1:284-432.

104. See Brown, *Gospel According to John* (note 102 above) on the Prologue Hymn, 1:xxiv-xl and pp. 18-23, and Rudolf Schnackenburg, *The Gospel According to St. John,* (New York: Herder and Herder, 1968), pp. 221-232.

105. Brown, *Gospel According to John,* vol. 2, "Appendix V: The Paraclete", pp. 1135-1144.

106. "Just as through one man's disobedience all became sinners, so through one man's obedience all shall become just" (Romans 5:19).

107. C. H. Talbert, "The Problem of Pre-existence in Phil. 2:6-11", *Journal of Biblical Literature* 86 (1967): 141-153 raised the question whether the Philippian hymn could not be interpreted as referring simply to the choice of the earthly Jesus to take the role of a servant rather than the kingly role of the first Adam, and a similar position has more recently been defended by Jerome Murphy-O'Connor, O.P., "Christological Anthropology in Phil. II, 6-11", *Revue Biblique* (83) 1976: 25-50. See the discussions of the question by T. F. Glasson, "Two Notes on the Philippian Hymn (ii:6-11)," *New Testament Studies* 21 (1974): 133-139; R. P. Martin, *Carmen Christi: Philippians ii 5-11 in Recent Interpretation and in the Setting of Early Christian Worship* (Cambridge: Cambridge University Press, 1976), and his commentary, note 43 above, and G. Howard, "Phil. 2:6-11 and the Human Christ", *Catholic Biblical Quarterly* 40 (1978): 368-387. On the authenticity of *Colossians* see Lohse, (note 43 above) pp. 177-183 (who rejects it) and Martin, pp. 32-40 (who defends it), and on that of *Philippians* see Martin, *Philippians,* New Century Bible (London: Aliphant, 1976), pp. 10-22 (favorable).

108. Brown, *Gospel According to John,* 1:3-37; Schnackenburg, *Gospel According to John* (note 104 above), pp. 232-235, and C. H. Dodd, *The Interpretation of the Fourth Gospel* (Cambridge: Cambridge University Press, 1968), pp. 292-296.

109. Since the Son as God is eternal, and eternity transcends time, it is not correct to speak of the Son as "pre-existing" his incarnation except in the sense that he did not *begin* to exist at the moment of the Incarnation but is present equally to every moment of human history, including the events that preceded the Incarnation. The theological question, therefore, is whether the eternal existence of the Son is independent of his Incarnation or dependent upon it. Some contemporary theologians seem to want to answer this question by saying that the distinction of the Son from the Father is dependent on the eternal divine decree that the Incarnation take place at a certain moment in created history. But it will not do to make the existence of the Son depend on a *free* act of the Father, since that would make the Son a contingent being and therefore a creature, as the Arians maintained.

110. See George Wesley Buchanan, *To the Hebrews,* Anchor Bible, (Garden City, N.Y.: Doubleday, 1972), pp. 256-268, who argues for dating it at Jerusalem before 70 A.D., and attributing it to a Jewish Christian. It has been attributed to Barnabas or to Apollos of Alexandria and dated after 70 but before 95 when it was used by *I Clement;* but the evidence of Essene influence resulting from comparison with the Dead Sea Scrolls has led to the view Buchanan supports.

111. See Chapter 9, pp. 381-384.

112. For the current difficulties about the Chalcedonic formula see Piet Schoonenberg, S.J., *The Christ,* (New York: Herder and Herder, 1971), pp. 51-66.

113. See Hill, *Three-Personed God* (note 21 above) pp. 262-272.

114. See Yves Congar, O.P., "Meditation théologique sur la troisième Personne", *Proche Orient Chrétien,* 29 (1979): 201-211.

115. "Dogmatic Reflections on the Knowledge and Self-Consciousness of Christ", *Theological Investigations,* (London: Darton, Longman and Todd, 1966), pp. 193-218.

116. For some speculations on the different modes of consciousness see Erich Neumann, *The Origins and History of Consciousness,* Bollingen Series no. 42, Princeton, N.J.: Princeton University Press, 1954, especially pp. 314-335; Lancelot Law Whyte, *The Unconscious Before Freud* (New York: Basic Books, 1960); Charles T. Tart, ed., *Altered States of Consciousness* (New York: John Wiley and Sons, 1969); Wilson Van Dusen, *The Natural Depth in Man,* (New York: Harper and Row, 1972); Robert E. Ornstein ed., *The Psychology of Consciousness* (San Francisco, W. H. Freeman, 1972), and Julian Jaynes, *The Origin of Consciousness in the Breakdown of the Bicameral Mind,* (Boston: Houghton Mifflin, 1976), especially pp. 130-145, where Jaynes argues for the very late development of language among the hominoids.

117. "The unbelief of Israel in Jesus as the Christ is not unbelief in God (for God could have worked in Jesus of Nazareth, all things being possible to God). It is only that the Jew — who is saved by God himself, being with him from his own birth — is not saved by him who came after for the sake of those who were born after." Arthur A. Cohen, *The Myth of the Judeo-Christian Tradition* (New York: Harper and Row, 1960), p. 41; for a more conciliatory view see Samuel Sandmel, *Two Living Traditions,* Detroit: Wayne State University Press, 1972. On the history of the division see Han Joachim Schoeps, *The Jewish-Christian Argument* (New York: Holt, Rinehart and Winston, 1963), especially "The Parting of the Ways" pp. 258-264.

118. *Declaration on the Relation of the Church to Non-Christian Religions (Nostra Aetate)* Oct. 28, 1965 and *Guidelines on Religious Relations with Jews* (Committee for Religious Relations with Jews) 1 Dec., 1964, in Austin Flannery, O.P., *Vatican Council II: The Conciliar and Post-Conciliar Documents* (Collegeville, Minn.: Liturgical Press, 1975), pp. 738-749.

119. see note 43 above for exegesis of *Philemon.*

120. On the notion of the Spirit in comparative religions see *Spirit and Nature: Papers from the Eranos Jahrbuch* (Bollingen Series xxx) (New York: Pantheon, 1954).

121. Recently there has developed a whole scientific discipline of *futurology* (or *futuristics*), of which an idea can be gathered from Victor C. Ferkiss, *Futurology: Promise, Performance, Prospects* (Beverly Hills, California: Sage Publications, 1977), Edward Cornish and members of staff of the World Future Society, *The Study of the Future,* (Washington, D.C.: World Future Society, 1977), and Magoroh Maruyama and Arthur M. Harkins, eds., *Cultures of the Future,* The Hague: Mouton, 1978 (who emphasize that anthropology is ethnocentrically western and that the future may not belong to the western world). Burnham P. Beckwith, *The Next 500 Years: Scientific Prediction of Major Social Trends* (New York: Exposition Press, 1967) is an interesting example by a futurologist who denies human free will, and predicts that in the future the believers in *any* religion will be reduced to between 10 and 20% of the global community.

122. "This is the salvation which the prophets carefully searched out and examined . . . They knew by revelation they were providing, not for themselves but for you, what has now been proclaimed to you by those who preach the gospel to you, in the power of the Holy Spirit sent from heaven. *Into these matters angels long to search,"* I Peter, 1: 10, 12. In *Revelations* 4:2-5, an angel asks who can open the sealed scroll of history, and "no one in heaven or on earth or under the earth could be found to open or examine the scroll" Only the victorious Lamb can do so.

Chapter 14

The Godliness Of Matter

I: How Matter Mirrors God

At the beginning of this book I raised the question "Can we create ourselves?" Humanists and Marxists say yes, that modern science and technology have made this a real possibility. I have tried to show that the Christian answer is also yes, but that human creativity is a stewardship of gifts given us by God. It is a participation in God's own creative activity in bringing the universe to its completion. Yet our share in this creativity goes beyond the mere execution of God's designs to a real share in determining what the course of history is to be.

I also asked, "How can we have an ethics based on human nature, if human nature itself is problematic?". The answer I gave to that question was grounded on the insight that our creativity is not autonomous, but a stewardship. If we misuse our God-given gifts, we will destroy our world and our own intelligence. Consequently, our use and re-making of our own bodies must respect their essential structure and basic needs.

This means, however, that ethics must be rooted in self-knowledge, not merely of our subjective selves, but of ourselves as bodies. Modern science is constantly increasing our understanding of our bodies and of

693

our environment, out of which we have emerged by evolutionary processes. Human thought is an activity which transcends the spatial limits of material processes, but because it is human it is essentially and substantially dependent on the brain and the sense organs, and these are dependent on the rest of the body for their life and activity and the execution of human creativity in the world.

Christian theology has always taught reverence for the human body as the temple of the human spirit and the Holy Spirit, but influenced by dualistic conceptions of the relation of mind to matter it has failed to take full advantage of our increasing scientific self-understanding. There remains the haunting idea that matter, although created by God, is somehow the source of the evils in the world. To overcome this notion, throughout the argument of this book I have tried to show how all human knowledge arises out of matter or is at least expressed only in terms derived from matter, even our loftiest ideas of God and the spiritual world.

To summarize all this, I want to state succinctly both the negative and the positive aspects of matter. Matter, in the radical sense of primary matter, is a negative principle in that it is *pure potentiality* which cannot actualize itself, but requires to be actualized by actually existing beings. Therefore, it exists only in existing natural units as their intrinsic capacity to be transformed into other actually existing units, and can never exist in its own right. Matter is never self-developing, and evolution cannot be explained by the analogy of a seed or an embryo, since matter is radically passive. Because all material things always retain this radical passivity in their materiality, they are all liable to be destroyed by exterior forces. All living things are eventually killed; all non-living things are eventually disintegrated by the action of external agents.

The total passivity of primary matter is also the reason that material things are separated from each other spatially, and act only through processes that are temporal and transitory, so that material units exist in a Now that is always perishing into the past and stretching forward into the future. Each material primary unit is a whole of few or many parts, and these parts are spatially and temporally exterior to each other so that its unity is tenuous. Only in the most complexly organized primary units, such as living animals, does this unity reach a high level.

Thus in the material world of our direct experience, everything is in process, everything eventually perishes, yet when things exist they preserve themselves by relative isolation, so that although they are always interacting, the universe as a whole is very spread out in time and place and has only a very loose unity. The ultimate outcome of all physical process is that material things tend to a state of total disorder or entropy,

694

when the unity of the universe will become minimal and all creative processes will subside.

These negative or limiting aspects of matter, however, are only the reflection of the positivity of matter which makes it a real principle of existing things and not merely, as Plato thought, sheer negation or lack of being. Oddly, it was Plato himself who grasped the positive aspect of real potentiality when he said that Socrates, although ugly himself, was a lover of beauty, and therefore potentially beautiful.[1] Aristotle corrected Plato's dualism by developing this insight of his teacher, and showed that primary matter, although it has nothing of actuality of its own, yet "loves" actuality, that is, it has a teleological "appetite" for form. In every primary unit, matter exists as a potentiality for new and perhaps more complex actualizations in other units. Matter is thus evolutionary in the radical sense, that it is always open to new and higher information. It is only when it reaches the upper limit of such organization in the human body informed by an intelligence which transcends the limitations of matter, that this thirst for form finally is satisfied. In using such metaphors as "love", "appetite", "thirst" there is no implication that matter has a psychic dimension as Julian Huxley, Teilhard de Chardin and others have thought,[2] but only that the potentiality of matter is *intrinsically related to actualization* by an unlimited variety of forms ranging from those of very simple units to the most complex.

This wonderfully protean positivity of matter, its openness to novelty and to the future, gives to our world the possibility of its dynamic, historical, dramatic, narrative development, of evolutionary movements arising out of the entropic decline of the universe which results from the mutual interaction and mutual frustration of its conflicting forces.

All that science has to tell us today about the vast variety of forms that this primary matter assumes ranging from the atoms to humans and all the transitional forms of the great "empty" spaces of our universe (which are actually material fields vibrant with radiant energy passing in every direction) — all this variety manifests the infinite openness of the material world, and this openness turns toward the Creator who draws forth these forms freely as He writes the history of the universe. At least we know this much about this history, that it has in fact led, not indeed by a straight path, but nevertheless surely, through billions of years and as many crises that might just as well have terminated in dead ends, to the emergence of intelligent life on our earth.

Thus in the physical universe, to the infinite creativity of God corresponds the infinite potentiality of matter. And in that ocean of matter, in the ever shifting and transient forms that cross it, we can see the face of God reflected, in what the medievals called the Mirror of God. Yet that metaphor of the mirror is too Platonic, because God's epiphany in the world is not through mere surface shadows, but is in the coming to be, development, and passing away to make room for novelty of primary natural units, each of which truly exists and acts in its own right and according to its own nature and structure for its time, and interacts with other units in a process of mutual actualization and eventual replacement. The physical universe is not a mere shadow, it is a drama of billions of actors, some minute, blindly moving atoms, some living plants and moving animals, and some intelligent body-persons enacting an evolutionary history whose scenario still remains open to the future.

II: The Human Body: God's Image

The outcome of this history, then, has been the human body, an intelligent body whose intelligence transcends the limits of the body yet remains necessarily rooted in it. As primary matter is infinitely open to form, so the human intelligence, although it is not material, is in pure potentiality to knowing the forms of matter, and through them by analogy divining something of the vaster spiritual universe which lies outside the range of direct human investigation by science. The human intelligence has a teleology corresponding to the teleology of matter. As matter lies open to receiving concrete existence under an unlimited variety of forms, so the human intelligence lies open to learning about these forms.

As our human intelligence investigates the world by ever more refined methods, it comes to see that the human body recapitulates the universe because it is the most complex and highly unified primary unit in the material universe, and the only one capable of knowing the universe. All the principles and forces that have built the universe in its vast variety are summed up in the human body, small and frail as it seems when it stands naked under the sun and the stars. It is the human intelligence itself which is the unifying principle in the human body, which is mostly deeply hidden from itself. We cannot look into the depths of our spirit, and must learn who we are from studying the world around us until we know our body better, and then from our body begin to know something about its soul. But what we *do* know is that the intelligence, which alone can understand and somewhat control through technology the varied forms of the universe, *can know and know that it knows*, and that it exceeds the limits of matter, which can only receive forms that are spread out in space and limited to the here and now, while the intelligence is present to itself and draws all the past together in memory as it reaches out to plan for the future, to create what has never been.[3]

696

If the matter of the universe "mirrors" God and the human body sums up the universe in this mirroring of God, and if the human intelligence uses this body to know the universe in which it begins to see God, and in seeing its own body as part of the universe begins to know its inner self, then indeed in its inner self it finds the image of God.

III: The Body Is God's Glory

The dignity of matter, therefore, is in its *humility*, its openness to God's creative love, his *agape* of generous giving love, to which it responds with *eros*, love that needs and receives the gift. The mystery of the Christian Gospel is that God himself is humble, willing to come down to what is humblest in the universe and dwell in its humility.

When the Son of God, who is the very *form* of the universe, the *Logos* or pattern after which the universe is formed, chose to become flesh, to become human, to enter into the world's history and to serve the poor, did the hidden teleology of the universe achieve its goal? Yes, if, as is in fact true, the purpose of the material universe was to evolve to the level of intelligence, and the purpose of the whole universe including the realms of the pure spirits was to come to know and praise the Creator. After the sin of free creatures had attempted to set up false gods, this purpose was finally realized in Jesus of Nazareth, the suffering servant. Thus the humility of matter, its faithful love, obtained for the universe what all the spirits, lofty as the seraphim, could not obtain, the entrance of the Son of God himself into creation as a human, bodily being, who has become the center of the total universe, visible and invisible, material and spiritual.

This humility of the material world is most perfectly personified in Mary, the New Eve, in whom was recapitulated all of history as *expectation*, prayer for the coming of God into the world. It was her faith, summing up the faith of all who had believed in God from the beginning, not only in Israel but in every land, which alone was able to believe that God would really come into the world. Feminists have resented the comparison of woman to matter, because it seemed to make woman inferior to the masculinity of God. This is the last blindness that male pride has imposed on women, to make them think that to be woman is to be inferior, that the humility of matter is inferiority rather than its greatest opportunity. Because of Mary's humility she, a creature in the material world, became the Mother of God from whom God, as it were, learned to be humble, and revealed to us what God really is, the God of love, love which is *eros* in openness, *agape* in fullness.

697

Mary is the Church, the beginning of the Reign of God. She is the Reign of God in which her Son will reign as Head of the body which will become the community of all humankind, and will be raised to the community of the spirits and of the Trinity itself. Mary has been assumed already with her son in the resurrection, the beginning of the redemption and resurrection of the universe, when God will be all in all. She is the New Jerusalem descending like a bride from heaven as the universe of time enters into eternity.

Even in the eternal Now the body of Jesus, the body of Mary, our bodies, the Body which is the Church, which is the material universe or temple of God, will remain forever, summing up the memory of our world, and the biographies of each of us, the Book in which the story of God's love for us was written and which will be the theme of our endless praise.

> Then I saw new heavens and a new earth. The former heaven and the former earth had passed away, and the sea was no longer. I also saw a new Jerusalem, the holy city, coming down out of heaven from God, beautiful as a bride prepared to meet her husband. I heard a loud voice from the throne cry out: "This is God's dwelling among men. He shall dwell with them and they shall be his people and he shall be their God who is always with them (Rv 21:1-4).

Notes

1. *Symposium*, 203, Diotima explains that Love is the child of Poverty and Plenty, and Alcibiades says (216) that Socrates looks like the ugly Silenus, yet seduces others by the beauty of his character.

2. "If, as is the case, mind and matter co-exist in the higher animals and man; and if, as seems certain, the higher animals and man are descended from lower animals, and these in turn from lifeless matter, then there seems no escape from the belief that all reality has both a material and a mental side, however rudimentary and below the level of anything like our consciousness that mental side may be." Julian Huxley, *Religion without Revelation* (London: Watts and Co., 1945), p. 27. For Teilhard see Chapter 6, pp. 237-238.

3. See Karl Rahner, S.J., "Theology of the Symbol", *Theological Investigations* (Baltimore: Helicon, 1966) vol. 4, pp. 245-252, section on "The Body as Symbol of Man." On the theme of the "convergence of spirit and matter" see R.C. Zaehner, *Matter and Spirit: Their Convergence in Eastern Religions, Marx, and Teilhard de Chardin*, New York: Harper and Row, 1963 and Chapter IV "Spirit and Matter" in his *Concordant Discord*, (Oxford: Clarendon Press, 1970), pp. 61-82.

700

Baudry, Leon, 188, 192
Bauer, Bruno, 221
Baumann, Urs, 409
Baumer, Franklin L. van, 94, 95
Baumgarten, Charles, 203
Baumgarten, A.G., 214
Baur, Ferdinand Christian, 221
Bautain, Louis, 223, 225
Bayer, Ronald, 475
Bayle, Pierre, 211, 349, 610, 629
Beauchesne, Richard J., 557
Beck, Hans-George, 142
Beck, J. Chr., 214
Beck, Lewis W., 241, 334, 685
Becker, Carl, 13
Becker, Joachim, 481
Beckwith, Burnham P., 692
Becque, L., 629
Becque, M., 629
Beelzebul, 352
Beethoven, Ludwig van, 318, 387
Begheyn, Paul, 202
Beierwaltes, Werner, 146
Beit-Hallahmi, Benjamin, 93
Bell, Alan P., 475
Bell, Norman W., 473
Bellah, Robert, 53, 92
Bellarmine, Robert, 185, 360
Belloc, Hilaire, 465, 482
Benaceraff, P., 343
Benedek, Therese, 469
Benedict XV, 244
Benedict, St., 521
Bengel, J.A., 214
Benoit, Pierre, 551, 620, 623, 624,
    626
Bentham, Jeremy, 5
Bercovitz, J. Peter, 477
Berdyaev, Nicholas, 337, 399
Berger, Klaus, 403
Berger, Peter, 52, 91
Bergson, Henri, 84-86, 99, 237-238,
    262, 285, 291
Berkeley, George, 214
Berkouwer, G.C., 408, 628
Berlin, Isaiah, 398
Bernadine of Siena, St., 198
Bernard of Clairvaux, St., 134, 150
Bernard of Thierry, 147
Bernini, Giovanni Lorenzo, 181

Bertalanffy, Ludwig von, 44, 291
Bertocci, Peter A., 412
Bessel, Friederich, 239
Best, Ernest, 628, 683
Bettoni, Efrem, 190, 191
Beveridge, W.I.B., 337
Beza, Theodore, 179
Bidney, David, 406
Bieler, Andre, 201
Bieler, Ludwig, 146
Black, Matthew, 683
Blackwell, Richard J., 289
Blanchet, Leon, 195
Blandino, Giovanni, 341
Bloch, Ernst, 511, 555
Bloch, Rene, 94
Bloesch, Donald, 247
Blondel, Maurice, 230
Bloom, E.D., 47
Blum, Harold F., 48, 49
Blumenthal, H.J., 135
Bock, Gisela, 195, 476
Bode, Edward L., 624
Boehme, Jakob, 178
Boehner, Philotheus, 192, 193, 336
Boer, Rebecca R. De, 341
Boethius, 150
Bogen, James, 334
Bohm, Walter, 187
Bohm, David, 48
Boismard, M.E., 551
Boissard, Edmond, 627
Bolotin, Susan, 474
Bonald, Louis de, 223
Bonasea, Bernardine M., 191
Bonaventure, St., 157, 159, 652
Bonelli, M.L. Righini, 239
Bonhoeffer, Dietrich, 199, 200, 234,
    406
Boniface VII, 164
Bonifazi, Conrad, 294
Bonner, Gerald, 145
Bonnetty, Augustin, 225
Borgese, G.A., 96
Borgnet, Auguste, 188
Boroja, Julio Caro, 689
Boros, Ladislas, 686
Boscovich, Joseph R., 262, 291
Bosquet, G.H., 626
Bossuet, Jacques Benigue, 223, 460

Dahood, Mitchell, 580, 622
Dalbiez, Roland, 97
Daly, Mary, 474, 479, 554
Daly, Robert J., 469, 621
Daniel-Rops, H., 242, 407
Danieli, Giuseppe, 550
Daniélou, Jean, 18, 112, 138, 139,
   140, 141, 142, 143, 406, 409,
   686, 689
Danilevsky, Nikolai, 399
Dante Alighieri, 157, 160, 587, 608
Danto, A., 48
Darwin, Charles, 20, 64, 72, 373
Davey, F.N., 622
David, 106, 111, 454, 490, 493,
   510, 526, 669
Davies, 409
Davies, Alan T., 554
Davies, J.G., 624
Davies, W.D., 408, 411, 570, 571,
   621, 628, 683
Dawkins, Richard, 400
Day, Sebastian J., 190
De Wulf, Maurice, 189
Debus, Allen G., 194
Dedek, John F., 403
Deeken, Alfons, 402
Deely, John N., 46, 247, 271, 275,
   294, 296, 341, 479
Deferrari, R.J., 142
Deidun, T.J., 628
Dekker, E., 146
De Koninck, Charles, 290, 480
Delehaye, Hippolyte, 628
Delling, Gerhard, 624
Deman, Th., 203
Demarest, Robert J., 12
Democritus, 116, 134, 207, 266,
   267, 277, 284, 288, 293
Denziger, H., 196
De Raeymaeker, Louis, 229
Derrett, J. Duncan M., 137, 476, 622
Derrida, Jacques, 81, 99
Descamps, A., 408, 557
Descartes, René, 59, 61-62, 70, 73,
   77-78, 129, 204-211, 215-219,
   222-231, 262-263, 269-278, 282,
   288-289, 302, 310, 317-323, 327,
   335, 633, 644
Descoqs, Pedro, 229

Desmond, Robert J., 43
Despland, Michael, 96
Deutsch, Diana, 342
Devilin, Robert M., 337
Dewey, John, 73, 99, 339
DeLumeau, Jean, 242
Dhanis, Edouard, 687
Diamond, J.M., 337
Diamond, Malcolm L., 683
Dickens, Charles, 81
Dickson, F.B., 293
Diderot, Denis, 63, 69, 96
Diebolt, J., 242
Dijksterhuis, E. J., 189, 240
Dillon, John, 140
Dilthey, Wilhelm, 357, 358, 401
Diodorus of Tarsus, 148
Dionysius the Areopagite (Pseudo),
   122, 133, 149, 152, 160, 185
Diotima, 544
Dobzhanzky, Theodosius, 41, 44,
   47, 400
Dodd, C.H., 401, 411, 622, 691
Dodds, E.R., 135, 476
Dolan, John, 196, 292, 203, 482
Dolger, F., 149, 187
Dollinger, J.J.I.von, 203
Domanski, Boleslaw, 142
Dominic of Flanders, 202
Donaldson, J., 143
Dondeyne, Albert, 93
Donfried, Karl P., 550, 559
Doniger, Simon, 407
Donington, Robert, 407
Dorner, J.A., 197, 241
Dorr, D.J., 405
Dorrie, H., 140
Douglas, Mary, 470
Douglass, James W., 471
Dourley, John R., 247
Doutreleau, Louis, 138
Downey, Edward A., Jr., 201
Dreano, Maturin, 240
Drey, Johann Sebastian von, 224,
   244
Dreyfus, Hubert L., 44
Drummond, Alexander L., 241
Drummond, Andrew L., 202
Drummond, Richard, 409
Dryden, John, 180

705

Feigenbaum, E.A., 342
Feigl, Herbert, 340
Feinberg, Gerald, 47
Feldman, G.J., 47
Feldman, J., 342
Fenelon, François Salignac de la Mothe, 223
Fenton, John Y., 472
Fenton, Norman E., 241
Ferder, Fran, 561
Ferkiss, Victor C., 692
Fernandez, Aurelio, 558
Ferrariensis, Francis Sylvester de Sylvestris, 202
Ferré, Frederick, 683
Ferris, Timothy, 47
Festugière, A.J., 15, 139, 140, 195
Feuerbach, Ludwig, 226, 233, 633, 635, 642, 684
Feuillet, André, 551, 555, 558
Feyerabend, Paul K., 340
Fichte, Johann Gottlieb, 73, 217, 218, 243
Ficino, Marsilio, 169, 206
Fiedler, Maureen, 561
Figgis, J.N., 398
Filas, F.L., 563
Findlay, John N., 140, 243, 401
Fiorenza, Elisabeth Schüssler, 137, 474, 561
Fiorenza, Francis, 245
Fischer, Alden L., 335
Fitzmyer, Joseph A., 411, 493, 501, 503, 504, 505, 535, 550, 551, 552, 553, 563, 620, 625, 626
Flannery, Austin, 136, 240, 249, 481, 557, 621, 687, 692
Flannery, Edward H., 399
Flaubert, Gustave, 81
Flavell, John H., 295
Fleming, David L., 202
Fletcher, Joseph, 368, 405
Flew, Anthony, 682
Flick, Maurizio, 408
Floeri, F., 145
Fodor, Jerry A., 341
Folcer, Louis, 242
Fonck, A., 244
Fonseca, Peter, 203
Fontaine, J., 140, 141, 477

Foot, Henry W., 410
Forest, Aimé, 229
Fortman, E.J., 628, 630
Foster, Frank Hugh, 201
Foucault, Michel, 479
Foucher, Louis, 244
Fouyas, Methodos, 17
Francis of Assisi, St., 420, 427, 491, 492
Francis of Marchia, 156
Francis of Vitoria, 182, 185, 360
Francke, A.H., 178
Francoeur, Robert T., 476
Frank, Erich, 139
Frankena, William K., 412
Franz, M.L., 43, 97
Fraser, J.T., 48
Frassen, Claude, 214
Frederick II (the Great) of Prussia, 63
Freeman, Derek, 406
Frend, W.H.C., 628
Freud, Sigmund, 21, 72, 226, 313, 429, 436-437, 440, 467, 469, 481
Freudenstein, Eric G., 469
Freyne, Sean, 554
Friedman, Morton P., 342
Froehlich, Karlfried, 469
Froidevaux, L.M., 138
From, Peter, 342
Fromlan, Victoria, 42
Fromm, Erich, 99, 471
Fuchs, Josef, 362, 367, 368, 402, 405, 409
Fulbert of Chartres, 134
Fuller, Reginald, H., 410, 552, 624, 684
Fulton, Robert B., 480
Furfey, Paul Hanley, 471
Furnish, Victor Paul, 410, 411
Fuse, Wolfram Malte, 192

Gadamer, Hans-Georg, 11, 17, 77, 98, 236, 248
Gaffney, James, 136
Gaith, Jerome, 141
Gal-Or, Benjamin, 47, 48
Gale, Richard, 48
Galen, 135

707

Galileo Galilei, 58, 61, 64, 156, 169, 170, 205-212, 282, 288, 301-304, 335, 345
Galot, Jean, 409, 550, 684
Galtier, Paul, 143
Gammie, John G., 137
Gamow, George, 48, 49
Gannon, M. Anna Ida, 144
Garces, Narcisus Garcia, 563
Gardet, Louis, 12
Gardiner, Anne Marie, 561
Gardner, E.C., 409
Garrigou-Lagrange, Reginald, 203, 229, 551
Gassendi, Pierre, 59, 94, 207
Gatje, Helmut, 684
Gaudemaris, André de, 146
Gaullier, Bertrand, 626
Gautama, Siddhartha, the Buddha, 387, 430, 463 659
Gauthier, Theodore, 79
Gay, Peter, 63, 69, 87, 94, 95, 241, 401
Geankopolis, Deno I., 187
Geary, Patrick J., 628
Geffchen, Johannes, 689
Geiger, Abraham, 15
Geiger, L.B., 229, 232
Gelin, Albert, 136, 137, 407, 408, 470
Gell-Mann, Murray, 47
George, F.H., 343
Gerhard, John, 178
Gersh, S.E., 143
Gersh, Stephen, 146
Gerson, John, 160
Getzels, Jacob W., 337, 338
Geyer, Hans-Georg, 624
Ghiselin, Brewster, 338
Giacon, Carlo, 202-203
Gibellini, Rosino, 247
Gibot, A., 47
Gibson, James, 342
Giet, S., 141
Gilbert de la Porée, 147
Gilbert, Neal W., 202
Gilby, Thomas, 191, 197
Gilchrist, John, 200
Gilkey, Langdon B., 248
Gillespie, Charles C., 291

Gillet, Robert, 146
Gillon, Louis, 409
Gilson, Étienne, 15-16, 144, 160, 187, 188, 189, 190, 191, 192, 194, 229, 230, 231, 240, 244, 627, 688
Gioberti, Vincenzo, 223
Giotto di Boudoni, 157
Glaab, Charles N., 99
Glasner, Peter F., 92
Gleven, Michael, 295
Globus, G.G., 341
Glock, C.Y., 16
Glover, Edward, 97
Glozik, Josep, 482
Glutz, Melvin A., 246, 336
Gödel, Kurt, 17, 67, 330, 331, 335, 342
Godfrey of St. Victor, 147
Goergen, Donald, 477
Gogarten, Friedrich, 91
Golden, Leon, 338
Gombrich, H., 342
Good, I.J., 343
Goodenough, E.R., 139
Gordis, Robert, 477, 622
Gordon, William J., 337
Görres, Joseph, 217
Gotti, Vincenzo Ludovico, 214
Gottlieb, Hans, 400
Gough, E.K., 473
Gould, Stephen Jay, 46
Grabmann, Martin, 188, 189
Graef, Hilda C., 563
Grane, Leif, 197
Grassi, Francesco, 227
Graves, Robert, 407
Greeley, Andrew M., 475
Green, Robert W., 200
Green, William M., 144
Greenstein, J.L., 48
Gregoras, Nicephoras, 149
Gregory of Nyssa, St., 119-123, 133, 135, 149, 611
Gregory of Valentia, 203
Gregory I the Great, 132-133
Gregory Nazianzen, St., 119, 145
Gregory XVI, 225
Gregory, Michael S., 289
Gregory, Palamas, St., 124, 591, 603

708

Gregory, Richard L., 342
Gregory, S., 400
Gregory, Tullio, 146
Grelot, Pierre, 408, 476, 550, 557, 563, 623
Gremillion, Joseph B., 401, 480
Grene, Marjorie, 98, 334, 337
Gressmann, H., 467
Grillmeier, Alois, 406, 684
Grillot, Bernard, 686
Grisar, Hartmann, 409
Grisez, Germain, 402, 412, 685, 686
Gritsch, Eric W., 198, 199, 410
Groote, Gerard, 160
Gross, Julius, 408
Grosseteste, Robert, 151, 156, 206, 627
Grotius, Hugo, 185
Grunbaum, Adolf, 48, 294
Gryson, Roger, 477, 560, 562
Guagliardo, Vincent A., 248
Guibert, J. De, 622
Guibert, Pierre, 624
Guilday, Peter, 481
Guilford, J.P., 338
Gundry, Robert H., 407, 555
Gunkel, H., 467
Gunther, Anton, 224, 225, 227
Gurr, John E., 241
Gustafson, James M., 15, 402, 409, 410
Gutierrez, Pedro, 561
Guyton, Arthur C., 46

Haag, Herbut, 408
Haber, F.C., 48
Habig, Marion H., 550
Hacker, Paul, 200
Hackett, Stuart C., 482, 629
Hadamard, Jacques S., 339
Haddad, Yvonne Yazbeck, 626
Hahn, Roger, 290
Hailman, Jack P., 45
Halacy, D.S., 14
Hall, Marie Boas, 241, 296
Halley, Edmond, 293
Hamer, Jerome, 246
Hamilton, Michael P., 561
Hammerschmidt, William W., 292
Hammersmith, Sue K., 475
Hammond, Phillip, 92

Handel, Georg Friedrich, 175, 180
Handy, Robert T., 246
Hanney, Alasdair, 340
Hanson, Norwood Russell, 48
Hanwalt, P.C., 47
Harakas, Stanley S., 402
Hardison, O.B., Jr., 338
Haring, Bernard, 361, 362
Haring, H., 555
Harkins, Arthur M., 692
Harkness, Georgia, 200, 201
Harl, Marguerite, 141
Harnack, Adolf, 471
Harned, David B., 92
Harre, Rom, 290
Harries, Karsten, 247
Harrington, Wilfrid J., 467
Harris, Marvin, 338, 400, 474, 479
Harris, R. Baine, 140
Hartley, Thomas J.A., 227, 244
Hartmann, Edwin, 341
Hartshorne, Charles, 14, 84, 238, 292, 683
Harvey, William, 58, 170, 205, 206
Hasel, Gerard, 467
Hassler, August Bernhard, 555
Hathaway, Ronald F., 142
Haughton, Rosemary, 476
Hausman, Carl R., 338
Hauy, Rene, 688
Hawkins, David, 45
Haya-Prats, Gonzalo, 684
Hayen, Andre, 229
Haynes, R.H., 47
Hazard, Paul, 94, 96, 242, 481
Hebbelthwaite, Brian, 683
Hecht, M.K., 44
Hegel, Georg W.F., 71, 82, 99, 217-222, 224-225, 228-230, 234, 310, 347, 358, 401
Hegel, Martin, 554
Hehir, J. Bryan, 472
Heidegger, Martin, 17, 76-77, 81, 99, 230, 235-236, 239, 248, 271, 274-275, 295, 341, 351, 360, 395, 405, 459, 633, 661-662, 689
Heidel, W.A., 139
Heinrich, Dieter, 96
Heisenberg, Werner, 48, 284, 293
Heller, Nicholas, 192

709

710

Matthew, St., 112, 392, 428, 431,
490, 492, 493, 497, 498, 502,
503, 506, 537, 538, 543, 565,
566, 589, 590, 607, 649
Matyniak, K.A., 45
Maupertuis, Pierre-Louis Moreau de,
63
Maximus the Confessor, St., 122,
123, 133, 135, 149
Maxwell, James C., 260,
Maxwell, Grover, 340, 341
Maxy, Carl E., 201
May, William E., 404
Maylender, M., 194
Mayr, Ernst, 46, 296
McAdoo, Henry R., 241
McCabe, Henry, 405
McCauley, Leo P., 145
McCool, Gerald A., 230, 231, 242,
244, 246
McCormick, Richard A., 362, 402,
403, 404
McCrea, William H., 48
McCulloch, Warren S., 43
McDonagh, Enda, 409
McEleney, Neil J., 552
McEvoy, James, 187
McFarland, William N., 45
McGhee, Paul A., 337
McGinn, Bernard, 147, 191, 398
McGregor, Geddes, 629
McHugh, John, 552
McInerny, Ralph M., 339, 690
McKane, William, 137
McKay, Donald, 343
McKenna, David L., 409
McKenna, Stephen, 143
McKenzie, John L., 136, 467
McKnight, Edgar V., 399
McLellan, David, 99, 244
McLelland, Joseph C., 140
McManner, John, 629
McManus, J., 472
MacMullin, Ernan, 47, 296, 335
McNeill, John J., 475
McNeill, John T., 199, 410
McNeill, William H., 459, 482
McReynolds, Paul, 338
Maine de Biran, 224, 244
Mead, Margaret, 473

Medawar, P.B., 100
Medici, Cosimo de, 150
Meer, Haye van der, 560
Meerloo, Jost A.M., 338
Meersseman, G., 188
Mehl, Roger, 15
Meier, John P., 410
Meijering, E.P., 138, 140
Melanchthon, Philip, 178
Melchizedek, 523, 574
Mendel, Gregor, 64
Mendelejeff, D.I., 651
Mendelssohn, Felix, 175
Mendelssohn, K., 50
Menninger, Karl, 470, 471
Meredith, J.C., 96
Meredith, James Creed, 242
Merlan, Philip, 139
Merleau-Ponty, Maurice, 75-76,
98, 337
Mersenne, Marin, 59, 94, 207
Merton, Robert K., 93
Merton, Thomas, 471
Messner, Johannes, 402
Metz, Johannes B., 234, 247,
402, 470
Meyendorff, John, 136, 142, 143
Meyer, Charles R., 558
Michaud-Quantin, Pierre, 203
Michel, A., 143, 631
Michel, Johann, 686
Michelangelo Buonarroti, 168, 181
Micheli, Guiseppe, 685
Miel, Jan, 203
Miguens, Manuel, 552, 558
Milgrom, Jacob, 469, 471
Milhaven, John G., 470
Millard, Richard M., 412
Miller, A.V., 243
Miller, B.F., 472
Miller, John H., 136
Milton, John, 180, 611
Minear, Paul S., 136
Minucius Felix, 125
Minus, Paul M., Jr., 18
Mirandola, Pico della, 401
Misciatelli, Pietro, 192
Mitchell, Nathan, 690
Mohammed, 387, 388, 659
Mohanty, J.N., 98

Mohl, Hans, 92
Mohler, James A., 557
Möhler, Johann Adam, 224, 228, 244
Moingt, Joseph, 623
Molesworth, Sir William, 340
Molina, Luis de, 183
Moller, J., 690
Moltmann, Jurgen, 14, 234, 247, 409
Mondin, Battista, 339, 685
Mongillo, D., 402
Monica, 506
Monod, Albert, 242
Monod, Jacques, 65, 95, 400
Monroe, James, 562
Montagu, Ashley, 41, 471
Montague, George T., 410, 684
Montaigne, Michel Eyquem de, 59,
    171, 208
Montefiore, Alan, 95
Monteleone, James A., 472, 474
Montgomery, Nancy S., 561
Moon, Rev. Mr., 388
Mooney, Christopher F., 248, 689
Moor, James H., 342
Moore, Charles A., 629
Moore, James R., 689
Moore, Peter, 561
Moorman, John R.H., 470, 550
Moraczewski, Albert S., 246, 474
Moraux, Paul, 187, 188
More, Henry, 212, 335
Moreau, Joseph, 141, 188
Moreno, Antonio, 407, 629
Morgan, John H., 560
Morgenbesser, S., 48
Morreall, John S., 339, 685
Morris, Aldyth V., 629
Morris, Desmond, 14, 45
Morris, Leon, 138
Moses, 111, 118, 168, 453, 462,
    497, 518, 520, 574, 581, 607
Mossner, Ernest C., 94
Most, William B., 554
Moule, C.F.D., 624
Mountford, Charles P., 407
Mouroux, Jean, 407
Mouy, Paul, 94, 240
Mowinckel, S., 481
Mozart, Wolfgang Amadeus, 387
Muck, Ott, 245

Mullahy, B.L., 239
Muller-Thym, Bernard J., 191, 342
Muller, Michael, 146
Muller, G.H., 48
Mulligan, Lotte, 94
Munitz, Milton K., 682
Murdoch, J.E., 239
Murdock, George Peter, 406, 473,
    474
Murphy-O'Connor, Jerome, 555, 691
Murphy, Richard T.A., 245
Murphy, Roland E., 137, 411, 553,
    564, 626
Murray, George B., 335
Murray, John, 683
Murray, Michael, 98, 247
Myers, J.M., 411

Nagel, Ernest, 48, 95, 290, 336, 342
Napoleon Bonaparte, 387
Nardi, Bruno, 192
Nash, George H., 93
Needham, Joseph, 688
Nehemiah, 454
Neiman, David, 141
Nelson, Benjamin, 97
Nelson, R.J., 340
Nemesius of Emesa, 121, 135, 141
Nestorius, 535
Neumann, Erich, 17, 338, 407, 692
Neumann, John von, 43 (10)
Neusner, Jacob, 402
Newman, James R., 95, 342
Newman, Jeremiah, 402
Newman, John Henry, 228, 350,
    465, 482, 538
Newton, Isaac, 58, 62-69, 212-219,
    242, 259-265, 291, 294, 299,
    335, 345
Nicholas of Autrecourt, 161
Nicholas of Cusa, 170, 171, 266,
    269
Nickelsburg, George W.E., Jr., 623
Nicodemus, 567
Nicolas, M.J., 550
Niebuhr, Reinhold, 233, 398
Nielsen, Kai, 70, 682
Niesel, William, 410
Nietzche, Friedrich, 42, 73, 76, 114,
    139, 226, 234-236, 633

717

719

723

724

Subject Index

Alienation from God, 611
Alimentary system, 23
Alteration, 301
American Psychiatric Association, 439
Amorphous substances, 280
Amphibians, 26
Anabaptists, 575
Analogia fidei, 505, 537
Analogy, 149, 234, 276, 317-318, 328, 536, 640, 644-645, 659
  of attribution, 317
  of proportionality, 317, 318
Analytic Philosophy, 67, 236, 320, 452
Anamnesis, 112, 423, 424, 524, 575
Anarchism, 516, 517
Anarchy, 453
Anawim, 109, 426
Androgyny, 381, 437, 451, 544
Angelology, 122
Angels, 264, 441, 451, 495, 500, 597, 600, 646, 647, 649, 650, 652, 656, 681, 696, 697
  and communication, 604, 655
  and ecumenism, 658, 659
  and evolution, 647, 650, 657
  and future, 653
  and history, 648, 655
  and idle speculation, 659
  and innate, ideas, 654
  and Islam, 650
  and Judaism, 650
  and magisterium, 648
  and mathematics, 652
  and natural laws, 654
  and non-Christian prophets, 658
  and pagan gods, 654
  and redemption, 659
  and space, 654
  and time, 653
  as messengers, 654
  as ministers, 654
  communication of, 653
  contemplation of, 653
  effect on humans, 654
  existence of, 646
  evil, 351, 352, 610, 654-655, 658-659
  fall of, 655

finitude of, 657
fixed choice of, 616
good, 648
hierarchy of, 653-654
language of, 653
model of, 652
of nations, 657
power over matter, 654
prayer of, 653
species of, 653
teaching by, 653
virtual presence of, 654
visions and miracles, 655
Angiosperms, 26
Animal creativity, 325
Animal knowing, 309, 311, 322
Animal learning, 325
Animal rights, 419-420
Animal society, 450
Animals:
  as machines, 209
  care of, 420
  worship, 420
Animus and *Anima,* 381, 437
Anisotropy of time, 35
Annihilation, 33
Annihilation of damned, 618
Annihilationism, 617, 618
Annunciation, 492
Anomie, 425
Anthropocentrism, 269
Anthropology, 119, 161, 164, 185, 231, 238, 361, 374, 386, 416, 461, 467, 582, 594, 646
Anthropology of Cappadocians, 120
Anthropology, Biblical, 372, 373
Anthropology:
  of Augustine, 130
  of Cappadocian Fathers, 114
  of Irenaeus, 114
  of Nemesius, 122
  of Origen, 114
  of Pseudo-Dionysius, 121, 122
  of St. Maximus, 121-123
Anti-matter, 30
Anti-Christ, 464, 681
Anti-Semitism, 510, 657, 678
Antinomianism, 425, 605
Antinomies, of Kant, 643
Antiochene School of exegesis, 354

Biography and body, 458
Biography, Christian, 587
Biologism:
   and Stoicism, 370
   in ethics, 369, 421
Biosphere, (see Ecosystem,)
Birds, 26 ff.
Birth, 283
Birth:
   physical, 111
   spiritual, 111
Bishop, plenitude of priesthood, 534
Bisociation, 314
Black-holes, 37
Blood: 418
   and life, 572
   and spirit, 572
   as sacrifice, 569
   circulation, 170
   feuds, 569
   not to be eaten, 572
   transfusion, 421
Blood, Eucharistic, 662, 664
Blood of Jesus, 112, 500
Body size, 301
Body, analogy to Church, 517
Body, awareness of, 321
Body of Christ: 524
   and sexual sin, 512
   corpus caeleste, 125
Body, Eucharistic, 662, 664
Body, Mystical, 106, 107, 511, 524,
   574, 578, 593, 608
Body, Phenomenal, 204
Body, Resurrected:
   activity of, 597
   and artistic activity, 597-598
   and communication, 597-598, 605
   and food, 597
   and memory, 602
   and metabolism, 598
   and sex, 597-598
   as light, 603, 604
   condition of, 598
   identity with earthly body, 604
   transformation of, 592, 603
Body, Spiritual, 592, 593
   location of, 609
Body:
   analogy of, 698

and biologism, 370
and breath of life, 581
and communication, 604
and ethics, 39
and freedom, 433
and historicity, 132, 345-346
and human nature, 356
and humanism, 88
and memory, 601
and mental illness, 576
and Merleau-Ponty, 75 ff.
and Palamism, 124
and St. Irenaeus, 114
and sociality, 458
and spirit, 333
and teleologism in ethics, 370
and Trinitarian image, 639
and violence, 433
as evil, 112
as free choice, 616
as garment, 103
as God's glory, 697
as image of God, 696
as machine, 209
as prison, 578
as sacred, 356
as sacrifice, 568, 569
as sexual, 437
as spouse of soul, 103
as tomb, 103
beauty of, 332
care of, 425
desacralization of, 185
expression of soul, 486
glorified, 632
health of, 425
hermeneutics and, 8
holiness of, 441
idealization of, 165
immortal, 579
in art and literature, 78-82
in heaven, 616
in hell, 616
Jesus in his bodily existence, 155,
   389, 506
language and, 8
limitations of, 596
mobility of, 28
nude in art, 79
nudity of, 166

of Risen Christ, 597
sense organs of, 28
spiritual, 121, 565, 588, 596, 604
spiritual, theories of, 595
transformation of, 585, 596
Bonding, Sexual, 435, 436
Bonds, chemical, 34
Book of Wisdom, 109
Boredom, 429
Brain, 22-23, 260, 320, 326-330,
434, 595, 694
bits of information in, 24
dependence of sensation on, 324
not sense organ, 323
number of neurons, 6, 22
unmoved mover, 24
Brain Hemispheres:
Male and female, 435
Breath of Life, 581
and blood, 572
Bride Symbol, 378, 379, 530, 535,
546, 571, 587, 698
Bridegroom Symbol, 110, 370, 530,
535, 546, 587
Brownian movements, 300
Buddhasatva, 636
Buddhism, 53, 426, 453, 459, 463,
602
and non-violence, 431
and personhood, 613
Hinayana and Mahayana, 68
Zen, 427
Bultmannians, 236
Byzantine Empire, 464, 465
Byzantine Theology:
and Latin Theology, 124
Calvinism, 210, 211, 221, 233, 426
and ethics, 176
and marriage, 177
and social reform, 177
and tyrant God, 610
Orthodox, 179
Cambridge Platonists, 212
Cana, miracle of, 110
Canaanites:
and immortality, 580
Candide, 433
Canon of Bible, 173, 416, 490,
505, 513
Canon within Canon, 391

Capital punishment, 429, 432
Capital Sins, Seven, 133
Capitalism, 165
and Humanism, 55, 61
Cappadocian Fathers, 119, 120, 124,
585
and creation, 121
and incomprehensibility of God,
603
and origin of sex, 121
Carbohydrates, 30
Carolingian Empire, 464, 465
Carolingian Renaissance, 133
Cartesian Christian Theology, 205
Cartesianism (see also Descartes), 52,
204-225, 269-281, 288-289, 310,
317-323, 327, 633, 642, 644
and Jesuits, 61
and structuralism, 78
in seminaries, 61
Caste system, 456
Castration, 443
Cat:
essence defined, 312
Catechisms, 181
Categorical Imperative, 224
Categories of Being-in-Process:
list of, 304-305
Categories: 216
and language, 298
and process, 298
and Ockhamism, 161
Aristotelian, 298
Kantian, 298
kinds of real, 305, 306
mental, 305
of Actions, 297-299
of Places, 299
of Qualities, 302-303
of Quantities, 301, 303
of Receptions, 297-299
of Relations, 297, 305-306
of Substances, 288, 289, 297
of Times, 300
Catharism, 113
Catholic Reaction, 223
Catholicism, Early, 391, 514
Catholicism, Liberal, 223

Causality, 161, 216, 281, 285-287, 306, 640, 643
  and evolution, 256
  observability of, 281
  simultaneity of, 281-282
Cause, First, 641, 644, 645
Cave art, 458
Celebration, 452
Celibacy: 110, 166, 440, 449, 515, 529, 647
  and basic human needs, 442
  and Eschatology, 443
  and feminism, 444
  and population, 681
  as charism, 442
  as witness, 442
  consecrated, 443
  morality of, 442, 443
  of laity, 444
  of Jesus, 443, 444, 448, 509, 528
  of St. Paul, 443, 528
  premarital, 443
Celibacy of clergy:
  in Eastern Church, 527
  in Western Church and Reformation, 528
  optional, 528
Celibate sexuality, 443
Cell, 261, 280
  human somatic, 28, 30
Cellulose, 26
Cenozoic, 27
Central American civilization, 458
Central organ, 320
Chalcedon, Council of, 123, 484-485, 493, 547, 674, 676
Chance, 65, 116, 256, 353
Change, 155, 268, 285, 287
Change:
  kinds of, 281
Chaos, 108-109, 378-379, 417-418, 430
Character, Sacramental, of Orders, 527
Charismatics, 526
Charisms, 594
Chartres, School of, 134, 151
Chastity:
  marital, 448
Chemical evolution, 257

Chemical radicals, 280
Chemistry, 34 ff., 155, 212
Child development, 435
Child-care, 436
Child-like innocence of Adam and Eve, 114
Child:
  as victim of sin, 385
  unbaptized, 600
China, 458, 462, 463, 466
  ethics of, 7
Chivalry, 438
Chordates, 26
Chosen People, 385, 392
Christendom, 465
Christian world-view: 8 ff., 51
  agreements with Humanism, 57
Christianity, 53, 453, 585
  and historicity, 348, 354
  scandals of, 351
Christology, 123-125, 221, 350, 386, 390, 493, 547, 565, 674-675
  and ethics, 393
  from above, 483
  from below, 483-484
  from within, 483-487
  from without, 483
  Küng's views, 484
  Logos Christology, 119
  of Origen, 119
  of St. Anthanasius, 119
  Schillebeeckx's view, 484, 487, 489
Chromosomes, 28
Church in the Modern World (Vatican II), 237
Church Dogmatics (Barth), 233
Church Order:
  in Early Church, 533
Church, as credible witness:
  apostolicity, 389
  catholicity, 389, 466
  holiness, 389
  unity, 389
Church: 511, 678, 698
  and historical experience, 389
  and Marian symbols, 539
  and papacy, 465
  and politics, 679
  and slavery, 679

Community, World, 453, 458, 576
Compassion:
  and Humanism, 86
  for damned, 613, 617
  of God, 614
Complementarity, equalizing, 530,
  544, 546
Complexity, infinite, 266
Compounds, 254
Compulsions, sexual, 447
Computors, 24, 322-330, 603
Concepts, 329
Conceptualism, 161
Concrescence, 307-308, 315
Concupiscentia, 130, 131
Concupiscible appetite, 429
Confession (Reconciliation):
  Sacrament of, 181
Confessions (Augustine), 130
Confirmation, 535
Confucianism, 463
Congregatio de auxiliis, 183
Connaturality, in knowledge, 270,
  314
Conscience, rights of, 519
Consciousness, 21 ff. 205, 270, 310,
  320, 321, 322, 323, 326, 327,
  328
  manifold, 548
  of Jesus, 675, 676
  stream of and self, 64
  unconsciousness, 21, 72
Consensus, 456
Consensus in Church Life, 516
Consequences of human act, 365
Consequentialism, 395
Conservation, 357
Conservatism, 75
Consistency theory of truth, 64, 69,
  207, 215, 298
Consolation of Theology (Gerson),
  160
Constantinian Establishment, 459,
  464, 465, 657
Consubstantiation, 666
Contact, 299
Contemplation, 77, 80, 117,
  123-124, 127, 133-134, 450, 681
Contemplation, aesthetic, 459-460
Contemplatives, 543, 576

Contentment, 428
Contingency, 210
Continuum, 208
Contraception, 361, 421-422, 434,
  440, 444, 445
Contradiction, 317
Contradictions, Inner:
  of Humanism, 82-85, 89
Convergence of religions, 466
Conversion, 127, 541, 571, 572,
  574, 575, 608, 611
  after death, 606-607, 616
  of St. Paul, 588
Cooperation:
  and Humanism, 86
Copernican revolution, 641
Corporate personality, 511, 573
Corpse, 110, 280
Corpse, care of, 421
Correspondence theory of truth, 64,
  69, 207, 215, 452
Cosmic Christ, 107, 237, 660, 661
Cosmic epoch, 35 ff.
Cosmic evil, 351, 649
Cosmic History, 35 ff.
Cosmic order, 416, 424, 653,
  656, 696
Cosmic rays, 36
Cosmic religion, 169
Cosmic sin, 619
Cosmic transformation, 107
Cosmology (see also Philosophy of
  Nature), 259-260, 269, 649-650
  of Cappadocians, 120
  Whiteheadean, 84 ff.
Cosmopolitanism, 117
Cosmos, 430
  as feminine, 379
  transformation of, 660
Council of Florence, 150
Councils, ecumenical, 104
Councils of Church, 539
Counter-Reformation, 180, 186
Covenant, 106-112, 163, 179, 513,
  570, 571-574, 668
Covenant Theology, (see Federal
  Theology)
Creation, 109, 113, 133, 153,
  283-284, 348, 374, 548, 633,
  638, 645, 656, 667, 671, 693

745

Jesuits, 181, 184, 210, 227
  and moral theology, 362
  and natural science, 206-207
Jewish sects, 112
Jewish Christianity, 112
Jewish-Christian theology: 585
  heavens in, 606-607
Jewishness of Jesus, 508, 510
Jewishness of Mary, 540
Jews and Hellenism, 112
Jews, election of, 462
Jews, persecution of, 432
Job, 109, 658
Johannine writings, 112
Jonah, 354
Judaism, 454, 658
Judaism:
  and Humanism, 510
  ethics of, 7, 9
Jude, Epistle of, 657
Judges in Israel, 454
Judgment:
  Last, 275, 327, 329, 450, 499
  on serpent, 383
  on Adam and Eve, 383
Judgment of others forbidden, 447,
  611
Judith, 354
Jungian psychology, 381
  theory of sexuality, 437
Justice, social:
  and Humanism, 54
Justice, Original, 594
Justice:
  Distributive, 455
Justification, 174

Kabbala, 168, 210
Kantian *a priori,* 643
Kantian antinomies, 643
Kantianism, 213-233, 264, 274, 276,
  281, 298, 361, 635, 641-642
  epistemology of, 68
  ethics, 368
  formalism, 368
Karma, 613
Kenosis, 509, 640
Kingdom of God, (see Reign of
  God).
Kings of Israel, 106, 109, 454

Kiss of peace, 599
Knowing:
  animal, 309
  definition of, 322
Knowledge of self, 176
Knowledge of Jesus:
  infused, 508
  experiential, 507

Laicism, 223
Laity:
  mission of, 525
  priesthood of, 524, 525
Lamentabili, 228
Language, 115, 237, 276, 418, 450,
  451, 452, 453
  and Love, 452
  and God, 634-635
  animal, 328
  Greek, 126, 132, 133
  human, 8, 20 ff., 328
  Latin, 466
  of bees, 325
  of primates, 325
  ordinary language philosophy (see
    Analytic Philosophy)
  origin of, 27, 376
  sacred, 521
Larva, 25
Last Supper, 499, 568, 571, 573,
  663
Latin America, 466
Law, 160-164, 177
Law, biblical
  and Faith, 392
  and Jesus, 392
  and Paul, 391-392
  and Luther, 391
  Antinomianism, 391
  Divine, 394
  in *James,* 392
  in *Matthew,* 392
  of love, 596
  of sin, 417
  Old, 106, 363, 416, 422-424,
    570, 582
  New, 363
  universalization of, 392
  uses of, 176
Law, international, 185, 360

Macroscopic Man, 267
Magic, 169, 172, 463, 580, 658
Magisterium, ordinary, 648
Magnificat, 540, 567
Majority rule, 456
Making of Man (Gregory of Nyssa),
   133, 135
Male, 434
Male and Female, Equality of, 438
Male dominance, 436, 448
Malum per se, 365, 367
Mammals, 26 ff.
Man symbol:
   as steward, 381
Man, (See Humanity)
Manicheism, 113, 126-128, 131, 425
Manna, 424
Manuals, moral, 184
Marcionism, 510
Marian privileges, 536
Mariological hermeneutic, 536, 543
Mariology, 538, 546
Marriage, spiritual, 546
Marriage: 110, 436-441
   and celibacy, 449
   and marital selfishness, 132
   and Augustine, 131, 132
   and Calvinism, 177
   as ministry, 444
   as sacrament, 532
   as sanctifying, 449
   compared to celibacy, 443
   in heaven, 597
   loveless, 445
   obligation to marry, 442
Martyrdom, 111, 430, 431
   of John the Baptist, 567
   and Jesus, 566
Martyrs: 106, 464, 573
   Maccabean, 581
   of Old Law, 608
   veneration of, 608
Marxism, 5, 7, 12, 53, 89, 234, 239,
   276, 285, 320, 346-347, 372,
   389, 432, 453, 456, 459-466,
   514, 566, 575, 587, 633-635,
   680, 693
   and asceticism, 426
   and women, 448
   and Jews, 511

critique of Humanism, 72, 82-83
decline of, 347
Humanist critique of, 85
origin of, 350
symbols of, 521
world view of, 19
Mary: (see Mariology)
   as Bride, 610
   as Church, 610
   as New Eve, 610
   veneration of, 608
Masculinity, 20, 273
   and aggression, 492
Masculinity of Jesus, 509
Masochism, 430
Mass as Sacrifice, 575
Mass, inertial, 31
Mass (see Eucharist)
Massacre of Innocents, 566
Masturbation, 370, 439, 445-447
Materialism, 73, 116, 132, 211, 212,
   222, 234
   of Tertullian, 125
Mathematical space, 299
Mathematicism, 205, 207, 212, 270,
   302
Mathematics, 115, 129, 156,
   169-170, 186, 206, 210, 218,
   303, 314, 331, 652
   and change, 268
   and logic, 67
   and physics, 66
   and the arts, 316
Mathematization, 268, 288, 304
Matrilineality, 436
Matter, 4, 123, 133, 155, 209,
   215-219, 254, 259, 262-264,
   279-280, 288, 693
   and anti-matter, 33
   and dynamism, 695
   and energy, 30 ff.
   and history, 695
   and potentiality, 284
   and Scotus, 158
   and sexism, 544
   appetite for form, 695
   Cartesian, 52
   Einstein's formula, 31
   humility of, 697
   negativity of, 694

Panpsychism, 308, 309, 320, 322, 695
Pantheism, 133, 134, 169, 210, 220, 224, 226
Papacy, 172, 465
Parable, 523
Parables: 419, 487, 517, 523, 600, 634
  and food, 424
  of Good Shepherd, 598
  of Lazarus and Rich Man, 427, 601, 607, 617
  of Sheep and Goats, 601
  of Talents, 109
  of Ten Virgins, 442
  of Two Sons, 386
  of Vineyard, 567
  of Wedding Feast, 567
Paraclete, 670
Paradigm shift, 266
Paradise, 461, 541, 580
Paradise, Earthly, 608
Paradox, of Olbers, 266
Paradox: 70, 275, 320
  in philosophy, 13
  in theology, 13
Parallelism, 320
Parenthood, Responsible, 422
Parenting:
  animal, 435
  human, 435
Parousia, 515, 609
Participation, (metaphysical), 153, 230
Participation in decisions, 457, 519
Participation in Church, 518, 526
Particles, elementary, of matter: 32-35, 66, 299
Particle and wave models, 35
Pascendi, 228
Paschalian wager, 644
Passion of Christ:
  and Stoicism, 117
  Jesus' foreknowledge of, 566-567
Passive resistance, 430
Passivity, 289, 694
Passover, 423, 569, 573, 592
Patriarchal society, 453
Patriarchalism, 450, 457, 532
Patriarchs, 461, 462

  souls of, 606
Pattern and ground, 324
Peace, 126, 433
Pedagogy, divine, 394
Pederasty, 132
Pedophilia, 445
Pelagianism, 162, 168, 172-173, 600
Penance, public 608
Pentecost, 670, 677
Perception, 324, 327, 329, 330
  fallibility of, 325
Performer, 333
Periodic Table of Elements, 651
Peripatetics, 115
Persia, 458
Person, Corporate, 511, 573
Person, Divine of Jesus:
  correct understanding of, 547
Person, human: 116, 154, 216, 278, 307, 310, 382, 450-451
  and community, 419, 455
  and dualism, 104
  and Humanism, 55
  as ends or means, 419
  autonomy of, 380
  bodiless, 646
  dignity of, 679
  discovery of, 380
  empirical, 613
  etymology of, 675
  identified with soul, 114
  identity of, 67 ff.
  in Eastern thought
  inner, 126
  integrity of, 421-422
  knowledge of, 450
  needs of, 467
  outer, 126
Personal Good, 455
Personhood, 544
Persons, Trinitarian, 638, 674-675
Peru, 458
Perversion, polymorphous, 439
Perversion, sexual, 109, 445, 447
Pessimism, 128, 211, 352
Petrine Office, 520
Phantom Limb, 325
Pharisees, 106, 454, 455, 571, 647, 669
  and immortality, 580

and resurrection, 581
authority of, 518
Phenomena, 161
saving of, 206
Phenomenology, 73-78, 90, 215,
230, 235-236, 275, 322, 367, 452
and ethics, 362, 369
phenomenological reduction, 74
Phenotype, 28
Philosopher:
Jesus as, 116
Philosophes, 63
Philosophy of Nature (of science,
cosmology): 231-232
and idealism, 72
and Humanism, 90
Philosophy of History, 350, 460
Philosophy:
and faith, 150
and language, 8
and religion, 219
and theology, 8
biblical, 109
British-American, 9
Christian choice of school, 115
Continental European, 115
Greek, 115
task of, 89
Photon, 32 ff., 280
Photosynthesis, 25
Phhagoreanism, 115
Phyla:
of animals, 26
of plants, 26
Physical Education, 425
Physicalism in ethics, see Biologism,
Physics, 169
Pietism, 179, 214, 221
Place, natural, 299
Planets, 37, 651
Plant Life, evolution of, 26
Plasma, cell, 30 ff.
Platonic Christian Theology: 103,
119, 133, 135, 159
Eastern, 112 ff.
Western, 125 ff.
Platonism, 103-104, 117-118,
124-126, 130-135, 150, 157, 160,
165, 167, 178, 185, 204-208,
211-212, 224, 227-231, 259,

269-271, 278, 287, 310-315, 321,
323, 327, 331, 379, 582, 584,
617, 696
and Christology, 485
and creation, 117
and immortality, 579
and objectification, 127
and spirituality, 585
body as prison, 578
continuity with Neo-Platonism,
120
errors of, 117
Middle, 118-119
of Aquinas, 153
preferred by Justin Martyr, 115
reconciled with Aristotelianism,
121, 123
varieties of, 134
why preferred by theologians,
116
Pleasure, 116, 270, 274, 324, 423,
426
Pleasure Principle, 429, 436
Plenum, 209
Pluralism, 236, 273
ontological, 656
religious, 514
Poetry, 11, 316, 452
Polarization, 178
within world-view, 68, 90
Politics:
and Church, 679
Pollution of environment, 7, 596
Polyandry, 436
Polygenism, 375
Polygyny, 384, 436, 448
Polytheism, 462, 463, 636
Polyvalence of texts, 353
Poor, option for, 111, 427-428
Pope, 104, 225
and social doctrine, 455
infallibility of, 105
Population:
and evolution, 29
explosion, 7, 439
control, 422, 681
Pornography, 446
Positivism, 73, 218, 222, 226
opposed by Process Philosophy,
84

Possiblity, 317
Potency, 153
Potentia absoluta and *ordinata*,
162
Potentiality, 209, 284, 285, 301,
694-695
Poverty, 117, 383, 426
and asceticism, 428
and Beatitudes, 428
and reign of God, 428
and technology, 427
and the arts, 427
as evil, 426
as injustice, 427
as mystique, 426, 427
as virtue, 426
not inevitable, 427
Power:
sharing of, 428
misuse, 457
Powerlessness, 427
Practical judgment, indeterminacy
of, 456
Pragmatism, 64, 69 ff., 212, 216,
276, 395, 642, 643
Praise of God, 633, 682
Praxis, 104, 276
Prayer: 182
for dead, 607, 609
for sinners, 613
intercessory, 574-576
vigils, 576
Pre-existence of souls, 115, 118,
121, 585
Pre-existence of Son, 671, 672
Pre-history, 459-461
Pre-understandings, 389
Pre-Adamites, 376-377
Pre-Socratics, 209
Preaching, 115, 175
Preaching of Jesus:
themes of, 390
Predestination, 150, 177, 179, 183,
612
Predestination, Double, 612
Prediction, 258, 265, 346, 612
Predictions of Cross, 497, 499
Prehension, 307, 315
Presbyterium, 534
Presence, Eucharistic, 663

Presence of Church in world, 514
Presence of God, 547
Presence, personal, 329, 549
Pride, 109, 462
Priest:
responsibility of, 257
Priesthood: 527, 572, 574, 578
and sinful domination, 530
as sign, 527
Christian, 520
commitment and, 527
in persona Christi, 527, 529-530
Jewish, 521, 523, 569
of Jesus, 522-524, 569
of laity, 524
pagan, 462
relation to Eucharist, 529
servanthood of, 529
threefold Ministry, 529
within not above church, 527
Priestly Document, 417
Primary matter, 694-695
Primary natural law, 281
Primary Natural Unit, 253, 267-268,
276-277, 281-284, 287-289,
297-299, 306-311, 319, 323, 327,
334, 434, 450, 458, 486, 650,
652, 656, 664, 694, 696
ambiguity of, 279
in process, 280
of minerals, 179
Primates, 27
Prime Mover(s), 153-155, 650-651
Primitive humans, 458
and history, 359
Primum cognitum, 274
Principle of
Causality, 281
De Broglie, 34
Double Effect, 367-368
Functionalism, 47
Indeterminancy, 34
Non-contradiction, 272, 274
Totality, 421-422
Subsidiarity, 457
Uncertainty, 34
Uniformity, 460
Principles of Church order, 516
Probabiliorism, 184
Probabilism, 64, 183-184

763

764

Theosis, 119, 595
Theotokos, 535
Thermodynamics, Second Law of, 38 ff.
Thermonuclear reactions, 36, 39
Thesis and antithesis, 225
Thief, Good, 580, 587
Thinking, 236
    human capacity for, 23
    in Heidegger, 77
Thomism, (and Neo-Thomism), 165, 182-186, 205, 211, 214, 229, 231, 239, 288, 317, 583
    and Aristotelianism, 230
    and ethics, 361, 362
    and existentialism, 230
    and history, 348
    and Kantianism, 230
    and natural law, 360
    and Platonism, 230
    and Proportionalism, 369
    decline of, 156
    participation and, 229
    revival of, 226-227
    Strict Observance, 229
    Transcendental, 229, 230, 231, 236
Timaeus (Plato), 3, 117, 134
Time, 120, 161, 255, 300, 306-307, 310, 694
Time:
    absolute, 265
    angelic, 653
    anisotropy of, 35
    Eschaton, 71
    infinite, 266
    in Process Philosophy, 84
    linear vs. cyclical, 71
    mathematical, 265
    natural, 300
    quantified, 308
Toilet training, 424
Tomb, Empty: 579, 589
    silence of St. Paul and, 590
Tools:
    invention of, 20, 27
Torah, 106, 108, 386, 416, 418, 420, 424, 462, 582
    and natural law, 392
    Jesus and, 109

Torture, 429, 432, 616
Totalitarianism, 347, 450, 455
Totality, (see Principle of Totality).
Touch, 208, 324, 329, 597
    and existence, 325
Tradition, 220, 461
Tradition, Sacred, 105, 118, 389, 537
    and Bible, 104
    and history, 388
Traditionalism, 223-224, 350
Traducianism, 125, 131
Trans-signification, 661
Trans-substantiation, 661-665
Transcendence, 311, 319, 321, 327, 334, 514
    of God, 6-7
Transcendental Thomism, (see Thomism).
Transcendentalism, 224
Transfiguration, 482, 496, 591, 592
    vision of, 497
Transformation:
    of earth, 619
    of human body, 585, 591, 596, 603
    of universe, 107, 619, 661, 677, 682
Transformism, (see Evolution).
Transitional states, 279
Translation problem (see Hermeneutics).
Transmigration of souls, 103, 110, 115
Transparency:
    of self-consciousness, 600
Transplantation of organs, 421
Tree Symbol, 107, 379-382, 417
Trent, Council of, 131, 180, 182
    and Eucharist, 661
Tribal society, 452
Trinity, 121, 134, 219, 351, 488, 493, 496, 632, 638-639, 669, 673, 682, 698
    and communication, 674
    and marriage, 443
    and personhood, 674-675
    Augustine's theory of, 128-129
    bodily image and, 639
    center of universe, 667

770